AF522658

SP
Pvt. Ltd.

ANIMAL PRODUCTS TECHNOLOGY

Editors

Prabhat Kumar Mandal
Ashim Kumar Biswas

2014

Studium Press (India) Pvt. Ltd.

ANIMAL PRODUCTS TECHNOLOGY

ISBN: 978-93-80012-62-9

Published by:

Studium Press (India) Pvt. Ltd.
4735/22, 2nd Floor, Prakash Deep Building
(Near Delhi Medical Association),
Ansari Road, Darya Ganj, New Delhi-110 002
Tel.: + 91-11-43240200-15 (15 lines); Fax: 91-11-43240215
E-mail: studiumpress@gmail.com

Printed at:

Akash Press New Delhi

List of Contributors

Anant Vishwanathrao Dhotre: *Assisstant Professor*, Department of Dairy Engineering, College of Dairy Technology, Warud (Pusad), Dist-Yavatmal, Maharashtra – 445 204, (Maharashtra Animal & Fishery Sciences University).

E-mail: anantdhotre@yahoo.com

Ashim Kumar Biswas: *Senior Scientist*, Division of Post-Harvest Technology, Central Avian Research Institute, Izatnagar, Bareilly – 243 122, (U.P.).

E-mail: biswaslpt@gmail.com

Atanu Jana: *Professor*, Department of Dairy Technology, SMC College of Dairy Science, Anand Agricultural University, Anand – 388 110, Gujarat.

E-mail: atanujn@gmail.com

Bhuvana Koodakandi Somappa: *Researcher*, Department of Livestock Products Technology, Rajiv Gandhi Institute of Veterinary Education and Research, Pondicherry – 605 009.

E-mail: bhuvana.vet@gmail.com

Chidanandaiah: *Assisstant Professor*, Department of Livestock Products Technology, Veterinary College, Vinobhanagar, Shimoga, Karnataka – 577 204, (Karnataaka Veterinary and Animal Sciences University).

E-mail: chidalpt2004@rediffmail.com

Gauri Jairath: *Researcher*, Department of Livestock Products Technology, College of Veterinary and Animal Sciences, Guru Angad Dev Veterinary and Animal Sciences University, Ludhiana, Punjab – 141 004.

E-mail: gauri.jairath625@gmail.com

Immanuel Prince Devadason: *Senior Scientist*, National Research Centre on Meat, Chengicherla, Boduppal Post, Hyderabad – 500 092.

E-mail: princenadar@yahoo.co.in

Kaushik Khamrui: *Senior Scientist*, Division of Dairy Technology, National Dairy Research Institute, Karnal – 132 001, Haryana.

E-mail: kkhamrui@gmail.com

Malay Kumar Sanyal: *Dean*, College of Dairy Technology, West Bengal University of Animal Fishery Sciences, Kolkata – 700 037.

E-mail: sanyal_mk@rediffmail.com

Mini K.P.: *Assisstant Professor (S.G.)*, Department of Pharmacology and Toxicology, Rajiv Gandhi Institute of Veterinary Education and Research Kurumbapet, Puducherry – 605 009, (Pondicherry University).

E-mail: minikpm@gmail.com

Muthupalani Muthukumar: *Scientist (Senior)*, National Research Centre on Meat, Chengicherla, Boduppal Post, Hyderabad – 500 092.

E-mail: muthukumar55@rediffmail.com

Nagappa S. Karabasanavar: *Assisstant Professor*, Department of Veterinary Public Health, Bombay Veterinary College, Parel, Mumbai – 400 012, (Maharashtra Animal and Fishery Sciences University).

E-mail: pub.nag@gmail.com

Najeeb Ayapurackal Pareeth: *Researcher*, Department of Livestock Products Technology, Rajiv Gandhi Institute of Veterinary Education and Research, Kurumbapet, Puducherry – 605 009, (Pondicherry University).

E-mail: najeebvet@gmail.com

Parminder Singh: *Researcher*, Department of Livestock Products Technology, College of Veterinary and Animal Sciences, Guru Angad DevVeterinary and Animal Sciences University, Ludhiana, Punjab – 141 004.

E-mail: pssandhulpt@gmail.com

Pavan Kumar: *Assistant Meat Technologist*, Department of Livestock Products Technology, College of Veterinary and Animal Sciences, Guru Angad DevVeterinary and Animal Sciences University, Ludhiana, Punjab – 141 004

E-mail: vetpavan@gmail.com

Prabhat Kumar Mandal: *Associate Professor*, Department of Livestock Products Technology, Rajiv Gandhi Institute of Veterinary Education and Research, Kurumbapet, Puducherry – 605 009, (Pondicherry University).

E-mail: mandalpkm@gmail.com

Ramesh Chandra Keshri: *Principal Scientist (Retired)*, Division of Livestock Products Technology, Indian Veterinary Research Institute, Izatnagar, Bareilly (UP) – 243 122.

E-mail: rckeshri@gmail.com

Shilpa Vij: *Senior Scientist*, Division Dairy Microbiology, National Dairy Research Institute, Karnal – 132 001, Haryana.

E-mail: shilpavijn@ yahoo.co.in

Subrota Hati: *Assisstant Professor*, Department of Dairy Microbiology, SMC College of Dairy Science, Anand Agricultural University, Anand – 388 110, Gujarat.

E-mail: subrota_dt@yahoo.com

Sudhakar Ganapati Bhandare: *Assistant Professor*, Department of Veterinary Public Health, Bombay Veterinary College, Parel, Mumbai – 400 012, (Maharashtra Animal and Fishery Sciences University).

E-mail: vetsudhakar@bristolalumni.org.uk

Sudip Kumar Das: *Researcher*, Department of Livestock Products Technology, Faculty of Veterinary Science, West Bengal University of Animal and Fishery Sciences, 37, Belgachia Road, Kolkata – 700 037.

E-mail: dr.sudipdas@yahoo.com

Sunil Kumar: *Associate Professor & Head*, Division of Livestock Products Technology, Faculty of Veterinary Sciences and Animal Husbandry, Sher-e-Kashmir University of Agricultural Sciences & Technology-Jammu, R.S. Pura, Jammu – 181 102.

E-mail: sunilskuast@gmail.com

Surajit Mandal: *Scientist*, Division of Dairy Microbiology, National Dairy Research Institute, Karnal – 132 001, Haryana.

E-mail: mandalndri@rediffmail.com

Uttam Kumar Pal: *Professor*, Department of Livestock Products Technology, Rajiv Gandhi Institute of Veterinary Education and Research, Puducherry – 605 009.

E-mail: paluttamkumar@gmail.com

Vikas Suryakant Wasker: *Professor*, Department of Veterinary Public Health, College of Veterinary and Animal Sciences, Udgir, Dist. Latur – 413 517 (Maharashtra), (Maharashtra Animal and Fishery Sciences University).

E-mail: vikaswaskar@gmail.com

Zuhaib Fayaz Bhat: *Assisstant Professor*, Division of Livestock Products Technology, Faculty of Veterinary Sciences and Animal Husbandry, Sher-e-Kashmir University of Agricultural Sciences & Technology, R.S. Pura, Jammu – 181 102.

E-mail: zuhaibbhatvet@gmail.com

About the Editors

Prabhat Kumar Mandal

Prabhat Kumar Mandal was born in 1967 in West Bengal, graduated as a Veterinarian from Bidhan Chandra Krishi Viswavidyalaya, Kolkata, in 1991 and did his MVSc & PhD majoring in Livestock Products Technology from Indian Veterinary Research Institute, Bareilly, India in the year 1993 and 1996, respectively. He is a recipient of *National Scholarship* from Govt. of India, (1984–1991) during High School and Under Graduate studies. He was awarded *Junior Research Fellowship* (1991–1993) and *Senior Research Fellowship* (1991–1996) during Master's and Doctoral studies by Indian Veterinary Research Institute. He worked for a short period (1996) as Scientist (ARS) in Indian Council of Agricultural Research, India and then joined as *Assisstant Professor* from September 1996, in Rajiv Gandhi College of Veterinary and Animal Sciences, Pondicherry. He was promoted as Associate Professor in March, 2005 and Associate Professor, Senior Grade from March, 2008. Dr. Mandal is engaged in teaching (UG & PG) and research on different areas of Meat Science and Technology/Livestock Products Technology since last 20 years. He is teaching several UG and PG courses related to LPT and guiding PG students. His research achievements include development of chicken meat balls from spent hen meat, utilization of blood protein isolates in sausages and cakes, development of value added poultry products utilizing edible byproducts, development of low fat restructured chicken block and using gizzard and natural preservatives. He worked as *Visiting Professor* in Hankyong National Univrsity, South Korea and Konkuk University, Seoul from March, 2008 to February, 2010. He has been involved in projects like "Purification and characterization of a protease as meat tenderizer", during his stay in South Korea. Presently he is working on development of functional meat foods with natural preservatives. Dr. Mandal has more than 100 publications in his credit. Presently Dr. Mandal is *Editor-in-Chief* of International Journal of Meat Science and *Editor*, Journal of Meat Science. He is *'Regional Editor'* of, Asian Journal of Animal Science, Asian Journal of Animal and Veterinary Advances, American Journal of Food

Technology and Asian Journal of Poultry Science. Dr. Mandal is founder member & Treasurer of Indian Meat Science Association. He was *Associate Editor*, Journal of Meat Science published by IMSA since 2003 -10, *Managing Editor* from December 2010 and Editor from July 2011. He was *Joint Organising Secretary* of the 1st National Symposium (2003) of IMSA.

Ashim Kumar Biswas

Ashim Kumar Biswas Associate Editor, Born on 05th March 1978 in West Bengal, Dr Ashim Kumar Biswas did his BVSc & AH from WBUA& FSc, Kolkata in 2000 and received his MVSc (in 2002) and PhD (in 2005) degrees in Livestock Products Technology (LPT) from Deemed University, Indian Veterinary Research Institute, Izatnagar, Bareilly. In 2005, he joined in the Division of Livestock Products Technology, SKUAST-Jammu as Assistant Professor and served for one and half years. On Feb 2007, he joined as Assistant Professor in the Department of LPT, GADVASU, PAU Campus, Ludhiana, India. Presently he is serving as Senior Scientist in Division of Post-Harvest Technology at Central Avian Research Institute (ICAR), Izatnagar. He has more than 150 publications in his credit. His main achievements in research relating to evaluation of export buffalo meat for quality, development of enrobed meat products, restructured buffalo meat blocks, functional based & shelf stable value added meat products. His research also focused on development and validation of HPLC methodologies for detection of pesticide and veterinary drug residues in meat and meat products. He also standardized molecular techniques (PCR) for the detection of bacterial pathogens in livestock products. He teaches several LPT courses at the University and guides many undergraduate and postgraduate students since 2006. He handled or currently handling several research projects and grants funded by MFPI, DST, RKVY (Govt. of India), ICAR, etc in the capacity of PI and Co-PI. Dr Biswas has association with several scientific bodies and organizations, He is acting as review editor in different peer-reviewed journals of national and international repute like Journal of AOAC-International, Journal of Food Science, Food Chemistry, LWT-Food Science and Technology, Meat Science, International Food Research Journal, Asian-Australasian Journal of Animal Sciences, Journal of Food Science and Technology, International Journal of Food Properties, Indian Journal of Animal Science, Indian Journal of Poultry Science etc.

Preface

Animal Products includes meat, milk, poultry, egg, fish etc. Animal Products Technology deals with the scientific harvesting, handling, processing, preservation, quality assurance, food safety and marketing of animal products. It is a discipline in which the knowledge of Chemistry, Biochemistry, Microbiology and Engineering is applied. Livestock Products Technology has emerged as an important academic discipline with the introduction of new Veterinary Curriculum by Veterinary Council of India in 1994. Apart from the veterinary curriculum, animal products technology related courses are also taught in several universities and national institutions under UG and PG programmes. Recently ICAR also introduced new postgraduate syllabus for the discipline of Livestock Products Technology. Keeping these facts in mind this book is brought out to cater the need of a reference book as an updated compilation for all concerns in this area.

This book is an important facet of animal agriculture in relation to processing, preservation, quality control, food safety and marketing of livestock products. Animal proteins are essential and inevitable input for life sustenance of mankind. With the heavy economic burden towards energy saving, improving socio-economic status and increasing number of working couples, the processed food industry is growing very fast and processed animal foods find their way in the day-to-day menu of the family. The use of food additives and some chemical and microbial contaminants in food is not new, but their excessive and unauthorized use or presence in recent years raised food safety issues that place certain responsibility upon the livestock products technologists and the government.

Processing of animal products is a utopian task and a great challenge to cater to the needs of essential nutrients to vast population. All animal origin foods are very rich in all the nutrients and highly perishable in nature. Their preservation and judicial use are highly important keeping in mind safety of finished products, energy conservation and waste utilization for better environment. The oldest methods of processing practiced by man are sun drying for meat, and fermentation for dairy products. The modern methods of food processing and preservation came up with the advent of chilling, freezing and other scientific advances. The demand for processed animal products has increased globally after GATT and WTO agreement. But imposition of Sanitary and Phyto-sanitatary (SPS) measures were the key issues for new pattern of food trade. However, processed animal products are mushrooming up in the major cities in the world including in India and contributing reasonable share in the super markets.

This advance reference book "Animal Products Technology" deals with the current topics of interest related to Processing, Preservation, Packaging, Quality Assurance and Food Safety of Animal Products. The book is divided into 28 Chapters spread over about 600 pages. *Chapter 1- 'Animal products and human health'* deals with the role of animal products viz. meat, poultry, eggs, milk and their products as component of a balanced diet and their implications on human health. *Chapter 2- 'Animal products processing and marketing'* covers the gist of some of the basic meat processing practices, methods for improving the tenderness, some recent innovations in the area of fresh meat technology, several novel techniques such as use of high pressure treatment, microwave heating, irradiation, microwave, ohmic heating and pulsed electric fields. Brief chemistry, characteristics, properties, uses, advantages and disadvantages of each type of packaging materials including modern packaging materials has been discussed in *Chapter 3 'Packaging materials for animal products'*. *Chapter 4 'Milk pasteurization and equipments'* deals with the principles and methods of milk pasteurization with special emphasis on HTST pasteurization along with understanding of the operation and equipments and their maintenance. In *Chapter 5 'Technology of comminuted meat products'* principles and technology of comminuted meat products for utilization of low value meat has been discussed in length with recent updates covering aspects of healthy meat products.

Chapter 6 'Quality assurance of meat' highlights the holistic approach for quality assurance of Indian meat with special emphasis on hygienic processing of meat, HACCP system, microbial decontamination, meat preservation systems, importance of Codex standards, ISO 9000 series including Food Safety and Standards Act, 2006 to meet the global challenges in meat food safety. *Chapter 7 'Processing of mozzarella cheese'* describes the principle and chemistry of mozzarella cheese making, legal standards, mozzarella cheese manufacture, factors affecting quality, composition and flavour of mozzarella cheese, equipment for mozzarella cheese manufacture and the recent developments. *Chapter 8 'Biotechnology for animal products'* discusses about the application of biotechnology in the food processing sector to improve the quality, safety and consistency of products. Use of genetic techniques including mutation and selection, gene transfer methods and genetic engineering, molecular techniques in the production of fermented foods, for understanding the genetics of microbial starter cultures and characterization of the culture strains, nutritional and genetic interventions for altering the milk composition.

HACCP concept, with specific plan for its application in meat processing and critical approach to control hazards in the food industry with systematic assembling of reliable data relating to the occurrence, elimination, prevention and reduction of hazards, the process of developing an HACCP program for specific food product and processing line is discussed in *Chapter 9 'HACCP for clean meat production'*. *Chapter 10 'Bio-preservatives for animal products'* deals with various natural preservatives of plant and animal origin their efficacy and safety for use in the food, bio-preservation systems that are optimum for different food and future of natural preservatives and regulatory aspects. *Chapter 11 'Functional starters for fermented dairy products'* covering about new starter cultures of lactic acid bacteria with an industrially important functionality that can contribute to the microbial safety or offer one or more organoleptic, technological, nutritional, or health advantages, insights into the metabolism of novel starter cultures and several health, marketing, and technological advantages. *Chapter 12 'Novel processing of animal products'* -in this chapter several emerging and novel technologies for processing and preservation of animal products including microwave/radio frequency heating, high pressure processing and pulsed electric fields, infra red heating, irradiation, U V processing, membrane technology, plasma and ultrasound treatment are discussed. *Chapter 13 'Enrobed meat products'* is about enrobing process, edible coating and various ingredients and food quality of the product, composition of coating material, proper frying medium for enrobed meat products, microbiological profile and lipid oxidation etc. In *Chapter 14 'Natural antioxidants for muscle foods'* deals with efficacy of naturally occurring compounds in plant leaves, fruits and herbs as antioxidants, their effects on the physicochemical, microbiological and sensory attributes of meat and meat products, to augment the shelf life of meat and meat products, to replace the synthetic antioxidants and as functional additives by providing adequate health promoting effects.

Chapter 15 'Processing and storage of meat soup' covering about types of soup and their chemical composition, methods of preparation of soups, role of ingredients, nutritive value of soup, packaging of soup. It also discusses about the changes in, physico-chemical, microbiological and sensory properties during storage and shelf life of soup. The pesticide residues and mycotoxins in animal products, rapid methods of analysis of samples including ELIZA, LFA, biosensors, molecular imprinted polymers (MIPS), microarray etc. immunoassay, multi-analyte detection, quantitative analysis of food contaminants, government regulations and surveillance are discussed in *Chapter 16 'Pesticides and mycotoxins in animal products'*. *Chapter 17 'Functional dairy beverages'*is about the fermented dairy-based beverages containing probiotics and / or prebiotics, milk based beverages, whey or soya based functional beverages, functional dairy beverages and health, marketing and technological advantages. *Chapter 18 'Modern packaging of meat and meat products'* presents about the different methods of packaging for meat and meat products, innovations and developments in new packaging technologies for meat and meat products including active packaging, antimicrobial packaging, intelligent packaging, their benefits, practical and economic viability. *Chapter 19 'Low-fat meat products'* describes role of fat in diet, rationale for reducing fat in meat products, fat replacer and their classification, strategies for fat reduction and low fat meat products, effect of fat reduction on product quality, procedure of reducing fat content. *Chapter 20 'High pressure processing of meat'* HHP processing to improve meat and seafood safety and quality, to control food-borne pathogens in meat products and combined HP-thermal treatment, its effect on nutritional and sensory qualities, development of high-efficiency HHP machines, HHP as a low-temperature treatment and regulations to recognize HHP.

Chapter 21 'Functional meat and meat products' focuses on strategies to improve the functional value of meat and meat products by genetic manipulation and by adding functional compounds in animal feeds to improve carcass composition and meat quality, incorporating functional ingredients into meat products during processing, reduction of specific unhealthy compounds, their generation during processing such as fermentation, curing and aging. *Chapter 22 'Dairy ingredients in meat products'* highlights role of dairy ingredients in meat products, function and use of whey proteins in meat products, use of caseinates, NFDM in meat products, use of dairy ingredients in restructured and reformed meats, miscellaneous dairy ingredients in meat products, benefits of using dairy ingredients in meats. *Chapter 23 'Nanotechnology for animal products'* is about potential of nanotechnology in the food industry for providing safer and more nutritious foods, nanotechnology in food packaging and efficient production, potential of nanotechnology in meat industry, uncertainties about the technology and its applications. *Chapter 24 'Natural preservation of meat'* explores the use of some natural antioxidants and antimicrobials like clove, *Ceratonia siliqua* essential oil, pomegranate extract, mint and curry leaves, celery powder, leafy green vegetable extracts, rosemary extract, citrus extracts, propolis etc. that can be used in meat. Detection of residues at very low levels, sample preparation by protein precipitation combined with ultra-filtration and quantitation, screening assays for licensed veterinary drugs, analysis of antibiotic residues using inhibitory assays, regulatory aspects and residue surveillance programs are reviewed in *Chapter 25 'Detection of drug residues in animal products'*.

Chapter 26 'Production of designer eggs' deals with the effect of consumption of eggs on blood cholesterol and heart disease, health benefits beyond basic nutrition, modification of the nutritional composition of eggs and reduction of cholesterol by feeding special diets. Dietary manipulation to increase the amount PUFAs directly or indirectly, enriching the level of antioxidant such as vitamin E in the egg. *Chapter 27 'Application of proteomics in meat science'* describes use of proteomics on skeletal muscle study and meat quality, proteome analyses in detection of changes in protein profiles during myogenesis, for biological markers of meat quality, particularly tenderness and of post-mortem muscle metabolism, effects of stress on meat quality traits or meat processing ability, recent technical developments in 2DE to MS technology. *Chapter 28 'Animal by-products utilization'* presents scope for processing and utilization of major by-products from slaughter and fallen animals. This includes processing and utilization of blood, bones, hides and skin, gut and intestine, glandular and other by-products.

Altogether the book covers the advance aspects of Animal Products Technology with contribution from eminent teachers, scientists and professionals recognized both nationally and internationally. It is assumed that this reference book will be of great value to the researchers, academicians, entrepreneurs, meat industry personnel, public health specialists and post graduate students related to production, processing, preservation, food safety & quality control and marketing of various animal products.

Editors

Table of Contents

List of Contributors v
Preface xi

1. **Animal Products and Human Health** 1
U.K. Pal and *P.K. Mandal*

2. **Animal Products Processing and Marketing** 18
P.K. Mandal and *U.K. Pal*

3. **Packaging Materials for Animal Products** 33
A.K. Biswas and *P.K. Mandal*

4. **Milk Pasteurization and Equipments** 51
Anant V. Dhotre

5. **Technology of Comminuted Meat Products** 79
M. Muthukumar and *I. Prince Devadasan*

6. **Quality Assurance of Meat** 102
P.K. Mandal and *U.K. Pal*

7. **Processing of Mozzarella Cheese** 133
Atanu Jana

8. **Biotechnology for Animal Products** 152
P.K. Mandal and *A.P. Najeeb*

9. **HACCP for Clean Meat Production** 174
S.G. Bhandare, V.S. Wasker and *S.K. Nagappa*

10. **Bio-Preservatives for Animal Products** 185
A.K. Biswas, P. Singh and *P.K. Mandal*

11. **Functional Starters for Dairy Products** 203
S. Mandal, S. Hati and *K. Khamrui*

12. **Novel Processing of Animal Products** 216
A.K. Biswas and *P.K. Mandal*

13. **Enrobed Meat Products** 254
Chidanandaiah and *R.C. Keshri*

14. **Natural Antioxidants for Muscle Foods** 270
A.P. Najeeb, P.K. Mandal and *K.P. Mini*

15. **Processing and Storage of Meat Soup** 289
Chidanandaiah and *M.K. Sanyal*

16. **Pesticides and Mycotoxins in Animal Products** 305
A.K. Biswas, P.K. Mandal and *K.P. Mini*

17. **Functional Dairy Beverages** 344
S. Hati, S. Mandal and *S. Vij*

18. **Modern Packaging of Meat and Meat Products** 354
A.K. Biswas and *P.K. Mandal*

19. **Low-fat Meat Products** 371
S.K. Das and *P.K. Mandal*

20. **High Pressure Processing of Meat** 387
A.K. Biswas, Gauri Jairath and *P.K. Mandal*

21. **Functional Meat and Meat Products** 404
Sunil Kumar, Z.F. Bhat, Pavan Kumar and *P.K. Mandal*

22. **Dairy Ingredients in Meat Products** 456
P.K. Mandal

23. **Nanotechnology for Animal Products** 472
A.K. Biswas, Gauri J. and *P.K. Mandal*

24. **Natural Preservation of Meat** 490
K.S. Bhuvana, P.K. Mandal and *K.P. Mini*

25. **Detection of Drug Residues in Animal Products** 510
A.K. Biswas, P.K. Mandal and *K.P. Mini*

26. **Production of Designer Eggs** 543
Z.F. Bhat, Sunil Kumar and *Pavan Kumar*

27. **Application of Proteomics in Meat Science** 569
A.K. Biswas and *P.K. Mandal*

28. **Animal By-Products Utilization** 586
P.K. Mandal

Subject Index 609

1

Animal Products and Human Health

UTTAM KUMAR PAL AND PRABHAT KUMAR MANDAL

ABSTRACT

Animal products play an important role in planning the meal for a balanced diet. Meat supplies many important nutrients viz. *high quality protein, most of the B-vitamins and vitamins A, D, E and K. Lean red meat is a vital source of highly bioavailable iron, zinc and many more micronutrients. Recently, many bioactive substances have been detected in various meats in different concentrations. These compounds have been shown to have antioxidative, antiatherosclerotic, antihypertensive and anticarcinogenic activities which make the lean red meat to be considered as a functional food. Besides the saturated fatty acids, meat supplies substantial quantity of monounsaturated fatty acids (MUFA) and good amount of polyunsaturated fatty acids (PUFA) which are considered to be healthy. Various studies have proved that meat alone cannot be specifically identified as an independent risk factor for coronary heart diseases (CHD) and colorectal cancer as these diseases are multifactorial in origin. Milk and dairy products supply high quality proteins, bone-forming minerals and health giving vitamins and furnish energy-giving lactose and milk fat with certain essential fatty acids, in an easily digestible and assimilable form. The time has come to change consumer's attitude about eggs to overall health and as a risk factor in heart disease. Scientists, medical professionals and communicators should change their attitudes and thinking about eggs. Furthermore, it is time that eggs be recognized as a nutritious food which also have functional health benefits beyond basic nutrition. In this article an attempt has been made to examine the role of animal products viz. meat, poultry, eggs, milk and their products as component of a balanced diet and their implications on human health in the light of existing dietary recommendations reviewing the available literature.*

1. INTRODUCTION

Foods are generally grouped in to five major categories *viz*. cereals and breads, protein foods, greens, vegetables and fruits and oil, fats and sugars. In order to meet our nutritional requirements, it is important to know the amounts of food from each group to be included in our diet. A balanced diet is one which provides all the nutrients in the amount and proportion required according to one's age, sex and activity. For good health, we require energy providing nutrients (carbohydrates, fats and proteins), vitamins, minerals and water. Human requirements for organic nutrients include nine essential amino acids, several fatty acids, glucose, four fat soluble vitamins, ten water soluble vitamins, dietary fiber and choline. Several inorganic substances, including four minerals, seven trace minerals, three electrolytes and some ultra trace elements must also be supplied.

Current recommendations for healthy diet call for at least 10–14% of calories from protein, no more than 30% calories from total fat. Saturated fat and trans fat should be limited to <10% of calories and polyunsaturated fats also to <10% of calories with monounsaturated fats comprising the remainder of fat intake. At least 45–55% of total calories should be derived from carbohydrates (Dwyer, 2012). The objective of this article is to examine the exact role of animal products including meat, poultry, eggs and meat in the human health system as a component of a balanced diet and its implications in light of the above recommendations. Further, an attempt has been made to develop a logical opinion about the effect of eating animal products on human health based on the available and accessible literature to the authors on the subject.

2. MEAT IN THE EVOLUTION OF HUMAN BEING

From time immemorial, meat has been an integral component of human diet. It has so become in the evolutionary development of mankind. Our dentition, gut structure, enzyme range, adaptability and our dependence on both plant and animal sources for our essential nutrients are all supportive evidence of our omnivorous status. In the evolution of mankind, we have adapted well as omnivore with the hunting of animals for food in addition to consumption of fruits and vegetables. Paleolithic men had a high reliance on animal foods rather than on plant foods. Their diet was high in animal protein and low in carbohydrates (Abraham, 2003). Biswas (2013) mentioned that since human beings diverged from chimpanzees a million years ago, the human gene called apoE had mutated, giving the human being "meat eating" habit.

A second mutation appeared around 226,000 years ago which helped in breaking down the fats and whisk cholesterol from human blood. The gene apoE probably also boosted our brains. To function properly, brain cells need to sheath their axons in myelin which acts like rubber insulation on wires and helps brain signals to travel much faster. Cholesterol is a major component of myelin and the version of apoE that helps whisk cholesterol from our blood also helps to distribute brain cholesterol where it is needed and thereby helps to prevent myelin deterioration. The ability to eat more meat was perhaps just a side benefit of boosting our brain power (Biswas, 2013).

3. MEAT AS A SOURCE OF NUTRIENTS

3.1. Protein

Meat and meat products are an important source of high quality protein which is essential for growth and development. Meat in general is a relatively concentrated source of protein of high quality (NPU 0.75–0.80). Meat proteins are 95–100% digestible, whereas, some plant proteins are only 65–75% digestible. Moreover, meat proteins supply all the essential amino acids in required proportion needed for human health. In particular, these proteins supply a relatively surplus of one essential amino acid, lysine which is deficient in most of the cereals. A small quantity of meat, egg, fish or milk in a predominantly cereal/legume based diet can markedly improve the biological value of vegetable origin proteins because of the complementarities of amino acids.

3.2. Vitamins

Besides protein, meat is a rich source of B-complex vitamins especially thiamine, riboflavin, niacin, biotin, vitamin B_6 and B_{12}, pantothenic acid and folacin. The latter two are especially abundant in liver which, together with certain other organs is rich in Vitamin A and supplies appreciable amounts of vitamins A, D, E and K. Thiamine is required as part of co-enzyme that converts carbohydrates and fats in to fuel to supply energy to the body. It also promotes normal appetite and contributes to normal nervous system function. A typical serving of pork will supply all the daily requirements of thiamine. Vitamin B_6 is a cofactor for several cellular enzyme reactions including those related to amino acid metabolism and inter-conversion (Abraham, 2003).

Vitamin B_{12} is exclusively supplied in human diet through foods of animal origin. This important vitamin is required to produce red blood cells and acts as a co-factor for many enzyme reactions and has an important role in the synthesis of nucleic acids. Deficiency of B_{12} leads to abnormal haemopoesis, resulting in megaloblastic anaemia. In addition, B_{12} deficiency may result in neurological manifestations such as sub-acute combined degeneration of spinal cord (Rao, 2010). Moreover, since B_{12} is required by active enzymes within methylation cycle, low intake of B_{12} as well as folate and B_6 have been associated with elevated homocysteine, which is a risk factor for cardiovascular disease (CVD) and stroke (Wagemakers, 2009). It has been shown that regular consumption of moderate amount of red meat (40–72 g/day) can help to reduce the risk of B_{12} deficiency (Prynne *et al.*, 2009).

Meat and meat products contain sufficient quantities of 25-hydroxy cholecalciferol, which has five times higher biological activity than cholecalciferol. Thus meat has now been recognized as the richest dietary source of vitamin D. Low intake of meat and meat products has emerged as an independent risk factor for rickets amongst the Asian population and pure vegetarianism in adults as a risk factor for Asian osteomalacia (Abraham, 2003).

3.3. Minerals

Red meat is one of the richest and best sources of iron in the human diet. Iron deficiency anaemia is a major nutritional deficiency affecting children and young women worldwide irrespective of their socio-economic status (McAfee *et al.*, 2010). Regarding high prevalence of anaemia among Indian population, Rao (2010) has suggested that anaemia cannot be fully corrected by treatment with iron alone; coexisting folate, vitamin B_{12}, vitamin A, riboflavin and vitamin B_6 deficiencies have to be corrected for optimum improvement in haemoglobin level. Under these circumstances, provision of moderate quantity of lean red meat in the diet of such subjects appears to be the most logical solution to combat this public health problem. Haem iron found in meat is more bioavailable than non-haem iron found in plant sources and for this reason, meat consumers maintain better iron status than vegetarians and vegans (Cosgrove *et. al.*, 2005). Moreover, apart from being the major source of haem iron, meat can influence and enhance the bioavailability of non-haem iron present in other foods.

Zinc is an essential trace element, being a component of a large number of enzymes. It is necessary for growth, wound healing, the immune system, reproduction and cognitive development. Zinc deficiency has been reported among different population groups of the world. Beef and lamb are considered to be rich sources of zinc as they contain 4.1 mg and 3.3 mg/100 g, respectively. Consumption of red meat even at the level of 41g/day is estimated to contribute 10.2–10.6 mg/day which can provide adequate zinc for optimum health (McAfee *et al.*, 2010). Recent scientific studies suggest that the selenium is a constituent of several enzyme systems, acts as an antioxidant and protects against coronary heart disease and certain cancers. Beef and pork are very good sources of selenium (Hazarika *et al.*, 2008). Choline is an essential nutrient that is necessary for central nervous system development, homocysteine metabolism and plays a role in immune system, fat metabolism and improves athletic performance. Beef, in particular liver is one of the richest sources of choline (Abraham, 2003).

4. MEAT FATS AND HUMAN HEALTH

4.1. Fats and Heart Health

Lipid content of muscle is extremely variable, ranging from approximately 1.5–13%. As meat plays the central role in planning of a meal in most of the industrialized countries, meat has come under fire as a source of saturated fatty acids which cause elevated blood cholesterol level leading to coronary heart disease (CHD). But as such, etiology of CHD is multifactorial which include increasing age, male sex, family history, genetic factors, cigarette smoking, high blood pressure, elevated serum cholesterol, diabetes, obesity, sedentary lifestyle including dietary habits and stress (Park, 2013).

4.2. Saturated Fats

Among the dietary factors, saturated fatty acids, mainly myristic (C:14) and palmitic (C:16) have been established as the most important of dietary risk factors in CHD. On analysis of fatty acid contents of intramuscular fats of various animals, myristic and palmitic fatty acids together were found to contribute about 25.6, 30.6, 31.4 and 31.5% in chicken, pork, beef and mutton fat, respectively. Unsaturated fatty acids contents were reported to be around 67.3, 53, 45.5 and 41.9% in chicken, pork, beef and mutton fat, respectively. Out of these, chicken, pork, beef and mutton fats contain around 58.7, 51.8, 44.4 and 39.7% mono unsaturated fatty acids (MUFA), mainly as oleic acid (C 18:1 n-9) (Hedrick *et al.,*1994) which is considered to be healthy and also found in olive oil which is associated with healthy Mediterranean diet. Besides these, meat also contains polyunsaturated fatty acids (PUFA), predominantly the essential n-6 and n-3 PUFA, linoleic (LA, C18:2) and alpha-linoleic acid (ALA, C18:3), respectively.

4.3. Cholesterol

Cholesterol is transported in plasma by very low density lipoprotein (VLDL), low density lipoprotein (LDL) and high density lipoprotein (HDL). High levels of HDL and low levels of LDL have been found to be associated with low incidence of coronary diseases and levels of both LDL and HDL cholesterols in blood are influenced by the consumption of PUFA in the diet. Meat and fish are the only significant sources of preformed very long chain n-3 PUFAs (eicosapentanoic acid, C20: 5n-3; docosahexanoic acid, C22:6n-3). It has been reported that moderate consumption of lean red meat was found to lower total cholesterol, LDL cholesterol and triglycerides (Beauchesne-Rondeau *et al.*, 2003). As per dietary guidelines, daily cholesterol intake through foods should be limited to 300 mg/day and lean meat contains only 60-80 mg cholesterol/100 g depending on its fat content.

4.4. Conjugated Linoleic Acid

One of the most important emerging health benefits of eating meat is its conjugated linoleic acid (CLA) content. It is composed of a group of positional and geometric isomers of octadecadienoic acid (Nagao & Yanagita, 2005). It is a fatty acid found in ruminant meats, with beef and mutton being the best dietary sources of CLA. It is more abundant in grass fed, grazing animals than in animals reared intensively with concentrate feeding. Recent epidemiological studies have suggested that intake of high fat dairy products and CLA may reduce the risk of colorectal cancer. In addition to its anticarcinogenic property, CLA has antioxidative, antiatherosclerotic and immunomodulative properties. CLA may also play a role in the control of obesity, reduction of the risk of diabetes and modulation of bone metabolism (Arihara & Ohato, 2008).

5. RED MEAT AND CANCER

High consumption of animal proteins and saturated fats have been implicated as a predisposing factor for cancers at various sites like pancreas, breast, colon, prostate and endometrium though there are many contradictory reports. Larsson *et al.* (2005) reported an elevated risk of colorectal cancer with high red meat intakes, whereas, Robertson *et al.* (2005) found no association between red meat intake and the incidence of colon cancer. They further argued that it was difficult to compare across studies owing to the many differences in study design such as differences in sample size or method of dietary assessment, to variation in end point measurement, whether it was cancer incidence or occurrence of adenoma. However, Larsson and Wolk (2006) commented that carcinogenicity of meat could be assigned to meat consumed (red or processed), the method of cooking, the quantity consumed and the individual genetic risk.

Two different groups of factors are responsible for carcinogenic effects. One group is intrinsic in meat such as fat and iron, and another group are those that are added to meat during preparation and processing such as salting, curing, smoking, very high temperature cooking etc. (Hazarika *et al.*, 2008). The most supported mechanism is the formation of mutagenic compounds in meat including heterocyclic amines and polycyclic aromatic hydrocarbons during high temperature cooking (Alaejos *et al.*, 2008). However, very few studies have assessed the method of cooking when measuring meat consumption (McAfee *et al.*, 2010). Approximately 80% of colon cancer cases are thought to be caused by modifiable diet and life style factors (Willet, 1995).

Based on the available data of various studies on meat and colon cancer, Williamson *et al.* (2005) noted that it was unlikely that red meat consumption could be assigned as an independent risk factor for colon cancer development owing to the complex nature of the disease and large number of interacting risk factors, including smoking and low physical activity. Furthermore, McAfee *et al.* (2010) has remarked that putting in context, it is unlikely that reducing meat consumption alone is sufficient to reduce risk, unless complete dietary balance has been addressed. In relation to other causes, WCRF (2007) has recorded that evidence associating either red meat or processed meat consumption with cancers of prostate, lungs, bladder, oesophagus or pancreas is limited and not thought to be convincing.

6. POULTRY MEAT AND HEALTH

Indian meat industry has made rapid progress during the last four decades and ranks 4th in the world with poultry meat production of 1.93 million tons annually (FAO, 2010). The poultry meat industry is witnessing the highest growth rate of 15 per cent per annum contributing 25 per cent of the total meat production in the country and accounts for Rs. 400 billion to the national GDP (GOI, 2010). Poultry meat is the fastest growing component of global meat production, consumption, and trade with developing economy playing a pivotal role in its growth due to its enormous potential to bring rapid economic growth with low investment.

Indians prefer to consume chicken rather than other species meat primarily due to its taste, health benefits, free from religious taboos, cost competitiveness and easy availability. Poultry meat is relatively rich in polyunsaturated fatty acids with comparable protein content and good balance of amino acids and better digestibility (Bourre, 2005). Consumer's consciousness about the impact of diet on health, life expectancy and well being leads to an increased demand for animal food products including chicken with added functional value. Consumption of chicken meat containing high amounts of polyunsaturated fatty acids (PUFAs) has greatly increased in the last two decades keeping in view the nutritionists' recommendations to reduce intake of saturated fatty acids.

High content of unsaturated fatty acids in diet may cause lipid oxidation which creates a risk to human health because of the possible formation of atherogenic, mutagenic and cytotoxic compounds (Fernandez *et al.*, 2005). Lipid oxidation in meat products is initiated when polyunsaturated fatty acids react with molecular oxygen, forming peroxides followed by development of oxidative rancidity which adversely affect the nutritional quality and safety of meat and meat products.

7. MEAT BASED FUNCTIONAL FOODS

Functional foods are the processed foods with disease preventing and health promoting benefits in addition to their nutritive values. Due to increasing concerns about health, in recent years, enhanced attention has been paid to the specialty (tertiary) functions of food which include role of food components in preventing diseases by offering enhanced physiological functions like anticarcinogenicity, antimutagenicity, antioxidative and antiaging activities (Dentali, 2002; Fernandez-Giniz *et al.*, 2005). Several meat based bioactive substances have been studied which are found to have promising functional properties.

ACE (Angiotensin-I converting enzyme) inhibitory peptides are the most extensively studied bioactive peptides. ACE, which is a dipeptidyl carboxy peptidase is present in vascular endothelial cells and plays an important physiological role in regulation of blood pressure. ACE inhibitory peptides have been shown to have antihypertensive effects and have been utilized for pharmaceuticals and physiologically functional foods. Several ACE inhibitory peptides have been derived from meat and meat products (Arihara & Ohata, 2008). The most abundant antioxidatives in meats are histidyl dipeptides namely, carnosine (β-alanyl-L-histidine) and anserine (N-β-alanyl-1-methyl-L-histidine). The concentration of carnosine in meat ranges from 500 mg/kg of chicken to 2700 mg/kg of pork shoulder and anserine is reported to be abundant in chicken muscle. These antioxidative peptides play a positive role in wound healing, recovery from fatigue and prevention of diseases related to stress (Arihara & Ohata, 2008).

Another important bioactive compound, L-carnitine (β-hydroxy-gamma-trimethyl amino butyric acid) has been detected in the skeletal muscle of various animals (Shimada *et al.*, 2005). Beef contains an abundant quantity of L-carnitine (1300 mg/kg). It assists in producing energy and in lowering the cholesterol level in human and helps the body to absorb calcium and selenium picolinate to improve the strength of the bones and to build lean muscle mass, respectively. Further, it has been demonstrated that L-carnitine blocked apoptosis and prevented skeletal muscle myopathy in heart failure. A product, containing a good amount of L-carnitine and carnosine as functional food ingredients derived from a by-product of corned beef has been developed and marketed in Japan (Arihara & Ohata, 2008).

Meat is a good source of glutathione (12–26 mg/100 g beef) and taurine (77 mg/100 g beef). Glutathione is an important antioxidative compound which provides cellular defense against toxicological and pathological processes, whereas, taurine is reported to protect our body from oxidative stress (Purchas *et al.*, 2004). Beef also contains coenzyme Q10 (ubiquinone) (2 mg/100 g) which is having antioxidant and creatine (350 mg/100 g) and creatine phosphate which play a critical role in muscle energy metabolism (Purchas & Busboom, 2005). Arihara and Ohata (2008) opined that choline, betaine, creatine, lipoic acid, putrescine, spermidine and spermine should also be listed as meat based bioactive compounds.

8. AVAILABILITY AND CONSUMPTION OF MEAT

Data related to meat consumption and animal protein intake in different regions of the world over a period of 1995–2005 have been presented in Table 1. World average of meat consumption in 2005 (g/head/day) on carcass weight equivalent is worked out to be 112.87 g against 224.93 g in developed countries, 36.45 g in Sub-Saharan African countries and 15.90 g in South Asian countries. Animal protein intake of South Asian and Sub-Saharan countries are much less than the optimal nutritional requirements of 20 g animal protein per person per day or 7.3 kg per year/head to efficiently combat rampant malnutrition and under nutrition in many developing countries in these regions. This can be achieved by an annual consumption of 33 kg lean meat or 45 kg fish or 60 kg eggs or 230 kg milk (FAO, 2009).

In the debate over meat eating and health, it definitely goes in favour of meat taking in to perspective the wide range of macro and micro-nutrients and beneficial bioactive compounds present in meat. Moreover, availability and consumption of meat is highly skewed in favour of industrialized countries of the world where comparatively more sedentary life style has resulted in increased incidences of various chronic diseases. On the other hand, a large percentage of population, especially the children and women in many countries of the developing world are suffering from malnutrition and under nutrition.

Table 1: Meat production, consumption and animal protein intake in developed and developing countries.

Country/Region	*1995*	*2005*	*Annual growth (%)*
Meat production (thousand tones)			
World	206 853	286 700	2.7
Developed Countries	99 572	110 250	0.9
Former Centrally Planned Economies	19541	18993	–0.2
Near East and North Africa	6610	9633	3.2
South Asia	7159	9353	2.3
Sub-Saharan Africa	7129	9291	2.2
Meat consumption (kg/person/year)			
World	35.7	41.2	1.5
Developed Countries	77.3	82.1	0.6
Former Centrally Planned Economies	50.6	51.5	0.2
Near East and North Africa	22.6	27.3	1.9
South Asia	5.6	5.8	0.3
Sub-Saharan Africa	12.4	13.3	0.7
Per capita protein intake from livestock products (meat, milk and eggs) (g/person/day)			
World	21.1	23.9	1.3
Developed Countries	47.1	49.8	0.6
Former Centrally Planned Economies	35.2	37.5	0.7
South Asia	8.6	9.4	0.9
Sub-Saharan Africa	8.1	8.6	0.6

Source: FAO (2012, 2013)

9. EGGS AND HUMAN HEALTH

Eggs have been described as nature's original functional food packed with thirteen important vitamins and minerals (Shallo, 2001). Eggs provide valuable nutrients, from the high-quality protein to significant levels of beneficial vitamins, antioxidants, and other healthful compounds. Egg components have been attributed to diverse biological activities, including antimicrobial and antiviral activity, protease inhibitory action, vitamin-binding properties, anticancer activity, and immunomodulatory activity (Li *et al.,* 1995).

9.1. Nutritional Value of Egg

It is well known that eggs contain a rich array of nutrients and are the least expensive. For years, eggs have been held up as a powerhouse of nutrition due to their exceptional nutrient profile containing high quality protein and a substantial amount of many essential vitamins and minerals (Applegate, 2000). They have been promoted for their high quality protein comparable to other animal protein sources. Egg white is considered an ideal protein because it contains all the essential amino acids in proper proportion for human nutrition. A typical egg would contribute 3–4% of an adult's average energy requirement and approximately 6.5 g of protein. The major albumen proteins are ovalbumin (54%), ovotransferrin (14%), ovomucoid (11%), ovomucin (3.5%), lysozyme (3.5%) and globulins (8%) (Li *et al.,* 1995).

Egg is a source of the B vitamins and vitamin A in particular and contain a significant amount of phosphorus, iodine, selenium and iron. Of the total fat in eggs, more than half is the unsaturated. Unlike many other foods that are high in cholesterol, such as fatty-meats and full-fat dairy products, eggs are relatively low in saturated fat and calories. Thus, eggs provide a high nutrient density in proportion to their energy content. The nutrient density of eggs makes them a valuable contributor to the overall nutritional balance of the diet and as an economical source of high quality protein.

Egg is a significant source of iron, riboflavin, folate and vitamins B_{12}, D, and E. The iron in egg yolks, like the iron in meat, is highly bioavailable, egg yolks may therefore be valuable in the diets of individuals who may need more iron, such as infants (Makrides *et al.,* 2002). An egg's contribution of vitamin D is noteworthy because the egg is one of the very few foods that supply this nutrient. Meyer *et al.* (2003) ranked eggs at third most important provider of n-3 PUFA (16%), behind seafood (71%) and meat (20%) in an Australian survey.

9.2. Facts about Egg and Health

Despite of many beneficial nutrients, eggs have been linked with adverse effects on blood cholesterol. A significant barrier to increasing egg consumption in western countries is the perception that egg consumption is associated with a rise in blood cholesterol levels (Yaffee *et al.,* 1991) and as a consequence is deleterious to health and life expectancy. The expert opinion on the role of dietary cholesterol in the development of heart diseases has changed recently, indicating that, for the general population, dietary cholesterol makes no significant contribution to atherosclerosis and risk of cardiovascular disease (McNamara, 2000a). The specific recommendations for strict limitations on egg or egg yolk intake/week have been dropped by American Heart Association although (AHA, 2000).

Experimental studies have shown that the consumption of an additional egg per day leading to a modest increase in average serum cholesterol and population studies examining the association between egg consumption and serum cholesterol levels showed no association (McNamara, 2000b). In a meta-analysis of 224 studies covering more than 30 years of research, it was demonstrated that dietary cholesterol has only a limited effect on plasma cholesterol (Howell *et al.,* 1997). Moreover, in the largest epidemiologic study conducted to date on the relationship between egg consumption and coronary heart disease, involving more than 118,000 men and women revealed that consumption of up to one egg per day did not have a substantial impact on the risk of coronary heart disease and stroke (Hu, 1999). Moreover, research demonstrated that persons who reported eating more than four eggs per week had significantly lower mean serum cholesterol than those who reported eating one or fewer eggs per week (Kerver, 2000).

The time has come to change consumer attitude about eggs to overall health and as a risk factor in heart disease. Scientists, medical professionals and communicators should change their attitude and thinking about eggs. Furthermore, it is time that eggs be recognized as a nutritious food which also has health benefits beyond basic nutrition. The concept of egg as a 'functional food' is new to many and requires a change in the perception. Finally, the evidence that eating eggs is unrelated to heart disease risk needs to be widely disseminated to health professionals and the public for including eggs in the diet (Applegate, 2000).

10. MILK AND MILK PRODUCTS IN HUMAN HEALTH

Milk is an almost ideal food having high nutritive value. It supplies high quality proteins, bone-forming minerals and health giving vitamins and provides energy-giving lactose and milk fat. Besides supplying certain essential fatty acids, it contains all the above nutrients in an easily digestible and assimilable form. All these properties make milk an important food for pregnant mothers, growing children, adolescents, adults, invalids, convalescents and patients alike (De, 1985). However, during last few decades, milk, particularly the milk fat has received adverse publicity for its suspected role in cardiovascular disease, although there is little evidence from the scientific standpoint that dairy fat contributes to an atherogenic blood profile when consumed in recommended amounts. Milk fats, in spite of containing unique fatty acids and other potential therapeutic micronutrients (conjugated linoleic acid, vaccenic acid, sphingolipids, butyric acid, β-carotene etc.) have been constantly discredited owing to health risks mainly associated with hydrogenated vegetable oil (Kansal, 2010).

10.1. Desi Ghee

Desi ghee, the clarified butter fat (99.5–99.8% fat) has been held in high esteem for its nutritional attributes, characteristic flavour and pleasant aroma. Desi ghee with its therapeutic and medicinal values is a vital ingredient of many Ayurvedic medicines used for treatment of various diseases and it is in use for this purpose for more than 2000 years. Bhatia (2005) reported that dairy ghee (cow and buffalo) when fed to rats, diminished diet induced hypercholesterolaemia. It was further noted that feeding on diet containing ghee increased plasma HDL levels and significantly lowered the contents of cholesterol and triglycerides in aortic tissue compared to soybean oil fed rats. Contrary to the popular belief that milk fat and other fat rich dairy foods is difficult to digest, in reality, milk fat is easily digestible and is a valuable dietary constituent with a therapeutic potential in the diseases of stomach, intestine, liver, kidney and gall bladder as well as in disorders of fat digestion. From a diet containing 50 g ghee, the cholesterol absorption is only 20–30 mg, which is negligible in view of 1–4 g cholesterol are synthesized in liver to meet body's requirement (Kansal, 2010). The same author reported that recent studies at National Dairy Research Institute, Karnal, India have demonstrated the protecting effects of cow ghee from the development of mammary and gastrointestinal tract cancers.

10.2. Fermented Milk Products

A variety of fermented milk products such as curd (*dahi*), *lassi, shrikhand,* yoghurt etc. have become highly popular among the consumers, not only because of their unique sensory properties but also due to their nutritional and therapeutic values. Fermented dairy products are excellent carriers of probiotic bacteria such as lactobacilli, bifidobacteria which are proven to provide multiple benefits to the gut health of the consumers. These foods not only provide nutrition but also confer a number of health benefits like immune system modulation, reduction in serum cholesterol, antihypertensive effects, anticancer effects, antidiabetic effects, etc. As fermentation reduces the lactose content of milk, such foods are beneficial to lactose intolerant population. Further, these products exhibit huge potential to act as a carrier of functional food (discussed in next para) ingredients like ω-3 fatty acids, bioactive peptides, phytosterols, dietary fibres, etc. which are now increasingly added to foods to make them healthier (Khetra *et. al.*, 2011).

10.3. Milk Based Functional Foods

Functional food is often described as the "food that may provide health benefit beyond basic nutrition." Milk itself may be considered as functional food, since it contains a variety of bioactive compounds. Many of the milk products have high contents of calcium, proteins, peptides, CLA, sphingolipids, butyric acid, etc. As mentioned in the previous section, different functional ingredients can be added to various milk products to enhance their functionality. Some of the recently developed probiotic functional dairy foods are *laban rayeb* (traditional fermented milk) of Egypt, *Bifidobacterium longum* containing Sri Lankan buffalo milk curd, functional cheeses such as cottage cheese with various LAB, white soft cheese with a new *Lactobacillus* strain and *Streptococcus thermophilus*. There are reports on the development of yoghurt fortified with PUFA and vitamin E, milk-cereal based functional foods, *lassi* fortified with hydrocolloids and dietary fibres, low fat *paneer* enriched with whey protein concentrate, low calorie *sandesh* fortified with macro and micronutrients and bifidus organisms, herbal *sandesh* containing fibres, antioxidants and probiotics (Bandyopadhyay & Sen, 2012).

10.4. Goat Milk

South Asian countries including India, Bangladesh and Pakistan are the major producers of goat milk, whereas, in Europe, the most developed dairy goat industry is in France, Greece, Spain and Italy. Goat milk contains higher amount of Ca, Mg and P than cow milk but vitamins D, B_{12} and folate contents are lower. Fat globules of goat milk are smaller in size and do not coalesce upon cooling because of lack of agglutinin, which is responsible for the aggregation of fat globules in cow milk (Pal *et al.,* 2011). Because of predominance of smaller fat globules in goat milk, it is easier to digest than cow milk which may be attributed to the faster lipase activity on smaller fat globules due to greater surface area (Chandan *et al.,* 1992). Hence goat milk is recommended for infants, old and convalescent people. Three medium chain fatty acids named after goat: caproic (C_6), caprylic (C_8) and capric (C_{10}) contribute to 15% of total fatty acids content in goat milk compared to 5% in cow milk and these fatty acids are reported to have great medicinal values for patients suffering from a variety of malabsorption, childhood epilepsy, cystic fibrosis and gallstones (Haenlein, 1992). Like cow milk, goat milk caseins have the same four species of αs_1, αs_2, beta and kappa but in different proportions. It has been reported that beta casein is the major component of goat milk casein, whereas, αs_1 is the major component of cow milk casein. Level of αs_2 casein is relatively higher in goat milk but total αs_1 and αs_2 casein fractions together in goat milk is lower than αs_1 fraction of cow milk. Such differences might contribute to soft curd forming properties, better digestibility and least allergenicity of goat milk to children (Chandan *et al.*, 1992).

10.5. Bovine Colostrum

Colostrum, the first mammary secretion after giving birth to an offspring, is emerging as one of the most natural immune booster for human beings. It is a rich source of immune components and growth factors which impart the immunity to the colostrum fed newborn. It contains protein, immunoglobulins (I_g), fat, minerals and vitamins in much higher concentration than normal milk. Some colostrum based products are being marketed for humans as dietary supplements. Evidences suggest that supplementation of bovine colostrum has beneficial effects on improving health, atheletic performance, diarrhoea in persons with immnodefficiency syndromes, NSAID (non-steroidal anti inflammatory drug) induced gastrointestinal disturbances and some aspects of the acute phase response that occurs secondary to surgery. Research has shown that specific hyperimmune bovine colostrum with neutralizing titre activity can be produced against *Cryptosporidia* spp., *Helicobacter pylori*, *Shigella* spp., measles and rota virus. This apparently has clinical utility in conditions associated with these infectious organisms. The growth factors present in colostrum have significant muscle and cartilage repair capacity. They promote wound healing in surgical patients. Colostral growth factors have multiple regenerative effects on all structural body tissues. In future, it is possible that colostrum based preparations may play a significant role in healthcare (*Das et al.*, 2013).

11. CONCLUSIONS

The most important debatable point about eating animal products is their contribution of saturated and unsaturated fat in the diet. If thought logically, then such false perception about meat may change. For example, a 100 g portion of lean meat can provide 2.5 g saturated fat considering the fat content of the lean at 5% level and calories derived from 2.5 g of saturated fat is calculated to be around 22.5 calories which is far below the recommended 10% of energy to be derived from saturated fats in the diet. Unfortunately, consumers often associate meat and meat products with a negative health image due to the contents of fat, saturated fatty acids, cholesterol and their association with chronic diseases which are mostly unwarranted as far as lean red meat is concerned. It would be highly beneficial for the developing countries to take adequate measures to increase the meat production so that common people can have access to this valuable food item at an affordable price which can help in combating most of the health problems related to malnutrition and under nutrition.

Milk and dairy products supply high quality proteins, bone-forming minerals and health giving vitamins and furnish energy-giving lactose and milk fat with certain essential fatty acids, in an easily digestible and assimilable form. All these properties make milk an important food for pregnant mothers, growing children, adolescents, adults, invalids, convalescents and patients alike (De, 1985). The time has come to change consumer attitudes about eggs to overall health and as a risk factor in heart disease. Scientists, medical professionals and communicators should change their attitudes and behaviors about eggs. Furthermore, it is time that eggs be recognized as a nutritious food which also has health benefits beyond basic nutrition. The concept of egg as a 'functional food' is new to many and requires a change in the perception.

FAO (2013) paper on Animal Production and Health has rightly pointed out that most people with micronutrient deficiencies live in low income countries and are typically deficient in more than one micronutrient. Highly nutritious foods such as meat, milk, poultry, eggs, fish and their products are particularly required for HIV/AIDS infected communities and also for women and children. The wide range of bioactive components present in animal products may have the ability to fight cancers, coronary heart diseases and enhance immunity and cognition, regulate body weight, improve bone health and may be used in sports medicine. More recently, the focus on the diets of our ancestors has helped in rediscovering the animal products as a mainstay of human diet evolution. Lean meat, chicken, low fat milk and eggs should be considered as natural functional foods which should be regularly consumed in moderate quantities along with lots of good vegetables and fruits as part of balanced diet.

REFERENCES

Abraham, J. (2003). Anxieties, concerns and facts about meat consumption and health. *In*: Compendium, National Symposium on "Impact of Globalization on Indian Meat Industry" (*Eds.*) Rao, V.K., Pal, U.K., Mandal, P.K. and Das, C.D. Held on 11–12 Dec., 2003 at Pondicherry, India. pp. 03–11.

AHA, (2000). AHA Dietary Guidelines. Revision: A statement for healthcare professionals from the nutrition committee of the American Heart Association. *Circulation,* 102: 2296–2311.

Alaejos MS, Gonzalez V and Afonso AM 2008. Exposure of heterocyclic aromatic amines from the consumption of cooked red meat and its effect on human cancer risk: A review. *Food Additives and Contamianants*, 25: 02–24.

Applegate, E. (2000). Introduction: Nutritional and functional roles of eggs in the diet. *J. Am. Col. Nutr.,* 19(5): 495S–498S.

Arihara, K. and Ohata, M. (2008). Bioactive compounds in meat. *In*: Meat Biotechnology (*Ed.*) Toldra, F. Springer Science and Business Media, New York. pp. 231–249.

Bandyopadhyay, S. and Sen, D.C. (2012). Recent developments in milk based functional foods. *Indian Dairyman,* 64(12): 52–57.

Beauchense-Rondeau, E., Gascon, A., Bergeron, J. and Jacques, H. (2003). Plasma lipids and Ipoproteins in hypercholesterolemic men fed a lipid-lowering diet containing lean beef, lean fish and poultry. *Am. J. Clin. Nutr.*, 77: 587–593.

Bhatia, E. (2005). Comparative studies on bio-protective properties of cow and buffalo ghee. Ph.D. Thesis. National Dairy Research Institute, Karnal, India.

Biswas, S. (2013). Meat and meat products in healthy diet. *In*: Compendium, National Symposium on "Emerging Technological Challenges to Meet the Demands of Domestic and Export Meat Sector" held on 7–9 Feb., 2013 at Hyderabad, India. pp. 72–83.

Bourre, J.M. (2005). Where to find omega-3-fatty acids and how feeding animals with diet enriched in omega-3-fatty acids to increase nutritional value derived products for human: What is actually useful? *J. Nutri. Health Aging,* 9: 232–242.

Chandan, R.C., Attaie, R. and Sahani, K.M. (1992). Nutritional aspects of goat milk and its products. Proceedings of 5th International Conference on Goats. March 2–8, New Delhi, India, pp. 399–420.

Cosgrove, M.F.A. and Kiely, M. (2005). Consumption of red meat, white meat and processed meat in Irish adults in relation to dietary quality. *Brit. J. Nutr.*, 93: 933–942.

Das, A., Seth, R., Kumar, N. and Pritee, J.A. (2013). Bovine colostrum: An emerging neutraceutical. *Indian Dairyman,* 65(2): 52–57.

De, S. (1985). Outlines of dairy technology. Oxford University Press, New Delhi, India, pp. 14.

Dentali, S. (2002). Regulation of functional foods and dietary supplements. *Food Technol.,* 56(6): 89–94.

Dwyer, J. (2012). Nutrition requirements and dietary assessment. *In*: Harrison's Principles of Internal Medicine, Vol. 1, 18th edition, (*Eds.*) Long, D.L., Fauci, A.S., Kasper, D.L., Hauser, S.L., Jameson, J.L. and Loscalzo, J., McGraw Hill Medical, New York. pp. 588–593.

Fernandez-Gines, J.M., Fernandez-Lopez, J., Sayas-Barbera, E. and Perez-Alvarez, J.A. (2005). Meat products as functional foods: A review. *J. Food Sci.*, 70: R37–R43.

FAO, (2010). Animal Production Year Book. Food and Agricultural Organization, Rome.

FAO, (2012). World Agriculture: Towards 2015/2030—An FAO Perspective. Food and Agriculture Organization www.fao.org/docrep/005/Y4252E/y4252e05b.htm accessed on 7th March, 2013.

FAO (2013). Animal Production and Health Year Book, Food and Agriculture Organization www.fao.org/ag/againfo/themes/en/meat/background.html accessed on 7th March, 2013.

Fernandez, J.M., Fernandez, L.J., Sayas, B.E. and Perez, J.A. (2005). Meat products as functional foods: A review. *J. Food Sci.*, 70: 37–43.

GOI (2010). Basic animal husbandry statistics. Department of Animal Husbandry, Dairying and Fisheries, Ministry of Agriculture, Government of India.

Haenlein, G.F.W. (1992). Role of goat meat and milk in human health. Proceedings of 5th International Conference on Goats. March 2–8, New Delhi, India, pp. 575–580.

Hazarika, M., Brahma, M.L. and Laskar, S.K. (2008). Meat and meat products as healthy food—certain facts and fallacies, *In*: Compendium, National Symposium on "Safe Meat for Good Health and Environment" (*Ed.*) Nadeem Fairoze, M. held on 4–5 July, 2008 at Bangalore, India.

Hedrick, H.B., Abrele, E.D., Forrest, J.C., Judge, M.D. and Merkel, R.A. (1994). Principles of Meat Science, 3rd edn, Kendall Hunt Publishing Co., New York. pp. 50–51.

Kansal, V.K. (2010). Reaffirming health benefits of desi ghee. *Indian Dairyman,* 62(11): 60–63.

Kerver, J.M., Bianchi, L.J., McNamara, D.J. and Song, W.O. (2000). The nutritional contributions of eggs to the American diet in the 90's. *FASEB J.*, 14: A 220 (161.1).

Khetra, Y., Raju, D.N., Hati, S. and Kanawjia, S.K. (2011). Benefits of traditional fermented milk products. *Indian Dairyman*, 63(9): 54–60.

Larsson, S.C., Rafter, J., Holmberg, L., Bergkvist, L. and Wolk, A. (2005). Red meat consumption and risk of cancers of proximal colon, distal colon and rectum: The Swedish Mammography Cohort. *Int J Cancer*, 113: 829–834.

Larsson, S.C. and Wolk, A. (2006). Meat consumption and risk of colorectal cancer: A meta-analysis of perspective studies. *Int. J. Cancer,* 119: 2657–2664.

Li-Chan, E.C.Y., Powrie, W.D. and Nakai, S. (1995). The chemistry of eggs and egg products. *In*: Egg Science and Technology. *Eds.* Stadelman, W.J. and Cotterill, O.J. the Haworth press Inc., New York.

Makrides, M., Hawkes, J.S., Newmann, M.A. and Gibson, R.A. (2002). Nutritional effect of including egg yolk in the weaning diet of breast-fed and formula-fed infants: A randomized controlled trial. *Am. J. Clin. Nutr.*, 75: 1084–1092.

McAfee, A.J., McSorley, E.M., Cuskelly, G.J., Moss, B.W., Wallace, W.W., Bonham, M.P. and Fearon, A.M. (2010). Red meat consumption: An overview of risks and benefits. *Meat Sci.,* 84: 1–13.

McNamara, D.J. (2000a). Dietary cholesterol and atherosclerosis. *Biochem. Byophys. Acta,* 2529: 310–320.

McNamara, D.J. (2000b). The impact of egg limitations on coronary heart disease risk: Do the numbers add up? *J. Am. Coll. Nutr.*, 19: 540S–548S.

Meyer, B.J., Mann, N.J., Lewis, J.L., Milligan, G.C., Sinclair, A.J. and Howe, P.R.C. (2003). Dietary intakes ad food sources of omega-6 and omega-3 polyunsaturated fatty acids. *Lipids*, 38(4): 391–398.

Pal, U.K., Mandal, P.K., Rao, V.K. and Das, C.D. (2011). Quality and utility of goat mulk with special reference to India: An overview. *Asian J Anim. Sci.*, 5: 56–63.

Park, K. (2013). Park's Textbook of Preventive and Social Medicine, 22nd edition. M/S Banarsidas Bhanot Publishers, Jabalpur, India. pp. 338–339.

Prynne, C.J., Wagemakers, J.J.M.F., Stphen, A.M. and Wadsworth, M.E.J. (2009). Meat consumption after disaggregation of meat dishes in a cohort of British adults in 1989 and 1999 in relation to dietary quality. *European J. Clin. Nutr.*, 63: 660–666.

Purchas, R.W. and Busboom, J.R. (2005). The effect of production system and age on levels of iron, taurine, carnosine, coenzyme Q10 and creatine in beef muscle and liver. *Meat Sci.*, 66: 629–637.

Rao, B.S.N. (2010). Nutrient requirement and safe dietary intake for Indians. *Bulletin Nutr. Found India*, 3(1): 01–08.

Shallo, H.E. (2001). Designer foods: Egg Products. Egg Nutrition Center, 1050 17th St NW, Suite 560, Washington, DC 20036, http://www.fass.org/fass01/pdfs/Shallo.pdf.

Shimada, K., Sakura, Y., Fukushima, M., Sekikawa, M., Kuchida, K. and Mikamai, M. (2005). Species and muscle differences in L-carnitine levels in skeletal muscles based on a new simple assay. *Meat Sci.*, 68: 357–362.

Willet, W.C. (1995). Diet, nutrition and avoidable cancer. *Environmental Health Perspectives*, 103: 165–170.

WCRF (2007). World Cancer Research Fund/American Institute for Cancer Research. Food, nutrition and the prevention of cancer: A global perspective. American Institute for Cancer Research, Washington, DC.

Yaffee, M., Schutz, H., Stone, J., Bokhari, S. and Zeidler, G. (1991). Consumer perception and utilization of eggs and egg products. *Poult Sci.*, 70(S1): 88.

2

Animal Products Processing and Marketing

P.K. Mandal and U.K. Pal

ABSTRACT

For centuries, great efforts have been devoted to find ways of preserving food and protecting it against physical damages, chemical changes and further microbial contamination and spoilage. Innovations include the use of chemical preservatives such as polyphosphates, sodium ascorbates or isoascorbates, antibiotics, bacterial peptides etc. The advancement in science and technology leads to development of new processes and brought about changes in the basic meat and dairy processing viz. hygienic production, handling, and preservation and further processing of animal products. The fresh meat after being subjected to any of the physical or chemical processes like grinding, curing, comminution, freezing, canning, fermentation etc. to modify its properties is termed as processed meat. In this chapter, an attempt has been made to present the gist of some of the basic meat processing practices. Different methods for improving the tenderness of meat from aged animals are discussed. Some of the recent innovations in the area of fresh meat technology viz. electrical stimulation, high pressure technology, use of calcium chloride for tenderization and their applications under Indian conditions are also discussed in this paper. Several novel techniques such as use of high pressure treatment, microwave heating, and irradiation have joined the array of food protection methods. New preservation technologies, such as high pressure processing, microwave, ohmic heating and pulsed electric fields offer advantages in meeting consumer demands of freshness, convenience and safety. There is no single process that will allow the high-quality production of every food product while ensuring safety; all of these processes, as well as thermal processing, have their own set of limitations and advantages.

1. INTRODUCTION

Animal products play a vital role in sustaining animal production and providing valuable animal protein for humanity. To create demand for meat and sustained growth in meat animal production, organized development of processed meat industry is important to realize the real benefits from meat and edible byproducts of slaughtered animals. Processing of meat to develop products has many advantages in terms of providing variety meat products, increasing demand and marketability, utilization of different low value cuts and by products gainfully, incorporation of non-meat ingredients for quality and economy, preservation and distribution to larger populations at an affordable price.

In India, only 1–2 percent of total meat is processed into products while in developed countries, it is above 60 percent. Bacon factories which were established in 1960s to popularize pork and pork products could not sustain largely due to inefficient governmental management system. Under the erstwhile Meat Food Products Order, 1973, licensing is done for processed meat units in the country and there are about 170 units under different categories of licensing. In view of considerable proportion of slaughter animals being aged and culled producing generally tough and low palatable meat, development of processed meat sector would be an advantage both to producers and consumers. Meat products are also produced, particularly pork and buffalo meat products, in the unorganized sector in a primitive way giving least consideration to hygiene. Public health risks of these products need to be assessed.

Processes such as portioning, deboning, size reduction, seasoning (addition of spices and condiments), tenderization, tumbling, reforming, emulsion preparation, battering, breading and a variety of cooking methods are utilized to produce a variety of value added products. Production of value added products is labour intensive. But recently, automatic machines are being developed in the country for a variety of products as the imported machines are very costly. Correct formulation, processing, package, storage stability, prevention of flavour and colour changes, nutritional value, labeling requirements, regulations etc are important factors in the success of the processed products. In the Indian situation, more production of value added products are required to meet the increased demand for meat products and to sustain prospects of animal production. The products should be developed and marketed as per the needs, expectations and acceptance of as large a population as possible.

The development of emerging technologies in food processing addresses specific consumer needs toward safe, healthy, and minimally processed foods. These innovative processes also lead to environment friendly and sustainable food manufacturing techniques with low energy requirements and reduced water use that overcome some limitations given by current food processing practices. Taking advantage of specific potentials and opportunities of these new processes, including the understanding and control of the complex process-structure-function relationships, offers the possibility for a science-based development of tailor-made foods.

2. MEAT PROCESSING AND VALUE ADDED MEAT PRODUCTS

Sustainability of animal production depends on availability of good productive breeds, feeds at an affordable cost, production efficiency and economic utilization of produce for value added products. The quality and economic success of processed meat sector depends on hygienic production, cost efficient processing technologies, creating sustainable demand for the products, better utilization of byproducts, providing positive brand image and innovative marketing.

Processed products provide increasing convenience to consumers through decreasing preparation time, minimizing preparation steps, allowing use of specific parts, thus taking risks out of kitchen. It also allows differentiation of products and increasing value of the products through provision of wide choice to the consumers. They protect and promote good health and meet the requirement of social, cultural and recreational activities. Value added products could be broadly classified based on processing, variety, convenience and function. The growth of processing industry assures the farmers a regular off-take of their produce at reasonable prices and generation of employment.

In order to compliment and supplement qualities and availability of different meats and their byproducts, a suitable combination of meats is desirable to produce value added products. Blends of mutton and chicken or chicken byproducts or chicken fat resulted in highly acceptable products such as nuggets, patties and sausages. Eggs possess several functional properties, which may compliment meat proteins in meat products. Highly acceptable meat products of relatively lower cost were produced incorporating eggs up to 30% in the formulation.

2.1. Value Added Products from Spent Animals

Meat from spent animals is fairly tough and less palatable. Value added products, particularly from minced meat or emulsion based, would be advantageous to incorporate other ingredients to improve palatability and reduce cost of production. A variety of emulsion based products such as sausages, patties, nuggets, chicken balls, slices, kababs, chicken vada etc have been developed utilizing meat and byproducts including spent hen meat which are byproducts of the egg industry.

2.2. Technology of Sausages and Nuggets

A simple technology has been developed to produce meat sausages and nuggets from different meats and their combination by forming a block using a mould and cutting the cooked meat block into nuggets of different sizes and shapes. This technology is of very low cost as compared to modern technology using forming machines and emulsion of any desired consistency could be used. Fried, smoked and pickled type variety meat nuggets could be produced. Economic formulations with a cost reduction of 30% have been developed to produce economy type nuggets with higher nutritive value.

2.3. Enrobed Meat Products

Enrobing/coating of meat products with edible materials in the form of batter using flours, whole egg liquid and other additives is a method of value addition which enhances the acceptability of meat products. Enrobing imparts the product a crispy texture and increases the pleasure of eating with more desirable colour. Products will be juicier as natural juices are retained. Enrobing is not only important for foods which are traditionally coated, such as sea foods and poultry including egg but has become a means of offering consumers a new product (Pandey, 2003). The variety of foods that can be enrobed has no limit and depends upon the imagination of the various technologists, restaurants and other food processing establishments.

A variety of enrobed products are frozen. Commercial enrobing provides an opportunity of value addition at a low cost to the processors. Because of the high cost of meat, smooth to crunchy coating can be economical to the consumer and profitable for the industry. The enrobed products can be formed into many shapes and sizes resembling chops, steaks, and other popular cuts/products. The characteristics of enrobed products important to consumer are appearance, colour, crispiness, flavour, juiciness and nutritional quality. The appearance includes colour and overall aesthetics, which are determined by the amount of coating material used. The composition of the enrobing medium especially of carbohydrate and protein sources are important for the final colour (Maillard reaction) development. The ingredients of coating function individually and collectively and they include polysaccharides, proteins, fats and hydrogenated oils, water and seasonings. Generally two types of batters, *i.e.*, wheat flour and corn flour based are used for enrobing the meat products. The corn flour based batters when used, needs continuous stirring to keep the solids in suspension as they tend to settle out easily.

2.4. Restructured Meats Products

The purpose of producing restructured products is to effectively market less valuable carcasses and carcass components. The basic processes include chunking, flaking, tearing, grinding, chopping, mixing, tumbling, emulsification, forming, freezing, tempering, pressing, slicing, cooking etc. Tumbling, massaging and blade tenderization facilitate production of high quality restructured products. The products include steaks, cutlets, chops, roasts, rolls and hams. When meat is restructured, its form is changed. Restructuring does not signify re-hydration, addition or removal of something. It is a processing or manufacturing step. It creates consumer ready products, which resembles an intact steak, chop or roast. These products differ in composition, particle size, non-meat ingredients and certain manufacturing procedures and in final appearance of the products.

Restructuring signifies the binding of smaller pieces of meat, using naturally occurring proteins, to form a meat product with many of the properties of steak and roast meat. Non-meat ingredients are also used for improving functional and organoleptic properties. A major advantage of restructured meat is that they can be adapted to consumer needs for convenience.

The tender, tasty and very palatable restructured product, which combines the important properties of muscle foods perceived by the consumer, offers a good future (Pearson & Gillett, 1997). The products are processed in a manner that causes the meat pieces to adhere to each other so that they resemble an intact cut in regard to consistency and appearance. The cohesive substance that binds the pieces or muscles together may be from non-meat additives or extraction of myofibrillar proteins derived from the chunks or pieces of meat.

2.5. Fermented Meat Products

Processing of traditional fermented meat and meat products as speciality items is a well developed industry in many European countries. Modern approach of product development and preservation including selection of cultures, controlled fermentation and drying and packaging have added new values to these products. Based on the moisture content, the fermented meat products are categorized into three major classes *viz.*, **moist** (50–60% moisture), **semi dry** (35–50 % moisture) and **dry** (20–35%moisture). Fermented sausages are one such product where preservation is usually achieved by a combination of fermentation along with reducing water activity, dehydration and addition of salt. Smoking can also be used to aid drying and deposit residues on the surface to inhibit the surface mould growth (Cambell-platt, 1995).

As mentioned earlier, Europe is the major producer and consumer of fermented meat products which are most commonly produced from pork and beef. Sometimes lamb is also fermented, while chicken, duck and buffalo meats are used in some products. Hygiene and preservation is very important in the preparation of fermented products. *Pediococcus acidilactici*, a commonly used bacterial starter culture for processing fermented products is reported to produce “pediocin”, a bacteriocin which inhibits the growth of *Listeria monocytogenes* while drying of the products. Encouraging works have been carried out on fermented meat products in India at Indian Veterinary Research Institute, Izatnagar (Mukhopadhyay, 1991; Salahuddin, 1992).

As the meat produced in India is mostly from spent animals, meats from such animals are generally tough and fibrous. So, there is need to treat such meats in a way which will make them tender. Indian workers have developed a number of methods for tenderization of spent meats (Mendiratta, 2003) which need to be transferred to the industry in acceptable form. Considering the present infrastructure facilities available for marketing of meat products, it is felt that development of shelf stable products in the form of intermediate moisture meats, fermented meat products, meat processed in thermostable pouches are few of the viable propositions.

2.6. Intermediate Moisture Foods

These are a heterogeneous group of foods that are based on hurdle concept (Leistner, 1992). Intermediate moisture meats are characterized by a moisture content of around 15–50%, being intermediate between fresh and dry meats and they are partially dehydrated and have a suitable concentration of dissolved solids to bind the remaining water sufficiently to inhibit the growth of bacteria, moulds and yeasts. IM meat products have a water activity (a_w) between 0.6 and 0.85. These products are shelf stable at ambient temperature for varying period of times and can be eaten without re-hydration.

Glycerol in view of its low flavour impact is undoubtedly the most widely advocated humectant for intermediate moisture meats. Taking into consideration the different hurdles in preservation of meat, a mixture of sodium chloride, glycine and lactic acid in combination with glycerol is a very effective humectant system (Jay, 1996). Suitable antioxidants like BHA or BHT are included to minimize the problem of auto oxidation. In some formulations, potassium sorbate is added as an antifungal agent. The IM meats developed using these systems need to be explored for being most suitable under Indian conditions where there is a lack of refrigeration facilities due to its huge cost.

2.7. Use of Enzymes

Proteolytic enzymes of plant origin such as papain, ficin and bromelin have been either injected into the animal immediately before slaughter or applied to the meat surface before cooking. Such enzymes have the greatest activity at higher temperatures and tenderization takes place during the early stages of cooking and the enzymes are subsequently inactivated. The amount of the enzymes used for these purposes varies between 300 and 500 ml as 5–10% solution. The dose should be carefully controlled according to the breed, age, sex and weight of the animal. After injection, the animal must be slaughtered within 10 to 15 minutes, this being the optimum time range if maximum tenderness of the meat is to be secured (Lawrie, 1998; Gracey, 1999).

3. ECONOMICS OF VALUE ADDED MEAT PRODUCTS

A number of factors affect the economics of value added meat products. Cost of raw materials, processing cost, scale of operation, infrastructure, manpower, transportation, finance and working capital costs contribute to cost of production while marketing and shipment costs and sale price determine the returns. Quantity of by-products produced and their disposal costs/returns is an important consideration. Availability of adequate raw materials in the area of production and distribution and marketing at minimum costs contribute favourably to the economics. Sale price depends on the type of customers the product is aimed at. A project appraisal on different aspects of the value added products is necessary before attempting the venture.

WTO provides a wealth of choices to the consumers. Hence, value added products need to be produced to satisfy the consumer demand and prevent adverse impact of large scale imports. Small scale ventures of value added products would go a long way in improving the economics of livestock production and meeting the consumer demands. Dollar break-up of a processed product in International Market (Vijay Sardana, 2003) is reported in percentage as, Labour-38, Farm value-20, Packaging-8, Depreciation-5, Transportation-4, Profile-4, Advertising-4, Rent-4, Energy-3, Business taxes-3, Interest-2, Repairs-1 and Other costs-4.

4. CONSTRAINTS IN MEAT PRODUCTS MARKETING

4.1. Government Policies

The present animal-slaughter rules and regulations do not favour producing high quality meat animals or good quality meat to produce products of international standards and quality. A central excise duty of 8% on packaged processed meats adversely affects the profits and has not favoured the development of processed meat sector.

4.2. Religious Taboos

Sausage products are the largest component of processed meats world over and produced using lean from beef trimming and fat from pork trimmings as pork fat being unsaturated is superior for emulsion production. However, in the Indian situation, beef/buffalo meat is not acceptable for certain communities and pork for some other communities, affecting production of good quality sausage products in the processed meat sector.

4.3. Meat Quality

In general, fresh meats used for processing meat products are of poor quality. Good quality fresh meat and edible byproducts are not available due to various constraints associated with meat animal production, handling and transport, slaughter and poor and unscientific handling of fresh meats. Adverse effects of PSE and DFD meats is well known. In the Indian situation, due to stress in the handling of meat animals before slaughter, incidence of PSE meat is a significant factor to affect processed meat yields and quality. Proper rigor development to realize full physico-chemical and functional properties is a necessity. Meat from aged and poor conformation animals form a considerable proportion of meat animals in India and is rich in connective tissue which is detrimental to processed meat production. Practice of effective culling of animals before the animals reach the status of poor conformation would be advantageous not only to obtain better returns for the farmers but also to produce processed meats of acceptable quality.

4.4. Food Additives and Machinery

Non-availability of good quality food additives required for meat products such as phosphate blends, soya products, milk proteins, emulsifiers etc. at affordable cost is another hindrance for producing high quality products. Further, non-availability of processing machinery adds to the misery of the processors. As the demand for meat processing equipment is not large due to poor development of processed meat sector, manufacture of processing equipment has not been taken up in the country unlike dairy sector. Therefore, import of machinery at higher cost and difficulty in maintenance and repair and servicing of these machines has deterred small scale entrepreneurs entering into processed meat sector. Packaging materials for packaging specialized and variety meat products are not easily available for small scale ventures at reasonable cost.

4.5. Cold Chains

Inadequate cold-chains for distribution and marketing of processed meat products which are highly perishable in nature is another constraint faced by the small scale processors. The acute shortage of power/electricity, particularly during summer, is posing serious economical problems to maintain the existing cold-chains and thus hamper the marketing of meat products. Refrigerated transport system has yet to be developed for economic transport of perishable meat products. Malpractice by some traders to put off their cold chains during night hours to save electricity without considering the adverse effect of temperature abuse on the quality and shelf life of perishable products is a serious problem which can be solved through proper guidance and counselling by the experts.

4.6. Consumer Attitude

Consumer attitude is not favourable towards marketing frozen meat and meat products as they have concern about the quality and temperature abuse in marketing channels. Hence, most of the common people prefer fresh products. Ethnic concerns and apprehensions of consumers about the possible mixing/ adulteration of processed meats with meats of not their choice should be addressed properly. Negative impact of unwarranted publicity about the unhygienic meat production in India by some vested interest groups is hampering the development of meat and meat products sector. Therefore, it is the responsibility of the processors and traders to overcome some of the above bottlenecks by building the confidence among the consumers about the wholesomeness and quality of "branded" meat products.

5. MODERN TRENDS IN MEAT PROCESSING

5.1. Electrical Stimulation

Among the techniques used, electrical stimulation plays an important role in tenderizing meat from older age animals (Gracey *et al.,* 1999). Electrical stimulation of carcass following the slaughter is a relatively new commercial method for meat tenderization being regularly applied in beef and sheep meat industry from mid seventies, although the method was developed and patented long back in 1951. It is the post slaughter application of an electrical current to the animal carcass. Electrical stimulation is used during processing of beef and lamb, but not usually in pork where there is a significant risk of developing pale, soft, exudative (PSE) condition. Electrical stimulation has also been applied to poultry destined for further processing. Electrical stimulation involves application of sufficient voltage to cause rapid glycolysis in muscle which produces both physical and biochemical effects in the muscle tissue. During electrical stimulation, pH fall by 0.7 units can occur in two minutes. Impulses of about 50 with 100 voltage electricity, given within 30 minutes of slaughter increased meat tenderness, organoleptic qualities, shelf life and altered muscle structure (Lawrie, 1998). Electrical stimulation is a new concept in meat processing and it greatly reduces the risk of cold shortening in muscles by hastening the development of rigor mortis. The process of electrical stimulation has been adopted by the meat industry as means of countering cold shortening in beef and lamb.

5.2. Use of Calcium Chloride

Calpain proteolytic system in post mortem tenderization revealed that elevated calcium concentration of muscle by infusing / injection of calcium chloride would result in accelerating the process of ageing in beef and lamb. Injection of 100 mM calcium chloride solution @10% (w/w) in sheep meat and 250 mM calcium chloride solution @10% (w/w) or marination in 200 mM calcium chloride solution for 24 h in buffalo meat could be successfully used to improve tenderness and overall palatability of meat without adversely affecting other characteristics (Mendiratta, 2003). This enhanced tenderness has been attributed to stimulation of calpain during subsequent postmortem conditioning (Whipple *et al.*, 1994). Besides tenderness, number of other advantages have also been reported by the use of calcium chloride. It can eliminate the requirement for post mortem storage beyond 24 h for ageing or conditioning. Due to the inbuilt mechanism, the meat is never over tenderized by even higher level of $CaCl_2$. Use of $CaCl_2$ is approved by FDA as GRAS (Generally recognized as safe) compound up to a maximum level of injection of 3% (w/w) 0.8 M $CaCl_2$ solutions. Meat, being a low source of calcium, the process of tenderization may serve as a mode of calcium fortification of meat. Thus, the use of $CaCl_2$ in meat industry seems to be very promising.

5.3. Development of Functional Meat Products

The functional meat products may provide added health benefits following the addition/concentration of a beneficial ingredient, or the removal/substitution of an ineffective or harmful ingredient. The increase in food related diseases, such as cardiovascular disease, diabetes, cancer and obesity has led consumers to demand food products that not only offer nutritional value but also have functional and added health benefits (Solan, 2008). The alteration of nutritional composition of meat products is possible either by direct addition of bioactive food ingredients or by the inclusion of bioactive compounds into animal diets. The latter technique has the advantage that the bioactive compounds would be biologically introduced into the food and thus would not have to be declared as food additives (Decker & Park, 2010). In recent years, more attention has been paid to develop meat and meat products with physiological functions to promote health conditions and prevent the risk of diseases. Value improvement can be realized by adding functional compounds including conjugated linoleic acid, vitamin E, n-3 fatty acids and selenium in animal diets to improve carcass composition and fresh meat quality. In addition, functional ingredients such as vegetable proteins, dietary fibers, herbs and spices, and lactic acid bacteria can be directly incorporated into meat products during processing to improve their functional value for consumers (Zhang *et al.*, 2010; Muthukumar *et al.*, 2012). Functional compounds, especially peptides, can also be generated from meat and meat products during processing such as fermentation, curing and aging, and enzymatic hydrolysis.

5.4. Low Fat Meat Products

The demand for low fat meat products has greatly increased in recent years as high fat intake, especially saturated fats, is associated with increased risk of chronic diseases of the circulatory system and some types of cancers. More than twenty government and health organizations from many parts of the world have formulated recommendations asking for a reduction in total fat and saturated fat from the diet (USDA, 1991). The American Heart Association and several other professional health organizations throughout the world have recommended lowering the daily intake of dietary fat to an average of 30% of total calories. Consumers are advised to choose lean cuts of red meat, use low-temperature cooking methods, and to eat moderate amount of meat (approximately 170 g/day, cooked), and remove visible fat from the meat cuts (Keeton, 1994). As a result, reduction of fat in food products is a growing area of interest to food manufacturers especially in the meat sector. Consumption of meat products is, therefore, under pressure and in order to maintain their market share of convenience foods, the meat industry must change the image and quality of traditional meat products by developing low-fat versions.

Since fat acts as a reservoir of flavour compounds and contributes to the texture of the product, reducing the fat content in meat products may alter product quality. Long chain polyunsaturated n-3 fatty acids (PUFA) have been implicated as critical nutrients for human health. In order to improve healthy lipid profile, reformulation is generally based on the partial replacement of the animal fat normally present in the product with another fat whose characteristics are more in line with health recommendations and if possible cholesterol free. A variety of non-meat fats of plants and marine origin have been added to different meat products as partial substitutes for meat fats (Cengiz & Gokoglu, 2007).

5.5. Reduced Sodium Meat Products

Since diet is the main source of sodium, awareness among people regarding its possible role upon health has driven demand for various low sodium foods including meat products. The association between excessive sodium intake and the development of hypertension has prompted public health and regulatory authorities to recommend reducing dietary intake of sodium to 2.4 g (6 g salt) per day. However, this is not an easy task as common salt affects taste and flavor, functional attributes, stability, and food safety of meat products. The various properties such as taste and flavor, binding, as well as microbiological characteristics should be given due care while developing low salt meat products. The major barriers to salt replacement are cost, as salt is one of the cheapest food ingredients available (Verma and Banerjee, 2012). Different approaches have been proposed to reduce the salt content in processed foods (Desmond, 2006).

5.6. Substitution of Nitrite in Meat Products

The ability of nitrite to inhibit *C. botulinum* combined with the improvement in aesthetic qualities of colour, texture and flavour contributed to the widespread use of nitrite in meat products. However, there has been some concern over the use of nitrites and nitrate in foods due to the formation of carcinogenic nitroso compounds. Reduction of allowable sodium nitrite levels in finished products from 200 to 10 or 20 ppm has been urged. Thus, efforts have been made to find out substitutes for nitrite. Several hundred compounds have been examined, but none has been a good substitute. Combinations of various additives have been attempted in order to provide several functions of nitrite. Among the additives, ascorbate or erythorbate, alpha tocopherol, sorbate, propylene glycol and lauricidin, *Lactobacillus fermentum* strain, monascus extract, nitrite ligands, short-chain alkynoic and alkenoic acids and esters and protoporphyrin have been tried as nitrite substitutes.

6. FUTURE FOOD PROCESSING

6.1. Irradiation

There are three main types of ionizing radiation-Gamma rays from a radioactive source such as Cobalt-60 and Cesium-137, high energy electrons produced by an electron beam accelerator and X-rays. When beef is irradiated, it is conveyed through a treatment chamber and exposed to high energy electrons. In the case of gamma facility, the food is exposed to an ionizing energy source contained in a stainless steel tube. In the case of linear accelerator facility, the food is exposed to electrons or X-rays, which is delivered in rapid pulses. The irradiator emits short wavelengths of energy that pass through the food. Some electron accelerators can be designed to generate only electrons and thus can be used as an online process inside a food processing plant with just a few inches of shielding surrounding the source. High energy ionizing radiation produces positive and negative charges that disable bacterial DNA, thereby eliminating the threat to human health. The process of food irradiation is often called "cold sterilization" because it kills most bacteria without increase in temperature. This means that food can be irradiated within its packaging and remain protected against recontamination until opened by user.

6.2. UV Processing

UV light has been applied for many years in food processing but is still considered as an innovative technology. UV light (wavelength of 254 nm) is predominantly used as a disinfection method to inhibit or inactivate food borne microorganisms, mainly in liquid food products. Fresh produce can be processed using UV light, which has a germicidal effect on many types of microorganisms like bacteria, viruses, protozoa, molds, and yeasts. However, the effect of UV light on microorganisms in liquids depends on variables such as density of the liquid, types of microorganisms. Although the use of UV light is well established for air and water treatment and surface decontamination, its use for treating liquid foods is still limited. Recently, interest in using UV has increased as a viable alternative to thermal pasteurization of liquid foods, fresh juices, soft drinks, raw milk, liquid eggs etc. Fruit and vegetable liquid products are generally very suitable for processing by UV light to reduce the microbial load by exposing to a suitable dose of UV radiation. The germicidal properties of UV irradiation are mainly due to DNA damage induced through absorption of UV light.

6.3. Microwave Heating

Microwave heating is the interaction between an electromagnetic alternating field and the dipoles and ionic charges within a food that enables the heating of the product. MW systems operate by forcing polar molecules, such as water, and ionic species to constantly realign by reversing an electric field around the food. This molecular movement is extremely fast due to the high frequency range from 300 to 3000 MHz. MW are capable of offering rapid, non-contact volumetric heating. They can be distinguished by the way the electric field is generated and depth of electromagnetic waves. Wavelength of the two frequency values commonly used for MW are 915 MHz and 2450 MHz. Microwaves are primarily a radiation phenomenon that usually takes place in an enclosed cavity, or in close vicinity to a waveguide applicator, where the electrical component of the microwave field approaches the food product from all directions. MW processing in preservation in general and pasteurization or sterilization in particular is a more homogeneous heat treatment at a faster rate than conventional method of heating.

6.4. Ohmic Heating

Research on the potential of ohmic heating on food processing and energy savings started in 1980s. In this method, an AC current is passed through a food sample which leads to generation of internal energy in foods. Ohmic heating is somewhat similar to microwave heating but with very different frequencies.

Ohmic heating is based on the principle that most food products have the ability to resist the passage of an electrical current. Heating occurs when an alternating electrical current is passed through a food due to the electrical resistance. The advantage of this technique is that it uniformly heats food. A quality product with minimal structural, nutritional and organoleptic changes can be produced. Potential application of this technique includes blanching, evaporation, dehydration, fermentation and extraction. It saves significant time and energy in hot air and freeze drying of foods. The ohmic heating is useful for value added food processing, and in wide variety of food processing operations involving heat and mass transfer.

6.5. Infrared Heating

The application of infrared radiation to food processing has gained momentum due to its inherent advantages over the conventional heating systems. Infrared heating has been applied in drying, baking, roasting, blanching, pasteurization, and sterilization of food products. Infrared heating occupies an important place in drying technology. Application of IR drying in the food industry is expected to represent a new process for the production of high-quality dried foods at low cost. The advantages of IR technology includes reduction in drying time, increased energy efficiency, uniform temperature during drying, better-quality finished products etc.

6.6. Ultrasound Treatment

Ultrasound is defined as sound waves with frequencies above the threshold for human hearing (>16 kHz). Generally, ultrasound equipment uses frequencies from 20 kHz to 10 MHz. The bactericidal effect of ultrasound is attributed to micro-mechanical shocks that disrupt cellular structure and result in cell lysis. Critical processing factors are: nature of the ultrasonic waves, exposure time, the type of microorganisms, the volume of food, the composition of food, and the temperature. High-intensity ultrasound is used to generate emulsions, disrupt cells, promote chemical reactions, inhibit enzymes, tenderize meat etc. Ultrasound technology has a wide range of current and future applications in the food industry, including inactivation of micro-organisms and enzymes. Combinations of ultrasound with other preservation processes (*e.g.*, heat and mild pressure) appear to have a great potential in food processing.

6.7. Pulse Electrified Field (PEF)

PEF is considered as one of the promising non-thermal technologies alternative to the traditional food pasteurization. PEF is employed with short bursts of electricity for microbial inactivation and causes minimal effect on food quality. PEF can be used for processing liquid and semi-liquid food products. PEF technology is successfully demonstrated for the pasteurization of juices, milk, yogurt, soups, and liquid eggs. Application of PEF processing is restricted to food products with no air bubbles and with low electrical conductivity. PEF is a continuous processing method, which is not suitable for solid food. The technology is based on the phenomena that biological membranes are punctured when an external electrical impulse is applied. This process is referred to as non-thermal as structural damage to membranes is realized at significantly low energy levels when compared to the process of heating. It is assumed that PEF has preservative effect by two fold-physical inactivation of bacteria and High Electrical field Pulses-improves the drying and re-hydration properties of dried vegetables to extend the shelf life. PEF technology seems to be better compared to thermal processing to avoid the detrimental effects on flavor, freshness and nutritional and functional components.

6.8. Membrane Technology

Membrane processing is a technique that permits concentration and separation without increase in temperature. Particles are separated on the basis of their molecular size and shape with the use of pressure and specially designed semi-permeable membranes. When a solution and water are separated by a semi-permeable membrane, water will move into the solution to equilibrate the system. If a mechanical force is applied to exceed the osmotic pressure (up to 700 psi), the water is forced to move down the concentration gradient *i.e.* from low to high concentration. The different membrane processes include reverse osmosis (RO), nanofiltration (NF), ultrafiltration (UF), microfiltration (MF) and electrodialysis (ED). Various improvements have been achieved by utilizing membrane technology in the dairy industry, *e.g.*, for bacteria removal (MF), whey treatment (UF, NF), demineralization (NF, ED), fractionation of milk and whey (UF, NF) and water recycling (RO). Among the different membrane systems, reverse osmosis (RO) and ultra-filtration (UF) are being used extensively in food and dairy industries. RO is a single phase concentration process which uses a pressure gradient across a semi permeable membrane to squeeze water through membrane. RO process is extremely energy efficient compared to both evaporation and freeze concentration.

6.9. High Pressure Technology

The application of pressure up to 150 Mpa to muscle and related systems has revealed responses of considerable magnitude and potential usefulness. The post-mortem glycolysis can be controlled to the extent that it can be greatly accelerated so that it is virtually complete within a few minutes. Meat of any desired pH value in the usual post mortem range can be produced. The pressure treatment also causes changes in other structures of muscle.

A procedure has also been developed for pressure tenderization of post-rigor muscle. It provides the only means by which toughness due to muscle contraction can be eliminated in intact post-rigor muscle but it is necessary to heat the meat to about 60°C for 30 minutes during treatment. This combined pressure and heat treatments have been shown to result in a substantial decrease in shear force value even in cold shortened meat (Bouton *et al.*, 1977).

The treated meat has a cooked appearance. Pressure also influences the action of salt in the swelling of muscle tissues and the dissolution of muscle proteins.

REFERENCES

Arihara, K. and Ohata, M. (2008). Meat biotechnology (*Ed.*) Toldra, F. Springer Science and Business Media, New York. pp. 231–249.

Cambell-platt, G. (1995). *In*: Fermented Meats (*Eds.*) Cambell-platt, G. and Cook, P.E. Blakie Academic & Professional, London.

Cengiz, E. and Gokoglu, N. (2007). Effects of fat reduction and fat replacer addition on some quality characteristics of frankfurter-type sausages. *Int. J. Food Sci. Tech.*, 42: 366–372.

Decker, E. and Park, Y. (2010). Healthier meat products as functional foods. *Meat Sci.*, 86: 49–55.

Dentali, S. (2002). Regulation of functional foods and dietary supplements. *Food Technology*, 56(6): 89–94.

Desmond, E. (2006). Reducing salt: A challenge for the meat industry. *Meat Sci.*, 74: 188–196.

Fernandez-Gines, J.M., Fernandez-Lopez, J., Sayas-Barbera, E. and Perez-Alvarez, J.A. (2005). Meat products as functional foods: A review. *J. Food Sci.*, 70: R37–R43.

Gracey, J.F., Collins, D.S. and Huey, R.J. (1999). Meat Hygiene 10th Edn., WBSaunders Co. Ltd., London.

Hedrick, H.B., Abrele, E.D., Forrest, J.C., Judge, M.D. and Merkel, R.A. (1994). Principles of Meat Science, 3rd edn, Kendall Hunt Publishing Co., New York. pp. 50–51.

Lawrie, R.A. and Ledward, D.A. (2006). Lawries Meat Science,7th edn, Woodhead and CRC press, Cambridge, ISBN: 13: 978-1-84569-159-2.

Muthukumar, M., Naveena, B.M., Sen, A.R. and Vaithiyanathan Sureshkumar, K. (2012). Effect of incorporation of *Moringa oleifera* leaves extract on quality of ground pork patties. *J. Food Sci. Technol.*, DOI 10.1007/s13197-012-0831-8.

Nollet, L.M.L. and Toldra, F. (2006). Advanced technologies for meat processing, 1st edn, Taylor and Francis, London, ISBN: 13: 978-1-57444-587-9.

Pandey, N.K. and Yadav, A.S. (2003). *Indian J. Poult. Sci.*, 38: 22.

Pearson, A.M. and Gillett, T.A. (1997). Processed Meats. CBS Pub., New Delhi.

Solan, A.E. (2008). The top ten functional food trends. *Food Technol.*, 62: 25–44.

USDA (1991). Labelling of low fat and reduced fat/salt products. *In*: Production and processing of healthy meat, poultry and fish products. Advances in meat research, Volume 11. Pearson, A.M. and Dutson, T.R. (*Eds.*), Blakie Academic and Professional London, pp. 48–52.

Verma, A.K. and Banerjee, R. (2012). Low-sodium meat products: Retaining salty taste for sweet health. *Crit Rev. Food Sci. Nutr.*, 52: 72–84.

Whippe, G., Koohmarie, M. and Arbone, J.R. (1994). *Meat Science*, 38: 133.

Wilkinson, B.H.P., Janz, J.A.M., Morel, P.C.H., Purchas, R.W. and Hendrik, W.H. (2006). The effect of modified atmosphere packaging with carbon monoxide on the storage quality of master-packaged fresh pork. *Meat Sci.*, 73: 605–610.

Zhang, W., Xiao, S., Samaraweera, H., Lee, E.J. and Ahn, D.U. (2010). Improving functional value of meat products. *Meat Sci.*, 86: 15–31.

3

Packaging Materials for Animal Products

ASHIM K. BISWAS AND PRABHAT K. MANDAL

ABSTRACT

Packaging materials and their properties play a very vital role in the delivery and protection of foods of animal origin. Commonly used food packaging materials are polymers viz., low density polyethylene (LDPE), high density polyethylene (HDPE), polypropylene (PP), and polyamide (PA), Polyesters (PET), polyvinyl chloride (PVC), polyvinylidene chloride (PVDC), polystyrene (PS), and ethylene vinyl acetate copolymer (EVAC). In this chapter the brief chemistry, characteristics, properties, uses, advantages and disadvantages of each type of packaging materials including modern packaging materials has been discussed.

1. INTRODUCTION

The important properties of plastic resins for meat packaging are presented in Table 1. The most commonly used polymers for food packaging are low density polyethylene (LDPE), high density polyethylene (HDPE), polypropylene (PP), and polyamide (PA). Polyesters (PET), polyvinyl chloride (PVC), polyvinylidene chloride (PVDC), polystyrene (PS), and ethylene vinyl acetate copolymer (EVAC) are also used with food. Each type of packaging material has advantages, disadvantages, consumer and marketing issues, environmental considerations and cost (Marsh & Bugusu, 2007). Plastic containers were introduced around 1950 extensively for packaging of food products. The first commercial film was introduced in 1924 as cellophane. In 1936, a copolymer of vinyl chloride and vinylidene chloride was prepared, also known as saran. Most polymers or plastics used for food packaging are thermoplastics *i.e.*, they can be softened by heating and hardened on cooling repeatedly provided that no chemical decompositions occurs. Thermosets (*i.e.*, bakelite-handle of cooker, frying pan etc.) cannot be softened or hardened repeatedly. They undergo an irreversible change on heating and do not soften. They get charred at higher temperature since their molecular structure is a complex three dimensional bonded network. Main uses of thermosets are bottle caps and in most cases contact with food does not occur due to use of liners as barriers.

Table 1: Physical properties of plastic films.

	Molecular wt	***Crystallinity (%)***	***Density (g/cm^2)***	***Softening point (°C)***	***Melting point (°C)***
LDPE	$1–3 \times 10^4$	55–70	0.915–0.935	86	112
HDPE	10^5	80–95	0.945	120–130	137
PP	10^5	65–70	0.90	150	168
PVC	10^4	<5	1.4	70–80	220
PVDC	–	–	1.6–1.7	100–150	–
PTFE	10^6	93–98	2.1–2.2	–	342
PS	$10^4–10^6$	–	1.04–1.05	78–103	240
Polyamide (nylon)	–	–	1.04	175	185
Polycarbonate	10^5	–	1.02	135–165	220–230

In food packaging application, thermoplastics are preferred since basic polymer has to undergo several heating and cooling cycles during its conversion into containers or films. Most thermoplastics are chemically regarded as derivatives of ethylene ($CH_2 = CH_2$). They are also referred as vinyl plastics since they contain vinyl groupings (CH_2 = CH–R) or polyolefin since all monomers contain an olefin linkage before polymerization.

2. EVOLUTION OF PACKAGING MATERIALS

The evolution of food packaging materials start from the period of Palaeolithic times-man used natural containers such as hollow tree trunks, gourds, hollow rocks, shells, leaves, and pieces of bark. Mesolithic man stored food surpluses in this type of basket. Neolithic man fashioned metal containers and discovered pottery. Glass container was known in Sumerian times and small glass jar or bottle was known in 1500 BC. These glass containers are believed to be offshoot of pottery. Wooden barrel was invented by Alpine tribes, while paper and cardboard were Chinese invention. First paper was made from mulberry bark about 200 BC. Modern food packaging is believed to have in the 19th century with the invention of canning by Nicholas Appert. Nicholas Appert though earned nobel prize in the year 1810 for his invention of canning process during 1809, large scale of canning process adopted in food industry only after establishment of the fundamental principles of bacteriology by Louis Pasteur followed by Samuel *C. Prescott* and *William L.* Underwood (Brody, 2010).

In the beginning of 20th century, 3 pieces tin-plated steel cans were used for food distribution. Aluminum had its start in 1825. In 1886, Heroult and Hall discovered the modern electrolytic process for extracting aluminum from alumina. Most tin alloys were made from tin and either lead, antimony, zinc, or copper. Pure tin is difficult to roll. Many packaging innovations occurred during the period between World War I and World War II, these include aluminum foil, electrically powered packaging machineries, plastics such as polyethylene and polyvinylidene chloride, aseptic packaging, flexographic printing, and flexible packaging. The development of polypropylene, polyester and ethylene vinyl alcohol polymers led the incredible move away from metal, glass and paperboard packaging to plastic and flexible packaging.

In the later of 20th century, innovations include active packaging and intelligent or smart packaging. Retort pouch packaging, stand-up flexible pouches, self-sealing flexible pouches, coextrusion films and bottles emerged as rigid and semi-rigid packaging. Multilayer barrier plastic cans, microwave susceptors, dispensing colours, gas barrier bags for prime cuts of meat, modified atmosphere packaging, rotogravure printed full-panel shrink film labels, and dual ovenable trays are examples of innovations for the convenience attributes that have propelled food packaging in 21st century. More recently, innovations are related to nanotechnology whose future may lie in improving barrier and structural/mechanical properties of packaging materials and development of sensing technology (Brody, 2010).

3. ETHYLENIC OR POLYOLEFIN THERMOPLASTICS

Monomer-single unit of a substance. Polymer-repeating unit of monomer results in to high molecular unit. Polymer is of mainly three types—(*i*) Homopolymer-polymer of only one substance; (*ii*) Copolymer-blend of two or more types of polymers; and (*iii*) Condensation polymer-chemical reaction in which two or more molecules combine with the separation of water.

3.1 Polyethylene

Polyethylene is polymer (by addition polymerization) of ethylene gas obtained from a by-product from petrochemical industries. Polyethylene is the largest volume film used in flexible packaging industry. They are manufactured by addition polymerization.

3.1.1. *Low density polyethylene (LDPE)*

LDPE is produced by polymerization of ethylene gas at high pressure (1000–3000 atmospheric) and at high temperature (150–200°C). This process results in a macromolecule with high degree of branching *i.e.*, a mixture of single long chain and chains with branches linked at right angles to the main chain. But when low temperature process is used, then formation of long parallel linear chains occurs. The molecular formula of polymer is $(CH_2)n$.

The carbon atoms are linked by strong covalent bonds at a distance of 1.54 A (15.4 nm) at an angle of (α) 109°.

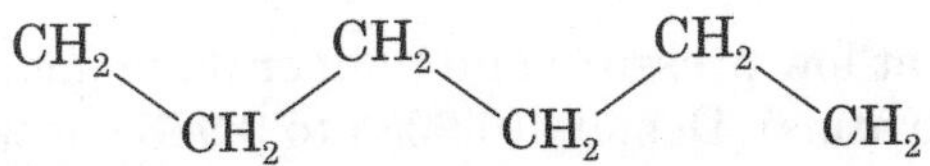

Fig. 1: Linear chain polyethylene macromolecule.

In Fig. 2 neighbouring chains are stabilized by a much weaker Vander Walls forces which accounts for much of the plasticity observed in LDPE.

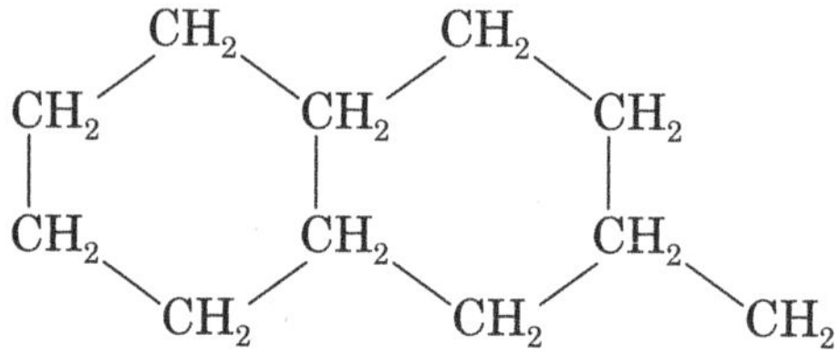

CH_2 CH_2
CH_2 CH_2 CH_2
CH_2 CH_2 CH_2
CH_2 CH_2 CH_2

Fig. 2: Secondary forces between linear chains.

Figure 3 shows the branch chain of LDPE. The branch chains prevent close packing of polymer chain and hence the finished polymer is of low density. The branching of chain also lowers crystallinity (regularly order regions) and softening point.

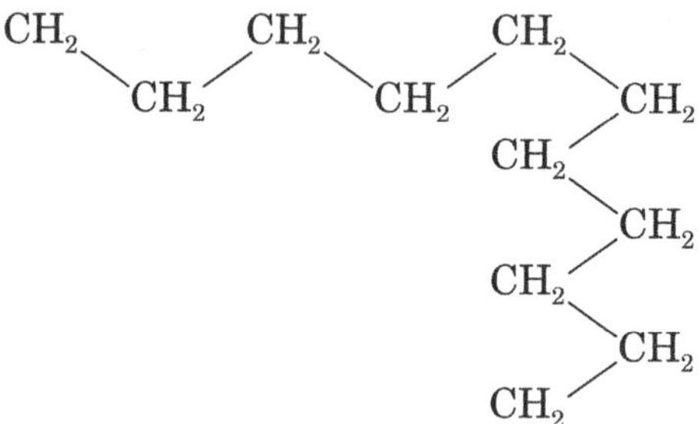

Fig. 3: Branch chain polymer of LDPE.

Properties of LDPE

(*i*) Tough, transparent and flexible.

(*ii*) Excellent resistance of most chemicals below 60°C but above this temperature the polymer becomes soluble in hydrocarbon solvents.

(*iii*) Good barrier to water vapour but less for gas such as O_2.

(*iv*) Lower melting (112°C) or softening point hence not suitable for packaging of food products which are to be heat sterilized.

(*v*) Because of high flexibility and heat sealability properties, it is easily converted into film for packaging of fresh and frozen foods.

(*vi*) Suitable for coating onto other materials such as paper and Al foil.

(*vii*) Moderate tensile strength.

3.1.2. *High density polyethylene*

HDPE is produced at low pressure and temperature (*i.e.*, 50–75°C & 10 atms or 60–160°C and 40 atms). Density (0.9384 to 0.9634 at 27°C).

Properties of HDPE

(*i*) Stiffer, harder, less transparent.

(*ii*) High melting (127°C) and softening point (increased stability to heat, hence steam sterilized).

(*iii*) Better resistance to oils and gases.

(*iv*) Lower permeability to water vapour and gases than LDPE.

Uses: Films, bottles, tubs, crates, trays etc.

3.2. Polypropylene (PP)

This is the polymer of propylene and the structural formula is as follows:

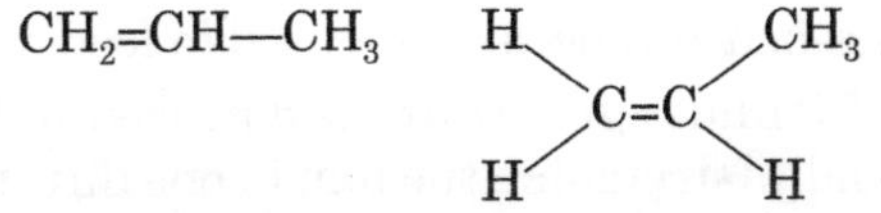

$CH_2{=}CH{-}CH_3$ H C=C CH$_3$ H H

Polypropylene molecule (monomer)

One of the hydrogen atom of ethylene molecule has been replaced by CH_3 (methyl) group. Polypropylene molecule exists in two forms *i.e.*, atactic and isotactic forms.

(*i*) *Atactic form:* (irregular arrangement of methyl group).

CH_2 CH CH_3 CH_2 CH CH_3 CH_2 CH_3 CH CH_3 CH_2 CH_3 CH CH_3 CH_2 CH_3

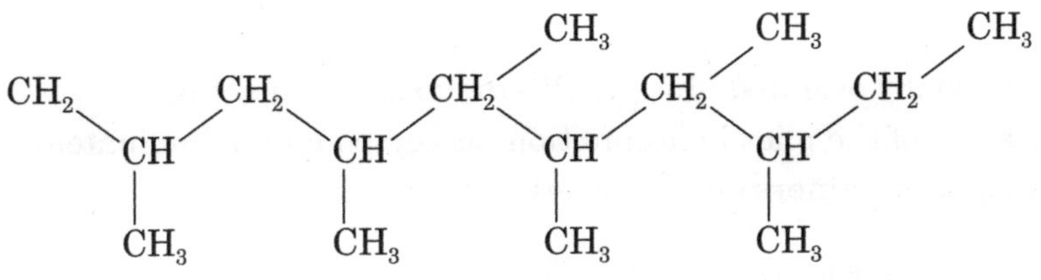

Atactic form of polypropylene

The methyl group is randomly distributed on either side of the main chain (*i.e.*, exists as irregular macromolecule of polypropylene).

(*ii*) *Isotactic form*: (regular arrangement of methyl group)

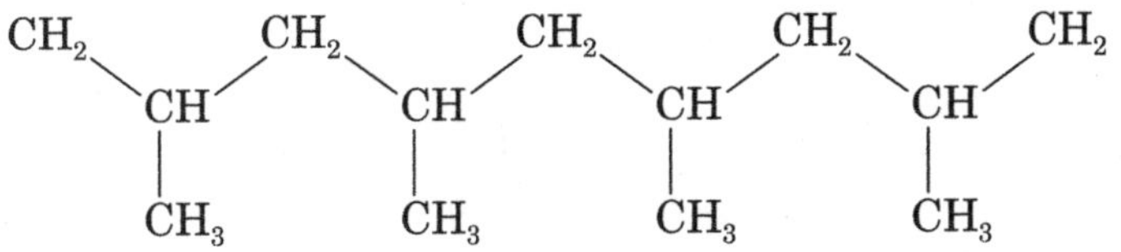

CH_2 CH CH_3 CH_2 CH CH_3 CH_2 CH CH_3 CH_2 CH CH_3 CH_2

Isotactic form of polypropylene

Isotactic form of polypropylene molecule is produced by using stereospecific catalysts and low pressure which results in regular arrangement of the methyl groups onto the same side of the polymer chain.

The isotactic form of polypropylene molecule has high crystallinity due to regular arrangement of methyl groups but not as high as HDPE. As a result this polymer (isotactic) is harder, and has high softening point than HDPE but low shock resistance, especially at low temperature.

Properties of PP

(*i*) PP is more rigid, stronger and lighter than PE.
(*ii*) The film has low WVTR, good grease resistance.
(*iii*) The film has high temperature resistance (melting point 168°C) and good gloss (hence good printability).

Types of PP

(*a*) Unoriented PP film—It has low temperature stability and has poor gas barrier properties hence generally not used in food packaging application.
(*b*) Oriented film—PP film is generally used in oriented form. By stretching the film monoaxially (stretching the film in one direction only) or biaxially (film is stretched in two directions at right angles under suitable temperature condition). The resulting film has
   (*i*) Good barrier properties to water vapour and gases.
   (*ii*) High tensile strength.
   (*iii*) High abrasion resistance.
   (*iv*) Good gloss and clarity (transparency).

Uses

(*i*) As biaxially oriented film (BOPP) in flexible packaging.
(*ii*) As a one of the plies in lamination (as component of laminates).
(*iii*) As rigid containers (Jars, bottles, trays etc.).

3.3. Polyvinyl Chloride (PVC)

It is the polymer of vinyl chloride. Polymerization is done in the presence of catalysts. By addition of plasticizers, flexible film is obtained. Structural formula is as follows:

$$CH_2{=}CH{-}CHCl \qquad \begin{matrix} H & & H \\ & C{=}C & \\ H & & Cl \end{matrix}$$

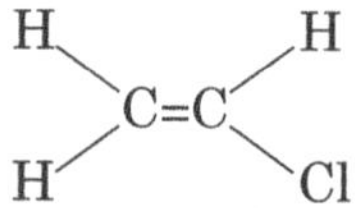

Vinyl chloride (monomer)

Here one of the hydrogen atoms in the ethylene molecule is substituted by halogen.

Properties of PVC

(*i*) The film is hard, stiff, clear and glossy surface.
(*ii*) Excellent moisture barrier.

(*iii*) Low gas permeability.
(*iv*) Good grease resistance.
(*v*) Low tensile strength.

Uses

(*i*) Rigid containers such as Jars bottles and trays etc.
(*ii*) Flexible film-in binding oriented form for shrink packaging of meat cheese.
(*iii*) Film used as copolymer (*i.e.*, component of laminates).

3.4. Polyvinylidene Chloride (PVDC)

Saran or cryo-vac is the copolymer of PVDC *i.e.*, PVDC /PVC. PVDC is the polymer of vinylidene chloride. The structural formula of vinylidene chloride is

$$CH_2{=}CH{-}CCl_2 \qquad \begin{matrix} H & & & & Cl \\ & \diagdown & & \diagup & \\ & & C{=}C & & \\ & \diagup & & \diagdown & \\ H & & & & Cl \end{matrix}$$

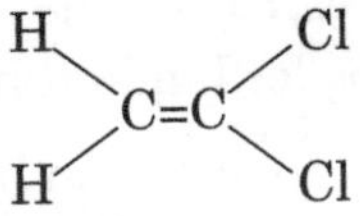

Vinylidene chloride (monomer)

Here two atoms of hydrogen in the ethylene molecule have been substituted by chlorine atoms-hence polymer has high degree of crystallinity. It is generally used as copolymer with PVC (13–20%). PVDC is also known as Saran film.

Properties of PVDC

(*i*) The film is clear and has good mechanical strength.
(*ii*) Low water vapour transmission rate, gas permeability.
(*iii*) Good grease resistance.
(*iv*) Insoluble in many organic solvents.

Uses

(*i*) Used as film (saran film) for packaging of meat.
(*ii*) It is a component of laminates.
Food packaging application: meat, sausage, cheese, dried fruits etc.

3.5. Polystyrene (PS)

It is the form of styrene. Styrene consists of an ethylene molecule in which one of the hydrogen atoms has been substituted by a phenol radical.

$$CH_2{=}CH{-}CCl_2 \qquad \begin{matrix} H & & & & Cl \\ & \diagdown & & \diagup & \\ & & C{=}C & & \\ & \diagup & & \diagdown & \\ H & & & & \text{phenol} \end{matrix}$$

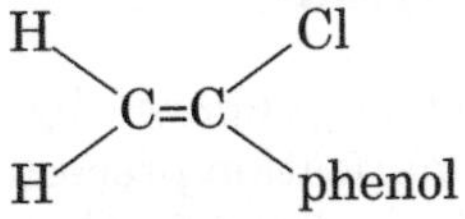

Styrene molecule (monomer)

Because of bulky nature of phenol radical, polymerization of styrene results in amorphous molecular chain. The phenyl radical during polymerization prevent close packing of polystyrene (PS) molecule.

Properties of PS

(*i*) PS polymer is transparent.
(*ii*) Possess good barrier to gases.
(*iii*) Poor barrier to water-vapour.
(*iv*) Soluble in alcohol, ketones, esters and aromatic and chlorinated hydrocarbons.

Disadvantages: It is brittle, hence low tensile strength and hence often copolymerized with their butadiene and/or acrylonitrile.

Uses

(*i*) As a moulded tubes, cups for dairy products (yoghurt, ice cream and cheese).
(*ii*) Polystyrene slabs for laminating card board boxes.

Polystyrene film is brittle and hence it is biaxially oriented to reduce the brittleness.

3.6. Polytetrafluoroethylene (PTFE)

It is polymer of tetrafluoroethylene. In this case, the ethylene molecule has been fully substituted with fluorine atoms.

$$F_2C=CF_2$$

Tetrafluoroethylene molecule (monomer)

The PTFE is highly crystalline and has very high molecular wt.

Properties of PTFE

(*i*) Extremely inert to chemicals.
(*ii*) Very low coefficient of friction producing non-stick surface and can withstand wide range of temperature.

Uses: Its principal use is as coating on frying pans and other cooking ware.

4. NON-ETHYLENIC THERMOPLASTICS

4.1. Polyamides (Nylons)

Polyamides are polymer prepared by the condensation polymerization (and not by addition polymerization process as mentioned previously) of diacid and diamine. Condensation polymerization occurs between two different monomers having different functional groups leading to foundation of polymer with the release of water molecule.

The reaction is presented below

$$\underset{\text{Hexamethylene diamine}}{NH_2(CH_2)_6NH_2} + \underset{\text{Diacid (adipic acid)}}{COOH(CH_2)_4COOH}$$

$$\rightarrow \underset{\text{Polyhexamethyl adipamide}}{[-NH(CH_2)_6NH\ CO(CH_2)_4COOH-]_n} + H_2O$$

The first compound prcduced was nylon 6. Nylon 6 made from adipic acid and hexamethyl diamine. Some polymers in their series are obtained by self condensation of one molecule only. For example—nylon 6 prepared from ω-aminocapnoic acid.

Properties of nylons

(*i*) Very strong and tough (excellent strength).
(*ii*) High melting point.
(*iii*) Low friction and hence high abrasion resistance.
(*iv*) Low gas and water-vapour permeability.

Uses

(*i*) Boil-in-bag type application for convenience foods.
(*ii*) As a components of laminates.

4.2. Polycarbonates

They are linear polymer of carbonic acid

$$H—[ORO—\overset{\overset{\displaystyle O}{|}}{C}—]_n OROH$$

Polycarbonates

Properties of polycarbonates

(*i*) Tough, stiff, hard and transparent (*i.e.*, high impact strength).
(*ii*) Withstand wide range of temperature (up to 300°C).
(*iii*) Soluble in chlorinated hydrocarbon solvents.
(*iv*) Film do not have good barrier to water vapour/gases.
(*v*) Resistant to weak acid and alkalis.

Uses

(*i*) As a flexible film, but not generally used due to cost prohibition.
(*ii*) As a rigid/semi rigid containers (thin walled bottle, baby feeding bottle, oil and fruit juices etc.).
(*iii*) Heat stable film.

5. MISCELLANEOUS FILMS

5.1. Polyester (Mylar/Terelyn)

It is the condensation polymer of polyalcohol with a diacid or its anhydride.

$$\underset{\text{Glycol (Dihydric alcohol)}}{CH_2(OH)\text{–}CH_2(OH)} + \underset{\text{Teraphthalic acid}}{\text{COOH-benzene ring-COOH}} \rightarrow \frac{\text{Heat}}{\text{polyester}} + \frac{H_2O}{\text{Catalyst}}$$

*Metalised polyester is done by coating the plastic film with small quantities of metal (aluminum) by vapourisation. The resulting film is transparent and has good resistant power.

Properties of mylar films

(*i*) Excellent tensile strength.
(*ii*) Not heat sealable.
(*iii*) Wide temperature resistance.

Uses: As a component of laminates in retortable pouches.

5.2. Rubber

Rubber is a polymer of isoprene (C_5H_8).It has two varieties. Natural rubber made from coagulation of latex obtained from rubber tree, contains 95% hydrocarbons with some proteins, lipids and inorganic salts *e.g.*, rubber hydrochloride (pliofilm). Synthetic rubber made from by-products of petroleum industry. Ethylene, propylene, butadiene, styrene and acrylonitrile are used as monomers.

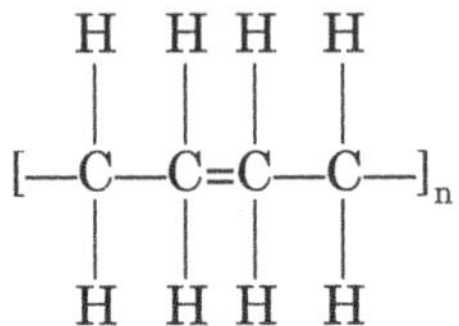

Isoprene unit (monomer)

5.3. Rubber Hydrochloride (Pliofilm)

It is made from rubbers by treatment of rubber by HCL.

Properties of pliofilms

(*i*) Stretchable.
(*ii*) Non-toxic.
(*iii*) Resistance to grease and oil.
(*iv*) Unaffected by weak acids and alkalis.
(*v*) Heat sealable but poor water vapour resistant.

Uses

(*i*) As a wrapper or bags for meat, cheese etc.

(*ii*) As a liner for thermoplastics.

5.4. Synthetic Rubbers

Synthetic rubber made from by-products of petroleum industry. Ethylene, propylene, butadiene, styrene and acrylonitrile are used as monomers. Synthetic rubbers are made not only to duplicate the properties of natural rubber but also to impart certain desired properties to give a range of materials with a wide variety of end users.

Types of synthetic rubber

(*i*) *Synthetic and butadiene rubber* (*SBR*): It has properties to natural rubber but has more abrasion resistance. Elasticity is slightly inferior to natural rubber but it possesses superior heat ageing characteristics.

(*ii*) *Butyl rubber*: They are resistance to vegetable oil/grease/solvent.

(*iii*) *Nitrile rubber*: It is copolymer of butadiene and acrylonitrile. Excellent resistant to solvents and oils.

(*iv*) *Silicon rubber*: Excellent resistant to oil and greases and also excellent temperature resistance (–75 to 260°C).

$$[-\underset{\underset{CH_3}{|}}{\overset{\overset{CH_3}{|}}{Si}}-O-]_n$$

Silicon Rubber

Uses: Rubber is not often used as food packaging material as such because, it may come in contact with food through its used as such as a seal, gasket or as one adhesive in food packaging machineries.

(*i*) Container lining.

(*ii*) Sealing rings for cans, glass bottles etc/pressure or retort cooker.

(*iii*) Stopper for bottle and flask.

(*iv*) Conveyor belts in food packaging/processing operation.

(*v*) Rubber aprons and gloves for working with foods.

5.5. Cellulose/Cellophane

Cellophane is derived from 'cello' (from cellulose) and 'phane' meaning transparent. It is manufactured from highly purified cellulose. The cellulose is treated with NaOH solution and carbon disulfide to produce xanthate which is dispersed in sodium hydroxide to produce viscose. The viscose is extruded through a slot into acid-salt bath to produce regenerated cellulose film. The cellulose content of regenerated cellulose film is usually 70% and 30% additives.

Types of cellophane

(i) *Plain and transparent cellophane (PT)*: The regenerated cellophane is washed, desulphated, bleached, soften[§], dried and wound as plain and transparent film. It is non-sealable and non-moisture proof hence use as wrapper for dry foods.

(*ii*) *MST cellophane (moisture proof, heat sealable and transparent)*: The cellophane film is coated on one or both sides with heat sealable, moisture resistant materials to make it moisture proof and capable of being sealed with heat. By application of various coating several grades of cellophane is available. Grades of MST/MSAT cellophane are 300, 400, 600 g/10 cm^2 of film.

Coated cellophanes are

(*i*) *Nitrocellulose coated cellophane*: Cellophane is coated with nitrocellulose. It has good water-vapour barrier property, flexible, heat sealable but not good barrier to oxygen.

(*ii*) *Saran coated cellophane*: It is coated with PVDC (saran). It has good water vapour barrier and gas property, flexible and heat sealable. Saran coated cellophanes are widely used for meat wrapping, as a component of laminate, and baked goods (cakes, bread, vegetables and confectionaries).

**Extrusion*: Method by which heated or unheated plastic is forced through a shaping orifice to become one continuously formed piece.

- Soften by use of glycerine, polyethylene glycol, urea etc.

5.6. Copolymer

Two or more repeating units of monomers produced either by addition polymerization or condensation polymerization. The properties of a copolymer often differ markedly from either of the homopolymers and depend on composition of the monomers use and amount of monomers along the chain.

5.6.1. *Ethylene-vinyl acetate (EVA) copolymers*

They are made from copolymerization of LDPE and vinyl acetate. Vinyl acetate may be up to 20% in the copolymer.

Properties

(*i*) More flexible than LDPE (more stretchable).

(*ii*) Not good barrier to water-vapour and gases.

(*iii*) Film is clear than LDPE.

(*iv*) Impact strength is excellent.

(*v*) Thermally instable at high temperature but very stable at lower temperature.

(*vi*) Can be heat sealed.

Disadvantages: EVA has tendency to block (due to polar nature of side chain in copolymer) that is high coefficient friction. Hence anti-blocking agents are added to control adhesion between touching layers of plastic during storage or use.

Uses

(*i*) As a coating to bags or industrial sacks.

(*ii*) Shrink film packaging.

(*iii*) Suitable for low temperature storage of food products.

5.6.2. *Vinyl chloride copolymers*

Vinyl chloride is copolymerized with

(*i*) Vinyl acetate (VA)—VC/VA for increase flexibility.

(*ii*) Vinylidene chloride (VDC)—VC/VDC for flexible packaging for better protection of product to water vapour and gases.

(*iii*) Polypropylene (PP)—VC/PP (up to 10%) for making bottles.

(*iv*) Acrylonitrile (AN).

5.6.3. *Polystyrene copolymers*

(*i*) *SAN (styrene and acrylonitrile)*: used for rigid containers (bottles, jars etc.).

$$\mathrm{H_2C{=}CH{-}CH{=}CH_2}$$

Butadiene monomer of ethylenic thermoplastics

(*ii*) *ABS (Acrylonitrile butadiene styrene)*: Contains three or more monomers. It is used for manufacture of tubes, trays, boxes etc and as a liner of refrigerator.

$$\mathrm{H_2C{=}CH(CN)}$$

Acrylonitrile monomer of ethylenic thermoplastics

These two copolymers posses much improved high impact resistance than polystyrene alone.

Table 2: Physico-chemical properties of some thermoplastics/paper/laminates.

Thermoplastics/ paper/laminates	***WVTR****	***Tensile strength (kg/15mm)***		***% Elon-gation***		***Bursting strength (kg/cm²)***	***Tear resista-nce (g)***	
	–	***MD***	***TD***	***MD***	***TD***	–	***MD***	***TD***
LDPE (100G)	10.7	3.7	2.0	300	175	0.98	27	32
LDPE (200G)	5.7	2.9	2.2	>250	>250	1.4	120	120
LDPE (300G)	4.1	1.8	0.90	>250	>250	2.2	>120	>120
HDPE (100G)	3.8	7.0	4.2	305	110	0.70	34	47
HDPE (200G)	1.8	5.2	4.7	270	240	1.54	38	95
HDPE (300G)	1.4	3.6	3.2	>250	190	1.97	52	52
PVC (120G)	27	2.7	2.6	5	3	2.7	4.4	6.4
Polystyrene (150G)	66	2.9	2.4	0.3	0.3	1.34	7.8	0.2
Polyester (100G)	13.1	8.3	6.7	58	81	4.64	20	30
PP(140G)	6.7	3.6	2.6	602	937	–	–	–
Kraft paper (124GSM)	4260	9.6	5.8	3.0	5.0	2.39	87	165
Kraft/LDPE (124 GSM/100G)	19.2	10.3	6.8	5.0	505	2.88	143	138
Glassine (40GSM)	1920	4.1	1.9	2.0	606	1.27	12	19
Glassine/LDPE (40 GSM/150G)	10.6	5.1	2.8	3.0	5.0	2.39	87	105
MST cello (100G)	27	6.7	2.9	16.5	58.5	–	–	–
PET/Al foil/PE (12μ/ 0.009″/50G)	Nil	11.6	11.1	56	60	–	–	–

*Unit: g/m²/24hrs at 38°C & 90% RH

5.6.4. *Polypropylene or ethylene copolymer*

Polypropylene is good barrier to water-vapour/gases and has low impact resistance at low temperature. Hence a copolymer made of PP/PE offers increased impact resistance to low temperature.

Other films

(1) *TPX*: Chemically closely related to ethylene. Manufactured from polymerization of 4-methylpentene-1. Transparent, low density compound to any plastic and has high melting point. Hence used for hot-filling and cooked food applications.

Disadvantages: It is highly permeable to water-vapour and gases.

(2) *Acrylic film (XT)*: excellent clarity, oil and grease resistance and gas permeability very low, good impact but poor resistance to aromatic and chlorinated hydrocarbons. They are used for manufactured of rigid containers.

(3) *Polyurethane (PU)*: They are tough, strong and abrasion resistant. Due to excellent oil and grease resistance-mainly used for motor oil packaging and boil-in-bag and flexible packaging of meat.

5.7. Laminates or Composite Film

Layers of individual materials bonded together by adhesive, coated or co-extrusion. No single polymer possesses all the desired properties which may be required for efficient packaging of particular foodstuff. In other words, mono-layer flexible packaging materials offer limited desired properties which may be achieved to a greater extent by the use of coated and laminated packaging materials consisting of two or more laminates depending upon the specific packaging needs to a products.

Types of laminates: All plastic laminates and laminate based on paper/Al foil/thermoplastics. Numerous possible combinations of individual layers of materials for production of laminate are possible which depends on the specific requirements of the food to be packaged.

Examples

Non-retortable pouch

(*i*) Glassine paper/LDPE.
(*ii*) Grease proof paper/LDPE.
(*iii*) Litho paper/LDPE.
(*iv*) Kraft paper/LDPE.
(*v*) LDPE/poster paper/LDPE.
(*vi*) Cellophane/LDPE.
(*vii*) PVDC/PVC copolymer.
(*viii*) Paper/Al foil/LDPE.
(*ix*) Polyester/Al foil/LDPE.

Retortable pouch

(*i*) Polyamide/Al foil/PP.
(*ii*) Polyester/Al foil/oriented nylon/PP.
(*iii*) PP/Al foil/PP.
(*iv*) Metalized polyester.

Tetra brick for milk, fruit juices and soft drinks

(*i*) Paper/Al foil/PE.

In India, the following plastic packaging materials are being manufactured-LDPE, linear LDPE, HDPE, High molecular HDPE, PP, PE, PS and PVC. Rigid plastic containers are made from HDPE, PP, PE and high impact PS.

Table 3: Oxygen transmission rate (cc/24 hrs/m^2 at 27°C at atmospheric pressure) of some thermoplastics/paper/laminates.

Thermoplastics/ paper/laminates	*WVTR**	*Resistance to acid*	*Resistance to alkalis*
LDPE (185G)	8181	Excellent	Excellent
LDPE (365G)	3380	Excellent	Excellent
HDPE (330G)	2228	Good	Good
PP(140G)	6682	Good	Good
PET/Al foil/PE (12µ/0.009″/50G)	Nil	–	–

Cryovac/saran (PVC)/PVDC—0.1–0.2 cc/100 sq inches/24 hrs at 72°C; Nylon (polyamide)—1.0 cc/100 sq inches/24 hrs at 72°C

5.8. PET Bottle/Jars/Containers

Polyethylene terephthalate containers are made by thermoforming or by co-extrusion process. This is food grade kitchen containers for storage of solid, semi-solid and liquids. PET containers are transparent, light weight (1/15th of the glass jars wt of same dimensions), non breakable, coloured (amber/green/ blue/clear transparent). PET bottle/jars are also made in different shape and sizes. PET containers have approval of FDA for food contact applications.

5.9. Boil-in Bag or Cook–in Bags

(*i*) The bags should withstand the cooking, frying, baking, roasting or oven simmering temperature.

(*ii*) Should reduce losses due to excessive dehydration during cooking.

(*iii*) Should require relatively shorter time to bring the food to desired temperature during cooking.

(*iv*) The package/bag should be strong to protect the food from the time of packaging, storage, handling and retail distribution.

Ideal packaging materials for boil-in bags are polyester (Mylar), polyamides (Nylon) and polyurethane film.

Grease resistance is expressed as transudation time in seconds.

Below 30 sec	:	D
Between 30 sec and 2 min	:	C
Between 2 min and 30 min	:	B
More than 30 min	:	A

Grease resistance of films/laminates

Cellophane/LDPE (150G)	:	A (More than 30 min)
HDPE (300G)	:	A

Table 4: Properties of major packaging materials used for meat and poultry.

Packaging material (0.025 mm thickness)	***Water vapour transmission rate, g/m2/24 h***	***O_2 transmission rate, cm3/m2/24 h***	***Tensile strength, MPa***	***Tear strength, g/mL***	***Impact strength, J/m***	***Haze, %***	***Light transmission, %***	***Heat seal temperature range, °C***
PVC	1.5–5	8–25	9–45	400–700	180–290	1–2	90	135–170
PVDC	0.5–1	2–4	55–110	10–19	–	1–5	90	120–150
PP	5–12	2000–4500	35.8	340	43	3	80	93–150
HDPE	7–10	1600–2000	38.2	200–350	373	3	–	135–155
LDPE	10–20	6500–8500	11.6	100–200	375	5–10	65	120–177
LLDPE	15.5–18.5	200	7–135	150–900	200	6–13	–	104–170
Ionomer	25–35	6000	24–35	20–40	150	–	–	107–150
EVAC	40–60	12500	14–21	40–200	45	2–10	55–75	66–177
EVAL	1000	0.5	8–12	400–600	–	1–2	90	177–205
Polyamide	300–400	50–75	81	15–30	50–60	1.5	88	120–177
PET	15–20	100–150	159	20–100	100	2	88	135–177
Polystyrene	70–150	4500–6000	45.1	2–15	59	1	92	121–177

PVC—Poly vinyl chloride; PVDC—Poly vinylidene chloride; PP—Polypropylene, LDPE—High density polyethylene; LDPE—Low density polyethylene, LLDPE—Linear low density poly ethylene; EVAC—Ethylene vinyl acetate; EVAL—Ethylene vinyl alcohol; PET—Poly ethylene terephthalate Cited from Scetar *et al.* (2010).

PVC (120G)	:	A
Polystyrene (100G)	:	B (Between 2 min and 30 min)
HDPE (100G)	:	B
Kraft paper/LDPE	:	B
Grease proof/LDPE	:	B
LDPE (100G)	:	C (Between 30 sec and 2 min)
Glassine	:	C

REFERENCES

Brody, A.L. (2010). Intelligent packaging: A powerful tool for food. *Food Technol.*, 5: 79–80.

Kerry, J.P., O'Grady, M.N. and Hogan, S.A. (2006). Past, current and potential utilisation of active and intelligent packaging systems for meat and muscle-based products: A review. *Meat Sci.*, 74: 113–130.

Marsh, K.S. and Bugusu, B. (2007). Food packaging- roles, materials and environmental issues. *J. Food Sci.*, 72(3): R39–R55.

Ozdemir, M. (2001). Food packaging lecture notes, Gebze Institute of Technology, Gebze-Kocaeli, Turkey.

Radhakrishnan, K.T., Venkataramanujam, V.V., Dushyanthan, K., Appa Rao, V. and Narendra Babu, R. (2002). Manual for short course on recent trends in packaging of meat and meat products. Department of Meat Science and Technology, TANUVAS, INDIA

Robertson, G.L. (1993). Food Packaging *Principles and Practice*'. Marcel Dekker, Inc., New York.

Sacharow, S. (1980). Principles of food packaging, AVI Pub. Co.

Scetar, M., Kurek, M. and Galic, K. (2010). Trends in meat and meat products packaging—A review. *Croat J. Food Sci. Technol.*, 2(1): 32–48.

4

Milk Pasteurization and Equipments

ANANT VISHWANATH DHOTRE

ABSTRACT

Pasteurization is a legally essential operation for commercial handling of milk in most countries. It ensures the safe status of milk for human consumption and prolongs keeping quality of milk moderately. The severity of any heat treatment depends upon temperature as well as time of exposure and therefore, various time temperature combinations have been proposed for pasteurization. Similarly the equipment of pasteurization can also be batch type or continuous type, later being more common and highly energy efficient. Good understanding of the operation and equipment of pasteurization can enable one for judicial use of utilities leading to economic operation. CIP of the pasteurization is very important aspect, improper cleaning is the root cause of many problems related to the unit. Hence it should be intelligently designed and religiously followed.

1. INTRODUCTION

1.1. History and Origin

The term 'pasteurization' has been named so after its inventor, Louis Pasteur—the well known French scientist. The pioneering investigations on such treatment were carried out in 1765 by Spallanzani (Silva & Gibbs, 2009). In the middle of nineteenth century some commercial wine makers encountered a problem of abnormal fermentation and subsequent spoilage of wines and beers. They approached Louis Pasteur to seek a solution. Pasteur studied the problem during 1864–65 and found that some microorganisms are the root cause of abnormal fermentation leading to early spoilage of wines and it can be prevented by heating the wine around 60°C, which kills these microorganisms. Later he reported that for spoilage of milk also the microorganisms are responsible. Milk being a complete food is a highly nutritious growth medium for micro organisms, which may include beneficial organisms (*e.g.*, lactic acid bacteria) as well as numerous spoilage organisms (*e.g.*, Gram-negative psychrotrophs) and bacterial pathogens (*e.g.*, *Salmonella, Listeria)*. Like wines milk too can be preserved longer by heat treatment. Following this, pasteurization of milk on commercial scale was started around 1890s, which spread all over the world.

Now, pasteurization stands as an essential milk processing operation worldwide for commercial milk handling.

1.2. Definition

By definition, 'milk pasteurization' refers to the process of heating every particle of milk to a specific temperature, holding it at the same temperature for specific duration (time) and followed by rapid chilling to less than 7°C. Accordingly, there can be numerous combinations of time and temperatures for heat treatment to get the desirable microbial or chemical effects like improvement in keeping quality, destruction of pathogens or destruction of all microorganisms. Higher temperature with longer holding time may give drastic reduction in the number of microorganisms but it also damages the nutritional constituents in milk, hence, it becomes necessary to select a combination of time-temperature, which would cause maximum microbial destruction with minimal nutritional loss. Thermization is such a treatment, in which milk is heated at 63°C for 15 seconds. It is milder than pasteurization and usually adopted to improve keeping quality rather than to eliminate pathogens. Whereas, sterilization involves heating milk above 100°C is much more severe than pasteurization and it kills all the living cells in milk. Pasteurization is considered as an optimized heat treatment as it kills all the pathogenic microorganisms and renders the milk safe for human consumption with minimum nutritional loss.

1.3. Public Health Aspects

It is mandatory by definition that the pasteurized milk, irrespective of the method of pasteurization employed, should be safe for human consumption. The criteria to test proper pasteurization include the negative alkaline phosphatase test that ensures the absence of all yet identified pathogens. Numerous incidences of illnesses due to milk borne pathogens have been reported worldwide; the intensity of such events decreased with the advance of time though. A brief summary of historical, current and uncommon/emerging public health concerns regarding milk and dairy products have been tabulated below (Table 1):

Table 1: Historical, current and uncommon/emerging public health concerns regarding milk and dairy products.

Historical	***Current***	***Uncommon/Emerging***
Diphtheria	*Bacillus cereus* poisoning	Aflatoxin
Poliomyelitis	Botulism	*Citrobacter freundii*

Table 1: *(Contd...)*

Table 1: *(Contd...)*

Historical	***Current***	***Uncommon/Emerging***
Scarlet fever Septic sore throat Tuberculosis	Brucellosis Campylobacteriosis Enteropathogenic	*Corynebacterium ulcerans,* Creutzfeldt-Jakob disease, Cryptosporidiosis Haverhill fever
Typhoid fever	*Escherichia coli* Enterohaemorrhagic	Histamine poisoning, Infectious hepatitis
	Escherichia coli	Johne's and Crohn's diseases
	Listeriosis Salmonelosis	Mycotoxins Q fever
	Staphylococcal poisoning	Shigellosis
	Yersiniosis	*Streptococcus zooepidemicus*
		Tick-borne encephalitis, Toxoplasmosis

Source: Encyclopedia of Dairy Sciences

2. METHODS OF PASTEURIZATION

There are four different methods of pasteurization as given below, the first and second being more common in literature than the third and fourth, which sometimes are not even referred as pasteurization.

2.1. Low Temperature Long Time (LTLT) Pasteurization

The LTLT pasteurization is also called as Batch Pasteurization or Vat Pasteurization as it is carried out in an agitated kettle in batch wise manner. It involves heating the milk up to temperature not less than 62.8°C and holding at that temperature for minimum 30 minutes. This was the first time-temperature treatment developed to render the milk free from active microbial pathogens especially *Mycobacterium tuberculosis,* the most heat-resistant pathogen of major public health concern in milk at that time. The LTLT pasteurization was widely used since inception of commercial milk pasteurization during 1940s. In 1940s due to the introduction of HTST pasteurization in the industry, use of LTLT pasteurization declined. Nevertheless, it is still used by small scale milk processors due to its low initial cost. Cheese manufacturers also use the LTLT pasteurization but they carry out the operation in Plate Heat Exchanger (PHE) rather than kettle to get the technological benefit of low temperature in cheese quality as well as benefit of regenerative heating in PHE.

2.2. Continuous (HTST) Pasteurization

The HTST pasteurization is also called as Continuous Pasteurization as it is carried out in continuous manner in plate heat exchanger with online detection and recirculation of insufficiently treated milk. It involves heating the milk up to temperature not less than 71.7°C and holding at that temperature for minimum 15 seconds. As mentioned earlier, it was commercially adopted in 1940s for the first time and since then it has been in vogue in dairy processing. The HTST process operationally comprises five stages *viz.*, regenerative heating, heating, holding, regenerative cooling and chilling in sequence. The HTST pasteurization is essentially accomplished in a specially designed PHE installed with additional components like booster pump, steam battery, holding section, flow diversion valve, back pressure valve, etc. The biggest advantage of HTST pasteurization is the regenerative heating/cooling of milk, which reduces the utility requirements by almost 80 to 90% as compared to batch pasteurization. Therefore, the operational cost of HTST pasteurization is very low, the initial cost is much higher though.

2.3. Flash Pasteurization

Flash pasteurization is the process of heating the milk up to 85 to 90°C and holding from 1 to 4 seconds (Kessler, 1981). It is severe treatment than HTST and commonly done by direct heating of the product through steam injection or infusion. After direct heating, the product needs to be throttled to remove the condensate in the form of flash. The name flash pasteurization might have come from this flashing or the second possibility is the holding time, which sometimes is just fraction of second referred as flash. The method of flash pasteurization is specially employed to the pasteurize cream during butter making (Eckles, 1993) and not commonly used for pasteurization of milk.

2.4. Ultra High Temperature (UHT) Pasteurization

The UHT is also elaborated as Ultra Heat Treatment and it refers to the heating of milk to not less than 135°C and holding at the same temperature for minimum 1.0 second followed by aseptic packaging. As the temperature of heating is higher than the normal boiling point of milk (100.16°C), a higher positive pressure of at least 3.2 bars is maintained in the milk line to attend the hundred plus temperature without evaporation (flashing). The method of heating can be direct or indirect or both. The most important advantage of UHT pasteurized milk is that it can be stored up to 3 to 6 months without refrigeration. Both LTLT and HTST pasteurizations destroy most of the vegetative bacterial cells but it can't destroy some thermophiles and spore forming organisms, and this necessitates prompt cooling and refrigerated storage of LTLT or HTST pasteurized product to arrest the growth of survivors. However, during UHT pasteurization, almost all microorganisms including spore formers are killed and therefore such treated product can be stored at room temperature without cooling unlike first three methods of pasteurization. Although the UHT pasteurization is severe, it is different from the milk sterilization, UHT milk shows negative turbidity test. It indicates less nutritional damage to UHT milk than sterilized milk.

In last few decades, the UHT processing has developed into a separate field of milk processing, its identity as method of pasteurization has seemed to have dropped. Due to its extra high (hundred plus) processing temperature, the designing and engineering aspects of UHT equipment differ remarkably from that of other pasteurizers. Similarly, the kinetics of the UHT process and technology of such processed milk also differ. For these reasons, UHT process and equipment has been kept out of the scope of this chapter, only is restricted to LTLT and HTST pasteurization only.

Pertinent literature reveals several other reported heat treatments such as Thermization, which refers to heating milk at 63°C for 15 seconds; High pasteurization, which refers to heating milk at 85°C for 20 seconds with objective to destroy Lactoperoxidase (EC:1.11.1.7); Sterilization, which is meant to kill all living cells including bacterial spores (Walstra, *et al.,* 1999) and Botulinum process involving heating temperature of 118–125°C (Holdsworth, 2009).

3. DESIGNING THE PASTEURIZATION PROCESS

The pasteurization process is identified by its specific temperature and time combination so designing of pasteurization process refers to the fixing of Time-Temperature combination of certain process. The basic objective while designing a pasteurization process can be described by the official definition of pasteurization given by International Dairy Federation (1980). It states that 'Pasteurization is a process applied to a product with the object of minimizing possible health hazards arising from pathogenic micro-organisms associated with milk by heat treatment which is consistent with minimal chemical, physical and organoleptic changes in the product'.

That is the temperature and time combination selected should be high enough to destroy all pathogenic microorganisms in milk. At the same time, it should be low enough to cause minimum possible damage to the heat sensitive milk constituents. Hence, the process designing aims to achieve an intelligent compromise to maximize microbial destruction and minimize the chemical changes in milk. It becomes necessary to identify the relationship between temperature and the rate of destruction.

When the decrease in the number of microorganisms or concentration of some constituent is in linear proportion with the time, the reaction is said to be of Zero Order. When the relationship is exponential *i.e.*, linear on semi-log plot, it is said to be of First Order and if the plot is hyperbolic, it is said to be of Second Order. Microbial destruction in milk under heating and most of other reactions involved in the food processing follow the First Order reaction (Toledo, 2000), the thermal destruction curve (plot of number of survivors vs heating time) is exponential and linear on semi-log plot as shown in Fig. 1. One has to have thorough understanding of following terms and its value before designing pasteurization (any thermal treatment) process.

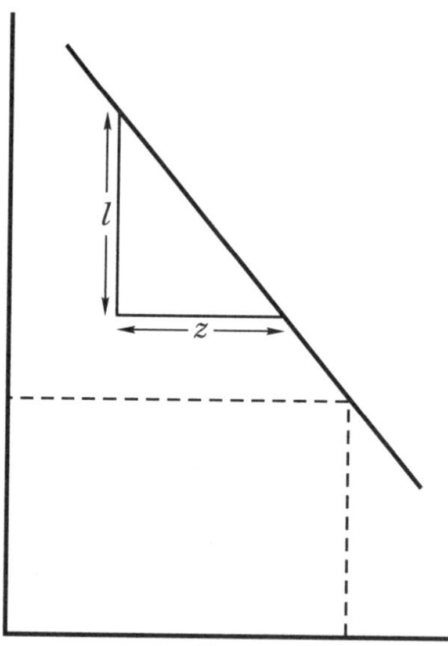

Fig. 1: Thermal Destruction Curve.

3.1. Thermal Death Time (TDT)

It is the time necessary to kill given number of organisms at a specified constant temperature. The TDT is one of the ways of indicating the thermal resistance of certain organisms. Alternatively, the Thermal Death Point, which is the temperature necessary to kill the given number of organisms in specific time, (often 10 min), is also used to indicate the same but it is not so common. Several methods have been reported to determine the TDT (Jay, 1996). The general method includes placing known numbers of specific organism or spore in test tube followed with instant sealing. These tubes are then put in water/ oil bath and heated at specific temperature for required time period followed by instant cooling. Such treated suspensions are then transferred to suitable growth medium and incubate at its' favourable temperature. Since dead organisms are unable to develop colonies, survivors counts is the number of colonies found.

3.2. D-Value

It is the time required to destroy 90% of the total number of microorganism initially present *i.e.*, to achieve one log cycle reduction in the microbial population. The D-value is also called as Decimal Reduction Time as after heating for the time equivalent to it, the number of microorganisms is reduced by one decimal. Mathematically, D-value (D) is the reciprocal of the survivor curve and numerically defined as,

$$\log\left(\frac{N_0}{N_t}\right) = -\frac{t}{D}$$

Where, N_0 is the number of microorganisms initially present (*cfu*) and N_t is the number of microorganisms survived after 't' minutes of heat treatment at some constant temperature (*cfu*). If survivors are exactly 10% of initial count, the right hand side assume unity and time 't' becomes the D-value. Negative sign indicates the decrease in N.

The D-value depends upon the type of micro organism, product in which the micro organism is present, pH, water activity and more prominently on temperature. Hence, while indicating the D-Value, the temperature is mentioned *e.g.*, $D_{82}^{\text{M.tuberculosis}}$=0.003 minutes, indicates that the D-value of mycobacterium tuberculosis at 82°C is 0.0003 minutes. When the temperature is specified as zero (D_0 value), it is 121.1°C or its equivalent in corresponding units. The D-values for some microorganisms of concern are given below (Table 2):

Table 2: D value of some microorganism

Sr. No.	*Organism*	*D-value (min)*	*Temperature*	*Reference*
1.	*S. aureus*	0.2–2.2	150°F	Jay, 1996
2.	*Coxiella burnetti*	0.5–0.6	150°F	Jay, 1996
3.	*Mycobacterium homnis*	0.2–0.3	150°F	Jay, 1996
4.	*Bacillus coagulens*	13.7	95°C	Jay, 1996

It is obvious that at higher temperature the D-value for given microorganism would be lower, provided all other factors are unchanged.

3.3. Z-Value

It is defined as the change in temperature required to change the microbial inactivation rate by tenfold. It can be simply defined as the change in temperature required bringing tenfold change in the D-value. The z-value is expressed in the unit of temperature and it describes the temperature dependence of D-value in quantitative terms. While D value reflects the resistance of a microorganism to specific temperature, the z-value provides information on the relative resistance of an organism to different destructive temperatures. Mathematically, z-value is

$$\ln\left[\frac{k_2}{k_1}\right] = \frac{T_2 - T_1}{z}\ln(10)$$

Where, k_1 and k_2 are the reaction rate constants at temperatures T_1 and T_2°C, respectively.

Alternatively, the equation relating D-values at any two temperatures (T_1 and T_2) and the z-value is

$$\frac{\log D_1 - \log D_2}{T_1 - T_2} = -\frac{1}{z}$$

The z-value facilitates to calculate the equivalent thermal processes at different temperature. If, for example, 15 sec at 72°C is an adequate process and z = 6°C, then 150 sec at 78°C and 1.5 sec at 66°C are equivalent process.

3.4. Q_{10} Value

It is defined as the change occurred in the rate of a reaction when its temperature is increased by 10°C and expressed as the ratio of the increased rate of reaction to the earlier rate of a reaction. In simple words it is defined as,

$$Q_{10} = \frac{k\,at(T+10)}{k\,at(T)}$$

Where, k is the reaction rate constant and T is the temperature.

That is, if the rate of any reaction doubles with a 10°C change in temperature, it has Q_{10} value of 2. Most of the enzymatic reactions like flavour change, browning, pigment formation, vitamin destruction, etc. exhibit Q_{10} value of 2. The chemical reaction rates are much less sensitive to changes in temperature than reaction rates for microbial destruction. The Z-values for spores always tend to be higher than those for respective vegetative cells (Lewis, 1994).

Effect of temperature on the reaction rate is given by the Arrhenius equation and that on the D-values is given by the Bigelow Model as follows;

Arrhenius equation

$$\frac{1}{(T+273.15)} - \frac{1}{(T_{ref}+273.15)}$$

Where, K_T is the rate of a reaction at any temperature T°C; K_{Tref} is the rate of a reaction at reference temperature T_{ref} °C; E_a is the activation energy (J/mol) and R is the universal gas constant (8.31434 J/mol/K).

Bigelow model

$$\log\left(\frac{D_T}{D_{T_{ref}}}\right) = \frac{T_{ref} - T}{z}$$

Where, D_T is the decimal reduction time (min) at any temperature T°C; D_{Tref} is the decimal reduction time (min) at reference temperature T_{ref}°C and Z is the z-value.

3.5. F-Value

It is the equivalent time in minutes at any process temperature required to achieve the destruction (lethality) which would have been achieved with 121.1°C as the process temperature. The F–value indicates the lethality of the process and is calculated as,

$$F_0 = D_0\,[\log(N_0)] - [\log(N_t)]$$

Where, N_0 is the initial number of microorganisms (*cfu*) and N_t is the number of microorganisms survived (*cfu*). The term (log N_0–log N_t) yields a number so the process can be identified as nD. Generally, for pasteurization 6D to 8D reduction is recommended (Ricardo, 2009).

Like D-values, F_0 also indicates the process time at 121.1°C, so for other process temperatures, the value is subscribed to F *e.g.*, F_{72}. If the z value is known, the equivalent F-value at any process temperature t is can be calculated by the following relationship.

$$F_t = F_0 \, 10^{\left(\frac{T_t - T_0}{z}\right)}$$

For all the thermal processes like pasteurization where heating and cooling times are relatively short the calculations are based on holding time and temperature only (Lewis, 2006). Hosahalli *et al.* (1997) have described standard methods for thermal process calculation. Based on this and the terminologies discussed, following protocol can be put forth to decide the time-temperature combination for pasteurization or other alike thermal treatments.

(*i*) Enumerate the possible micro flora for the given product.

(*ii*) Identify the most heat resistant pathogenic organism, among the micro flora, *i.e.*, target organism on the basis of TDT or D-values (from literature or through experimentation).

(*iii*) Decide the required lethality on the basis of the probability of occurrence and population of the target organism. For pasteurization, an average 7D reduction can be considered.

(*iv*) Find the Z-value for the target organism and calculate the F-values for a range of temperature to achieve the defined lethality, a curve of Temperature *Vs* Time, may be plotted.

(*v*) Identify the most heat sensitive but important constituent which is to be protected during the heating process and fix the degree to which the constituent can be sacrificed.

(*vi*) Calculate the active concentrations of the heat sensitive constituent remaining after treating at various F-values decided (used at 4) and select those F-values which cause the least (acceptable) degradation of the heat sensitive constituent.

(*vii*) Select any F-value *i.e.*, time & temperature from those selected at 6 on the basis of applicability and suitability for the process equipment.

(*viii*) Finally, it should be ensured that the selected time-temperature combination besides achieving the desirable microbial destruction must render the milk to satisfy all other requirements as laid down by the regulatory bodies.

The worldwide-accepted criteria to judge the aptness of pasteurization is negative alkaline phosphatase test. Additionally, the pasteurized milk should not have Total Viable Count, more than 30000 cfu/ml and coliform count more than 1 cfu/ml.

4. PASTEURIZATION EQUIPMENT

As discussed above, the equipment for batch and continuous units of pasteurization are different in construction as well as operation, albeit the objective is same. The milk pasteurization equipments have undergone rapid and notable changes in last century. Some of the distinguishable designs include coil vats, glass lined tanks, spray vats, Danish heater, drum or film heater, internal tubular heater and external tubular (surface) heater, which have been discussed in literature (Eckles, 1993).

4.1. Batch Pasteurizer

The batch pasteurizer is basically an agitated kettle; the jacket is used for circulation of the heating medium during heating and holding period. After holding period is over, chilling medium is circulated through the same jacket. In its simplest form, a batch pasteurizer consists of the following components as shown in Fig. 2.

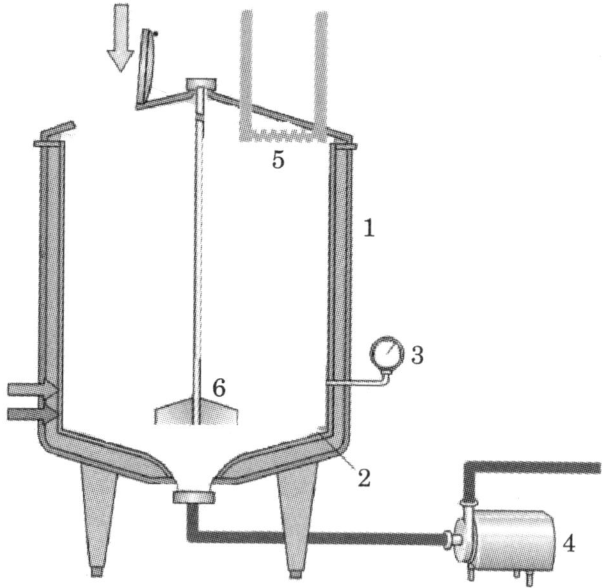

Fig. 2: Batch Pasteurizer.

4.1.1. *Jacketed vessel*

It holds the milk and has common inlet and outlet for milk being a batch type unit. The bottom is sloped towards outlet to ensure complete drainage. The dimensions of the vessel for the desirable throughput (Q) is calculated by,

$$Q = \frac{\pi}{4} d^2 h$$

Where, d is the inner diameter of the tank and h is the height to which milk would be filled. The actual dimensions can be kept on higher side for vapor space, etc. The ratio of d : h is adjusted on the basis of availability of floor space and height for small and marginal capacities. Ideally, it can be obtained by minimizing the area as the function of volume, diameter and height to save on material cost. Commonly, the h : d ratio of one is optimum.

The volume of the jacket is fixed primarily on the type of heating medium *i.e.*, steam or hot water. For steam the volume is calculated considering the rate of steam consumption and specific volume at the steam pressure. For hot water, consideration has to be given to a number of factors such as setting up the turbulent flow, required water velocity, overall heat transfer coefficient, and permissible temperature gradient from the view point of milk properties, etc. To boost the prospects of turbulence, water circulation may be guided within the jacket in swirling manner. An intelligent compromise between all these factors is the key to better design; it can be achieved with the experience. Some application softwares such as CFD that facilitate the simulation can be very useful in finding the most effective combination of the above set of parameters.

The jacket is often insulated with glass wool with an additional shell of mild steel or sometimes stainless steel as moisture barrier. It also adds to the aesthetic appeal of the equipment. Instruments like temperature sensor, often P_{T100}, bourdon tube pressure gauge, safety valve, etc are provided to the jacket.

4.1.2. *Agitator*

The propeller type agitator is preferred in batch pasteurizer and is mounted on top of the tank. The diameter of the propeller should be from 15 to 30% of the inner diameter of tank. The speed of rotation should be such that it does not create turbulence. Normally, a peripheral (linear) speed of 300 to 500 meter per min is expected, which can be calculated as,

$$v = r \times w = r \times (2\pi N) = \pi DN$$

Where, D is the diameter of agitator (D = 0.15 × tank diameter) and N is the rotational speed in rpm.

Power required to operate an agitator depends on several factors such as properties of milk, volume to be handled, size of the tank, type, size and speed of agitator, etc. For a propeller, which is mostly used in batch pasteurizer, power requirement (E) is calculated by,

$$E = N_e \cdot \rho \cdot n^3 \cdot d^5$$

Where; N_e is the Newton's number, ρ is density of liquid stirred, n is revolution per second of agitator, d is the diameter of agitator.

The value of N_e can be found from Reynold's number or can be read from the graph of Newton's Number *Vs* Reynold's Number supplied by the manufacturer (Kessler, 1981).

4.1.3. *Foam heater*

During heating and agitation foaming takes place at the surface of the milk due to some milk constituents like phospholipids. The foam contains air that is bad conductor of heat; hence albeit the milk is heated to specified temperature, the foam may remain under heated and may shelter microbes. To safeguard against this it is recommended that the foam be heated by 5 degrees above the set pasteurization temperature. It is accomplished with an additional steam heater or sometimes electric heater as shown in the Fig. 2.

4.2. Energy Requirement, Heating Time and Cooling Time for Batch Pasteurization

As discussed batch pasteurizer needs both thermal and electrical energies for fulfilling heating requirement and for pasteurizer accessories, respectively. The electrical energy requirement, being quite lower than that of thermal energy, is often neglected. The thermal energy required for the batch pasteurization of milk is supplied by the steam condensation or hot water circulation through the jacket. The requirement is influenced by number of factors;

(*a*) Initial temperature of milk.
(*b*) Desirable pasteurization temperature (few degrees above the ideal 63°C).
(*c*) Overall heat transfer coefficient, which again depends upon the degree of milk agitation, velocity of steam or hot water within jacket and conductance of the vessel wall.
(*d*) Heat losses to be overcome.

This thermal requirement, also called as the heat load, is given by,

$$Q = V_m \times \rho_m \times C_p \times (T_p - T_i)$$

Where, V_m is the volume of the milk (m^3), ρ_m is the mass density of milk (kg/m^3), C_p is the specific heat of the milk (kJ/kg–°C), T_p is the pasteurization temperature (°C) and T_i is the initial temperature of milk (°C).

In order to attain the pasteurization temperature, the heat load must be satisfied by the employed heating medium *i.e.*, steam or hot water. The quantities of steam (Q_s) or hot water (Q_w) required is estimated by,

$$Q_s = \frac{V_m \times \rho_m \times C_p \times (T_p - T_i)}{h_g - h_f}$$

$$Q_w = \frac{V_m \times \rho_m \times C_p \times (T_p - T_i)}{C_{pw} \times (T_i - T_o)}$$

Where, h_g is the enthalpy of steam entering the jacket (kJ/kg), h_f is the enthalpy of condensate leaving the jacket (kJ/kg), C_{pw} is the specific heat of the water (kJ/kg–°C), is the temperature of hot water at the inlet of jacket (°C), is the temperature of hot water at the outlet of jacket (°C).

If electrical foam heater is used, assuming it runs on single phase AC supply, its energy requirement is given by,

$$P_{FM} = V \times I \times \cos \varnothing = I^2 \times R \times \cos \varnothing$$

Where, V is the voltage available in the power line (Volts), I is the electric current flowing through the coil (Ampere), cos ∅ is the power factor.

The electrical energy consumption can be converted to its heat equivalent if desirable for comparison.

Though the effectiveness of a pasteurization process in terms of microbial lethality depends on the holding temperature and holding time, the sensory and chemical attributes of milk are greatly influenced by the rate of heating ($d\theta/dT$). It is very essential to predict the time that would be required to raise milk temperature from initial temperature to the pasteurization temperature, this time is known as the 'Come up time', often in context with in-container sterilization, here we can use 'heating time'. Accordingly the time required to decrease milk temperature after holding at pasteurization temperature to the desirable chilling temperature is called as 'come down time', often in context with in-container sterilization, here we can use 'heating time'. Due to change in the temperature profile of the process, unsteady state heat transfer occurs, which makes it difficult to precisely predict the heating and cooling times. Equating the rate of heat transfer from the heating medium and the rate at which the fluid absorbs energy is easy approach to decide the heating time. Thus, following equation is obtained:

Rate of heat transfer = Total heat to be transfered (Heat Load)

$$U \times A \times \Delta T_m = V_m \times \rho_m \times C_p \times \frac{dT}{dt}$$

On integration and simplification, it yields the time of heating $t_{heating}$ as

$$t_{heating} = \frac{V_m \times \rho_m \times C_p}{U \times A} \ln\left[\frac{T_h - T_i}{T_h - T_p}\right]$$

Where:n_m is the volume of milk (m^3), r_m is the mass density of milk (kg/m^3), C_p is the specific heat of the milk (kJ/kg–°C), U is the overall heat transfer coefficient (kW/m^2 K), A is the area of heat transfer (m^2), T_h is the temperature of heating medium (°C), T_i is the initial temperature of milk (°C), T_p is the Pasteurization temperature (°C).

Similarly, the cooling time can be obtained by

$$t_{cooling} = \frac{V_m \times \rho_m \times C_p}{U \times A} \ln\left[\frac{T_p - T_c}{T_f - T_c}\right]$$

Where: T_f and T_c are the final temperature of chilled milk and temperature of cooling medium, respectively (°C).

Both these equation are based on certain assumptions such as the values of overall heat transfer coefficient, specific heat of milk and density of milk are negligibly affected by change in temperature and the temperatures of heating/cooling media remain constant, which is possible only if the heating medium is condensing and the cooling medium is boiling/evaporating. When hot and chilled water is used as heating and cooling medium, respectively, sensible heat transfer occurs and the above equations would become futile due to change in temperatures. In such case following equations can be used.

$$t_{heating} = \frac{(V_m \times \rho_m \times C_p) \times e^{\left[\frac{UA}{m_h C_{ph}}\right]}}{m_h \times C_{ph} \times e^{\left[\frac{UA}{m_h C_{ph}}\right]}} \ln\left[\frac{T_h - T_i}{T_h - T_p}\right]$$

Where, m_h is the mass flow rate of heating medium and C_{ph} is the specific heat of the heating medium (Singh, 2003).

Similarly, time of cooling would be,

$$t_{cooling} = \frac{(V_m \times \rho_m \times C_p) \times e^{\left[\frac{UA}{m_c C_{pc}}\right]}}{m_c \times C_{pc} \times e^{\left[\frac{UA}{m_c C_{pc}}\right]}} \ln\left[\frac{T_p - T_c}{T_f - T_c}\right]$$

Where, m_c and C_{pc} are the mass flow rate and specific heat of the cooling medium respectively (Singh, 2003).

The overall heat transfer coefficient (U) is the function of milk side convective heat transfer coefficient (hi), jacket side convective heat transfer coefficient (ho) and conductance (l/k) of the heat transfer surface as follows;

$$U = \frac{1}{\frac{1}{h_i} + \frac{l}{k} + \frac{1}{h_o}}$$

4.3. Continuous Pasteurization

Any continuous (HTST) pasteurization unit should essentially consist of plate heat exchanger (PHE), holding tube, flow diversion valve, booster pump, heating mechanism, chilling mechanism and a back pressure valve. However, a typical commercial HTST pasteurizer would have few additional components from the view point of operational convenience, better quality of milk and energy efficiency. Figure 3 depicts a typical HTST pasteurization unit.

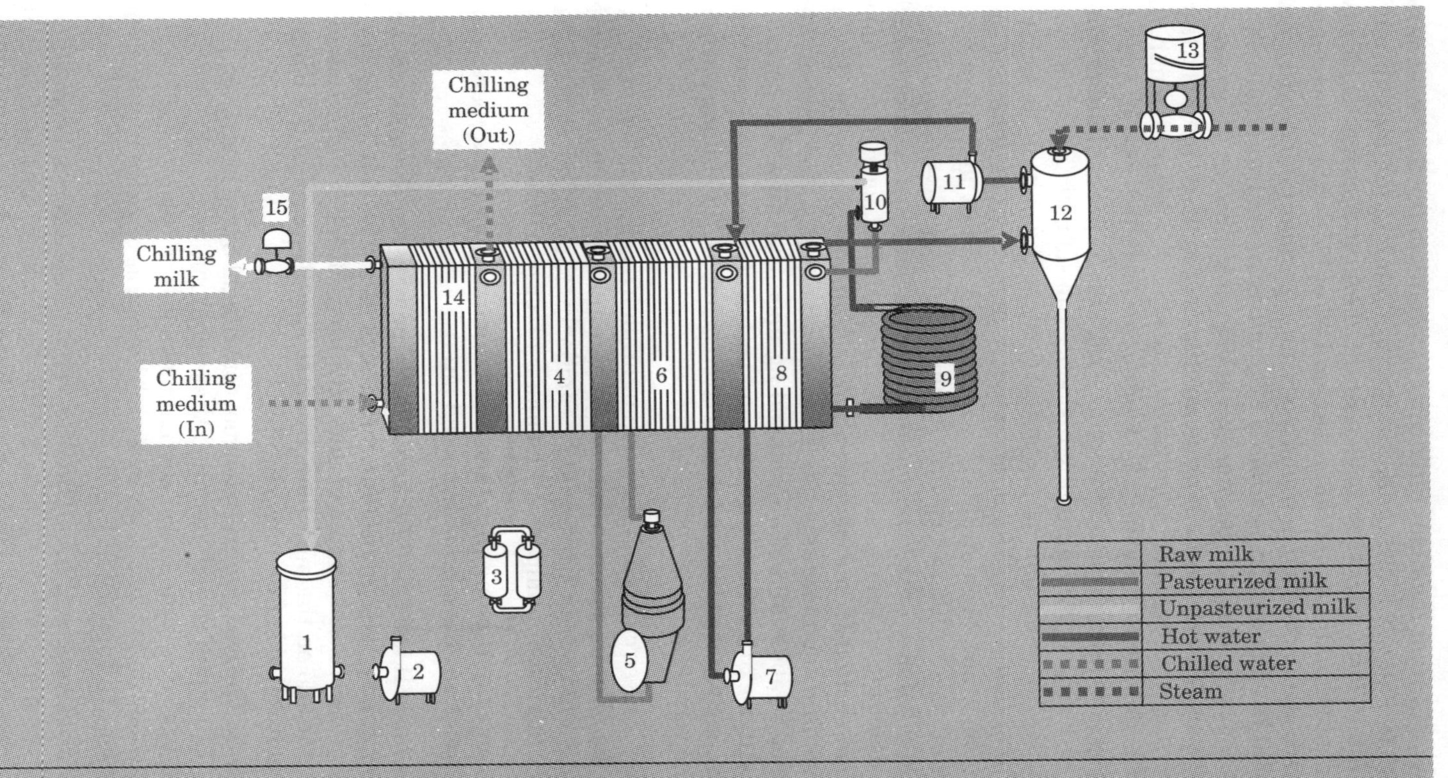

1. Balance tank; 2. Feed pump; 3. Duplex milk filter; 4. Regeneration section-I; 5. Cream separator; 6. Regeneration section-II; 7. Booster pump; 8. Heating section; 9. Holding tube; 10. FDV; 11. Hot water circulation pump; 12. Steam battery; 13. Steam valve; 14. Chilling section; 15. Back pressure valve

Fig. 3. Schematic diagram of HTST pasteurization unit.

The raw milk from the main tank comes initially to the float controlled balance tank (1). From balance tank, feed pump (2) forces the milk through duplex milk filter (3) in order to remove unwanted extraneous matter, if any. Filtered milk enters in the regeneration section-I (4) of PHE, where it is heated to about 50°C. So heated milk can be fed to cream separator (5) if desired, else it advances into regeneration section-II (6) and is heated to about 65°C. The milk then passes through booster pump (7), which boosts its pressure, to heating section (8). Whenever homogenization is required, the milk from (6) can be routed through homogenizer by-passing the booster pump. In heating section of PHE, the milk is heated to specified pasteurization temperature by hot water. The milk from heating section goes to holding tube (9), which is so designed that minimum residence time of milk in it is bit more than the specified holding time of pasteurization. At the exit of the holding tube, flow diversion valve (10) is situated, which decides whether the approaching milk has been properly pasteurized or not and allows the properly pasteurized milk to proceed further and unpasteurized milk is diverted back to balance tank (1) directly. Properly pasteurized milk, which is still at pasteurization temperature, flows through the downstream sides of regeneration section-II and regeneration section-I consecutively. It acts as heating medium for incoming raw milk. Consequently, the pasteurized milk gets cooled to about 10 to 15°C depending upon the initial temperature of raw milk. After exiting from regeneration section-I, the pasteurized milk enters into chilling section (14), where it is cooled below 5°C by chilling medium and goes to storage tank.

Hot water is re-circulated in the system through a compatible centrifugal pump (11) by maintaining its temperature through steam injection in a specially designed unit called steam battery (12). Similarly, the chilled water is re-circulated by centrifugal pump from ice bank tank or ice silo.

All the components along with their constructional as well as operational features and functions have been discussed later.

(a) *Balance tank*: It is a small tank that is placed between the milk silo and pump supplying milk to pasteurizer; it acts as a flow stabilizing agent and ensures constant supply of milk. It also nullifies the effect of milk head in the silo on the milk flow. By float the milk level in the balance tank is controlled between two limits.

(b) *Feed pump*: The feed pump is a normal centrifugal pump having sanitary design. The pump is such installed that negligible suction head will be exerted. It obviates the use of suction valve as the valve is undesirable from the view point of sanitary design. The details of pump construction accessories are depicted in Fig. 4.

(c) *Plate heat exchanger (PHE)*: The PHE is the most important component in HTST pasteurizer. In its simplest form, as shown in Fig. 5, a PHE consists of a *front base* also known as head terminal, a *back base* also known as end terminal, 4 bars known as guiding rail that connect the

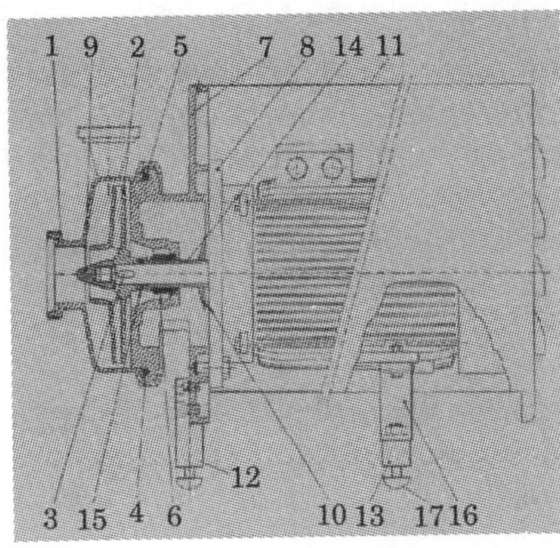

1. Casing
2. Impeller
3. Impeller key
4. Clamp
5. Clamp 'O' ring
6. Mechanical seal
7. Cover
8. Motor
9. Impeller nut
10. Thrower
11. Shroud
12. Front foot
13. Rear foot
14. Bush
15. 'O' ring
16. Rear foot frame
17. Level adjustable ball feet

Fig. 4: Milk Pump (Feed pump).

front and end terminals forming a frame like structure. Number of thin plates is fitted in the frame forming a horizontal stack, which may or may not fill the complete frame length. When the complete length of the frame is not filled with the plate stack, an additional thick plate known as pressure plate is included as the last supporting plate and is tighten with a central screw to provide the required contact pressure.

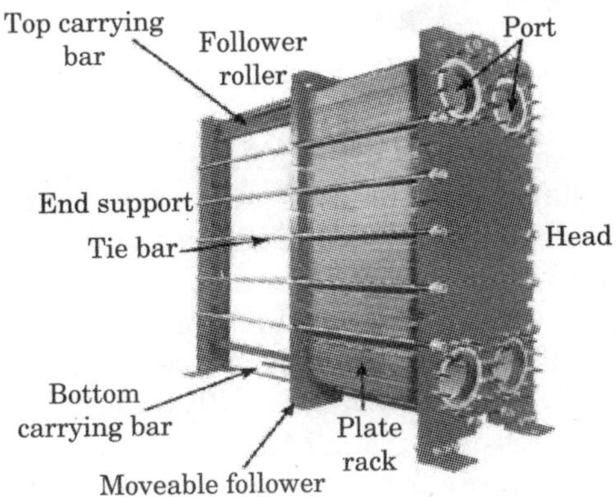

Fig. 5: Plate Heat Exchanger.

The plates are sealed against the two sides and exterior by flexible gaskets made of nitrile rubber. The structure of an individual plate as shown in Fig. 6 has four ports for the passage of liquid and two grooves for engaging the plate on the guiding rail. Two of the four ports act as inlet and outlet for one liquid (say milk) on that side of the plate and are called as open ports. Remaining two ports function as the bypass passages for other liquid (say water) to another side of the plate. These are called blind ports. The gasket lining isolates two ports and prevents the liquid from entering into one side so these are sometime called as blind ports or openings. Each plate is made of polished stainless steel and offers around 1 m^2 surface area, the thickness may vary from 0.5 to 1.25 mm. Each plate is given corrugations by stamping to narrow down the residence time and creating turbulence over the plate, which enhances the heat transfer (due to higher hi and ho). The corrugations also increase the actual heat transfer area improve mechanical bending strength of the plate. Different pattern of corrugations have been reported, the wash boards pattern and herringbone pattern, as depicted in Fig. 6, are common.

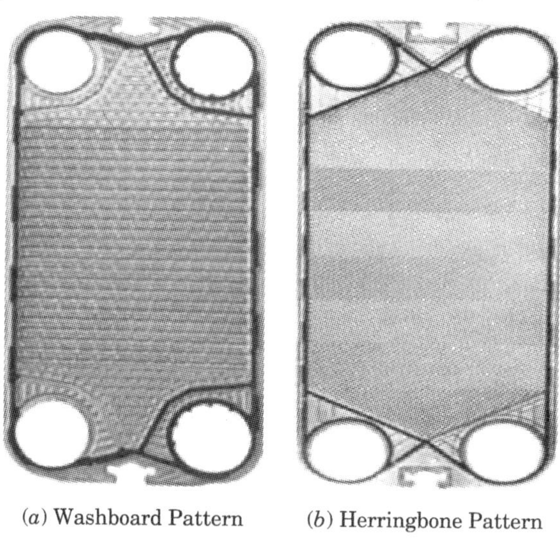

(*a*) Washboard Pattern (*b*) Herringbone Pattern

Fig. 6: Plate of PHE.

All the intermediate plates in the frame are divided into 3 to 4 groups by separating terminals also called as intermediate terminals. Each group is called a section; the terminals separate the plates as well as support the inlet and outlet passages of the section. This grouping is necessary because a narrow gap of around 3 to 6 mm is maintained between two plates to ensure the passage of milk in the form of thin film, consequently resulting into high degree of turbulence and large pressure drop. Total pressure drop over all the plates can't be overcome by pumping from single starting point; it would cause very high pressure in first few plates. Plate Heat Exchangers can handle the pressure up to 2 MPa only (Lewis, 2006) until specially designed for high pressures. Moreover, this grouping also makes it possible to draw the milk out of the PHE at intermediate temperatures between initial temperature and pasteurization temperature.

The total milk entering in the PHE simultaneously flows through many alternative plates, as shown in Fig. 7, downwards and then through same number of plates upward. It takes few passes within the plates' stack before coming out from the terminal. Likewise the heating / cooling medium also flows through many plates simultaneously depending upon its flow rate and takes few passes before reaching the terminal. Accordingly, the flow pattern in any

PHE is described by a ratio, the numerator describes the milk flow and denominator describes the heating or cooling medium (Das, 2005 and Kessler, 1981) flow as below;

Flow Pattern = [*Number of passes of milk* × *Number of plates (gaps) filled by milk per pass*]/(*Number of pass*)

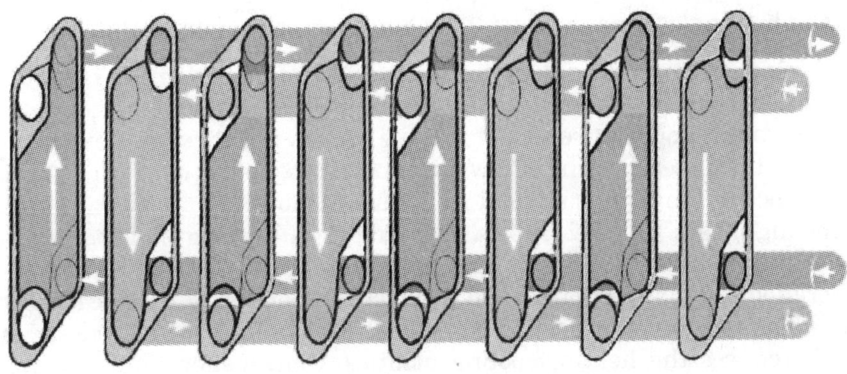

Fig. 7: Flow Pattern in Plate Heat Exchanger.

In the given figure the flow pattern is $\frac{1 \times 4}{1 \times 4}$ for the section shown, however it may differ with other units.

Example: A flow pattern of $\frac{6 \times 2}{3 \times 4}$ indicates, the milk takes six passes in all throughout the section, flowing through two plates at a time while the medium takes three passes throughout the section, flowing through four plates at a time.

This flow pattern depends upon the ratios of flow rates of respective fluids and the desired temperature profile in the heat exchanger. In typical milk pasteurizer the flow rate ratio of Milk to Hot Water to Chilled Water is 1:2:3 (De, 1982).

Normally, the PHE of HTST milk pasteurization assembly is divided into four sections *viz.*, chilling section, 1st regeneration section, 2nd regeneration section and heating section. Sometimes there can be three regeneration sections depending upon the process profile.

In regeneration section milk to milk heat transfer takes place, which is the characteristic feature of HTST pasteurization that decreases the utility cost by multiple times. Until few years back there used to be an additional cooling section to cool the pasteurized milk by ambient water before cooling by chilling agent. But now a days, raw milk is available at quite lower temperature so regenerative cooling obviates the use and possibility of ambient water cooling section. Further, the sizing of any section (*i.e.*) number of plates can be done by using the following relation,

$$N = \frac{V_m \times \rho_m \times C_p \times \Delta T}{U \times \Delta T_m \times A_p}$$

Where: V_m is the volumetric flow rate of milk (m^3/s), r_m is the mass density of milk (kg/m^3), C_p is the specific heat of the milk (kJ/kg–°C), ΔT is the desired rise or drop in temperature in the section (°C), U is the overall heat transfer coefficient (kW/m^2K), ΔT_m is the log mean temperature difference between the milk and heating/cooling medium (°C), A_p is the area of single plate (m^2).

During continuous pasteurization at any instance, there is raw milk at inlet which is to be heated and there is pasteurized milk after hold at pasteurization temperature which is to be cooled. So, these two are made to flow across the plates, heating of raw milk and consequent cooling of pasteurized milk will occur due to mutual heat exchange without use of external heating and cooling media. It gives substantial saving in heating and cooling loads since the same heat is regenerating again and again. This is called as heat regeneration and that portion of PHE in which it is done is called as regeneration section. The regeneration section can be sub divided in two or three sections. Since it reduces the heating/cooling utilities, the regeneration as much as possible is expected. The ratio of heat exchanged in regeneration to that required to accomplish the treatment is called as regeneration efficiency (η_{RE}) and is expressed in percent.

$$\eta_{RE} = \frac{\textit{Heat exchanged in regeneration section}}{\textit{Total heat exchanged required}} \times 100$$

This generalized equation can be modified to calculate the regeneration efficiency of heating *i.e.*, upstream regeneration efficiency and regeneration efficiency of cooling *i.e.*, downstream regeneration efficiency.

Upstream Regeneration efficiency $\eta_{\downarrow RE_{\downarrow heating}}$,

$$\eta_{RE_{heating}} = \frac{\textit{Heating of raw milk occured in regeneration}}{\textit{Maximum heating possible}} \times 100$$

$$\eta_{RE_{heating}} = \frac{m \times C_p \times (T_{hi} - T_i)}{m \times C_p \times (T_p - T_i)} \times 100 = \frac{(T_{hi} - T_i)}{(T_p - T_i)} \times 100$$

Downstream Regeneration efficiency $\left(\eta_{RE_{cooling}}\right)$,

$$\eta_{RE_{cooling}} = \frac{\begin{array}{c}\textit{Cooling of pasterized milk occured in}\\ \textit{regeneration section}\end{array}}{\textit{Maximum cooling possible}} \times 100$$

$$\eta_{RE_{cooling}} = \frac{m \times C_p \times (T_p - T_{ci})}{m \times C_p \times (T_p - T_f)} \times 100 = \frac{(T_p - T_{ci})}{(T_p - T_f)} \times 100$$

Where, T_i is the initial temperature of raw milk, T_{hi} is the temperature of milk after regeneration *i.e.*, at the inlet of heating section, T_{ci} is the initial temperature of pasteurized milk after regeneration in downstream side *i.e.*, at the inlet of chilling section, T_f is the final temperature of pasteurized chilled milk and T_p is the pasteurization temperature.

So the actual load on pasteurizer *i.e.*, $m.C_p.(T_p - T_i)$ would be reduced to $m.C_p.(T_p - T_{hi})$ due to regenerative heating from T_i to T_{hi}. Similarly, the cooling load *i.e.*, $m.C_p.(T_p - T_f)$ would be reduced to $m.C_p.(T_{ci} - T_f)$ due to regenerative cooling from T_p to T_{ci}. The reduced loads with known values of upstream and downstream regeneration efficiencies can be predicted by using following relations,

$$\text{Heating Load} = m \times C_p \times (T_p - T_i)\left(1 - \eta_{\text{RE}_{\text{heating}}}\right)$$

$$\text{Cooling Load} = m \times C_p \times (T_p - T_f)\left(1 - \eta_{\text{RE}_{\text{cooling}}}\right)$$

The heat lost by pasteurized milk will always be equal to that gained by raw. So using energy balance one regeneration efficiency can be easily calculated when another is known.

Existing generation of HTST pasteurizers can give up to 94% regeneration efficiency which means there thermal energy consumption is just 6% of that of batch units. The regeneration efficiency can be further increased by introducing additional plates in regeneration section but after certain number of plates it becomes uneconomical due to consequent pressure drop over the flow and thereby increasing the pumping requirement.

(d) *Holding tube*: The milk heated to pasteurization temperature must be held at the same temperature for specific duration of time to affirm proper pasteurization. This is accomplished in holding section or tube, the later being more common. Though ideally, the milk be heated to minimum 71.7°C and held for 15 seconds, the actual temperature and the minimum residence time of milk in the holding section should always exceed. This is due to the fact that some drop in temperature of milk while flowing through the holding tube is inevitable. Special emphasis is given while designing holding tube to keep this heat loss as low as possible. Ideally, plug flow is desirable in the holding tube to ensure uniform holding time (Brennan, 2006). But it is not practically possible to achieve perfect plug flow in long tube, laminar flow or turbulent flow can be achieved. Unlike plug flow, the velocity profiles in laminar and turbulent flows are curvilinear as shown in Fig. 8. Therefore, the residence time has to be set such that fastest moving particle should get the minimum required hold but during that time the slower moving particles get much longer hold.

In laminar flow the maximum velocity as shown in Fig. 8 is twice the average velocity. While in turbulent flow the maximum velocity is approximately 1.19 times the average velocity, so narrower distribution of residence time. However, in turbulent flow the pressure drop is higher than that in laminar flow, which necessitates the use of higher diameter holding tube. For practical implementation the average velocity (V_{avg}) can be calculated by using continuity equation *i.e.*, volumetric flow rate divided by cross sectional area of pipe and the maximum velocity (V_{max}) can be found out using dye method.

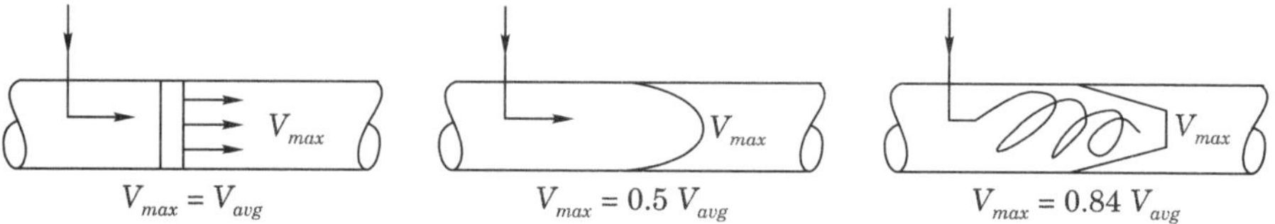

Fig. 8: (*a*) Plug flow, (*b*) Laminar flow and (*c*) Turbulent flow.

The selection is made as per requirement; however the turbulent is often preferred. Reynolds number is used to characterize the flow as given below,

$R_e = \dfrac{\rho v d}{\mu}$	Where; ρ is the density of milk, v is the average velocity, d is the pipe diameter and μ is the absolute viscosity of milk.

Reynolds numbers more than 4000 and less than 2100 indicate turbulent flow and laminar flows, respectively, while the intermediate value indicates transitional flow. The holding tube is generally spiral or helical with a continuous upward slope, which may range from 2.0 to 2.5 cm per meter of horizontal length. The length of the holding tube for certain capacity of pasteurizer is given by,

$$l = \frac{4Q.t_{hold}}{\rho d^2 \eta}$$

Where Q is the capacity of the pasteurizer in m^3 per second

L is the length of holding tube in meter

D is inside diameter of the tube in mm

t_{hold} is desirable the holding time in seconds

η is the efficiency factor.

The value of efficiency factor depends upon the flow type and tube design as discussed earlier. Often a value of 0.85 to 0.9 is used (Anon, 1995).

(e) *Flow diversion valve (FDV)*: FDV is the decisive component in HTST pasteurizer that ultimately decides whether the milk approaching it is properly pasteurized or not. Mostly, FDV is a pneumatic three way valve operated through an ON/OFF process controller, it diverts the milk flow to raw milk balance tank, if it is not properly pasteurized. As shown in Fig. 9, the FDV consists of three way diaphragm valve. There are two forces *viz.*, the air pressure and spring tension acting on either sides of the diaphragm. The imbalance between these two forces decides the valve position. Since the spring tension remains unaltered, it is mainly the air pressure that decides the valve position. When air pressure exceeds spring tension the valve stem is lifted upward and vice versa. The flow of air to the valve is controlled by a solenoid valve, which is activated through a process controller and the process controller is attached with a temperature sensor installed at the end of holding tube. The sensor senses the temperature of milk at specified time interval, which is commonly in milliseconds. The sensed temperature (Process Value, P_V) is compared with the set point (Pasteurization Temperature, P_s). If the sensed temperature is more than the set pasteurization temperature, it is an indication that the milk all along its flow through the holding tube had a temperature higher than the specified temperature, thus signify proper pasteurization. The air pressure is maintained. The moment sensed temperature is lower than the set pasteurization temperature, process controller deactivates the solenoid valve, the air pressure on the diaphragm is released and spring tension comes into action. The valve stem drops and the flow is diverted. To set the flow again when the temperature has been achieved air needs to be supplied. The default position of the FDV is diverted or closed; it needs constant air pressure to remain. This type of FDV is called as air-to-open FDV. Another type of FDV, which would normally remain open and needs air pressure to get diverted or closed, is known as air-to-close FDV. In fact these terminologies are used in context with any type of pneumatic valve.

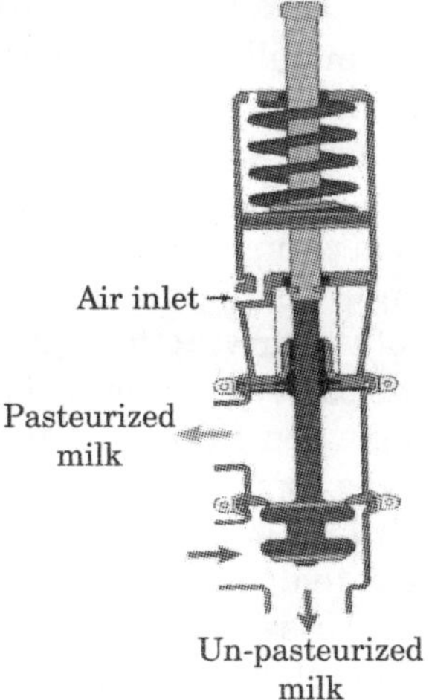

Fig. 9: Flow Diversion Valve.

If the distance between the outlet of holding tube is very less and the response of FDV is slow, it may happen that after sensing a lower temperature, by the time FDV diverts, the under pasteurized milk would have surpassed the FDV. Hence, in some pasteurizers, the temperature sensor is installed at the inlet of holding tube as it provide more time for reaction, ensuring that no under pasteurized milk can flow downstream the FDV. But this system does not account for the drop in temperature of milk that can while flowing through holding tube especially when it is not insulated. Or when accounted for the heat loss, the milk is slightly overheated to overcome the anticipated loss in temperature. It may again lead to nutritional losses and alterations to flavour and properties. Hence, by and large the former position (sensor at the end of holding) is thus preferred. In small and skid mounted HTST pasteurizers having throughput up to 5000 liter milk per hour, the solenoid operated FDV can also be installed.

(f) *Booster pump*: The booster pump is an ordinary centrifugal pump as that of feed pump. It is called so because it boosts the milk pressure flowing downstream side. During the continuous operation, there is possibility of generation of pin holes in the pasteurizer plates in the channeled portion, which can be easily identified. The occurrence of such holes would cause intermixing of the fluid on either sides of the plate and eventually lead to contamination of the pasteurized milk with raw milk, heating medium or chilling medium. It is not possible to frequently dismantle the plates or continuously monitor the plates for pin holes formation and prevent the contamination by replacing such plates. Hence, a booster pump is installed in the pasteurized milk line as an anticipatory remedial measure. The booster pump increases the pressure of the pasteurized milk by approximately 0.5 bar so in any incidence of pin hole leaking the pasteurized milk being under higher pressure will leak to the other side and its contamination would be prevented.

The booster pump can be installed before heating section, after holding section or after FDV. When installed before heating section the operating temperature is less and it prolongs the life of the pump as well as safeguard against the contamination with heating medium. When installed after FDV, it can also function as recirculation pump when the FDV goes in diverted condition *i.e.*, when FDV diverts the improperly pasteurized milk to balance tank, the pasteurized milk in absence of the driving force remain held up in the line. If by some valve arrangement the pasteurized chilled milk at the outlet of chilling section is fed to booster pump, when FDV is in diverted position, the booster pump can keep the pasteurized milk in circulation. It avoids the emptying and cleaning of complete plant after diversion and the process can resume immediately after the required temperature is attained. As an alternative to booster pump, the number of passes of pasteurized milk in the regeneration sections can be increased. The elongated path would thereby cause more pressure drop and exert more back pressure within the pasteurized milk section of the plant.

(g) *Back pressure valve*: Though it is assumed that boosting pressure with pump creates positive differential pressure in the pasteurized milk line, the exact value cannot be affirmed. The pressure drop occurs on both the side and due to deposits it may be more on the raw milk side and may be lower on the pasteurized milk side if the pasteurized milk silo is situated at lower level. Hence, to ensure the creation and maintenance of certain minimum positive differential pressure on the downstream side of the booster pump, a pressure relief valve of known pressure value is installed at the outlet of the chilling section. The valve exerts certain known back pressure required for its opening so is called as back pressure valve (BPV).

(h) *Heating medium mechanism*: In the heating section, milk is heated by hot water being circulated in a closed loop by hot water pump. The temperature of the hot water is maintained so that throughout heating section, it should be ~5°C above the set pasteurization temperature. Drop in hot water temperature due to heat transferred to milk is overcome by steam injection using steam battery. The steam condensate continuously overflows the hot water circulation loop. Alternatively, in some units brazed PHE are used if out flowing condensate is undesirable.

The required flow rate of heating medium (M_w) depends upon the heat load and drop in temperature between inlet and outlet heating section. Generally, it is twice the milk throughput but for specific unit it can be estimated by equating the heat gained by milk to heat lost by the, heating medium as follows

$$M_w = \frac{V_m \times \rho_m \times C_p \times \Delta T_h}{C_{pw} \times \Delta T_{hw}}$$

Where V_m is the volumetric flow rate of milk (m^3/s)

ρ_m is the mass density of milk (kg/m^3)

C_p is the specific heat of the milk (kJ/kg–°C)

ΔT_h is the desired rise in milk temperature across heating (°C) = $(T_p - T_{hi})$

ΔT_{hw} is difference between temperatures of hot water at inlet & outlet of heating section (°C)

(i) *Chilling medium mechanism*: The milk cooled in regeneration section is chilled to final storage temperature in separate chilling section using external chilling medium being circulated in a closed loop like heating section. Usually pasteurized milk is chilled below 7°C, if to be pouched and below 4°C, if to be transported in bulk container. The final chilled milk temperature in any case must not be set too close to initial freezing point of milk (about –0.55°C); else freezing of milk in chilling section may clog it and cause severe damage to the plates. Chilled water around 1–2°C is the most preferred chilling medium, however, brine or glycol solutions are also used as next alternatives in some units. Generally, the flow rate of chilling medium (M_c) is thrice the milk throughput. Like heating medium for specific unit, it can be estimated by equating the heat lost by milk to heat gained by the chilling medium as follows,

$$M_c = \frac{V_m \times \rho_m \times C_p \times \Delta T_c}{C_{pc} \times \Delta T_c}$$

Where V_m is the volumetric flow rate of milk (m^3/s)

ρ_m is the mass density of milk (kg/m^3)

C_p is the specific heat of the milk (kJ/kg–°C)

ΔT_c is the desired drop in milk temperature during chilling (°C) $= (T_{ci} - T_f)$

ΔT_c is difference in chilling medium temperature at inlet & outlet of chilling section (°C)

4.4. Process Control in Pasteurization

The main aim of process control system in HTST pasteurization is to ensure that the milk gets desired holding at specified temperature. It is likely to be influenced by several factors *viz.*, flow rates of milk and heating medium; their respective temperature gradients and heat transfer characteristics within the PHE.

The holding time is anticipated from the length of holding tube and minimum flow velocity in holding tube considering the prevailing flow type. In most types of HTST set ups temperature of milk after holding is measured. As long as it is higher than the set pasteurization temperature, FDV forwards the flow as discussed in earlier section. Otherwise the flow of milk coming out of the FDV is diverted. To avoid errors due to fluctuation in velocity, a flow controller is often installed in the downstream line of the feed pump or booster pump.

The temperature of circulating hot water is maintained by steam injection control. The steam flow to the battery is regulated by a PID operated Needle valve. In place of PID, fuzzy control is being recommended recently, however the former is preferred at industry. The temperature of circulating chilled water is maintained by regulating its flow rate.

A safety thermal limit recorder (STLR) is installed at control panel. It plots the temperatures of raw milk, temperature of milk after holding, inlet and outlet temperatures of hot water and chilled water over the 24 hour or given run time at frequent intervals. The plot is called as *thermogram*. At any instance, when the recorded temperature crosses the set safety thermal limit, an alarm is actuated automatically. Simultaneously, the process of flow diversion is also actuated through the process control loop.

4.5. Pasteurizer Trouble Shooting

Sr. No.	*Problem (Symptom)*	*Possible cause*	*Possible solution*
1.	• Decrease in capacity *i.e.*, flow rate Or	• Fouling of plates either on milk side	• Carry out complete CIP for de-scaling. If the problem still persists, dismantle the PHE and clean the plates manually.
	• Increase in pressure drop	• Milk freezing occurring in chilling section	• Check the temperature of outgoing chilling agent from chilling section and adjust it above 1°C.

(Contd..)

(Contd..)

Sr. No.	*Problem (Symptom)*	*Possible cause*	*Possible solution*
2.	• Visible leakage	• The gasket might have damaged	• Replace the gaskets
		• Operating pressure is more than the rated pressure	• Decrease the operating pressure. (Sometimes when booster pump is not installed in the system, decreasing operating pressure leads to low capacity, hence booster pump must be installed).
3.	• Low regeneration efficiency	• Less heat transfer area in regeneration section	• Increase the number of plates in regeneration section
		• Flow rates less than its rated values	• Increase the flow rates
4.	• Frequent flow diversion	• Insufficient plates in heating section	• Increase the number of plates in the heating section in accordance with the heat load
		• Drop in temperature of incoming hot water	• Check the working of steam injection valve.
		• Drop in flow rate of hot water	• Inspect the hot water recirculation pump for leakage and capacity.
5.	• Less MBRT for pasteurized milk	• Cross contamination in the pasteurizer due to pin holes on plates	• Inspect all the plates for pin holes by dye penetration method.
			• Install booster pump in the system to maintain higher pressure in the pasteurized milk stream.

5. FUTURE RESEARCH NEEDS

The technology of pasteurization has witnessed several advances like heat regeneration that decreased the energy consumption by almost 90%. The future research can be broadly focused on *de-fouling techniques or antifouling plates in PHE* and *narrowing residence time distribution in holding tube*. The non conventional technologies such as *microwave heating, ohmic heating, inductive heating, electric pulsed heating, irradiation, high pressure processing,* etc may be studied for their aptness as pasteurization.

REFERENCES

Anon, (1983). Pasteurizing Plant Manual, SDT.

APV, (2006). Heat Transfer Handbook.

APV, (2009). Heat Transfer Handbook.

Burton, H. (1988). UHT processing of milk and milk products, Elsevier Applied Science, London.

Das, H. (2005). Food processing operations analysis, Asian Books Pvt. Ltd., New Delhi.

De, S. (1981). Outlines of dairy technology, Oxford Press, London.

Eckles, C.H., Combs, W.B. and Macy, H. (1993). Milk and milk Products, 4th edition. Tata Mc Graw Hall Hill Publishing Co. Ltd, New Delhi.

Holdsworth, S.D. (2009). Principles of thermal processing: Sterilization. *In*: Engineering aspects of Thermal Food Processing, Simpson Ricardo (Ed), CRC Press, London, pp. 3–12.

Hosahalli, S.R. and Singh, R.P. (1997). Sterilization process engineering. *In*: Handbook of Food Engineering Practices, Valentas, K.J., Rostein, E. and Singh, R.P. (*Eds.*), CRC Press, London, pp. 37–70.

Jay, J.M. (1996). Modern Food Microbiology, 4th edition. Chapman & Hall Inc., New York.

Kelly, A.L. and Osheo, N. (2002). Pasteurizers, design and operation. *In*: Encyclopedia of Dairy Sciences, Rogonski H, Fuquay JW and Fox PF (Eds). Academic Press, London, Vol. 3, pp. 2237–2247.

Kessler, H.G. (1981). Food engineering and dairy technology, Verlag A. Kessler Publishing house, D-8050 Freising, Germany.

Lewis, M.J. (1994). Heat treatment of milk. *In*: Modern Dairy Technology, Robinson, R.K. (*Ed.*), Chapman & Hall Inc., London., pp. 1–59.

Lewis, M.J. (2006). Thermal processing. *In*: Food Processing Handbook, Brennan, J.G. (*Ed.*), WILEY-VCH Verlag GmbH & Co. KGaA, Weinheim. p. 33–70.

Mc Cabe, W.L., Smith, J.C. and Horriott, P. (1993). Unit operations of chemical engineering, 5th edition, Mc Graw Hill Inc, New York.

Riser, E.T. (2002). Pasteurization of liquid milk products. *In*: Encyclopedia of Dairy Sciences, Rogonski, H., Fuquay, J.W. and Fox, P.F. (*Eds.*). Academic Press, London, 3: 2232–2237.

Silva, F.V.M. da and Gibbs, P.A. (2009). Principles of thermal processing: Pasteurization. *In*: Engineering of Thermal Food Processing, Simpson, Ricardo. (Ed), CRC Press, London, pp. 13–39.

Singh, R.P. (2003). Heating and cooling processes for foods. *In*: Handbook of Food Engineering, Heldman, D.R. and Lund, D.B. (*Eds.*), CRC Press, London. pp. 9–22.

Singh, R.P. and Heldman, D.R. (2004). Introduction to Food Engineering, 1st Indian reprint, Academic Press, London.

Toledo, R.T. (2000). Fundamentals of Food Process Engineering, 2nd edition, Chapman & Hall Inc., New York.

Walstra, P., Geurts, T.J., Noomen, A., Jellema, A. and Van Boekel, M.A.J. Svan (1999). Dairy technology: Principles of milk properties and processes, Marcel Dekker, New York.

5

Technology of Comminuted Meat Products

M. MUTHUKUMAR AND I. PRINCE DEVADASON

ABSTRACT

New products are designed to provide variety to consumer, utilise various ingredients, improve quality of products, meet consumer demand and reduce cost of products. It is to be kept in mind that new product should be developed and marketed to the needs, expectation and acceptance of as large a population as possible. The advancement in the processed muscle foods development has enabled to produce varieties of meat products with better palatability and shelf life. The comminution process enabled commercialization of meat from spent birds and animals, meat with poor texture and appeal, which are previously unmarketable due to low consumer appeal and poor eating quality attributes. Further, the incorporation of non meat ingredients brings down the price of these products and makes them reachable to common consumers as well. The utilization of non meat ingredients is also important for meat industry, as increasing profit and decreasing the cost of production is required for their viability. Increasing health concerns of consumers lead to the development of functional foods. Proper and consistent application of sound manufacturing techniques is of great importance in the successful production of quality comminuted meat products.

1. INTRODUCTION

The processor's ability to produce food choices for consumer increases day by day due to constant emergence of new technologies, equipment and ingredients. The whole area of further processing is highly dynamic and products are constantly evolving. It is remarkable to see the switch from a basic array of raw agricultural commodities to so many ready to eat foods that has occurred in recent years (Fletcher, 2004). Convenience and variety are the foremost desirable characteristics of processed foods. Consumers demand for products will continue to be driven by safety, quality and consistency. Processed product can undergo different types of alteration, most of them aimed to improve the quality of product. However, food safety should be addressed properly in order to have a stable product without considerably altering its wholesomeness. Therefore, food processing is a complex art of scientific principles and good manufacturing practices. Increasing health concerns of consumers lead to development of new category of products that also have functional and health benefits. This chapter reviews the basic steps in comminuted meat products processing, ingredients used, types of comminuted meat products, their microbiological aspect and emerging trends in the comminuted products developments.

2. STEPS IN COMMINUTED MEAT PRODUCTS PROCESSING

A number of different principles may be applied to cut down meat into the required particle size. The nature of comminuted meat varies according to the method used and different methods of comminution may be used for different end products. However, most of them involve several common steps-comminution to reduce meat and fat particle size (grinding, mincing, chopping or flaking), mixing with ingredients, making into specific shape and finally packaging. The typical machines commonly employed for comminution are dicer, grinder and bowl chopper.

2.1. Dicing/Cubing

Meat cubes are used in many dishes and specialized sausages such Polish Krakowska sausage, which is done with visible chunks of meat (Marianski & Marianski, 2010). Meat cubes of uniform size are produced using dicer. Two main categories of machines are employed for making cubes. In one kind, machine pushes meat through a lattice, forming strips and cutting those strips into cubes with a rotating sickle blade. The other type machine cuts meat into strips, then use a multi-segment cutter head to produce cubes. The perfect cutting of meat products is a complex matter, since factors such as elasticity, fibre structure, dimensions, temperature and fat influence the cutting quality.

2.2. Flaking

In flaking, the frozen and tempered meat pieces are cut into flat slices (flakes). Block and impeller flakers are the two equipments commonly employed for flaking meat. In case of block flakers, meat block of specific size and temperature (–2 to –4°C) is essential for good quality flaking. The impeller flaker employs series of high speed rotating blunt blades or paddles to slice meat. Flakes of varying thickness could be produced by modifying the space between blades. The requirements of freezing and uniform block size are not required in impeller flaker. The flakes can be mixed with salt and other ingredients prior to being stuffed into a casing to form a log. After a period of frozen storage, the shape of the log can be modified by being processed in a hydraulic press that creates a particular shape. The log can then be cleaved into chops or steaks of various thicknesses (Claus *et al.*, 1994).

2.3. Grinding/Mincing

Grinding as the definition implies, grinds meat and pushes it along a worm-screw and then through the plate. A meat grinder is a machine used to force meat or meat trimmings by means of an auger under pressure through a horizontally mounted cylinder. At the end of the cylinder, there is a cutting system consisting of star-shaped knives rotating with the auger and stationary perforated discs (grinding plates). The perforations of the grinding plates normally range from 1 to 13 mm. The meat is compressed by the rotating feeding auger, pushed through the cutting system and extrudes through the holes in the grinding plates after being cut by the revolving star knives. The degree of mincing is determined by the size of the holes in the last grinding plate. If frozen meat and meat rich in connective tissue is to be minced to small particles, it should be minced first through a coarse disc followed by a second operation to the desired size. For non-emulsion type sausages, grinding is often the only form of comminution (FAO, 1991).

2.4. Chopping

Bowl cutter also known as silent cutter, cuts meat very finely and is a must have machine for commercial production of finely comminuted emulsified products. It is composed of revolving metal bowl that contains meat, while knife blades (set of 3–12) rotating at high speed on an axle, cut through the revolving meat mass. The meat is guided toward the knives by a plough fixed inside the bowl. Most of the large and high-speed bowl cutters are equipped with mechanical charger and discharger devices for loading and emptying the cutter, respectively. The meat can be chopped very finely, the degree largely depending on residence duration in the bowl. The speed of the knife, rotation speed of the bowl and sharpness of the blade determines the performance of chopper. The stainless steel bowl turns about 14–16 times per minute and the knives rotate about 3,000 times per minute. The resulting friction raises the temperature of meat mass during chopping by 10–20°C in 10–15 minutes of chopping. Modern bowl cutters are equipped with a thermometer for displaying the temperature of the meat mixture in the bowl during chopping. They also have devices to operate under a vacuum, which helps to improve colour and texture of the meat products by keeping oxygen out of the meat mixes and avoid air pockets (FAO, 1991).

2.5. Blending

As a separate processing step, blending refers to additional mixing for the purpose of solubilizing and swelling meat proteins before further processing.

By mixing meat and salt at the slowest speed, salt starts to extract proteins from broken muscle cells. Addition of ice water speeds up this process and controls the temperature rise. A separate blending step ensures a more uniform distribution of ingredients, especially of the cure and seasoning than could be achieved by grinding alone. Coarsely ground sausages are blended prior to being stuffed into casings. When grinding and mixing of raw materials occurs several hours before batter production, the process is known as pre-blending. This allows additional time for protein solubilization and swelling to occur. In addition, this time allows for sampling and analysis of protein, fat and moisture content of raw materials. Pre-blends of differing fat content can then be combined using knowledge of their composition to control precisely the composition of finished products.

2.6. Emulsifying

It is particularly useful in making fine emulsion type products. The emulsifiers can help the product manufacturer to achieve flexibility in raw material selection, excellent dispersibility in the meat batter, stable emulsions in products with high fat and water content, reduction of purge and cooking loss, and an optimized cost profile for sausages. The emulsion mill has a perforated plate, attached to which two edged extremely high speed revolving blades (rotor blade). The centrifugal pump of emulsifier forces the pre-ground meat through the perforated plate. Compared to the bowl cutter, the emulsifier operates at much higher speed, producing a finer emulsion like mix. The emulsifier is also perfectly suited to produce semi-processed products such as pig skin emulsions.

2.7. Moulding/Shaping

The simplest method of moulding patties/burgers is to shape in hand press. This method, however, is slow and only suited to small scale manufacture (Muthukumar and Kondaiah, 2011). Extrusion process is most widely used for large scale purposes. In case of sausage, the batter mix is transferred to the hopper of the filling machine/stuffer and fed to a nozzle by a piston pump. The casing is filled from the nozzle on a continuous basis and linked to form a string of individual sausages.

3. INGREDIENTS IN COMMINUTED MEAT PRODUCTS

In addition to meat, a host of ingredients like binders, spices, flavourings and additives are used to give varied meat products with distinctive flavour, taste and texture. Non-meat proteins are added to help to retain the moisture, provide additional emulsifying capacity, stability and texture (Zayas, 1997). Further, certain ingredients are incorporated into products to prevent quality loss during storage and improve shelf life.

3.1. Meat

The nature and amount of protein content in the meat determines its suitability in product preparation. A careful selection of meat ingredients is essential due to its role in the water binding, maintaining the fat content of the mixture and determining product cohesiveness. Animal tissues vary widely in moisture, protein and fat content, pigmentation and ability to bind water and fat. Thus, the knowledge about the properties and composition of various available meat tissues is essential to arrive at the correct formulation. Meat used for binding should have a sufficiently high protein content and the proteins should be readily extracted and form gels during cooking. The skeletal muscle has higher binding capacity, whereas high fat trimmings are generally poor binders. Head and cheek meats are intermediate in binding ability. Meats with high amount of connective tissue or large proportion of smooth muscle have low binding ability. Pre-rigor meat is superior to post rigor meat because more salt soluble proteins can be extracted. Filler meats are included in emulsion products formulations to "fill" the void space in emulsion meat products. Filler meats includes organs such as heart, glands, pork and beef tripe, pork/chicken skin, or snout and have poor water holding capabilities (Forrest *et al.*, 1975).

The meaty parts left attached to the bones after normal deboning are mechanically separated and widely used in meat processing. The use of such mechanically recovered meat (MRM) materials certainly contributes to the integration of all edible parts from carcasses into the food chain without wastage of valuable animal proteins. The meat can be recovered by grinding the bones into a paste and forcing it through very small holes in a drum. The pasty meat passes through the holes, whereas the hard bone particles are retained in the drum. Very high pressure is also employed to force the meat off the bones prior to separation. Very high quality standards of hygiene should be practiced while handling of the raw bones and production of MRM. Incorporation of MRM as raw material for meat products should be well balanced as it is prone to the development of rancidity (Pussa *et al.*, 2008).

3.2. Fat

Fat is an essential component of formulated meat products because it improves tenderness, juiciness and overall palatability. It also serves as the dispersed phase in meat emulsion. The fat content of meat ingredients varies more widely than the moisture/protein ratio and depends primarily upon the type of cut or trimmings as well as on the carcass grade. There is a relationship between the fat melting point and the degree of fat dispersion and absorption by protein film and fat water interface. Animal fats are sometimes separately emulsified in water by using sodium caseinate or by vegetable proteins. These emulsified fat batters may be directly incorporated into finely comminuted sausages or they may be chilled and diced into discrete particle sizes to create a desired visual effect (FAO, 1991).

3.3. Binders, Extenders and Fillers

These are another group of ingredients available for use in emulsion type products to enhance protein content, improve processing yields, sensory attributes and reduce formulation costs (Barbut, 1994; Naveena *et al.*, 2006). Binders are high level protein substances of animal or plant origin, which effect significant improvement in consistency and emulsifying and water binding capacities. Soy isolate, wheat gluten, caseinate, etc are some of classical examples for binders. Extenders and fillers are low in protein and not primarily intended for change of appearance or quality improvements but serve to add volume to the meat products. Their main purpose is to make meat products lower cost. Addition of starch and non-meat proteins improves fat coating of comminuted products, thereby enhances the textural properties of the products. Starch gelatinizes during cooking, increasing the emulsion viscosity and reducing fat globule mobility (Pearson & Gillett, 1997). Cereal flours, vegetable, roots and tubers, non fat dry milk, whey protein concentrate, sodium caseinate, wheat gluten, and polysaccharide gums at limited levels are used as extenders or fillers (Chin *et al.*, 1999). Cereal binders are also used to develop thermally processed emulsion type meat products due to their ability to improve the textural properties and sensory attributes (Prince Devadason *et al.*, 2010). The permitted maximum amount of meat extenders in sausage production is usually 3.5% and is strictly regulated by law in certain countries. Extenders and fillers are always used in combination with binders. Table 1 depicts the proximate composition of certain non meat proteins.

Table 1: Proximate composition of non meat proteins.

Non-meat proteins	*Moisture, %*	*Crude protein, %*	*Crude lipid, %*	*Ash, %*	*Carbohydrate,[a] %*
Whey protein concentrate	5.0	77.0	8.0	5.0	5.0
Soybean flour	9.4	37.4	16.7	4.3	32.2
Soybean protein isolate	6.0	84.6	1.0	4.5	3.9
Texturized soybean protein	10.0	46.8	1.2	4.2	42.0
Wheat gluten	8.0	83.0	1.5	1	6.5
Egg white powder	8.8	80.2	0.2	5.1	5.7
Gelatin	12.0	85.0	0.1	0.7	2.2

[a] Calculated by difference. *Source*: (Hsu & Lung, 2006)

3.3.1. *Soy proteins*

As a versatile food ingredient, soy proteins with its functional and nutritional properties greatly enhance the value of finished foods. These table proteins from soybeans are well known for their emulsifying and emulsion stabilizing properties. Soy proteins are available in the form of flours (full fat and defatted), grits, concentrates and isolates, texturized, untexturized, extruded and spun soy protein fibres. Soy protein is incorporated in meat products in a pre-hydrated form, at levels up to 4%, to improve product quality through its moisture and fat binding ability as well as through its emulsifying and structure forming potential (Rhee, 1994). Slightly bitter taste to meat products limits usage of both flour and grits to maximum of 2%. Soy protein concentrates are high protein products (not less than 70%). They possess good water absorption abilities and are used at 2.5–3.0% level in emulsion type sausages, luncheon loaves and meat patties (FAO, 1991). Soy protein isolates are the most refined form of soy proteins, without water soluble sugars and other insoluble components. Owing to their high protein content (minimum 97%) they have excellent dispersing, emulsifying, gelling and water and fat binding properties (Decker *et al.*, 1986). Textured soy proteins are available in a variety of shapes, sizes, textures, colours and flavours for example beef, pork, ham, chicken etc. They are used as extenders in comminuted sausages, meat patties and canned meat products.

3.3.2. *Milk proteins*

A number of milk protein products preparations like non fat dry milk (NFDM), calcium reduced non fat dry milk, dried whey, whey protein concentrate, casein, lactalbumin, lactoglobulin and other serum proteins are widely used in processed meat. NFDM increase water and fat binding ability, texture and flavour of meat products. It can be used up to 3.5 % level. Sodium and potassium caseinate are most widely used as an emulsifying agent in processed meats, while calcium and potassium caseinates are used when lower sodium formulations are required (Baardseth *et al.*, 1992). It is a fine white powder with neutral taste and a protein content of about 94%. Milk protein does not coagulate, shrink or form a gel while heating. If caseinate is added to meat before salt addition, it will surround free fat particles during comminution. Thus, the binding capacity of salt soluble meat proteins may be enhanced to form a more stable emulsion (Mittal & Usborne, 1985). Milk co-precipitates contain both casein and whey protein and used as good emulsifiers, water binder, gelling agent and thickeners, exist in a strongly folded and organized globular structure. Under appropriate heating conditions, they unfold and build intermolecular disulfide bonds resulting in a gel matrix whose capillaries entrap water. Hydrophobic and hydrophilic regions on the protein surface provide emulsifying properties.

3.3.3. *Egg proteins*

Egg proteins derived from whole egg, yolk or whites, are incorporated in processed meat for their good foaming, binding, thickening and emulsifying ability. They also possess good water retention capacity. Egg proteins have excellent nutritional properties (Colmenero, 2004).

3.3.4. *Flours and starch*

Flours of cereal grains (corn, wheat, rice) and tubers (potato, cassava) are used as extenders due to their high carbohydrate and low protein contents. Starch is a complex sugar of plant origin, used up to 4% in meat products. During heat treatment of product, added starch binds released water and juice, thus decreasing weight losses (Annor Frempong *et al.*, 1996). Wheat flour contains 65–75% starch, 8–14% gluten and 12–16% water. If flour is added in product formulation, care must be taken to prevent quick decomposition of the product mixture. Owing to the activity of flour enzymes, flour is commonly used up to 4% in low quality sausages (FAO, 1991).

3.3.5. *Hydrocolloids*

Carrageenans are sulfated linear polysaccharides of D-galactose and 3,6-anhydro-D-galactose extracted from red seaweeds. In the food industry, they have been used for their gelling, thickening, and stabilizing properties. Among the three basic types, Iota and kappa carrageenan are gelling agents and lambda is a non gelling thickener and posses very good water retention properties. Iota carrageenan will gel with calcium ions to form a syneresis free, clear plastic gel with good resetting properties after shear. Hydrocolloids are being tried as fat replacements in low-fat products. Reduction of fat in ground meat products results in a different mouth feel and dry taste, which consumers do not always accept. When carrageenans are incorporated along with phosphates, they improve the textural characteristics of the product by decreasing toughness and increasing juiciness (Trius & Sebranek, 1996). In cooked sliced meat products carrageenan is used to improve moisture retention, cooking yields, slicing properties, mouth feel and juiciness (Bater *et al.*, 1992).

3.4. Water

Added water accelerates curing process and helps to extract water soluble meat proteins influencing most of the functional properties and yield. Water addition aids in dissolving and uniform distribution of non meat ingredients, providing fluid condition during fat chopping and increasing the product yield. It is usual to add up to 50% of water as ice to minimize temperature rise and microbial growth during chopping. Texture and tenderness of the products are markedly affected by the added water content. Ice flakes and/ or chilled water added during chopping permits longer and efficient churning of meat mass without mechanical overheating. It reduces the products cost and lower product energy value and it has direct impact on product texture by increasing juiciness (Ockermann & Basu, 2004).

3.5. Salt

In a processed meat product, salt helps to bind the meat proteins together and acts as a binding agent between meat and fat. Salt assists in the extraction of soluble muscle proteins. It also increases the water binding capacity during cooking so that the final product has improved texture, tenderness, and palatability. Salt also imparts flavour and antimicrobial effects (Gillette, 1985). Levels of added salt in processed meat products vary between different types of products depending on recipes and formulations. Most of the emulsion meat products contain 1.5–2.0% added salt.

3.6. Sugar

Sugars (sucrose, dextrose or corn syrup) may be added at 0.3 to 0.5% level to meat products to provide specific flavor and counteract salty taste, lower the a_w value, which may be important for dried and canned products. Sugar also acts in dry fermented sausages and raw hams as a nutrient source for microbes, which convert sugars into organic acids (lactic, acetic) resulting in souring.

3.7. Nitrite

Nitrite in meat greatly delays development of botulinal toxin (botulism), develops cured meat flavour and color, retards development of rancidity, off-odors and off-flavours during storage, inhibits development of warmed-over flavour, and preserves flavours of spices, smoke, etc. Sodium nitrite, rather than sodium nitrate, is most commonly used for curing. In a series of normal reactions, nitrite is converted to nitric oxide. Nitric oxide combines with myoglobin, the pigment responsible for the natural red color of uncured meat. They form nitric oxide myoglobin, which is a deep red color that changes to the characteristic bright pink normally associated with cured and smoked meat when heated during the smoking process. Nitrite contributes flavour stability by complexing with heme iron, which if free could act as a potent catalyst in lipid oxidation. Sodium nitrite is commonly used, although in certain cases it may be substituted for by potassium nitrite. The nitrite is added at levels of 120–150 (Pegg & Shahidi, 2000). Nitrite is used in conjunction with the reducing agents ascorbate or erythorbate and phosphates.

3.8. Polyphosphate

Food grade phosphates are used in meat products for several reasons such as changing and/or stabilizing of pH value, increasing water holding capacity in order to lead to higher yields, decreasing losses of weight in cooking, improving texture and sensory properties like tenderness, juiciness, color and flavor, extending shelf-life, etc. The action of phosphate in raising the pH also improves the water binding capacity. Products formulated with phosphates retain more natural juices and added water during heat processing and subsequent reheating. The alkaline phosphate enhance the fat emulsifying capacity of myofibrillar proteins by solubilising and dissociating actomyosin into actin and myosin, which in their dissociated forms can bind more water (Kondaiah *et al.*, 1985; Anjaneyulu *et al.*, 1990). Phosphates may also help to stabilize flavour and colour in finished product, presumably by sequestering transitional metal ions (Fe and Cu), thus reducing oxidation. Phosphates used at a level up to 0.5% in finished products (Claus *et al.*, 1994). Some of the approved phosphates include sodium tri polyphosphate, sodium pyrophosphate, sodium hexametaphosphate and sodium acid pyrophosphate. If both salt and phosphates are present in meat products, they act synergistically to improve yield and the binding of meat particles (Huffman *et al.*, 1981).

3.9. Seasonings

Natural spices, herbs and vegetable bulbs are the main groups of seasonings used to flavour food. Mixtures of seasonings were developed in order to serve as flavouring agents for various meat products. Spices may be added as natural spice or spice extracts. Flavorings refer to extractives that contain flavor constituents from fruits, vegetables, herbs and roots. The commonly used herbs are sage, parsley and mint, while common spices include black pepper, paprika, mace, cinnamon, coriander and nut mug. To achieve uniformly and well flavoured comminuted meat products, spices must be properly sprinkled in the meat batter and are added after additives, but always before adding fatty tissue to avoid them sticking to it causing a non-uniform color (FAO, 1991). All spices for sausage formulation should be mixed before adding to the meat batter. Monosodium glutamate (MSG), inosine mono phosphate (IMP) and guanosine monophosphate (GMP) are some of the synthetic compounds used as potent flavor enhancers.

3.10. Additives

These are substances incorporated in product to provide technological and quality characteristics in foods. Additives preserve flavor or enhance products taste and appearance, shelf life and safety. The use of food additives has become more prominent in recent years due to the increased production of prepared, processed, and convenience foods. Food additives can be divided into several group *viz.* acidity regulators, antioxidants, emulsifiers, flavor enhancers, preservatives, etc.

(*a*) Ascorbic acid/sodium ascorbate—also known as Vitamin C, prevents oxidation that causes color change and spoilage. Ascorbic acid is commonly added to cured meats when nitrite is also used.

(*b*) Butylated hydroxyanisole (BHA), butylated hydroxytoluene (BHT)—antioxidants that protect natural nutrients like Vitamin A and prevent the fat in meat from developing off flavors and odors.

(*c*) Smoke flavoring—a condensed form of smoke, smoke flavoring is an alternative to natural wood smoking. Smoke flavoring gives products a smoky taste without a grill.

(*d*) Sodium erythorbate–derived from Vitamin C, erythorbate is an antioxidant that maintains the color of processed meats.

(*e*) Tocopherol—a form of Vitamin E that prevents fat in meat and poultry from becoming rancid.

(*f*) Glucono delta lactone (GDL) is a cure accelerator that produces an acid tangy flavor similar to flavor resulting from natural fermentation.

4. TYPES OF COMMINUTED MEAT PRODUCTS

Considerable variations in raw materials used and process involved in production of comminuted meat products, resulted in a wide range of products, which meant to cater to specific target market. Comminution can be coarse to produce non-emulsified meat products like hamburger, dry and semi dry fermented sausages and restructured products, or fine to produce emulsified sausages like frankfurters, bologna, etc.

4.1. Coarse Comminuted Meat Products

Hamburger, fresh pork sausage, semi dry and dry sausages are some of the important coarse ground products. These groups of products have coarse texture with small pieces of meat and fatty tissues visible. To make the connective tissues are less visible and more digestible, meat should be passed through a 15 to 25 mm grinder plate, while trimmings and fatty tissues need to pass through a 2 to 6 mm plate. Meat is passed through the grinder first, followed by the trimmings and finally the fatty tissues. This avoids fat separated from the fatty tissues greasing the surface of meat pieces, thereby decreasing their binding ability.

4.1.1. *Dry and semi dry sausages*

These types of sausages represent one of the oldest forms of meat preservation. Dry and semi dry sausages as their name imply differ in moisture content, which averages between 30 to 40% and 40 to 50%, respectively. These sausages differ from all other types of sausage in that they require microbial fermentation. During smoking, these sausages are deliberately held at temperatures, which encourage bacterial growth and fermentation. Dried sausages should never exceed 30°C at any stage as this would retard bacterial growth. The natural flora of meat, processing equipment and plant environment forms source of useful organisms responsible for desirable fermentation. To overcome the problems associated with natural bacterial fermentation, starter cultures of desired bacterial strains belonging to the genera *Lactobacillus* (*L. plantarum*, *L. pentosus*, *L. curvatus* and *L. sake*), *Pediococcus* (*P. pentosaceus* and *P. acidilactici*), *Micrococcus* (*Korucia varians*) and *Debaryomyces* (*D. hansenii*) are used in commercial production (Hugas & Monfort, 1997). This fermentation reduces the pH of the product to 5.0 to 5.3 for dry sausage and 4.6 to 5.2 in the case of semi-dry sausage. In addition, some of the moisture is removed by drying to produce a relatively low water activity (a_w) of approximately 0.85 in the case of dry sausage and 0.85 to 0.91 in the case of semi-dry sausage. The combined low pH, reduced a_w and high salt content contribute to the extended shelf life of the product. Several biochemical and physical changes occur during the ripening of fermented sausages that determine the flavour and odour of the end product (Ordonez *et al.*, 1999). One of the distinctive features of dry sausages is that they are processed uncooked and mostly smoked. The best known are dry pork sausages, dry beef sausages, mixed dry sausages, summer sausages and salamis.

4.1.2. *Burgers*

These were originally derived from hamburgers, a large beef sausage, which is cut into slices before cooking. They are made of minced or flaked beef, but nowadays other meats can also be used. The meat content varies from 100% including about 20% fat, to 80% or less with various additions of cereals, soya protein, salt, water, etc., to improve the cohesiveness of burgers after cooking and reduce the cost and fat content of the product (Varnam & Sutherland, 1995). However, in the USA, extenders are not permitted and hamburger meat must consist only of minced beef together with beef fat. The total fat content must also not exceed 30%, but is generally varies from 15 to 20%. Use of low quality meat with more connective tissue leads to burger of undesirable texture. The herbs and spices are less important in flavouring of burgers than that of sausage. However, novelty burgers are made with high level of flavouring ingredients like chilli, pepper and curry powder. Burger products with barbecue flavour are produced by incorporating liquid smoke (Warriss, 2000).

4.1.3. *Reformed and restructured steaks*

Restructuring has been defined as the use of manufacturing steps to create a consumer ready product, which resembles intact muscle, such as a steak, chop, etc (Pearson & Gillett, 1997). Restructuring has been developed as a method for transforming lower value cuts and quality trimmings into products of higher value. Flaked meat is used, rather than minced meat, because the particles bind together better and produce a better texture. Bind in restructured meat products may be achieved through the formation of heat or cold set gels. Conventional restructured meat products depend upon hot set binding of meat proteins extracted with the combined effects of salt, phosphate and mechanical action. This technology limits the marketing of the product to either the precooked or frozen state. However, high yields (25% above meat weight) are possible. In cold set binding, adhesion is promoted by the thickening agents like sodium alginate, Fibrimex (Harimex Inc., Alberta, Canada) are used to form a gel that binds the product together. The thermo-irreversible gel is formed by reaction of the sodium alginate with calcium ions. Fibrimex is a blood-based cold binding system that binds meat pieces based on the blood clotting action between brinogen and thrombin resulting in the conversion of brinogen to brin which cross-link with collagen and gel, thereby binding the meat pieces being restructured (Boles & Shand, 1998). Much of the success of a product depends on how closely this reaction can be controlled to produce the desired gel properties. Cold set binding have been developed to meet the demand for restructured meats that can be marketed in the raw, refrigerated state. Larger pieces of meat can also be formed into blocks of tissue using this technique (Farouk, 2010).

4.2. Fine Comminuted Meat Products

4.2.1. *Emulsion type meat products*

Emulsified meat products are those that have been prepared from finely comminuted meat. Hot dog, frankfurter, mortadella, bologna, liver sausage, pâté are typical examples. The amount of fat that can be incorporated in a stable emulsion depends on fat particle size, meat pH, temperature during emulsification and the amount and type of soluble proteins. It is very important to maintain low temperature during emulsion formation in order to avoid melting of fat particles, denaturation of soluble proteins and lowering of viscosity. As a general rule, it is recommended to attain the end point chopping temperature below 18, 12 and 8°C for beef, pork and poultry fat, respectively (Mandigo & Esquivel, 2004). If the temperature becomes too high, emulsion break down can occur during subsequent heat processing (Brown & Ledward, 1987). Addition of water as ice, properly sharpened knives and keeping the clearance between the knives and the bowl below 0.7 mm will reduce heating of the batter caused by friction.

Sausages are very popular products made from finely chopped or comminuted lean meat (45–60%) and fat (15–20%), mixed with salt, non meat binder, spices and other ingredients, and then filled into a container. The latter is often a natural casing, made from the cleaned intestines of cattle, sheep or pigs. Artificial casings made from reconstituted collagen derived from other parts of the animal, particularly the skin of cattle, and have the obvious benefit of uniformity of the product (Warriss, 2000).

4.2.2. *Canned/retort pouched comminuted meat products*

Corned beef, pork luncheon meat and chopped pork are the most important range of products in this category. The emulsion mixture must be sufficiently stable to withstand excessive fat cook-out under the required heat processing conditions. The mixture used for the high quality lean meat, part of which is emulsified with polyphosphates and other functional additives before addition of the remaining lean and fatty meats. In the case of luncheon meat—the ready to eat products made from finely chopped meat (pork/beef/veal/lamb), often contains high proportions of offals such as lungs and relies for their fat and water binding on additives such as potato starch, soya protein, etc. The heat processing given to this type of products is sub-botulinal as for whole cured meats, the aim being to achieve a centre temperature of about 105°C. Their stability relies on the interaction of the salt, nitrite and meat. Corned beef relatively a unique semisolid meat product traditionally manufactured as a byproduct of beef extract industry. However, with the decline of beef extract market, corned beef is now produced from uncooked beef which may be cured before packing or comminuted and cured by addition of coarse grains of salt—corns and nitrite before packing. Other comminuted meat products such as sausages, frankfurters and meat balls are also frequently canned. They are packed either in the dry form, in brine—as happens with frankfurters or in gravy or sauce. The formulations used are rather better than for the comparable open pack products with a higher lean: fat ratio to avoid excessive cook out (Wilson *et al.*, 1981).

Recently metal cans have been replaced by retort pouches *i.e.*, laminated multi-layer, flexible pouches. These laminated pouches must provide superior gases and moisture barrier properties, seal integrity, toughness, puncture resistance and must withstand higher pressure and cooking temperatures up to 135°C. Retort pouches usually consists of three-layer laminates composed of an outer layer of polyester film, a middle layer of aluminium foil and an inner layer of polypropylene. Foil laminated retortable pouches costs less, lighter in weight with faster processing time (Prince Devadason *et al.*, 2010).

5. MICROBIOLOGY OF COMMINUTED PRODUCTS

Comminuted meat products generally have a higher incidence of contamination with pathogenic microorganisms and are of poorer overall microbiological quality than whole meat (Varnam & Sutherland, 1995). This stems from three main factors: Firstly, the use of low quality ingredients, which have been subject to high levels of handling and, possibly, temperature abuse. The use of mechanically recovered meat has been of particular concern. Secondly, the thorough mixing of a wide range of ingredients leads to spread of contamination throughout the product. Thirdly, release of cellular constituents during comminution and subsequent operations, providing a readily available source of nutrients. At the same time, the surface area available for microbial growth is greatly increased and the organisms, originally on the surface, are distributed throughout the meat.

The meat to be processed should be at a temperature of 0–3°C and the bacterial count, whether fresh chilled or frozen thawed should be as low as possible, preferably between 10^2–10^4/g of meat. High counts will arise during comminution, if the temperature of the product mixture is not properly controlled. Addition of salt and nitrite, smoking and heat treatment will extend the shelf life. If the product mixture is kept at a relatively high temperature (above 20°C) before it is mould and cooked, vegetative bacteria will grow and cause deterioration. In order to kill most microorganisms present, emulsion type products must be heat treated to an internal temperature of 72–75°C (FAO, 1991). The product must be quickly chilled after heat treatment, in order to prevent growth of any surviving flora. Further, finished products must not come into contact with the floor or any other contaminated materials, especially raw meat.

Spoilage microfloras of fresh emulsion generally are similar to those found in ground meat. The type of meat, presence of preservatives, storage temperature and atmosphere of packing determine the predominant microbes. Products in air permeable packaging will have predominantly more *Pseudomonas* spp. during chilled storage and, Enterobacteriaceae during temperature abused conditions. Presence of these bacteria on fully cooked products is the result of post heating contamination, usually during casing removal, slicing and subsequent handling. Lactic acid bacteria and Micrococcaceae predominate in meats packaged with preservatives and *Brochothrix thermosphacta* in vacuum and modified atmosphere packaged products (Garcia Lopez *et al.*, 1998). In overall terms, uncooked comminuted meat products have a good safety record although the incidence of pathogens may be high. Products of this type are usually cooked at or shortly before, the point of consumption and food borne disease can result only from raw consumption, undercooking, or contamination of the cooked product. In recent years, infection with both *Salmonella* and *Escherichia coli* O157:H7 has occurred due to undercooking (Varnam & Sutherland, 1995).

6. EMERGING TRENDS AND RESEARCH NEEDS

6.1. Development of Healthier Meat Products

The increase in food related diseases, such as cardiovascular disease, diabetes, cancer and obesity has led consumers to demand food products that not only offer nutritional value but also functional and health benefits (Solan, 2008). In recent years, much attention has been paid to develop meat and meat products with physiological functions to promote health conditions and prevent the risk of diseases. The functional meat products may provide added health benefits following the addition/concentration of a beneficial ingredient, or the removal/ substitution of an ineffective or harmful ingredient. Value improvement can be realized by adding functional compounds including conjugated linoleic acid, vitamin E, n-3 fatty acids and selenium in animal diets to improve carcass composition and fresh meat quality. In addition, functional ingredients such as vegetable proteins, dietary fibers, herbs and spices, and lactic acid bacteria can be directly incorporated into meat products during processing to improve their functional value for consumers (Zhang *et al.*, 2010; Muthukumar *et al.*, 2012). Though the alteration of nutritional composition of meat products is possible either by direct addition of bioactive food ingredients or by the inclusion of bioactive compounds into animal diets, the later technique has the advantage that the bioactive compounds would be biologically introduced into the food and thus would not have to be declared as food additives (Decker & Park, 2010). Functional compounds, especially peptides, can also be generated from meat and meat products during processing such as fermentation, curing and aging, and enzymatic hydrolysis. The following approaches could be adopted to develop functional meat products (*a*) to raise the concentration of beneficial natural compound in the food up to the desired effective level (*e.g.*, increasing n-3 fatty acids or conjugated linoleic acid); (*b*) to add a compound that is not normally present in food but has proven benefits (*e.g.*, prebiotics and probiotics); (*c*) to replace a compound, usually a macronutrient (*e.g.*, fat); (*d*) to remove specific compounds, either naturally present or produced by processing to reduce adverse health effect (*e.g.*, saturated fatty acids, trans fatty acids); (*e*) to improve the bioavailability (*e.g.*, improved iron absorption) and (*f*) any combination of the preceding possibilities (Ashwell, 2002; Holm, 2003).

6.1.1. *Dietary fibre*

Dietary fibre consists of the remnants of edible plant cells, which include cellulose, hemicellulose, lignin, pectin and a variety of gums and mucilages. An adequate fibre diet optimizes gastrointestinal physiological function. Fibres mainly lower blood cholesterol levels by interfering with absorption of dietary cholesterol. Dietary fibre is thought to protect against breast cancer by altering estrogens metabolism. It may reduce the prostrate cancer by decreasing the amount of androgen available to the body by removing them from the enterohepatic circulation (Walker *et al.*, 1992; Stoll, 1996). Besides health benefit effects, dietary fiber supplementations increase the bulk and prevent cooking loss in meat products with no or fewer changes in textural parameters by enhancing water binding capabilities and carries great economical advantages for both the consumers and processors (Grigelmo Miguel *et al.*, 1999). Sugar beet, wheat, oat and pea fibre have been used in cooked meat products (Sen *et al.*, 2008). The functional properties of the fibre and the decrease in the product caloric content allows for marketing as healthy alternative, taking advantage of growing health concerns of the consumers. Limes, orange and lemons fibres have been effectively used in dry fermented meat sausage for increasing soluble fibre content. Dietary fibre from citrus fruits contain higher proportion of soluble dietary fibre (about 33%) compared to cereal source like wheat bran which is about 7% (Gorinstein *et al.*, 2001).

6.1.2. *Fat and fatty acids*

The demand for low fat meat products has greatly increased in recent years as high fat intake, especially saturated fats, is associated with increased risk of chronic diseases of the circulatory system and some types of cancers. Since fat acts as a reservoir for flavour compounds and contributes to the texture of product, reducing the fat content in meat products may alter product quality. Long chain polyunsaturated n-3 fatty acids (PUFA) have been implicated as critical nutrients for human health. The beneficial health effects of omega-3 PUFAs have been attributed to their ability to lower serum tri acylglycerol and cholesterol. In addition, omega-3-fatty acids are essential for normal growth and development and may also play a role in the prevention and treatment of hypertension, arthritis, other inflammatory and autoimmune disorder and cancer. In order to improve their healthy lipid profile, reformulation is generally based on the partial replacement of the animal fat normally present in the product with another fat whose characteristics are more in line with health recommendations *i.e.*, with smaller proportions of SFAs and larger proportions of MUFAs or PUFAs, better n-6/n-3 PUFA and PUFA/SFA ratios, and if possible cholesterol free. A variety of non-meat fats of plants and marine origin have been added to different meat products as partial substitutes for meat fats (Cengiz & Gokoglu, 2007). Fortification of meat with n-3 polyunsaturated fatty acids is an emerging area of commercial interest for development of functional meat products. Addition of omega 3 fatty acids into food can cause off flavour because these oils are prone to oxidation. However, several processes have been developed that microencapsulates the oil in a gelatine matrix with double shell protection for inclusion in foods. However, the potential losses of n-3 PUFAs due to meat processing and cooking could be significant. The use of antioxidants to maintain lipid stability in muscle food products without alteration of the fatty acid profile has been an effective means to delay lipid oxidation (Lee *et al.*, 2006).

Conjugated linoleic acid (CLA) is another fatty acid which has got much attention nowadays in meat and functional meat products. The CLA represent a heterogeneous group of positional and geometric isomers of linoleic acid, which are predominantly found in milk, milk products, meat and meat products of ruminants. Beef fat contains 3.1 to 8.5 mg CLA/g fat with the 9-cis and 11-trans isomers contributing 57–85% of the total CLA (Decker, 1995). The major health benefits attributed to CLA, includes anti-obesitic, anti-carcinogenic, anti-atherogenic, anti-diabetagenic, immunomodulatory, apoptotic and osteosynthetic effects (Benjamin & Spener, 2009). More recently, CLA has been investigated for its ability to change body composition, suggesting a role as a weight reduction agent. The CLA supplementation exhibited lower body fat and increased lean body mass possibly by reducing fat deposition and increasing lipolysis in adipocytes.

6.1.3. *Bioactive peptides*

Bioactive peptides have been defined as food derived components (genuine or generated) that, in addition to their nutritional value exert a physiological effect in the body. Food derived bioactive peptides have been shown to display a wide range of physiological functions including antihypertensive, antioxidative, opioid agonistic, immunomodulatory, antimicrobial, prebiotic, mineral binding, antithrombotic and hypocholesterolemic effects. Interestingly, within the parent protein sequence, the peptides are inactive and thus must be released to exert an effect (Ryan *et al.*, 2011). They are released by proteolytic enzymes. Primarily, meat proteins are degraded into peptides by endogenous enzymes during fermentation of meat products. Utilizing commercial proteases is an efficient way of releasing bioactive peptides from meat proteins. Carnosine, anserine, L-carnitine, taurine, histidyl dipeptides are some of the important meat based bioactive compounds (Quinn *et al.*, 1992). L carnitine helps the body to absorb calcium to improve skeletal strength (Arihara, 2004). Angiotensin converting enzyme (ACE) inhibitory peptides have attracted much attention because of their ability to prevent hypertension. These peptides could be used as potent functional food additives and could constitute a natural and healthier alternative to ACE inhibitory drugs. The commercial chicken-meat extract known as "Brand's Essence of Chicken" (BEC; Cerebos Pacific Ltd., Singapore) also claims to contain antihypertensive components (Slim, 2001). Major classes of bioactive components are listed in Table 2.

Table 2: Examples of major classes of bioactive food components used for the development of functional meat products.

Lipids and fatty acids	n-3 polyunsaturated fatty acids (eicosapentaenoic and docosahexaenoic acids), monounsaturated fatty acids, conjugated linoleic acids (CLA).
Proteins and peptides	Soy protein, bioactive peptides.
Prebiotic	Dietary fibres (Oats, soy, citrus, wheat bran etc.), Oligosaccharides.
Probiotic	Lactic acid bacteria, bifidibacteria.
Vitamins	Tocopherols, folic acid, ascorbic acid.
Minerals	Calcium, magnesium, selenium , zinc, iron.
Phytochemicals	Polyphenols, phytosterols, flavonoids etc.

Source: (Jimenez *et al.,* 2006)

6.1.4. *Probiotics*

Probiotics are defined as live micro organisms, which when administered in adequate amounts confer a health benefits on the host. The most common use of probiotic microorganisms is in fermented dairy products, although recently a new idea is developed to use probiotic strains in other foods such as fermented meat products. However, commercial application of probiotic microorganisms in fermented sausages is not common yet. There are both advantages and disadvantages connected to fermented meat matrices. They are adequate for the carriage of probiotic bacteria since they are usually not heated or only mildly heated and may promote the survival of probiotic bacteria in the gastrointestinal tract. In contrast, bacterial viability may be reduced due to the high content in curing salt and the low water activity and pH. Therefore, results are expected to be strain dependent. Candidate probiotic strains have been obtained through screening for technological requirements among bacteria that are naturally present in the meat or that originate from meat starter cultures. Alternatively, existing probiotic bacteria have been applied in meat products. Finally, the evaluation of the end-products needs to deal with both health effects through human intervention studies and technological characteristics by taste panels (Vuyst *et al.*, 2008).

6.1.5. *Vitamins and minerals*

Dietary antioxidants have been suggested to be beneficial to immune function, heart disease and cancer. Muscle foods could be even better sources of vitamin E through meat fortification or dietary supplementation with α tocopheryl acetate. Supplementation of livestock with α tocopheryl acetate in addition to improving nutritional quality of meat, preserve flavour and colour of meat by protecting it from lipid oxidation. The most commonly under consumed minerals *viz.*, calcium, potassium and magnesium could be incorporated into meat products without affecting sensory profile (Decker & Park, 2010).

6.2. Reduced Sodium Meat Products

As diet is the main source of sodium, awareness among people regarding its possible role upon health has driven demand for various low sodium foods including meat products. The association between excessive sodium intake and the development of hypertension has prompted public health and regulatory authorities to recommend reducing dietary intake of sodium to 2.4 g (6 g salt) per day. However, this is not an easy task as common salt affects taste and flavor, functional attributes, stability, and food safety of meat products. The various properties such as taste and flavor, binding as well as microbiological characteristics should be given due care while developing low salt meat products. The major barriers to salt replacement is cost, as salt is one of the cheapest food ingredients available (Verma & Banerjee, 2012). Different approaches have been proposed to reduce the salt content in processed foods (Desmond, 2006). This includes

6.2.1. *Use of salt substitutes*

Potassium chloride is probably the most common salt substitute in low or reduced salt/sodium foods. However, substitution of sodium chloride in meat products with KCl over 50% led to some degree of bitterness and loss of saltiness (Marsh, 1983). Also concerns were raised about the possible vulnerability of certain population subgroups to high potassium load from these salt substrates. The use of mineral salt mixtures is a good way to reduce the sodium content in meat products (Ruusunen *et al.*, 2007). Almost half of sodium is removed and replaced with potassium chloride, magnesium sulphate and the essential amino acid L-lysine hydrochloride. Replacements at 25–40% appear to be the range at which the flavour impact is most noticeable (Seman *et al.*, 1980).

6.2.2. *Use of flavour enhancers and masking agents*

Flavour enhancers such as yeast extracts, lactates, MSG and nucleotides, amongst others. These enhancers work by activating receptors in the mouth and throat, which helps compensate for the salt reduction (Brandsma, 2006).

Bitter bhosepers [Betra TM-adenosine 5-monophosphate (AMP)] can be used to improve the taste of NaCl/KCl mixtures. The AMP works by blocking the activation of the gustducin in taste receptor cells and thereby preventing taste nerve stimulation.

6.3. Substitution of Nitrite in Meat Products

The ability to inhibit *C. botulinum* combined with the improvement in aesthetic qualities of colour, texture and flavour contributed to the widespread use of nitrite in meat products. However, there has been some concern over the use of nitrites and nitrate in foods due to the formation of carcinogenic nitroso compounds. Reduction of allowable sodium nitrite levels in finished products from 200 to 10 or 20 ppm has been urged. Thus, there have been efforts to find out substitutes for nitrite. Several hundred compounds have been examined, but none has been a good substitute. Combinations of various additives in an attempt to provide the several functions of nitrite have been attempted. Among nitrite substitutes, ascorbate or erythorbate, alpha tocopherol, sorbate, propylene glycol and lauricidin, *Lactobacillus fermentum* strain, monascus extract, nitrite ligands, short-chain alkynoic and alkenoic acids and esters and protoporphyrin have been tried as nitrite substitute.

REFERENCES

Anjaneyulu, A.S.R., Sharma, N. and Kondaiah, N. (1990). Specific effect of phosphate on the functional properties and yield of buffalo meat patties. *Food Chem.*, 36: 149–154.

Annor Frempong, I.E., Annan Prah, A. and Wiredu, R. (1996). Cassava as a non-conventional filler in comminuted meat products. *Meat Sci.*, 44: 193–202.

Arihara, K. (2004). Functional foods. *In*: Encyclopedia of meat science. Vol I. Chemistry and physics of comminuted products. Devine, C. and Dikeman, M. (*Eds.*), Oxford, Elsevier, pp. 492–499.

Ashwell, M. (2002). Concepts of functional foods. International Life Sciences Institute, Washington DC, Europe Concise Monograph Series.

Baardseth, P., Naes, T., Mielnik, J., Skrede, G., Holland, S. and Eide, O. (1992). Dairy ingredients effects on sausage sensory properties studied by principal component analysis. *J. Food Sci.*, 57: 822–828.

Barbut, S. (1994). Protein ultrastructure and functionality. *In*: Protein functionality in food systems, Hettiarachchy, N. and Ziegler, G. (*Eds.*), Marcel Dekker, New York, pp. 383–433.

Bater, B., Descamps, O. and Maurer, A.J. (1992). Quality characteristics of hydrocolloid added oven roasted turkey breasts. *J. Food Sci.*, 57: 1068–1070.

Benjamin, S. and Spener, F. (2009). Conjugated linoleic acids as functional food: An insight into their health benefits. *Nutr. and Metab.*, 6: 36.

Boles, J.A. and Shand, P.J. (1998). Effect of comminution method and raw binder system in restructured beef. *Meat Sci.*, 49: 297–307.

Brandsma, (2006). Reducing sodium: A European perspective. *Food Technol.*, 60: 25–29.

Brown, S. and Ledward, D.A. (1987). Effect of temperature of comminution on the stability and eating quality of English sausages. *Meat Sci.*, 20: 97–105.

Cengiz, E. and Gokoglu, N. (2007). Effects of fat reduction and fat replacer addition on some quality characteristics of frankfurter-type sausages. *Int. J. Food Sci. Tech.*, 42: 366–372.

Chin, K.B., Keeton, J.T., Longnecker, M.T. and Lamkey, J.W. (1999). Utilization of soy protein isolate and konjac blends in a low fat bologna (model system). *Meat Sci.*, 53: 45–57.

Claus, J.R., Colby, J.W. and Flick, G.J. (1994). Processed meats/poultry/seafood. *In*: Muscle foods, Donald, M., Anthony, W. and Burdette, C.B. (*Eds.*), Chapman & Hall, New York, pp. 106–162.

Colmenero, F.J. (2004). Non meat proteins. *In*: Encyclopedia of meat science. Vol I. Chemistry and physics of comminuted products. Devine, C. and Dikeman, M. (*Eds.*), Oxford, Elsevier, p. 276.

Decker, C.D., Conley, C.C. and Richert, S.H. (1986). Use of isolated soy protein in the development of frankfurters with reduced level of fat, calories, and cholesterol. *Proc. European Meeting of Meat Research Workers*, 7: 333–349.

Decker, E. and Park, Y. (2010). Healthier meat products as functional foods. *Meat Sci.*, 86: 49–55.

Decker, E.A. (1995). The role of phenolics, conjugated linoleic acid, carnosine, and pyrroloquinoline quinine as nonessential dietary antioxidants. *Nutr. Rev.*, 53: 49–58.

Desmond, E. (2006). Reducing salt: A challenge for the meat industry. *Meat Sci.*, 74: 188–196.

FAO, (1991). Guidelines for slaughtering, meat cutting and further processing. FAO Animal production and health paper 91. Food and Agriculture Organization of The United Nations, Rome.

Farouk, M.M. (2010). Restructured whole-tissue meats. *In*: Hand book of meat processing. Blackwell publishing, USA.

Fletcher, D.L. (2004). Further processing of poultry. *In*: Poultry meat processing and quality. Woodhead Publishing Ltd., Cambridge, England.

Forrest, J.C., Aberle, E.D., Hedrick, H.B., Judge, M.D. and Merkel, R.A. (1975). Principals of Meat Science, WH Freeman and Co, San Francisco.

Garcia Lopez, M.L., Preito, M. and Otero, A. (1998). The physiologic attributes of Gram negative bacteria associated with spoilage of meat and meat products. In: The microbiology of meat and poultry. Davies, A. and Board, R. (*Eds.*), Blackie Academic and Professional, New York, pp. 1–28.

Gillette, M. (1985). Flavour effects of sodium chloride. *Food Technol.*, 39: 47–52.

Gorinstein, S., Martin-Belloso, O., Park, Y.S., Haruenkit, R., Lojek, A., Ciz, M., Caspi, A., Libman, I. and Trakhtenberg, S. (2001). Comparison of some biochemical characteristics of different citrus fruits. *Food Chem.*, 74: 309–315.

Grigelmo, M.N., Abadfa, S.M.I. and Martin B.O.A. (1999). Characterisation of low fat high dietary fibre frankfurters. *Meat Sci* 52: 247–256.

Hamm, R. (1986). Functional properties of the myofibrillar system and their measurements. In: Muscle as food. Bechtel, P.J. (*Ed.*), Academic Press, Orlando, FL, pp. 135–199.

Holm, L. (2003). Food health policies and ethics: Lay perspectives on functional foods. *J. Agr. Environ. Ethics*, 16: 531–544.

Hsu, S.Y. and Lung, Y.S. (2006). Comparisons on 10 non-meat protein fat substitutes for low fat Kung-wans. *J. Food Eng.*, 74: 47–53.

Huffman, D.L., Cross, H.R., Campbell, K.J. and Cordray, J.C. (1981). Effect of salt and tri-polyphosphate on acceptability of flaked and formed hamburger patties. *J. Food Sci.*, 46: 34–36.

Hugas, M. and Monfort, J.M. (1997). Bacterial starter cultures for meat fermentation. *Food Chem.*, 59: 547–554.

Jimenez, C.F., Reig, M. and Toldra, F. (2006). New approaches for the development of functional meat products. *In*: Advanced technologies for meat processing. Nollet, M.L., Toldra, F. (*Eds.*), CRC Press, Boca Raton, USA, pp. 275–299.

Kondaiah, N., Anjaneyulu, A.S.R., Rao, V.K., Sharma, N. and Joshi, H.B. (1985). Effect of salt and phosphate on the quality of buffalo and goat meats. *Meat Sci.*, 15: 183–192.

Lee, S., Faustman, C., Djordjevic, D., Faraji, H. and Decker, E.A. (2006). Effect of antioxidants on stabilization of meat products fortified with n-3 fatty acids. *Meat Sci.*, 72: 18–24.

Mandigo, R.W. and Esquivel, O. (2004). Emulsion and batters. *In*: Encyclopedia of meat science. Vol I. Chemistry and physics of comminuted products. Devine, C. and Dikeman, M. (*Eds.*), Oxford, Elsevier, pp. 266–271.

Marianski, S. and Mariañski, A. (2010). Home production of quality meats and sausages. Book Magic, LLB, USA.

Marsh, A.C. (1983). Processes and formulations that affect the sodium content of foods. *Food Technol.*, 37: 45–49.

Mittal, G.S. and Usborne, W.R. (1985). Meat emulsion extenders. *Food Technol.*, 39: 121–130.

Muthukumar, M. and Kondaiah, N. (2011). Technology of comminuted meat products. *In*: Technology of muscle foods. Studium Press (India) Pvt. Ltd., New Delhi. pp. 179–214.

Muthukumar, M., Naveena, B.M., Sen, A.R. and Vaithiyanathan, Sureshkumar, K. (2012). Effect of incorporation of Moringa oleifera leaves extract on quality of ground pork patties. *J. Food Sci. Technol.*, DOI 10.1007/s13197-012-0831-8.

Naveena, B.M., Muthukumar, M., Sen, A.R., Babji, Y. and Murthy, T.R.K. (2006). Quality characteristics and storage stability of chicken patties formulated with finger millet (*Eleusine coracana*). *J. Muscle Foods*, 17: 92–104.

Naveena, B.M., Muthukumar, M., Anjaneyulu, A.S.R., Sen, A.R. and Kondaiah, N. (2011). Value added chicken products: An entrepreneur guide. Hind Publications, Hyderabad, India.

Ockermann, H. and Basu, L. (2004). Chemistry and physics of comminuted products. *In*: Encyclopedia of meat sciences. Vol. I. Devine, C. and Dikeman, M. (*Eds.*), Oxford, Elsevier.

Ordonez, J.A., Hierro, E.M. and Bruna, J.M. (1999). Changes in the components of dry fermented sausages during ripening. *Crit. Reviews Food Sci. Nutr.*, 39: 329–367.

Pearson, A.M. and Gillett, T.A. (1997). Processed meats. CBS Publishers and Distributors, New Delhi, India.

Pegg, R.B. and Shahidi, F. (2000). Nitrite curing of meat. Trumbull, Conne, Food and Nutrition Press.

Prince Devadason, I., Anjaneyulu, A.S.R. and Babji, Y. (2010). Effect of different binders on the physico-chemical, textural, histological, and sensory qualities of retort pouched buffalo meat nuggets. *J. Food Sci.*, 75: S31–S35.

Pussa, T., Pallin, R., Raudsepp, P., Soidla, R. and Rei, M. (2008). Inhibition of lipid oxidation and dynamics of polyphenol content in mechanically deboned meat supplemented with sea buckthorn (*Hippophae rhamnoides*): Berry residues. *Food Chem.*, 107: 714–721.

Quinn, P.J., Boldyrev, A.A. and Formazuyk, V.E. (1992). Carnosine: Its properties, functions and potential therapeutic applications. *Mol. Aspects Med.*, 13: 379–444.

Rhee, K.C. (1994). Functionality of soy proteins. *In*: Protein functionality in food systems, Hettiarachchy, N. and Ziegler, G. (*Eds.*), Marcel Dekker, New York, pp. 311–324.

Ruusunen, M. and Puolanne, E. (2005). Reducing sodium intake from meat products. *Meat Sci.*, 70: 531–541.

Ruusunen, M., Sarkka Tirkkonen, M. and Puolanne, E. (2007). The effect of salt reduction on taste pleasantness in cooked bologna-type sausages. *J. Sensory Stud.*, 14: 263–270.

Ryan, J.T., Ross, R.P., Bolton, D., Fitzgerald and Stanton, C. (2011). Bioactive peptides from muscle sources: Meat and fish. *Nutrients*, 3: 765–791.

Seman, D.L., Olson, D.G. and Mandigo, R.W. (1980). Effect of reduction and partial replacement of sodium on bologna characteristics and acceptability. *J. Food Sci.*, 45: 1116–1121.

Sen, A.R., Naveena, B.M. and Muthukumar, M. (2008). Effect of wheat fibre on sensory, physic-chemical traits and storage stability of low fat mutton nuggets. *Fleischwirtschaft Int* ., 24: 71–74.

Slim, M.K. (2001). Cardiovascular actions of chicken-meat extract in normo- and hypertensive rats. *Br. J. Nutr.*, 86: 97–103.

Solan, A.E. (2008). The top ten functional food trends. *Food Technol.*, 62: 25–44.

Stoll, B.A. (1996). Can supplementary dietary fibre suppress breast cancer growth? *Br J Cancer*, 73: 557–559.

Trius, A. and Sebranek, J.G. (1996). Carrageenans and their use in meat products. *Crit. Rev. Food Sci. Nutr.*, 36: 69–85.

Varnam, A.H. and Sutherland, L. (1995). Meat and meat products - technology, chemistry and microbiology. Chapman and Hall, London.

Verma, A.K. and Banerjee, R. (2012). Low-sodium meat products: Retaining salty taste for sweet health. *Crit. Rev. Food Sci. Nutr.*, 52: 72–84.

Vuyst, L.D., Falony, G. and Leroy, F. (2008). Probiotics in fermented sausages. *Meat Sci.*, 80: 75–78.

Walker, A.R., Walker, B.F., Tsotetsi, N.G., Sebitso, C., Siwedi, D. and Walker, A.J. (1992). Case control study of prostate cancer in black patients in Soweto, South Africa. *Br. J. Cancer*, 65: 438–441.

Warriss, P.D. (2000). Meat science—An introductory text. CABI Publishing, New York, USA.

Wilson, N.R.P., Dyett, E.J., Hughes, R.B. and Jones, C.R.V. (1981). Meat and meat products. Factors affecting quality control. Applied Science Publishers, London.

Xiong, Y.L. (1997). Structure functionality relationships of muscle proteins. *In*: Food proteins and their applications, Damoradan, S. and Paraf, A. (*Eds.*), Marcel Dekker, Inc., New York, pp. 341–392.

Xiong, Y.L. (2000). Meat processing. *In*: Food proteins: Processing applications, Nakai, S. and Modler, H.W. (*Eds.*), Wiley VCH, New York, pp. 89–145.

Zayas, J.F. (1997). Functionality of proteins in food. Springer, New York.

Zhang, W., Xiao, S., Samaraweera, H., Lee, E.J. and Ahn, D.U. (2010). Improving functional value of meat products. *Meat Sci.*, 86:15–31.

6

Quality Assurance of Meat

P.K. MANDAL AND U.K. PAL

ABSTRACT

Quality and safety of meat foods have become vital to cater to the needs of consumers, to regain the market and confidence of consumers. There is a need for a holistic approach in the reorganization and strict implementation of quality control systems for safety of meat and meat products. Meat food chain consists of animal farms, transport of animals, processing in the abattoir, deboning, packaging, preservation, further processing, retailing and preparation and serving at home. Risks of microbial contamination exist at all these stages. The Hazard Analysis and Critical Control Points (HACCP) system and ISO 9000 series are recognized worldwide as quality control measures in meat food. Many undesirable substances enter into meat and meat products from the animal or during production and processing and cause health hazards in human beings. These include pesticides, heavy metals, animal drugs, antibiotics, antiparasitic drugs, hormones and growth promoters, tranquilisers, non-steroidal and antinflammatory drugs, carcinogenic substances produced during processing (heterocyclic amines, polycyclic aromatic hydrocarbons and nitrosamines). Minimum use of drugs and antibiotics with a safe withdrawal period, avoidance of pesticides, hormones and heavy metals and application of optimized processing conditions during product preparation are to be advocated. Application of Codex standards would minimize the residues in meat. This chapter highlights the holistic approach for quality control of Indian meat and meat products with special emphasis on hygienic processing of meat, HACCP system, microbial decontamination of meat by irradiation and organic acids, meat preservation systems, importance of Codex standards, ISO 9000 series including Food Safety and Standards Act, 2006 to meet the global challenges in meat food safety.

1. QUALITY OF FRESH MEAT

Meat has occupied a unique position in human diet, as it is one of the most nutritious and nutrient dense food used for human consumption. It is an excellent source of high quality protein and contains large numbers of important minerals in considerable quantity and serves as a good source of essential vitamins. Therefore, it is imperative that the meat supplied to the consumer should be of good quality, which requires a high standard in production, processing and handling of meat starting from the place of production to the consumers' table. The quality of meat obtained from a particular animal can vary considerably depending on the various direct or indirect factors that influence the life of the animal.

The consumer demands that a meat product should be tasty, tender, appealing to the eye, wholesome and safe for consumption, nutritious and should be available at a price that he can afford. The consumer's preference for taste, tenderness, etc., depends on the area/country/region, ethnic group, age group, sex, income or standard of living, attitude towards a food item and eating habits. A quality attribute which is acceptable in one society, may not be acceptable in another. Moreover the values change with the passage of time.

The control of quality standards in meat and meat products is changing day by day with the expansion of markets. The selection of the procedures which provide a maximum retention of taste and visual attractiveness, with the minimum of risk to the health of consumers has been a criterion in the development of quality control. The food industries unlike other industries must maintain a rigid quality control of their product not only to improve standards of the finished product but also to achieve the highest standards of hygiene during the processing. The prime target of the producer is to achieve the basic quality demanded by the consumer. It must be accepted that not all consumers have a liking for the same taste and what is appealing to one may not be appealing to another. A compromise must therefore be made by selecting the taste and quality which will appeal to the majority.

The quality of meat is influenced by a number of independent but interacting variables, which can be classified into production and post production factors affecting meat quality.

2. PRODUCTION FACTORS AFFECTING MEAT QUALITY

The body composition of an animal is always in a dynamic state. It changes in response to variations in internal and external environment. Qualitative and quantitative changes in the physico-chemical and microstructural characteristics are taking place continuously during the life span of an animal.

2.1. Age

It should be pointed out that most of food animals produced for meat are slaughtered at a young age before the animal reaches physical maturity. Age affects the body composition of animals, particularly the amount of fat and proportion of muscle to bone. Feeding livestock to an optimum age will increase the proportion of fat in the carcass. Meat from aged animal tends to be tougher than meat from younger animals, which is due to changes in the chemical nature of connective tissue in older animals and not due to the amount of this tissue. Meat tends to be darker in older animals due to the deposition of pigments in muscle and greater amounts of myoglobin. The flavour of meat appears to intensify with increasing age of live animal.

2.2. Sex

Body composition varies according to the sex of the animal and one of the most noticeable differences is in fat content. Carcasses from male animals have a lower proportion of fat and a higher proportion of lean meat than those of female animals of similar slaughter weight. This is also observed in "entire males" as compared to castrated males. In some instances, bull meat is darker in colour (dark cutting beef) than meat from steer and heifers particularly in older animals and this colour is considered detrimental in the meat trade. Occasionally, the meat from young boars that have reached sexual maturity may have an objectionable odour and this is a common feature of old boars (bore taint). However, it has been shown by a number of workers that meat from young boars of normal slaughter weight has no such odour or flavour and is of high quality.

2.3. Genetics and Breeding

A number of conditions that occur in livestock with detrimental effects on meat quality and a few of which may be advantageous are due to heredity. One of the most serious amongst them is "Stress susceptibility" (SS) in pigs, which in its extreme form results in sudden death at slaughter hall, while in less severe form manifests "Pale Soft Exudative" (PSE) condition. The SS condition is considered to have a genetic basis and occurs most commonly in Pietrain breed, Poland China cross and certain lines of Landrace but to a lesser extent in Large White Yorkshire. The prevalence of this condition seems to be related to selection of certain heavily muscled strains, in which the change in a particular proportion of muscle types leads to the condition. Asymmetric hind quarter syndrome and myopathy such as black muscle necrosis are other detrimental conditions in pigs which have a genetic basis.

Muscular steatosis, which is replacement of muscle by fat is considered to be a hereditary condition and is occasionally observed in cattle, sheep and pigs. Hereditary pectoral myopathy in turkeys, also known as green muscle disease, arises due to interference of blood supply to deep pectoral muscles resulting in greenish appearance. Affected breasts have to be removed at processing because of its objectionable appearance. Certain breeds mature early such as Aberdeen Angus in cattle and Southdown in sheep both of which tend to lay down fat at a relatively young age, hence tend to yield greater dressing percentage, which however is due to greater amount of fat in the carcass. Fat is typically yellow coloured in certain breeds of cattle such as Guernsey, Jersey and South Devon. Double muscling or dopplender cattle, a condition in which there is an increase in both number and size of cells, is also hereditary and most common in Belgian Blue breed of cattle. The advantages include more muscle, less fat, better muscle to bone ratio.

2.4. Housing and Meat Quality

Type of housing practised for animals prior to slaughter can affect meat quality. These effects are mainly due to lack of proper rest or the level of stress inflicted on the animal due to improper rearing environment. If animals are housed in conditions that result in lower rates of gain, then animals may be slightly older and may not have the same levels of lean mass and fatness as their counterparts which are reared in more ideal conditions. Free range animals also have the potential to have access to higher variety of feedstuffs prior to slaughter that may affect the flavour of the meat. But grass fed beef has been found to produce high quality lean meat. In general, confinement feeding has minimal to no effects on meat quality. If animals are overcrowded there may be limited access to feed and water and animals exhibit undesirable social behaviours such as fighting, chewing and inability to rest properly. In these situations, animal growth rate will be affected and meat obtained from them will be lower in overall fatness and may have a higher incidence of quality problems related to stress during slaughter.

2.5. Feeding

Feeding of high energy carbohydrate diets leads to faster growth and fat deposition in all livestock. Tainting of meat is a problem associated with feeding feeds such as fish meal, certain plants such as sting weed, certain strains of cloves and other legumes. Use of animal manure as a source of organic nitrogen in feed stuff may lead to tainting of meat if the concentration of manure is high. The color of veal can be manipulated by feeding. A very pale color veal is produced by feeding calves with diet deficient in iron, which would impair myoglobin formation without adversely affecting haemoglobin formation and either appetite or growth. Liver abscesses might result from high-energy diet containing little roughage. Carbohydrate rich diet leads to abnormally high acidity in the rumen ultimately causing ruminitis and ulceration. Several agents have been employed as growth promoters which not only increase a weight and enhance FCR but also increase the lean meat and reduce the fat content. Commonly used growth promoting agents include anabolic steroids, feed antibiotics and mineral preparations but recently there is an increased concern among the health conscious consumers about their use in animal feed.

2.6. Veterinary Practices

Poor castration methods might lead to infected castration wound at slaughter. Accident, improper injections of vaccines into muscles might lead to abscess formation. Certain substances such as iron injections, oxytetracycline etc. cause discoloration. Care must be exercised to provide adequate withdrawal period after medication prior to slaughter. Volatile chemicals such as turpentine, linseed oil, gas ammonia lead to tainting of meat, which can be managed by chilling for 24 hours. Surgical operations might result in downgrading of carcasses due to bruising, adhesions or local abscesses. Peritonitis may result from improper use of trochar and canula leading to spillage of rumen content.

3. POST PRODUCTION FACTORS AFFECTING MEAT QUALITY

3.1. Transport

The quality of the final products produced by the processing plants is largely determined by the quality of the live products supplied. As a result of improper loading and transport, a first class animal can be reduced to a product of poor quality. It is therefore best to proceed with extreme care along the route from shed to shackle. The link between the growing farm and the processing plant should offer the best possible conditions for a product of perfect quality for processing and a guarantee for optimal quality of the final produce. Thus the transport of animals plays an important role in producing quality products.

Animals may be moved from place of production to the abattoir by: 1. Driving on the hoof, 2. Road, 3. Rail and 4. Sea transport. In India, the animals are transported mostly by hoof to the markets and from there by trucks to urban abattoirs. All these movements will have stressful effects upon animals resulting in poor quality of carcass and loss of weight, referred as "Shrinkage". The excretory shrinkage results from elimination of body waste while the tissue shrinkage is due to respiratory activities and catabolic activities of the body. The shrinkage in fat lambs after one hour in transit will be 1.35 per cent and it is increased to 8.33 per cent after 35 hours. Average shrinkage on feeder lambs after one hour in transit was one percent which increased to 11.49 per cent after 41 hours.

3.1.1. *Long term stress and meat quality*

Animals exposed to long-term pre-slaughter stress have reduced glycogen supplies in muscles at slaughter and upon onset of rigor mortis pH decline does not proceed at a normal rate and the ultimate pH is higher than normal (>6.00) resulting darker colour of meat, firm texture, higher water holding capacity and less drip loss popularly defined as Dark Firm Dry (DFD) meat. When DFD meat is cooked it is often described as juicy, tender and of intense flavour. DFD meat while perfectly fit for food has a less desirable colour and flavour and lower keeping quality and resisted by the meat trade for use as normal meat cuts. Additionally due to high pH, DFD meat will spoil more rapidly. Conditions that induce long-term stress are long transit times, exposure to extremes of temperature (hot or cold), extended periods without food, bulls expressing sexual behaviour or improper handling prior to slaughter. It is obvious that conditions imposing long-term stress should be avoided.

3.1.2. *Short term stress and meat quality*

Short-term stress results in Pale Soft and Exudative (PSE) meat having a lower pH than normal. The meat is pale in colour with very poor water holding capacity and watery appearance. During cooking, PSE meat will lose a high amount of moisture and resulting in drier, tougher and not less flavourful meat. Improper handling including rough handling, mixing of animals from different pens during transportation or at the lairage, poorly designed holding and handling facilities at the slaughter plant and other conditions that induce stress just immediately prior to slaughter can result in PSE meat.

The following procedures will help to reduce the incidence of PSE and DFD meat:

(*i*) Animals should he handled gently and quietly at the farm and at the meat plant.

(*ii*) Loading ramps should be provided at the farm and efficient unloading facilities at the meat plant.

(*iii*) Stock should be kept in their original social groups as far as possible and there should be no mixing within the last vital 24–48 hour period before slaughter.

(*iv*) Lairage pen design and race arrangements should allow for easy and efficient movement of stock.

(*v*) Water should be provided at all times in the lairage, and food when necessary.

(*vi*) The use of sticks and electric goads should be avoided.

(*vii*) Aggressive animals should be isolated in the lairage as should females in oestrus.

(*viii*) The use of overhead fine sprays of water in lairages has been shown to reduce the incidence of PSE. This is thought to be due to the cooling effect.

(*ix*) Careful selection of stock for slaughter is important so that the period of time animals spend in the lairage is the optimum for the particular species, sex and time of year.

(*x*) Feeding of sugar in liquid form to pigs in the lairage is recommended to replace depleted liver glycogen reserves.

3.2. Effect of Stunning and Slaughter on Meat Quality

The type of stunning method used can affect meat quality, either by inducing short-term pre slaughter stress or by affecting the process of bleeding or exsanguination. During electrical stunning, kicking can occur during the clonic phase as the brain's inhibitory influence on the spinal cord is reduced. This can result in increased time for shackling, increased worker risks and less effective exsanguination. Pigs immobilized with carbon-di-oxide are more relaxed, and the incidence of ecchymoses reduced and can remain motionless for upto 60 seconds. The process is reported to reduce the incidence of PSE and reduced stress resulting in better meat quality. The methods of converting the live animal in to a carcass and subsequent conversion of muscle to meat have great impact on meat quality. Stress of either long or short term induced before slaughter will directly influence the course of rigor mortis.

3.3. Contamination During Slaughter

The methods usually employed for the slaughter can provide entry for micro-organisms into the tissues. Stunning instruments and knives are potential sources of contamination. Bacteria present on captive bolt pistol have been recovered from spleen of cattle, and those present on the pithing rod were found in the muscles of the neck and the flank. Organisms from the knife used for cutting the throat of the sheep were isolated from the heart, spleen, kidneys and liver though rarely from the muscles. Scalding water used for pigs is also another source of contamination, especially if the scalding is done in tank which is very common. The significance of contamination during slaughter depends on the number of bacteria introduced, and the extent to which they are eliminated by the defence mechanisms. With reasonable hygienic precautions, contamination from this source should not pose a problem.

3.4. Fabrication

Meat undergoes extensive handling during boning and cutting operations and fresh surfaces are exposed, making it more susceptible to contamination. The extent of contamination depends on the surrounding environment, the throughput of the boning room, the temperature, duration of time the meat is present in the room and the cleanliness of the cutting tables, conveyor belts, knives and other equipments and personnel hygiene observed by the meat handlers. In a properly managed boning or cutting room where temperature is maintained at 10°C or below, microbial growth and aerial contamination should not contribute significantly to the microbial flora of the meat. The major source of contamination is likely to be the incoming meat, and the extent to which the contamination is spread on freshly cut surfaces by contact with work surfaces, equipments and hands. Products that undergo extensive handling are likely to be contaminated with bacteria of human origin particularly *Staphylococci*. Hot boning increases water holding capacity and is claimed to have a number of advantages in relation to meat quality but concerns have been expressed about the microbial quality of hot boned meat, as hot boned meat if packed prior to cooling. The packaged, hot boned meat appears to present a less hostile environment to micro-organisms and undesirable microbes would grow rapidly, unless it is chilled rapidly.

3.5. Effect of Storage on Meat Quality

Storage of meat can affect the quality both positively and negatively. The positive effect of meat storage influences meat tenderness, also referred to as meat ageing or conditioning. During refrigerated post-mortem storage, meat tenderness improves. The major factor responsible for post-mortem increase in meat tenderness is degradation or proteolysis of muscle proteins. Proteolysis of muscle post-mortem has mainly contributed to sarcoplasmic calcium dependent proteases, the calpains, and the level of their inhibitor, calpastatin. The physiological changes in post-mortem muscle have been associated with degradation of Z-lines, troponin-T, nebulin, and desmin and the appearance of a 95,000 dalton and 30,000 dalton components. During post-mortem storage, the major improvements in tenderness occur within the first 7–14 days but degradation continues with increased storage, but at a slower rate. The negative effect of meat storage on meat quality is mainly due to microbial growth and/or lipid oxidation. Both these processes result in reducing the shelf life of meat.

3.6. Colour

The colour of fresh meat is of utmost importance to the meat industry. The physical appearance of a retail cut is the most important factor determining retail selection. The consumers equate the colour of meat to freshness and rely on colour as a criterion of quality. The colour of the meat, as perceived by the consumer, is primarily determined by factors such as the concentration and chemical form of myoglobin, morphology of the muscle structure and the ability of the muscle to absorb or scatter light.

Discoloration of packaged fresh meat in the retail case is known as "Loss of bloom" in meat industry. Though many factors are responsible for this loss of bloom, the consumers ordinarily relate this to bacterial growth which is not always the case. Discoloration is also important to the retailer because the areas of discoloured lean on primal cuts must be trimmed, which can result in substantial economic losses. The colour intensity is influenced by factors such as species, stress, sex and age of the animal, rate of post-mortem pH decline and ultimate pH. The colour intensity differences between species are primarily caused by differing concentrations of myoglobin.

The colour of meat is further influenced by a number of independent but interacting variables. The animal husbandry practice, feeding regime employed, the pre-slaughter treatments which might induce stress in animal and the way animals are maintained in lairage can markedly influence the initial colour of the meat. The process of distribution of meat for sale, the chilling rate, temperature and duration of chilled storage for ageing the carcass or primal cuts for optimum tenderness and possible use of frozen storage affect the stability of meat colour at retail. The major variables at retail are temperature of the display cabinet and the surrounding gaseous atmosphere in which the meat is packed.

Bureau of Indian Standards (BIS) has formulated several standards for the meat industry. These standards are prepared keeping in mind the need of industry, protecting the interests of both producers and consumers under Indian conditions and are reviewed periodically. BIS specifications are voluntary but an adherence to these guidelines improves the quality of processed products. However adherence to Meat Food Products Order which exclusively deals with quality control of processed meat products is mandatory. Microbial standards help to improve the plant sanitation, ensure safety of products and prevent losses due to microbial spoilage. The concept of quality control at few points in the entire production chain is now giving way to integrated quality management systems of which HACCP and ISO 9000 series are of particular interest to meat processors.

4. MEASURES TO IMPROVE FRESH MEAT QUALITY

(*i*) As the quality control starts from the farm itself, all possible steps should be taken to rear the animals under strict hygienic conditions.

(*ii*) During transportation of animals, all steps should be taken to transport the animals without much undue stress.

(*iii*) All animals should be given proper rest before slaughter.

(*iv*) Steps should be taken to remove the extraneous dirt from the skin/fleece by spraying water.

(*v*) Water used for all slaughter operations should be of potable quality.

(*vi*) Humane methods of slaughter should be practised.

(*vii*) Simple and cost effective decontamination techniques like washing of carcasses with chlorinated water (10–20 ppm) should be practised.

(*viii*) All personnel engaged in slaughter and processing should follow hygienic standards and they should be screened periodically for health status.

(*ix*) Training should be imparted for butchers/slaughter house personnel on scientific flaying, slaughter and dressing operations for wholesome meat production.

(*x*) Meat should be hygienically packed preferably under modified atmospheric conditions.

(*xi*) Carcasses, meat and meat products should be transported only in covered and refrigerated vehicles.

(*xii*) All retail outlets should maintain strict hygienic measures by way of providing fly proof enclosures and low temperature display cabinet and storage facilities.

(*xiii*) Consumer awareness should be created about the importance of hygienic measures on meat quality through print and electronic media.

5. MICROBIOLOGICAL QUALITY OF MEAT

Meat had been an integral component of human diet since time immemorial. Demand for safe and quality foods is growing day-by-day and poultry and poultry products are no exception to this trend. Meat and meat products would continue to remain as important foods for the human beings due to cultural habits. In the recent past, the world has witnessed meat crisis due to Bovine Spongiform Encephalopathy (BSE), Foot and Mouth Disease (FMD), Avian Influenza, *E. coli* (O 157:H7), Classical Swine Fever and Dioxin poisoning. These trends have reduced the confidence among consumers on the quality and safety of meat and meat products with adverse impact on global meat economy. It has become a challenging task world over to rebuild the quality assurance system in meat food chain from farm-to-fork to produce healthy and safe meat and meat products free from pathogenic microorganisms and parasites, toxins, drug residues, pesticides and hormones. Quality and safety of meat foods have become vital to regain the market and confidence of consumers.

Today, food safety is an important global issue related with international trade and public health implications. Moreover, in the process of globalization, our country is a signatory on agreements, namely, Sanitary and Phytosanitary (SPS) measures, Technical Barriers to Trade (TBT) etc. which prescribed a number of quality requirements. It is felt world over that there is a need for a holistic approach in the re-organization and strict implementation of quality control systems for safety of meat and meat products involving farm, abattoir, processor, further processor, retailer, government agencies, promotional bodies and consumer forum.

The Hazard Analysis and Critical Control Points (HACCP) system and ISO 9000 series are the powerful tools recognized by FAO, Codex Alimentarius, Food and Drug Administration (FDA) and World Trade Organization (WTO) for maintenance of quality control in meat food chain. Meat production and consumption are on the increase in India. The foreign exchange revenue from meat exports is close to Rs 20,000 million and over Rs 100,000 million from animal byproducts. Globalisation has further provided potential opportunities in meat business for domestic and export markets but quality control is the key word in meat food safety system to reach out to the global markets. Quality control systems need to be strengthened to improve the safety status of meat and meat products in the country.

In the recent years, poultry meat has been regarded as the healthy meat because of its high polyunsaturated fatty acid contents, unique flavour, texture, easy digestibility, dense nutrient contents and moreover, acceptability to all sections of population due to easy availability and lower price compared to sheep and goat meat. Therefore, food safety and quality control has assumed tremendous importance in food production chain including poultry and poultry products. Food safety hazards in relation to poultry and poultry products include pathogenic and spoilage microorganisms, biogenic amines, veterinary drug residues, mycotoxins, chemicals etc. of which microorganisms are of paramount importance and hence, will be discussed in detail in this chapter. Strict implementation of quality control systems for safety of poultry and poultry products needs co-operation and similar involvement of all the concerned sections of the production and marketing chain as mentioned above.

6. MICROBIOLOGICAL ISSUES

Meat food chain consists of animal farms, transport of animals, abattoir, deboning, packaging, preservation, further processing, retailing and home. Microbial contamination risks exist at all the stages.

6.1. Pathogenic Microorganisms Associated with Meat and Poultry

The major groups of pathogenic microorganisms associated with meat and poultry are *Salmonella* spp, *E. coli* (0157:H7), *Staphylococcus aureus*, *Listeria monocytogenes*, *Campylobacter* spp, *Yersinia enterocolitica*, *Aeromonas hydrophila*, *Bacillus cereus*, *Clostridium perfringens*, *Clostridium botulinum* and *Aspergillus flavus* which are responsible for majority of the meat borne disease outbreaks world over. *Salmonella* was recorded up to 9% in meat and up to 10% in poultry meat in India. The incidence of *E. coli* (0157:H7) was 3.7% in beef, 1.5% in pork, 2.0% in lamb and 1.5% in poultry meat in USA. The incidence of *Salmonella* spp in frozen chicken was 41% while chilled chicken showed 33% in UK in 1994. In Scotland, frozen chicken and chilled chicken carried 22% and 16% of *Salmonella* spp, respectively. An EU survey revealed 21% of *Salmonella* spp and 0–60% incidence of *Campylobacter* in poultry meat. The incidence of *L. monocytogenes* in chicken meat varied between 12 to 27% in UK.

Table 1: Microbiological standards in meat and meat products

Product type	*Microbiological Counts*	*Standards suggested in USA*	*Microbial profiles recorded*	*Suggested standards in CFTRI*
Fresh meat	Total plate count	15,000,000/g	1000–10,000/g	5,000,000/g
	E. coli.	50/g	10/g	50/g
	Staphylococci (coagulase positive)	100–200/g	25/g	100–200/g
	Clostridium perfringens	–	1000–2000/g	1000–5000/g
	Salmonella spp.	Absent	Absent	Absent
Cooked meat products	Total plate count	100.000/g	100–1000/g	100,000/g
	E. coli	Absent	Absent	Absent
	Salmonella spp.	Absent	Absent	Absent
Canned meat	Total plate count	100/g	10/g	100/g
	Anaerobic bacteria	Absent	Absent	Absent
	Salmonella spp.	Absent	Absent	Absent

Source: CFTRI, 199

6.2. Spoilage Microorganisms Associated with Meat and Poultry

Spoilage microorganisms seriously affect the marketability and economics of meat and poultry products as they disturb the preservation system and reduce the shelf life of the products. Thus, spoilage microorganisms are another group that needs attention from the point of quality maintenance of the products. The important spoilage microorganisms that affect poultry are *Pseudomonas* spp., *Micrococcus* spp., *Lactic acid bacteria*, *Brochothrix thermosphacta*, *Acinetobacter* spp., *Moraxella* spp., *Enterobacteriaceae*, *Aspergillus* spp., *Penicillium* spp., *Thamnidium* spp., *Cladosporium* spp., *Sporotrichum* spp., *Candida* spp., *Torulopsis* spp., *Rhodotorula* spp. and *Debaromyces* spp.

6.3. Foot and Mouth Disease (FMD)

FMD outbreak in UK in 2001 touched the headlines of news in the world. FMD has thrown UK meat industry into chaos. This has resulted in the slaughter of millions of beef animals in UK, ban on imports and exports of animals and posed problems related to safety of meat and meat products. The situation has been attributed to wrong vaccination policy, not stopping of animal movements and delays in slaughter. FMD has hit beef industry in South America also. Import bans were imposed on Uruguayan beef because of FMD and the country's exports were crashed down by 45%. Argentina too suffered the drop in exports of beef due to FMD. FMD has become the subject of active research world over. World regulations indicate that immunological and prophylactic approaches are ideal in order to eradicate FMD. Virologists, veterinarians, meat scientists and meat technologists have to play a vital role to free meat and meat products from FMD virus. Standard methods of chilling, freezing and curing would not eliminate FMD virus from meat and meat products. Chemical treatments such as acidification of meat to pH 2.5 would eliminate FMD virus but the product is unacceptable sensorily. High dose of irradiation to the tune of 3.0 Mrad is necessary to inactivate FMD virus in meat. However, high thermal processing temperatures would inactivate FMD virus easily. The FMD virus did not survive in ground beef when processed to internal temperatures of 90–95°C. FMD virus was inactivated in beef frankfurters subjected to steam cooking for 133 minutes at minimum temperature of 92.2°C. It was reported that FMD virus did not survive in beef burgers grilled for 3–5 min at 208°C followed by cooking in the steam oven. The practicality of these heating systems is questionable. Application of hurdle technology is likely to provide solutions for the elimination of FMD virus in meat and meat products.

6.4. Bovine Spongiform Encephalopathy (BSE)

BSE is a worldwide problem though it has created havoc in livestock industry (especially cattle and sheep) and caused enormous economic losses in the EU. It has been established that BSE has similarities with CJD in man and scrapie in sheep and goat. It is evident that BSE is transmissible to human beings, produces new variant CJD and damages central nervous system. When inadequately heated meat meal or bone meal made from sheep offals is fed to cattle, prion proteins present in meat meal or bone meal enter central nervous system (CNS) of cattle and multiply and cause damages to the CNS. The most widely accepted hypothesis is that a distorted prion molecule combines with the normal prion in the brain and acts as a three dimensional template, causing the normal protein to adopt a similar shape. The newly converted molecule, in turn, is able to act as a template and repeat the process, resulting in an exponentially increasing amount of distorted prion. Brain, spinal cord, bone and retina contain infective prions in BSE infected cattle but beef muscle and milk do not carry the infective prions.

The incidence of BSE decreased from 37,487 in 1992 to 3,980 in 1997 and serious attempts are being made to bring the epidemic to an end. European Commission (EC) has banned SRMs (Specified Risk Materials) (brain, bone, spinal cord, retina and lymphatic system) as a health safeguard against BSE and the human form of CJD. Ten countries have been declared as BSE—free by EC and they are Australia, Argentina, Botswana, Chile, Namibia, New Zealand, Nicaragua, Paraguay, Uruguay and Switzerland. These countries are allowed to export beef to EU without removing SRMs. EC accepts that this SRMs rule should not be applied to countries where the presence of BSE is very unlikely. This would facilitate the export of buffalo meat from India and other Asian Countries as BSE is very unlikely among buffaloes since these animals are not given feeds containing meat meal or bone meal made from sheep offals.

Tests for detection of BSE in live animals are application of monoclonal antibody techniques. To detect BSE in meat, fluorescent ELISA, PCR, DNA hybridization (southern blotting) and iso-electric focusing are suggested. Regulations on control of BSE to protect consumers are: ban on feeding cattle with feed containing meat meal, bone meal and offal meal prepared from sheep/goat, slaughter and disposal of infected cattle, identification and disposal of infected carcasses, prohibition of eating meat suspected of BSE, prohibition of use of brain, spinal cord, retina, bone, tonsils, offals, milk and meat as human food from suspected BSE cattle. Methods for reducing the risk of infection with the BSE prion in meat industry personnel include use of protective clothing, practices related to hygiene, sanitation and disinfection, decontamination of equipments and careful handling of SRMs.

7. MEAT PRESERVATION SYSTEMS

7.1. Chilling

Chilling is the most commonly used method for low temperature preservation of carcasses immediately after slaughter, during transport and storage and for packed meat. Cold shortening can be prevented by cooling the meat to about 15°C and keeping it at this temperature to allow the onset of *rigor mortis*. Electrical stimulation has been found effective in the prevention of cold shortening of carcasses. Chilled carcasses are then fabricated in to wholesale cuts, retail cuts or converted into mince. Though chilling is an ideal method of preservation, it cannot prevent the growth of psychrotrophs on meat surfaces eventually resulting in meat spoilage. Vacuum packaging and modified atmosphere packaging are in use for increasing the shelf life of refrigerated meat for extended periods. Carcass chilling rooms are normally operated in the temperature range of 2 to 4°C with relative humidity of 88–92%. Shrinkage loss from carcass is in the range of 1–2%. Accelerated cooling is achieved by using low temperature (–15°C) or by spraying with or immersion in cryogenic liquids (liquid nitrogen or solid carbon dioxide) for a very short period of time.

7.2. Freezing

Freezing is an effective method of storing cuts of large carcasses, retail cuts or minced meat in fresh state for a longer period. Marketing of frozen meat in retail outlets is unsuccessful in India due to the resistance from the consumers. Consumers prefer to see the appearance of the product. Frozen meat will not give the appearance of fresh meat due to the ice crystal formation on the meat surface. Meat freezes between –1.5 and –2°C. The recommended storage temperature for frozen meat is –18°C. A lower temperature maintains quality better but this should be balanced against cost. Meat must be chilled properly before it is subjected to the process of freezing. Films impermeable to oxygen and water need to be used for packaging of meat. Freezing must be rapid. Blast freezers and contact plate freezers are commonly used. Individual quick freezing is gaining importance for certain small sized meat cuts. Most rapid freezing can be achieved by cryogenic agents (liquid nitrogen or solid carbon dioxide). Rapid freezing produces smaller ice crystals and damage to the meat tissues is less.

Frozen storage of meat is very critical. Due to fluctuations in temperature, ice crystals grow in size and damage the meat tissue during frozen storage. 'Freezer burn' is a surface dehydration due to moisture loss from the unprotected meat surface, which appears dry. It occurs if proper packaging material is not used. Enzymatic and chemical reactions proceed at a slower rate even in frozen meat. Lipid oxidation is common but very slow. Thawing is a process of bringing back the meat to its original chilled state. Immersion in water or microwave method is advocated for thawing. Thawing methods are to be standardized from the point of microbiological safety. The lesser the damage to the tissues during freezing and frozen storage, the lesser is the drip loss. Drip losses were less in blast frozen samples. Plate frozen meat samples scored higher for texture, juiciness and aroma. However, both the plate and blast frozen meat samples were similar in overall quality according to taste panel results. It was reported that improvements were observed in the quality of frozen ground buffalo meat by blending with natural antioxidants and vacuum packaging.

7.3. Packaging

Packaging protects meat from moisture loss, contamination by dirt, dust and microorganisms, changes in colour and physical damage. Colour of meat can be displayed in a very appealing way through packaging to the consumers. Packaging imparts attractiveness to the product. Packaging of fresh meat varies from simple wrapping to advanced systems like vacuum packing (VP) and modified atmosphere packaging (MAP). Fresh retail meat cuts are commonly packed in polyethylene pouches. These pouches allow oxygen transmission, which maintains the bright red colour of meat and reduces moisture loss. Shelf-life of these meat cuts varies between 3 and 5 days at 4°C. Exclusion of oxygen prior to packing is the principle in the vacuum packing of meat. Oxygen impermeable films are used. The growth of aerobic microorganisms is prevented in meat under vacuum and the growth of anaerobes is inhibited since vacuum packed meat is stored below 4°C. Vacuum packaging provides at least three weeks shelf-life for the product under adequate refrigeration, but the product looks dark. When the pack is opened and exposed to air, the meat regains its bright colour. MAP means alteration of atmospheric gas concentrations in the package prior to packing. The three principal gases used in MAP are carbon dioxide, nitrogen and oxygen. Carbon dioxide exerts inhibitory effects on bacterial and mould growth. Nitrogen can inhibit oxidation of fats, rancidity and pack collapse and reduce the possibility of mould growth. Oxygen prevents anaerobic spoilage. A gas mixture containing 10% carbon dioxide, 5% oxygen and 85% nitrogen was very effective for extending the shelf-life of meats in zero barrier film.

In India, there is no proper packing of meat in the traditional system of meat handling and marketing. The common practice at retail level is to keep the meat in a black polyethylene bag to hand over to customers or to keep meat in the containers brought by the customers. In metro cities, there is a shift from traditional meat handling to real packing of meat. In export houses, low and high-density polyethylene films are used for packing buffalo meat. VP and MAP have not yet been applied to buffalo meat in India.

8. QUALITY MONITORING AND CONTROL SYSTEMS

Assurance of safety and quality poultry and poultry products to the consumers requires continuous monitoring of all operations involved in raising the birds, their transport, slaughter, processing, distribution, marketing, handling in retail outlets, restaurants and homes.

8.1. At Farm Level

Broiler chickens should be placed in sheds which have been thoroughly cleaned and disinfected. The surrounding area and the equipments in the sheds must be cleaned and disinfected. Adequate floor space for each bird is required for its growth, health, quality and general well being. Supply of adequate, well balanced feed and clean drinking water must be ensured. All poultry farms should have contract with veterinarians who will advise on the health and management of the flock including vaccination and medication, whenever required. Adequate withdrawal periods are essential for antibiotics, anti-coccidial and other drugs used to maintain the flock health. The withdrawal of feed prior to loading for transport can also help to reduce the level of contamination by reducing the crop and intestinal tract contents.

8.2. At Slaughter, Dressing and Processing Level

The chances of eliminating microbial contamination from poultry meat will be improved if measures are taken to ensure that birds entering the slaughter and processing chain are free from infection or are identified as contaminated and treated accordingly during ante-mortem inspection. Dressing and processing should be carried out by trained personnels maintaining adequate hygiene at different stages of dressing and processing.

8.3. At Retail, Restaurant and Home Levels

Cold chain should be maintained strictly. Proper handling of poultry products at retail level is very important to maintain the quality of the product. Products must be sold before the expiry of 'best before' date for use. Strict separation of raw and cooked products is a must at restaurant kitchen and at home for food safety. At restaurant, food handlers and the utensils are the main sources of contamination. Therefore, care should be taken towards personal hygiene of the food handlers and proper washing and disinfection of utensils. At home, raw chicken should be washed with potable water before cooking. Cooked poultry products should be served after thorough cooking. Left-over, if any, and in case of delay between preparation and consumption, product should be kept under refrigeration. Before serving, the stored left-over should be heated thoroughly. Raw chicken, if stored at home, preferably be kept frozen.

8.4. Hygienic Processing of Meat

The quality of meat produced in the existing traditional systems in India is far from satisfactory. Few modern abattoirs that exist are facing the problems related to maintenance. Export houses (buffalo processing units) have modern abattoirs. Modern abattoir is an essential unit for hygienic processing of meat animals. It provides facilities for ante- and post-mortem inspection, humane slaughter, clean meat production, proper recovery of by-products and waste treatment. The essential steps involved in modern abattoir practice are: (*i*) holding the animals in lairage, (*ii*) ante-mortem inspection, (*iii*) stunning, (*iv*) slaughter and bleeding on the cradles/rail, (*v*) skinning, dressing and evisceration on the overhead rail, (*vi*) post-mortem inspection, (*vii*) washing the carcasses, (*viii*) dispatch of carcasses for marketing, (*ix*) holding the carcasses in chill room (optional) and (*x*) disposal of by-products and wastes and follow scientific method of handling of meat. The CFTRI has developed modern abattoir models for hygienic processing of meat. The CFTRI studies revealed low microbial profile (Table 2) in meat processed in a modern abattoir as compared to traditional units.

Table 2: Microbial counts on sheep carcasses.

	Mean counts (log cfu/cm²)	
Microbial counts	***Mysore city abattoir***	***CFTRI modern abattoir***
Total plate counts	6.2	4.0
Coliforms	4.4	1.6
Staphylococci	5.2	2.1
Enterococci	4.4	1.6
Psychrotrophs	3.5	2.6

8.5. Good Manufacturing Practice (GMP)

Practice of GMP is vital in the industry before aiming at application of HACCP system. This has to be carried out as per the Codex Principles of Food Hygiene. Training of Production Managers on GMP is the key issue to avoid contamination of product with microorganisms.

8.6. Cold Water Sprays or Dips

The final step in slaughter operations is the washing of carcasses by cold water. Carcasses are either sprayed or dipped before they are sent to chill room. Bacterial reductions are possible up to one log total plate count (TPC). Removal of bacteria is greater if carcasses are washed at each processing point.

8.7. Hot Water Dips or Sprays

The effect of hot water dips or sprays on the elimination of microorganisms from carcasses or meat cuts is shown in Table 3. The data revealed that the reduction in the levels of microorganisms can be more than 2.0 log TPC of pathogenic bacteria. Hot water sprays are used in abattoirs for decontamination of carcasses in USA.

Table 3: Decontamination of carcasses or muscles by using water.

Species	***Conditions***	***Effect***
Beef	80°C	2.64 log *E. coli* reduction
		2.21 log *Salmonella* reduction

Table 3: (*Contd...*)

Table 3: (*Contd...*)

Species	*Conditions*	*Effect*
Mutton	80°C	3.32 log *E. coli* reduction 3.32 log *Salmonella* reduction
Beef	Commercial carcass washing unit	1.07 log TPC reduction
Beef	Hand wash with hose	0.99 log APC reduction
Beef	550 kPa (80 psi)	1–1.9 log reduction of *E. coli* (0.157:H7)
Sheep Carcass	Spray wash	0.2–0.6 log reduction in TPC
Buffalo meat cuts	80°C	1.6–1.8 log reduction TPC 1.6 –1.8 reduction in *Enterobacteriaceae*

8.8. Acid Dips or Sprays

Acetic acid and lactic acid have been extensively evaluated as decontaminating agents in meat and poultry (Table 4). In general, bacterial counts will be reduced by 1–2 log cfu/g or /cm^2 by acid treatments. It is the undissociated acid molecule that is effective against microorganisms. Further, Gram negative bacteria are more susceptible to organic acids than Gram positive organisms. There will be a residual antimicrobial effect of these acids and the shelf life of product is extended during storage. Acetic and lactic acids are inexpensive and these acids have GRAS (generally recognised as safe) status. Acid treatment has been approved for decontamination of carcasses in USA and in Canada but not in EU.

Table 4: Decontamination of carcasses or muscles by using organic acids.

Species	*Concentrationof acid*	*Effect*
Beef	3%	1.8 log TPC reduction
Beef	2%, 55°C	>2.0 log reduction in *Salmonella*
Beef/pig	1%, 55°C	1.0 log reduction in TPC
Sheep Carcass	2%	1.5–2.0 log reduction in TPC and 1.0 log reduction *S. aureus*
Beef	1, 3, 5%	1.75 log reduction in *E. coli* 0157:H7
Beef	1, 3, 5%	3.5 log reduction in *P. fluorescense*

8.9. Gamma Irradiation

Extensive studies were conducted on sterilization or pasteurization (radurisation) of foods by gamma irradiation. Gamma irradiation has been proved as a potential tool in inactivating parasites and to greatly reduce the microbial levels (pathogenic and spoilage) and extending shelf life of meat and poultry. World Health Organization (WHO) clarified in 1980, the medical acceptability of irradiated foods and stated that no health hazard could result from consuming any food irradiated up to a dose of one mega rad (1 Mrad or 10 kGY). A dose of 2–3 kGY is adequate to decontaminate meat and poultry from Salmonella, *Campylobacter* spp., *E. coli* (0157:H7), *Listeria monocytogenus* and *Pseudononas* spp. A dose of 0.3–1 kGY eliminates parasites from meat. The sale of irradiated foods has been permitted in many countries. The USA permitted irradiation in pork and poultry. The UK has permitted this in poultry. It has strongly been said that efforts on education of public on the benefits of irradiation should be strengthened since public protests exist in USA and UK on the use of irradiation in foods wrongly considering it to be radioactive. South Africa has permitted irradiation in cold meat, chicken, sausages and bacon and smoked salami. Detailed studies were conducted on the irradiation of meat, fish and poultry at Bhabha Atomic Research Centre, Mumbai. CFTRI-BARC collaborative studies demonstrated the potentiality of irradiation for extending the shelf-life of intermediate moisture meat products at ambient temperatures for longer periods. Irradiation has been permitted in meat, fish and poultry recently in India.

9. HAZARD ANALYSIS AND CRITICAL CONTROL POINTS SYSTEM (HACCP)

Safe meat production starts from healthy animal reared at farm. Assessment of risks (microbial, toxins and residues of animal origin) that cause hazards in human health is vital during production of animals at farm. The critical aspects that affect the quality management of animals are diseases, contamination of animal feed with soil and excreta, environment and treatment of animals with medicines. Though microorganisms play a major role in causing health hazards among humans, toxins, drug residues, pesticides and hormones are also contributing factors. HACCP offers good potential in risk assessment and improvements in the production of healthy animal at farm. HACCP has not yet been implemented effectively at farm production world over due to lack of correct scientific data documentation with regard to the practices of animal production and management. Farm production practices are to be understood thoroughly for application of HACCP system.

Application of the HACCP system has been made mandatory in EU Directive 93 /43 and the USDA FSIS Pathogen Reduction HACCP system. The HACCP system is often applied qualitatively and subjectively. Subjective assessment of critical control points would mislead the interpretation of process control system during slaughter and dressing operations. A quantitative approach of the HACCP system can provide ideal criteria for critical process steps for administering control measures and optimizing processes to a particular risk. Baseline data on the microbiota of meats collected by the USDA Food Safety and Inspection Service (FSIS) revealed that indicator organisms [viable aerobic bacteria, total coliforms and *E.coli* (Biotype I)] are the better predictors for process control during implementation of HACCP system.

The National Advisory Committee on Microbiological Criteria for Foods in the USA standardized the HACCP principles as follows:

1. Conduct a hazard analysis.
2. Determine the critical control points.
3. Establish critical limits.
4. Establish monitoring procedures.
5. Establish corrective actions.
6. Establish verification procedures.
7. Establish record-keeping and documentation procedures.

HACCP is a proactive method of quality assurance and is must in modern food processing units to ensure food safety. At international level, Codex Alimentarius Commission has recommended the application of HACCP as the baseline for consumer protection under the Agreement on the Application of Sanitary and Phytosanitary Measures. In simple words, HACCP is a tool to assess hazard and establish control systems that focus on prevention rather than relying on end product testing. Hazard has been defined as "the unacceptable contamination, unacceptable growth and/or unacceptable survival by microorganisms of concern to safety or spoilage and/or the unacceptable production or persistence of products of microbial metabolism (*e.g.*, toxins, enzymes, biogenic amines) in foods' (ICMSF, 1988).

The purpose of hazard analysis is to identify potential problems which could occur in an operation. The definition of critical control points (CCP) is of practical importance to the meat and poultry processors as it points out the limits of what should be achieved when a HACCP program is established. The ICMSF definition for a critical control point is a "location, practice, procedure or process at which control can be exercised over one or more factors which, if controlled could minimize or prevent a hazard." CCPs are of two types based upon the level of confidence that hazards can be prevented. CCP1 will assure hazard control and in some cases elimination; while CCP2 will minimize hazard but cannot assure control/elimination of a hazard.

Some of the important examples of CCP1 in poultry processing are

- Cooking to 63°C or its equivalent at a lower temperature for a specified time for the assurance of destruction of *Salmonella* in meat.
- Chilling of dressed carcasses to 5°C or below to check the growth of spoilage microorganisms and proteolytic bacteria such as *Clostridium* spp.
- Acidification to pH < 4.6 prevents growth of pathogens in shelf stable pickled meats and poultry.
- Drying to water activity < 0.86 prevents growth of pathogenic bacteria in snack chicken products.

Some of CCP2 in poultry processing are as follows

- Change of scalding water at regular interval reduces the risk of cross contamination.
- Proper care in evisceration reduces the incidence of *Salmonella* spp and *Campylobacter* spp on fresh poultry.
- Use of potable quality water in carcass washing reduces surface microbial load.
- Separating raw meat from cooked meat reduces risk of contamination of pathogenic microorganisms on cooked products.
- Correct personal hygiene practices reduce the risk of product contamination.
- Refrigeration of cooked products reduces risk of growth of pathogenic bacteria between cooking and serving.

This HACCP system ensures hygiene, safety, better use of resources and timely response to food safety problems in poultry processing as well as in other food processing plants. It is recognized as one of the most effective means of minimizing the occurrences of identifiable food borne biological, chemical and physical hazards. It is a rational, systematic, cost-effective and documented technique which can be used for organizing and implementing entire quality assurance program.

10. BEYOND HACCP

Though HACCP system is a good process control tool, it is not a total answer to eliminate risks challenging the safety of food including meat. Two new concepts have emerged for managing the risks after HACCP application—(*i*) Kill steps and (*ii*) Due diligence.

Kill Steps are mainly intended to kill residual bacteria that survive the HACCP technique. These can be used in combination with HACCP system. Kill steps are application of heat in the form of cooking or pasteurization. Electron beam acceleration can also be used. Here, electrically generated electrons are concentrated on the surface of meat products. Irradiation is a non–thermal technique that is widely accepted world over for elimination of pathogenic and spoilage microorganisms in meat and meat products. Thus irradiation serves as an important kill step in meat production and processing in combination with HACCP systems.

Due diligence concept is mainly to control an individual or organization that exhibits negligence in maintaining the minimum standards in the production of meat. The concept provides flexibility to the producer even if the manufactured food causes illness. He will not be punished provided he does not neglect to maintain minimum standards during production of food. In other words, the concept puts responsibility on the producer to improve the product quality.

11. RESIDUES IN MEAT AND MEAT PRODUCTS

With the advent of intensive production systems to meet the demands of the growing population, the present day concern is on the occurrence of substances not naturally present, which the biological systems are unable to handle in their metabolism resulting in their accumulation in the food chain. The contaminants can occur at any point in the food chain from the field to the point of consumption. Processing may even concentrate the contaminants in particular fractions. Once contamination has occurred, it is difficult to remove. In case of animal products, traditional meat inspection procedures cannot guarantee the absence of the residues. Majority of farm animals receive medication and are exposed to many environmental contaminants. Presently, more than 300 feed additives, antimicrobials and hormone type agents are used in animal production. Further, a large variety of chemicals have been in use in modern agro-horticultural and animal husbandry practices and may be hazardous to the consumers. The concept of *"plough to plate"* is now being increasingly recognized.

11.1. Antibiotics

Antibiotics are primarily used from the view point of animal welfare and well-being on production grounds and public health stand point. Occurrence of residues is dependent on many factors such as age, disease condition and husbandry practices. One of the major concerns with antibiotic residues is *allergic reactions*. Antibiotic resistance in the gut bacteria of human being is another malady, which is caused by multi drug resistant *Pneumococci*, glycopeptide resistant *Enterococci* and Gram negative bacteria with extended spectrum and beta lactamases. Aplastic anaemia with chloramphenicol and carcinogenic effects on laboratory animals by sulfa drugs and nitrofurans are also of public health importance. Penicillins and cephalosporins used to treat mastitis are responsible for residues. Hypersensitivity reactions to cephalosporins have been reported in people who are allergic to penicillins. Amino glycosides effective against Gram-negative bacteria may persist in tissues and not eliminated. Macrolides used as feed additives effective against Gram positive and Gram-negative organisms develop cross-resistance with other antibiotics. Lincosamides used to treat Gram-positive aerobes and anaerobes cause hypersensitivity.

11.2. Hormones

The United States Food and Drug Administration (FDA) has approved six growth hormones *viz.* estradiol, progesterone, and testosterone (natural ones) and synthetic hormones zeranol, trenbolone acetate and melengestrol acetate (MGA). In 1980, the presence of hormone Diethyl Stilbesterol Acetate (DES), that promotes weight gain and muscle development in livestock sparked a health scandal in Italy when it was reported that children who had eaten the hormone treated meat had developed breasts and infant girls had begun menstruating. In response to these events, EC adopted Directive 81/602 in 1981 to prohibit the use of hormones. A ban on zeranol and trenbolone was imposed in 1985 on the grounds that their safety has not been conclusively proven. EU believes that these hormones cause cancer, whereas, increased levels of estrogens results in male infertility. Subsequently, EU also banned meats treated with thyrostatic, estrogenic, androgenic or estrogenic hormones. In 1991, Codex decided that maximum residue limits (MRLs) were not necessary for three natural hormones and adopted MRLs for trenbolone and zeranol.

Hormones used to promote growth and lactation performance include Zeranol, trenbolone acetate, PST, BST, estrogens and progesterones that are permitted to use. These are chemically identical to the natural hormones. However, some countries do not promote them. DES being a strong carcinogen is banned from use. Similarly anabolic steroids such as non-testosterone and testosterone are not permitted to use for producing animals. Beta agonists such as salbuterol, clenbuterol and climaterol though increase lean content in meat in animals, they also cause toxicity and poisoning in human. Post-mortem inspection should now include specific examination for the presence of implants and the lesions thereof at normal site (base of the ear) or at unusual sites like legs, brisket, or tail. Abscesses may be found, as the implantation does not take place under sterile condition. In case of synthetic or xenobiotics, the concentrations occur in urine, bile, or faeces and more stable than in liver, kidney, and muscle. In case of some drugs of beta agonists, residues persist in the retina of the eye for very long periods.

11.3. Heavy Metals

Heavy metals are widely distributed in the environment. Most of them eventually accumulate in soft and hard tissues of wildlife, animals and humans. Feed stuffs such as mineral supplements and fish meals contaminated with heavy metals result in residues in livestock products. Moreover, high concentration of heavy metals is found in tissues of animals which are grown in the vicinity of metal processing unit, tannery, refinery, asbestos product units which are major sources of atmospheric emission of heavy metals. Unfortunately most of the countries have no regulations on these contaminants. MFPO (Amendment) (1993) has prescribed maximum permissible limits of heavy metals in meat products (Table 5).

Table 5: Level of permissible limits for some heavy metals (MFPO, 1973)

S. No.	*Name of the heavy metal*	*ppm by weight (max. permissible limit)*
1.	Lead	2.5
2.	Copper	20
3.	Arsenic	1
4.	Tin	250
5.	Zinc	50

11.4. Residues in Poultry Meat and Products

A residue is defined in EC directive 96/23/EC as "a residue of substances having a pharmacological action, of their metabolites and of other substances transmitted to animal products and which are likely to be harmful to human health". They have been classified into seven categories:

1. Antimicrobials *e.g.*, sulphonamide, tetracycline etc.
2. Growth promoters *e.g.*, oestradiol, nortestosterone etc.
3. Pesticide *e.g.*, DDT, lindane etc.
4. Heavy metals *e.g.*, lead, arsenic etc.
5. Natural toxins *e.g.*, aflatoxins, ochratoxins etc.
6. Preservatives *e.g.*, sulphur dioxide, sodium nitrite etc. (Table 6)
7. Other substances *e.g.*, fluoride etc.

11.5. Natural Toxins

Mycotoxins are metabolic products of fungi produced in foods and feeds. Risk to human health due to mycotoxins is much higher from direct consumption of grains than that from animal products. Residues of aflatoxins have been reported in some edible tissues and meat (Jemmali & Murty, 1974). Ochratoxin residues may accumulate in kidney and present a higher risk to the consumers. Aflatoxin B1 and M1 present in milk are very carcinogenic. Zearalenone has estrogenic effect in swine and suspected to be endocrine disruptor.

11.6. Processing Contaminants

A more systematic approach is needed to identify processing contaminants. However, there are few established examples such as carcinogenic N-nitroso-ammines in fermented foods and cured meats, carcinogenic polycyclic aromatic hydrocarbons in smoked foods, and 3-Monochloropropane-1, 2-diol (3-MCPD) and ethyl carbamate and dioxins in fish. Chemical migration into the foods from packaging materials is another area, which needs investigation.

Illegal practices in commerce have been known to occur in the preparation of certain meat foods such as hamburgers and hot dogs where worst leftovers of slaughter house (ears, snouts, and hooves) may be incorporated in the form of mince. To keep the false freshness of such meats, red dyes may also be added. Frying or baking may also release substances such as acryl imides. Pepperoni and sausage toppings are high-risk processed meats, which are added with very high quantities of nitrates, chemicals, preservatives and saturated fats that are harmful to health. Monosodium glutamate, the most commonly employed flavouring agent has also been reported to cause headaches and allergic reactions in few individuals.

Table 6: Level of permissible limits of some chemical preservatives.

S. No.	*Name of the preservative*	*ppm by weight (max. permissible limit)*
1.	Sulphur dioxide	450
2.	Sodium & Potassium nitrite (calculated as sodium nitrite)	200
3.	Commercial salt petre (calculated as sodium nitrate)	500

11.7. Pesticide Residues

Chemical residues in meat and milk could result from indirect chemical contamination if animals are fed grain or forage that has been treated with pesticides or by accident (Table 7). Organochlorine pesticides such as dieldrine, DDT or heptachlor, heavy metals such as cadmium, lead or mercury, naturally occurring toxins such as aflatoxins and alkaloids and environmentally occurring compounds such as dioxins enter the livestock products causing serious health hazards. Organochlorine (OC) pesticides like dieldrine, heptachlor and DDT were banned from use on pastures, agricultural purposes and termite control. However, they persist in the environment through horticultural crops, cotton or sugar cane fields. Livestock graze on paddocks spread with organochlorine compounds, hay or grain stored in contaminated areas act as sources for residues.

Under drought conditions, there is increased reliance on purchased fodder, and temptation to feed unusual materials. Loss of pasture cover forces animals to eat soil leading to increased levels of soluble chemical residues. Risks can be regulated by ensuring that feeds purchased do not contain unacceptable residues, exercising extreme care over any decision to use unconventional stock feeds, using chemicals strictly in accordance with label directions, keeping a safety margin to withholding periods and keeping away livestock from farm building or other sites where chemicals especially organo-phosphorous chemicals have been stored (APHS, 2003).

Approximately 2.5 million tones of pesticide chemicals are applied throughout the World. WHO (1992) reports that 3.0 million pesticide poisonings occur annually resulting in 2,20,000 deaths globally. Litovitz *et al.* (1990) reported that 67,000 pesticide poisoning cases occurred in US of which 27 were accidental fatalities, which costed heavily. Approximately 80% pesticides produced in the world are used in developed world but less than half of all pesticide induced deaths occur in these countries whereas the proportion of pesticide poisonings and deaths that occur in developing countries were very high. Average pesticide residue levels in food are often higher in developing countries than developed countries. A study in Egypt reported that majority of assayed milk samples when tested for 15 different pesticides, contained residue levels between 60–80% (Dogheim *et al.*, 1990) whereas 50% of milk samples analyzed in a US milk study had traces well below EPA and

FDA regulatory limits (Trotter & Dickerson, 1993). Since farmers and farm workers directly handle 70–80% of pesticides, they are at the greatest risk of exposure.

Epidemiological evidence suggests a significantly higher rate of cancer incidence among farmers and farm workers in US and Europe than non-farm workers in some areas (Cantor *et al.*, 1992). Based on animal studies, it was found that sufficient evidence exists for carcinogenicity for 18 pesticides (WHO/ UNEP, 1989) and increased prevalence of cancer in farmers (Cantor *et al.*, 1992). One percent human cancers are attributed to pesticides in US. Recently banned pesticide dibromo-chloro-propane (DBCP), which is used for plant pathogen control, was found to cause testicular dysfunction in animals and infertility in human and sperm counts decreased by 50% in males. Decreased immunity has been observed due to exposure to aldicarb (Fiore *et al.*, 1986). Reproductive organs have become small in male alligators due to chronic exposure to pesticides (Colborn *et al.*, 1996).

Some pesticides such as organochlorine insecticides (aldrine, dieldrine, gamma-BHC, DDT, endrin, endosulfan, hepachlor) though used to control ectoparasites in aniamls, their destruction in the environment is very slow and accumulate highly in fat tissues of wildlife and livestock. Of these, pyrethride methomyl, dieldrin, gamma-BHC, DDT, endrin, endosulfan, hepachlor, thoxlyatenonylphenol, butyl adipate and dioctyl phthalate are endocrine disruptors. Endosulphan used for controlling cotton insects cause tissue residues in beef.

Table 7: Joint FAO/WHO recommended limits for pesticides and drug residues.

Residue	***ADI μg/kg***	***MRL in Poultry meat***	***MRL in Egg***
Abamectin	0–11	–	–
Albendazole	0–50	–	–
Benzylpenicillin	30	50	–
Chlortetracycline/ Tetracycline	0–3	100	200
Cypermethrin	0–50	200	100
Alphacypermethrin	0–20	100	50
Diclazurial	0–20	500	–
Dihydro streptomycin & Streptomycin	0–30	500	–
Flubendazole	0–12	200	400
Levamisole	0–6	10	–
Neomycin	0–60	500	500
Spectinomycin	0–40	300	–

Table 7: (*Contd...*)

Table 7: *(Contd...)*

Residue	***ADI μg/kg***	***MRL in Poultry meat***	***MRL in Egg***
Spiromycin	0–50	200	–
Sulphadimidine	0–50	100	–
Thiamphenicol	0–6	40	–
B-Adrenoceptor Blocking Agents			
Clenduterol	0–0.004	–	–
Anthelmentic Agents			
Abamectin	0–1	–	–
Insecticide			
Cypermethrin	0–50	200	100
Alpha cypermethrin	0–20	100	50

ADI—Acceptable daily intake, MRL—Maximum residue limit. *Source*: Thulasi, G (2006).

12. INTERNATIONAL REGULATIONS

12.1. European Union (EU)

HACCP has received greater consideration in food policy as far as food safety is concerned. However regulatory acceptance is absent in HACCP from EU food law applicable to foods of animal origin. EU has legislation on hygiene regulations for meat and stressed the need for training personnel in meat hygiene and safety including the principles of HACCP. To ensure food safety, EU regulations include (1) general food hygiene directive (93/43/EEC), (2) Fresh meat hygiene directive (Section 2, (3) Poultry meat (71/118/EEC and 92/116/EEC) (4) Rabbit and game meat (91/495/EE#C and 92/45/EEC), (5) Minced meat and meat preparations (94/65/EEC) and (6) Meat products (77/99/EEC and 92/5/EEC).

12.2. USA

In 1985, National Academy of Sciences published two reports—(1) The report on Meat and Poultry Inspection. The National Program endorsed the implementation of HACCP system for production of safer meat and poultry products. (2) The other report on 'An Evaluation of the Role of Microbiological Criteria for Foods and Food Ingredients'—emphasized on the application of HACCP system especially for high-risk foods. It was in July, 1996, HACCP system was made mandatory for meat and poultry industry and implemented through the Pathogen Reduction HACCP system. These regulations are inspected by Food Safety and Inspection Service (FSIS) and similar state agencies. Each processing plant will have its HACCP plan for prevention of meat food safety hazards and is fully responsible for implementation of the system. HACCP plan of each processing unit will be audited regularly by independent experts but it is the FSIS which is endorsed with the power of accepting the HACCP plan. With regard to the import of meat and poultry products into USA, the exporting countries should meet the requirements of HACCP principles.

12.3. Codex Alimentarius Commission (Codex)

Codex Alimentarius is a common platform for all the countries to discuss on the meat food safety issues based on HACCP principles. Codex was formed by FAO and WHO to implement their food standards and harmonize international food standards and food hygiene requirements to protect consumers. HACCP system has become part of the texts of food hygiene of Codex, which also includes meat and meat products.

12.4. World Trade Organization

WTO covers the Sanitary and Phytosanitary (SPS) Agreement that avoids the international trade barriers and protect consumers world over. SPS agreed that Codex Standards are to be considered by all the countries. Article 5 says that measures stress on an assessment of risk, with particular internationally accepted assessment and control techniques. HACCP is such a technique, applicable in the realm of food safety.

12.5. ISO 9000 Series

ISO 9000 is one of the quality management systems to be followed by the industries for production of wholesome and safe food products and to create confidence among consumers on the quality of food products. ISO 9000 certification means that the industry is committed to follow the quality control procedures, standards and specifications set by international agencies. Globalization of Indian meat trade would insist on the implementation of ISO 9000 series to reach out the global markets.

13. INDIAN SCENARIO

Systems in operation for production of safe meat and meat products for domestic market are:

(*i*) *Food safety and standards act, (2006)*: Under this, all the legal provisions and acts related to quality control has been repealed. Under this act, a Food Safety and Standards Authority of India has been constituted which has formulated food safety and standard regulations which is already effective from November, 2011. Under this Act, specific provisions are given for the quality control of different meat and meat products.

(*ii*) *Export quality control and inspection act (1963)*: This Act deals with the promotion of export of safe food products to global markets. The Export Inspection Council has been established to ensure compulsory quality control and inspection of various commodities including their manufacturing set-ups.

(*iii*) Agriculture and Processed Food Products Export Development Authority (APEDA) operates under Ministry of Commerce, Government of India, covering the export of agro and processed food products including meat and meat products. It looks after the development of export trade of food products and strengthens the capabilities related to the quality control systems. It conducts training programs in Quality Management Systems such as HACCP, Total Quality Management (TQM) and ISO 9000 Series.

(*iv*) Bureau of Indian Standards is a national agency that set standards for a variety of foods including meat and meat products. It operates quality system certification in India.

(*v*) *Ministry of food processing industries (MFPI)*: The Ministry is concerned with the development of food industries including meat in the country. It extends financial support for upgrading infrastructural facilities in meat industry and support R & D programs in improving the quality of meat and meat products.

14. CONCLUSIONS

Good hygienic practice can go a long way towards minimizing chemical and microbial contamination of foods. Farmers and processors who follow such practices enjoy a better market share due to their high quality products and the consumers are assured about the quality of such products. In India, data on occurrence of residues in animal products is sporadic and national surveys are required to facilitate international trade. Residue regulating strategies include Integrated Pest Management (IPM) and monitoring and surveillance of livestock products, fertilizers for residues and certain innovative detection methods. The international harmonization control includes Codex Committee on Pesticide Residues (CCPR) and Joint meetings on Pesticide Residues (JMPR), Codex Committee of Veterinary Drugs in Foods (CCRVDF) and the Joint Expert Committee on Food Additives and Contaminants (JECFA) and Codex Committee on Food Additives and Contamination (CCFAC).

15. ACTION PLAN FOR QUALITY CONTROL AND QUALITY MANAGEMENT OF MEAT AND MEAT PRODUCTS

(*i*) Quality Control at farm production.

(*ii*) Popularization of modern abattoir concepts for production of hygienic meat and preservation systems.

(*iii*) Implementation of Quality Control Measures.

(*iv*) Establishment of Regional Quality Control Labs.

(*v*) Establishment of Export Inspection Agency for meat and meat products.

(*vi*) Surveillance of the incidence of pathogenic microorganisms, and parasites, toxins, residues, pesticides and heavy metals in meat and meat products.

(*vii*) Training of manpower in quality control and quality management.

(*viii*) Long-term policy for a holistic approach for Quality Control in meat food chain from Farm-to-Fork.

(*ix*) Establishment of Meat Food Safety Authority.

(*x*) Government's firm commitment on the organization of Quality Control Systems for meat food safety.

(*xi*) Implementation of food safety and standards regulation as specified by Food Safety and Standards Authority of India.

(*xii*) Establishment of Export Inspection Council for meat and meat products.

(*xiii*) Implementation of Codex Standards, ISO 9000 series.

(*xiv*) Involvement of Indian Meat Science Association in the organization of Quality Control and Management of meat food chain.

REFERENCES

APHS, (2003). Monitoring and regulating chemical residues. The Queens land state department of primary industries.

Cantor, K.P., Blair, A., Everelt, G., Gibson, R., Burmeister, L.F., Brown., L.M., Schuman, L. and Dick, F.R (1992). Pesticides and other agricultural risk factors for non-Hodgkins's lymphoma among men in Iowa and Minnesota. *Cancer Research*, 52: 2447–2455.

Chen, J. and Brody, A.L. (2013). Use of active packaging structures to control the microbial quality of a ready-to-eat meat product. *Food control.*, 30: 306–310.

Chen, J.H., Seow, R.J., Liu, T., Bang, W.S. and Yuk, H.G. (2012). Intervention technologies for ensuring microbiological safety of meat: current and future trends. *Comprehensive Reviews in Food Science and Food Safety*, 11: 119–132.

Colborn, T., Myers, J.P. and Dumanoski, D. (1996). Our stolen future: how we are threatening our fertility, intelligence and survival: A scientific detective story. New York.

Debis Book Science (2003). The Hazards of America's 10 most popular Foods, Boca News Article. Fit America MD.

Dogheim, S.M., Nasr, E.N., Almaz, M.M. and El-Tohamy, M.M. (1990). Pesticide residues in milk and fish samples collected from two Egyptian Governorato. *J. Asso. Off. Anal. Chem.*, 73: 19–21.

FAO (2000). Food safety management in developing countries, Proc. of the International workshop, Hanak, E., Boutrif, E., Fabre, P. and Pineiro, M. (*Eds.*).

Fiore, M.C., Anderson, H.A., Hong, R., Golubjatnikov, J.E., Seiser, J.D., Nordstrom, D., Hanrahan, L and Belluck, D. (1986). Chronic exposure to aldicarb-contaminated ground water and human immune function. *Environ. Res.*, 41: 633–645.

Jenson, I. and Summer, J. (2012). Performance Standards and meat safety-developments and direction. *Meat Sci.*, 92: 260–266.

Litovitz, T.L., Schmitz, B.F. and Bailey, K.M. (1990). Annual report of the american association of poison control centers national data collection system. *Am. J. Emergency Med.*, 8: 394–442.

Nollet, L.M. (2012). Handbook of meat, poultry and seafood quality, 2nd edn, John Wiley & Sons, Inc.

Tiwari, B.K., Valdramidis, V.P., O'Donaell, C.P., Muthukumarappan, K., Bourke, P. and Cullen, P.J. (2009). Application of natural antimicrobials for food preservation. *J. Agric. Food Chem.*, 57: 5987–6000.

Trotter, W.J and Dickerson, R (1993). Pesticide residues in composited milk collected through the US. Pasteurized milk network. *J. AOAC Int.*, 76: 1220–1225.

Vandekinderen, I., Develieghere, F., De Meulenaer, B., Ragaert, P. and Van Camp, J. (2009). Optimization and evaluation of a decontamination step with peroxy acetic acid for fresh-cut produce. *Food Microbiol.*, 26: 882–888.

Varnam, A.H. and Sutherland, J.P. (1995). Meat and meat products—technology, chemistry and microbiology, Chapman and Hall.

WHO, (1992). Our planet, our health Report of the WHO commission on health and environment. Geneva: World Health Organization.

WHO/UNEP (1989). Public health impact of pesticides used in agriculture, Geneva: World Health Organization/United Nations Environment Programme.

7

Processing of Mozzarella Cheese

ATANU JANA

ABSTRACT

Mozzarella cheese is a soft, unripened cheese variety of the Pasta-filata family, which had its origin in Italy. Though it is traditionally made from buffalo milk, it is made all over Italy, in other European countries and USA from cow milk, however the process needs to be modified accordingly. The finished cheese, lightly salted, is white, soft with a very lively surface sheen and has unique property of stretchability. Mozzarella cheese owes its characteristics mainly to the action of lactic acid on dicalcium-para-caseinate. Pure white color is the basic requirement of Mozzarella cheese, because of which buffalo milk is preferred over cow milk. This chapter is about the principle and chemistry of mozzarella cheese making, legal standards, mozzarella cheese manufacture, factors affecting quality of mozzarella cheese, composition of mozzarella cheese, value-added ingredients as additive, equipment for mozzarella cheese manufacture, flavour of mozzarella cheese and the recent developments in mozzarella cheese manufacture. The microstructure and rheological characteristics of cheese, functional properties of mozzarella cheese for use on pizza, browning / blistering of mozzarella cheese, packaging, shelf life and storage changes of mozzarella cheese have also been discussed

1. INTRODUCTION

Mozzarella cheese is a soft, unripened cheese variety of the Pasta-filata family, which originated in the Battipaglia region of Italy. Mozzarella cheese was originally manufactured from high fat buffalo milk. However, it is made all over Italy, in other European countries and USA from cow milk with certain modifications (Ghosh *et al.*, 1990). The finished cheese, lightly salted, is white, soft with a very lively surface sheen and has unique property-stretchability. This makes it suitable for the preparation of lasagna, veal cutlet *alla* parmagiana and as a topping on pizza (Kosikowski, 1982; Jana, 2001).

Mozzarella cheese is used as a topping on pizza pie owing to its inherent stretching qualities. Its production has boosted due to the emergence of the successful business enterprise, the pizza parlour. In Italy, Mozzarella and pizza cheeses accounts for 78.0% of the total Italian cheese product (Merrill *et al.*, 1994). In 'Pasta filata' principle of cheese making, the smooth texture and grain of the cheese is achieved through a skillful stretching of curd in hot water. Mozzarella cheese owes its characteristics mainly to the action of lactic acid on dicalcium-para-caseinate. At a pH of 5.2 to 5.4, much of this compound gets converted to mono-calcium paracaseinate, which provides the strings and sheen to the cheese (Kosikowski, 1986).

In India according to Food Safety and Standards Act (FSSA), there are two variants *viz.*, Mozzarella and Pizza cheeses whose requirements are depicted in Table 1. The microbial requirements of FSSA are given in Table 2. Based on moisture and fat-on-dry matter (FDM) content, four classes of Mozzarella cheese has been suggested in Table 3.

Table 1: FSSA requirements for Mozzarella and Pizza cheese.

Cheese type	***Moisture*** (%)	***Fat-on-dry matter*** (%)
Mozzarella	Max. 60.0	Min. 35.0
Pizza	Max. 54.0	Min. 35.0

Table 2: Microbial specifications for Mozzarella and Pizza cheese as per FSSA.

Microbial profile	***Requirements***
Total plate count/g, Max.	50,000
Coliform count	Absent in 0.1 g
E. coli	Absent in 1 g
Yeast and Mould count	Absent in 1 g
Anaerobic spore count	Absent in 1 g
Salmonella and Shigella	Absent in 25 g
Staph. aureus	Absent in 1 g
Listeria monocytogenes	Absent in 25 g

Table 3: Classification of Mozzarella cheese.

Type of cheese	***Moisture*** (%)	***Fat-on-dry matter*** (%)
Mozzarella (Regular)	52.0–60.0	Min. 45.0
Mozzarella (Part skim)	52.0–60.0	30.0–45.0
Mozzarella (low moisture)	45.0–52.0	Min. 45.0
Mozzarella (low moisture, part skim)	45.0–52.0	30.0–45.0

The low moisture, part skim Mozzarella (LMPSM) is considered suitable for use as pizza topping. New York state in USA specifies 52.0–58.0% moisture and a minimum fat content of 18.0% (for whole milk product) and a moisture content of 45 to 52% moisture for Part skim Mozzarella cheese (Kosikowski, 1986). A maximum moisture content of 58.0% and minimum FDM of 28.0% is prescribed for such cheese in Sau Paulo state (Schiftan & Komatsu, 1979). In Italy, a minimum of 50.0% FDM and a maximum of 65.0% moisture are specified for buffalo whole milk Mozzarella cheese (Cortesi & Maranelli, 1982).

2. QUALITY DETERMINANTS FOR MOZZARELLA CHEESE

Several factors affects the quality of Mozzarella cheese such as the type of milk used, the pre-treatment meted out to cheese milk, the type of starter culture or the rennet used.

2.1. Milk as Raw Material

Pure white color is the requirement of Mozzarella cheese, which is lacking in cow milk cheese due to presence of carotene. Kosikowski (1982) suggested use of neutralizing dyes (*e.g.*, chlorophyll), benzyl peroxide and titanium dioxide (@ 0.03%) to cheese milk, to overcome this problem. Buffalo milk is reported to be more suitable than cow milk for Mozzarella cheese due to high yield, its characteristic aroma and physical attributes (Bonassi *et al.*, 1982).

Bonassi *et al.* (1982) did not find any difference in taste, aroma or texture of Mozzarella cheese made from cow milk, buffalo milk or their combination (1:1). The cheese yield and fat content were significantly higher for buffalo milk product, but the difference in pH, titratable acidity (TA), moisture, protein, soluble nitrogen and ash content were not significant. Singh and Ladkani (1984) standardized the method for Mozzarella cheese from cow and buffalo milk, with and without cheddaring. The stretchability was relatively inferior in buffalo cheese. Patel *et al.* (1986) standardized the starter culture (SC) and direct acidification (DA) method for the manufacture of Mozzarella cheese from buffalo milk.

An initial fat level of 4.0% fat in buffalo milk gave best result; 0.75–0.80% lactic acid was found to be optimal for stretching operations yielding excellent melt and chewiness in the resultant product. Criteria such as fat content, C_{14}/C_{12} fatty acid ratio and electrophoretic mobility of αs_1-casein has been suggested to detect adulteration of genuine buffalo milk with cow milk for mozzarella cheese making. Buffalo milk mozzarella had higher protein content than cows' milk cheese. Rheological properties of cows' milk cheese were superior to those made from buffaloes' milk or mixture of cow and buffalo milk (1:1) (El-Batawy *et al.*, 2004a).

The meltability of ripened cheeses was alike when the fat content of bovine and caprine milks was standardized. Bovine cheese exhibited more free oil compared to caprine cheese. Bovine cheese had a greater structural degradation of cheese matrix than caprine cheese during storage (Imm *et al.*, 2003).

2.2. Pre-treatment of Cheese Milk

2.2.1. *Standardization of milk*

Satisfactory quality Mozzarella cheese can be made from cow or buffalo milk standardized to 3.0–6.0% fat (Kosikowski, 1982; Patel *et al.*, 1986). Use of lower fat content resulted in a harder texture and the cheese lost some of its milky, nutty flavor (Patel *et al.*, 1986). Sundar and Upadhyay (1991) obtained Mozzarella cheese with best sensory quality having superior attributes as a topping on pizza, when buffalo milk was standardized to casein/fat ratio of 0.7. The fat content, yield, meltability and fat leakage of cheese increased with increasing milk fat levels, whereas, moisture and protein content tended to decrease. The pH and stretchability of cheese were not influenced by the fat level in milk. A milk fat level of minimum 2.5% was considered optimum for pizza making (Valle *et al.*, 2004).

2.2.2. *Milk pasteurization*

Traditionally, mozzarella cheese was manufactured from raw milk. However, pasteurization of milk is recommended for mozzarella cheese that is to be consumed fresh, because the plasticizing process does not always destroy pathogens. Heat treatment of milk (72°C, no hold) meant for mozzarella cheese making improved protein and total solids (TS) recovery, but decreased the fat recovery, gave soft-bodied cheese, improved flavor score and keeping quality and ensured public health safety (Patel *et al.*, 1986). Sharma (1988) made satisfactory quality mozzarella cheese from pasteurized and/or UHT treated milks with increased yield compared to the one made from raw milk. UHT heat treatment of milk resulted in greater recovery of the milk proteins in directly acidified mozzarella cheese leading to 3.4% increase in cheese yield (Partridge *et al.*, 1982).

2.2.3. *Lactose hydrolysis of milk*

Lactose hydrolysis of milk (> 60.0%) in the manufacture of mozzarella cheese resulted in reduction in the period for acid development, curd treatment and ripening as well as improved the curd characteristics and yield.

2.2.4. *Homogenization of cheese milk*

Homogenization of milk in mozzarella cheese manufacture led to whiter appearance and improved flavor, reduced fat losses in whey and moulding water, increased cheese yield and reduced fat leakage during baking of cheese on pizza. However, it necessitated some modifications in the cheese making procedure and impaired the curd stretching as well as melting quality during baking trials (Breene *et al.*, 1964; Jana & Upadhyay, 1992a, b). Use of higher homogenization pressure (> 50 kg/cm^2) resulted in greater fat recovery in cheese, while protein recovery and curd stretching property were adversely affected. Homogenization (25 kg/cm^2, 60°C) of cheese milk led to increased cheese moisture, TA, total volatile fatty acid (TVFA) and cohesiveness, but decreased stretchability, meltability, hardness, gumminess and chewiness. The flavour development was rapid in cheeses prepared from homogenized milk (El-Batawy *et al.*, 2004b). Homogenization of cheese milk greatly reduced free oil; homogenization of skim milk alone had no effect. Free oil from homogenized mozzarella melted at a lower temperature (Tunick, 1994).

2.3. Starter Cultures

Manufacture of high-moisture mozzarella entails use of starters like *S. lactis*, *S. durans* or *S. faecalis*, whereas, for low-moisture mozzarella the starters recommended are *S. thermophilus*, *L. bulgaricus* and/or *L. helveticus*. The starter used for fresh Italian mozzarella is only *Streptococcus thermophilus*. To achieve consistent quality in the end product, the parameters to be controlled include avoiding batch-to-batch variation in the starter strain ratio and constant acidification. According to Ferris (1981), direct-to-vat starters such as 'Superstart' are used where cooking temperature does not exceed 39.4°C. For those using starters composed of *S. thermophilus* and *L. bulgaricus*, which are necessary when cooking at temperatures of ≥ 40°C, the popularity of the bacteriophage-inhibiting medium 'Thermostar' is advisable.

2.4. Renneting of Cheese Milk

Various types of rennet *i.e.*, of animal origin—calf, bovine, porcine pepsin, of vegetable origin—*Irpex lacteus*, and of microbial origin fungal rennet *i.e.*, Meito, Marzyme and Surecurd have been used successfully to produce mozzarella cheese. According to Quarne *et al.* (1968b) cheese made with porcine pepsin or fungal rennet gave sensorily superior product than veal rennet. Micketts and Olson (1974) found that cheese made with porcine pepsin was slightly softer than the one made using calf rennet. Identical results have been reported using Marzyme and calf rennet in Mozzarella cheese making. The extract of dried berries of *Withania coagulans* has been successfully used in preparation of Mozzarella cheese (Nawaz *et al.*, 2011).

Recovery of TS of cheese decreased when using *Mucor miehei* rennet, while the recovery of protein and fat content in cheese were nearly the same as for calf rennet. *Mucor miehei* rennet exhibited greater rise in soluble nitrogen (SN) content than other coagulants. TVFA content increased with storage period, and with increased fat content (1.5 *vs.* 3.0%) of the milk. The coagulant types did not affect the SN or TVFA. The type of coagulant had negligible effect on meltability, fat leakage and oiling-off of cheese. Mozzarella cheese made with *Mucor miehei* had the highest firmness. *Mucor miehei* rennet exhibited highest rate of proteolysis during storage and such cheeses were preferred over cheese made by calf rennet or chymosin (Ahmed *et al.*, 2011).

Mozzarella cheese prepared by starter culture method, using *E. parasitica* protease was softer in unmelted cheese texture, was more meltable and had lower apparent viscosity and exhibited more free oil release on melting than the cheese made using chymosin and *Mucor miehei* protease. Decreasing the concentration of chymosin coagulant by up to 40.0% had limited impact on the composition, proteolysis and functional characteristics during refrigerated storage of LMPSM cheese (Kindstedt, 1991); the free oil formation was significantly affected.

LMPSM cheeses were made with single culture (SC) of *Streptococcus thermophilus* or mixed culture (MC) of *Streptococcus thermophilus* and *Lactobacillus helveticus* using chymosin or *Cryphonectria parasitica* (CP) coagulant. The melt area increased by ~2 times in cheeses made with SC as against 3-4 times for cheeses made with MC. The melt characteristics of 7 day old cheeses prepared with MC almost corresponded to that of 30 day old cheeses made with SC, suggesting faster ripening using MC. The degradation of total αs-casein was higher in the chymosin cheeses and that of α_s-casein in the CP cheeses. Meltability of Mozzarella cheese was better correlated to hydrolysis of β-casein, but least to α_s-casein (Dave *et al.*, 1998).

2.5 Salting of Cheese

The cheese maybe salted by any of the three methods *viz.*, (*a*) dry salting, (*b*) brine salting and (*c*) hot brine salting. Direct salting of cheese involves stretching the cheese in hot water, draining the water followed by sprinkling salt (~1.6%) over the cheese during mixing (Ferris, 1981). Direct injection of salt solution (~6 MPa hydraulic pressure) into cheese has been attempted. Use of dry salting, rather than brine salting (23.0% NaCl, pH 5.15, 0.02% calcium) or combined salting (dry + brine), resulted in a significantly higher cheese moisture and yield (Guinee *et al.*, 2000). Brine-salted cheeses usually develop reasonably smooth and predictable gradients of decreasing moisture from center to surface. Cold brining resulted in less moisture loss from the cheese surface to the brine. Reducing the salt content of cheese led to slight decrease in the meltability and free oil of Mozzarella cheese.

3. QUALITY OF MOZZARELLA CHEESE

The factors that determine the quality of resultant Mozzarella cheese, over and above the ones discussed above are as follows:

3.1. Influence of pH/Acidity of Whey at Draining

The ash, calcium and phosphorus concentration of Mozzarella cheese prepared by starter culture method decreased as the whey pH was lowered from 6.2 to 5.3. The cheese yield and calcium recovery were the highest in cheese with pH of 5.3.

Reducing the calcium content of cheeses increased the meltability of cheese. Fresh cheeses made by draining the whey at pH 5.6 were preferred (Yazici & Akbulut, 2007). The cheese made using 0.40% LA whey acidity was superior in terms of flavour and texture score; it had maximum springiness. Cheese made from 0.50% whey acidity at draining was least acceptable (Sundar & Upadhyay, 1992).

3.2. Effect of Coagulant Concentration and Composition

Mozzarella cheeses with lower fat content had the lowest moisture-in-nonfat-substance (MNFS) content which led to slower rate of proteolysis. Cheeses became softer during storage, and the decrease in modulus was influenced by the level of coagulant. High fat content and high coagulant level promoted more melting of cheese (Dave *et al.*, 2003).

3.3. Effect of Cooking Temperature

Higher cooking temperature (44°C *vs.* 38–41°C) produced cheese with lower moisture and decreased proteolysis during refrigerated storage. The apparent viscosity of melted cheese was higher when employing higher cooking temperature. Cooking temperature did not significantly affect the meltability and free oil of cheese (Yun *et al.*, 1993a).

3.4. Effect of Milling pH

Milling pH (5.1–5.4) affected the cheese pH and TA. The proteolysis during refrigerated storage of cheese was unaffected by differences in the milling pH (Yun *et al.*, 1993b).

3.5. Effect of Curd Acidity at Stretching

The optimum curd TA for cheese stretching was 0.8% LA. The sensory score of cheese increased with increasing curd acidity from 0.5 to 0.8% LA. Rise in curd acidity led to a decrease in cheese moisture and TS loss during stretching, while melting quality improved.

4. COMPOSITION OF MOZZARELLA CHEESE

The composition of Mozzarella cheese depends on the type of milk used for cheese making and the procedure used for cheese manufacture. The composition of Mozzarella cheese reported by few workers is depicted in Table 4.

5. VALUE-ADDED INGREDIENTS AS ADDITIVE

5.1. Concentrated Buttermilk

Cheese milks were prepared by adding 0–6.0% (w/w) of condensed sweet cream buttermilk (CSCB) (34.0% TS, 9.0% casein) to part-skim milk, and cream was

Table 4: Composition of Mozzarella cheese.

Constituents	*El-Owni and Oswan (2009)*	*Jana (1992)*	*Pizaia et al. (2003)*
Moisture, %	48.59	50.51	49.30
Fat, %	27.25	26.34	25.85
Protein, %	20.06	20.48	20.83
Lactose, %	1.59	–	–
Ash, %	2.25	2.00	3.62
Salt (NaCl), %	–	0.90	2.17
Acidity (% LA)	0.66	0.61	0.76

added to obtain milk with 11.2–12.7% TS and casein/fat of 1. Use of CSCB resulted in a significant increase in cheese moisture. Fat and nitrogen recoveries in CSCB cheeses were slightly lower and higher, respectively, than the control cheeses. Cheeses containing CSCB (4.0–6.0%) exhibited less melt and stretch than control cheese. Addition of CSCB significantly lowered free oil at 1 week. Addition of CSCB at low (*i.e.*, 2.0%) levels improved cheese yield without affecting the compositional, rheological and sensory properties of cheese (Govindasamy *et al.*, 2007).

Cheese milks of casein/fat of 1.0 and casein content of 2.7% were obtained by blending cold ultrafiltered (UF) sweet cream buttermilk (SCB) retentate (19.9% solids), Reverse osmosis (RO)-SCB retentate (21.9% solids), or evaporation (EV)-SCB retentate (36.6% solids) with partially skimmed milk and cream (34.6% fat). SCB-fortified washed, stirred-curd pizza cheeses had 4.0–5.0% higher moisture (51.0–52.0%) than control cheese without SCB (47.0%). Fat recoveries were significantly lower in RO- and EV-SCB cheeses than in control or UF-SCB cheeses. The SCB cheeses had lower meltability compared to control. On baked pizza, cheese made from milk fortified with UF-SCB tended to have least free oil, but flavor attributes of all cheeses were similar (Govindasamy *et al.*, 2007).

5.2. Milk Protein Concentrate

Addition of milk protein concentrate (MPC) improved the yield of cheese from 10.34% for control (C) to 14.50 and 16.65% for cheeses made from whole milk + MPC + culture (MP culture) and whole milk + MPC + direct acid (MP acid) respectively. TS recoveries were in the order MPculture > MPacid > C; fat and protein recoveries were not significantly affected. MPculture cheese had the highest Ca and lowest lactose content. MPacid cheese had the maximum meltability. Control cheese initially had superior meltability than MPculture; however, the difference became insignificant after 28 days of storage at 4°C. MP directly acidified cheese had soft texture and produced large-sized blisters when baked on pizza. The lowest and highest levels of proteolysis were found in MPculture and MPacid cheeses, respectively (Rehman *et al.*, 2003). Cold dispersible MPC containing 83.0% protein obtained by ultrafiltration process was used to standardize protein content of cheese milk to 4.0% (fat/protein = 1.0). There was an increase in the yield of cheese up to 16.7% (*vs.* 13.8% for control) as a result of high recovery of TS and protein and slightly higher cheese moisture content (Francolino *et al.*, 2010).

5.3. Whey Proteins

Heated (95°C at pH 4.6) whey protein homogenized dispersions added to standardized milk and reduced fat milk was used to enhance the yield of Mozzarella cheese. Homogenization of whey proteins improved fat retention and yield. The lower yield of reduced fat cheese was partially offset by adding whey proteins. Physical and sensory properties of reduced fat cheeses containing such whey proteins were similar to the full fat control. Inclusion of 0.1 or 0.2 w/w of heat denatured whey protein had no significant effect on the functionality of cheese compared to control, but helped in improving the water binding capacity of cheese.

5.4. Ultrafiltration Retentates

Pizza cheese was manufactured from two ultrafiltration retentates *i.e.*, 13.5% and 15.2% TS fortified milk. Cheese milks were obtained by blending cold processed UF retentate with partially skimmed milk. Gels made from UF-fortified milk had similar clotting times, but clotted faster than control milk. Fat recoveries in the cheese increased in the order: UF (15.2% TS) < control < UF (13.5% TS). Nitrogen recoveries were lower in control than in UF-fortified cheese. Standardization of cheese milk with cold UF retentates increased the cheese yield, without adversely affecting its functionality.

6. METHODS OF MANUFACTURING MOZZARELLA CHEESE

The methods of manufacture of Mozzarella cheese varies considerably according to the market of the cheese.

6.1. Starter Culture Methods

The stepwise description for preparation of Mozzarella cheese by traditional technique (starter culture technique) has been reported by few workers (Ohashi 1981; Patel *et al.*, 1986). The flow chart for preparation of Mozzarella cheese by Starter culture method is depicted in Fig. 1.

Pasteurized milk is incubated at 33°C with rennet and a starter comprising of lactobacilli and streptococci. The separated curd (pH 5.2) is heated in water at 80°C, kneaded and formed into 150–250 g balls, which are cooled in running water at 10–12°C for 30 min followed by immersion in chilled (5°C) brine for 30 min.

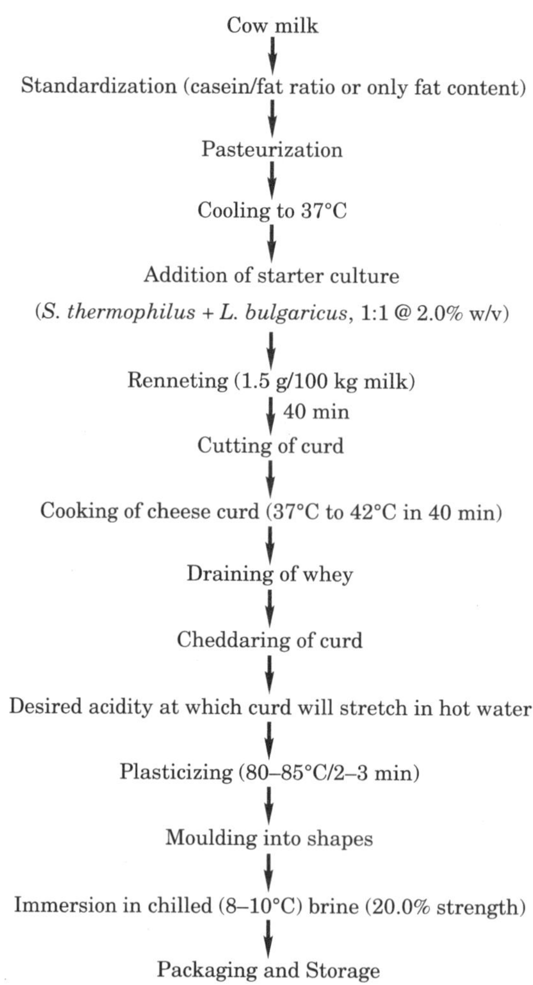

Fig. 1. Flow chart for preparation of Mozzarella cheese.

6.2. Direct Acidification Methods

Direct acidification (DA) technique has gained considerable commercial interest, as it does not rely on starter performance (unpredictable, risk of phage infection, milk contaminated with antibiotics) and helps towards mechanization of production. The manufacture of Mozzarella cheese using DA has been attempted by several workers (Ohashi, 1981; Demott, 1983; Singh & Ladkani, 1984). The different types of acids employed in DA technique include hydrochloric acid, phosphoric acid, lactic acid, acetic acid, malic, citric acid and glucono-delta-lactone (Quarne *et al.*, 1968a, b; Jensen *et al.*, 1987). Pizza cheese manufactured using citric acid as pH regulator had pronounced meltability and stretchability. No significant differences were noted with regard to fat and free oil contents when using acids such as citric, lactic acid, acetic acid, phosphoric and hydrochloric (Zhang *et al.*, 2011).

The two methods employed utilizing DA techniques are:

1. Direct acidification quiescent storage method (Breene *et al.*, 1964; Patel *et al.*, 1986; Singh & Ladkani, 1984), and
2. Direct acidification continuous agitation (DACA) method (Micketts & Olson, 1974; Kim & Yu, 1988).

DACA method involves use of continuous mixing, cooking, drawing, plasticizing, moulding, salting/brining and packaging machines.

When comparing Pizza cheese made by conventional SC and DA method, the recovery of fat and cheese yields was greater at pH 5.6 and 5.4 than at 5.9 and 5.2. The recovery of SNF and cheese hardness decreased, while moisture content, meltability and fat leakage increased as the pH decreased. Meltability of DA cheeses was lower than that of SC, except for cheeses made with citric or malic acid at pH 5.2 and 5.4. DA cheeses were most acceptable at pH 5.4 and 5.6, and the cheese made with lactic or acetic acid were superior compared to use of phosphoric, hydrochloric, malic or citric acids (Kim & Yu, 1988).

6.3. Other Cheese Making Methods

6.3.1. *Mozzarella cheese from reconstituted dried milk/skim milk*

Cheese made with 40.0–60.0% reconstituted skim milk (10.0% solids) as blends with cows' fresh whole milk, had the best shredding, melting and stretching characteristics without oiling-off. Higher substitution levels gave significantly higher moisture content and rigid curd. Such LMPSM cheeses had lower meltability but better shredding and stretching properties with longer strings than full-fat milk cheese, when baked.

Mozzarella cheese has been made from reconstituted 'low-heat' dried skim milk and cream. Use of low-heat powder avoided a weak curd with poor stretch and melt characteristics. The cheese made by DA from reconstituted milk had higher protein and TS recoveries than that made from fluid milk (Demott, 1983). Mozzarella cheese made from recombined milk does not exhibit the characteristic stretch and melt behaviour of fresh milk cheeses. Such problem can be overcome either by using low homogenization pressures or by coating the fat in the recombined milk emulsion with phospholipid and homogenizing at the usual pressure.

6.3.2. *Modified method*

An alternative method for stretching the Mozzarella curd involved addition of 1.0–1.5% of emulsifying salts comprising of tri-sodium citrate and disodium hydrogen phosphate; 1:1 or 2:1 w/w to curd, before heating to 65–70°C for 10–15 min, followed by moulding. Such cheese had improved melting characteristics, flavour and texture and recovery of milk constituents compared to control cheese made by stretching in hot moulding water (Patel & Upadhyay,

1999). Shah *et al.* (2008) prepared pasteurized Mozzarella cheese (negative alkaline phosphatase test) from raw milk (unpasteurized) just by employing boiling water for plasticizing the cheese curd.

7. MECHANIZATION OF MOZZARELLA CHEESE MANUFACTURE

The mechanization in the manufacture of Mozzarella cheese includes the development of machines for forming and shaping cheese balls, moulding, automatic plasticizing and stretching of curd, salting, automatic moulder-cooling, moulding and hardening, cheese compacting, automatic brining and chilling of cheese and shredding (Muzzarelli, 1985; Mongiello, 1986).

Mechanical mixers with single screw or twin screws coupled with steam injection system are used for heating and stretching of the cheese. Typically, the water temperature used in cooker-stretcher ranges from 60–85°C, and the cheese temperature as it exits the mixer ranges from 50–65°C (Renda *et al.*, 1997). DACA method of Mozzarella cheese manufacture greatly facilitates the mechanization and continuous production of curd. The process mainly involves use of continuous mixing, cooking, drawing, plasticizing, moulding, salting/brining and packaging machines.

8. FLAVOUR OF MOZZARELLA CHEESE

Mozzarella cheese has lower free fatty acids (FFA) concentration and milk flavor and except for C_4, the role of FFA in the characteristic flavor of this cheese was not apparent. It is considered that branched short chain fatty acids are important for the flavor of Mozzarella cheese (Woo & Lindsay, 1984).

9. RECENT DEVELOPMENTS IN MOZZARELLA CHEESE MANUFACTURE

9.1. Membrane Process in Cheese Production

Reviews pertaining to the manufacture of Mozzarella cheese from ultrafiltrated milk are given by Jensen *et al.* (1987) and Jana (1992). Some manufacturing lines like Firma Pasilac, Ridgeway Foods, Alfa-Laval Alcurd continuous coagulator and Maubois, Mocquot and Vassal (M.M.V.) process (Kosikowski, 1986) utilizes the ultrafiltration (UF) process in the manufacture of Mozzarella cheese on commercial basis.

The principle behind the manufacture involves separation of milk, UF of skim milk, cream concentration and mixing retentate with the cream before culturing and renneting. UF of milk is carried out to 39.0–40.0% TS, followed by conventional evaporation to 52.0–54.0% TS. Supplementation of cheese milk with retentate and hot brine stretching; addition of freeze dried retentate to fresh liquid retentate; and diafiltration with brine during UF with/without simultaneous fermentation of the retentate gave good result (Covacevich, 1981, Fernandez and Kosikowski, 1986). UF cheeses exhibited lower stretchability, higher oiling off and greater increase in hardness during storage than non-UF cheeses (El-Batawy *et al.*, 2004a).

UF whole milk retentates Volume Concentration Ratio (VCR of 1.4–2.0:1) were made into Mozzarella cheese by DA with 10.0% glacial acetic acid. Low concentrate retentates showed improved physical properties over that of non-retentate control whole milk. Yield efficiency, based on casein recovery, was higher in retentate cheese than in controls.

Retentate supplementation (4.5:1 retentate supplementing milk to 1.2:1 fat and 1.4:1 protein) improved LMPSM cheese yield. Increased output per vat, higher yield efficiency and good melting properties were observed in retentate cheese made using thermophilic starter over control; optimum retentate VCR was 1.75:1. Mozzarella cheese prepared from UF retentate (VCR of 2.3:1) had higher total protein, ash and pH, but lower fat, FDM, salt and TA than control cheese (Pizaia *et al.*, 2003). Impaired stretching properties and low meltdown characteristics in the resultant cheese are the difficulties encountered.

9.2. Low-Fat Mozzarella Cheese

Owing to the awareness among people regarding fat intake (especially milk fat) and health, the dairy industry has shown increasing interest in producing low-fat and fat-free dairy products, including Mozzarella cheese. Low fat Pasta filata cheese has been prepared from milk standardized to 1.6 and 3.2% milk fat by DA method.

Low-fat Mozzarella cheeses containing 6.0 % fat were made by pre-acidification of milk with citric acid to pH 6.1, using encapsulated ropy exopolysaccharide (EPS) producing *Streptococcus thermophilus*. Pre-acidification of milk helped in reducing the hardness and increase meltability of low fat cheeses. Use of capsular and ropy-EPS aided in increased moisture retention (2.0% higher moisture) in cheese. On baking, the cheeses containing EPS showed improved shred fusion, meltability and a reduction in surface scorching (Zisu & Shah, 2007). Low-fat (6.0%) cheese made with the EPS starter cultures exhibited slightly higher meltability than control cheese made using non EPS starter (Perry *et al.*, 1998).

10. TEXTURAL CHARACTERISTICS OF CHEESE

The effects of type of milk, standardization of milk for fat or casein/fat (C/F) ratio, heat treatment of milk, homogenization, milk coagulant, method of manufacture, type of acid and pH at coagulation whey draining acidity, composition and proteolysis during storage on the textural properties of Mozzarella cheese has been reported (Kindstedt, 1991; Tunick *et al.*, 1991). Cheese made using the lowest C/F ratio milk (0.50) had lower hardness, cohesiveness, springiness, chewiness and gumminess than cheeses made from milk with higher C/F ratios (up to 0.90) (Sundar & Upadhyay, 1991).

Mozzarella cheese is a visco-elastoplastic material (semisolid) at room temperature, which becomes visco-elastic (as liquid) at about 60°C. Mozzarella cheese exhibited poor meltdown, fat leakage, acid flavor, free surface moisture, poor cohesiveness and bleached discolouration after thawing. On decreasing the fat content (3.5 to 2.0%) of the cheese milk, the cohesiveness, gumminess and chewiness of the cheese increased, whilst elasticity decreased (Valle *et al.*, 2004).

11. FUNCTIONAL PROPERTIES OF MOZZARELLA CHEESE FOR USE ON PIZZA

The important characteristics of Pizza cheese are: (*i*) it should possess a moderate toughness and an adequate stringiness, (*ii*) it should shred, grind and slice with a minimum of matting. Mozzarella cheese manufacturers considered meltability, stretchability, elasticity, stringiness and shredability to be important for customer satisfaction (Mc Mahon *et al.*, 1993; Jana, 2001).

Freshly made Mozzarella melts to a tough, extremely elastic, and somewhat granular consistency with limited stretch. However, during the first few weeks of refrigerated storage, a transformation occurs as the unmelted cheese becomes softer and the melted cheese becomes more viscous, less elastic, and highly stretchable with optimal functionality for pizza (Kindstedt, 1991). With ageing of cheese there was an increase in the stretchability and flowability of cheese. On prolonged storage, shredded cheese developed susceptibility to clumping/balling and baked cheese tended to exude excess free oil and lost its desired level of chewiness. Melting increased during storage (Dave *et al.*, 2003).

11.1. Browning/Blistering of Mozzarella Cheese

Galactose accumulated in Mozzarella when either strain (non-galactose (Gal^-) and galactose fermenting (Gal^+) strains) of *Str. thermophilus* was used. The fermentation of accumulated galactose is as a result of metabolism by *Lactobacillus helveticus*. There was a positive correlation between galactose content and brown color intensity in heated Mozzarella cheese (Johnson & Olson, 1985; Matzdorf *et al.*, 1994). Mozzarella cheese obtained by conventional method was more prone to blistering than cheese made from UF retentate during the first month of storage; both cheeses exhibited similar behavior during subsequent storage (Pizaia *et al.*, 2003).

12. PROCESSED MOZZARELLA CHEESE AND PACKAGING

Processed cheese has been made using Mozzarella cheese curd (Kosikowski & Silverman, 1957). The young Cheddar cheese component of the processed cheese blend could be substituted by Mozzarella cheese up to an extent of 25.0–40.0% level, without adversely affecting the composition and sensory characteristics of the product (Patel *et al.*, 1986).

Anti-caking agents (cellulose, potato starch or mixtures including dextrose) are utilized in the production of shredded cheese for reducing clumping and increasing the appearance of shreds. Meltability and stretchability of Mozzarella cheese remained unaffected. The free oil formation was significantly influenced by the varying anticaking agents applied.

The antimicrobial (*viz.*, lemon extract combined with brine and gel solution made of sodium alginate) packaging system led to an increase in the shelf life of active packaged Mozzarella cheeses. Mozzarella cheese has been successfully packaged in polyethylene (300 gauge) and Cryovac packed in polyvinylidene chloride (PVDC) film (150 gauge) with or without vacuum packaging (Ghosh & Singh, 1992a, b).

13. STORAGE AND MICROBIAL PROFILE

A decrease of about 2 log of *Listeria monocytogenes* was observed after stretching of the curd in hot water (95°C), followed by complete elimination of Listeria after 24 h of cheese storage in conditioning liquid (Villani *et al.*, 1998). Stretching of Mozzarella cheese curd at 66°C for 5 min or 77°C for 1 min can effectively control *L. monocytogenes* during Mozzarella cheese production; brining too had a lethal effect (Kim *et al.*, 1998). The Mozzarella cheese when packaged in cryovac package in PVDC film and stored under refrigeration (5°C) and deep freeze, kept well for 42 and 90 days respectively (Ghosh & Singh, 1992a). A lactic acid/chitosan solution (0.075% chitosan strength) when added to the starter in Mozzarella cheese making inhibited the growth of spoilage microorganisms (*i.e.*, coliform, *Pseudomonas* species).

Storage of Mozzarella cheese at 4°C over 4 weeks decreased the apparent viscosity; the rough fibrous texture changed to a smooth one and changes also occurred in cheese melting and stretching characteristics (Kindstedt & Guo, 1997). Patel *et al.* (1986) reported that cheeses made by DA method had lower shelf life (1 week) as compared to those prepared by starter method (> 2 weeks), when stored at 8°C. The melting and fat leakage characteristics increased, while stretch tended to decrease with advancement in storage period of Mozzarella cheese (Ghosh & Singh, 1992a). During refrigerated (4°C) storage of Mozzarella cheese for 50 days the meltability increased, apparent viscosity of melted cheese decreased, and free oil formation from melted cheese increased (Yun *et al.*, 1993b). The Mozzarella cheese obtained from UF retentate exhibited lower proteolysis and melt capacity during storage than the one made from normal unconcentrated milk (Pizaia *et al.*, 2003).

REFERENCES

Ahmed, N.S., Abd El-Gawad, M.A.M., El-Abd, M.M. and Abd-Rabou, N.S. (2011). Properties of buffalo Mozzarella cheese as affected by type of coagulant. *Acta Scientiarum Polonorum—Technologia Alimentaria* 10: 339–357.

Asperger, H. and Brandl, E. (1982). Microbiology of Mozzarella cheese. 21st Intl. Dairy Congr., 1. *Book* 2: 274.

Breene, W.M., Price, W.V. and Ernstrom, C.A. (1964). Manufacture of Pizza cheese without starter. *J. Dairy Sci.*, 47: 1173–1180.

Bonassi, I.A., Carvalho, J.B. De C. and Villares, J.B. (1982). Use of buffaloes' milk for Mozzarella cheese. *Arch. Latinoam Nutr.*, 32: 903–12.

Cortesi, M.L. and Maranelli, A. (1982). Fior di Lette and Mozzarella cheese: Hygiene and standards. *Industri. Alimentari.*, 21: 528–531.

Covacevich, H.R. (1981). Cited by Jensen, L.A., Johnson, M.E. and Olson, N.F. (1987). Composition and properties of cheeses from milk concentrated by ultrafiltration and reverse osmosis—A review of literature. *Cult. Dairy Prod. J.*, 22: 10.

Dave, R.I., Sharma, Pragati and Mc Mahon, D.J. (1998). Melt and rheological properties of Mozzarella cheese as affected by starter culture and coagulating enzymes. *J. Dairy Sci.*, 81: 563–566.

Dave, R.I., Mc Mahon, D.J., Oberg, C.J. and Broadbent, J.R. (2003). Influence of coagulant level on proteolysis and functionality of Mozzarella cheeses made using direct acidification· *J. Dairy Sci.*, 86: 114–126.

Demott, B.J. (1983). Recovery of milk constituents in a Mozzarella-like product manufactured from non-fat dry milk and cream by direct acidification at 4° and 35°C. *J. Dairy Sci.*, 66: 2501–2506.

El-Batawy, M.A., Galal, E.A., Morsy, M.A. and Abbas, K.A.A. (2004a). Utilization of ultrafiltration technique in making Mozzarella cheese from different kinds of milk. *Egyp. J. Dairy Sci.*, 32: 303–314.

El-Batawy, M.A., Galal, E.A., Morsy, M.A. and Abbas, K.A.A. (2004b). Effect of homogenization on some properties of Mozzarella cheese. *Egyp. J. Dairy Sci.*, 32: 315–326.

El-Owni, O.A.O. and Osman, S.E. (2009). Evaluation of chemical composition and yield of Mozzarella cheese using two different methods of processing. *Pakistan J. Nutr.*, 8: 684–687.

Fernandez, A. and Kosikowski, F.V. (1986). Low moisture Mozzarella cheese from whole milk retentates of ultrafiltration. *J. Dairy Sci.*, 69: 2011–2017.

Francolino, S., Locci, F., Ghiglietti, R., Lezzi, R. and Mucchetti, G. (2010). Use of milk protein concentrate to standardize milk composition in Italian citric Mozzarella cheese making. *LWT-Food Sci. Technol.*, 43: 310–314.

FSSA (2006). Food Safety and Standards Act www.mofpi.nic.in/foodsafety.htm

Ghosh, B.C. and Kulkarni, S. (1996). Low cholesterol Mozzarella cheese—Technology standardization. *J. Food Sci. Technol.*, 33: 488–492.

Ghosh, B.C. and Singh, S. (1992a). Storage studies of Mozzarella cheese. 1. Sensory and rheological characteristics. *Indian J. Dairy Sci.*, 45: 199–202.

Ghosh, B.C. and Singh, S. (1992b). Storage studies of Mozzarella cheese. 2.Chemical changes. *Indian J. Dairy Sci.*, 45: 203–208.

Ghosh, B.C., Singh, S. and Kanawjia, S.K. (1990). Rheological properties of Mozzarella cheese. A Review. *Indian J. Dairy Sci.*, 43: 70–79.

Govindasamy, L.S., Lin, T., Jaeggi, J.J., Martinelli, C.J., Johnson, M.E. and Lucey, J.A. (2007). Impact of type of concentrated sweet cream buttermilk on the manufacture, yield and functionality of pizza cheese. *J. Dairy Sci.*, 90: 2675–2688.

Guinee, T.P., Mulholland, E.O., Mullins, C. and Corcoran, M.O. (2000). Effect of salting method on the composition, yield and functionality of low moisture Mozzarella cheese. *Milchwissenschaft*, 55: 135–138.

Imm, J.Y., Oh, E.J., Han, K.S., Oh, S., Park, Y.W. and Kim, S.H. (2003) Functionality and physico-chemical characteristics of bovine and caprine Mozzarella cheeses during refrigerated storage. *J. Dairy Sci.*, 86: 2790–2798.

Jana, A.H. (1992). Mozzarella cheese from concentrated milk—A review. *Egyp. J. Dairy Sci.*, 20: 201–218.

Jana, A. (2001). Mozzarella cheese and pizza—the compatible partners. *Beverage and Food World*, 28: 14–19.

Jana, A.H. and Upadhyay, K.G. (1992a). Homogenization of milk for cheese making—A review. *Aust. J Dairy Technol.*, 47: 72–79.

Jana, A.H. and Upadhyay, K.G. (1992b). Process standardization for manufacture of Mozzarella cheese from homogenized buffalo milk. *Indian J. Dairy Sci.*, 45: 256–260.

Jensen, L.A., Johnson, M.E. and Olson, N.F. (1987). Use of glucono-delta-lactone as an acidulant for low moisture part skim Mozzarella cheese produced from ultrafiltered milk. *J. Dairy Sci.*, 70: 66.

Kim, Y.H. and Yu, J.H. (1988). A study on the manufacture of pizza cheese by direct acidification continuous agitation. *Korean J. Dairy Sci.*, 10: 21–33.

Kim, J., Schmidt, K.A., Phebus, R.K. and Jeon, I.J. (1998). Time and temperature of stretching as critical control points for *Listeria monocytogenes* during the production of Mozzarella cheese. *J. Food Prot.*, 61: 116–118.

Kindstedt, P.S. (1991). Functional properties of Mozzarella cheese on pizza: A review. *Cult. Dairy Prod. J.*, 26: 27–31.

Kindstedt, P.S. and Guo, M.R. (1997). Recent developments in the science and technology of pizza cheese. *Aust. J. Dairy Technol.*, 52: 41–43.

Kosikowski, F.V. (1982). *In*: Cheese and Fermented Milk Foods. 2nd Edn., Edwards Brothers Inc., Am. Arbor., Michigan, pp. 153–167, 178.

Kosikowski, F.V. (1986). New cheese making procedures utilizing ultrafiltration. *Food Technol.*, 40: 71–77.

Kosikowski, F.V. and Silverman, G.J. (1957). Processed Mozzarella cheese. *Milk Prod. J.*, 48: 10–11,52.

Matzdorf, B., Cuppett, S.L., Keeler, L. and Hutkins, R.W. (1994). Browning of Mozzarella cheese during high temperature pizza baking. *J. Dairy Sci.*, 77: 2850–2853.

Mc Mahon, D.J., Oberg, C.J. and Mc Manus, W. (1993). Functionality of Mozzarella cheese. *Aust. J. Dairy Technol.*, 48: 99–104.

Merrill, R.K., Oberg, C.J. and Mc Mahon, D.J. (1994). A method for manufacturing reduced fat Mozzarella cheese. *J. Dairy Sci.*, 77: 1783–1789.

Micketts, R. and Olson, N.F. (1974). Manufacture of Mozzarella cheese by direct acidification with reduced amounts of rennet and pepsin. *J. Dairy Sci.*, 57: 273–279.

Mongiello, A.A. (1986). Improved cheese manufacturing method and apparatus. European Patent No. 0 174 637 A2. *Cited in Dairy Sci. Abstr.*, 49: 229.

Muzzarelli, G. (1985). Continuous kneading machine for the production of plastic curd cheeses. European Patent 0 146 765 A2. *Cited in Dairy Sci. Abstr.*, 47: 7593.

Nawaz, M.A., Masud, T. and Sammi, S. (2011). Quality evaluation of Mozzarella cheese made from buffalo milk by using paneer booti (*Withania coagulans*) and calf rennet. *Intl. J. Dairy Technol.*, 64: 218–226.

Ohashi, T. (1981). Technology of cheese manufacture in the US: With special reference to a utilization of lactic starter cultures. *Jap. J. Dairy Food Sci.*, 30: 237–239.

Partridge, J.A., Nilson, K.M., Atherton, H.V. and Duthie, A.H. (1982). Quality and yield of Mozzarella cheese manufacture from pasteurized milk in prolonged storage. *Cult. Dairy Prod. J.*, 17: 6–10.

Patel, H.G. and Upadhyay, K.G. (1999). An alternate technology for manufacture of buffalo milk pizza cheese. *J. Food Sci. Technol.*, 36: 235–238.

Patel, G.C., Vyas, S.H. and Upadhyay, K.G. (1986). Evaluation of Mozzarella cheese made from buffalo milk using direct acidification technique. *Indian J. Dairy Sci.*, 39: 394–403.

Perry, D.B., Mc Mahon, D.J. and Oberg, C.J. (1998). Manufacture of low fat Mozzarella cheese using exopolysaccharide producing starter cultures. *J. Dairy Sci.*, 81: 563–566.

Pizaia, P.D., Spadoti, L.M., Narimatsu, A., Dornellas, J.R.F. and Roig, S.M. (2003). Mozzarella by ultrafiltration: Composition, proteolysis, melting capacity and blisters formation. *Ciencia e Tecnologia de Alimentos*, 23: 485–491.

Quarne, E.L., Larson, W.A. and Olson, N.F. (1968a). Recovery of milk solids in direct acidification and traditional procedures of manufacturing Pizza cheese. *J. Dairy Sci.*, 51: 527–530.

Quarne, E.L., Larson, W.A. and Olson, N.F. (1968b). Effect of acidulants and milk-clotting enzymes on yield, sensory quality and proteolysis of Pizza cheese made by direct acidification. *J. Dairy Sci.*, 51: 848–852.

Rehman, S.U., Farkye, N.Y. and Yim, B. (2003). Use of dry milk protein concentrate in pizza cheese manufactured by culture or direct acidification. *J. Dairy Sci.*, 86: 3841–3848.

Renda, A., Barbano, D.M., Yun, J., Kindstedt, P.S. and Mulvaney, S.J. (1997). Influence of screw speeds of the mixer at low temperature on characteristics of Mozzarella cheese. *J. Dairy Sci.*, 80: 1901–1907.

Schiftan, T.Z. and Komatsu, I. (1979). Composition of Mussarela cheese consumed in the city of Sao Paulo. *Revista do Instituto de Laticinios Candido Tostes*, 34: 29–34.

Shah, R.D., Jana, A.H. and Solanky, M.J. (2008). Use of plasticizing treatment in producing pasteurized Mozzarella cheese from raw milk. *J. Food Sci. Technol.*, 45: 275–278.

Sharma, S.K. (1988). A study on manufacture of Mozzarella cheese from buffalo milk heated at various time-temperature combination. M.Sc. Thesis, Gujarat Agricultural University, Anand, India.

Singh, S. and Ladkani, B.C. (1984) Standardization of manufacturing technique of Mozzarella cheeses. Annual Report, National Dairy Research Institute, Karnal, p. 82.

Sundar, M.R. and Upadhyay, K.G. (1991). Influence of casein/fat ratio of milk on baking, rheological and sensory characteristics of buffalo milk Mozzarella cheese. *J. Food Sci. Technol.*, 28: 98–100.

Sundar, M.R. and Upadhyay, K.G. (1992). Influence of whey acidity at draining on the manufacture of buffalo milk Mozzarella cheese—baking, rheological and sensory characteristics. *Indian J. Dairy Sci.*, 45: 261–267.

Tunick, M.H. (1994). Effects of homogenization and proteolysis on free oil in Mozzarella cheese *J. Dairy Sci.*, 77: 2487–2493.

Tunick, M.H., Mackey, K.L., Smith, P.W .and Holsinger, V.H. (1991). Effects of composition and storage on the texture of Mozzarella cheese. *Neth. Milk Dairy J.*, 45: 117–125.

Valle, J.L.E.do, Silva, C.S.D.da., Yotsuvanagi, K. and Souza, G.de. (2004). Influence of the fat level on the functional properties of Mozzarella cheese. *Ciencia e Tecnologia de Alimentos*, 24: 669–673.

Vernon, H.R. (1972). Non dairy cheese-a unique reality. *Food Prod. Dev.*, 6: 22–26.

Villani, F., Pepe, O., Mauriello, G., Moschetti, G., Sannino, L. and Coppola, S. (1998). Behaviour of *Listeria monocytogenes* during the traditional manufacture of water buffalo Mozzarella cheese. *J. Food Prot.*, 61: 116–118.

Woo, A.H. and Lindsay, R.C. (1984). Concentration of major free fatty acids and flavor development in Italian cheese varieties. *J. Dairy Sci.*, 67: 960–968.

Yazici, F. and Akbulut, C. (2007). Impact of whey pH at drainage on the physicochemical, sensory and functional properties of Mozzarella cheese made from buffalo milk. *J. Agric. Food Chem.*, 55: 9993–10000.

Yun, J.J., Kiely, L.J., Barbano, D.M. and Kindstedt, P.S. (1993a). Mozzarella cheese: Impact of cooking temperature on chemical composition, proteolysis and functional properties. *J. Dairy Sci.*, 76: 3664–3673.

Yun, J.J., Barbano, D.M. and Kindstedt, P.S. (1993b). Mozzarella cheese: Impact of milling pH on chemical composition and proteolysis. *J. Dairy Sci.*, 76: 3629–2638.

Zhang, Jian-Qiang, Shi, Hui-Fang and Zhao, Li-Qin (2011). Effects of acidifying agent on the functional properties of pizza cheese. *Packaging Food Machinery*, 29: 1–4.

Zisu, B. and Shah, N.P. (2007). Texture characteristics and pizza bake properties of low-fat Mozzarella cheese as influenced by pre-acidification with citric acid and use of encapsulated and ropy exopolysaccharide producing cultures. *Intl. Dairy J.*, 17: 985–997.

8

Biotechnology for Animal Products

P.K. MANDAL AND A.P. NAJEEB

ABSTRACT

Biotechnology provides powerful tools for the sustainable development of food industry. Biotechnology in the food processing sector targets the selection and improvement of microorganisms to improve the quality, safety and consistency of products. Starter culture improvement and development of advanced tools has played a pivotal role in the production of fermented foods. Genetic techniques including mutation and selection, gene transfer methods such as transduction, conjugation and transformation and genetic engineering etc. Molecular techniques include temperature gradient gel electrophoresis (TGGE), single strand conformation polymorphism (SSCP), repetitive bacterial DNA elements (Rep)-PCR etc. Mutation and selection techniques are being used to improve the starter cultures used in fermented dairy and meat products. Natural gene transfer systems in bacteria facilitated the understanding of the genetics of microbial starter cultures. Genetic engineering provided the necessary tools for transfer, maintenance and optimal expression of cloned DNA in microbial starter cultures. The extensive development of molecular methods resulted in new tools for molecular characterization of the culture strains. Nutritional and genetic interventions have altered the milk composition. Introduction of cultured meat is the contribution of genetic engineering.

1. INTRODUCTION

Food technology and biotechnology are applied sciences and are closely related to adequate and economically sound processing of food, prolongation of shelf life by preservation, optimization of storage and handling, improvement of safety and nutritive value, adequate and appropriate packaging (Grba *et al.*, 1994). Therefore, application of biotechnology in food-processing targets selection and manipulation of micro-organisms with the objectives to improve quality, safety, consistency and yield of food products (FAO, 2004). Biotechnology in food processing makes use of micro-organisms for the preservation and production of foods including a range of products such as enzymes, flavour compounds, vitamins, microbial cultures and food ingredients.

Genetic techniques include mutation and selection, use of natural gene transfer methods such as transduction, conjugation and transformation and genetic engineering etc. Mutation and selection techniques are being used to improve the metabolic properties of microbial starter cultures used in food fermentations. Natural gene transfer systems in bacteria facilitated the understanding of the genetics of microbial starter cultures and have been used for strain improvement.

The extensive development of molecular methods resulted in new tools that could be used for molecular characterization of the strains isolated by culture-dependent methods. Molecular techniques like temperature gradient gel electrophoresis (TGGE), or single strand conformation polymorphism (SSCP), repetitive bacterial DNA elements (Rep)-PCR, randomly amplified polymorphic DNA (RAPD)-PCR, and enterobacterial repetitive intergenic consensus (ERIC)-PCR are common practices. Genome sequence information coupled with the support of highly advanced molecular techniques have augmented the scientists to study the mechanisms of action of various host-defensive pathogen and aid them to produce more healthy and safer food by optimizing or eliminating the effect of harmful bacteria .

2. BIOTECHNOLOGY IN FERMENTED FOODS

Fermentation is a bio-processing which makes use of microbial inoculants for enhancing properties such as the taste, aroma, shelf-life, safety, texture and nutritional value of foods. Microbes associated with the raw food material and the processing environment serve as inoculants in spontaneous fermentations while inoculants containing high concentrations of live micro-organisms referred to as starter cultures are used to initiate and accelerate the rate of fermentation processes in non-spontaneous or controlled fermentation processes. Microbial starter cultures vary widely in quality and purity. Starter culture development and improvement is the subject of research both in developed and developing countries. Considerable work on genetically modified starter culture development is ongoing at the laboratory level in developed countries. In 1990, the United Kingdom became the first country to permit the use of a live organism (GMO) in food.

The development and improvement of microbial starters has been a driving force for the transformation of traditional food fermentations in developing countries from an "art" to a science. Microbial starter culture development has also been a driving force for innovation in the design of equipment suited to the hygienic processing of traditional fermented foods under controlled conditions in many developing countries. Starter cultures improvement together with the improvement and development of advanced technological tools for the control of fermentation processes in developed countries has played a pivotal role in the production of high value products such as enzymes, microbial cultures and functional food ingredients. These products are increasingly produced in more advanced developing countries and are imported by less advanced developing countries as inputs for their food processing operations.

One of the developments in biotechnology of food fermentation is the use of defined starter culture. "Defined starter cultures" consist of single or mixed strains of micro-organisms. They may be incorporated in adjunct culture preparations that serve a food-safety and preservative function. Adjunct cultures do not necessarily produce fermentation acids or modify texture or flavour but are included in the defined culture owing to their ability to inhibit pathogenic or spoilage organisms. Their inhibitory activity is due to the production of one or several substances such as hydrogen peroxide, organic acids, diacetyl and bacteriocins. Defined cultures are produced by pure culture maintenance and propagation under aseptic conditions. They are generally marketed in a liquid or powdered form or else as a pressed cake. Loog-pang is a defined culture marketed in Thailand in the form of a pressed rice cake consists of *Saccharomyces cerevisiae*, *Aspergillus oryzae* or *Rhizopus* sp. and Mucor. Defined starter cultures are widely imported by developing countries for use in the commercial production of dairy products such as yogurt, kefir and cheese. The use of DNA-based diagnostic techniques for strain differentiation is being used for the tailoring of starter cultures to yield products with specific flavours and/or textures. Random amplified polymorphic DNA (RAPD) techniques have been applied in the molecular typing of bacterial strains especially for flavour development during the production of the fermented pork sausage like *nham* (Wiriyacharri, 1990).

Another milestone in food biotechnology of fermentation is the genetically modified starter culture. Rennet which is widely used as a starter in cheese production across the globe is produced using GM bacteria. Thailand currently makes use of GM *Escherichia coli* as inoculants in lysine production. Many industrially important enzymes such as α-amylase, gluco-amylase, lipase, pectinase and bio-based fine chemicals such as lactic acid, amino acids, antibiotics, nucleic acid and polysaccharides are produced in China using GM starter cultures. Other developing countries which currently produce enzymes using recombinant micro-organisms include Cuba, Brazil, India and Argentina.

Biotechnology can be of help for the improvement of fermented foods at three levels:

- *Raw materials*: Exploitation of biotechnological tools for the improvement of animal production.
- *Fermentation engineering*: Recent developments in biotechnology have given rise to great innovations in bioreactor designs. Starter cultures for fermented foods are today developed mainly by design rather than by screening. The design principles are based on knowledge of bacterial metabolism and physiology as well as on the interaction with the food product. The design tools available are food grade tools for genetic, metabolic and protein engineering and an increased use of laboratory automation and high throughput screening methods. Most of these designs deal with liquid reaction media. Even though a great number of fermented foods are produced through solid-substrate fermentation in which the

fermenting paste is frequently hand mixed. Bioreactors to simulate such a process are needed for the modernization of such traditional fermented foods (Hansen, 2002).

- *Microbiology and enzymology*: There are many opportunities for biotechnological innovations in the microbiology of fermented foods. First, all the microorganisms involved should be isolated, characterized, and preserved. Second, the metabolic role of each of the strains involved should be clearly identified and their full potential should be studied. Use of monoclonal antibodies for the characterization of different strains of the same species can be of great help in this area. Many of these organisms have the enzyme complement to produce vitamins and amino acids in fermented foods. This potential can be improved via recombinant DNA technology to produce strains which are capable of producing and releasing the required amino acid or vitamin into the food. To avoid food losses due to spoilage organisms and to avoid possible development of food-poisoning microbes, it is possible to use genetically engineered strain as a pure culture. Such a strain may bring about all the desired changes required in the food and grow at a convenient temperature.

3. BIOTECHNOLOGY FOR STARTER CULTURE

3.1. Mutation and Selection

Mutation and selection techniques have been used to improve the metabolic properties of microbial starter cultures used in food fermentations. In nature, mutations (changes in the chromosome of an organism) occur spontaneously at very low rates. The mutation rate can be dramatically increased by the exposure of microorganisms to mutagenic agents such as ultraviolet light or various chemicals which induce changes in the deoxyribonucleic acid (DNA) of host cells. These techniques are having several limitations. Mutagenic agents cause random mutations, thus specificity and precision are not possible. Potentially deleterious undetected mutations can occur, since selection systems may be geared for only the mutation of interest. Additionally, traditional mutation procedures are extremely costly and time-consuming and there is no opportunity to expand the gene pool. In spite of these limitations, mutation and selection techniques have been used extensively to improve industrially important microorganisms (Harlander, 1992).

3.2. Natural Gene Transfer Methods

The discovery of natural gene transfer systems in bacteria facilitated the understanding of the genetics of microbial starter cultures and has been used for strain improvement. Genetic exchange in bacteria can occur naturally by three different mechanisms: transduction, conjugation and transformation. Numerous genera of bacteria harbor plasmid DNA. Lactic acid bacteria naturally contain one to more than ten distinct plasmids. Metabolically important traits including lactose-fermenting ability, bacteriophage resistance and bacteriocin production have been linked to plasmid DNA. Conjugation is being used to transfer these plasmids into recipient strains for the construction of genetically improved commercial dairy starter cultures (Harlander, 1992).

3.3. Electroporation

It involves the application of high-voltage electric pulses of short duration to induce the formation of transient pores in cell walls and membranes. Under appropriate conditions, DNA present in the surrounding medium may enter through the pores. Applications of this method have been limited by the relatively low expression levels of the transferred gene. Electroporation is the method of choice used for strains that are recalcitrant to other gene transfer techniques. The applicability of *in vivo* electroporation or gene transfer into muscle has been investigated using plasmid DNA expressing interleukin-5 as the vector (Aihara & Miyazaki, 1998).

3.4. Genetic Engineering

Genetic engineering provides an alternative method for improving microbial starter cultures. It revolutionizes the science of strain improvement and is destined to have a major impact on the food fermentation industry. Plasmid-free, genetically characterized and highly transformable hosts coupled with multifunctional expression vectors provide the necessary tools for transfer, maintenance and optimal expression of cloned DNA in microbial starter cultures. Once gene(s) have been identified and cloned into the appropriate vector, they must be introduced into a viable host. Several properties could be enhanced using genetic engineering. Transfer of bacteriocin production to microbial starter cultures could improve the safety of fermented products. Increasing the number of copies of the genes that code for the enzymes involved in acid production might increase the rate of acid production ensuring that the starter will dominate the fermentation and rapidly destroy less-aciduric competitors. Certain enzymes are critical for proper development of flavor and texture of fermented foods. Lactococcal proteases slowly released within the curd are responsible for the tart flavor and crumbly texture of aged cheddar cheese. Cloning of additional copies of specific proteases involved in ripening could greatly accelerate the process. Main limitation of this technique is that genetic improvement of microbial starter cultures requires sophisticated equipment and expensive biological materials that may not be available in developing countries (FAO, 2004).

Genetic engineering is often used in the production of sausage and ham. Many additives commonly found in sausage and hams are often produced using methods involving genetically modified microorganisms. With the development of genetic engineering, it became possible to insert gene from the calf stomach cell (which makes the cell produce the enzyme, rennin) into certain bacteria, fungi or yeasts to make them produce chymosin. Chymosin produced by genetically modified organisms was the first artificially produced enzyme to be registered and allowed by the US Food and Drug Administration. Genetically modified microbial rennet is used more often in industrial cheese making because it is less expensive than animal rennet. 'Chymax' is commercially available genetically engineered rennet. The most widely used genetically engineered rennet is produced by the fungus *Aspergillus niger*. Often a mixture of genetically engineered chymosin and natural pepsin is used to imitate the complexity of natural rennet and to get the same results in coagulation and in development of flavour and taste. Microbial rennet produced by genetically engineered bacteria has proven suitable substitutes for animal rennet, but increasing attention has been directed toward natural rennet extracted from plants (Ahmed *et al.*, 2009).

Introduction of cultured meat is another contribution of genetic engineering in the field of food biotechnology. Cultured meat is meat produced *in vitro* in a cell culture rather than from an animal. The production of cultured meat begins by taking a number of cells from a farm animal and proliferating them in a nutrient-rich medium. Cells are capable of multiplying so many times in culture. A single cell could be used to produce enough meat to feed the global population for a year. After the cells are multiplied, they are attached to a sponge-like scaffold and soaked with nutrients. They may also be mechanically stretched to increase their size and protein content. The resulting cells can then be harvested, seasoned, cooked and consumed as a boneless processed meat such as sausage, hamburger or chicken nuggets. The cells that can be used to produce cultured meat are muscle and stem cells from farm animals. It is possible that genetically modifying a muscle cell would allow it to proliferate number of times in culture and thus make cultured meat production more economical.

4. MOLECULAR TECHNIQUES

With the advancement of molecular biology and the invention of the polymerase chain reaction (PCR), a new range of techniques were developed that help to understand the microbial complexity in natural ecosystems. As a consequence, the traditional microbiological techniques based on plating, isolation and biochemical identification are now supported by new methods that rely on the analysis of the nucleic acids for detection, identification and characterization of the microorganisms. In this context, several groups of food scientists started to apply molecular methods to various food fermentations. Viable but not culturable (VNC) cells that are not able to form colonies on agar plates but possess metabolic activity can be better identified as isolated by various molecular techniques.

Generally they are based on the extraction of the nucleic acids directly from the samples. The nucleic acids are then amplified by PCR and subjected to either cloning or sequencing or profiling techniques, such as denaturing gradient gel electrophoresis (DGGE), temperature gradient gel electrophoresis

(TGGE) or single strand conformation polymorphism (SSCP). An alternative culture-independent approach that does not rely on extraction of nucleic acids from the sample matrix is fluorescence *in situ* hybridization (FISH). The extensive development of molecular methods resulted in new tools that could be used for molecular characterization of the strains isolated by culture-dependent methods. Randomly amplified polymorphic DNA (RAPD)-PCR, repetitive bacterial DNA elements (Rep)-PCR, and enterobacterial repetitive intergenic consensus (ERIC)-PCR are common practices in almost all the laboratories that deal with microbial ecology and diversity.

4.1. Culture-dependent Techniques

4.1.1. *Strain identification*

The identification of strains isolated from sausage fermentation is most commonly achieved by targeting the ribosomal RNA operon (rRNA). In case of the bacteria, 16S rRNA gene is the most used target for molecular applications (Collins *et al.*, 1991), while D1-D2 loops of the 26S RNA gene is a generally accepted target for yeast identification (Kurtzman & Robnett, 1997). The process of identification of microbial strains by targeting rRNA usually consists a step in which the specific gene (16S or 26S rRNA), or portion of it is amplified by PCR that subsequently subjected to DNA sequencing. It is possible to achieve the identification by alignment of the sequence in the gene bank using available software programmes online.

This experimental tactic is not applicable for ecological studies because of the high number of isolates. Moreover, it is not feasible and economically convenient to sequence hundreds of strains. For this reason, researchers developed species-specific PCR and ribosomal RNA probes as fastest ways to obtain species identification. Ribosomal RNA probes (Nissen & Dainty, 1995), species specific PCR primers (Berthier & Ehrlich, 1998; Blaiotta *et al.*, 2003; Marot-Bizot *et al.*, 2003; Rossi *et al.*, 2001) and multiplex PCR (Corbiere *et al.*, 2004) for identification of microorganisms in traditional meat products have been used in the field of fermented sausages for the identification of lactic acid bacilli and coagulase negative cocci isolated during fermentation.

The techniques that allow strain grouping based on differences in a DNA sequence have been applied in the field of fermented sausages are the restriction fragment length polymorphism (RFLP) analysis of the 16S rRNA gene (Lee *et al.*, 2004; Sanz, *et al.*, 1998) and denaturing temperature gradient gel electrophoresis (DTGGE) (Blaiotta, *et al.*, 2003; Cocolin, *et al.*, 2001; Ercolini, *et al.*, 2001).

With the RFLP approach, the differences in the DNA sequence are identified by using a restriction endonuclease that will cut the DNA in specific restriction sites giving a specific pattern for a given species. The DTGGE methods are used to detect the differences in the sequences by analyzing their denaturation behavior. The change in the DNA conformation determines a reduction in the migration rate of the molecules and thereby DNA with different nucleotide sequences will show differential electrophoretic patterns.

4.1.2. *Strain characterization*

Strains isolated during the fermentation can also be subjected to molecular characterization in order to understand the dynamics and diversity within the same species by either PCR amplification or non-PCR methods. Randomly amplified polymorphic DNA PCR has been applied to examine the diversity of *Lactobacillus sakei* in naturally fermented Italian sausages and has the ability to take over fermentation compared with other strain (Urso *et al.*, 2006). Regarding the molecular methods that do not exploit an amplification step used in the field of fermented meats, pulsed-field gel electrophoresis (PFGE) technique is considered the "gold standard" for the characterization of strains since it is very precise, reproducible and reliable.

4.1.3. *Culture-independent techniques*

The culture-independent methods are able to profile the microbial populations in complex microbial ecosystems without any cultivation. The strategy based on which these methods work is the analysis of the nucleic acids extracted directly from the matrix. Once the DNA and RNA are available, they can be subjected to several types of analysis that can be preceded by a PCR step or not.

The culture-independent method that has been more extensively applied to sausage fermentation is denaturing gradient gel electrophoresis. This method is able to differentiate DNA molecules based on their denaturation behavior. When this method is used for microbial population profiling, PCR is carried out with universal primers able to prime amplification for all the microbes present in the sample. After this step, a complex mixture of DNA molecules will be obtained that can be differentiated if separated in gels with denaturant gradients. Every single band visible in denaturing gradient gel electrophoresis represents a component of the micro-biota. The more bands are visible, the more complex is the ecosystem. By using these methods, it is possible not only to profile the microbial populations but also to follow their dynamics.

The use of species-specific primers in the PCR amplification analyzing the DNA extracted directly from the sample is considered as another culture independent method. This approach is being used to identify lactic acid bacilli and coagulase negative cocci species in fermented sausages (Aymerich *et al.*, 2003).

A promising culture-independent method that has not yet been efficiently exploited to study the microbial diversity in fermented sausages is the fluorescence *in situ* hybridization technique. In this technique a set of specific probes are used to target different microorganisms directly in the sample. The probes are labeled with different fluorophores thereby allowing the detection of several species simultaneously. Since the probes are generally designed on the ribosomal RNA, only alive cells are detected by fluorescence *in situ* hybridization technique (Bottari *et al.*, 2006). One of the most fascinating features of fluorescence *in situ* hybridization technique is the possibility to localize the microorganisms directly into the food matrix. This has been applied usually in dairy products.

5. BIOTECHNOLOGY FOR FERMENTED MEAT PRODUCTS

Research on the use of starter cultures in meat products began in USA in 1940s by inoculating the raw material in fermented sausages using lactobacilli with the aim to govern and accelerate the fermentation. Natural diversity of different strains of micro organism occur in foods is being explored (Fig. 1) via comparative genomics, microarray analysis, transcriptomics, proteomics and metabolomics tools in biotechnology. Genetic studies of the main bacterial species *L. sakei*, *S. xylosus*, and *S. carnosus* have provided an understanding of the metabolic activities and stress resistance functions. The era of molecular genetics of lactic acid bacteria was opened in 1982 by the development of a protocol for transforming DNA into *Streptococcus lactis* (Kondo & McKay, 1982).

It has been found that the presence of methylglyoxal synthase in lactic acid bacilli utilizes a pathway to counteract frequent glucose starvation and modulate metabolism of alternative carbon sources (Chaillou *et al.*, 2005). The catalase gene, *kat*A of *S. xylosus* has been studied in detail (Barriere *et al.*, 2001). Catalases help to dissociate hydrogen peroxide to oxygen and water. Hydrogen peroxide produced as a result of metabolism by lactic acid bacilli usually interferes with the sensory properties of meat and causes discoloration of nitroso-heme pigments in meat products. *Staphylocuccus* and *Kocuria* have a role for their nitrate reductase and antioxidant activities (Talon *et al.*, 1999, 2002). These microorganisms reduce nitrate to nitrite which is important for the formation of nitrosylmyoglobin, the compound responsible for the characteristic red color of fermented meats.

The use of starter cultures with negative decarboxylase activity was shown to prevent the growth of biogenic amines producers and leads to end-product nearly free of biogenic amines. This negative-decarboxylate strain can decrease the pH quickly during the fermentation step and predominant throughout the process thus preventing the growth of bacteria that can produce biogenic amines (Bover *et al.*, 2001, Gonzalez *et al.*, 2003).

The evaluation of proteolysis by lactobacilli using muscle sarcoplasmic and myofibrillar proteins as substrates have been carried out by incubation of these proteins with whole-cells and cell free extracts of different lactobacilli strains (*L. casei* CRL705, *L. curvatus* CECT904, *L. plantarum* CRL681, and *L. sakei* CECT4808) isolated from sausages. The addition of only cell-free extracts of *L. casei*, *L. curvatus*, and *L. sakei* results in a net increase in free amino acids but a decrease was reported when cell-free extracts of *L. plantarum* were incorporated (Champomier *et al.*, 1999). This indicated that this strain has higher intracellular metabolic activity for amino acid degradation. The combination of both cell-free extracts and whole-cell suspensions especially of *L. sakei* and *L. plantarum* results an increase in free amino acid concentration. When only *S. xylosus* was used as starter without lactobacilli, the observed proteolysis was identical to that in non-inoculated control sausages reinforcing the importance of proteolysis contribution by lactobacilli (Casaburi *et al.*, 2008).

Berthier and Ehrlich (2000) developed the pairs of PCR primers to detect *Lb. curvatus* and *Lb. sakei*. They also evaluated the phenotypic and genotypic diversity among 165 isolates and compared the results of biochemical assays and randomly amplified polymorphic DNA data and found that *S. xylosus*, *S. equorum*, and *S. carnosus* are frequently involved in fermentation process. Simonova *et al.* (2006) employed primers targeting *sodA* genes to detect *S. xylosus* and *S. carnosus* strains in different types of Slovak traditional sausages for the selection of potential starter cultures used in sausage processing.

Denaturing gradient gel electrophoresis PCR was also used to investigate the yeast populations in fermented sausages. Yeast isolates were subjected to PCR denaturing gradient gel electrophoresis and the strains giving identical migration patterns were grouped together and amplified with the primers. It has been found that *D. hansenii*, is the dominant species at the initial stage of fermentation (Kurtzman & Robnett, 1998). This species was usually accompanied by *Candida zeylanoides* and only at the end of the fermentation by *Metschnikowia pulcherrima*. At the start of the fermentation a high biodiversity was observed. About five species could be identified apart from *D. hansenii*, includes *Pichia triangularis*, *Candida parapsilosis*, *Saccharomyces cerevisiae*, and *Sterigmatomyces elviae* (Andrade *et al.*, 2006; Martorell *et al.*, 2005; Ramos *et al.*, 1998).

The diagnostic role of biotechnology in starter culture development improves the value of *Nham*. *Nham* is traditionally made from fresh lean pork that is trimmed, minced, mixed thoroughly with salt, potassium nitrate, cooked rice and seasonings and packed in either banana leaves or cylindrical plastic bags. The addition of *L. plantarum* to the *nham* accelerated very distinctly the decrease in the pH of *nham* (Elmer *et al.*, 1992). *L. plantarum* inoculation had a very distinct effect in terms of firmness and colour development when it was used together with *P. cerevisiae* and *M. varians* (Wiryacharee, 1990).

6. BIOTECHNOLOGY IN FOOD SAFETY

6.1. Biocontrol of Pathogens in Meat

Salmonella, *Campylobacter jejuni* and Enterohaemorraghic *Escherichia coli* (EHEC; predominately serotype O157:H7) and *Listeria monocytogenes* are the most predominant food borne bacterial pathogens reported in the developed countries (United States Department of Agriculture, 2001). Pathogens in food animals such as livestock and poultry can lead to both direct and indirect contamination of raw and processed meats. Inadequate processing techniques allow increase in the number of pathogens especially in foods of poor microbiological quality. Numerous methods have been reported for the reduction of food pathogens at the pre-harvest stage (in the animal) and post-harvest and processing stages of the meat chain. These approaches include the use of organic compounds, vaccines and bacteriophages as well as antagonistic bacteria (Callaway *et al.*, 2004; Joerger, 2003; Koomaraie *et al.*, 2005; Le & Wetzel, 2007).

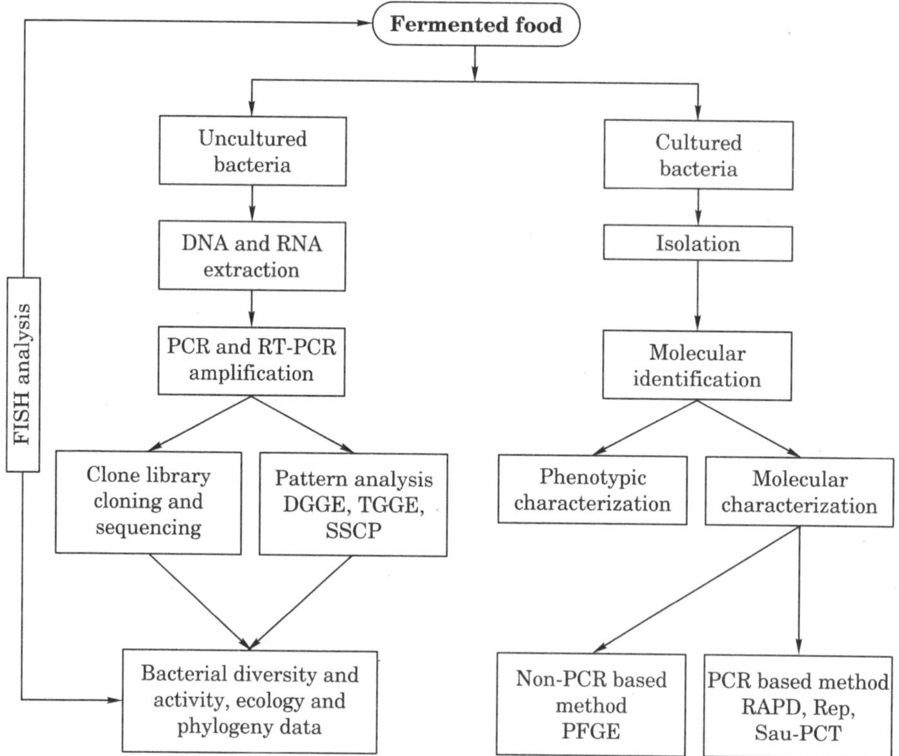

Fig. 1: Methods to study the microbial ecology of fermented foods (Fidel *et al.*, 2007).

6.2. Competitive Exclusion Technology

Competitive exclusion (CE) technology involves the addition of a non-pathogenic bacterial culture to the intestinal tract of food animals in order to reduce colonization or decrease populations of pathogenic bacteria in the gastrointestinal tract (Callaway *et al.*, 2004). A competitive exclusion culture usually composed of one or more strains or species of bacteria. The use of competitive exclusion technology is similar to using probiotics which are defined as a preparation of a product containing viable defined microorganisms in sufficient numbers which alter the microflora of the host and exert beneficial health effects (De Vrese & Schrezenmeir, 2002). Some of the bacteria may produce antimicrobial compounds such as acids or bacteriocins to eliminate species competing within the same niche. Some of these antimicrobial compounds have been specifically investigated to use as biopreservatives for food safety applications in meat and meat products. One of the limitations of competitive exclusion technology is that, the viable bacteria may transfer virulence genes to the animal or human microbiota resulting in enhanced antimicrobial resistance (Wagner, 2006).

6.3. Application of Bacteriocins in Meat Products

Bacteriocins are potentially valuable biological tools to improve food preservation and food safety by reducing the prevalence of undesirable spoilage microorganisms or food borne pathogens in a food product (Deegan *et al.*, 2006). The application of bacteriocins is not recommended as a primary processing step or barrier to prevent the growth or survival of pathogens but should form part of a system with multiple hurdles. Bacteriocins can be incorporated into food as an ingredient in the form of a purified/semi-purified bacteriocin preparation. Alternatively, bacteriocin producing bacteria can be introduced as 'protective cultures' in the form of a live culture which produces the bacteriocins *in situ* in the food (Table 1). A major criterion for a protective culture in meat products is to inhibit pathogens and/or prolong shelf life while not changing the sensory properties of the product. The use of purified bacteriocins is not always attractive as some can be inactivated in meats and may also require regulatory approval or be labeled as an additive on the meat product (Aasen *et al.*, 2003). Bacteriocins generally act by creating pores in their membrane of their target cells and may be cytotoxic to the target cells as a result of disturbance of the bacterial inner or outer membranes. Different mechanisms of action by bacteriocins have been described but interaction with the bacterial membrane is an important requirement for most of bacteriocins (Cleveland *et al.*, 2001).

Table 1: Use of bacteriocin-producing cultures to control pathogens in meat products (Fidel *et al.*, 2007).

Culture	*Meat product*	*Target pathogen(s)*
Lb. sakei CECT 4808 and/or *Lb. curvatus* CECT904	Pork sausage	*L. monocytogenes*
Enterococcus faecalis A-48–32	Dry-fermented sausage	*L. monocytogenes*
Lactococcus lactis subsp. LMG21206 and *Lb. curvatus* LBPE	Raw sausage (mergeuz)	*L. monocytogenes lactis*
Leuconostoc carnosum 4010	Pork meat sausage slices	*L. monocytogenes*
Lb. sakei TH1	Beef meat	*L. innocua B. thermosphata*
Lb. casei CRL705	Poultry breast	*L. monocytogenes*
Lb. curvatus CRL705	Pork adipose tissue	*B. thermosphacta*

Most commonly used bacteriocin in foods including meat products is nisin (produced by *Lactococcus lactis* subsp. *lactis*). Nisin is one of the commercially available bacteriocins with US Food and Drug Administration (FDA) approval. It inhibits the growth of wide range of Gram-positive bacteria including *L. monocytogenes*. Nisin has been shown to be effective in number of food systems but is predominantly used in canned foods and dairy products particularly in cheese production where it protects against heat-resistant, spore forming organisms such as *Bacillus* and *Clostridium* spp (Cotter *et al.*, 2005; Deegan *et al.*, 2006).

One of the main limitations is the narrow spectrum of activity of bacteriocins. They are active against Gram-positive organisms but are not as effective against Gram-negative food borne pathogens such as *E. coli* (particularly *E. coli* O157:H7) or *Salmonella* spp which are a great concern in meat products. Another potential problem associated with using bacteriocins in foods is the development of resistant populations of problematic bacteria. Some bacteriocin-producing bacterial strains may spontaneously lose their ability to produce bacteriocin(s) due to genetic instability (Holzapfel *et al.*, 1995; Riley & Wertz, 2002) and can also become ineffective if the cell membrane of a target organism changes in response to a particular environmental condition.

6.4. Bioprotective Cultures

Antimicrobial metabolites have been termed as bioprotective or biopreservative culture. Antagonistic cultures are added to foods to inhibit pathogens and/or extend the shelf life while changing the sensory properties as little as possible is called protective cultures (Lucke, 2000). The main objective of this concept is the storage life extension as well as the enhancement of food safety. Lactic acid bacilli have a major potential for use in biopreservation because they are safer for consumption. During storage, they naturally will dominate over the micro-biota of many foods. In addition, antimicrobial peptides produced by lactic acid bacilli can be easily broken down by digestive proteases so it will not disturb the gut micro-biota. Lactic acid bacilli can exert a bioprotective or inhibitory effect against other microorganisms as a result of competition for nutrients by the production of bacteriocins or other antagonistic compounds such as organic acids, hydrogen peroxide and enzymes. A distinction can be made between starter cultures and protective cultures in which metabolic activity (acid production, protein hydrolysis) and antimicrobial action constitute the major objectives (Fidel, 2008).

6.5. Bacteriophage

Bacteriophages are viruses that infect and kill bacterial cells by reproducing within the bacteria and disrupting the host metabolic pathways causing the bacterium to lyse without infecting the mammalian cells. When bacteriophage was applied to chicken skin inoculated with *C. jejuni* and stored at 4 and 20°C, a reduction of approximately 1 log 10 CFU/ml was obtained (Atterbury *et al.*, 2003). They also observed a reduction in Campylobacteraceae on the frozen samples by 2 log 10 CFU/ml after day 1, and remained at similar levels thereafter. It also has been reported that bacteriophage preparation consists of six *Listeria* specific bacteriophages, combined to reduce the possibility of *L. monocytogenes* in meat product. Listex TMP100 is a commercially available preparation of a bacteriophage used for the control of *L. monocytogenes* in meat and cheese products (Carlton *et al.*, 2005).

6.6. Limitations of Bacteriophage

Adsorption of a bacteriophage to receptors of bacterial cell occurs due to brownian motion. Host bacterial cells may be protected by biofilms present in a food environment as bacteriophage would be incapable of penetrating and accessing the bacterial cell within the biofilm. Many meat products are distributed and stored at refrigeration temperatures under which many pathogens may not grow. This poses a problem with bacteriophage use, as replication cannot occur under conditions where the host is not dividing. Bacteriophages may be destroyed in food processing environments as they are effectively cleaned and sanitized. There are reports of dairy bacteriophages being destroyed by sodium hypochlorite (100 ppm free chlorine) and per acetic acid (Quiberoni *et al.*, 1999). Some pathogens can survive at low pH while bacteriophages are unstable at such a low acidic environment. Bacteriophages also have the capability to transfer unfavorable genes from one bacterium to another; these could be virulence genes or antibiotic resistance genes (Alisky *et al.*, 1998; Figueroa & Bossi, 1999; Schmeiger & Schicklmaier, 1999; Mirold *et al.*, 2001).

6.7. Use of Biosensors for Safer Meat Products

Meat products can be contaminated from a number of sources including the environment, the animal itself, fecal-contaminated water or from human contact during the conversion of meat into food products. *Escherichia coli* O157:H7 and *Salmonella species* are among the deadliest of the food borne pathogens in the United States. *E. coli* O157:H7 may be introduced into food through meat grinders, knives, cutting blocks and storage containers. Quick detection of target biomolecules can be achieved by technologies like polymerase chain reaction (PCR), enzyme linked immunosorbent assay (ELISA), flow cytometry, electronic noses and biosensors. A biosensor can detect pathogens within minutes, allow the food processors to take quick corrective action when pathogens are detected, resulting in minimized product recalls and lawsuits for the industry and savings from medical costs for consumers.

Biosensors are analytical instruments possessing a bio-molecule as a receptor or reactive surface in close proximity to a transducer which converts the binding of the target analyte with capturing bio-molecule or receptor into a measurable signal (Muhammad & Alocilji, 2003; Tumer & Newman, 1998; Mandal *et al.*, 2011). The capturing bio-molecule can be in the form of an antibody, enzyme, nucleic acid or cell. The transducer can be based on electrochemical (potentiometric, voltammetric, and conductometric), optical, thermal or surface acoustic waves. Electrochemical biosensors often referred to as amperometric, conductometric or impedimetric have the advantage of being highly sensitive, rapid, inexpensive and are highly amenable towards microfabrication. They measure the change in electrical properties of electrode structures as cells become entrapped or immobilized on or near the electrode (Ghindilis *et al.*, 1998; Gau *et al.*, 2001; Muhammad & Alocilja, 2004; Radke & Alocilja, 2004; Pal *et al.*, 2007).

Biosensors using antibodies as receptors are often referred to as immunosensors. In immunosensors development, antibody immobilization is the vital step. The immobilization method must preserve the biological activity of antibody and enable efficient binding. These methods can be grouped into three main categories: adsorption; immobilization via entrapment or immobilization via cross-linking agents (Vikholm, 2005; Ye & Letcher, 1997). The antibodies used are selective to the epitope on the antigen. If the epitope is present on a living or dead microorganism, the antibody will capture the antigen and register a positive signal. In case of bacterial food borne pathogens that must be ingested by the host to cause disease, a positive result from non-viable cells may raise false alarm (Mandal *et al.*, 2011).

7. DESIGNER MILK IN DAIRY BIOTECHNOLOGY

Nutritional and genetic interventions to alter the milk composition for specific health and/or processing opportunities are gaining importance in dairy biotechnology. The composition of milk can be altered by several ways. For diet and human health measures, the actions that would be beneficial are: (*i*) generate a greater proportion of unsaturated fatty acids (USFA) in milk fat, (*ii*) reduce lactose content in milk in order to cater to persons suffering from lactose intolerance and (*iii*) remove *b*-lactoglobulin from milk. From a technological point of view, there exist vast opportunities in: (*a*) alteration of primary structure of casein to improve technological properties of milk, (*b*) production of high-protein milk, (*c*) engineering milk meant for cheese manufacturing that leads to accelerated curd clotting time, (*d*) increased yield and/or more protein recovery, (*e*) milk containing nutraceuticals and (*f*) replacement for infant formula (Latha, 2004).

Transgenic cows secreting elevated levels of *b*-casein (8–20%) and *k*-caseins (two fold) have been produced by genetic engineering which is the most abundant milk protein involved in calcium phosphate binding and thus controlling milk calcium levels (Brophy, 2003). Higher *k*-casein content in milk is linked to smaller micelles, better heat stability and improved cheese-making properties. Edible casein is used in vitamin tablets, instant drinks and infant formulas. Caseins particularly *b*-caseins, αs_1 and αs_2-caseins being easily digestible are quite sensitive to plasmin, a serine protease occurring naturally in milk along with plasminogen. Plasmin activity leads to limited proteolysis in milk. This offers a dual disadvantage of decreasing the curd yield in cheese and inducing organoleptic defects and gelation of ultrahigh temperature treated milk. Milk augmented with specific inhibitors of either plasmin or plaminogen activator would therefore be a boon to the process industry.

The 'ideal' milk fat for human health would contain <10% poly unsaturated fatty acids (PUFA), <8% saturated fatty acids (SFA), and >82% mono-unsaturated fatty acids (MUFA). It has been found that desaturase gene contributes to the production of milk with decreased levels of SFA (California Dairy Research Foundation, 2004).

Alpha-lactalbumin (*a*-LA) is one of the major milk proteins present in almost all mammalian milk. It interacts with *b*-lactalbumin and 4-UDP galactosyltransferase (UDP-gal) to modify substrate specificity of this enzyme, virtually creating a unique binding site for glucose and leads to the synthesis of lactose (Vilotte, 2002). The methodology of reducing lactose involves the removal of *a*-LA or introducing the lactase enzyme into milk. They reduce the overall sugar content of milk resulting in highly viscous milk.

Studies on mice have revealed that reduction of lactose via *a*-LA deletion was inappropriate because it impaired milk volume regulation. The milk of such mice was highly viscous with high protein (88%) and fat (60%), no *a*-LA and no lactose. Knocking out the UDP-gal gene in mice also produced milk with no lactose, but of high viscosity. An alternative to produce low-lactose milk is over expression of *b*-gal in milk. Jost *et al.* (1999) explained an *in vivo* technique for low-lactose milk production. They generated transgenic mice that selectively produced a biologically active *b*-gal in their milk. In these transgenic mice, the lactose content of the milk was at least halved, even though the *b*-gal expression levels were relatively low.

7.1. Humanization of Bovine Milk

Lactoferrin (LF) is an iron-binding protein, has antimicrobial properties and may also mediate some effects of inflammation and have a role in regulating various components of the immune system. Its level in human milk is about 1g/L (in human colostrum about 7g/L). The level of lactoferrin in cow's milk is only about one-tenth that in human milk. This has caught the attention of those involved in designing human milk replacement formulas. Pharming NV (Leiden, Netherlands) developed the first transgenic bull in the late 1980s and a line of transgenic cows to produce several proteins including human LF (hLF) (Subramanian, 2004).

Human milk contains 0.4 g/L of lysozyme (LZ), an enzyme that provides antibacterial activity. Active human lysozyme (hLZ) has been produced in the milk of transgenic mice at concentrations of 0.78 g/L (Maga & Anderson, 1995). The expression of lysozyme in milk results in a reduction of rennet clotting time and greater gel strength in the clot. Yet another application of transgenic technology is the production of human lipase. The lipase thus produced could be used as a constituent of formulas to increase the digestibility of lipids especially in premature infants who have low *b*-gal activity (Lonnerdal, 1996).

8. BIOTECHNOLOGY IN FLAVOR PRODUCTION

Studies of flavour genesis led to the identification of the contribution of the enzyme activities present in the meat matrix (Hammes & Hertel, 1998). Microbial enzyme or their metabolic product is mostly flavor generating compounds and this characteristic is used in food industry for the production of desired aroma. Microbial lipases are used for ester hydrolysis, esterification (alcohol and acid), transesterification (ester and alcohol), interesterification (ester and acid) and transfer of acyl groups from esters to other nucleophiles like amines or thiols. Metabolic products of all these reactions are flavor compounds. *Mucor miehei* lipase brings lipolysis and cause the production of short chain fatty acids including acetic acid, propionic acid, butyric acid etc which are flavor generating compounds.

Lipolysed milk fat products can serve as cream-like/butter-like flavouring agents (Chen & Yang, 1992). Stereoselectivity of lipases is often used to yield pure optically active flavour compounds from racemic precursors. Lipases from *Penicillium*, *Rhizopus*, *Trichoderma* and various bacteria cause hydrolysis of methyl esters (Fukui, 1981). Lipase-catalysed esterification and transesterification reactions have a wide range of applications in the synthesis of aroma compounds. *Candida antarctica* lipase catalyses the esterifacation reaction and the subsequent production of methyl octanoic acid contributing to the goat like aroma in cheese prepared from goat milk.

Esters of acids from acetic acid to hexanoic acid and alcohols from methanol to hexanol, geraniol and citronellol have been synthesized using lipases from *Mucor miehei*, *Aspergillus* sp., *Candida rugosa*, and *Trichosporum fermentans*. Butyrolactone produced intramolecular transesterification of 4-hydroxycarboxylic esters with lipase from *Mucor miehei* is also a flavor enhancing compound. Flavorzyme is a commercially available proteolytic enzyme preparation that can be used to obtain meat-like flavor from defatted soybean meal (Dubal *et al.*, 2008).

9. FUTURE RESEARCH NEED

Biotechnology provides powerful tools for the sustainable development of food industry. It comprises mainly of both genetic and molecular techniques. Future research may be focused in the following aspects. Natural gene transfer systems in bacteria for understanding the genetics of microbial starter cultures and for strain improvement. Genetic engineering for the use of plasmid-free genetically characterized and highly transformable hosts to provide the necessary tools for transfer, maintenance and optimal expression of cloned DNA in microbial starter cultures. Nutritional and genetic interventions to alter the milk composition for specific health and/or processing in dairy biotechnology.

Extensive research work is needed for development of molecular methods for characterization of the strains isolated by culture-dependent methods. Genome sequencing with advanced molecular techniques to study the mechanisms of action of various host-defensive pathogen and aid them to produce more healthy and safer food by optimizing or eliminating the effect of harmful bacteria. More extensive studies are needed for cultured meat with the application of genetic engineering in the field of food biotechnology.

REFERENCES

Aasen, I.M., Markussen, S., Moretro, T., Katla, T., Axelsson, L. and Naterstad, K. (2003). Interactions of the bacteriocins and nisin with food constituents. *Int. J. Food Microbiol.*, 87: 35–43.

Achi, O.K. (2005). The potential for upgrading traditional fermented foods through biotechnology. *African .J Biotechnol.*, 4(5): 375–380.

Ahmed, I.A.M., Morishima, I., Babiker, E.E. and Mori, N. (2009). Characterisation of partially purified milk clotting enzyme from *Solanum dubium* fresen seeds, *Food Chem.*, 116: 395–400.

Aihara, H. and Miyazaki, J. (1998). Gene transfer into muscle by electroporation *in vivo. Nature Biotechnol.*, 16: 867–870.

Alisky, J, Iczkowski, K, Rapoport, A and Troitsky, N (1998). Bacteriophages show promise as antimicrobial agents. *J. Infection.*, 36: 5–15.

Andrade, M.J., Rodrýguez, M., Sanchez, B., Aranda, E. and Cordoba, J.J. (2006). DNA typing methods for differentiation of yeasts related to dry cured meat products. *Int. J. Food Microbiol.*, 107: 48–58.

Atterbury, R.J., Connerton, P.L., Dodd, C.E., Rees, C.E. and Connerton, I.F. (2003). Application of host-specific bacteriophages to the surface of chicken skin leads to a reduction in recovery of *Campylobacter jejuni. Applied Environmental Microbiol.*, 69: 6302–6306.

Aymerich, T., Martin, B., Garriga, M. and Hugas, M. (2003). Microbial quality and direct PCR identification of lactic acid bacteria and nonpathogenic staphylococci from artisanal low-acid sausages. *Applied Environmental Microbiol.*, 69: 4583–4594.

Barriere C, Centeno, D., Lebert, A., Leroy-Setrin, S., Berdague, J. and Talon, R. (2001). Role of superoxide dismutase and catalase of *Staphylococcus xylosus* in the inhibition of linoleic acid oxidation. *FEMS Microbiol. Letters*, 201: 181–185.

Berthier, F. and Ehrlich, S.D. (1998). Rapid species identification within two groups of closely related lactobacilli using PCR primers that target the 16S/23S rRNA spacer region. *FEMS Microbiol Letters*, 161: 97–106.

Berthier, F. and Ehrlich, S.D. (1999). Genetic diversity within *Lactobacillus sakei* and *Lactobacillus curvatus* and design of PCR primers for its detection using randomly amplified polymorphic DNA. *Int. J. Systematic Bacteriol.*, 49: 997–1007.

Blaiotta, G., Pennacchia, C., Ercolini, D., Moschetti, G. and Villani, F. (2003). Combining denaturing gradient gel electrophoresis of 16S rDNA V3 region and 16S-23S rDNA spacer region polymorphism analyses for the identification of staphylococci from Italian fermented sausages. *Systematic and Applied Microbiol.*, 26: 423–433.

Bottari, B., Ercolini, D., Gatti, M. and Neviani, E. (2006). Application of FISH technology for microbiological analysis: Current state and prospects. *Applied Microbiol. Biotechnol.*, 73: 485–494.

Bover, C.S., Hugas, M., Izquierdo-Pulido, M. and Vidal-Carou, M. (2000). Reduction of biogenic amine formation using a negative amino acid decarboxylase starter culture for fermentation of Fuet sausages. *J. Food Prot.*, 63: 237–243.

Brophy, B., Smolenski, G., Wheeler, T., Wells, D., Huillier, P.L. and Laible, G. (2003). Cloned transgenic cattle produce milk with higher levels of beta-casein and kappa-casein. *Nature Biotechnol.*, 21: 157–162.

California Dairy Research Foundation (2004). Investigators aim to improve milk composition for increased utilization. *Pre-harvest Research*, www.cdrf.org/newsletter.

Callaway, T.R., Anderson, R.C., Edrington, T.S., Genovese, K.J., Harvey, R.B. and Poole, T.L. (2004). Recent pre-harvest supplementation strategies to reduce carriage and shedding of zoonotic enteric bacterial pathogens in food animals. *Animal Health Research Reviews*, 5: 35–47.

Carlton, R.M., Noordman, W.H., Biswas, B., de Meester, E.D. and Loessner, M.J. (2005). Bacteriophage P100 for control of *Listeria monocytogenes* in foods: Genome sequence, bioinformatic analyses, oral toxicity study and application. *Regulatory Toxicol Pharmacol.*, 43: 301–312.

Chaillou, S., Champomier-Verges, M.C., Cornet, M., Crutz-Le Coq, A.M., Dudez, A.M. and Martin, V. (2005). The complete genome sequence of the meat borne lactic acid bacterium *Lactobacillus sakei* 23K. *Nature Biotechnol.*, 23: 1527–1533.

Champomier, V.M.C., Zuniga, M., Morel-Deville, F., Martýnez, G.P., Zagorec, M. and Ehrlich, S.D. (1999). Relationships between arginine degradation, pH and survival in *Lactobacillus sakei. FEMS Microbiol Letters*, 180: 297–304.

Chen, J.P. and Yang, B.K. (1992). Enzyme and flavour biotechnology. *In*: Flavours and frangrances: Chmeistry, bioprocessing and sustainability, (*Ed.* Bergger, R.G.) Springer Publications, Newyork., pp: 155–160.

Cleveland, J., Montville, T.J., Nes, I.F. and Chikindas, M.L. (2001). Bacteriocins: Safe, natural antimicrobials for food preservation. *Int. J. Food Microbiol.*, 71: 1–20.

Cocolin, L, Manzano, M., Cantoni, C. and Comi, G. (2001). Denaturing gradient gel electrophoresis analysis of the 16S rRNA gene V1 region to monitor dynamic changes in the bacterial population during fermentation of Italian sausages. *Applied Environmental Microbiol.*, 67: 5113–5121.

Collins, M.D., Rodriguez, U., Ash, C., Aguirre, M., Farrow, J.E. and Martinezmurcia, A. (1991). Phylogenetic analysis of the genus *Lactobacillus* and related lactic acid bacteria asdetermined by reverse-transcriptase sequencing of 16S ribosomal-RNA. *FEMS Microbiol, Letters*, 77: 5–12.

Corbiere Morot-Bizot, S., Talon, R. and Leroy, S. (2004). Development of a multiplex PCR for the identification of *Staphylococcus* genus and four staphylococcal species isolated from food. *J. Applied Microbiol.*, 97: 1087–1094.

Cotter, P.D., Hill, C. and Ross, R.P. (2005). Bacteriocins: Developing innate immunity for food. *Nature Reviews Microbiol.*, 3: 777–788.

De Vrese, M. and Schrezenmeir, J. (2002). Probiotics and non-intestinal infectious conditions. *British J Nutr.*, 88: S59–S66.

Deegan, L.H., Cotter, P.D., Hill, C. and Ross, R.P. (2006). Bacteriocins: Biological tools for bio-preservation and shelf-life extension. *Int. Dairy J.*, 16: 1058–1071.

Dubal, S.A., Yogesh, P.T., Momin, S.A. and Borkar, I.V. (2008). Biotechnological routes in flavour industries, *Advanced Biotech.*, 8: 20–31.

Elmer, L., Gaden Bokanga, M., Harlander, S., Clifford, W.H. and Steinkraus, H.K. (1992). Applications of biotechnology to traditional fermented foods, National Academy Press, Washington, D. C.

Ercolini, D., Moschetti, G., Blaiotta, G. and Coppola, S. (2001). Behavior of variable V3 region from 16S rDNA of important lactic acid bacteria in denaturing gradient gel electrophoresis. *Current Microbiol.*, 42: 199–202.

FAO, (2010). http://www.fao.org/biotech/forum, asp.

Fidel, T. (2008). Meat Biotechnology. 1st edn. Springer Science Publishing Ltd, Newyork.

Figueroa, B.N. and Bossi, L (1999). Inducible prophages contribute to *Salmonella* virulence in mice. *Molecular Microbiol.*, 33: 167–176.

Fukui, S. (1981). Biotechnological routes in flavour industry. *In*: Flavours and frangrances: Chmeistry, bioprocessing and sustainability, (*Ed*. Bergger, R.G.) Springer Publications, Newyork., pp: 135–139.

Gau, J.J., Lan, E.H., Dunn, B., Ho, C.M. and Woo, J.C.S. (2001). A MEMS based amperometri detector for *E.coli* bacteria using self-assembled monolayers. *Biosensors and Bioelectr.*, 16: 745–755.

Ghindilis, A., Atanasov, P., Wilkins, M. and Wilkins, E. (1998). Immunosensors: Electrochemical sensing and other engineering approaches. *Biosensors and Bioelectr.*, 13: 113–131.

Gonzalez, F.C., Santos, E., Jaime, I. and Rovira, J. (2003). Influence of starter cultures and sugar concentrations on biogenic amine contents in chorizo dry sausage. *Food Micro biol.*, 20: 275–284.

Hamm, W.P. and Hertel, C. (1998). New developments in meat starter cultures, *Meat Sci.*, 49: 125–138.

Hansen, E.B. (2002). Commercial bacterial starter cultures for fermented foods of the future. *Int. J. Food Microbiol.*, 78: 119–131.

Harlander, S. (1992). Genetic improvements of Microbial Starter culture, *In*: Application of Biotechnology in Traditional Fermented Foods, 1st edn. National Academic Press, Washington, pp: 20–27.

Harshman, M.R. (1997). Food biotechnology: Food industry, nutrition and public health, *Nutrition Soci.*, 56: 845–854.

Holzapfel, W.H., Geisen, R. and Schillinger, U. (1995). Biological preservation of foods with reference to protective cultures, bacteriocins and food-grade enzymes. *Int. J. Food Microbiol.*, 24: 343–362.

Joerger, R.D. (2003). Alternatives to antibiotics: Bacteriocins, antimicrobial peptides and bacteriophages. *Poultry Sci.*, 82: 640–647.

Jost, B., Vilotte, J.L., Duluc, I., Rodeau, J.L. and Freund, J.N. (1999). Production of low-lactose milk by ectopic expression of intestinal lactase in the mouse mammary gland. *Nature Biotechnol.*, 17: 160–164.

Kondo, J.F. and McKay, L.L. (1982). Transformation of *Streptococcus lactis* protoplast by plasmid DNA. *Appl. Environ. Microbiol.*, 43: 1213–1215.

Koohmaraie, M., Arthur, T.M., Bosilevac, J.M., Guerini, M., Shackelford, S.D. and Wheeler, T.L. (2005). Post-harvest interventins to reduce pathogens in beef. *Meat Sci.*, 71: 79–91.

Kurtzman, C.P. and Robnett, C.J. (1997). Identification of clinically important ascomycetes yeast based on nucleotide divergence in the 5' and large-subunit (26S) ribosomal DNA gene. *J Clinical Microbiol.*, 35: 1216–1223.

Latha, S. (2004). Designer milk—An imminent milestone in dairy biotechnology. *Current Sci.*, 87: 10–18.

Le, J.J.T. and Wetzel, A.N. (2007). Pre-harvest control of *Escherichia coli* O157 in cattle. *J Animal Sci.*, 85: 73–80.

Lee, J., Jang, J., Kim, B., Kim, J., Jeong, G. and Han, H. (2004). Identification of *Lactobacillus sakei* and *Lactobacillus curvatus* by multiplex PCR-based restriction enzyme analysis. *J Microbiological Methods.*, 59: 1–6.

Lonnerdal, B. (1996). Recombinant human milk proteins—An opportunity and a challenge. *Animal J. Clinical Nutri.*, 63: 622–626.

Lucke, F.K. (2000). Utilization of microbes to process and preserve meat. *Meat Sci.*, 56: 105–115.

Maga, E.A. and Anderson, G.B. (1995). The effect of mammary gland expression of human lysozyme on the properties of milk from transgenic mice. *J Dairy Sci.*, 78: 2645–2652.

Mandal, P.K., Biswas, A.K., Choi, K. and Pal, U.K. (2011). Methods for rapid detection of foodborne pathogens: An overview. *Am. J Food Technol.*, 6: 87–102.

Martýn, A., Cordoba, J.J., Aranda, E., Cordoba, M.G. and Asensio, M.A. (2006). Contribution of a selected fungal population to the volatile compounds on dry cured ham. *Int. J. Food Microbiol.*, 110: 8–18.

Mirold, S., Rabsch, W., Tschape, H. and Hardt, W.D. (2001). Transfer of the *Salmonella* type III effector *sopE* between unrelated phage families. *J Molecular Biol.*, 312: 7–16.

Morot, B.S., Talon, R. and Leroy, S.S. (2003). Development of specific primers for a rapid and accurate identification of *Staphylococcus xylosus*, a species used in food fermentation. *J Microbiological Methods*, 55: 279–286.

Muhammad, T.Z. and Alocilja, E. (2004). A disposable biosensor for pathogen detection in fresh produce samples. *Biosystems Eng.*, 88: 145–151.

Muhammad, T.Z. and Alocilja, E.C. (2003). A conductimetric biosensor for biosecurity. *Biosensors and Bioelectronics*, 18: 813–819.

Nissen, H. and Dainty, R. (1995). Comparsion of the use of rRNA probes and conventional methods in identifying strains of *Lactobacillus sake* and *L. curvatus* isolated from meat. *Int. J. Food Microbiol.*, 25: 311–315.

Pal, S., Alocilja, E.C. and Downes, F.P. (2007). Nanowire labeled direct-charge transfer biosensor for detecting bacillus species. *Biosensors Bioelectronics.*, 22: 2329–2336.

Quiberoni, A., Suarez, V.B. and Reinheimer, J.A. (1999). Inactivation of *Lactobacillus helveticus* bacteriophages by thermal and chemical treatments. *J. Food Prot.*, 62: 894–898.

Radke, S. and Alocilja, E. (2004). Design and fabrication of an impedimetric biosensor. *IEEE Sensors J.*, 4: 434–440.

Ramos, J.P., Valente, P., Hagler, A.N. and Leoncini, O. (1998). Restriction analysis of ITS region for characterization of the *Debaryomyces* species. *J. General and Applied Microbiol.*, 44: 399–404.

Riley, M.A. and Wertz, J.E. (2002). Bacteriocins: Evolution, ecology and application. *Annual Review of Microbiol.*, 56: 117–137.

Rossi, F., Tofalo, R., Torriani, S. and Suzzi, G. (2001). Identification by 16S-23S rDNA intergenic region amplification, genotypic and phenotypic clustering of *Staphylococcus xylosus* strains from dry sausages. *J Applied Microbiol.*, 90: 365–371.

Saiyavit, V., Sujin, S. and Manop, S. (2000). Production of meat like flavor. *Science Asia.* 26: 219–224.

Sanz, Y., Hernandez, M., Ferrus, M.A. and Hernandez, J. (1998). Characterization of *Lactobacillus sake* isolates from dry-cured sausage by restriction fragment length polymorphism analysis of the 16S rRNA gene. *J. Applied Microbiol.*, 84: 600–606.

Schmieger, H. and Schicklmaier, P. (1999). Transduction of multiple drug resistance of *Salmonella enterica* serovar *typhimurium* DT104. FEMS Microbiol. Letters, 170: 251–256.

Simonova, M., Strompfova, V., Marcinakova, M., Laukova, A., Vetserlund, S. and Moratella, S. (2006). Characterization of *Staphylococcus xylosus* and *Staphylococcus carnosus* isolated from Slovak meat products. *Meat Sci.*, 73: 559–564.

Subramanian, S. (2004). Transgenic therapeutics—medicines in milk. *Biotechnology and Soci.*, 12: 18–24.

Talon, D., Deliere, E. and Bertrand, X. (2002). Characterization of methicillin resistant *Staphylococcus aureus* strains susceptible to tobramycin. *Int. J. Antimicrobial Agents*, 20(3): 174–179.

Talon, R., Walter, D., Chartier, S., Barriere, C. and Montel, M. (1999). Effect of nitrate and incubation conditions on the production of catalase and nitrate reductase by *Staphylococcus aureus. Int J Food Microbiol.*, 52: 47–50.

Turner, A.P. and Newman, J.D. (1998). An introduction to biosensor. *In*: Biosensor for Food Analysis, (*Ed.* Gateshead, T.W.). Athaenaeum Press Ltd. UK. pp: 13–27.

United States Department of Agriculture (2001). Service Economics of foodborne disease: Estimating the benefits of reducing foodborne disease. Washington, DC.

Urso, R., Comi, G. and Cocolin, L. (2006). Ecology of lactic acid bacteria in Italian fermented sausages: Isolation, identification and molecular characterization. *Systematic Applied Microbiol.*, 29: 671–680.

Vikholm, I. (2005). Self assembly of antibody fragments and polymers on gold for immunosensing. *Sensors and Actuators B-Chemical*, 106: 311–316.

Vilotte, J.L. (2002). Lowering the milk lactose content *in vivo*: Potential interests, strategies and physiological consequences. *Repr. Nutri. Development*, 42: 127–132.

Wagner, R.D. (2006). Efficacy and food safety consideration of poultry competitive exclusion products. *Molecular Nutri. Food Res.*, 50: 1061–1071.

William, J.D. (1991). Applications of biotechnology and separation technology in dairy processing. *Int. J Dairy Technol.*, 44(3): 67–72.

Wiriyacharee, P. (1990). The Systematic Development of a Controlled Fermentation Process Using Mixed Bacterial Starter Cultures using mixed starter cultures 127 for *Nham,* a Thai Semi-dry Sausage. Ph. D. thesis, Massey University, New Zealand.

Wu, Y.F. and Cadwallader, K. (2002). Enzyme and flavour biotechnology. *In*: Flavours and frangrances: Chmeistry, bioprocessing and sustainability, (*Ed.* Bergger, R.G.). Springer Publications, Newyork. pp: 125–128.

Ye, J.M., Letcher, S.V. and Rand, A.G. (1997). Piezoelectric biosensor for detection of *Salmonella typhimurium. J. Food Sci.*, 62: 167–174.

9

HACCP for Clean Meat Production

S.G. BHANDARE, V.S. WASKER AND S.K. NAGAPPA

ABSTRACT

The traditional quality control measures and microbiological testing are fraught with limitations, as the reports of food borne illnesses remain high. The Hazard Analysis and Critical Control Point (HACCP) concept, identifies specific hazards and control measures to ensure the safety of food. A HACCP plan is specific to the particular food and processing application. HACCP offers a more comprehensive and science based alternative for controlling food safety hazards compared with traditional sanitation programmes based upon good manufacturing practices. The HACCP provides a more specific and critical approach to control hazards in the food industry than traditional inspection and quality control approaches. Controlling hazards under HACCP programme requires a systematic assembling of reliable data relating to the occurrence, elimination, prevention and reduction of hazards. The process of developing an HACCP program for specific food product and processing line requires technical expertise and considerable time. Though the food safety management is meticulously followed in the developed countries the necessary know how about the implementation of performance criteria like HACCP is imperative for developing countries.

1. INTRODUCTION

Meat has been used by man as food since prehistoric time and still continues as an important segment of our diet, mainly providing essential amino acids which man cannot synthesize. The consumer expects the meat he purchases should be safe and wholesome. Wholesome meat is the one which is produced hygienically, pathogen free, retains its natural state and nutritive value, endowed with optimum fat and unconditionally acceptable to the consumer. Meat is vulnerable for bacteria and other infections by its very nature and serves as an ideal medium for bacterial multiplication which induces deteriorative changes in meat. The risk of contamination exists from the point of entry of animal to the slaughter floor till the time of consumption of meat. It is, therefore, very essential to take utmost care and precautions at every stage of its production, processing, transport, storage and till it reaches to the consumer.

Even today, in most of the developing countries, the slaughtering, dressing and subsequent handling of the carcasses is done under poor sanitary conditions in an unorganised sector by local butchers. Meat is sold in small quantities from hung carcasses throughout the day in retail outlets, which is in freely exposed state while being produced and is thus highly prone to contamination. Several factors like slaughtering on the floor, use of non potable water, washing of guts side by side with meat cutting, neglect of personal hygiene etc. contribute to the unhygienic production of meat in unorganised sector. For these reasons more emphasis should be given to the centralized slaughtering at organised sector having skilled butchers and hygienic environment with implementation of Hazard Analysis and Critical Control Point Approach (HACCP).

Most of the slaughterhouses in the developing world are in bad shape having temporary or old civil structure with leaking roof, poor ventilation and inadequate light and water supplies. The lairages are not available and clean as well as unclean operations are carried out in the same hall. Butchers/workers engaged in the slaughterhouses and meat-processing plants are not properly trained. Refrigerated transportation of meat is the basic requirement for the transport of meat, which is virtually non-existent in the developing countries. Much of the meat by-products are not utilized for edible or inedible purposes, contributing to huge losses. Due to lack of effective legislation and a regulated meat inspection system it is difficult to supervise private slaughterhouses as well. Even the animals are not raised especially for meat purpose with proper feeding and management practices as like in the developed countries. Most often, rejected and culled animals are slaughtered for meat purpose, which give less than 50% carcass-yield.

The present system of meat hygiene programme of examination of grossly apparent abnormality, which do not recognize microbiological contamination, or traditional meat hygiene system, which concentrates on routine ante and post mortem inspection, is now considered as an inadequate effort. The traditional quality control measures and microbiological testing are widely applied either individually or in combinations, but they are fraught with limitations, as the reports of food borne illnesses remain high. This necessitates rational approach based on Hazard Analysis and Critical Control Point (HACCP) concept, which identifies specific hazards and control measures to ensure the safety of food. A HACCP plan is specific to the particular food and processing application. The HACCP system is capable of accommodating changes such as advances in equipment design, new information concerning health hazards or risks, new processing procedures or technological developments. HACCP offers a more comprehensive and science based alternative for controlling food safety hazards compared with traditional sanitation programmes based upon good manufacturing practices.

Controlling hazards under HACCP programme requires a systematic assembling of reliable data relating to the occurrence, elimination, prevention and reduction of hazards. Superficially, the development of a HACCP program appears to be quite simple. However, the process of developing an HACCP program for specific food product and processing line requires technical expertise and considerable time. The time and effort are well spent, because HACCP provides a more specific and critical approach to control hazards in the food industry than traditional inspection and quality control approaches. Though the food safety management is meticulously followed in the developed countries the necessary know how about the implementation of performance criteria like HACCP is imperative for developing countries.

2. HISTORY AND BACKGROUND

The concepts for HACCP originated from work done in 1960's by the National Aeronautics and Space Administration (NASA) and the U.S. Military Natick Laboratory to establish the risk of *Salmonella* infection in food and on quality assurance systems. The HACCP system as is seen today, took shape in the 1971 National Conference of Food Protection, where risk assessment was combined with the Critical Control Point (CCP) concept. Recognizing the importance of HACCP to food control, the 20th session of Codex Alimentarius Commission held in Geneva, Switzerland adopted Guidelines for the application of Hazard Analysis Critical Control Point (HACCP). Since then, HACCP has been recognized internationally as a logical tool for adapting traditional inspection methods to a modern, science-based, food safety system.

Based on risk-assessment, HACCP plans allow both industry and government to allocate their resources efficiently in establishing and auditing safe food production practices. In 1994, the organization of *International HACCP Alliance* was established initially for the US meat and poultry industries to assist them with implementing HACCP and now its membership has been spread over other professional/industrial areas. FAO/WHO has published a guideline for all governments to handle the issue in small and less developed food businesses.

While the application of HACCP to all segments and sectors of the food chain is possible, it is assumed that all sectors should be operating according to Good Manufacturing Practices (GMPs) and the Codex General Principles of Food Hygiene. The ability of an industry segment or sector to support or implement the HACCP system depends on the degree of its adherence to these practices. The successful application of HACCP requires the full commitment and involvement of management and the workforce. The top management should express an overall intentions and direction for quality in their organization, which is called as Quality Policy. It is a statement of their commitment to the customers for the provision of quality products. The three-level structure of execution is shown in Fig. 1.

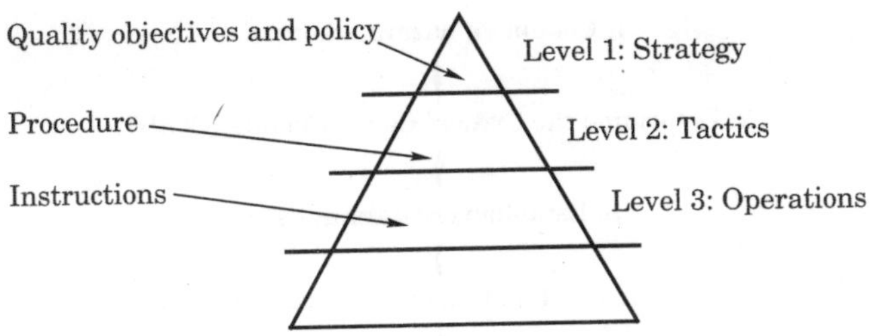

Fig. 1: The three-level structure of execution.

Implementation of HACCP requires a multi disciplinary approach which should include, as appropriate, expertise in agronomy, veterinary health, microbiology, public health, food technology, environmental health, chemistry, engineering, etc. according to the particular situation. The application of the HACCP system is compatible with the implementation of Total Quality Management (TQM) systems such as the International Organisation for Standardization (ISO) 9000 series.

Of late, HACCP is considered to be a part of more integrated food safety management called as ISO 22000. This is a management system having an interactive communication throughout the food chain to identify the food hazards and control them, based upon the HACCP principles. This approach is well summarized in a statement 'Farm to fork approach'. Thus, showing the coordinated efforts both backward *i.e.*, towards the farm; where raw material is produced and forward *i.e.*, towards the consumers; by whom the commodity is utilized.

3. APPLICATION OF HACCP

In order to apply HACCP to the meat processing facility, it requires thorough planning and work out. The HACCP system works on seven principles (Fig. 2) which depict the step by step work out and implementation procedures for effective application of HACCP for clean meat production. The multidisciplinary team work is executed as per the guidelines of the seven principles of HACCP.

The HACCP principles are applied and executed in the following manner:

3.1. HACCP Team Assembly

For the development of an effective HACCP plan, the food operation should assure that the appropriate product specific knowledge and expertise is available. This is accomplished by assembling a multidisciplinary team. Expert advice should be obtained from other sources, when such expertise is not available on site. The team should identify the scope of the HACCP plan which should describe the segment of the food chain involved and the general classes of hazards to be addressed.

1. Conduct a hazard analysis.

↓

2. Determine the Critical Control Points (CCPs).

↓

3. Establish critical limit(s).

↓

4. Establish a system to monitor control of the CCP.

↓

5. Establish the corrective action to be taken when monitoring indicates that a particular CCP is not under control.

↓

6. Establish procedures for verification to confirm that the HACCP system is working effectively.

↓

7. Establish documentation concerning all procedures and records appropriate to these principles and their application.

Fig. 2: Flowchart showing seven principles of HACCP.

3.2. Product Description

Description of the product should include composition, physical/chemical structure (including A_w, pH, etc.), bactericidal/bacteriostatic treatments (heat-treatment, freezing, brining, smoking, etc.), packaging, durability and storage conditions and method of distribution.

For example the product description for frozen chicken can be as below:

Product: Raw frozen chicken (broiler) parts packed in polypropylene; product to be fully cooked before consumption.

Formal product name: Frozen product.

*Alternate name(*s): Frozen poultry parts.

Product holding temperature: –28°C

Preparation and serving

Procedures: Defrost in refrigerator overnight, cook properly and serve.

Packaging used: Polypropylene and carton box.

Transportation and distribution: In frozen condition.

Intended suppliers: Supermarkets.

Intended consumers: General population.

Label warnings required: Keep frozen all the time.

3.3. Identification of Intended Use

The expected uses of the product by the end user or consumer, is the intended use of that particular product. Even institutional feeding can be an intended use.

3.4. Construct Flow Diagram

All steps in the operation should be covered in the flow diagram. Steps preceding and following the specified operation should be considered when applying HACCP to a given operation. The flow diagram for frozen chicken meat production is depicted in Fig. 3.

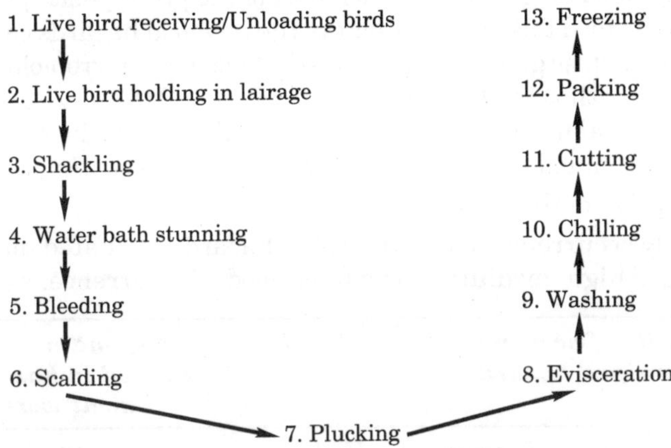

Fig. 3: The flow diagram for frozen chicken meat production.

3.5. On-site Confirmation of Flow Diagram

The HACCP team should confirm the processing operation against the flow diagram during all stages and hours of operation and amend the flow diagram where appropriate.

3.6. Hazard Analysis

All the hazards that occur at each step from primary production, processing, manufacture, and distribution until the point of consumption should be listed.

The HACCP team should next conduct a hazard analysis to identify for the HACCP plan which hazards are of such a nature that their elimination or reduction to acceptable levels is essential to the production of a safe food.

In conducting the hazard analysis, wherever possible the following should be included:

- The likely occurrence of hazards and severity of their adverse health effects
- The qualitative and/or quantitative evaluation of the presence of hazards
- Survival or multiplication of microorganisms of concern
- Production or persistence in foods of toxins, chemicals or physical agents and
- Conditions leading to the above.

The results of the Hazard Analysis should be documented into the Hazard Identification & Hazard Assessment Worksheets maintained for each process step.

Steps involved in Hazard Analysis are:

- Identification of all potential hazards associated with that process step.
- Classification of hazard based on type (whether physical, biological or chemical) and category (into which part of the prerequisite program (PRP) does this hazard fall). In meat plants the physical hazards can the foreign materials such as glass, metal or plastic, biological/microbiological hazards can be the pathogens like *Salmonella*, *E. coli O157*, *Campylobacter* etc., while chemical hazards can be the chemicals used in the processing plant such as chlorine, lubricants, cleaners, paints and coatings.
- A description of the hazard.
- The likely occurrence of the identified hazard, evaluated quantitatively, in terms of high, medium or low likelihood of occurrence, as follows:

Probability of occurrence of identified hazard	***Level***	***Corresponding number to be entered in hazard assessment worksheet***
Very high chances of occurrence *e.g.*: Occurrence once in every months	High	10
Moderate chances of occurrence *e.g.*: Occurrence once in 4 to 6 months	Medium	5
Marginal chances of occurrence *e.g.*: May not occur or once in 5 years	Low	3

- The severity of the hazard's adverse health effects, evaluated quantitatively, in terms of critical, major or minor impact severity, as follows:

Severity of hazard in case of occurrence	***Level***	***Corresponding number to be entered in hazard assessment worksheet***
Will definitely result into unsafe product *e.g.*: Death or permanent damage	Critical	10
May result into unsafe product *e.g.*: Hospitalization may result in an unsafe product	Major	5
Will not result in unsafe product *e.g.*: Stomach pain	Minor	3

- The risk rating is done, in quantitative terms (arrived at by multiplying the probability of occurrence of identified hazard with the severity of hazard in case of occurrence) and the classification of the control measures is done, based on the risk rating. A risk rating of 30 or above indicates that the hazard is significant:

	Severity of impact		
Likelihood of occurrence	***Critical (10)***	***Major (5)***	***Minor (3)***
High (10)	Significant	Significant	Significant
Medium (5)	Significant		
Low (3)	Significant		

3.7. Determine Critical Control Points

For each food safety hazard that is controlled by the HACCP plan, CCPs are identified for the control measures identified by the method described above.

If a hazard has been identified at a step where control is necessary for safety, and no control measure exists at that step, or any other, then the product or process should be modified at that step, or at any earlier or later stage, to include a control measure.

For example in the frozen chicken production process the evisceration process may be become a Critical Control Point (CCP) for a biological hazard of food borne pathogens based on the hazard analysis by seeing the likelihood of occurrence and severity of hazard.

3.8. Establish Critical Limits for each CCP

The critical limit(s), action limit(s) and target value for each parameter should be documented in the HACCP Plan for CCPs Worksheet. The rationale for the chosen critical limits shall be documented.

In some cases more than one critical limit will be elaborated at a particular step. Criteria often used include measurements of temperature, time, moisture level, pH, A_w, available chlorine, and sensory parameters such as visual appearance and texture.

Suppose our CCP is evisceration process as discussed earlier then the critical limits for this CCP can be maximum of 2% of carcasses with visible intestinal contamination should be permitted, sanitization of instruments with 80°C hot water for 2 minutes should be compulsory, minimum of 10 ppm free chlorine be used in wash water and minimum 1 liter of water per carcass is pre-requisite.

3.9. Establish a Monitoring System for Each CCP

It is an observation of a CCP relative to its critical limits. This monitoring activity detects any deviation observed, so that steps can be taken to prevent violation of the critical limits. Data derived from monitoring must be evaluated by a designated person with knowledge and authority to carry out corrective actions when indicated. If monitoring is not continuous, then the amount or frequency of monitoring must be sufficient to guarantee the CCP is in control.

Most monitoring procedures for CCPs will need to be done rapidly because they relate to on-line processes and there will not be time for lengthy analytical testing. Physical and chemical measurements are often preferred to microbiological testing because they may be done rapidly and can often indicate the microbiological control of the product. All records and documents associated with monitoring CCPs must be signed by the person(s) doing the monitoring and by a responsible reviewing official(s) of the company.

As to our CCP of evisceration process may require monitoring like critical inspection of each carcass for intestinal contamination and recording of washing water temperature, it's flow as well as it's chlorine concentration continually. This monitoring is recorded in the respective check lists without fail by particular online supervisors.

3.10. Establish Corrective Actions

Corrective action is undertaken in a timely manner to prevent further occurrence of nonconformity. Any corrective action plan relating to safety, legality or quality is agreed by the personnel who have defined responsibility and accountability for the area/s of control. These personnel, along with the Food Safety Team Leader, are also responsible for verifying that the corrective action plan has been completed satisfactorily.

The actions must ensure that the CCP has been brought under control. Actions taken must also include proper disposition of the affected product. Deviation and product disposition procedures must be documented in the HACCP record keeping.

For example the deviation response for evisceration CCP would be removal of contaminated carcasses from line; if more than 2% are visibly contaminated adjusting eviscerating machine or slow line speed if necessary, increase in water temperature if it drops below 80°C, increase in chlorine level if it drops below 10 ppm and increase in water flow if it is below 1 liter per carcass.

3.11. Establish Verification Procedures

Verification and auditing methods, procedures and tests, including random sampling and analysis, are required to know whether we are doing right thing and HACCP system is working correctly. The frequency of verification should be sufficient to confirm that the HACCP system is working effectively. Examples of verification activities include:

- Review of the HACCP system and its records;
- Review of deviations and product dispositions;
- Confirmation that CCPs are kept under control.

Where possible, validation activities should include actions to confirm the efficacy of all elements of the HACCP plan.

All monitoring records are verified every hour by the superiors. Suppose the monitoring is recorded by a supervisor then it's verification is done by the officer immediate senior to him/her. Periodically, the records are verified by plant managers and the validation should be carried out by external auditors.

3.12. Establish Documentation and Record Keeping

It is a pre-requisite to maintain efficient and accurate record keeping for the application of a HACCP system. Every HACCP procedure should be well documented. Documentation for Hazard analysis; CCP determination; Critical limit determination and record for CCP monitoring activities; Deviations and associated corrective actions; Modifications to the HACCP system etc should be invariably maintained.

4. CONCLUSIONS

The meat processing Industry is growing at very faster rate in the developing countries. In order to remain competitive in the international markets, the food safety and quality standards should be maintained at the international levels. The present infrastructure and facilities in the developing countries are inadequate and outdated leading to poor sanitary conditions. Thus, the prevalence and emergence of food borne pathogens are encouraged. A lot has to be dealt with in the quality assurance arena; though there is a good future given the present attention by the Governments and private entrepreneurs for rapid progress in meat processing sector. HACCP, which can be used as a due diligence by the meat industry, is a beneficial tool for producers as well as consumers providing a win-win situation for the meat sector.

REFERENCES

Anyawu, R.C. and Jukes, D.J. (1990). Food safety control systems for developing countries. *Food Control*, 1: 17–26.

Bhandare, S.G. (2010). Food borne diseases and food safety management: an Indian perspective. *Asian Meat: Processing and Marketing*, pp. 28–29.

Bhandare, S.G., Sherikar, A.T., Paturkar, A.M., Waskar, V.S. and Zende, R.J. (2007). A comparison of microbial contamination on sheep/goat carcasses in a modern Indian abattoir and traditional meat shops. *Food Control*, 18: 854–858.

Codex Alimentarious. (1997). *Food Hygiene Basic Text*. Food and Agriculture Organisation, Rome.

Huss, H.H. (1994). *Assurance of seafood quality*. FAO Fisheries Technical Paper Series: T334, p. 169.

Kaitazov, G. and Stoyanchev, T. (2008). Modified Hazard Analysis. Basic programs of the modified HACCP for Poultry Industry. *Bulgarian Journal of Agricultural Science*, 14: 466–469.

Mortimore, S.E. (2001). How to make HACCP really work in practice. *Food Control*, 12: 209–215.

Panisello, P.J., Rooney, R., Quantick, P.C. and Stanwell-Smith, R. (2000). Application of foodborne disease outbreak data in the development and maintenance of HACCP systems. *Int. J. Food Microbiol.*, 59: 221–34.

Simonsen, B., Bryan, F.L., Christian, J.H.B., Roberts, T.A., Tompkin, R.B. and Silliker, J.H. (1987). Prevention and control of food-borne salmonellosis through application of Hazard Analysis Critical Control Point (HACCP). *Int. J. Food Microbiol.*, 4: 227–247.

10

Biopreservatives for Animal Products

A.K. Biswas, Parminder Singh and P.K. Mandal

ABSTRACT

Each preservation method has its inherent advantages and disadvantages. The most acceptable method is one which should not induce alterations in chemical and nutritional constituents as well as maintain all the sensory attributes of meat and meat products. Traditional food preservatives are an important tool in preserving the foods from the hazardous and detrimental effects of microorganisms. Among various natural preservatives, only few bacteriocins are used. This may be partially due to the fact that newly discovered broad-spectrum bacteriocins have yet to be fully characterized and officially approved. Natural preservatives are not only effective, but are also safe for use in the food supply. Moreover, it appears that their pertinent exploitation will produce bio-preservation systems that are optimum for each food. In future, natural preservatives are going to play a role in food preservation. However, it is quite doubtful that any newly discovered synthetic antimicrobials will be approved for use in foods by worldwide regulatory agencies. Hence, the future of natural antioxidants and antimicrobials lies in novel uses.

1. INTRODUCTION

Preservation of meat is a complex process of treatment in various ways so as to halt its spoilage, preserving the nutritious and sensory attributes thereby increasing the shelf-life and making it safe (in terms of physical, chemical and biological hazards) for consumption. Preservation of meat and meat products could be attempted by various ways *viz.*, temperature treatment, chemical treatment, irradiation and use of natural and synthetic preservatives. All these preservation techniques have their own problems and concerns but the basic idea is that one should select such a technique of preservation which should not affect the nutritional quality and sensory attributes of the product. Out of all the basic techniques of meat preservation, the use of preservatives is preferred due to cost, infrastructure, shelf life and quality concerns.

In meat and meat products, lipid oxidation and growth of microbes (bacteria, yeast and molds) are the two major bottle necks which should be combated to improve its shelf life and its acceptability amongst the masses. For this purpose, a large number of chemicals are being used these days. However, the use of synthetic chemicals is always a concern of debate due to their ill effects on human health such as heart attacks, cancer, stroke, brain damage etc. Therefore, modern trends toward the natural preservatives have obliged the meat industry to include the same in poultry meat for their antioxidant and antimicrobial action. Hence, the use of natural preservatives derived from plants, animals and certain species of microbes is warranted for preservation of meat and meat products.

Meat is a rich source of proteins, lipids, vitamins and minerals. Out of these, lipids are considered to be the integral components of muscle foods. Hence, all the muscle foods are prone to lipid oxidation which is the major cause of deterioration of quality of stored meat and meat products. However, among all the meat products, poultry meat *i.e.*, white meat is considered to be more prone to oxidative rancidity than the red meats. There is strong explanation for this fact as poultry meat is rich in phospholipids and poly-unsaturated fatty acids which are integral components of cell membranes (Kemin-Europa, 2009). Some of the tissue lipid per-oxidants are ferric heme pigments, non-heme iron, ascorbic acid, sodium chloride etc. In cured meat due to formation of stable pink colour by ferrous form of pigment, there is less oxidation as compared to cooked uncured meat (Love & Pearson, 1971).

Antioxidants are those chemical compounds or molecules which when present significantly inhibit or delay the oxidation, while often being oxidized themselves. There are two principle mechanisms of action for antioxidants. Primary antioxidants halt the free radical chain reactions just by donating a single electron to the free radical present in the system. Since donation of an electron to a lipid radical produces a new radical and is more stable than the initial one. However, secondary antioxidants act by various ways including quenching chain-initiating catalysts, binding of metal ions, scavenging of oxygen, converting hydroperoxides to non-radical species, absorbing UV radiation, deactivating singlet oxygen, inhibition of peroxidases etc. Sometimes two or more antioxidants show their synergistic effect when present together in a medium thereby greatly enhancing the antioxidative effect. Such types of antioxidants are referred to as co-antioxidants and process is known as co-antioxidation.

2. CLASSIFICATION OF NATURAL PRESERVATIVES

Briefly, the natural preservatives have been classified into three broad categories depending upon their origin. They may be of plant origin (Table 1), animal origin (Table 2) or microbial origin (Table 3).

2.1. Natural Preservatives of Plant Origin

Natural antimicrobials from plant origin include Phytoalexins, organic acids, essential oils, phenolic acids, polyphenols etc.

2.1.1. *Phytoalexins*

Phytoalexins are host synthesized low molecular weight broad spectrum antimicrobial compounds whose synthesis from distant precursors is induced in plants in response to microbial infection or treatment of plant tissues with a range of naturally occurring or synthetic, artificial compounds. The molecules that signal the plants to begin the process of phytoalexin synthesis are called elicitors.

2.1.2. *Organic acids*

Organic acids such as acetic acid, lactic acid, propionic acid, citric acid, succinic acid, malic acid, tartaric acid, benzoic acid, sorbic acid and their derivatives are commonly used as antimicrobial agents. As far as their spectrum of action is concerned, they are almost equally effective in inhibiting the growth of gram positive and gram negative bacteria. During fermentation of various foods, lactic and propionic acids are two normally produced acids. Their antimicrobial activity has been well documented. Cell walls, cell membranes, metabolic enzymes, protein synthesis systems and genetic material are the main targets of their action against a wide range of microorganisms. The antimicrobial activity of organic acids is pH dependent and as the pH increases, their activity decreases. The reduction of intracellular pH of microorganisms changes the ability of cells to maintain the homeostasis and thus affecting the substrate transport and metabolic pathways of microorganisms. Organic acids are commonly used in the form of washes and sprays for disinfection of animal and poultry carcasses.

2.1.3. *Essential oils*

Essential oils are nothing but true spices which are obtained from roots, bark, buds, seeds or fruits of plants growing in tropical countries. The phenolic compounds which commonly occur in natural foods have been divided into three groups *viz*. simple phenols and phenolic acids (*e.g.*, *p*- cresol, 3-ethylphenol, hydroquinone, vanillic, gallic, syringic, ellagic), hydrocinnamic acid derivatives (*e.g.*, *p*-coumaric acid, caffeic, ferulic, sinapic) and flavonoids. Proteins, fats, salts, pH and temperature are factors which affect the antimicrobial activity of phenolics. Their basic classification, active principles and use in poultry meat has been enlisted in the Table 1.

Table 1: Natural preservatives of plant origin, their active principles and use in meat.

Preservative (active principles)	***Use in poultry meat***
Clove (Eugenol, thymol, carvacrol)	1 and 2% clove oil as antimicrobial in chicken frankfurters (Mytle *et al.*, 2006); 0.2 and 0.5% clove oleoresin in chicken meat marination (Carlos & Harrison, 1999); clove powder as phytopreservative and antimicrobial in chicken nuggets (Kumar & Tanwar, 2011). Hao *et al.* (1998) applied eugenol to beef slices or cooked chicken and it was proved that eugenol inhibited the growth of *A. hydrophila* and *L. monocytogenes*.
Ginger (6-gingerol and its derivatives)	2.5% ginger extract as tenderizing, antioxidant and antimicrobial in smoked spent hen meat (Naveena *et al.*, 2001); 2%, 4% and 6% ginger paste as an antioxidant in spent hen meat balls (Rongsensusang *et al.*, 2005).
Garlic (allicin, diallyl sulfide, allyl sulfide, diallyl trisulfide and propyl sulfide)	Garlic powder and garlic oil as antioxidant and antimicrobial in raw chicken sausage (Sallam *et al.*, 2004); garlic extract as growth inhibitor of facultative aerobic, mesophilic, and faecal coliforms on the surface of poultry carcasses (Oliveira *et al.*, 2005).
Onion (Quercetin)	Onion controlled oxidative rancidity in cooked ground turkey meat (Younathan *et al.*, 1980); dried onion flesh @1.6% showed antioxidant activity in cooked chicken meat (Karastogiannidou, 1999); onion extract 50% significantly reduced lipid oxidation in cooked turkey breast rolls (Tang & Cronin, 2007).
Nutmeg (macelignan, d-camphene)	Nutmeg showed bactericidal activity against *E. coli* O157:H7 in Barbecued Chicken (Shekarforoush *et al.*, 2007a).
Green tea (tea catechins, epicatechins, epigallocatechin, epigallocatechin gallate and epicatechin gallate)	Tea catechins at 300 ppm level as antioxidant in duck, ostrich and chicken meat under refrigerated storage (Tang *et al.*, 2001); tea catechins @ 0.01% inhibited oxidation in chicken nuggets (O'Sullivan *et al.*, 2004), supplementation of poultry diets with natural tea catechins @ 300 mg/kg feed delayed lipid oxidation in chicken meat (Tang *et al.*, 2000 and Maher *et al.*, 2002).
Oregano (rosmarinic acid, carvacrol, thymol and glycosides)	Oregano oil in chicken breast meat reduced the microbial growth during refrigerated storage (Burt, 2004; Chouliara *et al.*, 2007); Oregano oil @ 200mg inhibited the development of lipid oxidation in breast and thigh turkey meat during refrigerated storage (Govaris *et al.*, 2004).
Honey	5% in ground turkey as antioxidant (McKibben & Engeseth, 2002); 10, 20 & 30% for marination of roasted chicken (Hashim *et al.*, 1999).

Table 1: *(Contd...)*

Table 1: (*Contd...*)

Preservative (active principles)	***Use in poultry meat***
Dried plums (flavonoids and polyphenols)	Plum extract @ >2% significantly retarded oxidation in turkey breast rolls (Lee & Ahn, 2005).
Grapes (Grape seed extract, catechins, epicatechins and reserveratol)	Grape seed extract @ 1.0 and 2.0% effectively inhibited the development of TBARS in dark poultry meat patties (Lau & King, 2003).
Sage (carnosol, rosmanol, rosemadiol, epirosmanol, isorosmanol, galdosol, methyl carnosate and carnosic acid)	Sage extract alone or in combination with sodium isoascorbate resulted in reduced mesophilic bacteria and coliforms counts in raw vacuum-packaged turkey meatballs (Karpinska-Tymoszczyk, 2007); In high-pressure processed chicken meat, sage inhibited lipid oxidation during chilled storage for 2 weeks (Mariutti *et al.*, 2008).
Orange	Addition of 2% orange fibre to bologna sausages significantly retarded the development of TBARs over 28-days of storage at 4°C (Fernández-López *et al.,* 2004).
Cinnamon (Cinnamldehyde, Eugenol, cinnamicaldehyde)	Cinnamon powder @ 2% and 4% (w/w) level extended the shelf life of minced chicken meat at room temperature and refrigeration temperature (Yadav *et al.,* 2002).
Caraway (carvone and limonene)	Caraway as antioxidant in chicken meat stored under frozen conditions (El-Alim *et al*, 1999).
Dry soya sprouts	@ 30 g/ kg in chicken patties (Romero *et al.*, 2008).
Pomegranate (tannins, antho-cyanins and flavonoids)	0.02% phenolics solution as antioxidant and antimicrobial in chicken meat (Vaithiyanathan *et al.*, 2011); Pomegranate juice @ 10 mg equivalent phenolics/100 g meat in chicken meat patties (Naveena *et al.*, 2008).
Rosemary (carnosol, carnosic acid, rosmanol, isorosmanol, rosmariquinone, rosmaridiphenol and rosmary-diphenol)	Used @ 400 ppm in cooked chicken slurry as antioxidant (Beltran *et al.*, 2004); rosemary extract for antioxidant (20%w/w) and antimicrobial activity (4%w/w) in chicken frankfurters (Riznar *et al.*, 2006).
Rice hull extract	Rice hull extract @ 0.2% (w/w) in cooked turkey meat (Nam *et al.*, 2004).
Tomatoes (Lycopene)	Use of tomato juice in low-fat cooked sausages (Yilmaz *et al.*, 2002).
Seabuckthorn (flavonols)	As antioxidant in mechanically deboned broiler meat (Pussa *et al.*, 2008).
Capsaicin	Capsaicin @ 100ppm in precooked chicken patties (Emrick *et al.*, 2005).
Vitamin E	Spraying of Vit E directly on turkey meat resulted in a lower myoglobin oxidation (Sante & Lacourt, 1994).

Besides these some other natural preservatives derived from plants such as Broccoli, carrot, Brussel sprouts, cabbage, cucumber, mint (flavonoids and polyphenols), oats, olive leaf extract, prune, cocoa leaves (polyphenols), Thyme (1,8 cineole, alpha-terpineol, linalool, carvacrol and thymol), Sesame oil, Cherry (flavones, isoflavones, anthocyanins, anthocyanidins and phenolic glycosides), Rapeseed (α-tocopherol, sinapine and sinapic acid), Black seed, coriander, Potato peel (Phenolics, catechins and chlorogenic acid), Evening Primrose (phenolics), Marjoram (T3b), Bamboo leaves (flavones, lactones and phenolic acids), Beetroot, Cloudberry, Willow herb, Lutein, Raspberry and strawberry (Ellagic acid), Vitamin E (α-tocopherol), Ascorbic acid, Black pepper, Red pepper etc. have also been used successfully in various meat and meat products.

2.2. Natural Preservatives of Animal Origin

Preservatives of animal origin includes mainly lactoferrin, lactoferricin, Lysozyme, Bone protein hydrolysates, Lactoperoxidase, chitosan, Avidin, carnosine, bioactive milk pepetides, Defensins, Cathepsins G, Insect, Attacins, Cecropins, Magainins etc.

2.2.1. *Lactoferrin*

Lactoferrin is an iron chelating bacteriostatic protein and it is a component of phagosome. Lactoferricin also inhibited the growth of *Klebsiella pneumoniae*, *Pseudomonas aeruginosa*, *Listeria monocytogenes* and *Staphylococcus aureus*. Lactoferricin B causes the release of lipopolysaccharides from the cell envelopes of gram negative bacteria, the death of a range of bacteria and candida albicans as well as the inhibition of glucose uptake by dermatophytes, microfungi and yeasts.

2.2.2. *Lactoperoxidase*

Lactoperoxidase system in presence of thiocyanate and hydrogen peroxide shows a strong antimicrobial action. Hydrogen peroxide may be added either directly or indirectly into the system. The hypothiocyanate ion is the major intermediate oxidation product of thiocyanate and the prinicipal antimicrobial agent generated by lactoperoxidase in milk.

2.2.3. *Lysozyme*

Lysozyme was discovered by Alexander Flemming in 1922. It is β-1, 4-N-acetyl-muramidase. It is found in many mammals, birds, fish, insects, plants and viruses. It constitutes 3–4% of albumen proteins. The concentration of lysozyme in hen, duck and goose egg is 3000, 1000 and 1200 μg/g. Hen egg white lysozyme is classified as Generally Recognised as Safe (GRAS) in the United states and is approved for use in some foods in the European Union, Japan and Australia (Losso *et al.*, 2000). Lysozyme from hen egg is extracted commercially by using cation exchange chromatography. Mostly it causes the cell death of gram positive bacteria. The enzyme makes contact with the peptidoglycan in the cell wall, the polymer gets cleaved which in turn leads to osmotic uptake of water and eventual rupturing of cytoplasmic membrane. Since its main cite of action is peptidoglycan in the cell wall, so gram negative bacteria are resistant to it. It has been documented that in combination with a cation binding agent the antimicrobial activity of lysozyme gets increased. The use of lysozyme as a food preservative requires Recombinant DNA technology and protein engineering so as to reduce the production costs and improve its activity. Arima *et al.* (1997) has revealed that upto 500 mg/L of recombinant lysozyme can be produced but this is still very low amount for its commercial use at industry level. Lysozyme activity is affected by various conditions like type of food, processing and storage conditions. Lysozyme has also been evaluated for its use as an antimicrobial packaging component.

2.2.4. *Avidin*

Avidin protein constitutes about 0.05% of the total proteins in the albumen of hen's eggs. Avidin is a glycoprotein whose one molecule binds to four molecules of vitamin biotin. Hence, it inhibits the growth of some bacteria and yeasts that have a requirement for biotin, with the primary mechanism being the nutrient deprivation. According to another scientific study, it has been suggested that it can also bind the porin proteins of *E. coli*. Thus it can also inhibit microbes *in vitro* by interfering with transport through porins.

2.2.5. *Chitosan*

Chitosan is a natural polymer obtained from exoskeletons of crustaceans and also from cell walls of fungi and insects but commercially it is produced from deacetylation of chitin. Chitin, chitosan and their modified derivatives have been applied in medicine, agriculture, food, and non-food industries. In food preservation, chitosan inhibits the growth of food borne bacteria, yeasts and molds. The minimum inhibitory concentration (MIC) for bacteria and yeasts vary widely depending upon the molecular weight of the polymer, degree of acetylation, pH, temperature and presence of interfering substances such as fats and proteins. Usually in combination with the other antimicrobials like benzoates, the MIC of chitosan can be reduced.

2.2.6. *Chitinases*

Chitinases and other hydrolytic enzymes are also having the antimicrobial activity. These enzymes are produced by plants, insects, protozoa and microorganisms. Plants produce endo-acting chitinases, whereas, bacteria produce exo-acting chitinases. The antifungal activity of chitinases and glucanases is well known. The use of animal origin preservatives in poultry meat has been enlisted in the Table 2.

Table 2: Preservatives of animal origin and their use in poultry meat.

Preservative	*Use in poultry meat*
Carnosine	Carnosine @ 1.5% inhibited lipid oxidation in chicken meat patties (O'Neill *et al.*, 1999).
Bioactive milk peptides	Casein peptides @ 20 mg ml^{-1} effectively inhibited lipid peroxidation in ground beef homogenates and mechanically deboned poultry meat (Rossini *et al.*, 2009).
Lysozyme	Pre-surface application of nisin and lysozyme reduced the growth of *Listeria monocytogenes* during refrigerated storage of low-fat turkey bologna (Mangalassary *et al.*, 2008); used as natural antimicrobial in combination with rosemary, oregano oil and EDTA in semi-cooked chicken meat fillets (Ntzimani *et al.*, 2010); Effect of lysozyme action on microbiological stability of confectioned raw poultry meat was investigated (Kijowski *et al.*, 2002); treatment of chicken skin with with 2–100 Mm TSP and lysozyme (10 μg/ml) reduced viable numbers of attached *Campylobacter jejuni*, *E. coli*, *Pseudomonas fluorescens*, and *S. enteritidis* by up to 99.9% depending upon target organism and temperature (Carneiro de Melo *et al.*, 1998); Samuelson *et al.* (1985) also found that combination of lysozyme and EDTA inhibited serovar typhimurium on poultry.

2.3. Natural Preservatives of Microbial Origin

Bacteria produce some inhibitory compounds which can inhibit the growth of potential spoilage or pathogenic microbes. These mainly include bacteriocins, hydrogen peroxide, CO_2, diacetyl, organic acids, reuerin etc. Preservation of meat and meat products by these methods is called as Biopreservation. Biopreservation of meat is of increasing interest for industry and consumers these days. It is a type of preservation in which natural micro flora (mostly lactic acid bacteria) and or their antimicrobial peptides are used so as to enhance the shelf life and or safety of meat and meat products.

2.3.1. *Bacteriocins*

The term "Bacteriocins" was first of all given by Jacob *et al.* (1953) to describe the inhibitory activity of certain microbes against closely related species. Bacteriocinogenic lactic acid bacteria and/or their isolated bacteriocins are considered safe additives (GRAS), useful to control the frequent development of spoilage and pathogenic microbes. Bacteriocins are produced by strains of all genera of lactic acid bacteria of relevance to meats and act against bacteria closely related to the producer organisms. The source of bacteriocins has been given in Table 3.

Table 3: Antioxidants and antimicrobials of microbial origin and their use in poultry meat.

Bacteriocin	***Producing organisms***
Lactocidin, Acidophilin, Lactacin B	*Lactobacillus acidophilus*
Acidocin B, Acidolin J	*Lactobacillus acidophilus*
Plantacin B and N, Plantaricin C	*Lactobacillus plantarum*
Plantaricin SA6, Plantaricin LP84	*Lactobacillus plantarum*
Nisin, Lacticin, Lactococin G	*Lactococcus lactic* subsp. *lactis*
Diplococcin, Lactococcin	*Lactococcus lactic* subsp. *cremoris*
Lactostrepcin 5	*Lactococcus lactic* subsp. *cremoris*
Sakacin, Sakacin P	*Lactobacillus sake*
Bulgaricin	*Lactobacillus delbruceckii* subsp. *bulgaricus*
Curvacin	*Lactobacillus curvatus*
Helveticin J	*Lactobacillus helveticus*
Leucocin	*Leuconostoc gelidum*
Lactacin F	*Lactobacillus johnsonii*
Bifidin	*Bifidobacterium bifidum*
Leuconocin S	*Leuconostoc paramesenteroides*
Caseicin A	*Lactobacillus caseii*
Pediocin AcH, Pediocin PA-1	*Pediococcus acidilactici*
Propionicin, Propionicin PLG-1	*Propionibacterium thoenii*

2.3.2. *Classification of bacteriocins*

Bacteriocins can be classified by various ways in different classes. However, generally they are classified into three broad categories:

(*i*) *Class Ia*: These are small peptides (< 5 kDa) called as lantibiotics, with some unusual amino acids like lanthionine, methyl-lanthionine, dehydrobutyrine and dehydroalanine. *e.g.*, nisin, etc. Lantibiotics are also produced by the genera other than LAB.

(*ii*) *Class Ib*: These are small globular peptides with no net charge or net negative charge. *e.g.*, mersacidin etc.

(*iii*) *Class IIa*: These are small heat stable peptides (non lantibiotics) which are synthesized in a form of precursor which is processed after two glycine residues and are mainly active against *Listeria monocytogens. e.g.*, pediocin PA-1, sakacins A and P, leucocin A etc.

(*iv*) *Class IIb*: Two component system: two different peptides required to form an active poration complex. *e.g.*, Some other bacteriocins like Lactococcins G and F, lactacin F, Plantaricin FE & JK.

(*v*) *Class III*: These are large heat sensitive peptide molecules. *e.g.*, Helveticins J and V-1829, lactacins A and B, acidophilucin A etc.

(*vi*) *Class IV*: These are the peptide molecules having lipid and carbohydrate moieties. *e.g.*, Leuconocin, lactacin 27, Pediocin SJ-I.

2.3.3. *Mechanism of action of bacteriocins*

(*i*) Lantibiotics destroy the bacterial cells by forming pores in the cell membrane thereby depleting the transmembrane potential and causing the leakage of cellular materials (Okereke & Montville, 1992).

(*ii*) Some bacteriocins cause membrane depolarization.

(*iii*) Class II bacteriocins inhibit septum formation in susceptible bacteria (Martinez *et al.*, 2000). *e.g.*, Lactococcin 972.

The efficacy of a bacteriocin differs in different Gram positive bacteria due to differences in membrane composition and fluidity (Bennik *et al.*, 1997). Gram-negative bacteria are protected by their outer membrane, which prevents bacteriocins from reaching the cytoplasmic membrane (Abee *et al.*, 1995). The above mechanisms of action need further scientific investigation.

2.3.4. *Methods of addition of bacteriocins to meat and meat products*

Broadly there are three ways in which bacteriocins can be incorporated into a food to improve its safety:

(*i*) Using a purified/semi-purified bacteriocin preparation as an ingredient in food.

(*ii*) Incorporating an ingredient previously fermented with a bacteriocin-producing strain.

(*iii*) Using a bacteriocin-producing culture to replace all or part of a starter culture in fermented foods to produce the bacteriocin *in situ*.

2.3.5. *Effect of bacteriocins in poultry meat*

The action of a bacteriocin in a food system depends upon its chemical composition and the physical condition. The action of bacteriocins may be reduced by the binding of bacteriocin molecules to food components, by the destabilizing action of proteases and other enzymes and by an uneven distribution in the food matrix (Schillinger *et al.*, 1996). Use of bacteriocins in meat and meat products is somewhat difficult than the liquid foods because of different reasons such as uneven distribution, bacteriocins can partition to the fat phase or may bind to meat proteins. Since lactic acid bacteria (LAB) are commonly found fauna in meat, hence the antimicrobial roles of bacteriocins have been well documented. All bacteriocins act well in meat systems but the role of nisin is well studied (Cleveland *et al.*, 2001).

Nisin was first discovered in England in 1928 as a result of difficulties experienced during cheese making. Nisin may be ineffective in meat due to high pH (Rayman *et al.*, 1983) or improper distribution and interference by meat components such as phospholipids (de Vuyst & Vandamme, 1994). Nisin in combination with chelating agents showed potent bactericidal activity against gram negative bacteria including *Salmonella* and *E. coli*, has led to the investigation whether this combination can can be applied to meat and poultry carcasses (Blackburn *et al.*, 1989; Delves-Broughton, 1993 and Cutter & Siragussa, 1995a). It may also inhibit *Brochothrix thermosphacta* in cold meat-binding system (Cutter & Siragusa, 1998). The use of nisin in spray form for decontamination of meat contact surfaces along the processing lines has also been documented. The nisin concentrate "Nisaplin" is used commercially as a preservative in dairy products, canned foods, cured meats and other segments of fermentation industry. While working with turkey carcasses, Mahadeo and Tatini (1994) documented that nisin at 2.5 μg/L provided synergistic effect on kill of *L. monocytogenes* by heat (52°C for 3 mins) in the scald water.

Shefet *et al.* (1995) successfully used a combination mixture of nisin (100 μg/ml) with chelating agents (EDTA, citric acid) and a surfactant (Tween 20) to reduce *S. typhimurium* contamination on chicken drumsticks and to extend the shelf-life of refrigerated chicken drumsticks. Bell and De Lacy (1986) documented that degree of nisin recovery was poor from minced meat and meat emulsions and it was affected by fat content in it. The use of nisin in raw meat systems is quite limited due to their limited activity against gram negative bacteria. The use of nisin is better understood in case of sausages and pate meats where homogenization of meat allows its even distribution in the meat system. Davies *et al.* (1999) found that nisin was more effective in lean meat and the type of phosphate used in meat affected the activity of nisin against lactic acid bacteria. Nisin at the level of 6.25–25 μg/g effectively inhibited the growth of a mixed lactic acid bacteria and extended the shelf-life of vacuum packaged sausages.

Other bacteriocins such as Leucocin A, enterocins, sakacins and the carnobactericins A and B also prolong the shelf life of fresh meat. Pediocins are more effective than nisin in inhibiting bacteria in meat and are even more effective in dairy products. Dipping meat in 5,000 A.U. of crude Pediocin PA-1 per ml decreases the viability of *L. monocytogenes* 100 to 1000 fold. Pediocin AcH at 1350 A.U/ml reduces listeria populations in ground beef and sausages by 1 to 7 logs. Pediocin AcH is more effective at 4°C than at 25°C against *L. monocytogenes* in wiener exudates. Pediocin AcH (PA-1) successfully controlled the growth of *L. monocytogenes* in raw chicken (Goff *et al.*, 1996). Pediocin may be used in chicken before cooking for maximum effectiveness. Pediocin AcH inhibits *L. monocytogenes* in chicken summer sausages. Besides this, Pediocin PA-1 has also been expressed in the yeast *Saccharomyces cerevisiae* to improve preservation of wine, bread and other food products where yeast is used.

2.4. Bacteriophages

Bacteriophages are the viruses infecting the bacteria. In 1960s it was shown that bacteriophages can destroy their host cells recovered from fish, meats and skim milk. Goode *et al* (2003) documented that application of specific phages on excised chicken skin can reduce *S. enteritidis* and *Camplobacter jejuni*. Atterbury *et al.* (2003) demonstrated the efficacy of host specific phages to reduce the numbers of *C. jejuni* on artificially contaminated chicken skin. Some important limitation of this technique are a limited host range, the requirement for threshold numbers of the bacterial targets, phage-resistant mutants, and the potential for the transduction of undesirable characteristics from one bacterial strain to another. Phages have been shown to reduce the numbers of *E. coli* O157:H7 and Salmonellae on fresh poultry. Kennedy *et al.* (1984) reported the apparent reduction of numbers of viable *E. coli* from fresh poultry where coliphages were quite common.

2.5. Hydrogen Peroxide

Hydrogen peroxide (H_2O_2) is Generally Recognised as Safe (GRAS) for use in food products as a bleaching, oxidizing, reducing and antimicrobial agent. The effectiveness of H_2O_2 to destroy the microorganisms depends on several factors including pH, temperature and environmental conditions. Under normal conditions, H_2O_2 itself gets decomposed into oxygen and water thereby leaving no toxic residues in it. It has bactericidal or bacteriostatic effect which is based on the formation of radicals that damage nucleic acids, proteins and lipids. Hydrogen peroxide @ 0.5% w/v in water acts as poultry carcass decontaminant (Bolder, 1997). A combination spray of hydrogen peroxide and sodium bicarbonate resulted in increased shelf-life of poultry (Fletcher *et al.*, 1993). The application of hydrogen peroxide seems to be an effective and safe method for carcass decontamination to control the spread of pathogens.

3. REGULATORY ASPECTS OF NATURAL PRESERVATIVES

According to FDA regulations, there should not be residual H_2O_2 present in the foods under any circumstance. LABs are GRAS for the production of fermented foods. GRAS status was conferred by U.S. Food and Drug Administeration (FDA) is desirable because it allows a compound to be used in a specific application without additional regulatory approval. Nisin has been approved for use in all the foods internationally. The Joint Food and Agriculture/ World Health Organization accepted nisin as a food additive in 1969 and set the maximum intake level as 33,000 I.U. per kg body weight. That's why many countries allow nisin in a variety of products, sometimes with no restrictions as to maximum level. Recently FDA has given the approval of Listeria-specific bacteriophage preparations for food preservation. The maximum tolerance of some GRAS chemical food preservatives has been given in Table 4.

Table 4: Summary of some GRAS chemical food preservatives.

Preservatives	***Maximum tolerance***	***Organisms affected***	***Foods***
Propionic acid/ propionates	0.32%	Molds	Bread, cakes, some cheeses, rope inhibitor in bread dough
Sorbic acid/ sorbates	0.2%	Molds	Hard cheeses, Figs, syrups, salad dressings, jellies, cakes.

Table 4: (*Contd...*)

Table 4: *(Contd...)*

Preservatives	***Maximum tolerance***	***Organisms affected***	***Foods***
Benzoic acid/ Benzoates	0.1%	Yeasts and Molds	Margarine, pickle relishes, apple cidar, soft drinks, salad dressings.
Parabens*	0.1%**	Yeasts and Molds	Bakery products, soft drinks, pickles, salad dressings.
SO2/ sulfites	200–300 ppm	Insects, micro-organisms	Molasses, dried fruits, wine making, lemon juice (not to be used in meats or other foods recognized as sources of thiamine).
Ethylene/ propylene oxides***	700 ppm	Yeasts, molds, vermin	Fumigant for spices, nuts
Sodium diacetate	0.32%	Molds	Bread
Nisin	1%	Lactics, clostridia	Certain pasteurized cheese spreads
Dehydroacetic acid	65 ppm	Insects	Pesticide on strawberries, squash
Sodium nitrite	120 ppm	Clostridia	Meat curing preparations
Caprylic acid	—	Molds	Cheese wraps
Sodium lactate	Up to 4.8%	Bacteria	Pre-cooked meats
Ethyl formate	15–220 ppm****	Yeasts and Molds	Dried fruits, nuts

Note: GRAS (Generally recognized as Safe) per section 201[32] (s) of the U.S. Food, Drug, and Cosmetic act as amended.
*Methyl-, propyl-, and heptyl- esters of p-hydroxybenzoic acid.
**Heptyl-ester-12 ppm in beers; 20 ppm in noncarbonated and fruit based beverages.
***May be involved in mutagenesis and/or carcinogenesis.
****As formic acid.

4. HEALTH BENEFITS OF NATURAL PRESERVATIVES

Natural antimicrobials though primarily used for killing or inhibiting number of bacterial species, have several beneficial effects. The use of LAB as dietary adjunct to ameliorate lactose intolerance is well documented. LAB in fermented foods in addition to their antimicrobial activity, produce desirable changes in texture, flavour and taste of the foods. Since these are commonly present in various fermented foods it is assumed that these bacteria do not pose any health risk to man and are designated as "GRAS" (Jeevaratnam *et al.*, 2005). Curry leaves commonly use as flavouring agent in many traditional dishes are gaining attention because of its medicinal values such as anticarcinogenic (Chevallier, 1996), antiplatelets and vasorelaxing effects (Wu *et al.*, 1998). Prochaska and Talalay (1992) demonstrated that phenolics and their derivatives present in green and black tea have protective effects against liver cirrhosis, emphysema and in cancer. Nonetheless, addition of plant phenols susceptible secondary amines and amides to form potent carcinogenic nitrosamines and nitrosamides, in-vivo as well as in-vitro during different processing aids of cured and smoked meat products. Similarly hypertensive effects of oleuropein in olive leaves and fruits were also reported.

Addition of bacteriocins to meat and milk products has been permitted by FDA and WHO. Moreover, nisin, pediocin and lactococcin are more active and stable at low pH and high temperature, respectively. So they have been given GRAS (generally regarded as safe) status and are naturally present in many food products. Lycopene helps in the prevention of cardiovascular disease and cancers of the prostate or gastrointestinal tract (Clinton, 1998). Ginger is anti-inflammatory and prevents against ovarian and colorectal cancer. Garlic reduces the risk of cancer, stroke and aging. Bioactive milk peptides may act as opioid agonists, opioid antagonists, antihypertensive agents, immunostimulants, antithrombotic, mineral carriers etc (Shah, 2000). Carnosine, vitamin E, Ascorbic acid are potent antioxidants. Chitosan acts as a fat binder and thereby helps to lose body weight. It is also shown to have beneficial effect on cholesterol, ulcers, osteoporosis and blood pressure.

REFERENCES

Abee, T., Krockel, L. and Hill, C. (1995). Bacteriocins: Mode of action and potential in food preservation and control of food poisoning. *Int. J. Food Microbiol.*, 28: 169–185.

Arima, H., Ibrahim, H.R., Kinoshita, T. and Kato. A. (1997). Bactericidal action of lysozymes attached with various sizes of hydrophobic pepetides to the C-terminal using genetic modification. *FEBS Lett.*, 415: 114–118.

Atterbury, R.J., Connerton, P.L., Dodd, C.E.R., Rees, C.E.D. and Connerton, I.F. (2003). Isolation and characterization of campylobacter bacteriophages from retail poultry. *Appl. Environmental Mirobiol.*, 69: 4511–4518.

Bell, R.G. and De Lacy, K.M. (1986). Factors influencing the determination of nisin in meat products. *J. Food Technol.*, 21: 1–7.

Bennik, M., Verheul, A., Abee, T., Naaktgeboren-Stoffels, G., Gorris, L.G. and Smid, E.J. (1997). Interactions of nisin and pediocin PA-1 with closely related lactic acid bacteria that manifest over 100-fold differences in bacteriocin sensitivity. *Appl. Environmental Microbiol.*, 63: 3628–3636.

Blackburn, P., Polak, J., Gusik, S. and Rubino, S.D. (1989). Nisin compositions for use as enhanced, broad range bacteriocins. International application No. PCT/US89/02625; Applied Microbiology Inc., New York, USA.

Bolder, N.M. (1997). Decontamination of meat and poultry carcasses. *Trends Food Sci. Technol.*, 8: 221–227.

Burt, S. (2004). Essential oils: Their antibacterial properties and potential applications in foods-A review. *Int. J. Food Microbiol.*, 94: 223–253.

Carlos, A.M.A. and Harrison, M.A. (1999). Inhibition of selected microorganisms in marinated chicken by pimento leaf oil and clove oleoresin. *J. Appl. Poult. Res.*, 8: 100–109.

Carneiro de Mel, A.M.S., Cassar, C. and Miles, R.J. (1998). Tri-sodium phosphate increases sensitivity of gram negative bacteria to lysozyme and nisin. *J. Food Prot.*, 61: 839–844.

Chouliara, E., Karatapanis, A., Savvaidis, I.N. and Kontominas, M.G. (2007). Combined effect of oregano essential oil and modified atmosphere packaging on shelf-life extension of fresh chicken breast meat stored at 4°C. *Food Microbiol.*, 24: 607–617.

Chevallier, A. (1996). The encyclopedia of medicinal plants. Darling Kindersley, London, UK.

Cleveland, J., Montville, T.J., Nes, I.F. and Chikindas, M.L. (2001). Bacteriocins: Safe, natural antimicrobials for food preservation. *Int. J. Food Microbiol.*, 71(1): 1–20.

Clinton, S.K. (1998). Lycopene: Chemistry, biology, and implications for human health and disease. *Nutr. Rev.*, 56: 35–51.

Cutter, C.N. and Siragusa, G.R. (1998). Incorporation of nisin into a meat binding system to inhibit bacteria on beef surfaces. *Lett. Appl. Microbiol.*, 27: 19–23.

Cutter, C.N. and Siragusa, G.R. (1995a). population reductions of gram negative pathogens following treatments with nisin and chelators under various conditions. *J. Food Prot.*, 58: 977–983.

Delves-Broughton, J. (1993). The use of EDTA to enhance the efficacy of nisin towards Gram negative bacteria. *Int. J. Biodeterioration and Biodegradation*, 32: 87–97.

de Vuyst, L. and Vandamme, E. (1994). Nisin, a lantibiotic produced by *Lactococcus lactis* subsp. lactis: Properties, biosynthesis and applications. In: L. de Vuyst and E. Vandamme, Editors, *Bacteriocins of Lactic Acid Bacteria Microbiology, Genetics* and *Applications*, Blackie Academic and Professional, London (1994), pp. 151–221.

El-Alim, S.S.L.A., Lugasi, A., Hovari, J. and Dworschak, E. (1999). Culinary herbs inhibit lipid oxidation in raw and cooked minced meat patties during storage. *J. Sci. Food Agri.*, 79(2): 277–285.

Emrick, M.E., Penfield, M.P., Bacon, C.D., Van Laack, R.V. and Brekke, C.J. (2005). Heat intensity and warmed-over flavor in precooked chicken patties formulated at 3 fat levels and 3 pepper levels. *J. Food Sci.*, 70: S600–S604.

Fernández-López, J., Fernández-Ginés, J.M., Aleson-Carbonell, L., Sendra, E., Sayas-Barberá, E. and Pérez-Alvarez, J.A. (2004). Application of functional citrus by-products to meat products. *Trends Food Sci. Technol.*, 15: 176–185.

Fletcher, D.L., Russell, S.M., Walker, J.M. and Baily, J.S. (1993). An evaluation of a rinse procedure using sodium bicarbonate and hydrogen peroxide on the recovery of bacteria from broiler carcasses. *Poult. Sci.*, 72: 2152–2156.

Frankel, E.N. (1984). Lipid oxidation: Mechanism, products and biological significance. *J. AOCS*, 61: 1908–1917.

Frankel, E.N. (1985). Chemistry of autoxidation: Mechanism, products and flavor significance, D.B Min, T Smouse, Editors , Flavor chemistry of fats and oils, American Oil Chemist Society Press, Champaign, Illinois (1985), pp. 1–37.

Goode, D., Allen, V.M. and Barrow, P.A. (2003). Reduction of experimental Salmonella and Campylobacter contamination of chicken skin by application of lytic bacteriophages. *Appl. Env. Mirobiol.*, 69: 5032–5036.

Goff, J.H., Bhunia, A.K. and Johnson, M.G. (1996). Complete inhibition of low levels of *Listeria monocytogenes* on refrigerated chicken meat with Pediocin AcH bound to heat-killed *Pediococcus acidilactici* cells. *J. Food Prot.*, 59: 1187–1192.

Govaris, A., Botsoglou, N., Papageorgiou, G., Botsoglou, E. and Ambrosiadis, I. (2004). Dietary versus post-mortem use of oregano oil and/or á-tocopherol in turkeys to inhibit development of lipid oxidation in meat during refrigerated storage. *Int. J. Food Sci. Nutri.*, 55(2): 115–123.

Gray, J.I. (1978). Measurement of lipid oxidation: A review. *J. AOCS.*, 55: 539–546.

Hao, Y.Y., Brackett, R.E. and Doyle, M.P. (1998). Efficacy of plant extracts in inhibiting Aeromonas hydrophila and Listeria monoctyogenes in refrigerated, cooked poultry. *Food Microbiol.*, 15: 367–378.

Hashim, I.B., McWatters, K.H. and Hung, Y.C. (1999). Marination method and honey level affect physical and sensory characteristics of roasted chicken. *J. Food Sci.*, 64:163–166.

Jacob, F., Lwoff, A., Siminowitch, A. and Wollman, E. (1953). Definition de quelques termes relatives a la lysogenie. *Annales des Institute Pasteur Paris*, 84: 222–224.

Jeevaratnam, K., Jamuna, M. and Bawa, A.S. (2005). Biological preservation of foods-Bacteriocins of Lactic acid bacteria. *Indian J. Biotechnol.*, 4: 446–454.

Karastogiannidou, C. (1999). Effects of onion quercetin on oxidative stability of cook-chill chicken in vacuum-sealed containers. *J. Food Sci.*, 64(6): 978–981.

Karpinska-Tymoszczyk, M. (2007). Effects of sage extract (*Salvia officinalis* L.) and a mixture of sage extract and sodium isoascorbate on the quality and shelf life of vacuum-packaged turkey meat balls. *J. Muscle Foods*, 18: 420–434.

Kennedy, J.E.Jr, Oblinger, J.L. and Bitton, B. (1984). Recovery of coliphages from chicken, pork sausage and delicatessen meats. *J. Food Prot.*, 47: 623–626.

Kijowski, J., Marciszewska, C. and Cegielska-Radziejewska, R. (2002). Quality and microbiological stability of chilled chicken breast muscles treated with lysozyme solution. *Pol. J. Food Nutr. Sci.*, 11(2): 47–54.

Kumar, D. and Tanwar, V.K. (2011). Utilization of clove powder as phytopreservative for chicken nuggets preparation. *Journal of Stored Products and Postharvest Research*, 2(1): 11–14.

Ladikos, D. and Lougovois, V. (1990). Lipid oxidation in muscle foods: A review. *Food Chem.*, 35: 295–314.

Lau, D.W. and King, A.J. (2003). Pre- and post-mortem use of grape seed extract in dark poultry meat to inhibit development of thiobarbituric acid reactive substances. *J. Agric. Food Chem.*, 51: 1602–1607.

Lee, S.C., Nam, K.C. and Ahn, D.U. (2004). Antioxidant properties of far infrared-treated rice hull extract in irradiated raw and cooked turkey breast. *Iowa State University Animal Industry Report.*

Losso, J.N., Nakai, S. and Charter, E.A. (2000). Lysozyme. *In*: *Natural Food Antimicrobial Systems* (*Ed*. Naidu, A.S.), CRC Press, Boca Raton, pp. 185–210.

Mahadeo, M. and Tatini, S.R. (1994). The potential use of nisin to control *Listeria monocytogenes* in poultry. *Lett. Appl. Microbiol.*, 18: 323–326.

Maher, M., O'Grady, M.N., Buckley, D.J., Troy, D., Moloney, A.P. and Kerry, J.P. (2002). Effect of dietary supplementation and direct addition of tea catechins and rosemary on the oxidative stability of beef. In *Proceedings of the 48th international congress of meat science and technology* (pp. 500–501), 25–30 August 2002, Rome, Italy.

Mangalassary, S., Han, I., Rieck, J., Acton, J. and Dawson, P. (2008). Effect of combining nisin and/or lysozyme with in-package pasteurization for control of *Listeria monocytogenes* in ready-to-eat turkey bologna during refrigerated storage. *Food Microbiol.*, 25(7): 866–870.

Martinez, B., Rodriguez, A. and Suarez, J.E. (2000). Lactococcin 972, a bacteriocin that inhibits septum formation in lactococci. *Microbiol.*, 146: 949–955.

Mariutti, L.R.B., Orlien, V., Bragagnolo, N. and Skibsted, L.H. (2008). Effect of sage and garlic on lipid oxidation in high-pressure processed chicken meat. *European Food Research and Technology*, 227: 337–344.

McKibben, J. and Engeseth, N.J. (2002). Honey as a protective agent against lipid oxidation in ground turkey. *J. Agric. Food Chem.*, 50(3): 592–595.

Mytle, N., Anderson, G.L., Doyle, M.P. and Smith, M.A. (2006). Antimicrobial activity of clove (*Syzgium aromaticum*) oil in inhibiting *Listeria monocytogenes* on chicken frankfurters. *Food Control*, 17(2): 102–107.

Nam, K.C., Kim, J.H., Ahn, D.U. and Lee, S.C. (2004). Far-infrared radiation increases the antioxidant properties of rice hull extract in cooked turkey meat. *J. Agric. Food Chem.*, 52(2): 374–379.

Naveena, B.M. and Mendiratta, S.K. (2001). Tenderisation of spent hen meat using ginger extract. *Brit Poult. Sci.*, 42(3): 344–49.

Naveena, B.M., Sen, A.R., Vaithiyanathan, S., Babji, Y. and Kondaiah, N. (2008). Comparative efficacy of pomegranate juice, pomegranate rind powder extract and BHT as antioxidants in cooked chicken patties. *Meat Sci.*, 80(4): 1304–1308.

Ntzimani, A.G., Giatrakou, V.I. and Savvaidis, I.N. (2010). Combined natural antimicrobial treatments (EDTA, lysozyme, rosemary and oregano oil) on semi cooked coated chicken meat stored in vacuum packages at 4°C: Microbiological and sensory evaluation. *Innovative Food Science and Emerging Technologies*, 11(1): 187–196.

Okereke, A. and Montville, T.J. (1992). Nisin dissipates the proton motive force of the obligate anaerobe *Clostridium sporogenes* PA 3679. *Appl. Environ. Microbiol.*, 58: 2463–2467.

Oliveira, K.A.M., Santos-Mendonca, R.C., Gomide, L.A.M. and Vanetti, M.C.D. (2005). Aqueous garlic extract and microbiological quality of refrigerated poultry meat. *J. Food Proc. Preserv.*, 29(2): 98–08.

O'Neill, L.M., Galvin, K., Morrissey, P.A. and Buckley, D.J. (1999). Effect of carnosine, salt and dietary vitamin E on the oxidative stability of chicken meat. *Meat Sci.*, 52(1): 89–94.

O'Sullivan, C.M., Lynch, A.M., Lynch, P.B., Buckley, D.J. and Kerry, J.P. (2004). Use of antioxidants in chicken nuggets manufactured with and without the use of salt and/or sodium tripolyphosphate: Effects on product quality and shelf-life stability. *Int. J. Poultry Sci.*, 3(5): 345–353.

Prochaska, H.J. and Talalay, P. (1992). Phenolic antioxidants as inducers of anticarcinogenic enzymes. *In*: Phenolic compounds in Food and Their Effect on Health I, Analysis, Occurance, and Chemistry (*Eds.* Chi-Tang, Ho., Lee Chang, Y. and Mou-Tuan, Huang.). ACS symposium series, Wsahington, DC.

Püssa, T., Pällin, R., Raudsepp, P., Soidla, R. and Rei, M. (2008). Inhibition of lipid oxidation and dynamics of polyphenol content in mechanically deboned meat supplemented with sea buckthorn (*Hippophae rhamnoides*) berry residues. *Food Chem.*, 107(2): 714–721.

Rayman, K., Malik, N. and Hurst, A. (1983). Failure of nisin to inhibit outgrowth of *Clostridium botulinum* in a model cured meat system. *Appl. Environ. Microbiol.*, 46: 1450–1452.

Riznar, K., Celan, S., Knez, Z., Skerget, M., Bauman, D. and Glaser, R. (2006). Antioxidant and antimicrobial activity of rosemary extract in chicken frankfurters. *J. Food Sci.*, 71(7): C425–C429.

Romero, A.M., Doval, M.M., Romero, M.C., Sturla, M.A. and Judis, M.A. (2008). Antioxidant properties of dry soya sprout hydrophilic extracts. Application on cooked chicken patties. *Electronic Journal f Environmental, Agricultural and Food Chemistry*, 7(8): 3196–3206.

Rongsensusang, S.R.K., Kondal, R.K. and Dhana, L.K. (2005). Effect of ginger on quality of frozen spent hen meat balls. *J. Food Sci. Technol.*, 42(6): 534–39.

Rossini, K., Noreña, C.P.Z., Cladera-Olivera, F. and Brandelli, A. (2009). Casein peptides with inhibitory activity on lipid oxidation in beef homogenates and mechanically deboned poultry meat. *LWT—Food Sci. Technol.*, 42(4): 862–867.

Sallam, K.I., Ishioroshi, M. and Samejima, K. (2004). Antioxidant and antimicrobial effects of garlic in chicken sausage. *Lebensmittel-Wissenschaft und-Technologie*, 37(8): 849–55.

Samuelson, K.J., Rupnow, J.H. and Froning, G.W. (1985). The effect of lysozyme and EDTA on Salmonella on broiler parts. *Poult. Sci.*, 64: 1488–1490.

Sante, V.S. and Lacourt, A. (1994). The effect of dietary tocopherol supplementation and antioxidant spraying on color stability and lipid oxidation of turkey meat. *J. Sci. Food Agri.*, 65(4): 503–507.

Schillinger, U., Geisen, R. and Holzapfel, W.H. (1996). Potential of antagonistic microorganisms and bacteriocins for the biological preservation of foods. *Trends Food Sci. Technol.*, 7: 158–164.

Shah, N.P. (2000). Effects of milk-derived bioactives: An overview. *Brit. J. Nutri.*, 84(1): S3–S10.

Shefet, S.M., Sheldon, B.W. and Klaenhammer, T.R. (1995). Efficacy of optimized nisin based treatments to inhibit Salmonella typhimurium and extend shelf-life of broiler carcasses. *J. Food Prot.*, 58: 1077–1082.

Shekarforoush, S.S., Nazer, A.H.K., Firouzi, R. and Rostami, M. (2007a). Effects of storage temperatures and essential oils of oregano and nutmeg on the growth and survival of *Escherichia coli* O157:H7 in barbecued chicken used in Iran. *Food Control*, 18(11): 1428–33.

Tang, S., Sheehan, D., Buckley, D.J., Morrissey, P.A. and Kerry, J.P. (2001). Antioxidant activity of added tea catechins on lipid oxidation of raw minced red meat, poultry and fish muscle. *Int. J. Food Sci. Technol.*, 36: 685–692.

Tang, S.Z., Kerry, J.P., Sheehan, D., Buckley, D.J. and Morrissey, P.A. (2000). Dietary tea catechins and iron-induced lipid oxidation in chicken meat, liver and heart. *Meat Sci.*, 56(3): 285–290.

Tang, X. and Cronin, D.A. (2007). The effects of brined onion extracts on lipid oxidation and sensory quality in refrigerated cooked turkey breast rolls during storage. *Food Chem.*, 100: 712–718.

Vaithiyanathan, S., Naveena, B.M., Muthukumar, M., Girish, P.S. and Kondaiah, N. (2011). Effect of dipping in pomegranate (*Punica granatum*) fruit juice phenolic solution on the shelf life of chicken meat under refrigerated storage (4°C). *Meat Sci.*, 88(3): 409–414.

Yadav, A.S., Pandey, N.K., Singh, R.P. and Sharma, R.D. (2002). Effect of garlic extract and cinnamon powder on microbial profile and shelf-life of minced chicken meat. *Indian J. Poult. Sci.*, 37: 72–77.

Yilmaz, I., Şimşek, O. and Işikli, M. (2002). Fatty acid composition and quality characteristics of low-fat cooked sausages made with beef and chicken meat, tomato juice and sunflower oil. *Meat Sci.*, 62: 253–258.

Younathan, M.T., Marjan, Z.M. and Arshad, F.B. (1980). Oxidative rancidity in stored ground turkey and beef. *J Food Sci.*, 45: 274–275, 278.

Wu, T.S., Chan, Y.Y., Liou, M.J., Lin, F.W., Shi, L.S. and Chen, K.T. (1998). Platelet aggregation inhibitor from. *Murraya keonigii. Phytotherapy Research*, 12: 380–382.

11

Functional Starters for Dairy Products

S. Mandal, S. Hati and K. Khamrui

ABSTRACT

The production of fermented foods is based on the use of starter cultures, for instance lactic acid bacteria that initiate rapid acidification of the raw material. Recently, new starter cultures of lactic acid bacteria with an industrially important functionality are being developed. The latter can contribute to the microbial safety or offer one or more organoleptic, technological, nutritional, or health advantages. Examples are lactic acid bacteria that produce antimicrobial substances, sugar polymers, sweeteners, aromatic compounds, vitamins, or useful enzymes, or that have probiotic properties. Novel insights into the metabolism of LAB offer perspectives for the application of a new generation novel starter cultures. Functional LAB starters may offer several health, marketing, and technological advantages. However, fundamental and applied research is still needed to optimally implement functional starter cultures in the existing production technology and to obtain quantitative data. Mathematical analysis of the kinetics of functional starter cultures may yield precious information about the relationship between the food environment and bacterial functionality, and may contribute to optimal strain selection and process design. This may result in better process control, enhanced food safety and quality, and reduction of economic losses.

1. INTRODUCTION

Lactic acid bacteria (LAB) are the most important bacteria used in food fermentations. Apart from general demands for starter cultures from the view of safety, technological effectiveness and economics, numerous specific aspects have to be considered when selecting strains for the different food fermentations. Therefore, selection criteria for LAB depend on the type and the desired characteristics of the final product, the desired metabolic activities, the characteristics of the raw materials and the applied technology. A starter culture can be defined as a microbial preparation of large numbers of cells of at least one microorganism to be added to a raw material to produce a fermented food by accelerating and steering its fermentation process. The group of lactic acid bacteria (LAB) occupies a central role in these processes, and has a long and safe history of application and consumption in the production of fermented foods and beverages. They cause rapid acidification of the raw material through the production of organic acids, mainly lactic acid. Also, their production of acetic acid, ethanol, aroma compounds, bacteriocins, exopolysaccharides, and several enzymes is of importance. In this way they enhance shelf life and microbial safety, improve texture, and contribute to the pleasant sensory profile of the end product. Lactic acid bacteria, propionibacteria, surface-ripening bacteria, yeasts, and moulds are used as starter cultures for manufacturing of various fermented milk products. Starter cultures have a multifunctional role in dairy fermentations. The production of lactic acid, by fermenting lactose is the major role of dairy starters. The acid is responsible for development of characteristic body and texture of the fermented milk products, contributes to the overall flavour of the products, and enhances preservation. Diacetyl, acetaldehyde, acetic acid, also produced by the lactic starter cultures, contribute to flavour and aroma of the final product. Carbon-di-oxide produced by some hetero-fermentative lactic acid bacteria involves in very characteristics texturization in some fermented dairy products, *viz.*, "eye" formation in cheeses. Development of flavour and changes in texture during ripening of cheeses is associated with enzymes originating from bacterial and fungal cultures, depending on the cheese variety. Dairy starters are also having some direct or indirect functional health promoting attributes, such as live probiotics, prebiotic exopolysaccharides and oligosaccharides, bioactive peptides and lipids, etc. Most of the cultured dairy products are produced using commercial starter cultures that have been selected for a variety of desirable properties in addition to rapid acid production. These may include flavour production, lack of associated off flavours, bacteriophage tolerance, ability to produce flavour during cheese ripening, salt tolerance, exo-polysaccharide (EPS) production, bacteriocin production, temperature sensitivity, etc.

2. NOVEL DAIRY STARTERS AND THEIR APPLICATION

Milk contains many health promoting constituents including immunoglobulin, bioactive fatty acids and peptides etc. The healthy image of milk has resulted in dramatic growth in the diversification of dairy products in recent years and in huge increase in the varieties of products such as dairy desserts, flavoured milk drinks, cheeses, yoghurt etc. Apart from the milk components, the health attributes are associated with fermented and probiotic milks and dairy products. Milk has been preserved by fermentation through the action of lactic acid bacteria (LAB), which convert lactose to lactic acid and other organic acids, thereby lowering the pH and subsequently inhibiting the growth of pathogenic and spoilage bacteria. Moreover, these LAB produce a range of secondary metabolites that can influence the products' flavour, aroma and texture as well as antimicrobial peptides. These bacteria also possess a diverse complement of proteases and peptidases that aid in digestion of milk proteins. In addition, many bifidobacteria and lactobacilli are increasingly exploited in probiotic dairy products such as cheeses, yoghurt, milk drinks etc.

A novel trend in the food industry is to explore the use of functional starter cultures of LAB for the manufacture of fermented foods. Functional starter cultures are defined as starters that possess at least one inherent, functional property, aimed at improving the quality of the end product. Functional starter cultures of lactic acid bacteria are defined as strains that are used as starter cultures for food fermentation, and that are able to express certain functional properties which give an added value to the end-product (*e.g.*, bacteriocin producers, exopolysaccharide producers, probiotic strains etc.). This functionality can contribute to microbial safety, or offer one or more organoleptic, technological, or nutritional and health advantages to the food. Promising examples are LAB that produces antimicrobial substances such as bacteriocins to assure food safety, sugar polymers to improve texture, desirable aromatic compounds to enhance taste properties, or strains that display probiotic effects (Table 1). To develop such cultures, the biodiversity of traditional fermented foods and spontaneous fermentation processes is analyzed.

2.1. Lactose-Negative Starters

In yoghurt production, lactose is converted by a yoghurt culture into lactic acid until a final pH of 4.2–4.5 is achieved. Upon storage, the pH can decrease below 4.0. This undesirable post-acidification, ascribed to *Lb. delbrueckii* subsp. *bulgaricus*, leads to an acid and bitter taste. Lactose-negative mutants of *Lb. delbrueckii* subsp. *bulgaricus* enable production of mild yoghurts since such cells can, given their proto-co-operation; only grow in the presence of actively lactose fermenting *S. thermophilus*.

Table 1: Starter cultures for fermented foods and beverages.

1. Hard cheeses without eyes	*Lactococcus lactis* ssp. *lactis, L. lactis* ssp. *cremoris*
2. Cheeses with small eyes	*Lactococcus lactis* ssp. *lactis, Lactococcus lactis* ssp. *lactis* var. *diacetylactis, L. lactis* ssp. *cremoris, Leuconostoc mesenteroides* ssp. *cremoris*
3. Swiss and Italian type cheeses	*Lactobacillus delbrueckii* ssp. *lactis, L. helveticus, L. casei, L. delbrueckii* ssp. *bulgaricus, Streptococcus thermophilus*
4. Butter and buttermilk	*Lactococcus lactis* ssp. *lactis, Lactococcus lactis* ssp. *lactis* var. *diacetylactis, L. lactis* ssp. *cremoris, Leuconostoc mesenteroides* ssp. *cremoris*
5. Yoghurt	*L. delbrueckii* ssp. *bulgaricus, Streptococcus thermophilus*
6. Fermented, probiotic milk	*L. casei, L. acidophilus, L. rhamnosus, L. johnsonii, Bifidobacterium lactis, B. bifidum, B. brevis*
7. Kefir	*L. kefir, L. kefiranofacies, L. brevis*
8. Kumiss	

3. METABOLITES PRODUCED BY LAB

LAB has relatively simple homo- or hetero- fermentative metabolism. These bacteria rely on lactose as their main carbohydrate source. Dairy LAB includes members of the genera *Lactobacillus, Lactococcus, Leuconostoc* and *Sterptococcus*. LAB fermentation yields primarily lactic acid, which plays a vital function in safeguarding food products. LAB metabolism beneficially affects the texture and flavour of fermented foods. The viscosity and texture of fermented dairy products can be greatly enhanced by the production of polysaccharides by LAB, while compounds such as diacetyl, ethanol, acetaldehyde, etc play vital roles in flavour development (Table 2).

In addition, many LABs produce compounds of human nutritional value as regular end products in their metabolisms, including some B-vitamins. Many strains of LAB and bifidobacteria produce other metabolites that promote human health. Bioactive peptides generated from milk proteins as a results of their proteinase and peptidases activity and production of CLA from linoleic acid by strains of lactobacilli and bifidobacteria. For instance, the meat isolates *Lb. sakei* CTC 494 and *Lb. curvatus* LTH 1174 turned out to be promising novel bacteriocin-producing starters for sausage fermentation, whereas *Lb. amylovorus* DCE 471 seemed more suitable as a starter for industrial sourdough fermentation. *E. faecium* RZS C5 has potential to be used as a bacteriocin-producing coculture for food fermentations. Moreover, modelling indicated how exopolysaccharide production by *S. thermophilus* LY03 in yoghurt may be optimised by adjusting the process conditions.

Table 2: Novel starter cultures and their applications.

Starter bacteria	***Functionality***	***Benefits***
Lactococcus lactis ssp. *lactis, L. lactis* ssp. *cremoris, Enterococcus* spp., *L. carvatus, L. sakei, P. acidilactici, E. faecium, L. plantarum, L. ruteri, Streptococcus thermophilus*	Bacteriocins production	For bio-preservation of foods
Several EPS producing lactic acid bacteria (lactobacilli and Streptococci, Lactococci)	Exo-polysaccharides production	Good body and texture of low fat fermented dairy products Probiotics for probiotics Health beneficial attributes
Galactose fermenting lactobacilli and streptococci	Good galactose utilization	Low level of galactose in fermented milk products Low browning in mozzarella cheese
Lactose negative *L. delbrueckii* ssp. *bulgaricus*	Prevention of over acidification in yoghurt	Good body and texture Curd syneresis prevented

Table 2: *(Contd...)*

Table 2: *(Contd...)*

Starter bacteria	***Functionality***	***Benefits***
Autolysing lactic acid bacteria	Enhanced proteolytic and lipolytic activities	Acceleration of ripening of cheeses Production of bioactive milk peptides
Mannitol, sorbitol producing lactic acid bacteria (*Leuconostoc*)	Production of low calorie sugars in fermented milks	Reduced calorie misthi dahi Health beneficial attributes
Probiotic lactic acid bacteria	Different functional attributes	Probiotic functional foods and dairy products
Vitamins producing lactic acid bacteria (*Strepotococci*) and propionibacteria	Improved the vitamin content in fermented dairy products	Good health to the consumers Improve vitamin malnutrition

3.1. Production of Aroma and Flavour

LAB contributes to the aroma and flavour of fermented products. They acidify the food, resulting in a tangy lactic acid taste, frequently exert proteolytic and lipolytic activities, and produce aromatic compounds. Wild strain starter cultures and NSLAB play an important role in flavour formation because they have a high biosynthetic capacity and produce aromatic compounds. The addition of NSLAB as adjunct cultures for cheese manufacturing increases the level of free amino acids, peptides, and free fatty acids, leading to flavour intensity and accelerated cheese ripening. Homofermentative LAB converts the available energy source (sugar) almost completely into lactic acid via pyruvate to produce energy and to equilibrate the redox balance. However, pyruvate can lead to the generation of many other metabolites such as acetate, ethanol, diacetyl, and acetaldehyde. In this way, LAB produce volatile substances that contribute to the typical flavour of certain fermented products, such as sourdough (determined by the lactate/acetate ratio), kefir and kumis (ethanol), butter and buttermilk (diacetyl), and yoghurt (acetaldehyde).

3.2. EPS Production and Texture Improvement

Long-chain, high-molecular-mass polymers that dissolve or disperse in water to give a thickening or gelling properties are indispensable tools in food products formulation. Such polymers are also used for secondary effects that include emulsification, stabilization, and suspension of particulates, control of crystallization, and inhibition of syneresis, encapsulation, and film formation in foods. Most of the biothickeners in current use by the food industry are polysaccharides from plants (*e.g.*, starch, pectin, locust bean gum, guar gum) or seaweeds (*i.e.*, carrageenan, alginate). The animal proteinaceous hydrocolloids gelatin and casein are also used. The functional properties of these polymers in foods are determined by quite subtle structural characteristics. However, these polysaccharides may not always be readily available in the quality needed or their rheological properties may not exactly match those required. Alternative biothickeners are microbial exopolysaccharides, either associated with cell surface in the form of capsules or secreted extracellularly in the form of slime, referred to as capsular or slime exopolysaccharides, respectively. Microbial exopolysaccharides are biothickeners that can be added to a wide variety of food products, where they serve as viscosifying, stabilizing, emulsifying or gelling agents. Many food grade microorganisms also produce exopolysaccharides, in lactic acid bacteria, propionibacteria and bifidobacteria.

The quest to find food ingredients with valuable bioactive properties has encouraged interest in exopolysaccharides from LAB. Functional food products that offer health and sensory benefits beyond their nutritional composition are becoming progressively more important to the food industry. The sensory benefits of exopolysaccharides are well established, and there is evidence for the health properties that are attributable to exopolysaccharides from LAB. However, there is a wide variation in molecular structures of exopolysaccharides and the complexity of the mechanisms by which physical changes in foods and bioactive effects are elicited.

To give a desired texture and mouth feel to yoghurt, skim-milk powder or whey is frequently added to the milk. Although the consumer does not consider this as unnatural, it represents an extra cost for the producer. In some countries, however, gelatine or plant (*e.g.*, starch, pectin, guar gum, and alginate) and microbial polysaccharides (*e.g.*, xanthene and gellan) are added. Polysaccharides increase the viscosity and firmness, improve the texture, reduce susceptibility to syneresis, and contribute to the mouth feel of low-fat products. Some polysaccharides, *e.g.*, plant carbohydrates, xanthene and gellan, have the additional advantage of being suitable to chemical modification to improve their rheological properties. However, the modified molecules are perceived by the consumer as unnatural and the food regulation of several European countries forbids the use of additives in, for instance, yoghurt. The *in-situ* production of natural texture-improving sugar polymers (exopolysaccharides) by LAB for the manufacturing of yoghurts, of sour cream and whipped toppings of ice cream and of low-fat Mozzarella are being explored. Functional, exopolysaccharide-producing starters of *Lb. delbrueckii* sub sp. *bulgaricus* or *S. thermophilus* are promising. Current research is investigating the biodiversity of exopolysaccharides produced by LAB from artisan yoghurts, fermented milks, vegetables, and cereals, the conditions for optimal production, and their technological implementation in the industrial production of fermented foods.

3.3. Bacteriocin Production

Functional starters' could protect consumers from harmful bacteria either by a rapid acidification or by the production of antimicrobials (bacteriocins). Bacteriocins, ribosomally-synthesized peptides or proteins with antimicrobial activity, are produced by different groups of bacteria. Many lactic acid bacteria produce bacteriocins with broad spectra of inhibitions that offer potential applications in food preservation. The use of these bacteriocins in food industry can reduce the dependency on chemical preservation as well as the intensity of heat treatments, resulting in naturally preserved foods rich in organoleptic and nutritional value. The live bacteriocin-producing LAB incorporation into a food gives it its own built in bio-preservation, thereby returning to a more natural method of shelf-life extension and improving the safety of food. However, there may be a lack of compatibility between the bacteriocin-producing strain and other cultures used in the fermentation. This was found to be the case for nisin-producing cultures, which inhibit the starter cultures required for cheese making. In an attempt to overcome this problem, the viability of using nisin-producing starters was assessed, but unfortunately these starters suffered from poor acid development, possessed inadequate proteolytic activity and often displayed an enhanced susceptibility to bacteriophage attack. The use of cultures to produce bacteriocins *in situ* as a means of bio-preservation has received a great deal of interest in recent times. The inhibition of *L. monocytogenes* in dry-fermented sausage using lyophilised bacteriocin-producing cultures as a protective adjunct has been investigated. To reflect realistic pathogen contamination levels, sausage meat was artificially spiked with a four strain cocktail. *Lb. curvatus* was more successful than *Lc. lactis*, following 4 h fermentation. The utilisation of *Corynaebacterium* strains to combat listerial contamination in cold-smoked salmon (CSS) has been studied and has shown potential, but due to the emergence of resistance to Class II bacteriocins there has been an increased interest in the use of bacteriocin-negative *Corynaebacterium* strains as protective cultures. Recently, a food-grade strain has been developed to produce both lacticin 3147 and lacticin 481. This strain addresses both the food safety and food improvement aspects of bacteriocin production. Significantly, the killing effect of this double producer was more pronounced, when tested against *Lb. fermentum* and *L. monocytogenes*, than either bacteriocin producer alone. The use of bacteriocin-producing cultures in foods can be very beneficial. While the choice of bacteriocin-producing culture may have to vary to meet with the particular microenvironment associated with certain foods, the net result of this technology can be the production of foods that have a built-in preservation system.

4. ACCELERATED RIPENING OF CHEESES

The ripening of cheese is a slow and consequently an expensive process. The economic advantage of rapid development of more intense cheese flavour in shorter periods of time would be substantial. Lactic acid bacteria play a key role during ripening and can therefore be used as accelerating agents. This review describes the different strategies where lactic acid bacteria or their enzymes were used to reduce the ripening time of cheese. The advantages, limitations and technical feasibility as well as the commercial potential of the different approaches are also considered. During cheese maturation, several aromatic compounds are generated due to the action of endogenous milk enzymes as well as the proteolytic and lipolytic activities of LAB present in the cheese. The maturation stage is time and space consuming so that accelerated maturation techniques are being sought. Besides rational selection of the LAB starter and co-cultures, and the application of process conditions for optimal activity of the endogenous enzymes, the addition of exogenous enzymes (enzyme-modified cheese) as well as the increased *in situ* autolysis of the LAB, represent alternative solutions. Autolysis of the starter cells is followed by the release of intracellular peptidases in the curd. Bacteriocins (see above) may cause bacteriolysis, for example by inducing autolysins or by deregulating enzyme action due to energetic deficiency, which will result in degradation of the cell wall and consequent cell lysis. Furthermore, genetic strategies exist for the induction of phage holins and lysins to promote lysis of the starter cells during maturation.

5. REDUCTION OF TOXIC OR ANTINUTRITIVE FACTORS

The fermentative action of specific LAB stains may lead to the removal of toxic or antinutritive factors, such as lactose and galactose from fermented milks to prevent lactose intolerance and accumulation of galactose. Other examples are the removal of raffinose, stachyose, and verbascose from soy to prevent flatulence and intestinal cramps, proteinase inhibitors from legumes and cereals to prevent maldigestion, phytic acid and tannins from cereals and legumes to increase mineral bioavailability, and natural toxins such as cyanogenic glucosides from cassava as well as biogenic amines from traditional fermented foods.

6. PRODUCTION OF NUTRACEUTICALS

Nutraceuticals are food components that, through specific physiological action, contribute to the health of the consumer. Several nutraceuticals from bacterial origin have been added to food. Through strain selection and process optimisation, the activity of LAB can be modified to increase the content of nutraceuticals in fermented foods such as fermented dairy products. As an example, fermented milks can be produced with LAB starter strains that produce high amounts of low-calorie polyols so as to reduce the sugar content. Also, the use of oligosaccharide-producing LAB that produce sugar polymers with a controlled structure and chain length (and hence molecular mass) may yield fermented products with health applications. Health effects of such oligosaccharides are ascribed to their low-calorie character, their fibre-like nature, and their bifidogenic effect. Certain LAB, such as the yoghurt bacteria *Lb. delbrueckii* subsp. *bulgaricus* and *S. thermophilus* are able to produce vitamins such as folate. A controlled use of these bacteria may lead to dairy products with increased folate content. The proteolytic system of LAB can contribute to the liberation of health-enhancing bioactive peptides from milk. The latter may improve absorption in the intestinal tract, stimulate the immune system, exert antihypertensive or antithrombotic effects, display antimicrobial activity, or function as carriers for minerals, especially calcium.

6.1. Low Calorie Sugar Production

Mannitol is a polyol or sugar alcohol that is produced by several organisms. Mannitol is assumed to have several beneficial effects, as an antioxidant (protection against oxidative damage by oxygen radicals) and as a non-metabolizable sweetener. Mannitol can therefore be applied to foods leading to health-promoting effects (functional foods). Mannitol-producing lactic acid bacteria may directly be applied in the manufacture of foods and this may lead to fermented food products with an extra nutritional value. Mannitol is a non-toxic, non-hygroscopic sweetener estimated to be about half as sweet as sucrose. Human are incapable of metabolizing mannitol entirely and therefore do not suffer from hyperglycaemia. Many lactic acid bacteria can synthesize mannitol and the mode of synthesis is depended on the pathway used for hexose fermentation. These are *Leuconostoc mesenteriodes, Leuconostoc pseudomesenteriodes, Lactobacillus plantarum* etc. Mannitol production by lactic acid bacteria and other food-grade microorganisms offer several important advantages. Firstly, food-grade microorganisms and their products are directly applicable in food products, without any restriction. Secondly, there is no need for a careful separation of products and microorganisms, which would be the case if microorganisms are not of food grade. Thirdly, some lactic acid bacteria are claimed as beneficial in the gastrointestinal tract. Several lactic acid bacterial strains are known to produce mannitol. Some homofermentative lactic acid bacteria were found to produce small amounts of mannitol. In the presence of large amounts of glucose or sucrose, cell suspensions of *Streptococcus mutans* produce approximately 0.1–0.3 mg mannitol. *Lactobacillus leichmanii* also showed mannitol production. Since in homofermentative lactic acid bacteria the carbon flow from carbohydrates is directed mainly to lactate production, the formation of other fermentation products like mannitol is often only possible when strains are more or less hampered in the lactate production pathway. A lactate dehydrogenase-negative mutant of *Lactobacillus plantarum* produced 6 mM mannitol from 50 mM glucose whereas the wild type was not able to produce any mannitol from glucose. The best mannitol producing lactic acid bacteria show a heterofermentative metabolism. In the presence of fructose or sucrose *Leuconostoc (pseudo) mesenteroides* produces high levels of mannitol. Two other heterofermentative lactic acid bacteria, *Lactobacillus* sp. and *Leuconostoc* sp., also produced mannitol from fructose and sucrose. The fructose is bio-converted to mannitol. Under optimized conditions *Leuconostoc mesenteroides* converted up to 95% (mol/mol) of fructose consumed into mannitol.

6.2. Vitamins Production

Folate the salt of folic acid is vital component of the human diet and involve in the biosynthesis of nucleotides and co-factors in many metabolic reactions. As mammalian cells cannot synthesize folate, the folate synthesizing LAB can be used to augment the folate content in fermented dairy products. Milk contains 20-50 microgram per liter folate and fermented milk contains higher amounts up to 150 microgram per litre in yoghurt. Some starter cultures have been shown to synthesize folate and secreted into the medium during fermentation. These are *Bifidobacterium longum, Lactobacillus acidophilus, L. delbrueckii* ssp. *bulgaricus*, and *Streptococcus thermophilus* strains. Vitamin B12 is an important co-factor for the metabolism of fatty acids, amino acids, carbohydrates and nucleic acids. This vitamin is synthesized by propionibacteria. Vitamin K, an essential co-factor, is involved in the post-translational modification of glutamic acid residues in certain proteins, such as blood clotting proteins and proteins involve in tissue calcification. Genera *Lactococcus, Lactobacillus, Leuconostoc, Bifidobacterium* and *Enterococcus* synthesize the vitamin K. Such strains may be useful as dairy starters to increase the vitamin K content in fermented dairy products.

6.3. Bioactive Peptide Production

Bioactive peptides have been identified within the amino acid sequences of milk proteins and may be released through the action of digestive enzymes in the intestine, added proteases or lactic cultures. Such peptides potentially have a wealth of health-related activities such as antimicrobial, opiate, antithrombic, antihypertensive, mineral binding properties, etc. Milk proteins have received increasing attention as potential ingredients of health-promoting functional foods targeted at diet-related chronic diseases, such as cardiovascular disease, diabetes type two and obesity. To this end, growing interest has been focused on physiologically active peptides derived from milk proteins. These peptides are inactive within the sequence of the parent protein molecule and can be liberated by gastrointestinal digestion of milk, fermentation of milk with proteolytic starter cultures or hydrolysis by proteolytic enzymes. Milk protein-derived peptides have been shown under *in vitro* and *in vivo* conditions to exert a number of activities affecting the digestive, endocrine, cardiovascular, immune and nervous systems. A great variety of naturally formed bioactive peptides have been found in fermented dairy products, such as yoghurt, sour milk and cheese. Recently, industrial-scale technologies suitable for the industrial production of bioactive milk peptides have been developed. In addition, a few commercial food products supplemented with milk protein-derived bioactive peptides have been launched on limited markets. Some of these products carry clinically documented benefits, in particular for reduction of mild hypertension. The multifunctional properties of milk peptides appear to offer considerable potential for the development of many similar products in the near future. Fermentation of milk with single industrial dairy cultures generated antioxidant activity in the whey fraction. The activity correlated positively with the degree of proteolysis suggesting that peptides were responsible for the antioxidative property. Fermentation of milk with a commercial starter culture mixture of five lactic acid bacteria (LAB) strains followed by hydrolysis with a microbial protease increased ACE inhibitor activity of the hydrolysate. Several novel ACE-inhibitory peptides in milk fermented with *Enterococcus faecalis* strains isolated from raw milk. The proteolytic activity of several dairy LAB cultures and probiotic strains (*Lactobacillus acidophilus, Bifidobacterium lactis* and *Lactobacillus casei*) as determinant of growth and *in vitro* ACE-inhibitory activity in milk fermented with these single cultures. All the cultures released ACE inhibitor peptides during growth with a *Bifidobacterium longum* strain and the probiotic *Lb. acidophilus* strain showing the strongest ACE-inhibitory activity. ACE-inhibitory peptide from milk fermented with *Lactobacillus delbrueckii* ssp. *bulgaricus* was identified and sequenced as Ser-Lys-Val-Tyr-Pro-Phe-Pro-Gly-Pro-Ile. A hypotensive peptide structure with a sequence of Ser-Lys-Val- Tyr-Pro was obtained from beta-casein by *Streptococcus salivarius* ssp. *thermophilus* and *Lactococcus lactis* biovar. *diacetylactis*.

6.4. Production of Conjugated Linoleic Acid (CLA)

CLA refers to mixture of positional and geometrical isomers of linoleic acid. A number of these isomers have attracted considerable attention due to their potentially beneficial effects, including anticarcinogenic and anti-inflamatory activities and inhibition of the onset of diabetes. A number of LAB can also produce CLA and CLA are higher in fermented dairy products than in unprocessed milk. Although, CLA production is not a general feature of LAB and bifidobacteria, strains are now available that can efficiently convert exogenous linoleic acid to CLA and such strains can be used to increase the CLA content in fermented dairy products. The productivity of conjugated linoleic acid synthesis by bacteria of the *Lactobacillus* and *Bifidobacterium* genera is a strain-specific trait. *Bifidobacterium* synthesize a higher number of different CLA isomers compared to *Lactobacillus*.

7. PROBIOTICS FOR FUNCTIONAL FOODS

Specially-selected cultures may also provide probiotic benefits, and, if properly modified, they may even be endorsed with nutraceutical traits. Probiotics are products aimed at delivering living, potentially beneficial, bacterial cells to the gut ecosystem of humans and other animals, whereas, prebiotics are non-digestible carbohydrates delivered in food to the large bowel to provide fermentable substrates for selected bacteria. Strains of LAB are the most common microbes employed as probiotics. Two principal kinds of probiotic bacteria, members of the genera *Lactobacillus* and *Bifidobacterium*, have been studied in detail. Most probiotic strains belong to the genus *Lactobacillus*. Probiotics have been evaluated in research studies in animals and humans with respect to antibiotic-associated diarrhoea, travellers' diarrhoea, pediatric diarrhoea, inflammatory bowel disease, and irritable bowel syndrome. It is possible that, in the future, probiotics will be used for different gastrointestinal diseases, vaginosis, or as delivery systems for vaccines, immunoglobulins, and other therapies.

The basic notion behind probiotics is pretty straightforward: restore the depleted ecology of the intestine with new, healthful bacteria. Many different microorganisms are added to dairy products for their probiotic potential. Probiotic microorganisms designed for delivery in food or dairy products, via supplementation or fermentation. These include Lactobacilli, such as *Lactobacillus acidophilus*, *L. casei*, *L. delbrueckii* ssp. *bulgaricus*, *L. reuteri*, *L. brevis*, *L. cellobiosus*, *L. carvatus*, *L. fermentum* and *L. plantarum*; Gram-positive cocci, such as *Lactococcus lactis* ssp. *cremoris*, *Streptococcus thermophilus*, *Enterococczus faecium*, *S. diacetylactis* and *S. intermedius* and Bifidobacteria, such as *B. bifidum*, *B. adolescentis*, *B. animalis*, *B. infantis*, *B. longum* and *B. thermophilum*. Nonpathogenic microorganisms that occupy important niches in the host gut or tissues, such as yeasts, enterococci and *Enterobacteriaceae,* are used as human and animal probiotics. Though, *Lactobacillus* and *Bifidobacterium* are the most common species of bacteria used as probiotics for the production of fermented milks and other dairy products.

Human probiotic species and strains

- *Bifidobacterium breve* Yakult
- *Bifidobacterium lactis* (BB12)
- *Bifidobacterium longum* (SBT2928, BB536)
- *Lactobacillus acidophilus* (NCFM, SBT2062)
- *Lactobacillus casei* (Shirota, CRL431, DN014001, immunits)
- *Lactobacillus delbrueckii* ssp. *bulgaricus* (2038)
- *Lactobacillus fermentum*
- *Lactobacillus johnsonii* (La1, Lj 1)
- *Lactobacillus paracasei* (CRL431, F19)
- *Lactobacillus plantarum* (299V)
- *Lactobacillus reuteri* (SD 2112)
- *Lactobacillus rhamnosus* (GG, 271, GR1)
- *Lactobacillus salivarius* (UCC118)
- *Streptococcus thermophilus* (1131)

8. SELECTION AND MAINTENANCE OF SUITABLE STRAINS

Selecting for strains with interesting properties to be used as new, functional starter cultures may lead to an improved fermentation process and an enhanced quality of the end product. However, as it has been shown for bacteriocin-producing LAB, the success of using functional starter cultures in a particular food is strongly strain dependent. Rational selection of appropriate strains is crucial. Because of legislation and marketing reasons, the industrial application of carefully selected natural food isolates with functional properties seems attractive. Selection of starter cultures must not only aim at expressing functional properties, but also at eliminating undesirable side effects, such as the formation of D-lactic acid or a race mate of lactic acid (DL), or the formation of biogenic amines. One of the challenges of using wild-type strains will be to allow large-scale production of fermented foods without losing their unique flavour and other traits. In the new generation of fermented foods, LAB with diverse physiological and metabolic traits is combined. Their metabolic and technological properties are often different from those of traditional starter cultures, so that appropriate production processes must be developed. The positive results observed with functional starters under optimal laboratory conditions are not necessarily obtained in practice, *i.e.*, in the actual food matrix.

REFERENCES

Cogan, T.M. and Hill, C. (1993). Cheese starter cultures, Ch.6 *In*: P.F. Fox, ed., Cheese: Chemistry, Physics and Microbiology, Vol. 1, General Aspects, 2nd ed., pp. 193–206. Chapman and Hall, London.

Heap, H.A. and Lawrence, R.C. (1976). The selection of starter strains for cheese making. *New Zealand J. Dairy Technol.*, 11: 16.

Tamime, A.Y. and Robinson, R.K. (1999). Preservation and production of starter cultures, *In*: Yoghurt, Science and Technology pp. 486–514. CRC Press, New York and Woodhead Pub. Ltd., Cambridge, UK.

DE VUYST, L. (2000). Application of functional starter cultures. *Food Technol. Biotechnol.*, 38(2): 105–112.

12

Novel Processing of Animal Products

Ashim K. Biswas and Prabhat K. Mandal

ABSTRACT

Improving food processing technology not only improves health, but reduces poverty. When food products are safe, nutritious, well marketed, and competitively priced thanks to efficient manufacturing, they attract consumers. Rising consumer demand, in turn, expands a nation's entrepreneurial base in food products, creating jobs and raising family incomes. Larger family food budgets then contribute to a further drop in malnutrition. New processing and preservation technologies, offer advantages in meeting consumer demands of freshness, convenience and safety. There is no single process that will allow the high-quality production of every food product while ensuring safety; all of these processes, as well as thermal processing, have their own set of limitations and advantages. In this chapter several emerging and novel technologies for processing and preservation of animal products are discussed. They include as microwave/radio frequency heating, high pressure processing and pulsed electric fields, infra red heating, irradiation, UV processing, membrane technology, plasma and ultrasound treatment.

1. INTRODUCTION

The development of emerging technologies in food processing addresses specific consumer needs toward safe, healthy, and minimally processed foods. These innovative processes also lead to environmentally friendly and sustainable food manufacturing techniques with low energy requirements and reduced water use that overcome some limitations given by current food processing practices (Toepfl *et al.*, 2006). Taking advantage of specific potentials and opportunities of these new processes, including the understanding and control of the complex process-structure-function relationships, offers the possibility for a science-based development of tailor-made foods.

High hydrostatic pressure (HHP), pulsed electric fields (PEFs), ultrasound (US), infrared (IR) heating, plasma treatment (PT), membrane processing, UV treatment and oscilitating magnetic field (OMFs) are used to exemplify scalable and flexible food manufacturing techniques. Significant, science-based achievements have been made to better understand the basic principles underlying HHP and PEF processing (Raso & Heinz, 2006). The food and beverage industry offers many possibilities for the use of US. The basic phenomena, such as micro streaming and cavitations, and the resulting hydrodynamic shear-forces, make US an alternative technology for the realization of homogenization, dispersion, and emulsification, as well as for the disintegration of tissue to enhance mass transfer processes.

PT based surface treatments, such as ultrafine cleaning, etching, surface fictionalisation, and thin film deposition, and environmental applications, such as exhaust air cleaning, are proven techniques and already applied using industrial scale plasma systems (Foest *et al.*, 2005). A strong increase of plasma applications in medical device technology and therapeutic medicine is currently taking place, including applications such as plasma decontamination, and research is focusing on the interaction between plasma and biological cells and tissue as well as on the plasma diagnostics with regard to the understanding and control of the complex behaviour of cold plasma (Daeschlein *et al.*, 2010, Weltmann *et al.*, 2009). Understanding the impact and potential of such technologies on food systems at the cellular level will enable the design of tailor-made foods and to establish process-structure-function relationships. Based on this knowledge, completely new process design and the incorporation of HHP, PEFs, US, and PT in traditional processes, as well as the generation of improved equipment design, will be possible.

2. NOVEL THERMAL PROCESSING

2.1. Microwave (MW) and Radio Frequency (RF)

Dielectric heating implies the interaction between an electromagnetic alternating field and the dipoles and ionic charges contained within a food product that enables the volumetric heating of the product. RF and MW systems operate by the same principle, forcing polar molecules, such as water, and ionic species to constantly realign themselves by reversing an electric field around the food product. This molecular movement is extremely fast due to the high frequency of the field, which in RF can range from 1 to 300 MHz and in MW from 300 to 3000 MHz. Despite that RF and MW are both capable of offering rapid, non-contact volumetric heating, they can be distinguished itself by the way that electric field is generated and depth of electromagnetic waves. Wavelength at the RF frequencies is up to 360 times greater than the wavelength corresponding to the two frequency values commonly used for MW (915 MHz and 2450 MHz), which allows RF energy to penetrate foods more deeply than MW.

Since heating applications in food processing require a uniform temperature in the food being heated, achievement of this using MW can be a difficult task that makes this technology profitable only for small sized foods (Marra *et al.*, 2007). In RF heating, the electrical field is generated in a directional manner between a pair of capacitor plates; the food material to be heated is placed between the two plates, where it plays the role of a dielectric. Microwaves are primarily a radiation phenomenon that usually takes place in an enclosed cavity, or in close vicinity to a waveguide applicator, where the electrical component of the microwave field approaches the food product from all directions. RF and MW processing in preservation in general and pasteurization or sterilization in particular is to deliver a more homogeneous heat treatment at a faster rate than conventional method of heating.

2.1.1. *Principles of heating*

The molecular friction produced by dipole rotation and by the migration of ionic species under the influence of the oscillating electromagnetic field, generates heat inside the food by energy dissipation. RF and MW are considered non-ionizing radiation because they have insufficient energy (<10 eV) to ionize atoms. Microwaves cause oscillation of molecules in the food system. The molecules oscillate about their axes while attempting to go proper positive and negative poles; intermolecular friction is created and manifested as a heating effect. Microwave frequencies *viz.*, 915 and 2450 megacycles are commonly used with wavelength of 32.8 and 12.25 cm, respectively. Although most commonly available microwave ovens operate at 2450 MHz, there is great interest in the 915 MHz frequency. This frequency produced two peaks one at the surface and other at the center, while 2450 MHz frequency produces only the surface peak.

In conventional microwave oven power tube in which electrical energy is converted to an oscillating electromagnetic field is known as magnetron. These molecular frictions would generate heat and subsequently lead to temperature rise. There are two major mechanisms, dipolar and ionic interactions. In *dipolar interaction*, once microwave energy is absorbed, polar molecules such as water molecules inside the food will rotate according to the alternating electromagnetic field. The water molecule is a "dipole" with one positively charged end and one negatively charged end. Similar to the action of magnet, these "dipoles" will orient themselves when they are subject to electromagnetic field. The rotation of water molecules would generate heat for cooking.

The *ionic interaction* works through acceleration of ionic compounds present in dissolve state in food by the electromagnetic field and collided with other molecules to produce heat. Hence, the composition of a food is very important in microwave processing of foods. Food with higher moisture content will be heated up faster because of the dipolar interaction. As the concentration of ions (*e.g.*, dissolved salts increase, the rate of heating also increases because of the ionic interaction with microwaves. Even though oil molecules are much less polar than water molecules and are non-ionic, food products with high oil content has a fast heating rate because the specific heat of oil is about less than half that of water. However, there is significant variation of cooking of food in conventional and microwave oven.

The promise of rapid and volumetric heating has called the attention of food industry for the potential use of dielectric systems. Conventional oven is heated by surrounding hot air, whereas, food cooked in microwave oven is heated as a result of the alternating electromagnetic field. The electromagnetic field generated is not uniformly distributed inside the cooking cavity and hence it leads to uneven heating of food. The uneven heating process may produce food items with hot and cold spots, which in turn may give rise to microbiological food safety concern. Nowadays, there have been improvements in the design of microwave oven such that the problems of uneven heating are minimized. Nevertheless, standing of food item for a while before consumption may allow time for heat re-distribution to other parts of the food item.

Another difficulty with microwave cooking is that microwave oven cannot induce browning or crisping of food. However, this can be resolved by using microwave packaging materials called microwave susceptors when heating food. Typically susceptors are made of polyester film with particulate aluminium and paper or board. The aluminium layer would absorb microwaves and get hot quickly to make the food crispy. In addition, some models of microwave oven have been specially designed with the addition of a grill heating system. Food would get crispy and brown faster when cooked in an oven with the combination of microwave cooking and grill heater. However, in microwave cooking, heat is produced inside the food and there is less energy loss. On an average most microwave oven takes only about 20% of time required by conventional oven and saves at least 20% of energy depending on the food type.

RF systems could be particularly suitable for heat processing of cured whole meat products such as ham (Lyng *et al.*, 2006) while, RF drying of foods ingredients, snack foods, potato products and pasta products are well established applications. Radio frequency heating systems for the purpose of food pasteurization or sterilization are still being investigated due to the capability for rapid and uniform heating, and it is anticipated that RF will be capable of producing shelf-stable food with higher quality for civilian and military use. However, before fully commercial implementation of RF technology for pasteurization and sterilization of packaged foods, potential problems such as dielectric arcing and thermal runaway heating need to be addressed (Zhao, 2006). The knowledge of dielectric properties of foods to be heated is a key factor for an effective RF and MW heating. Much in the same way, the factors that influence dielectric properties of the foods during heating, such as mass, shape, water content, chemical composition, temperature and frequency of treatment, must be fully understood for further development and industrial setup of dielectric heating systems in food processing.

2.1.2. *Risks associated with microwave cooking*

Cooking processes, especially the high temperature ones (*e.g.*, grilling, baking, etc.) are known to induce the production of potential carcinogens. There have been concerns that microwave cooking may also increase the production of carcinogens or mutagens in foods. But, there is no such scientific evidence that the production of any carcinogenic substances would increase upon the application of microwave heating. Among the carcinogens, the formation of the chemicals heterocyclic amines (HCAs), polyaromatic hydrocarbons (PAHs) and nitrosamines are of particular concern. The HCAs are a group of compounds that are present in cooked muscle meat after high-temperature cooking such as grilling/barbecuing, broiling or pan-frying. The International Agency for Research on Cancer (IARC) of the World Health Organization has classified one of the HCAs, 2-amino-3-methylimidzo [4, 5-f] quinoline, as probable human carcinogen (Group 2A), whereas, nine other HCAs as possible human carcinogens (Group 2B).

Cooking methods like microwaving and boiling can result in food with lesser amounts of HCAs. In a study it has been revealed that the formation of certain types of HCAs in chicken legs could be reduced by microwave cooking when compared with frying. Precooking of meat in microwave oven before barbecuing would significantly reduce the formation of HCAs. PAHs are referred to a large group of organic chemicals containing two or more fused aromatic rings made up of carbon and hydrogen atoms. Incomplete combustion could lead to formation of PAHs during cooking of foods using roasting, grilling, barbecuing and smoking. Nitrosamines are formed by a reaction between nitrite and a secondary or tertiary amine. The reaction may take place in certain types of foods as a result of curing, drying or cooking. In conclusion, microwave cooking did not produce significant amount of HCAs, nitrosamines or PAHs in meat products.

The use of microwave cooking to precook meats before grilling or barbecuing has in fact been recommended so as to minimize the formation of HCAs and PAHs. It is probably due to the lower cooking temperature and shorter cooking time of microwave cooking. The risk related to food-borne infection through consumption of meat prepared by microwave cooking is negligible as the effectiveness of microwave cooking in killing microorganisms and spores is comparable with conventional methods provided that appropriate temperature and time are reached. It is generally advised that raw animal food should be heated to a temperature of at least 75°C for 15 seconds to kill any food borne pathogen that may be present in the food. But opposite results can be attributed to uneven heating of the foods which could have been avoided by ways including covering the food during microwave cooking.

2.1.3. *Effect on food constituents*

The degradation rates of nutrients depend on the heating time and temperature. On heating protein undergoes modification as unfolding of protein molecules take place but the nutritive value of proteins in foods treated by conventional and microwave heating remain same. Lipid components including triglycerides, saturated and unsaturated fatty acids, as well as cholesterol undergo various decomposition reactions during heating in the presence of oxygen. The increase in fat oxidation products is of particular health concern. But studies suggested that the stability of lipids from soya, egg yolk and meats; fatty acid profiles in chicken and beef patties, chicken fat, beef tallow, bacon fat, rainbow trout and peanut oil did not result in significantly more chemical modifications upon microwave cooking. The retention of vitamin in micro waved foods is equal or better than conventionally prepared foods because of the shorter heating time of microwave cooking. Minerals are generally not destroyed during cooking including microwave cooking. However, they might be lost in cooking water or meat drippings. Therefore, the nutritional values of food cooked by microwave would be comparable with those by conventional methods.

2.1.4. *Interaction with food contact materials*

The common materials for packaging or containing foods are plastic, paper, glass, ceramics and metal. However, not all of these materials are suitable for microwave cooking. Materials like plastics, paper, glass and ceramics are generally transparent to microwaves. Nevertheless, some of them may absorb certain amount of microwave energy and hence reduce the amount of energy to be absorbed by food. On the other hand, there have been concerns on the possibility of chemical migration from such food contact materials into food during microwaving.

It is assumed that all types of plastic materials are not suitable for microwave cooking. Even though high density polyethylene can be used for foods with high water content, it cannot be used for foods with high fat or high sugar content as these foods may reach temperature above 100°C during microwave cooking. Among plastic materials, the most commonly used ones for microwave cooking are polypropylene and crystalline polyethylene terephthalate (CPET), which have melting points of 210–230°C. There have been concerns regarding the potential carcinogenicity of DEHA which may migrate into foods. Paper and board can also absorb some microwave energy so, it is not ideal for micro wave cooking.

When food is microwaved in glass container, heat is also retained in the glass. The degree of energy absorption depends on the types of glass. Moreover, microwave energy can be superimposed at the centre after passing through the glass containers, particularly the ones with small radius. Hence, the temperature at the centre might be much higher than the outside. Ceramics/metals are not suitable for microwave cooking because sparks caused by electric arcing. The arcing effect was resulted from reflection or bouncing-off microwaves from the metallic components. Then the air between two metallic components nearby would become ionized and luminous electric current would in turn pass across the gap between the two components. Extensive arcing is undesirable because it would damage the magnetron inside the microwave oven.

The use of microwave oven provides a convenient way to thaw, cook or reheat foods nowadays. Many studies have been conducted to assess the safety as well as possible nutrient loss associated with microwave cooking. The best available evidence supports that the use of microwave cooking resulted in foods with safety and nutrient quality similar to those cooked by conventional methods, provided that the consumers followed the given instructions.

2.2. Ohmic Heating

Ohmic heating (OH), also called Joule heating, electrical resistance heating or electro conductive heating, is based on the principle that most food products have the ability to resist to the passage of an electrical current. Heating occurs when an alternating electrical current is passed through a food resulting in the internal generation of heat, due to the electrical resistance of the food. This technology has been around since early 1900s. The food processing researcher, however, began investigating the potential of ohmic heating on food quality and cost and energy savings in 1980s. In this method an AC current is passed through a food sample which leads to generate internal energy in foods. Ohmic heating is somewhat similar to microwave heating but with very different frequencies.

The advantage of this technique is that it uniformly heats food with different densities such as chicken soup. The quality product with minimal structural nutritional and organoleptic changes can be produced. Potential application of this technique includes blanching, evaporation, dehydration, fermentation and extraction. It saves significant time/energy in hot air and freeze drying of foods and enhances extraction/yields in some processing operations. The parameters used during ohmic heating such as frequency of alternating current applied voltage and the temperature to which the sample was heated have a significant effect on its success. The electrical conductivity is also a significant factor. The ohmic heating is useful for value added processing, and it has great potential for use in wide variety of food processing operations involving a heat and mass transfer.

Ohmic heating is currently used in Europe, Asia and North America to produce a variety of high quality low and high acid products containing particulates. Electrical resistance heating allows particles and liquids to heat at the same rate and permits the rapid heating of mixtures of high solid fractions. The technique has been applied to a number of food processes, and has recently been developed into a commercial process for the sterilization of food mixtures. Ohmic heating occurs, when an electric current is passed through an electrically conducive product. Low frequency current from domestic supply could be effectively used for ohmic heating. The ohmic heating has many advantages over conventional heating. Continuous processing is possible without any heat transfer surface. Liquid-solid mixture can be rapidly and evenly heated with minimal heat damage and residence time difference.

2.2.1. *Effect on microbial inactivation*

The principal mechanisms of microbial inactivation in OH are thermal in nature. Recent research indicates mild non-thermal cellular electroporation mechanism may occur during ohmic heating similar to the effects utilized in PEF processing (Lebovka *et al.*, 2005; Kulshrestha & Sastry, 2006). The principal reason for the additional microbial inactivation effect to heating of ohmic treatment may be its low frequency (50–60 Hz), which allows cell walls to build up charges and form pores. This is in contrast to high-frequency methods such as microwave or radio frequency heating, where the electric field is essentially reversed before sufficient charge build up occurs at the cell walls. Nevertheless, temperature is the principal critical process factor in ohmic heating. As in conventional thermal processes, the key issue is identifying the slowest heating zone. Fundamentally, there is only one critical factor: the temperature-time history of the coldest point.

Several factors significantly affect the temperature within an ohmic process. The critical parameters in continuous flow ohmic heating systems include electrical conductivities of the respective phases of the food, temperature dependence of the electrical conductivity, design of the heating device (*e.g.*, location and orientation of the electrodes), extent of interstitial fluid motion, residence time distribution, thermal properties of the food, and electric field strength.

As a main consequence of this effect, the D value observed for the microbial inactivation under ohmic heating is reduced when compared to traditional heating methods. This reduction has been observed for *Bacillus licheniformis* and *Escherichia coli* (Pereira *et al.*, 2007), *Bacillus subtilis* (Cho *et al.*, 1999), *Streptococcus thermophilus* (Sun *et al.*, 2008) and *Byssochlamys fulva*. The thermal treatment required for microbial inactivation in biomaterials and foods could be potentially reduced if there exist any sublethal injury or additional lethal effect due to electric current (Palaniappan & Sastry, 1992). Pereira *et al.* (2007) have reported lower D and z values for the inactivation of *E. coli* and *B. licheniformis* when submitted to ohmic heating. In this research, comparing conventional against ohmic heating, the thermal history of the samples analyzed was adjusted to match.

The D values observed for *E. coli* at 65°C were 3.5 ± 0.2 and 0.86 min for conventional and ohmic process, respectively. The z values were also reported as 23.1° and 8.4°C, respectively. The observed results indicate that the electric current may have affected the microbial death rate (Pereira *et al.*, 2007). Similar observations were obtained in the same study for the spore inactivation of *B. licheniformis*. Considering both microorganisms strain, ohmic heating presented a lower D value. This fact indicates that an additional nonthermal lethal effect occurred under ohmic heating, due to the presence and effects of the electric current over vegetative cells of *E. coli* and bacterial spores of *B. licheniformis* (Pereira *et al.*, 2007).

Additionally to the heating promotion, research data strongly suggests that the applied electric field under OH causes electroporation of cell membranes. The cell electroporation is defined as the formation of pores in cell membranes due to the presence of an electric field and as consequence, the permeability of the membrane is enhanced and material diffusion throughout the membrane is achieved by electro-osmosis. It is assumed that the electric breakdown or electroporation mechanism is dominant for the non-thermal effects of OH.

2.2.2. *Effect on enzyme inactivation*

Electrical fields, applied during ohmic heating of lipoxygenase and polyphenoloxidase, caused their faster inactivation than during conventional heating (Castro *et al.*, 2004). Similarly, ohmic heating was found to be more efficient for the required micrcbial and pectin esterase inactivation due to a shorter residence time when released flavour compounds were not degraded as quickly as during conventional pasteurisation (Leizerson & Shimoni, 2005). Peroxidase in pea puree was also inactivated in a shorter processing time, by ohmic heating, than by conventional heating (Icier *et al.*, 2006). Moreover, ohmic heating caused less browning. The effects of voltage gradient, temperature and holding time on the polyphenol-oxidase activity were investigated for grape juice ohmic heating (Icier *et al.*, 2008). The inactivation kinetics of this enzyme was described by several mathematical models but simple first-order kinetics was found to be the best.

2.2.3. *Limitations of ohmic heating*

The main limitation of ohmic heating is the heterogeneous nature (in composition) of the food products and their corresponding electrical conductivities that leads to differences in the conversion of the electrical current into thermal energy. As in microwave heating, in ohmic heating, thermal runaway can also occur, because electrical conductivity, which is the property that influences electrical energy dissipation, usually increases with increasing temperature. Therefore, especially in stationary (*i.e.*, not moving in a stream) solid products, there may be areas that are very hot (usually areas close to the electrodes), which in some instances may even be burned, while in other areas (with initially lower electrical conductivities, or farther away from the electrodes) almost no heating occurs. Uniform heating with ohmic processing is theoretically possible, but at the same time challenging due to the various factors impacting on the slowest heating zone and the time—temperature history throughout the products.

3. NOVEL NON-THERMAL PROCESSING

3.1. High Pressure Processing

HPP has demonstrated wide applicability for producing high quality foods. HPP has become accepted as an attractive alternative to traditional preservation methods utilizing preservatives or thermal processing. HPP is commonly referred to as a nonthermal process of liquid and solid foods through application of high pressure in the order of 100 to 1000 MPa and holding times of several minutes. In agreement with the isostatic principle, during HHP pressure is applied uniformly and instantaneously through a food material (with or without packaging), independently of its mass, shape and composition. Under pressure, biomolecules obey the Le Chatelier-Braun principle and reactions that result in reduced volume will be promoted. Such reactions affect the structure of large molecules, such as proteins.

HPP causes a partial unfolding of proteins that can promote covalent and non-covalent interactions during and upon release of pressure, thus triggering their denaturation. These results in the inactivation of microorganisms and enzymes (Hendrickx *et al.*, 1998) which offers opportunities for increased shelf life, and can also promote changes in the rheological properties of the food products (Diels & Michiels, 2006; Knorr *et al.*, 2006; Sharma & Yadav, 2008). Small molecules that have little secondary, tertiary and quaternary structures, such as amino acids, vitamins and flavor and aroma components contributing to the sensory and nutritional quality of food, remain unaffected. Process temperature during HHP can be specified from below 0°C (to minimize any effects of adiabatic heat) to above 100°C and exposure times can range from a millisecond pulse to a treatment time of over 20 min.

HP offers several interesting applications for the food industry. For example, HP-treated starches have reduced digestibility and might be used to substitute fat in dietary foods (Zhang *et al.*, 2008). Furthermore, the inactivation behavior of enzymes under pressure could be used as an indicator for reaching HP pasteurization or sterilization conditions and to detect temperature differences in the pressure vessel during processing (Grauwet *et al.*, 2010). A further possible application of HP is its combination with temperatures below 0°C for subzero storage of food in the liquid state or a cold denaturation of proteins to form gels with unique properties. Moreover, HP could be used to reduce the thawing time for frozen products such as fish (Schubring *et al.*, 2003).

3.1.1. *Effect on microorganisms*

Pressure treatments do not necessarily weaken biological cells. Compression will increase the temperature of foods approximately 3°C per 100 MPa and may also shift the pH of the food as a function of imposed pressure. Pressure pasteurisation is feasible also at room temperature and energy saving as compared to heat treatment. At low pressure levels, increased resistance of microbial cells was observed, and pressure-induced thermo-tolerance of lactic acid bacteria occurred after HP treatment between 100 and 200 MPa (Ananta & Knorr, 2003). As a result of this phenomenon, pressure-induced stress response was found to offer promising processing options such as pre-treatment of lactic acid bacteria before drying or freezing for the purpose of starter culture production. Microbial inactivation of more than five-log cycles in food products is reported by several authors to occur at pressures between 300 and 800 MPa (Ananta *et al.*, 2005). The special shape of isokinetic lines at higher pressure for the inactivation of microorganisms shows the synergisms between pressure and temperature. This behaviour is typical for vegetative cells but was also observed for bacterial spores (Margosch *et al.*, 2006), viruses (Isbarn *et al.*, 2007), and proteins (Smeller, 2002).

By increasing the process temperature, it is possible to decrease the applied pressure, but unwanted reactions that would lead to quality losses also have to be taken into account. Water activity and pH are critical process factors in the inactivation of microbes by HPP. An increase in food temperature above room temperature and to a lesser extent a decrease below room temperature in some cases increases the inactivation rate of micro-organisms during HPP treatment. Temperatures in the range of 45–50°C appear to increase the rate of inactivation of food pathogens and spoilage microbes. Temperatures ranging from 90 to 110°C in conjunction with pressures of 500–700 MPa have been used to inactivate spore-forming bacteria such as *Clostridium botulinum*. Current pressure processes include batch and semi-continuous systems.

Bacterial spores have a higher barotolerance than vegetative bacteria and survive pressures above 1,200 MPa at ambient temperatures (Margosch *et al.*, 2006). Early approaches toward spore inactivation aimed at a pressure-induced germination at moderate pressure. However, combination processes with spore germination at pressures below 200 MPa and an additional moderate heat treatment could not guarantee sufficient inactivation because a small population of spores could not be germinated and remain in the dormant state (Heinz & Knorr, 1996).The HP inactivation of bacterial spores is not yet fully understood and is still highly relevant in modern HP sterilization research. A detailed discussion of HP-related spore inactivation mechanisms is given by Mathys (2007).

3.1.2. *Effect on enzymes*

HP is regarded as a mild process by which the primary structure of proteins is not affected. However, it could have an impact on hydrophobic interactions, which stabilize the quaternary and tertiary structure through reversible unfolding, and the secondary structure through irreversible unfolding. Generally, pressure-induced changes in proteins and enzymes between 100 and 300 MPa at ambient temperature are reversible, whereas, an increase of pressure above 400 MPa could cause an irreversible unfolding, leading to inactivation of the enzyme. Pressure also favours an unfolding of protein chains as well as a dissociation of oligomeric proteins. Owing to conformational changes, unfolding of an enzyme can alter its functionality and result in a decreased or increased biological activity and could even change its substrate specificity.

The pressure stability of enzymes can vary significantly ranging from pressure-sensitive enzymes such as phosphohexose isomerase from bovine milk ($p < 400$ MPa) (Rademacher & Hinrichs, 2006) to extreme pressure resistant enzymes such as horseradish peroxidase ($p > 700$ MPa) (Smeller & Fidy, 2002). However, a categorization of enzymes as a result of their pressure stability is not appropriate, because there is structural variability among enzymes catalyzing the same reaction. It has been observed that the inactivation of 90% of different polyphenol oxidases (PPO) after 10 min in the pressure and temperature landscape. Furthermore, the pressure-temperature resistance of enzymes shows a significant dependence on matrix conditions such as the pH value (Riahi & Ramaswamy, 2004), whereas, even isoforms of an enzyme from the same origin can vary in physical stability between several hundred MPa (Rodrigoa *et al.*, 2006).

3.1.3. *Effect on food texture*

HP could well be used for food texture engineering, owing to its influence on the properties of food ingredients. The primary structure of low molecular weight molecules such as peptides, lipids, vitamins, and saccharides is rarely affected by isostatic HP because of the very low compressibility of covalent bonds at pressures below 2 GPa (Oey *et al.*, 2006). Conversely, HP can change the native structure of macromolecules such as starch similar to thermal-induced structure changes. Potato starch was found to be less affected by pressure treatment than the other starches such as tapioca or wheat starches. Several studies revealed that the swelling of starch granules as measured by loss of birefringence depends on pressure treatment conditions as well as on the type of starch granules. Furthermore, damaged starch granules are less pressure resistant than intact starch granules (Bauer & Knorr, 2005). Another possible HP application is the modification of proteins. Dumay *et al.* (1998) observed gelatinization by HP treatment of solutions of a β-lactoglobulin isolate. The pressure-induced gel matrix offered small particles in highly packed β-lactoglobulin, unlike the gels after heating, which exhibit fine-stranded aggregates (Dumay *et al.*, 1998). Furthermore, Zeece *et al.* (2008) reported that an HP treatment from 400 MPa to 800 MPa at 20°C improved considerably the digestibility of β-lactoglobulin.

3.1.4. *Limitations of HPP*

Although great progress has been made in the development of economically viable high - pressure applications, the limitation of HPP to date mainly lies in the limited throughput and, relative to heat processing, the high cost of equipment, labour, and maintenance. High maintenance costs are caused mainly by the extreme processing conditions. Furthermore, there are only a few large-scale commercial high-pressure equipment suppliers worldwide that have expertise in the food industry, including Avure Technologies, Inc. (Kent, WA), Kobelco (Kobe Steel Ltd., Kobe, Japan), and NC Hyperbaric

(Burgos, Spain). Treatment non-uniformities can occur during HPP not only as a result of different compressibility of the various substances in the food product, including trapped air, but also because of the food packaging material. In addition, if the purpose of the process is the inactivation of the vegetative microorganisms, a non-uniform treatment can occur because some microorganisms are more resistant to the pressure when embedded in a fat matrix. Foods with higher fat or oil content may, therefore, protect the microorganisms in some areas in the food where fat is contained.

In addition to pressure, temperature is an important process variable. In heterogeneous food materials, with the contents exhibiting differences in compression heating, temperatures may not be uniformly distributed in the food products. Furthermore, the packaging material, the material of the product carrier, and the steel of the high-pressure vessel are not heated to the same extent as the food; therefore, temperature gradients are developed throughout the system, leading to heat flux from the products to the cooler areas. These spatial temperature heterogeneities increase over the process time. Although, theoretically, the preheated product heats up uniformly during compression to sterilization temperatures, during pressure holding time temperatures may decrease in certain areas of the vessel. This can affect spore inactivation, and spores may survive the process if temperature loss is not prevented. Product carriers have been developed as a means of retaining heat throughout the vessel during both pressure come-up and holding times.

3.2. Pulse Electrified Field (PEF)

PEF is considered as one of the most promising non-thermal food processing technologies alternative to the traditional food pasteurization. Pulsed electric field (PEF) employed with short bursts of electricity for microbial inactivation and causes minimal or no detrimental effect on food quality attributes. PEF can be used for processing liquid and semi-liquid food products. Application of PEF technology has been successfully demonstrated for the pasteurization of foods such as juices, milk, yogurt, soups, and liquid eggs. Application of PEF processing is restricted to food products with no air bubbles and with low electrical conductivity. The maximum particle size in the liquid must be smaller than the gap of the treatment region in the chamber in order to ensure proper treatment.

PEF is a continuous processing method, which is not suitable for solid food products that are not pumpable. The technology is based on the phenomena that biological membranes are punctured when an external electrical impulse is applied. This process is referred to as non-thermal as structural damage to membranes is realized at significantly low energy levels when compared to the process of heating. It is assumed that PEF has preservative effect by two fold-physical inactivation of bacteria and High Electrical field Pulses-improves the drying and re-hydration properties of dried vegetables to extend the shelf life. Application of PEF technology into food processing may avoid the detrimental effects on flavor, freshness, nutrition and bioactive functional components (Jia *et al.*, 1999). PEF may be helpful to increase the market share of the products due to the higher survival rate of functional components, such as IgG and Vitamin C, fresher and the better maintenance of nutrition of the food products compared to those processed with thermal processing. PEF is also applied to enhance extraction of sugars and other cellular content from plant cells, such as sugar beets. PEF also found application in reducing the solid volume (sludge) of wastewater.

3.2.1. *Mechanisms of PEF for inactivation of microorganisms*

The mechanisms involved in the inactivation of microorganisms have been investigated by many researchers. Under common conditions without any external electric field applied, the transmembrane potential across the cellular is about 10 millivolt, depending on different membrane composition and the environments where the cells stay. When a suspension of cells subjects to an external electric field, a potential difference across the membrane will be induced accordingly. More importantly, this across membrane potential difference, when it reaches to certain level, can dramatically increase the conductivity and permeability of the cell membranes. When this potential difference exceeds a critical value of 1 Volt in most cellular systems, membrane destruction or irreversible membrane changes can be observed, and, as a result, cause strong release of cytoplasm and the death or stress of the cells. Sale and Hamilton (1967) first reported this phenomenon with *E. coli* cell suspension at 10 kV/cm electric field strength. This phenomenon is nominated as "dielectric breakdown" by Zimmermann *et al.* (1980) and explained by the formation of pores within the cellular membranes due to the local electromechanical instability of cell membranes.

Sale and Hamilton (1968) reported that in microorganisms the structural changes due to a potential difference of about 1 volt gives rise to irreversible loss of membranes' function as the semi permeable barrier between the cell and its environment. Membrane damage is believed to be the number one cause responding to the inactivation effects on cells by PEF treatment. Some experiments also demonstrated that the cause of cell inactivation may be accomplished by electric fields. The correlation between the effects of the pulse treatment on cell inactivity and membrane damage, which was measured by poor to no spheroblast formation, suggested that cell inactivation and membrane damage have a very high positive association. This implies that membrane damage is a direct cause responding to the death of cells treated.

3.2.2. *Effects on microbes*

The factors that affect microbial inactivation during PEF treatment include, microbial factors such as type, shape, size, concentration, and growth stage of microorganism, and media factors such as pH, antimicrobials, ionic compounds, electrical conductivity, and medium ionic strength. The process factors such as electric field intensity, pulse width and shape, treatment time and PEF on the selected microorganisms are highly positive and, as a result, stimulated the studies on the inactivation of microorganisms by PEF, the novel nonthermal food pasteurization. It has been observed that electric pulses of high voltage electric fields up to 25kV inactivate *Escherichia coli*, *Staphylococcus aureus*, *Sarcina lutea*, *Bacillus subtillus*, *B. cereus*, *Clostridium welchii*, *Saccharomyces cerevisiae*, and *Candida utilis*. Electrical fields up to 25 kV/cm were applied as a series of direct current pulses from 2 to 20ms to the suspensions of microorganisms. The temperature increase during the electric field treatment during 10 pulses of 20 ms at 19.5 kV was 10°C so that we can think there was no thermal inactivation contributing to the total inactivation observed.

Electric field strength and the total treatment time are the two most important parameters responsible to the inactivation of microorganisms. However, as mentioned previously different cell structural properties, such as the size of the cell, the properties of the cell membranes, the growth phases of the microorganisms and the influences from media in which the cells grow are critical factors that show significant influences on the inactivation effects by PEF. Generally cells with larger size, such as yeast and mold vegetative cells are more sensitive to the electric fields than their smaller peers like bacterial cells and virus. Gram negative microorganisms are more sensitive to PEF treatment than Gram positive microorganisms. Some authors argued that the inactivation of bacterial cells is caused by the lyses of protoplasts, leakage of intracellular contents and loss of β-galactosidase activity. Castro *et al.* (1993) argued that PEF treated food retains "fresh" physical, chemical, and nutritional characteristics and possesses a satisfactory ambient shelf life.

Introduction of this new technology has the potential to provide consumers with microbiologically safe, minimally processed, nutritious, and fresh-like foods. These authors suggested that high intensity PEF is potentially the most important nonthermal pasteurization/sterilization technology available to replace or complement thermal processes. Jayaram *et al.* (1992) applied PEF to inactivate *Lactobacillus brevis* cells and reached the conclusion that the cell destruction was primarily due to the field-induced rupture of the cell wall and not to ohmic heating. Pothakamury *et al.* (1995) applied PEF to the cultures of *E. coli, S. aureus*, and *Bacillus subtillis*. PEF treatment at 16 kV/cm with 60 pulses at a pulse duration ranging from 200 to 300 ms inactivates 4 to 5 log cfu microbial population in model foods such as simulated milk ultra-filtrate. The cell suspension temperature was maintained below the lethal temperature. The results suggest that the microbial inactivation effect of PEF treatment is not due to thermal effect. Zhang *et al.* (1994) applied stepwise PEF treatment on *E. coli* at electric field strength ranged from 35 to 75 kV/cm with a pulse duration time of 2 ms and reported a 9 log-cycle reduction in the microbial population.

Many other microbial inactivation experimental results were reported by Qin *et al.* (1999). Pothakamury *et al.* (1996) reported their studies about the influences of microorganism growth stage and the processing temperature on the inactivation effects of *E. coli* by PEF. The authors reported that, at 3° to 40°C processing temperature, square 30 wave showed higher inactivation effects on *E. coli* suspended in simulated milk ultra filtrate (SMUF) than the exponential decaying wave pulses at same electric field strength of 36 kV/cm. Logarithmic-phase cells were more sensitive than stationary and lag-phase cells. This result implies that when we treat food samples not only the PEF parameter should be counted as the influential factors for the inactivation effects but the microorganisms themselves.

3.2.3. *Effect on enzymes*

The evaluation of the effect of PEFs on enzymes is complex. Available reports on mechanisms are limited, and different experimental setups and processing parameters make them difficult to compare. In general, most enzymes are not affected by PEF. The fact that the maximum temperature reached is lower than in thermal pasteurization means that some of the flavours associated with the raw material are not destroyed. PEF side effects such as changes of pH at the electrode surface due to electrochemical reactions (Saulis *et al.*, 2005), as well as the occurrence of temperature hot spots due to ohmic heating effects within a non uniform electric field (Jaeger *et al.*, 2009), may contribute to the observed overall enzyme inactivation during PEF processing. Although enzymes do not contain membrane structures, which are the target for an inactivation based on electroporation, the possible impact of PEF side effects indicates that process modifications toward the inactivation of microorganisms and enzyme structures are also possible The enzymes means that products have to be refrigerated after passing through PEF processing in order to slow the action of the enzymes. PEF alone is generally not capable of producing ambient shelf-stable products. However, acidic well-packaged products may have a useful ambient shelf life. Further a combination of mild heat and PEF might also be helpful to achieve sufficient enzyme inactivation to avoid the necessity of refrigerated storage.

3.2.4. *Effect on food constituents*

The influences of PEF on the main attributes of the food products treated with PEF have been under investigation. Jia *et al.* (1999) reported that the average loss of flavour compounds in orange juice processed by 240, 480 microseconds at 30 kV/cm with square waveform and 90°C 1 min thermal processing are 3.0, 9.0 and 22.0%, respectively. The results revealed that PEF causes minor flavor compound loss during pasteurization compared to its thermal peers. Yeom *et al.* (2000) claimed that PEF treatment inactivated 88% of pectin methyl esterase (PME) activity. No restoration of PME activity was observed during the storage time at 4°C and 22°C. PME causes cloud loss of orange juice by de-esterification of pectin (Kimball, 1991). Commercial thermal pasteurization of orange juice normally denatures 90~100% PME activity (Irwe and Olsson,

1994). The PEF induced enzyme activity changes are reported to be due to the conformational change of enzyme (Vega *et al.*, 1995).

Heat load during pasteurization can adversely impact the fresh flavor of orange juice (Ekasari *et al.*, 1988). Less heat encountered during PEF treatment thus causes less negative impacts on the fresh flavor of orange juice. Vitamin C is sensitive to thermal processing, particularly in low acid environment and with the present of oxidative substances. High temperature during processing and storage can cause loss of vitamin C concentration. The concentration of vitamin C in both PEF (35 kV/cm–59 ms) treated orange juice and heat treated (94.6°C for 1 minute) orange juice had almost same vitamin C concentration. The concentration of vitamin C were 53 mg/100 ml for PEF treated orange juice and 51 mg/100 ml for the heat treated ones, and very close to control samples. However, vitamin C in the PEF treated orange juice decreased significantly slower when stored at 4°C than that in the heat treated orange juice.

Among the main nutrients of food proteins, carbohydrates, lipids and vitamins are the ones that are influenced significantly by thermal pasteurization. Thermal processing causes denaturation of proteins and loss of functionalities of proteous bioactive components like IgG. However, PEF treatment of the immunoglobulin enriched soymilk at 41.1 kV/cm for 54 microseconds leads to no significant change in binding activity of bovine IgG (Li *et al.*, 2003). At this condition, PEF can inactivate about 5.3 logs of natural flora existing in the soymilk and lead to significant extension of shelf life of the product when stored at 4°C (Li *et al.*, 2003).

Nevertheless, Biedinger *et al.* (1991) reported that electromanipulation such as electrofusion and electropermeabilization can cause degradation of membrane lipids in plant cell protoplasts (*Vicia faba*). Addition of α-tocopherol and ascorbic acid can result in a decreased ethane formation of 22% in fused protoplasts as compared to 18% in non-fused ones. Slight changes can take place in lipid properties. So far no direct information about the PEF impacts on carbohydrates in food is available. However, Yeom *et al.* (2000) reported that the browning index of PEF treated orange juice is relatively lower than the heat treated orange juice. The theory behind this is not clear yet and worth further investigation. PEF impacts on minerals are negligible under PEF pasteurization conditions. Changes in electric conductivity after PEF treatment are very minor and negligible. Release of electrode materials into the food products is negligible. Furthermore, the release of electrode materials into the food products can be minimized by modifying the design of pulse generator (OSU-6), extending electrode life-time and minimizing the contacting area between electrodes and food.

3.2.5. *Economics*

The major costs of the PEF process are determined by the initial investment in equipment. This means that the major component in the additional costs per unit is determined by depreciation. Stork Food and Dairy systems have calculated that the total additional cost for PEF treatment for orange juice is 0.01 Euro per litre when compared to heat pasteurization. In the case of PEF large scale operation to a level of 5,000 to 10,000 L/hour is required to supply the markets of interest to a significant level. Large-scale operation of cost effective systems to a scale of 2000 L/hour has been demonstrated. Assuming a profit margin on freshly squeezed orange juice (currently sold at 2.80 EU/litre) of 0.04 Euro per litre this means that a return on an investment is achieved within 2 to 3 years. This is more than reasonable, taking into account that the costs of equipment and required modifications to the process are relatively small when PEF is implemented in an existing production line. Therefore the financial risks are acceptable when PEF treated products are introduced to the market.

3.2.6. *Limitations*

PEF treatment is effective for the inactivation of vegetative bacteria only. Treatment is not effective towards bacterial spores. Applications will therefore be found in the range of acid products and products that are distributed in the refrigerated chain. A second limitation is that treatment does not inactivate enzymes. After treatment products are therefore in general subject to enzymatic spoilage. Despite these limitations, PEF treatment has considerable added value for specific product ranges. For example, in the production of cheeses derived from raw milk PEF treatment is a possible solution to eliminate pathogens found in milk. A second example is the extension of the shelf life of stable acid products, *e.g.*, fruit juices obtained from concentrate, while maintaining the initial quality.

The technology that is required for the mild preservation of foods by PEF treatment has reached the stage of commercialization. PEF treatment is an alternative to continuous heat pasteurization of bulk product. The technical and economical feasibility of large-scale applications has been demonstrated. To ensure food safety, a risk assessment is required that deals with chemical safety of PEF treated products in relation to public health. Methods to quantify minute changes by electrochemical processes have been developed and are required for application within the framework of the EU Novel Food Regulation. Applications for bulk products such as dairy and fruit juices are most likely the first products that will be introduced to the market.

3.3. Ultrasound Treatment

Ultrasound is defined as sound waves with frequencies above the threshold for human hearing (>16 kHz) and in its most basic definition, refers to pressure waves with a frequency of 20 kHz or more (Butz *et al.*, 2002). Generally, ultrasound equipment uses frequencies from 20 kHz to 10 MHz. Higher-power ultrasound at lower frequencies (20 to 100 kHz), which is referred to as "power ultrasound," has the ability to cause cavitations (implosion of gas bubbles).

The bactericidal effect of ultrasound is attributed to intracellular cavitations, that is, micro-mechanical shocks that disrupt cellular structural and functional components up to the point of cell lysis.

Critical processing factors are the nature of the ultrasonic waves, the exposure time with the microorganisms, the type of micro-organism, the volume of food to be processed, the composition of the food, and the temperature. Low-intensity ultrasound provides information about physico-chemical properties, while high-intensity ultrasound is used to alter, either physically or chemically, the properties of foods, *e.g.*, to generate emulsions, disrupt cells, promote chemical reactions, inhibit enzymes, tenderize meat, and modify crystallization processes. Although ultrasound technology has a wide range of current and future applications in the food industry, including inactivation of micro-organisms and enzymes, presently, most developments for food applications are non-microbial. There are not many data on inactivation of food micro-organisms by ultrasound. Research activities centred on the combination of ultrasound with other preservation processes (*e.g.*, heat and mild pressure) which appear to have the greatest potential for industrial applications.

3.3.1. *Mechanisms of ultrasound effect*

When sound energy passes to the medium resulting in a continuous wave-type motion, longitudinal waves will be generated with the result that the motion creates alternative compression and rarefaction of the medium particles. In dependence of the frequency used and the sound wave amplitude applied, a number of physical, chemical and biochemical effects can be observed which enables a variety of applications. For food processing purposes, it is important to address the generation of heat due to ultrasound applications and the related cavitations caused by a rapid change of heating to 550°C and pressure increase to 50 Mpa. The temperature and pressure indicated are generated during very short periods of time at the point were cavitation occurs with an order of temperature variation of 109°C/s. Shock waves are generated due to cavitation, which contribute to the ultrasound effect. Formation and behavior of cavitation bubbles upon the propagation of the acoustic waves constitute the essential events, which induce the majority of the acoustic effects. The amount of energy released by cavitation depends on the kinetics of the bubble growth and collapse of the bubbles. This energy should increase with increasing surface tension at the bubble interface and decrease with increasing vapour pressure of the liquid. In particular, hydrated foods have a comparatively high surface tension, so it can be a very effective medium for cavitation (Save *et al.*, 1997).

Ultrasound is known to disrupt biological structures and when applied with sufficient intensity has the potential to cause cell death (Malo *et al.*, 2005). The bactericidal effect of ultrasound is attributed to intracellular cavitation; that is, micromechanical shocks that disrupt cellular structural and functional components up to the point of cell lysis. Critical processing factors include the nature of the ultrasonic waves; the exposure time with the microorganisms; the type of microorganism; the volume of food to be processed; the composition of the food, and the temperature. The lethal effect of ultrasound on some microorganisms was demonstrated first by Jacobs and Thornley (1954); thus, ultrasound has been proposed as a means of sterilization of liquid foods. The effects, however, are not severe enough for a sufficient reduction of microorganisms by ultrasound alone so most applications use combinations with other preservation methods. Because of the complexity and sometimes protective nature of the food, the singular use of ultrasound as a preservation method is impracticable. Although ultrasound technology has a wide range of current and future applications in the food industry, including inactivation of microorganisms and enzymes, presently, most developments for food applications are non-microbial.

3.3.2. *Effects on microbes*

Bactericidal effect of ultrasound was first reported in the 1920s. Ultrasound induced cell damage is primarily explained by cavitation phenomena such as shear disruption, localized heating, and free radical formation (Hughes & Nyborg, 1962). Cell wall and membrane of biological cells can be damaged by surface rubbing, leading to fracture and leakage (Kinsloe *et al.*, 1954). Equally, separation of the cytoplasmic membrane due to ultrasonic treatments has been reported, and free radicals are assumed to cause DNA damage (Hughes & Nyborg, 1962). Several studies showed an additive or even synergistic effect when US was combined with other lethal effects such as elevated temperatures (thermosonication), pressure (manosonication), or both (manothermosonication) (Knorr *et al.*, 2004), which has been explained by the weakening of the cell wall, making it more susceptible to the effects of cavitations. Ultrasonic pasteurization carried out at 50°C could present a high potential in preserving physicochemical properties, color, and flavor compared with conventional thermal pasteurization techniques. In combination of treatments *Escherichia coli* coud be reduced up to five-log cycle.

3.3.3. *Effect on enzymes*

The application of low energy ultrasound could have positive impact on enzyme activity. Substrates can be made available in large amounts, and the transport of substrate toward immobilized enzymes is increased by micro streaming. The use of thermosonication, monosonication and monothermosonication improves the inactivation of enzymes for food preservation than sonication alone. Nevertheless, the sensitivity can be very different from one enzyme to another. Positive results have been reported for tomato pectic enzymes as well as for α-chymotrypsin and porcine lipase, whereas phospholipase A2 was nearly insensitive to monothermosonication, and the sensitivity of trypsin was found to be temperature dependent (Vercet *et al.*, 2002).

3.3.4. *Effect on food constituents*

Sonication can modify proteins and product structure, improve flow behavior and increase heat transfer as it was exemplified with continuous formation of an edible coating of raw sausages (Knorr *et al.*, 2004). A heated tubular sonotrode was used to form a skin of denatured proteins around the raw emulsion inside. It has been observed that the physical properties of ultrasonic treated soy proteins showed significant texture changes in model food systems (Jambrak *et al.*, 2009). The application of ultrasonic treatment for the preservation of fruit juices was considered promising, as sonication resulted in only minor colour changes and retention of 94% of the anthocyanins in blackberry juice (Tiwari *et al.*, 2009).

Soy protein isolate creamed during the treatment with an ultrasonic probe (20 kHz, 15 min) and soy protein concentrate showed changes in conductivity, increased solubility, and increased specific surface area, which is of importance for food texture and functionality. Contrarily, sonication did not improve emulsifying or foaming properties. Extrusion processes can be improved by the ultrasonic excitation of a metal tube or extrusion dye and ultrasonic vibrations can lead to a reduction in drag resistance and improved flow characteristics (Mousavi *et al.*, 2007). In meat processing, sonication has been successfully applied to improve binding strength, water holding capacity, product color, and yield of processed meat (Vimini *et al.*, 1983), and was reported to be beneficial in the meat tenderization process (Roberts, 1992). Ultrasonic pretreatments of vegetables and fruits led to cavitation damage, an increased distortion of cells and the creation of microscopic channels (Fernandes *et al.*, 2009). These changes in tissue structure were shown to have potential for increasing mass transfer processes.

3.4. Infrared Heating

Food systems are complex mixtures of different biochemical molecules, biological polymers, inorganic salts, and water. The infrared spectra of such mixtures originate with the mechanical vibrations of molecules or particular molecular aggregates within a very complex phenomenon of reciprocal overlapping (Halford, 1957). Amino acids, polypeptides, and proteins reveal 2 strong absorption bands localized at 3 to 4 and 6 to 9 μm. On the other hand, lipids show strong absorption phenomena over the entire infrared radiation spectrum with 3 stronger absorption bands situated at 3 to 4, 6, and 9 to 10 μm, whereas carbohydrates yield 2 strong absorption bands centered at 3 and 7 to 10μm (Sandu, 1986; Rosenthal, 1992). Among all IR radiations [near- infrared (NIR), mid-infrared (MIR), and far-infrared (FIR), corresponding to the spectral ranges of 0.75 to 1.4, 1.4 to 3, and 3 to 1000 μm, respectively], FIR radiation is matter of interest for food processing because most food components absorb radioactive energy in the FIR region (Sandu, 1986).

The application of infrared radiation to food processing has gained momentum due to its inherent advantages over the conventional heating systems. Infrared heating has been applied in drying, baking, roasting, blanching, pasteurization, and sterilization of food products. Infrared heating provides an imperative place in drying technology and extensive research work has been conducted in this area. Application of FIR drying in the food industry is expected to represent a new process for the production of high-quality dried foods at low cost. The use of IR radiation technology for dehydrating foods has numerous advantages including reduction in drying time, alternate energy source, increased energy efficiency, uniform temperature in the product while drying, better-quality finished products, a reduced necessity for air flow across the product, high degree of process control parameters, and space saving along with clean working environment (Mongpreneet *et al.*, 2002).

Therefore, FIR drying operations have been successfully applied in recent years for drying of fruit and vegetable products such as potatoes (Afzal & Abe, 1998), sweet potatoes (Sawai *et al.*, 2004), onions (Sharma *et al.*, 2005), kiwifruit (Fenton & Kennedy 1998), and apples (Nowak & Levicki, 2004). It has been observed that solid materials absorb infrared radiation in a thin surface layer. However, moist porous materials are penetrated by radiation to some depth and their transmissivity depends on the moisture content. Energy and mass balance developed by Ratti and Mujumdar (1995) accounts for the shrinkage of the heated particle and absorption of infrared energy. Theoretical calculations showed that intermittent infrared drying with energy input of 10 W/m^2 becomes equivalent to convective drying in which the heat transfer coefficient would be as high as 200 W/m^{2}K.

3.4.1. *Inactivation mechanism of IR heating*

Inactivation of M/Os by IR heating may include inactivation mechanism similar to that of ultraviolet light (DNA damage) and microwave heating (induction heating) in addition to thermal effect, as infrared is located between ultraviolet and microwave in the electromagnetic spectrum. Thermal inactivation can damage DNA, RNA, ribosome, cell envelope, and proteins in microbial cell. Sawai *et al.* (1995) investigated the inactivation mechanism of *E. coli* treated with infrared radiation in phosphate buffer saline. They proposed that sublethally injured cells will become more sensitive to an inhibitory agent which has an inhibitory action on the damaged portion of the cell. Four inhibitory agents, namely, penicillin, chloramphenicol, rifampicin, and nalidixic acid, were used for the enumeration of pathogens. An 8 min infrared radiation at a wattage of 3.22 kW/m^2 resulted in approximately 1.8, 1.9, 2.7, and 3.2 $\log_{10}$ reduction of *E. coli*. When no inhibitory agents were present, a 1.8 log reduction was obtained. With conductive heating, similar damages were observed; however, RNA, protein, and cell wall showed more vulnerability to IR heating than conductive heating. The order of magnitude of infrared damages was as follows: protein> RNA> cell wall> DNA. RFP inhibits RNA polymerase in *E. coli* and CP binds ribosomal sub units and inhibits peptidyltransferase reactions (Sawai *et al.*, 1995).

3.4.2. *Effect on microbial inactivation*

IR heating can be used to inactivate bacteria, spores, yeast, and mold in both liquid and solid foods. Efficacy of microbial inactivation by infrared heating depends on the following parameters: infrared power level, temperature of food sample, peak wavelength, and bandwidth of infrared heating source, sample depth, types of microorganisms, moisture content, physiological phase of M/Os (exponential or stationary phase), and types of food materials.

Increase in the power of infrared heating source produces more energy and thus total energy absorbed by microorganisms' increases, leading to microbial inactivation. It has been found that surface temperature increased rapidly as infrared rays directly heated the surface without any need for conductors. Therefore, irradiating powers of 0.5, 1.0, 1.5, and 2.0 kW resulted in 60, 80, 125, and 195°C inside the experimental device, and 45, 65, 95, and 120°C on the surface of wheat stack, obtaining 0.83, 1.14, 1.18, and 1.90 log10CFU/g total bacteria after a 60 s treatment, respectively. Dry heat inactivation *of B. subtilis* spores by infrared radiation was investigated. D values of*B. subtilis* at 120, 140, 160, and 180°C were 26 min, 66, 9.3, and 3.2 s, respectively.

Shorter treatment time was enough to inactivate pathogens at higher temperatures and the estimated Z value was 23°C. *E. coli* population was reduced by 0.76, 0.90, and 0.98 $\log_{10}$ after 2 min exposure to IR radiation when the temperature of the bacterial suspension was maintained at 56, 58, and 61°C, correspondingly (Sawai *et al.*, 2003). As indicated earlier, food and microbial components absorb certain wavelengths of infrared radiation. Jun and Irudayaraj (2003) utilized selective infrared heating in the wavelength range of 5.88 to 6.66 μm using optical band pass filters for inactivation of *Aspergillus niger* and *Fusarium proliferatumin* corn meal. The selected wavelength denatures the protein in microorganisms, leading to a 40% increase in inactivation of*A. niger* and *F. proliferatum* compared to normal IR heating. For instance, a 5 min treatment with nonselective and selective heating resulted in approximately 1.8 and 2.3 $\log_{10}$CFU/g reduction of *A. niger*. The sample temperatures after selective or nonselective infrared heating were identical, absorption of energy by fungal spores increased in selective heating, leading to a higher lethal rate (Jun & Irudayaraj, 2003).

Total energy decreases as the peak wavelength increases. Therefore, NIR radiation with short wavelength has a relatively higher energy level than FIR radiation with longer wavelength. Hamanaka *et al.* (2006) studied the inactivation efficacy of *Bacillus subtilis* treated with 3 infrared heaters (A, B, and C) having different peak wavelengths (950, 1100, and 1150 nm) and radiant energies (4.2, 3.7, and 3.2 $\mu W/cm^2/nm$), respectively. It has been reported that pathogen inactivation was higher with heater A than those of heaters B and C, although temperature was the same for all the heaters. Therefore, it is obvious that inactivation efficiency is associated with the radiation spectrum. As the penetration depth of IR radiation is very low, an increase in the sample depth slows down the bulk temperature increase of the food sample. A 90% reduction in IR power was observed within a thin layer of 40 μm in bacterial suspension.

Therefore, the effect of IR radiation on the microbial inactivation diminishes as the sample thickness increases. Decreasing the sample depth also accelerates the inactivation of spores and *E. coli* and *S. aureus* (Hashimoto *et al.*, 1992). Further, resistance of bacteria, yeasts, and molds to infrared heating may be different due to their structural and compositional differences. In general, spores are more resistant than vegetative cells. When *Bacillus subtilis* spores in physiological saline were exposed to infrared heating, a spore population increased up to 5 times in the first 2 min, followed by subsequent exponential reduction, resulting in shoulder and tailing effects. Upon infrared heat treatment, vegetative cells were inactivated followed by activation of spores. Then vegetative cells formed from spores will be activated and thus spores will be inactivated. Again water molecules inside microorganisms readily absorb infrared radiation. These water molecules are attached to polar groups such as $-NH_2$, –COOH, and –COO within the cell (Hamanaka *et al.*, 2006). State and amount of water inside spores, bonding conditions of water molecules, and location of water molecule within M/Os affect their responses to infrared heating. Maximum D values of *B. subtilis* spores inactivated by IR heat differed with initial water activities ranging from 0.6 to 0.9. As the peak wavelength of the IR heating was short, the initial values of water activity leading to maximum D values for bacterial spores also increased.

3.4.3. *Effect on enzyme inactivation*

Infrared heating can be effectively used for enzyme inactivation. Lipooxygenase, an enzyme responsible for deterioration in soybeans, was inactivated 95.5% within 60 s of IR treatment (Kouzeh *et al.*, 1982). Certain enzyme reactions involving action of lipases and α amylases were affected by infrared radiation at a bulk temperature of 30 to 40°C (Sawai *et al.*, 2003). FIR radiation for 6 min resulted in a 60% reduction in lipase activity, while thermal conduction resulted in 70% reduction. It has been reported FIR treatment could successfully inactivate enzymes responsible for the development of off flavors in peas prior to the freezing process (van Zuilichem *et al.*, 1986). Galindo *et al.* (2005) investigated the application of IR heating of carrot slices prior to freezing as compared to blanching in terms of carrot cell and tissue damage. Carrot slices heated by FIR radiation contained damaged cells only in the first half millimeter from the surface and exhibited the texture characteristic of the raw tissue, thus providing the potential of FIR energy technology in the frozen carrots industry.

3.4.4. *Effect of IR on food constituents*

Application of infrared heating in a stepwise manner by slowly increasing the power, with short cooling between power levels, resulted in less color degradation than with intermittent infrared heating (Chua & Chou, 2005). Reductions in overall color change of 37.6 and 18.1% were obtained for potato and carrot, respectively. However, the quality of beef produced by infrared dehydration was similar to that of conventional heating as indicated by surface appearance and taste tests (Burgheimer *et al.*, 1971). Longer infrared heat treatments may darken the color of onion due to browning (Gabel *et al.*, 2006). Hebbar *et al.* (2003) suggested that 3 to 4 min infrared heat treatment was adequate for commercially acceptable products, with reduction in yeast cells and acceptable changes in hydroxymethylfurfural and diastase activity. Infrared heating raised the internal temperature of the strawberries not above 50°C, while the surface temperature was high enough to effectively inactivate microorganisms.

Therefore, infrared heating can be used for surface pasteurization of pathogens without deteriorating the food quality (Tanaka *et al.*, 2007). The evaluation of full-fat flour made from IR-heat treated soybeans maintained freshness similar to fresh flour for one year. However, untreated samples resulted in rancidity development (Kouzeh *et al.*, 1982). Compared to regular freeze drying, IR-assisted freeze-drying of yam brought about lower color differences as well as faster dehydration. Furthermore, infrared heating leading to a higher dehydration ratio implies that infrared heating reduces serious product shrinkage (Lin *et al.*, 2007). IR heat treated lentils were found to be darker than raw lentils, though there was no visible indication. Cell walls of lentils were less susceptible to fracture after infrared heat treatment, in addition to having a more open microstructure, thus enhancing the rehydration characteristics. It has been observed that pungency of onions following infrared radiation decreased with reduction in moisture (Gabel *et al.*, 2006). Infrared heating of carrots provided less damage to the tissue than blanching, as observed by lower relative electrolyte leakage values and microscopic observations (Galindo *et al.*, 2005). Infrared-treated carrots had higher tissue strength while effectively inactivating the enzymes on carrot surface.

Although infrared heat-treated turkey samples were slightly darker than the controls after treatments, refrigerated storage for an hour resulted in no significant difference in color values. Bitterness and protein solubility of peas were reduced after IR heat treatment (McCurdy, 1992). Furthermore, chlorophyll content of dehydrated onions treated by infrared increased with an increase in irradiation power (Mongpreneet *et al.*, 2002). Infrared heating provided a more appealing brown color and roasted appearance to deli turkey, in addition to effectively pasteurizing the surface (Muriana *et al.*, 2004). Infrared heating and jet impingement of bread resulted in rapid drying and enhanced color development, compared to conventional heat treatment (Olsson *et al.*, 2005). Though the thickness of bread crust increased faster, a short IR treatment time enabled the formation of thinner crust.

3.5. Plasma Treatment

Plasmas can be described as quasineutral particle systems in the form of gaseous or fluid-like mixtures of free electrons and ions, frequently containing neutral particles, with a large mean kinetic energy of the electrons and/or all of the plasma components and a substantial influence on the charge carriers and their electromagnetic interaction on the system properties. According to Fridman *et al.* (2005), plasma systems are two major categories: thermal and non-thermal plasmas. Thermal plasmas usually arcs or radiofrequency (RF) inductively coupled plasma discharges are associated with Joule heating and thermal ionization, and enable the delivery of high power (to over 50 MW per unit) at high operating pressures. Very high gas temperatures limit their applicability to food systems.

In non-thermal plasmas, the electron temperature is much higher (several to ten thousands K) than the bulk gas temperature (below 40°C). The non-thermal plasmas may be produced by a variety of electrical discharges at different pressure levels. The most suitable system for food processing is an atmospheric-pressure plasma device in which no extreme conditions are required and low temperatures can be realized. Atmospheric pressure plasma is commonly generated by corona discharge, dielectric barrier discharge (DBD), or plasma jet (Keener, 2008). For the treatment of non-uniformly shaped products, the application of plasma jets offers advantages due to various options regarding design and construction (Foest *et al.*, 2005).

3.5.1. *Effect on microorganisms*

Inactivation of food-related microorganisms by plasma treatment is commonly conducted using model systems. The antimicrobial activity of non-thermal plasma against Gram-negative and Gram-positive bacteria, yeast and fungi, biofilm formers, and endospores was reported (Brandenburg *et al.*, 2007). However, the inactivation mechanisms of plasma, is not yet fully understood. According to Moisan *et al.* (2001) three basic mechanisms are involved in plasma inactivation: (*a*) ultraviolet (UV) irradiation of genetic material, (*b*) intrinsic photo-desorption, and (*c*) etching. Among these UV plays a minor role in the inactivation of microorganisms at atmospheric pressure, and the inactivation process is controlled by chemically reactive species.

However, it was shown that in some cases UV photons can play a role in the inactivation process of microorganisms at atmospheric pressure (Boudam *et al.*, 2006). Moreau *et al.* (2008) compared plasma inactivation effects to the effects of micropulses. Similar to the effects of micropulses, the cell membranes of microorganisms are perforated after plasma treatment. Besides the perforation of the cell membrane, the inactivation effect of plasma is induced by the bombardment of the cell membrane by radicals (OH or NO). These radicals are absorbed onto the bacteria surface, and volatile components are formed and eliminated from the cells (etching). Two mechanisms of plasma inactivation described by Gaunt *et al.* (2006) are the electrostatic disruption of cell membranes and lethal oxidation of cellular components.

3.5.2. *Effect on enzymes*

It has been reported that the enzymatic activity of food substrate could be decreased after treatment with plasma generated in radio frequency driven glow discharge with the highest decrease in activity within the first 10 minutes of treatment (Dudak *et al.*, 2007). They showed plasma-chemical oxidation as well as fragmentation of the proteins. Changes in the secondary protein structure were also detected. A fragmentation of proteins was also found after dielectric barrier discharge treatment (Deng *et al.*, 2007) where, a atomic oxygen was shown to play a dominant role in destruction and degradation reactions. Oxygen plasma generated by RF discharge led to a reduction of C–H and N–H bonds in casein protein and to a modification of the secondary protein structure.

3.5.3. *Effect on food constituents*

There is very little information currently known about the effect of plasma treatment on bioactive plant substances. The degradation of mycotoxins in microwave-induced atmospheric pressure argon plasma has recently been shown by Park *et al.* (2007). Plasma treatment resulted in a significant time-dependent decrease in aflatoxin B_1, deoxynivalenol, and nivalenol, coming along with a dose-dependent reduced cytotoxity. A time and dose-dependent degradation has also been observed for flavonoids, known for their high antioxidant activity, which protects cells against the damaging effects of reactive oxygen species. The degradation rate strongly depended on the polyphenolics substitution pattern. Although glycosidic flavonoids showed a rather inert behaviour throughout plasma treatment, aglycoside derivates were quickly degraded. The potential of non-thermal plasma surface treatment to decontaminate food surfaces is investigated in various studies. However, organoleptic analysis of plasma-treated nut samples likewise showed no relation between treatment and perceptual sensory character (Basaran *et al.*, 2008). On a molecular level, it has been reported that cabbage and lettuce species subjected to plasma treatment may lead to changes of plant surface hydrophobic wax layers (Grzegorzewski *et al.*, 2010). On the other hand, Ragni *et al.* (2010) observed no negative effects of plasma treatment on egg quality.

3.6. Membrane Technology

Membrane processing is a technique that permits concentration and separation without the use of heat. Particles are separated on the basis of their molecular size and shape with the use of pressure and specially designed semi-permeable membranes. When a solution and water are separated by a semi-permeable membrane, the water will move into the solution to equilibrate the system. This is known as osmotic pressure. If a mechanical force is applied to exceed the osmotic pressure (up to 700 psi), the water is forced to move down the concentration gradient *i.e.*, from low to high concentration. Permeate designates the liquid passing through the membrane, and retentate (concentrate) designates the fraction not passing through the membrane. The different membrane processes include reverse osmosis (RO), nanofiltration (NF), ultrafiltration (UF), microfiltration (MF), electrodialysis (ED), and membrane reactions (MR).

Various improvements have been achieved by utilizing membrane technology in the dairy industry, *e.g.*, for bacteria removal (MF), whey treatment (UF, NF), demineralization (NF, ED), fractionation of milk and whey (UF, NF) and water recycling (RO). More and more, membrane technologies are refining or completely replacing conventional process lines. This results in reduced operating costs and increased capacities of existing plants. UF units are most effectively used to improve the separation of proteins, fats and solids and to yield products that are more uniform in appearance and composition. NF devices are most effective in demineralization processes. Following will be a description of membrane applications in the dairy industry, and an evaluation of an exemplary case study from a medium sized dairy company.

Among the different membrane systems, reverse osmosis (RO) and ultra-filtration (UF) are being used extensively in food and dairy industries. RO is a single phase concentration process which uses a pressure gradient across a semi permeable membrane to squeeze water through membrane. RO process is extremely energy efficient compared to both evaporation and freeze concentration. Ultra filtration uses much lower pressure 1 to 10 bars and much more open membranes, which pass salts, sugars and organics in the molecular range typically from 5,000 to 1,00,000 depending on the membrane type. It is limited by osmotic pressures, since the sugars are not concentrated. Both RO and UF have promising uses in fruit and vegetable juice industry as a unit operation for concentration or aroma recovery and clarification of juices respectively.

The primary goal of UF in fruit industry is to replace the holding filtration and decantation steps of traditional process. Enzyme treatment is required to reduce viscosity of juice by partially hydrolyzing. It is a clarification process to remove pectin, enzyme and other fibrous components, constituting the clear juice. For separation of molecules, semi permeable membrane is used at a temperature of 50–55°C (high) on 10–15°C (low), depending on the type of juice and sensitivity. Tubular modules can be used for viscous partially depectinized juice, whereas, prefiltration of juice is necessary when this channel on boiled fiber UF system is used. Many case studies have done on apple juice. UF system generates cost savings and manpower reduction, uses only electrical energy to raise the pressure of juice feed, operating costs are typically 5-10 times lower than normal operations, process control is simple, no cooling water equipment is needed and products have better flavour. In nutshell, UF has become economical viable alternative for clarification of juice in comparison to conventional method of clarification. RO is a well established process for concentration/pre concentration of raw and clear depectinized juice from fruits and vegetable. It consumes 10 times less energy for renouncing water when compared with conventional evaporators.

Microfiltration (MF) designates a membrane separation process similar to UF but with even larger membrane pore size allowing particles in the range of 0.2 to 2 micrometers to pass through. The pressure used is generally lower than that of UF process. The membrane configuration is usually cross-flow. MF is used in the dairy industry for making low-heat sterile milk as proteins may pass through but bacteria do not.

Electrodialysis is used for demineralization of milk products and whey for infant formula and special dietary products. Also used for desalination of water. Under the influence of an electric field, ions move in an aqueous solution. The ionic mobility is directly proportioned to specific conductivity and inversely proportioned to number of molecules in solution. $\sim 3\text{–}6 \times 10^2$ mm/sec. Charged ions can be removed from a solution by synthetic polymer membranes containing ion exchange groups. Anion exchange membranes carry cationic groups which repel cations and are permeable to anions, and cation exchange membranes contain anionic groups and are permeable only to cations. Electrodialysis membranes are comprised of polymer chains-styrene-divinyl benzene made anionic with quaternary ammonium groups and made cationic with sulphonic groups. 1–2V is then applied across each pair of membranes.

3.7. UV Processing

UV light for food processing has been investigated for many years but is still considered as an innovative technology in food processing. In this technology, UV-C light (wavelength of 254 nm) is predominantly being used as a disinfection method to inhibit or inactivate food-borne microorganisms, mainly in liquid food products. Fresh produce can be processed using UV light, which has a germicidal effect on many types of microorganisms like bacteria, viruses, protozoa, molds, and yeasts. However, the effect of UV light on microorganisms in liquids depends on variables such as density of the liquid, types of microorganisms. Although the use of UV light is well established for air and water treatment and surface decontamination, its use for treating liquid foods is still limited.

Recently, interest in using UV has increased as a viable alternative to thermal pasteurization for a range of liquid foods and ingredients like fresh juices, fruit purees, soft drinks, raw milk, liquid eggs, liquid sugars and sweeteners, etc. (Koutchma, 2009). Pumpable fruit and vegetable products are generally very suitable for processing by UV light to reduce the microbial load as long as sufficient fluid mixing allows the entire product to be exposed to a certain required dose of UV radiation. The germicidal properties of UV irradiation are mainly due to DNA damage induced through absorption of UV light by DNA molecules. This mechanism of inactivation results in a sigmoidal curve of microbial population reduction. UV treatment can be used for primary disinfection or as a backup for other purification methods such as carbon filtration, reverse osmosis, or pasteurization.

As UV has no residual effect, the best position for a treatment system is immediately prior to the point of use. This ensures that any incoming microbiological contaminants are destroyed and that there is little chance of post-treatment contamination. In addition to UV light with wavelengths other than 254 nm can also be used as a radiation source to inactivate microorganisms in foods. In general, wavelengths ranging from 100 to 400 nm are suitable for UV light processing (Sastry *et al.*, 2000). UV disinfection has many advantages over alternative methods. Unlike chemical treatment, UV does not introduce toxins or residues into the process and mostly does not alter the chemical composition, taste, odour, or pH of the water or liquid being disinfected. As a physical method, UV irradiation has a positive consumer image and is of interest to the food industry as a low cost non-thermal method of preservation. Recent advances in the science and engineering of UV light irradiation have demonstrated that this technology holds considerable promise as an alternative method to traditional thermal pasteurization for liquid foods and ingredients, fresh juices, soft drinks, and beverages.

3.8. Irradiation

Preserving food to maintain its quality and safety is challenging task by the meat industry. For centuries, great efforts have been devoted to finding ways of preserving food and protecting it against physical damages, chemical changes and further microbial contaminations and spoilage. Drying was one of the first techniques developed. Chilling, freezing, heating including canning, curing and smoking also have long histories of use in food preservation. Later several innovations include the use of chemical preservatives such as polyphosphates, sodium ascorbates or isoascorbates, antibiotics, bacterial peptides etc. Now several novel techniques such as use of high pressure treatment, microwave heating, and irradiation have joined the array of food protection methods.

There are three main types of ionizing radiation-Gamma (γ) rays from a radioactive source such as Cobalt-60 and Cesium-137, high energy electrons produced by an electron beam accelerator and X-rays. When beef is irradiated, it is conveyed through a treatment chamber and exposed to high energy electron. In the case of gamma facility, the food is exposed to an ionizing energy source contained in stainless steel tube. In the case of linear accelerator facility, the food is exposed to electrons or X-rays, which is delivered in rapid pulses. The irradiator emits short wavelengths of energy that pass through the food. Some electron accelerator can be designed to generate only electrons and thus can be used as an online process inside a food processing plant with just a few inches of shielding surrounding the source. High energy ionizing radiation produces positive and negative charges that disabled bacterial DNA, thereby eliminating the threat to the human health. The process of food irradiation often called "cold sterilization" because it kills most bacteria without the use of heat. This means that food can be irradiated within its packaging and remain protected against recontamination until open by user.

3.8.1. *The process of food irradiation*

Food irradiation is a controlled treatment of food with ionizing energy, a part of the electro-magnetic energy spectrum that also includes radio or television waves, microwaves, infrared radiation, visible light and ultraviolet radiation. Ionizing radiation has a shorter wave length and consequently higher energy than other radiation in the spectrum. Levels of absorbed radiation are measured in kilo grays (kGy). The scientific community has defined three levels of food irradiation. Again the FDA has approved medium dose treatments for fresh and frozen meat.

3.8.2. *Irradiation and food safety*

Substantial information has been collected on the chemical changes that occur when foods are irradiated. Sensitive analytical techniques have identified many of the radiolytic products, and most are proved to be familiar substances that exists in non-irradiated foods or are produced in food are processes such as cooking. No truly unique substances have been identified and no significant evidence of hazard to human health has been found. FDA scientists concluded that irradiation reduces or eliminates pathogenic bacteria, insects and parasites. It also reduces spoilage and, in certain fruits and vegetables, inhibits sprouting and delays the ripening process. It does not make food radioactive, compromise nutritional quality or noticeably change food taste, smell texture or appearance as long as it is applied properly to a suitable product.

3.8.3. *The FDA decisions*

On December 2, 1997, the FDA approved the irradiation of meat products (fresh and frozen beef, lamb and pork) for controlling disease-causing microorganisms such as *Escherichia coli* O157:H7, *Salmonella*, *Listeria* and other food borne pathogens. The ruling FDA amended the food additive regulations to provide for the safe use of a source of ionizing radiation to treat refrigerated (4.5 kGy maximum permitted dose) or frozen (7.0 kGy maximum permitted dose) uncooked meat, meat by-products and certain meat food products (*e.g.* ground beef and hamburger) to control food borne pathogens and extend product shelf life. In 1992, USDA approved fresh and frozen poultry irradiation at doses from 1.5 kGy to 3.0 kGy.

3.8.4. *Impact of food irradiation on human health*

Based on chemical analyses animal feeding trails and consumption by human volunteers, sound scientific studies indicate that the radiolytic products produced by irradiation pose no risk to humans. In fact, people requiring the safest food-hospital patients in USA and the UK receiving bone marrow transplants are given irradiated foods at higher doses to ensure product safety. Analyses of irradiated meat products show that a toxicological hazard due to the consumption of irradiated meat is highly unlikely because no substance resulting from irradiation has been found at levels that justify the toxicological concern.

3.8.5. *Impact of irradiation on food constituents*

Studies have shown that low- and medium dose irradiation treatments such as the type used for meats, do not cause significant decreases in the nutritional quality of foods. High dose treatments cause smell, but measurable losses of some vitamins; however, these losses are similar to those caused by other processing techniques, such as canning, drying, and heat pasteurization, which produced a similar degree of preservation. Such losses therefore are not considered a drawback to the use of irradiation for food preservation.

3.9. Oscillating Magnetic Fields

The strong static (SMF) or oscillating (OMF) magnetic fields (5–50 Tesla) have the potential to inactivate vegetative microorganisms. The impulse duration is between 10 ms and several milliseconds. The frequencies are maximally 500 MHz, because above that value the items begin to warm up noticeably. Preservation of foods with OMF involves sealing food in a plastic bag and subjecting it to 1–100 pulses in an OMF at temperature of 0 to 50°C for a total exposure time ranging from 25 to 100 ms (Pothakamury *et al.*, 1993). However, the effects of magnetic fields on microbial populations have produced controversial results (Tsuchiya *et al.*, 1996). So, before considering this technology for food preservation purposes consistent results concerning the efficacy of the method are needed.

REFERENCES

Afzal, T.M. and Abe, T. (1998). Diffusion in potato during far infrared radiation drying. *J. Food Eng.*, 37(4): 353–365.

Ananta, E. and Knorr, D. (2003). Pressure-induced thermotolerance of *Lactobacillus rhamnosus* GG. *Food Res. Int.*, 36: 991–997.

Ananta, E., Heinz, V. and Knorr, D. (2005). Assessment of high pressure induced damage on *Lactobacillus rhamnosus* GG by flow cytometry. *Food Microbiol.*, 21: 567–577.

Basaran, P., Basaran-Akgul, N. and Oksuz, L. (2008). Elimination of *Aspergillus parasiticus* from nut surface with low pressure cold plasma (LPCP) treatment. *Food Microbiol.*, 25: 626–632.

Bauer, B.A. and Knorr, D. (2005). The impact of pressure, temperature and treatment time on starches: Pressure induced starch gelatinisation as pressure time temperature indicator for high hydrostatic pressure processing. *J. Food Eng.*, 68: 329–334.

Biedinger, U., Bickert, C., Youngman, R.J. and Schnabl, H. (1991). The formation of free lipid radicals during the electromanipulation of protoplasts (*Vicia faba*). *Botanica Acta*, 104(1991): 217–221.

Boudam, M.K., Moisan, M., Saoudi, B., Popovici, C., Gherardi, N. and Massines, F. (2006). Bacterial spore inactivation by atmospheric-pressure plasmas in the presence or absence of UV photons as obtained with the same gas mixture. *J. Phys. D-Appl. Phys.*, 39: 3494–3507.

Brandenburg, R., Ehlbeck, J., Stieber, M., von Woedtke, T. and Zeymer, J. (2007). Antimicrobial treatment of heat sensitive materials by means of atmospheric pressure rf-driven plasma jet. *Contrib. Plasma Phys.*, 47: 72–79.

Burgheimer, F., Steinberg, M.P. and Nelson, A.I. (1971). Effect of near infrared energy on rate of freeze-drying of beef spectral distribution. *J. Food Sci.*, 36(l): 273–276.

Butz, P. and Tauscher, B. (2002). Emerging technologies: Chemical aspects. *Food Res. Int.*, 35(2/3): 279–284.

Castro, A.J., Barbosa–Canovas, G.V. and Swanson, B.G. (1993). Microbial inactivation of foods by pulsed electric-fields. *J. Food Process Preserv.*, 17(1): 47–73.

Castro, I., Teixeira, J.A., Salengke, S., Sastry, S.K. and Vicente, A.A. (2004). Ohmic heating of strawberry products: electrical conductivity measurements and ascorbic acid degradation kinetics. *Innovative Food Sci. Emerging Technol.*, 5: 27–36.

Cho, H.Y., Yousef, A.E. and Sastry, S.K. (1999). Kinetics of inactivation of *Bacillus subtilis* spores by continuous or intermittent ohmic and conventional heating. *Biotechnol. Bioengineer*, 62(3): 368–372.

Chua, K.J. and Chou, S.K. (2005). A comparative study between intermittent microwave and infrared drying of bioproducts. *Int. J. Food Sci. Technol.*, 40: 23–39.

Daeschlein, G., Woedtke, T.V., Kindel, E., Brandenburg, R., Weltmann, K.D. and Junger, M. (2010). Antibacterial activity of an atmospheric pressure plasma jet against relevant wound pathogens *in vitro* on a simulated wound environment. *Plasma Process Polym.*, 7: 224–230.

Deng, X.T., Shi, J.J., Chen, H.L. and Kong, M.G. (2007). Protein destruction by atmospheric pressure glow discharges. *Appl. Phys. Lett.*, 90: 501–504.

Diels, A.M.J. and Michiels, C.W. (2006). High-pressure homogenization as a non-thermal technique for the inactivation of microorganisms. *Crit. Rev. Microbiol.*, 32: 201–216.

Dudak, F.C., Kousal, J., Seker, U.O.S, Boyaci, I.H., Choukourov, A. and Biederman, H. (2007). Influence of the plasma treatment on enzyme structure and activity. Presented at ICPIG, 28th, Prague.

Dumay, E.M., Kalichevsky, M.T. and Cheftel, J.C. (1998). Characteristics of pressure-induced gel of beta-lactoglobulin at various times after pressure release. *Lebensmittel-Wiss Technol.*, 31: 10–19.

Ekasari, I., Jongen, W.M.F., Vermunt, A.E.M. and Pilnik, W. (1988). Measurment of heat load in orange juices: Use of microbiological methods. *Food Technol.*, 42(5): 124–128.

Fernandes, F.A.N., Gall˜ao, M.I. and Rodrigues, S. (2009). Effect of osmosis and ultrasound on pineapple cell tissue structure during dehydration. *J. Food Eng.*, 90: 186–190.

Foest, R., Kindel, E., Ohl, A., Stieber, M. and Weltmann, K.D. (2005). Non-thermal atmospheric pressure discharges for surface modification. *Plasma Phys. Control Fusion*, 47: B525–536.

Fridman, A., Chirokov, A. and Gutsol, A. (2005). Non-thermal atmospheric pressure discharges. *J. Phys. D. Appl. Phys.*, 38: R1–R24.

Gabel, M.M., Pan, Z., Amaratunga, K.S.P., Harris, L.J. and Thompson, J.F. (2006). Catalytic infrared dehy- dration of onions. *J. Food Sci.*, 71(9): E351–357.

Galindo, F.G., Toledo, R.T. and Sjoholm, I. (2005). Tissue damage in heated carrot slices. Comparing mild hot water blanching and infrared heating. *J. Food Eng.*, 67: 381–385.

Gaunt, L.F., Beggs, C.B. and Georghiou, G.E. (2006). Bactericidal action of the reactive species produced by gasdischarge nonthermal plasma at atmospheric pressure: A review. *IEEE Trans. Plasma Sci.*, 34: 1257–1269.

Grauwet, T., Van der Plancken, I., Vervoort, L., Hendrickx, M.E. and Van Loey, A. (2010). Protein-based indicator system for detection of temperature differences in high pressure high temperature processing. *Food Res. Int.*, 43(3): 862–871.

Grzegorzewski, F., Rohn, S., Kroh, L.W., Geyer, M. and Schluter, O. (2010). Surface morphology and chemical composition of lamb's lettuce (*Valerianella locusta*) after exposure to a low-pressure oxygen plasma. *Food Chem.*, 122: 1145–1152.

Halford, R.S. (1957). The influence of molecular environment on infrared spectra. *Ann. New York Acad Sci.*, 69: 63–9.

Hamanaka, D., Uchino, T., Furuse, N., Han, W. and Tanaka, S. (2006). Effect of the wavelength of infrared heaters on the inactivation of bacterial spores at various water activities. *Int. J. Food Microbiol.*, 108: 281–285.

Hashimoto, A., Sawai, J., Igarashi, H. and Shimizu, M. (1992). Effect of far-infrared irradiation on pasteurization of bacteria suspended in liquid medium below lethal temperature. *J. Chem. Eng. Japan*, 25(3): 275–81.

Hebbar, H.U., Nandini, K.E., Lakshmi, M.C. and Subramanian, R. (2003). Microwave and infrared heat processing of honey and its quality. *Food Sci. Technol. Res.*, 9: 49–53.

Hendrickx, M., Ludikhuyze, L., Van den Broeck, I. and Weemaes, C. (1998). Effects of high pressure on enzymes related to food quality. *Trends Food Sci. Technol.*, 9(5): 197–203.

Heinz, V. and Knorr, D. (1996). High pressure inactivation kinetics of *Bacillus subtilis* cells by a three-state-model considering distribution resistance mechanisms. *Food Biotechnol.*, 10: 149–161.

Hughes, D.E. and Nyborg, W.L. (1962). Cell disruption by ultrasound: Streaming and other activity around sonically induced bubbles is a cause of damage to living cells. *Science*, 138: 108–114.

Icier, F., Yildiz, H. and Baysal, T. (2006). Peroxidase inactivation and colour changes during ohmic blanching of pea puree. *Journal of Food Engineering*, 74(3): 424–429.

Icier, F., Yildiz, H. and Baysal, T. (2008). Polyphenoloxidase deactivation kinetics during ohmic heating of grape juice. *J. Food Engineer*, 85(3): 410–417.

Irwe, S. and Olsson, I. (1994). Reduction of pectinesterase activity in orange juice by high pressure-treatment. In *Minimal Processing of Foods and Process Optimization*. Singh. R.P. and Oliveira, F.A.R. *Eds.* CRC Press, Ann Arbor, MI, pp. 35–42.

Isbarn, S., Buckow, R., Himmelreich, A., Lehmacher, A. and Heinz, V. (2007). Inactivation of avian influenza virus by heat and high hydrostatic pressure. *J. Food Prot.*, 70: 667–673.

Jaeger, H., Meneses, N. and Knorr, D. (2009). Impact of PEF treatment inhomogeneity such as electric field distribution, flow characteristics and temperature effects on the inactivation of *E.coli* and milk alkaline phosphatase. *Innov. Food Sci. Emerg. Technol.*, 10: 470–480.

Jacobs, S.E. and Thornley, M.J. (1954). The lethal action of ultrasonic waves on bacteria suspended in milk and other liquids. *J. Appl. Bacteriol.*, 17: 38–56 (from Pagan *et al.*, 1999).

Jambrak, A.R., Lelas, V., Mason, T.J., Kresic, G. and Badanjak, M. (2009). Physical properties of ultrasound treated soy proteins. *J. Food Eng.*, 93: 386–393.

Jayaram, S., Castle, G.S.P. and Pargaritis, A. (1992). Kinetics of sterilization of *Lactobacillus brevis* cells by the application of high voltage pulses. *Biotech. Bioeng.*, 40: 1412–1420.

Jia, M.Y., Zhang, Q.H. and Min, D.B. (1999). Pulsed electric field processing effects on flavour compounds and microorganisms of orange juice. *Food Chemistry*, 65: 445–451.

Jun, S. and Irudayaraj, J. (2003). A dynamic fungal inactivation approach using selective infrared heating. *Trans ASAE*, 46(5): 1407–1412.

Keener, K.M. (2008). Atmospheric non-equilibrium plasma. *Encycl. Agric. Food Biol. Eng.*, 1: 1–5.

Kimball, D.A. (1991). Juice cloud, color of citrus juices and citrus microbiology. *In: Citrus processing-Quality Control and Technology*, Van Nostrand Reihold: New York, pp. 117–243.

Kinsloe, H., Ackermann, E. and Reid, J.J. (1954). Exposure of microorganisms to measured sound fields. *J. Bacteriol.*, 68: 373–380.

Knorr, D., Zenker, M., Heinz, V. and Lee, D.U. (2004). Applications and potential of ultrasonics in food processing. *Trends Food Sci. Technol.*, 15: 261–66.

Knorr, D., Heinz, V. and Buckow, R. (2006). High pressure application for food biopolymers. *Biochim. Biophys. Acta*, 1764: 619–631.

Koutchma, T. (2009). Advances in ultraviolet light technology for non-thermal processing of liquid foods. *Food Bioprocess Technol.*, 2(2): 138–155.

Kouzeh, K.M., van Zuilichem, D.J., Roozen, J.P. and Pilnik, W. (1982). A modified procedure for low-temperature infrared radiation of soybeans.II. Inactivation of lipoxygenase and keeping quality of full fat flour. *Lebensm. Wiss. Technol.*, 15(3): 139–142.

Li, S.Q., Zhang, Q.H., Lee, Y.Z. and Pham, T.V. (20030. Effects of pulsed electric fields and thermal processing on the stability of Bovine Immunoglobulin G (IgG) in enriched soymilk. *J. Food Sci.*, 68(4): 1201–1207.

Lin, Y.P., Lee, T.Y., Tsen, J.H., King and An-Erl, V. (2007). Dehydration of yam slices using FIR-assisted freeze-drying. *J. Food Eng.*, 79: 1295–301.

Leizerson, S. and Shimoni, E. (2005). Stability and sensory shelf life of orange juice pasteurized by continuous ohmic heating. *J. Agric. Food Chem.*, 53: 4012–4018.

Malo, A.L., Palou, E., Fernandez, M.J., Alzamora, S.M. and Guerrero, S. (2005). Multifactorial fungal inactivation combining thermosonication and antimicrobials. *J. Food Engineer*, 67: 87–93.

Marra, F., Lyng, J., Romano, V. and McKenna, B. (2007). Radio-frequency heating of foodstuff: Solution and validation of a mathematical model. *J. Food Engineer*, 79: 998–1006.

Margosch, D., Ehrmann, M.A., Buckow, R., Heinz, V., Vogel, R.F. and G¨anzle, M.G. (2006). High pressure-mediated survival of *Clostridium botulinum* and *Bacillus amyloliquefaciens* endospores at high temperature. *Appl. Environ. Microb.*, 72(5): 3476–3481.

Mathys, A., Chapman, B., Bull, M., Heinz, V. and Knorr, D. (2007). Flow cytometric assessment of *Bacillus* spore response to high pressure and heat. *Innov. Food Sci. Emerg. Technol.*, 8: 519–527.

McCurdy, S.M. (1992). Infrared processing of dry peas, canola, and canola screenings. *J. Food Sci.*, 57(4): 941–944.

Moisan, M, Barbeau, J., Moreau, S., Pelletier, J., Tabrizian, M. and Yahia, L.H. (2001). Low-temperature sterilization using gas plasmas: A review of the experiments and an analysis of the inactivation mechanisms. *Int. J. Pharmaceutics*, 226: 1–21.

Mongpreneet, S., Abe, T. and Tsurusaki, T. (2002). Accelerated drying of welsh onion by far infrared radiation under vacuum conditions. *J. Food Eng.*, 55: 147–156.

Moreau, M., Orange, N. and Feuilloley, M.G.J. (2008). Non-thermal plasma technologies: New tools for biodecontamination. *Biotechnol. Adv.*, 26: 610–617.

Mousavi, S.A.A.A., Feizi, H. and Madoliat, R. (2007). Investigations on the effects of ultrasonic vibrations in the extrusion process. *J. Mater Process Technol.*, 187–188: 657–661.

Muriana, P., Gande, N., Robertson, W., Jordan, B. and Mitra, S. (2004). Effect of prepackage and post- package pasteurization on postprocess elimination of Listeria monocytogenes on deli turkey products. *J. Food Prot.*, 67(11): 2472–2479.

Nowak, D. and Levicki, P.P. (2004). Infrared drying of apple slices. *Innov. Food Sci. Emerg. Technol.*, 5: 353–360.

Oey, I., Verlinde, P., Hendrickx, M. and Van Loey, A. (2006). Temperature and pressure stability of 1-ascorbic acid and/or [6s] 5-methyltetrahydrofolic acid: A kinetic study. *Eur. Food Res. Technol.*, 223: 71–77.

Olsson EEM, Tragardh AC and Ahrne LM. 2005. Effect of near-infrared radiation and jet impingement heat transfer on crust formation of bread. *J. Food Sci.*, 70(8): E484–91.

Palaniappan, S. and Sastry, S.K. (1992). Effects of electroconductive heat treatment and electrical pretreatment on thermal death kinetics of selected microorganisms. *Biotechnol Bioengineer*, 39: 225–232.

Park, B.J., Takatori, K., Sugita-Konishi, Y., Kim, I.H. and Lee, M.H. (2007). Degradation of mycotoxins using microwave-induced argon plasma at atmospheric pressure. *Surf. Coatings Technol.*, 201: 5733–37.

Pereira, R., Martins, J., Mateus, C., Teixeira, J.A. and Vicente, A.A. (2007). Death kinetics of Escherichia coli in goat milk and Bacillus licheniformis in cloudberry jam treated by ohmic heating. *Chemical Papers*, 61(2): 121e126.

Pothakamury, U.R., Barbosa-Ca Novas, G.V. and Swanson, B.G. (1993). Magnetic-field inactivation of microorganisms and generation of biological changes. *Food Technol.*, 47(12): 85–93.

Pothakamury, U.R., Monsalve-Gonzalez, A., Barbosa-Canovas, G.V. and Swanson, B.G. (1995). Inactivation of *Escherichia coli* and *Staphylococcus aureus* by pulsed electric field technology. *J. Food Res. Int*. Unpublished manuscript, University of Washington, Pullman.

Pothakamury, U.R., Vega, H., Zhang, Q.H., Barbosa-Canovas, G.V. and Swanson, B.G. (1996). Effect of growth stages and processing temperature on the inactivation of *E.coli* by pulsed electric field. *J. Food Prot.*, 59(11): 1167–1171.

Qin, B., Pothakamury, U.R., Vega, H., Martin, O., Barbosa-Cannovas, G.V. and Swanson, B.G. (1995). Food pasteurization using high-intensity pulsed electric fields. *Food Technology*. 49(12): 55–60.

Rademacher, B. and Hinrichs, J. (2006). Effects of high pressure treatment on indigenous enzymes in bovine milk: Reaction kinetics, inactivation and potential application. *Int. Dairy J.*, 16: 655–661.

Ragni, L., Berardinelli, A., Vannini, L., Montanari, C. and Sirri, F. (2010). Non-thermal atmospheric gas plasma device for surface decontamination of shell eggs. *J. Food Eng.*, 100: 125–32.

Raso, J. and Heinz, V. (2006). Pulsed electric fields technology for the food industry. Heidelberg: Springer Verlag. p. 245.

Ratti, C. and Mujumdar, A.S. (1995). Infrared drying. *In*: Mujumdar AS, editor. Handbook of industrial drying. New York: Marcel Dekker. pp. 567–88.

Riahi, E. and Ramaswamy, H.S. (2004). High pressure inactivation kinetics of amylase in apple juice. *J. Food Eng.*, 64: 151–60.

Roberts, R.T. (1992). High intensity ultrasonics. *In*: The Chemistry of Muscle-based Foods, *Eds*. Johnston, D.E., Knight, M.K. and Ledward, DA pp. 287–97. Cambridge, UK: R. Soc. Chem.

Rodrigoa, D., Cortesb, C., Clynenc, C., Schoofsc, L., Van Loeya, A. and Hendrickx, M. (2006). Thermal and high-pressure stability of purified polygalacturonase and pectinmethylesterase from four different tomato processing varieties. *Food Res. Int.*, 39: 440–48.

Rosenthal, I., Rosen, B. and Berstein, S. (1996). Surface pasteurization of cottage cheese. *Milchwiss,* 51(4): 198–201.

Sale, A.J. and Hamilton, W.A. (1967). Effect of high electric fields onmicro-organisms. I. Killing of bacteria and yeast. II. Mechanism of action of the lethal effect. *Biochim. Biophys. Acta*, 148: 781–800.

Sandu, C. (1986). Infrared radiative drying in food engineering: A process analysis. *Biotechnol. Prog.*, 2(3): 109–19.

Sastry, S.K., Datta, A.K. and Worobo, R.W. (2000). Ultraviolet light. *J. Food Sci.*, 65(12): 90–92.

Save, S.S., Pandit, A.B. and Joshi, J.B. (1997). Use of hydrodynamic cavitation for large scale microbial cell disruption. *Trans Chem. Eng.*, 75: 41–49.

Sawai, J., Sagara, K., Igarashi, H., Hashimoto, A., Kokugan, T. and Shimizu, M. (1995). Injury of Es-cherichia coliin physiological phosphate buffered saline induced by far-infrared irradiation. *J. Chem. Eng. Jap.*, 28(3): 294–299.

Sawai, J., Sagara, K., Hashimoto, A., Igarashi, H. and Shimizu, M. (2003). Inactivation characteristics shown by enzymes and bacteria treated with far-infrared radiative heating. *Int. J. Food Sci. Technol.*, 38: 661–667.

Sawai, J., Nakai, T., Hashimoto, A. and Shimizu, M. (2004). A comparison of the hydrolysis of sweet potato starch with β-amylase and infrared radiation allows prediction of reducing sugar production. *Int. J. Food Sci. Technol.*, 39: 967–974.

Sharma, G.P., Verma, R.C. and Pathare, P.B. (2005). Thin-layer infrared radiation drying of onion slices. *J. Food Eng.*, 67: 361–6.

Schubring, R., Meyer, C., Schluter, O., Boguslawski, S. and Knorr, D. (2003). Impact of high pressure assisted thawing on the quality of fillets from various fish species. *Innovative Food Sci. Emerg. Technol.*, 4: 257–67.

Saulis, G., Lape, R., Praneviciute, R. and Mickevicius, D. (2005). Changes of the solution pH due to exposure by high-voltage electric pulses. *Bioelectrochemistry*, 67: 101–108.

Sharma, A. and Yadav, B.S. (2008). Resistant starch: Physiological roles and food applications. *Food Rev. Int.*, 24: 193–234.

Smeller, L. (2002). Pressure-temperature phase diagram of biomolecules. *Biochim. Biophys. Acta*, 1595: 11–29.

Smeller, L. and Fidy, J. (2002). The enzyme horseradish peroxidase is less compressible at higher pressures. *Biophys. J.*, 82: 426–436.

Sun, H.X., Kawamura, S., Himoto, J.I., Itoh, K., Wada, T. and Kimura, T. (2008). Effects of ohmic heating on microbial counts and denaturation of proteins in milk. *Food Sci. Technol Res.*, 14: 117–123.

Tanaka, F., Verboven, P., Scheerlinck, N., Morita, K., Iwasaki, K. and Nicolaý, B. (2007). Investigation of far infrared radiation heating as an alternative technique for surface decontamination of strawberry. *J. Food Eng.*, 79: 445–452.

Tiwari, B.K., O'Donnell, C.P. and Cullen, P.J. (2009). Effect of sonication on retention of anthocyanins in blackberry juice. *J. Food Eng.*, 93: 166–171.

Toepfl, S., Mathys, A., Heinz, V. and Knorr, D. (2006). Review: Potential of high hydrostatic pressure and pulsed electric fields for energy efficient and environmentally friendly food processing. *Food Rev. Int.*, 22(4): 405–423.

Tsuchiya, K., Nakamura, K., Okuno, K., Ano, T. and Shoda, M. (1996). Effect of homogeneous and inhomogeneous high magnetic fields on the growth of *Escherichia coli*. *J. Fermentation Bioengineer*, 81(4): 343–346.

van Zuilichem, D.J., Vant Reit, K. and Stolp, W. (1986). An overview of new infrared radiation processes for various agricultural products. *Food Eng. Process Appl.*, 1: 595–610.

Vega, H.M., Powers, J.R., Canovas, G.V. and Swanson, B.G. (1995). Plasmin inactivation with pulsed electric fields. *J. Food Sci.*, 60: 1132–1136.

Vercet, A., Sanchez, C., Burgos, J., Montanes L. and Lopez Buesa, P. (2002). The effects of manothermosonication on tomato pectic enzymes and tomato paste rheological properties. *J. Food Eng.*, 53: 273–278.

Vimini, R.J., Kemp, J.D. and Fox, J.D. (1983). Effects of low frequency ultrasound on properties of restructured beef rolls. *J. Food Sci.*, 48: 1572–1573.

Weltmann, K.D., Kindel, E., Brandenburg, R., Meyer, C. and Bussiahn, R. (2009). Atmospheric pressure plasma jet for medical therapy: Plasma parameters and risk estimation. *Contrib. Plasma Phys.*, 49: 631–640.

Yeom, H.W., Streaker, C.B., Zhang, Q.H. and Min, D.B. (2000). Effects of Pulsed electric fields on the quality of orange juice and comparison with heat pasteurization. *J. Agric. Food Chem.*, 48(10): 4597–4605.

Zeece, M., Huppertz, T. and Kelly, A. (2008). Effect of high pressure treatment on *in vitro* digestibility of betalactoglobulin. *Innovative Food Sci. Emerg. Technol.*, 9: 62–69.

Zhang, Q., Qin, B.L., Barbosa-Canovas, G.V. and Swanson, B.G. (1994). Inactivation of *E. coli* for food pasteurization by high-intensity short-duration pulsed electric fields. *J. Food Proc. Preserv.*, 19: 103–118.

Zhang, G., Sofyan, M. and Hamaker, B.R. (2008). Slowly digestible state of starch: mechanism of slow digestion property of gelatinized maize starch. *J. Agric. Food Chem.*, 56: 4695–4702.

Zhao, Y. (2006). Radio frequency: Dielectric heating. *In*: Sun, D. (*Ed.*), Thermal food processing: Newtechnologies and quality issues. Boca Raton, USA: Taylor and Francis.

Zimmermann, U., Pilwat, G. and Riemann, F. (1974). Dielectric breakdown in cell membranes. *Biophys. J.*, 14: 881–899.

13

Enrobed Meat Products

CHIDANANDAIAH AND R.C. KESHRI

ABSTRACT

Enrobing is a process in which foods are traditionally coated with edible coating materials in the form of batter which preserves and enhances food quality. Enrobing improves product appearance, colour, crispness, flavour, juiciness, nutritive value and also the microbiological profile of the product. The amount of coating and the composition of the coating material influence the colour and appearance and overall aesthetics. Coatings act as sealants, preventing natural juices from flowing out. They act as moisture barrier, prevent moisture and weight loss, improve the juiciness and tenderness. Use of proper frying medium is important to the quality of fried foods. Proper and useful handling of frying medium will assure a more economic operation and higher quality products. The amount of cooking medium that is absorbed into the food should be replaced by fresh material for every 8 hr. Improvement in the microbiological standards of enrobed products is done by using good quality ingredients and processing under good hygienic conditions. Enrobed meat products are more susceptible to microbial contamination due to its increased handling and addition of various ingredients. Lipid oxidation is reduced in breaded samples and such products have higher overall ratings. In unbreaded sample oxidation would be double after frying.

1. INTRODUCTION

Enrobing is a process in which foods are traditionally coated with edible coating materials in the form of batter which preserves and enhances food quality. It provides water vapour tightness to prevent product desiccation, good barrier against microbial invasion to reduce the spoilage and decay of foods. Traditionally coated poultry and sea foods offer consumers new items that are promotable features in super markets. In restaurants coated vegetables (potato finger chips, batter coated chillies), nuts (groundnut) and fruits are extremely popular. Variety of enrobed/coated foods depends on imagination of various companies and food preparation establishments.

The examples of enrobed food items are:

1. Meat based	Pork patties, beef patties, pork and beef nuggets.
2. Poultry products	Precooked cutup poultry parts, raw cutup poultry parts, poultry patties, giblets and poultry nuggets etc.
3. Sea foods	Coated shrimp, natural and formed fillets, fish sticks etc.
4. Egg products	Egg bonda
5. Dairy products	Paneer, cheese nuggets, cheese sticks and cubes etc.
6. Vegetables	Onion rings, cauliflower, potatoes, mushrooms, chillies (enrobed shimla mirch or choti mirch, samosa, bread etc.
7. Nuts	Groundnut and peanut etc.
8. Fruits	Pine apple and bananas etc. (Cunningham, 1989)

Enrobing provides value addition to the product (Jessup, 1981). Small improvement in product quality fetches high price. In foreign countries most of the enrobed products are sold in frozen form (Anon, 1975). Shrimp is the first food item enrobed in the USA and achieved phenomenal success. The second item achieved success was enrobed fish sticks in 1951 and other frozen enrobed sea foods soon followed. The consumers demand for greater convenience and quest for new food items, along with renewed interest from a nutritional point of view, has led the industry to enrobe vegetables, some fruits and a wide variety of meat items.

2. ADVANTAGES OF ENROBED PRODUCTS

2.1. Improvement in Appearance/Colour

Enrobing improves product appearance, colour, crispness, flavour, juiciness, nutritive value and also the microbiological profile of the product. The amount of coating and the composition of the coating material influence the colour and appearance and overall aesthetics. Mixture of thickening agents offer the most promise for achieving the combination of attributes desired *viz.*, good adhesion, desirable thickness, crispness, colour and attractive appearance (Hanson & Fletcher, 1963). Colour of the product is influenced by thickening agents. Waxy rice flour and waxy corn starch produces glossy brown coatings, wheat flour a greenish brown colour, waxy corn starch and corn starch, a very light brown colour, potato flour, a golden brown and yellow corn flour, a greenish yellow colour. Coated fries and frozen product from wheat flour based coatings after baking had good appearance and organoleptic properties (Bernacchi *et al.*, 1990). Coating of chicken pieces by using coating mix containing malt dextrin, amylose starch, powdered fat and pre-gelatinized starch simulate the appearance and texture of chicken skin after baking (Fioriti & Ackermann, 1996). Batter containing gelatine and egg albumin improves the coating adhesion. The colour of the enrobed foods was dependent upon cooking time, cooking temperature, enrobing material composition and cooking oil characteristics (Landes & Blacksher, 1971). Batter containing pre-gelatinized or modified corn flour produced coating with acceptable colour (Baker & Scott, 1988).

2.2. Improved Adhesion/Coating

One of the important characteristics of enrobed products include adhesion of coating. The poultry skin ultra structure plays role in batter and breading adhesion (Suderman & Cunningham, 1980). Freezing of poultry prior to breading improves the coating adhesion (Suderman & Cunningham, 1981). Adhesion is simply the chemical and physical binding of a coating with itself and the food products it coats. Thinner coating reduces the peeling tendency. Waxy cereal thickening agents produce coatings more elastic and less tough than those made from the common cereals and they tended to peel in pieces, in contrast to the intact peeling of tough coatings. Coatings containing waxy or common starch have better adhesive qualities than those containing other cereals (Hanson & Fletcher, 1963). Egg yolk content of coating influences colour but have little effect on adhesion. Enrobed products cooked in soya bean oil were superior to those cooked in peanut oil. Among gums, carboxymethylcellulose is significantly better in improving batter adhesion to the chicken skin (Suderman *et al.*, 1981). Protein pre-dust material before battering the chicken produces crusts with better adhesion than starches, gums and alginates. High adhesion scores result in low cooking losses and high yields (Baker *et al.*, 1972).

2.3. Improved Acceptability

Coated uncooked nuggets maintain sensory quality for 4 months at –17.8°C. Afman *et al.* (1999) described breaded bone in poultry products with better acceptability which involve pre dusting, coating in batter, breading and pan frying. Preparation of battered and breaded chicken having appearance, taste and texture similar to fried foods without frying is patented by Samson and Bangs (1995). Coating mix which is effective in imparting the texture, taste and appearance of deep-fat fried foods to fresh, uncooked poultry meat, fish and vegetables is described by Rispoli *et al.* (1987). Evans *et al.* (1980) patented coating composition for moistened fowl, which after baking imparts a crisp, uniform and adhesive coating with good taste, texture and appearance.

2.4. Moisture and Fat Content

Batter with high solid content tends to absorb less fat. Batter containing pregelatinized corn flour and relatively low solids level produce the coatings with high moisture content which absorb little fat (Baker & Scott, 1988). Breading reduced the weight loss on subsequent deep fat frying. At the same time deep fat frying reduced the moisture content of both breaded and unbreaded samples. The fat content of the fried samples depends on the fat content before frying. Increase in fat content during frying depends on type of breading (Pikul & Wojciechowska, 1994). Coating generally reduces the moisture loss (Proctor & Cunningham, 1983). Crude fat was highest in uncoated thigh meat, cooked by deep fat frying.

2.5. Lipid Oxidation

In unbreaded sample TBA value would be doubled after frying. The fat oxidation was less in breaded samples and such products have higher overall ratings (Pikul & Wojciechowska, 1994). On contrary Chidanandaiah (2003) observed higher TBA value in deep fat fried enrobed buffalo meat patties when compared to broiled patties. Microbial analysis showed intermediate counts in raw and cooked product for standard plate count, psychrotrophs, coliforms and yeast and mold and *Salmonella* counts were negative. Coating of cooked meat, poultry and vegetables can stay without freezing or use of preservatives (Wilkening, 1991).

2.6. Coating Pickup/Yield

No significant correlation was found between size of the chicken cut up parts and percent coating pick up of the batter (Proctor & Cunningham, 1984). They also reported that correlation between original weight of a broiler drumstick and cooking loss was highly significant. Similarly relation between percent crumb loss and original weight of a drumstick was highly significant ($P<0.01$). Percent overall yield of the coated product was dependent on original weight (Proctor & Cunningham, 1984). Batter composition had a major influence on the amount of coating pick up by the chicken nuggets. Percentage coating pick up increases as the hydrocolloid concentration increases in the batter. Cunningham and Tiede (1981) indicated that as batter viscosity increased, the amount of breading pickup also increased. Temperature of coating either refrigerated or room does not have effect on percent coating pickup of drumsticks. But drumsticks coated at 110°C had lower percentage coating pickup, possibly due to a drier skin surface. Coating done at frozen temperature (–15.5°C) had least coating pickup, at the same time such product had higher cooking losses. Percent cooking loss is minimum when chicken cutup parts coated at room and refrigeration temperature. Percent crumb loss is also dependent on temperature (Proctor & Cunningham, 1984). Broiler drumsticks with skin had significantly increased coating pickup and decreased crumb loss over drumsticks without the skin.

3. TYPES OF ENROBING INGREDIENTS

Most of the ingredients for batter preparation are sold individually in the market or any stationery stall. The ingredients consist of flour (Bengal gram flour, wheat flour, corn flour etc.) mixed with variety of seasonings and converted into wet batter at the time of use by adding water. Wheat flour based batter mixes easily with water and apparently remains in solution for long periods of time. The corn flour based batter settles out easily and would change batter thickness. This result in uneven coating pick up. Continuous mixing to keep solids in suspension is necessary when using corn based batters. There are two types of enrobing materials (Johnson & Hutchinson, 1983). The ingredients can be classified into five categories (Cunningham, 1989).

3.1. Polysaccharides

Polysaccharide containing ingredients are starch (corn or rice), corn flour, wheat flour, modified derivatives of cellulose and gums. Polysaccharides interacting with themselves or other ingredient components such as proteins provide stability, lipid provide viscosity and sometimes emulsifying action. Saccharides have control over the form, texture and shelf life of enrobed foods (Suderman, 1979). Solids in the corn starch batter tended to settle rapidly. This is not a problem in mixture of corn starch with other thickening agents. Use of wheat flour based batter for coating poultry, fish, red meat, egg products, vegetables, fruits and cheeses was reported by Bernacchi *et al.* (1990). The dough coating containing fine ground rye flour and wheat flour for coating cooked meat, poultry, vegetables etc. is patented by Wilkening (1991). Pikul and Wojciechowska (1994) used potato flour based with fresh egg for breading chicken meat. Incorporation of corn flour, rice flour and bajra flour alone or in combination with Bengal gram flour improved the sensory scores particularly for crispness and adherence of enrobed buffalo meat patties (Chidanandaiah, 2003).

3.2. Proteins

Protein ingredients include milk powders and milk protein fractions, egg albumen, cereal flours, seed proteins and single cell proteins. Although their functions in relation to food coatings have not been fully established, their role in the other food product is well known. Non-fat dry milk (NFDM) has been used to improve the water absorption capacity of flour, which increases the viscosity of the system, baking quality of wheat flours improved, strengthen product structure and improves texture, retards the moisture loss and enhances crust colour and flavour development. Egg proteins act as thickeners. The functional properties of soya proteins include emulsification capacity, emulsion stabilization and water absorption.

Egg yolk content of coating influences colour but have little effect on adhesion. Higher yolk content increases the fat content of the coatings. Protein pre-dust materials generally improved adhesion of batter coating more than starches and gums. Gelatine and egg albumen most effectively improve the adhesion than NFDM, soya proteins and whey proteins. Baker *et al.* (1972) reported higher adhesion scores with egg albumen than with soya concentrate or NFDM. Bengal gram flour alone with other ingredients used as batter mix for enrobing increased the overall physico-chemical and sensory quality of buffalo meat patties (Chidanandaiah, 2003).

3.3. Fats and Hydrogenated Oils

Fats and hydrogenated oils are known to include carriers of fat soluble vitamins and to contribute to food flavour and palatability as well as to the feeling of satiety after eating. Shortening contribute to the tender quality or shortness of a baked product by their lubricating action. Mono- and di-glycerides and lecithin serve as emulsifiers and staling inhibitors.

3.4. Water

Water is used to form batter suspensions and aid in the adhesion of breading to the surface of the food product. Adhesion and viscosity are affected by the ratio of solids to water. This in turn affects the amount of coating on the food, its thickness, machinability and drying time. Water reacts with proteins and polysaccharides, acts in a structure forming role.

3.5. Seasonings

Seasonings include sugar, salt, pepper, paprika and other spices etc. Sugar improves both colour and flavours, when sugar combines with amines gives a brown colour (Maillard browning) and caramelized flavours. Salt and spices enhance the flavour of enrobed products.

4. FUNCTIONS OF INGREDIENTS

Coatings are applied to food products for many reasons. They enhance the food quality and storage characteristics of some vegetables, fruits, meat including chicken, fish and cheese products (Rizzoti & Tilly, 1988). They remove the monotony of some foods and increase their acceptance by giving the food product a more attractive appearance and a better taste (Elston, 1975). Nutritional value of the food product improved by enrobing (Kroger & Igoe, 1971), by the added ingredients and by protecting the nutrients within the food product from direct contact with the cooking medium and from moisture loss where nutrients can be leached out.

Coatings act as sealants, preventing natural juices from flowing out. They act as moisture barrier, prevent moisture and weight loss, improve the juiciness and tenderness (Dawson *et al.*, 1962; Suderman, 1979). Morgon (1971) reported that coatings enhanced food appearance, reinforced food structure, inhibited oxidation by excluding oxygen from the food product and extended shelf life. Composition of batter influences the adhesive qualities, toughness, crispness, elasticity, colour, smoothness and handling characteristics. These properties are important factors in acceptability of the product. Coating thickness can be varied with several thickening agents by changing the solid to moisture ratio in the batter (Hanson & Fletcher, 1963). Increasing the proportions of thickening agent increases the thickness of the coating. The ingredients function individually and collectively in many types of coating associated with different food products.

5. SYSTEMS OF ENROBING

Foods can be enrobed in many ways in order to provide acceptable adhesion or desired texture and functional properties. Automated systems for enrobing chicken parts have been designed to duplicate hand operation. According to Johnson and Hutchinson (1983), there are four different types of enrobing systems.

(*i*) *Single Pass System*: This method batters and breads the product only once and even with a pre-dust, coating pickup is around 30%.

(*ii*) *Double Pass System*: The system repeats the process of batter and breading and coating pickup can range from 30 to 50%.

(*iii*) *Batter Fry System*: In this method, a dry pre-dust is followed by enrobing the chicken in a viscous batter and coating pickup can vary from 30 to 55%.

(*iv*) *Tempura-Japanese*: In this method chicken is enrobed with a tempura batter and coated with Japanese breading. Coating pickup can range from 36 to 55%. There is slightly higher yield for fried chicken parts prepared in the following manner: (*a*) flour pre-dust, (*b*) batter, (*c*) a second flour dust and (*d*) use of batter-breading method (Yang & Chen, 1979).

6. COOKING MEDIA

Use of proper frying medium is important to the quality of fried foods. Proper and useful handling of frying medium will assure a more economic operation and higher quality products (Jacobson, 1967; Thorner, 1973). Type of fat did not significantly affect the fat uptake (Bognar, 1998). The cooking operations play a vital role in determining the quality of finished food products. Depending upon the physical form of the medium at room temperature, the cooking media have been described in various terms such as oils, shortenings and fats which are commonly used.

The cooking medium acts as a heat transfer medium and then as a food ingredients. Heat is transmitted through cooking medium to the food and then the medium itself is absorbed into the food to become a part of the edible item. The absorption or adsorption of reused corn oil by chicken pieces would undoubtedly affect product stability. Broken down frying medium is dark and viscous with a burnt odour and flavour tends to smoke and foam easily. The more unsaturated fats and oils developed colour of foods faster than did the harder shortening during frying (Weiss, 1970). In general, presence of milk solids or egg significantly darkened the shortening (Waimaleongora & Chen, 1983).

The amount of cooking medium that is absorbed into the food should be replaced by fresh material for every 8 hr. Cooking temperatures have a drastic effect on the shelf-life of the oil. Particles of food coatings, which are left in the cooking medium, accelerate the lipid degradation. Continuous filtration is the best way to reduce the degradation. Proper selection and use of cooking medium will make a vital difference to the quality of the final, cooked enrobed products. Fat uptake by foods during frying apparently increases up to a saturation limit depending on food and coating type (Bognar, 1998).

7. COOKING METHODS

7.1. Deep Fat Frying

Various combinations of cooking temperature and time have been suggested by different researchers. Weiss (1970) recommended that large pieces of chicken coated with batter or breading be given 10–15 min of frying at 176°C. Small pieces needed 9–10 min of frying at 185°C. The American Hospital Association suggested 8–14 min at 174°C for large chicken pieces and 7–10 min at 184°C for small pieces. Flour/batter/flour method coated broiler breast and thigh are deep fat fried at 168.3°C for 12 min (Waimaleongora & Chen, 1986). Frying in rape seed oil at 180°C to an internal temperature of approximately 79°C required 11 min 16 sec and 7 min 45 sec for breaded thighs and drumsticks, respectively as reported by Pikul and Wojciechowska (1994).

Fat uptake by meat, poultry and fish was generally lower for deep fat frying than for pan frying. Frying resulted in retention of almost all proteins, carbohydrates and minerals and vitamins B_1, B_2, B_3 and C when compared to boiling, steaming or stewing (Bognar, 1998). Chicken frying process after pre-dusting, battering and breading with minimum oil absorption is described by Babka *et al.* (1993). Dusting and batter coating of poultry product and flash frying is described by Cahill (1999).

7.2. Simmering Method

The raw parts without batter are first simmered in water for 14 min. After simmering the parts are cooled at 2°C, coated and deep fat fried at 193°C until brown.

7.3. Pan Baking

Breaded and coated poultry drumsticks are baked at 205°C in a rotary oven for 40 min. Then they are removed and allowed to cool at room temperature for 20 min.

8. PACKAGING OF ENROBED PRODUCTS

Majority of enrobed foods are absorbed into frozen food industry (Cunningham, 1989). During frozen storage value added products undergo quality changes *viz.,* desiccation, discolouration, development of rancidity etc. Proper packaging of the products retards these changes and enhances the shelf life. Besides conventional flexible plastic packaging materials, thermoformed containers are also preferred. For thermoformed trays high density polyethylene (HDPE), high impact polystyrene (HIP) and polyvinyl chloride (PVC) are commonly used (Srinivas Gopal & Ravi Shankar, 2001). Eyas (2001) used low density polyethylene (LDPE) packaging material for refrigerated (4 ± 1°C) and frozen (–18°C) stored enrobed buffalo meat cutlets.

As per USDA regulation the amount of battering with breaded material should not exceed 30% of the weight of the finished poultry product. Product with more than 30% breading based on meat weight must be labeled "Fritter" (Cunningham, 1989; Smith, 1999). Wehr (1978) recommended that the aerobic count should be within the limit of 4 log cycle and coli form, *Staphylococcus* sp. and *E. coli* should be absent in a fried breaded fish.

9. QUALITY DURING STORAGE

9.1. Physico-Chemical Quality

9.1.1. *pH and titratable acidity*

The increase in pH is caused by microbial production of alkalinizing substances especially ammonia and is related as a consequence of amino acid degradation (Jay, 1972). When meat is spoiled by gram positive bacteria, pH value decreased due to production of lactic acid (Shelef, 1975). A non-significant change in pH during refrigerated storage (5–7°C) upto one week was reported by Sunki *et al.* (1978). The pH values were fairly stable in low fat Chinese style sausages at 4°C for 12 weeks (Lin & Chuang, 1999). Eyas (2001) reported that pH values of the enrobed buffalo meat cutlet were lower than that of control sample on day 0, which later on increased more than that of control on day 10. However, there was no significant difference in pH values of control and enrobed cutlets after day 10. Biswas (2002) observed that pH values of pork patties were in proximity with control and enrobed pork patties at each interval of storage up to 35 days and overall pH value was also not much variable. Southerland *et al.* (1975) reported that the acidity decreased as the storage period increased in vacuum packaged beef. During storage at 4 ± 1°C, both pH and titratable acidity increased in enrobed buffalo meat patties at regular interval of 0, 7, 14 and 21 days (Chidanandaiah, 2003).

9.1.2. *TBA value*

Determination of thiobarbituric acid reacting substance (TBARS) number measures the amount of malonaldehyde, a water soluble secondary product obtained mainly from the oxidation of poly-unsaturated fatty acids in food sample. It is indicative of oxidative deterioration of lipids in muscle food (Pearson *et al.*, 1983). Lipid oxidation impairs the nutritive value of fats and lowers their utility. Oxidative rancidity occurs more rapidly in cooked meat than in raw meat due to denaturation of proteins in cooked meat (Yamuchi, 1973). Storage temperature, rate of cooking, addition of various ingredients *viz.*, Polyphosphates, dry spices etc. affect the TBARS number in meat and meat products (Wang *et al.*, 1996; Abd EL-Alim *et al.*, 1999).

Alginate coatings can provide protection against lipid oxidation of food ingredients as in case of precooked ground pork patties. Coated samples did not develop warmed over flavour and exhibited better texture when compared to uncoated samples (Krochta *et al.*, 1994). Calcium alginate coating in pork patties improved the sensory properties, eliminated warmed over flavour and decreased the thiobarbituric acid value (Wanstedt *et al.*, 1981). Alginate coating of buffalo meat cut reduced the thiobarbituric acid value during frozen storage (El-Ebazary *et al.*, 1981). Alginate coating application constitutes an oxygen barrier thus retards the oxidation and also retains the flavour (Earle & McKee, 1985). Wheat gluten, soya protein, carrageenan coatings and carrageenan films are very effective in reducing TBA reactive substances values during storage of precooked beef patties (Wu *et al.*, 2000). Low oxygen permeability and good continuity of the whey protein isolate (WPI) coatings to walnuts effectively prevents the oxidative rancidity of walnuts (Mate & Krochta, 1997).

Application of WPI powder to frozen king salmon delays the onset of lipid oxidation and reduces the peroxide value during storage (Stuchell & Krochta, 1995). Gelatin coating to frozen pork reduces the TBA and peroxide value during storage with improved retention of sensory qualities. Coatings on roasted and salted nut meats reduced their susceptibility to oxidative rancidity and extended their shelf life at least 3–5 times (Alikonis & Cosler, 1961). Lower lipid oxidation has been reported in breaded chicken (Pikul & Wojciechowska, 1994). WPI coatings plasticized with glycerol coatings to roasted peanut delayed the oxygen uptake thereby delaying the rancidity process (Alcantara & Krochta, 1996). Methyl cellulose coating had better moisture and fat barrier properties (Mallikarjunan *et al.*, 1997). During storage (at 4 ± 1°C) of enrobed buffalo meat patties TBA value increased up to day 14 and latter decreased on day 21(Chidanandaiah, 2003).

9.1.3. *Tyrosine value*

The extent of proteolysis in meat is measured by tyrosine and tryptophan content in non-protein extract of meat (Strange *et al.*, 1977). Determination of tyrosine complex has been used as one of the methods for detecting microbial spoilage in meats, poultry and sea foods (Jay, 1996). Tyrosine value in minced meat is in the range of 26–85 mg/100 g of meat (Daly *et al.*, 1976). Higher tyrosine, ammonia and non-protein nitrogen (NPN) values occur due to microbial activity only during advanced stage of spoilage (Dainty *et al.*, 1975). This was due to the synthesis of proteolytic enzymes by bacteria in late logarithmic phase of growth. A steady increase in the tyrosine value was observed during refrigerated storage of 9 days in both antibiotic treated and control buffalo meat samples (Agnihotri, 1988). Significant correlation between tyrosine value and length of storage period in untreated meat sample was noticed (Morrissey *et al.*, 1980). In enrobed buffalo meat cutlet inconsistent changes in tyrosine value was observed during refrigerated storage

(Eyas, 2001). During storage at 4 ± 1°C, both non protein nitrogen (NPN) and tyrosine value increased in enrobed buffalo meat patties at regular interval of 0, 7, 14 and 21 days (Chidanandaiah, 2003).

9.2. Microbiological Quality

Microbial growth in meat products depends on the extent of microbial contamination, initial microbial load, storage conditions and chemical properties of the product. Spoilage changes in meat become visible when the total number of bacteria per gram reached 10^8–10^9. A change from mesophilic to psychrotrophic dominance was observed during the course of commercial chill storage conditions (McDowell *et al.*, 1986). Total mesophilic aerobes in beef carcasses ranged from log 2.88 to 6.32/cm^2 (Kotula *et al.*, 1975). Anjaneyulu *et al.* (1988) reported that aerobic plate count of ground buffalo meat ranged from log 5.34 to 5.77 cfu/g. Microbial count of log 4.6 cfu/g for psychrophils and log 5.3 cfu/g for total plate count is considered to be indicative of unacceptability of cooked meat products (Cremer & Chipley, 1977).

Enrobed meat products are more susceptible to microbial contamination due to its increased handling and addition of various ingredients. Improvement in the microbiological standards of enrobed products is done by using good quality ingredients and processing under good hygienic conditions. The food before and after enrobing should be properly refrigerated (Cunningham, 1989). Reddy *et al.* (1990) reported high aerobic counts in fish finger after breading and battering. This increase was possibly due to addition of various ingredients. Deep fat frying of patties to a core temperature of 100°C resulted in the complete removal of pathogenic micro flora with more than one log cycle/g (Anand *et al.*, 1991). Reddy *et al.* (1990) reported that fish finger contained higher aerobic counts after breading and battering than control. Deep fat frying of the product eliminates pathogenic microorganisms (Anand *et al.*, 1991). Refrigerated chicken patties showed linear increase and frozen stored showed a decline in total plate count (Nath & Mahapatra, 1995). Mesophilic count in enrobed buffalo meat cutlet increased up to midway of storage period and then declined as compared to control (Eyas, 2001).

Acetylated monoglyceride coating of beef steak maintains the microbial counts (aerobic and coli form bacteria) during storage at 2 ± 2°C. But after 49 days, *Lactobacillus cellobiosis* was the dominant microorganism on treated and control steaks (Keeton *et al.*, 1988). Sodium alginate coating significantly reduces the microbial flora of lamb carcass (Lazarus, 1977). Alginate coating on frozen buffalo meat reduces the bacterial count during storage (El-Ebazary *et al.*, 1981). Surface microbial growth of intermediate moisture food can be stabilized by application of edible coatings (Vojdani & Torres, 1990). Total plate count, psychrophilic count, yeast and mould count, staphylococcal count increased linearly and enterobacteriaceae organisms could not be detected at each interval of storage at refrigeration temperature in enrobed buffalo meat patties (Chidanandaiah, 2003).

9.3. Sensory Quality during Storage

Coating of frozen pork with 10% gelatine improves the retention of flavour, taste and aroma. Acetylated monoglyceride coating of beef steaks had little effect on lean colour, surface discolouration, overall appearance, moisture, pH and sensory characteristics during storage at 2 ± 2°C for one week. However, it maintained acceptable fat colour and reduced the purge (Keeton *et al.*, 1988). Sensory attributes of enrobed patties *viz.*, colour and appearance, texture, flavour and overall acceptability decreased at each intervals of storage at 4 ± 1°C (Chidanandaiah, 2003). Batter mix prepared from CF either alone or in combination with BGF increased the sensory scores for crispness and adherence of batter resulting in improved overall palatability. Incorporation of rice flour in batter mix improved crispness and adherence of the product when compared to enrobing with BGF alone (Chidanandaiah & Keshri, 2006; 2007).

Alginate coated ground pork patties did not develop warmed over flavour and exhibited better texture when compared to uncoated samples (Krochta *et al.*, 1994). Coating of beef steaks with calcium alginate significantly decreased shrinkage loss, drip and degree of off odour after 96 hr storage and coating significantly prolonged the muscle colour (Williams *et al.*, 1978). Chicken patties coated with calcium alginate were more desirable in texture, flavour, juiciness and overall palatability when compared to uncoated controls (Roland *et al.*, 1981). Calcium alginate coating in pork patties improved the sensory properties and eliminated warmed over flavour (Wanstedt *et al.*, 1981). Frozen stored alginate coated buffalo meat retained texture and colour during storage (El-Ebzary *et al.*, 1981). Hydrocolloid coatings (alginate, gellan and K-carrageenan) to cheese reduced the weight loss and improved gloss and colour and produced cheeses of softer and less brittle texture without affecting flavour of cheese negatively during storage (Kampf & Nussinovitch, 2000). They concluded that hydrocolloid coatings improved the physical and sensory properties of white brined cheese.

10. FUTURE RESEARCH NEED

Enrobing provide processors with their greatest 'value added' opportunity at the low cost. The amount and composition of enrobing materials influence colour, appearance and overall aesthetics. The preferential appeal and desired texture of enrobed products create new opportunities, up scaling foods or making them more home style, recreating traditional foods. The crispness and adhesion of coating are the important sensory attributes of enrobed products for saleability and preservation of products. Traditionally coated products are commonly available around us for savour lack systematic study. The consumers demand for greater convenience and quest for new food items, along with renewed interest from a nutritional point of view, has led the industry to enrobe vegetables, some fruits and a wide variety of meat items. In view of discussion mentioned above, there is immense scope for scientific research needed particularly to improve crispness, texture, appeal, satiety etc.

REFERENCES

Afman, B., Hope, J.L., Fischer, R.W., Girdhar, B.K. and Tometich, P.E. (1999). Method of producing fully cooked and breaded bone in poultry product. United States Patent.

Agnihotri, M.R. (1988). Studies on the keeping quality and spoilage pattern of buffalo meat. Ph.D. Thesis. Indian Veterinary Research Institute, Izatnagar, U.P., India.

Alcantara, C.R. and Krochta, J.M. (1996). Mechanical properties of whey-protein isolate coated freeze dried chicken white meat dice. 1996, IFT annual meeting: Book of abstracts, p. 179, ISSN 1082–1236.

Alikonis, J.J. and Cosler, H.B. (1961). Extension of shelf-life of roasted and salted nuts and peanuts. Pea nut *J. Nutr. World*, 40(5): 16–20.

Anand, S.K., Pandey, N.K., Mahapatra, C.M. and Verma, S.S. (1991). Microbiological quality and shelf-life of chicken patties stored at -18^{0}C. *Indian J. Poultry Sci.*, 26(2):105–108.

Anjaneyulu, A.S.R., Sharma, N. and Kondaiah, N. (1988). Effect of salt and phosphate preblending on the quality of buffalo meat during frozen storage and acceptability of patties. *Indian J. Meat Sci. Technol.*, 1(2): 136–144.

Anon, (1975). Breading used for 1.5 billion pounds of fi; increased 342% in 12 years. Quick Frozen Foods. pp. 18–24, 51, 64–66.

Babka, J.R., Kvedoras, D.P. and Park, C.Y. (1993). Chicken frying process. United States Patent, US 5 262 185, US 845735 (19920302).

Baker, R.C., Darfler, J.M. and Vadehra, D.V. (1972). Prebrowned fried chicken. II. Evaluation of predust materials. *Poultry Sci.*, 51(4): 1220–1222.

Baker, R.C. and Scott, K.D. (1988). Development of a high protein coating batter using egg albumen. *Poultry Sci.*, 67(4): 557–564.

Bernacchi, D.B., Loewe, R.J. and Immel, D.L. (1990). Process for preparing storage stable, readily reconstituted frozen comestibles and frozen comestibles produced there by. United States patent, US 4 948 603, US232910 (19880816).

Biswas, A.K. (2002). Studies on enrobing of pork patties. M.V.Sc. Thesis, IVRI, Izatnagar, U.P., India.

Bognar, A (1998). Comparative study of frying to other cooking techniques influence on the nutritive value. *Grasas-Y-Aceites*, 49(3/4): 250–260.

Cahill, D. (1999). A method for preparing a poultry product. European Patent Application.

Chidanandaiah, (2003). Studies on the effect of edible coatings and enrobing on the quality buffalo meat patties. Ph.D Thesis, IVRI, Izatnagar, U.P., India.

Chidanandaiah Keshri, R.C. (2006). Bengal gram and rice flour in batter mix on quality of enrobed buffalo meat patties. *Indian Vet. J.*, 83(10): 1092–1095.

Chidanandaiah Keshri, R.C. (2007). Quality of buffalo meat patties enrobed with batter mix containing Bengal gram flour, finger millet flour and/or corn flour. *J. Food Sci. Technol.*, 44(3): 307–309.

Cremer, M.L. and Chipley, J.R. (1977). State lite food service system: Time and temperature and microbiological and sensory quality of precooked frozen hamburger patties. *J. Food Prot.*, 40: 603–607.

Cunningham, F.E. (1989). Developments in enrobed products. *In*: Processing of poultry. Mea, G.C. (*Eds.*), Elsevier Sci. Pub. Ltd., NewYork, USA p. 325.

Cunningham, F.E. and Tiede, L.M. (1981). Influence of batter viscosity on breading of chicken drumsticks. *J. Food Sci..*, 46: 1950–1952.

Dainty, R.H., Shaw, B.G., Boer, K.A. and Scheps, E.S.J. (1975). Protein changes caused by bacterial growth on beef. *J. Appl. Bacteriol.*, 39(1): 73–81.

Daly, M.C., Morrissey, P.A .and Buckley, D.J. (1976). Quality of raw minced beef. *Irish. J. Agric. Res.*, 15(3): 283–289.

Dawson, L.E., Zobik, M. and Sobel, N. (1962). An edible coating for poultry meat preservation. *Poultry Sci.*, 41: 1640.

Earle, R.D. and McKee, D.H. (1985). Coated food product and method of making same. *US Patent*, 4: 504, 502.

El-Ebzary, M.M., Askar, A.A., El-Dashhouty, M.S. and El-Baki, M.M.A. (1981). Evaluation of alginate coatings for frozen stored buffalo meat cut. *Gordian*, 81(12): 287–292.

Elston, E. (1975. Why fish fingers top the market. Fishing News International. pp. 14–30.

Evans, D.N., Moore, W.W. and Ng, J.L. (1980). Poultry batter coating. United States Patent.

Eyas, A.M. (2001). Studies on development of enrobed buffalo meat cutlets. M.V.Sc. Thesis, IVRI, Izatnagar, U.P., India.

Fioriti, F.R. and Ackermann, K.R. (1996). Coating mix for simulating chicken skin and method of use. *United States Patent*, US. 5:562 940, US 397513 (950301).

Hanson, H.L. and Fletcher, L.R. (1963). Adhesion of coatings of frozen fried chicken. *Food Technol Champaign*, 17: 793–795.

Jacobson, G.A. (1967). Quality control of commercial deep fat frying. *Food Technol.*, 21: 147–152.

Jay, J.M. (1972). Mechanism and detection of microbial spoilage in meats at low temperatures: A status report. *J. Milk Food Technol.*, 35: 467–471.

Jay, J.M. (1996). Modern Food Microbiology, 5th ed. Chapman and Hall, New York.

Jessup, L. (1981). Your lowest value added cost. *Quick Frozen Foods*, 44(4): 70–72.

Johnson, R.T. and Hutchinson, J. (1983). Batter and breading processing equipment. *In*: Batter and Breading Technology (*Eds.* Suderman, D.R. and Cunningham, F.E.), AVI Publishing Co., Westport, Connecticut, USA, pp. 120–154.

Kampf, N. and Nussinovitch, A. (2000). Hydrocolloid coating of cheeses. *Food Hydrocolloids*. 14(6): 531–537.

Keeton, J.T., Leu, R., Vanderzant, C., Bohae, J.J., Griffin, D.B., Savell, J.W. and Cross, H.R. (1988). Evaluation of fresh vacuum-packaged beef steaks coated with an acetylated monoglyceride. *J. Food Sci.*, 53(3): 701–704.

Kotula, A.W., Lusby, W.R. and Course, J.D. (1975). Variability in microbiological counts on beef carcasses. *J. Anim. Sci.*, 40: 834–837.

Krochta, J.M., Baldwin, E.A. and Nisperos-Carriedo, M.O. (1994). Edible coatings and films to improve food quality. Technomic publishing Co-Inc. Lancaster, Pennsylvania, USA, p. 379.

Kroger, M. and Igoe, R.S. (1971). Edible containers. *Food product Developments*, 5: 74–76.

Landes, D.R. and Blacksher, C.D. (1971). The effects of different cooking oil on flavour and colour of fried chicken breading material. *Poultry Sci.*, 50: 894–897.

Lazarus, C.R. (1977). The development of microbial decontamination and moisture loss control procedures for beef, pork and lamb carcasses. *Dissertation Abstracts Int B.*, 37(10): 4890, Order No. 77-819.

Mallikarjunan, P., Chinnan, M.S., Balasubramaniam, V.M. and Phillips, R.D. (1997). Edible coatings for deep fat frying of starchy products. *Lebensmittel-Wissen-Schaft-Und-Technologie*, 30(7): 709–714.5.

McDowell, D.A., Hobson, I., Strain, J.J. and Owens, J.J. (1986). Bacterial micro flora of chill stored beef carcasses. *Environ Health*. 94: 65–68.

Morrissey, P.A., Buckley, D.J. and Daly, M.C. (1980). Effect of four species of bacteria on minced beef stored at 7oC. *Irish J. Food Sci. Technol.*, 4: 1–11.

Nath, R.L. and Mahapatra, C.M. (1995). Effect of levels of chicken fat on the quality and storage life chicken patties. *Indian J. Poult. Sci.*, 30(1): 50–57.

Pearson, A.M., Gray, J.J., Wolzak, A.M.. and Horenstein, N.A. (1983). Safety Implication of oxidized lipids during in muscle foods. *Food Technol.*, 37: 121–129.

Pikul, J. and Wojciechowska, K. (1994). Effect of breading and deep fat frying of chicken meat on oxidation of lipids during refrigerated storage. *Gospodarka Miesna.*, 46(2), 27–31 (c.f. FSTA, 1994, 6S 141).

Proctor, V.A. and. Cunningham, F.E. (1983). Composition of broiler meat as influenced by cooking methods and coating. *J. Food Sci.*, 48(6): 1696–1699.

Proctor, V.A. and Cunningham, F.E. (1984). Effect of weight, temperature and skinning of broiler drumsticks on batter and breading adhesion. *J. Food Quality*, 6(4): 315–327.

Reddy, L., Shetty, T.M.R. and Dora, K.C. (1990). Utilization of low value fish preparation of fish fingers from croaker and perches. *Fishery Tech.*, 27(2): 133–137.

Rispoli, J.M., Sergenian, H.H., Topalian, H., Rogers, M.A. and Swartley, J.S. (1987). Coating mix containing fried component and process therefor. European patent.

Rizzoti, R. and Tilly, G. (1988). Edible coating agent, process for coating of foods and application to the preservation of foods. European Patent Application.

Roland, L.M., Seideman, S.C., Donnelly, L.S. and Quenzer, N.M. (1981). Physical and sensory properties of chicken patties made with varying proportions of white and dark spent fowl muscle. *J. Food Sci.*, 46(3): 834–837.

Samson, A.D. and Bangs, W.E. (1995). Coating composition for food products. *UK patent Application*, GB 2 283 655A, GB 9323230 (19931110).

Shelef, L.A. (1975). Microbial spoilage of fresh refrigerated beef liver. *J. Appl. Bact.*, 39: 273–280.

Smith, D. (1999). Breading: The golden process. *Meat Processing*, 38(12): 26–30.

Southerland, J.T., Gibbs, P.A. and Muray, J.G. (1975). Some metabolic and biochemical characteristics of representative microbial isolates from vacuum packaged beef. *J. Appl. Bacteriol.*, 39: 239–249.

Srinivas Gopal, T.K. and Ravishankar, C.N. (2001). Packaging of value added fish products. *Indian Food Industry*, 20(3): 64–67.

Strange, E.D., Benedit, R.C., Smith, J.C. and Swift, C.E. (1977). Evaluation of rapid tests for monitoring alterations in meat quality during storage. I. Intact meat. *J. Food Protect*, 40: 843–847.

Stuchell, Y.M. and Krochta, J.M. (1995). Edible coatings on frozen king salmon: Effect of whey protein isolate and acetylated monoglyeride on moisture loss and lipid oxidation. *J. Food Sci.*, 60(1): 28–31.

Suderman, D.R. (1979). Factors affecting the adhesion of coating to poultry skin. Ph.D. Dissertation, Kansas State University, Kansas, USA.

Suderman , D.R. and Cunningham, F.E. (1980). Factors affecting the adhesion of coating to poultry skin. Effect of age, method of chilling and scald temperature on poultry skin ultrastructure. *J. Food Sci.*, 45: 444.

Suderman, D.R. and Cunningham, F.E. (1981). Effect of freezing broilers drumsticks on breading adhesion. *J. Food Sci.*, 46: 1953–1955.

Suderman, D.R., Wiker, J. and Cunningham, F.E. (1981). Factors affecting adhesion of coating to poultry skin: Effects of various protein and gum sources in the coating composition. *J Food Sci.*, 46(4): 1010–1011.

Sunki, G.R., Annapureddy, R. and Rao, D.R. (1978). Microbial, biochemical and organoleptic changes in ground rabbit meat stored at 5 to 7°C. *J. Anim. Sci.*, 46: 584–588.

Thorner, M.E. (1973). Convenience and Fast Food Hand book. Avi Publ. Co. Westport, CT.

Tims, M.J. and Watts, B.M. (1958). Protection of cooked meat with phosphates. *Food Technol.*, 12: 240–243.

Vojdani, F. and Torres, J.A. (1990). Pottasium sorbate permeability of methylcellulose and hydroxypropyl methyl cellulose coatings: Effect of fatty acids. *J. Food Sci.*, 55(3): 841–846.

Waimaleongora-Ek, C. and Chen, T.C. (1983). Effect of shortening colour, frying temperature and coating ingredients on colour of fried chicken parts. *Poultry Sci.*, 62: 793–797.

Waimaleongora-Ek, C. and Chen, T.C. (1986). Effect of frying shortening quality, holding conditions and reheating on selected flavour volatiles of deep fat fried chicken parts. *Poultry Sci.*, 65(11): 2043–2050.

Wang, C.C., Kuo, C.C., Tsai, T.C. and Chi, S.P. (1996). Effect of smoke on quality of marinated chicken drumsticks during storage at room temperature. *Food Sci. Taiwan*, 23(2): 228

Wanstedt, K.G., Seideman, S.C., Donnelly, L.S. and Quenzen, N.M. (1981). Sensory attributes of precooked, calcium alginate-coated pork patties. *J. Food Protect.*, 44: 732–735.

Wehr, H.M. (1978). Attitudes and policies of state governments. *Food Technol. Champaign.*, 24: 189–199.

Weiss, T.J. (1970). Food oils and their uses. AVI Publishing Co. Weetpoint, Connecticut, USA.

Wilkening, H. (1991). Natural food product and method for its manufacture. German-Federal-Republic Patent-Application. DE39 25 055 A1 DE 3925055 (19890728).

Williams, S.K., Oblinger, J.L. and West, R.L. (1978). Evaluation of a calcium alginate film for use on beef cuts. *J Food Sci.*, 43: 292–296.

Wu, Y., Rhim, J.W., Weller, C.L., Hamouz, F., Cuppett, S. and Schnepe, M. (2000). Moisture loss and lipid oxidation for pre-cooked beef patties stored in edible coating and films. *J. Food Sci.*, 65(2): 300–304.

Yamuchi, K. (1973). Effect of heat treatment on the amount of free lipid released from isolated mitochondrial fraction of skeletal muscle tissue. *Jap J. Zootech. Sci.*, 44: 201.

Yang, C.S. and Chen, T.C. (1979). Yields of deep fat fried chicken parts as affected by preparation, frying conditions and shortening. *J. Food Sci.*, 44: 1074–1076.

14

Natural Antioxidants for Muscle Foods

A.P. Najeeb, P.K. Mandal and K.P. Mini

ABSTRACT

Consumers' interest in meat products formulated with natural antioxidants has motivated researchers to evaluate the effectiveness of naturally occurring compounds in plant leaves, fruits and herbs for functional purposes. Leaves and fruits are rich in polyphenolic compounds and possess significant DPPH, ABTS and superoxide anionic scavenging activity. So addition of leaves and fruits in different modes such as paste, extract and powder have potent natural preservative effects to augment the shelf life of meat and meat products and could replace the synthetic antioxidants effectively. Addition of these leaves and fruits were found to have substantial desirable effects on the physicochemical, microbiological and sensory attributes of meat and meat products. Hence, utilization of these natural preservatives in meat and meat products not only will increase the storage stability, but also will replace the toxic effect of chemical preservatives. These can work as functional additives by providing adequate health promoting effect in the human body system too.

1. INTRODUCTION

Indian meat industry has made rapid progress during the last four decades and ranks 4th in the world with poultry meat production of 1.93 million tons annually (FAO, 2010). The poultry meat industry is witnessing the highest growth rate of 15 per cent per annum contributing 25 per cent of the total meat production in the country and accounts for Rs. 400 billion to the national GDP (GOI, 2010). Poultry meat is the fastest growing component of global meat production, consumption, and trade with developing economy playing a pivotal role in its growth due to its enormous potential to bring rapid economic growth with low investment. Indians prefer to consume chicken rather than other species meat primarily due to its taste, health benefits, free from religious taboos, cost competitiveness and easy availability.

The strategies for producing healthier meat and meat products involve modification both at the farm level and meat product processing level.

Consumers' consciousness about the impact of diet on health, life expectancy and well being leads to an increased demand for animal food products even with added value. Considering the need of extending the storage stability of poultry meat in perspective of its marketing potentiality, various additives with antioxidant and antibacterial properties have been tried with varying degrees of success.

Functional food is the part of everyday diet and is demonstrated to offer health benefits by reducing the risk of chronic diseases beyond the widely accepted nutritional effects. The functional foods comprise of (*i*) conventional foods containing naturally occurring bioactive substances (*e.g.*, dietary fiber), (*ii*) foods enriched with bioactive substances (*e.g.*, probiotics, antioxidants) and (*iii*) synthesized food ingredients introduced to traditional foods (*e.g.*, prebiotics). At the moment, the most important and frequently used functional food additives are probiotics, prebiotics, plant antioxidants, vitamins and calcium.

Different reformulation strategies are being carried out to confer meat as functional food by modifying its lipid and fatty acid content, and/or by incorporating a series of functional ingredients like fiber, vegetable proteins, phytochemicals, prebiotics and other natural antioxidants (Fernandez *et al.*, 2004). Natural antioxidants are known to protect cells from damage induced by oxidative stress due to the production of free radicals, which are generally considered to be the cause of aging, degenerative diseases and cancer. Based on growing interest in free radical biology and the lack of effective therapies for most chronic diseases, the usefulness of antioxidants in protection against these diseases is warranted. Putting emphasis on these facts, researches on natural antioxidants are the current trends in meat and allied industry.

2. MECHANISM OF LIPID OXIDATION

Lipid oxidation is a highly complex process in which polyunsaturated fatty acids after reacting with molecular oxygen via a free radical chain mechanism form fatty acyl hydroperoxides which are referred to as either simply peroxides or primary products of oxidation (Gray, 1978). This primary auto-oxidation is followed by a series of secondary reactions thereby causing degradation of lipids and development of oxidative rancidity (Ladikos & Lougovois, 1990). As already mentioned, being a highly complex process, the entire process of lipid oxidation has been divided into three phases: Initiation phase, Propagation phase and Termination phase.

2.1. Initiation Phase

During this phase, initiation occurs by the removal of a hydrogen from a methylene carbon in a fatty acid to form an alkyl radical (L•). Being easier to remove hydrogen from a methylene carbon, it increases the number of double bonds in the fatty acids (Frankel, 1984 & Frankel, 1985). That's why phospholipids which are rich source of poly-unsaturated fatty acids are particularly susceptible to auto-oxidation. Hence, for lipid oxidation, muscle foods are in particular susceptible. This phase of oxidation is catalysed by a large number of molecules such as reactive oxygen species (OH•) and iron-oxygen complexes (*e.g.*, ferryl or perferryl radicals).

2.2. Propagation Phase

This phase involves a reaction between the alkyl radical (L•) oxygen to form a peroxyl radical (LOO•) which then extracts a hydrogen from another fatty acid and forms hydroperoxides (LOOH). These hydroperoxides are formed continuously via a free radical mechanism pathway. Formation of hydroperoxides is affected by various conditions including light, temperature, pH, presence of metal ions or iron, pigments, enzymes etc.

2.3. Termination Phase

During this phase, continued generation of hydroperoxides occurs by secondary reaction pathways thereby leading to their accumulation. Later on decomposition of these hydroperoxides leads to the formation of numerous volatile and non volatile compounds (Frankel, 1984). These compounds are mainly aldehydes, ketones and furans which affect the colour, flavour, texture and nutritional profile of meat and meat products. The aldehydes formed including hexanal, malondialdehyde and 4-hydroxynonenal can react readily with amino groups of proteins to yield Maillard-type complexes. These types of complexes affect the colour of meat and meat products.

Iron in muscle foods exists in the form of various oxidation states and the dominated states are Fe^{2+} and Fe^{3+}. Iron has unique feature to catalyze the detrimental oxidation of biomolecules such as DNA, protein, lipid etc. Therefore, iron metabolism *in vivo* should be tightly regulated by iron binding proteins so as to ensure the absence of free forms of iron existing.

3. OXIDATION OF MEAT

Meat is a rich source of proteins, lipids, vitamins and minerals. Out of these, lipids are considered to be the integral components of muscle foods. Hence, all the muscle foods are prone to lipid oxidation which is the major cause of deterioration of quality of stored meat and meat products. However, among all the meat products, poultry meat *i.e.*, white meat is considered to be more prone to oxidative rancidity than the red meats. There is strong explanation for this fact as poultry meat is rich in phospholipids and poly-unsaturated fatty acids which are integral components of cell membranes (Kemin-Europa, 2009). Some of the tissue lipid per-oxidants are ferric heme pigments, non-heme iron, ascorbic acid, sodium chloride etc. In cured meat, due to formation of stable pink colour by ferrous form of pigment, there is less oxidation as compared to cooked uncured meat (Love & Pearson, 1971).

Poultry meat is relatively rich in polyunsaturated fatty acids (Bourre, 2005). Consumption of meat containing high amounts of polyunsaturated fatty acids (PUFAs) has greatly increased in the last decade keeping in view the nutritionists' recommendations to reduce intake of saturated fatty acids. High content of unsaturated fatty acids in diet may cause lipid oxidation which creates a risk to human health because of the possible formation of atherogenic, mutagenic and cytotoxic compounds (Fernandez *et al.*, 2005). Lipid peroxides and their decomposition products may cause damage to proteins, membranes and biological components thus affecting vital cell functions (Frankel, 1984).

Lipid oxidation in meat products is initiated when polyunsaturated fatty acids react with molecular oxygen, via free radical chain mechanism, forming peroxides followed by a series of secondary reactions leading to the degradation of lipids and development of oxidative rancidity which adversely affect the nutritional quality and safety of meat and meat products (Gray, 1978). Lipid oxidation is influenced by the composition of phospholipids, amount of polyunsaturated fatty acids, presence of metal ions, oxygen, haeme pigments, mechanical processes and addition of salt during processing. The changes in meat products quality caused by lipid oxidation are manifested by adverse changes in colour, flavour, nutritive value and production of toxic compounds (Jensen *et al.*, 1998). Hence, it is radical to control these changes for better storage stability and acceptability of meat and meat products.

The most common strategies for preventing lipid oxidation are the use of antioxidants and restriction of access to oxygen (Tang *et al.*, 2001). Antioxidants are those compounds which help to delay or inhibit lipid oxidation when added to foods, thereby minimizing rancidity, retarding the formation of toxic oxidation products, helping to maintain the nutritional quality and increasing the shelf life of food products (Fukumoto & Mazza, 2000). Antioxidants can bind to metals, scavenge reactive oxygen species that initiate or perpetuate oxidation, quench high-energy oxygen species thus prevent formation of peroxides or decompose lipid peroxides. The synthetic compounds with antioxidant properties like butylated hydroxyanisole (BHA) and butylated hydroxytoluene (BHT) etc. are not encouraged in meat and allied industry due to their toxic potential and carcinogenic effect (Jayaprakasha *et al.*, 2003).

In view of present national and international barriers regarding the use of chemical food additives in food processing and preservation, search for biological and plant origin food additives have notably increased for the production of high-quality, ready-to-cook or ready-to-eat products in recent years. Over and above, growing awareness among consumers towards the health aspect has increased the interest in food additives with natural preservatives. Consumers' interest in meat products formulated with natural antioxidants has motivated researchers to evaluate the effectiveness of naturally occurring compounds in plant leaves and herbs for functional purposes. Utilization of natural preservatives offers adequate microbiological safety, replaces the toxic effect of chemical preservatives and contributes health promoting effect, hence making the food functional.

4. OXIDATION OF RESTRUCTURED MEAT PRODUCTS

In its broadest sense, any meat products that are partially or completely disassembled and then reformed into the same or different forms are called restructured products. Meat restructuring involves the assembling of meat pieces into a cohesive product which aims to simulate or retain the texture of high quality muscle (Sheard & Jolley, 1988). Cohesion is developed during cooking by the gelation of meat proteins solubilised during processing by the action of salt. Many technological advantages on processing combination have been used in the preparation of restructured meat products to acquire sensory characteristics somewhere between ground meat and intact muscle steaks. Lipid oxidation is one of the major problems in the development of new convenient meat products (Gray & Pearson, 1987).

The increasing demand for convenient foods has led to the expansion in the variety of ready-to-eat meat products including restructured meat products. Restructuring has been developed as a method for transforming lower value cuts and trimmings into products of higher value and considered as a way to expand the potential of muscle foods in the market. Advancement in technology and mechanization makes the carcass meat to be disassembled and reassembled in such a way that it gives a texture similar to steaks, chops and roasts.

Restructured meat products are commonly prepared using different methods of size reduction *viz.*, chunking and forming, flaking and forming or sectioning and forming or the combination of them (Boles & Shand, 1998). Chunking is accomplished by passing the meat through a coarse grinder plate leading to decrease in the particle size not greater than one and a half inch cubes. This technique increases the surface for the extraction of myosin and aids in better binding during mixing. High speed dicing or slicing machine is being used for flaking and reforming of restructured meat products. Fine flakes produce more acceptable appearance, increase tenderness and decrease shear force value. Sectioned and formed meats are primarily composed of intact muscle or section of muscle that are bound together to form a single piece (Pearson & Gillet, 1996).

In restructuring of meat, tumbling and massaging are the two techniques used quiet commonly. Tumbling utilizes impact energy and massaging utilizes frictional energy. The action of tumbling and massaging not only aids in better extraction of proteins but also improves the speed of curing by increasing salt absorption. A tumbler consists of a stainless steel tank mounted on a set of rotating wheels to agitate the contents while spinning. Massager is a slow mixer designed to stir or agitate large chunks of meat (Hedrick *et al.*, 1994).

Oxidation and colour fading are two major problems in retail acceptance of the restructured meat (Pearson & Dutson, 1987). A highly acceptable restructured chicken slice without addition of extra fat has been developed as a novel meat product (Mandal *et al.*, 2002). But increase in TBA and tyrosine values of this product were observed under refrigerated storage. The product was found to be acceptable only up to 10th day of storage under refrigeration. On 15th day, product had pale appearance and unpleasant odour, so it was not subjected for sensory evaluation. Gizzard was successfully added in the restructured chicken block up to 50% without any adverse effect (Mandal *et al.*, 2011). The gizzard added product was also prone to oxidation both under refrigerated and frozen storage (Sudheer *et al.*, 2011a). Sensory scores of the restructured chicken block decreased significantly ($P<0.05$) on the 10th day of storage (Sudheer *et al.*, 2011b). The gizzard added product also had a storage life only upto 10th day under refrigerated storage. The product was not subjected to sensory evaluation on 15th day due to oxidative changes. Use of natural antioxidants has been found to have an ameliorating effect on these undesirable changes produced via oxidation.

5. ANTIOXIDANTS AND THEIR MECHANISM OF ACTION

Antioxidants are those chemical compounds or molecules which when present, significantly inhibit or delay the oxidation, while often being oxidized themselves. There are two principle mechanisms of action for antioxidants. Primary antioxidants halt the free radical chain reactions just by donating a single electron to the free radical present in the system, since donation of an electron to a lipid radical produces a new radical and is more stable than the initial one. However, secondary antioxidants act by various ways including quenching chain-initiating catalysts, binding of metal ions, scavenging of oxygen, converting hydroperoxides to non-radical species, absorbing UV radiation, deactivating singlet oxygen, inhibition of peroxidases etc. Sometimes two or more antioxidants show their synergistic effect when present together in a medium thereby greatly enhancing the antioxidative effect. Such types of antioxidants are referred to as co-antioxidants and process is known as co-antioxidation.

6. USE OF NATURAL ANTIOXIDANTS OF PLANT ORIGIN IN MEAT

The antioxidant properties of natural food additives are being studied intensively due to the growing demand from the pharmaceutical and food industries. Plant sources may bring new natural products into the food industry with safer and better antioxidants that provide ample protection against oxidative damage occurring both in the human body and in the processed food. The use of natural preservatives to increase the shelf-life of meat products is a promising technology since many herbs, plants, fruits, and vegetable extracts or powders have antioxidant and antimicrobial properties (Biswas *et al.*, 2012). Synthetic antioxidants like butylated hydroxy toluene (BHT) and butylated hydroxy anisole (BHA) have successfully been used to prevent lipid oxidation in meat. But due to their adverse side effects on human health, an effective alternative particularly from natural sources is necessary to replace these synthetic chemicals.

Presently, the use of natural antioxidants in meat systems is limited. This is due to the lack of availability, extra production cost, inadequate knowledge about their active ingredients and molecular composition, and non-availability of relevant data on toxicity (Shahidi *et al.*, 1994). In traditional Indian preparations, leaves and fruits are added to meat as flavour enhancers either in raw chopped form or as ground paste with other spices. The total phenolics and free radical scavenging activity of these ingredients retained even after heating indicated that the active constituents were resistant to thermal denaturation and hence could be used as antioxidants in meat products processing (Kannat *et al.*, 2005). This combination treatment could have an important economic impact as it might reduce losses attributed to spoilage and also allow the products to reach new and distant markets.

6.1. Use of Leaves as Natural Antioxidants

6.1.1. *Drumstick leaves (Moringa oleifera L.)*

The leaves are rich in proteins, essential amino acids such as methionine, cystine, tryptophan and lysine, vitamin C, beta-carotene, riboflavin, nicotinic acid, pyridoxine, potassium, copper, calcium, iron and phosphorus. It is reported to be a potent source of natural antioxidants like ascorbic acid, alpha tocopherol and other flavonoids and carotenoids, and thus enhancing the shelf life of fat containing foods (Gupta *et al.*, 1989; Dillard & German, 2000). The presence of these substances in *drumstick* leaves reveals its importance as a functional dietary supplement (Makkar & Becker, 1996).

Antioxidant activity of *drumstick* leaves has been reported in meat and meat products. The crude extract of *drumstick* leaf can actively scavenge free radicals and thus prevent cellular damage (Sreelatha & Padma, 2009). The *drumstick* leaf extract contains polyphenols that have antioxidant effects. Polyphenolic compounds act as chain-breaking peroxyl radical scavengers which lead to the inhibition of lipid peroxidation and also prevent low density lipoprotein oxidation (O'Byme *et al.*, 2002). Hazra *et al.* (2011) reported a significantly ($P<0.05$) lower thiobarbituric acid value in cooked ground buffalo meat treated with *drumstick* leaves extract (1.5%).

Significant differences ($P<0.01$) in the pH values were observed among the restructured chicken slices formulated with leaf powders at 1% (*drumstick*, mint and *curry* leaves) and BHT (200 ppm) including control during 20 days of storage at refrigerator. But no significant differences were observed among the leaf powders added products throughout the storage period (Najeeb, 2012).The cooked ground buffalo meat treated with 1.5% crude extract of *drumstick* leaves had significantly ($P<0.05$) higher pH value compared to other products.

Caceres *et al.* (1991) reported that juice of fresh *drumstick* leaves is found to have antimicrobial properties. It has been noticed that the total plate counts of ground buffalo meat were significantly ($P<0.05$) lowered by the incorporation of *drumstick* leaf extract (Hazra *et al.*, 2011). The fresh leaf juice, powder from fresh leaves and cold water extract of fresh leaves from *drumstick* have displayed a potential antibacterial activity against both gram negative and gram positive bacteria as determined by disc diffusion and minimum inhibitory concentration method (Rahman *et al.*, 2009).

A significant (P<0.05) reduction in cooking loss was also observed in 1.5% and 2% *drumstick* leaves treated samples compared to control (Hazra *et al.*, 2011). Cooked ground buffalo meat formulated with *drumstick* leaves extract upto 1.5% did not show any significant difference in its moisture, total protein and ether extract content compared to control. However, the products treated with 2% crude extract had higher protein and fat content in comparison to other treated products. This might be due to the increase in solubility of the total protein and a higher moisture from the 2% treated cooked products. No significant differences were observed for both cooking yield and proximate composition (moisture, protein, fat and total ash %) of restructured chicken slices incorporated with drumstick leaf powders (1%) and BHT (200 ppm) throughout the storage period (Najeeb, 2012).

Addition of *drumstick* leaves extract also improved the quality of meat by enhancing its tenderness, juiciness and preventing discoloration (Hazra *et al.*, 2011). The appearance, flavor, texture, juiciness and overall acceptability scores of the restructured chicken slices containing leaf powders (*drumstick*, mint and *curry* leaves) and BHT were not affected significantly even up to 20 days of storage under refrigeration and scores remained above 6 indicating that all the products were well acceptable (Najeeb, 2012).

6.1.2. *Mint leaves (Mentha spicata. L.)*

Mint leaves have potent antioxidant and antiperoxidant properties due to the presence of eugenol, caffeic acid, rosmarinic acid and α-tocopherol (Marinova & Yanishlieva, 1997; Zheng & Wang, 2001; Dorman *et al.*, 2003). There are only few reports on the use of mint as an antioxidant in food. Marinova and Yanishlieva (1997) reported that *Mentha spicata L.* was able to retard the autooxidation of sunflower oil at 100°C. Decrease in enzymatic lipid peroxidation has also been reported by aqueous and alcoholic extracts of mint (Shobana & Naidu, 2000).

Application of antioxidant activity of mint leaf has been reported in meat and meat products. The suitability of mint leaf extract (0.1%) as an antioxidant was determined in irradiated lamb meat (Kannat *et al.*, 2005). Treatment with mint leaf extract reduced lipid oxidation significantly for all irradiated products throughout the storage period. Thiobarbituric acid values of irradiated meat containing mint leaf extract (0.1%) stored at chilled temperatures were significantly (P<0.05) lower than control. After four weeks of chilled storage, TBARS in irradiated meat containing mint leaf extract (0.1%) was half of that present in untreated irradiated meat and it was comparable to that of BHT treated products (Kannat *et al.*, 2005). No significant differences were observed among the TBA values of restructured chicken slices added with BHT (200 ppm), curry leaf and mint leaf powder (1%) during 20 days of storage under refrigerator. However, all the products incorporated with leaf powders and BHT were significantly (P<0.01) lower than control products (Najeeb, 2012).

Raw ground pork meat treated with mint leaf extract showed a significantly lower TBA value than control during refrigerated (4 ± 1°C) storage (Biswas *et al.*, 2012). They also reported a significantly lower pH value in raw ground pork meat treated with mint leaf extract stored at refrigeration (4 ± 1°C) temperature than control.

Crude extract of *Mentha spicata* was found to have better antibacterial activity against *Escherchia coli*, *Bacillus subtilis*, *Shigella flexeneri*, *Staphylococcus aureus*, *Pseudomonas aeruginosa* and *Salmonella typhi*. Among them, *E. coli* and *Pseudomonas* were shown to be the most sensitive on minimum inhibitory concentration method (Naseem *et al.*, 2011). It was observed that the essential oil from *M. spicata* has a dose dependent activity against bacteria as determined by disc diffusion assay and among which gram positive organisms were more sensitive compared to gram negative bacteria. It was reported that essential oil from *M. spicata* possessed better antibacterial activity than the standard drugs like streptomycin and penicillin (Habiba *et al.*, 2011). Restructured chicken slices prepared with leaf powders (*drumstick*, mint and *curry* leaves) showed significantly (P<0.01) lower microbial counts compared to both control and BHT (200 ppm) added products (Najeeb, 2012).

Antifungal assay has been carried out to evaluate the antifungal activity of mint leaves extract. It has been observed that mint leaves extract has potent antifungal activity against *Trichophyton longifusus*, *Microsporum canis*, *Candida albicans*, *Aspergillus flavus*, *Fusarium solani* and *Candida glaberata* (Naseem *et al.*, 2011). Yeast and mould counts were not detected in restructured chicken slices added with leaf powders (*drumstick*, mint and *curry* leaves), BHT (200ppm) and control products throughout the storage period (Najeeb, 2012).

6.1.3. *Curry leaves (Murraya koenigii. L.)*

Phytochemical studies on leaves, stem, bark and root of this plant revealed that the presence of carbazole alkaloids such as murrayanine, girinimbine, mahanimbine, murrayafoline-A and triterpene had vasorelaxing, anticarcinogenic and antiplatelet activity in addition to antioxidant and antimicrobial properties (Chakraborty *et al.*, 1978; Ito *et al.*, 1993; Baker *et al.*, 2007). Antioxidant property of *curry* leaves is attributed mainly to the presence of mahanimbine, murrayanol and mahanine (Tachibana *et al.*, 2003; Ningappa *et al.*, 2008). Mahanine is reported to have high antioxidant activity compared to BHA and alpha tocopherol (Gupta *et al.*, 2011).

Use of *curry* leaves as antioxidant in muscle foods on oxidative stability has been reported only exiguously. *Curry* leaf powder at a concentration as low as 0.2% is found to be a very effective inhibitor of primary and secondary oxidation products in raw ground and cooked goat meat patties stored under refrigeration (4 ± 1°C). It was found that raw ground goat meat product was not acceptable after three days while *curry* leaf powder added product was acceptable up to seven days in refrigerated storage (Das *et al.*, 2011).

Biswas *et al.* (2012) reported a significantly (P<0.05) lower TBA value in raw ground pork meat treated with *curry* leaf extract stored at refrigeration (4 ± 1°C) temperature than control. Reduction in TBARS values were observed in chicken patties treated with *curry* leaf extract (2%) compared to control during eight days of storage under refrigeration. Interestingly, it was reported that chicken patties treated with BHT had a higher TBARS values than curry leaves extract treated samples (Devatkal *et al.*, 2011).

No significant differences were observed on the tyrosine values of restructured chicken slices added with mint and *curry* leaf powders (1%) throughout the storage days at refrigerator (Najeeb, 2012). Raw goat meat with *curry* leaf powder (0.2%) had significantly (P<0.05) lowered free fatty acid content than control during nine days of refrigerated storage (Das *et al.*, 2011). Restructured chicken slices prepared with *drumstick* leaf and *curry* leaf powders (1%) did not show any significant differences in the FFA values during the entire storage period.

Cooking loss and pH were not affected by the addition of *curry* leaf powder (0.2%) in raw ground goat meat patties (Das *et al.*, 2011). Biswas *et al.* (2006) observed that pH was not affected due to the addition of *curry* leaf powder in ground poultry meat. But a significant (P<0.05) reduction in pH values were reported in raw ground pork meat treated with *curry* leaf extract than control stored at refrigeration (4 ± 1°C) temperature (Biswas *et al.*, 2012).

The protein designated as APC (antioxidant protein from *curry* leaves) isolated from *curry* leaves showed highly significant broad spectrum antibacterial activity by producing a clear zone of inhibition (20–25 mm) against *E. coli* and *Staphylococcus aureus*, *Vibrio cholerae*, *Klebsiella pneumoniae*, *Salmonella typhi* and *Bacillus subtilis* and was as potent as that of chloramphenicol, streptomycin and gentamicin (Ningappa *et al.*, 2008). It is also found that methanolic extract of *curry* leaves had potent antibacterial activity (Gupta *et al.*, 2011). But Das *et al.* (2011) reported that addition of *curry* leaf powder (0.2%) in ground goat meat did not show any significant effect on bacterial growth during chilling storage. Similarly, Biswas *et al.* (2006) did not find any antimicrobial effect of *curry* leaf powder (0.2%) in fresh and cooked chicken meat.

While considering the sensory attributes of meat products formulated with *curry* leaves, it was reported that there was no significant difference in colour and odour score of ground goat meat patties treated with *curry* leaves powder (0.2%) and control. But an improvement in the odour score has been noticed with *curry* leaves treated samples (Das *et al.*, 2011).

6.2. Use of Fruits as Natural Antioxidants

6.2.1. *Red grapes (Vitis vinifera)*

Grapes are a rich source of polyphenolic compounds. It mainly includes anthocyanins, flavonols, stilbenes and phenolic acids. Flavonoids and other plant phenolics have been reported to have multiple biological effects including inhibition of platelet aggregation (Kanner *et al.*, 1994), anti-inflammatory action (Moroney *et al.*, 1988), antioxidant activity (Frankel *et al.*, 1995) and antimicrobial activity (Renaud & Longeril, 1992). As the most notable bioactivity of phenolic compounds from grapes is the antioxidant activity, its characteristics like scavenging of free radicals, inhibition of lipid oxidation, reduction of hydroperoxide formation has been widely studied (Meyer *et al.*, 1997). By exploiting its antioxidant and other functional properties, it is being used as a natural preservative and functional additive in meat and meat products.

In raw meat, grape seed extract was found to be effective in reducing the amount of primary and secondary lipid oxidation products in beef (Banon *et al.*, 2007), chicken (Lau and King, 2003) and pork (Carpenter *et al.*, 2007). It was observed that 0.1% grape seed extract was as effective as BHT/BHA and completely inhibited the formation of TBARS and lipid hydroperoxides in cooked beef, pork, and chicken after a period of seven days under refrigerated storage (Brannan & Mah, 2007). The mechanism of the protective effect on lipid oxidation may be due to the fact that the grape seed extracts obtained from wine and juice processing are rich in proanthocyanidins, which have multiple mechanisms for its antioxidant activity *viz.* ability to sequestrate free radicals, chelate metals and synergise with other antioxidants (Lu & Foo, 1999).

Chicken patties formulated with grape seed extract (0.1%) did not show any significant changes in cooking yield. Similarly, moisture content of patties during refrigerated storage at four different RH (relative humidity) levels was not significantly affected by the period of storage. It was reported that addition of grape seed extract even at 0.1% in chicken patties had lowered its pH compared to control (Brannan, 2008). Cooking yield and proximate composition (moisture, protein, fat and ash %) of restructured chicken slices formulated with different fruit powders (red grapes, gooseberry and tomato) at 1% and BHT (200 ppm) did not vary significantly on the 1st and 20th day of storage (Najeeb, 2012).

Addition of grape extract (60 mg total phenolic compounds/kg of meat) provided greater stability to the chicken meat balls with regard to red discolouration indicating highest a* value and showed significantly darker colour compared with BHT treated samples (highest L* value) which were determined by an instrumental colorimeter using the L* (lightness), a* (redness) and b* (yellowness) parameters of the CIELAB system. Changes in L*, a* and b* values may be due to the colour interference of the grape residues (Selani, *et al.*, 2010). Rojas and Brewer (2007) reported that grape seed extract as low as 0.02% improved the sensory and chemical shelf life of refrigerated cooked beef and pork patties without affecting its colour.

It was found that addition of grape seed extract did not affect the initial changes in flavor scores in both irradiated and non-irradiated whole chicken breasts (Rababah *et al.*, 2005). It also reduced the warmed over flavor in cooked beef during three days of refrigerated storage (Ahn *et al.*, 2002) and rancid and wet-cardboard off-odour scores in beef and pork meat without affecting colour during eight days of refrigerated storage (Rojas & Brewer, 2007). All parts of grapes are having potent antimicrobial activity. The ascending order of the antimicrobial activity was in flesh, whole fruit grape extracts, fermented pomace, skin, leaf and seed (Xie *et al.*, 2010). Anastasiadi *et al.* (2009) suggested that high concentration of flavonoids and their derivatives such as stilbenes and phenolic acids in grapes were responsible for its antimicrobial activity.

Jayaprakasha *et al.* (2003) observed that grape seed extracts inhibited bacterial growth at 340–390 mg GAE/L and 475–575 mg GAE/L for gram positive and gram negative bacteria, respectively. Rotava *et al.* (2009) reported that phenolic compounds from defattened grape (*Vitis vinifera*) seed extract inhibited the growth of *Staphylococcus aureus* and *Escherichia coli*, while they showed no effect on *Salmonella* sp. Xie *et al.* (2010) reported that phenolic compounds in grape like resveratrol displayed potent antifungal activity against the *Candida albicans*. All these findings suggested the potential applications of phenolic compounds to maintain shelf life and improve safety of ready-to-eat food products. No significant differences were observed on the SPC values of restructured chicken slices added with red grapes and tomato powders (1%) throughout the storage period (Najeeb, 2012).

6.2.2. *Gooseberry (Emblica officinalis)*

Gooseberry is the richest source of antioxidants like vitamin C, emblicanin A and B, punigluconin, pedunculagin, superoxide dismutase, catalase, gluthathione peroxidase (Bhattacharya *et al.*, 2000; Rajak *et al.*, 2004), tannin, trigalloyl, polyphenol, flavonoids, ellagic acid and phyllembic acid (Anilakumar *et al.*, 2004). Gooseberry has also been reported to possess antifungal, antibacterial and antiviral activities (Godbole & Pendse, 1960; Dutta *et al.*, 1998; Rani & Khullar, 2004).

Restructured chicken slices added with gooseberry powder (1%) showed significantly ($P<0.01$) lower pH and higher tyrosine values compared to all other products prepared with BHT (200 ppm), red grapes and tomato powders (1%) throughout twenty days of storage (Najeeb, 2012). Reddy *et al.* (2004) reported that extracts from *drumstick* leaves (1%) and gooseberry (1%) were more effective compared to BHA for controlling lipid oxidation of biscuits during storage at ambient temperature. Biscuit formulated with gooseberry showed a lower FFA value and it was comparable to that of *drumstick* leaves treated samples.

Gooseberry is reported to possess high antibacterial activity against *E. coli*, *Klebsiella pneumoniae*, *Pseudomonas aerugenosa* and *Salmonella paratyphi* (Saeed *et al.*, 2007). Antimicrobial activity of aqueous and methanolic extract of gooseberry was tested against different microorganisms by Vimela *et al.* (2009). It was found that methanolic extract showed a zone of inhibition with maximum activity against *Geotricum* species and *Fusarium* species and minimum activity against *Salmonella* species.

6.2.3. *Tomato (Solanum lycopersicum)*

Tomato and tomato based products contain phytochemicals such as lycopene, folate, vitamin C, phenolics and flavonoids which are having health beneficial effects (Beecher, 1998; Agarwal & Rao, 2000). Lycopene is the major carotenoid present in tomato and a highly potent antioxidant that provides protection against cellular damage caused by reactive oxygen species and it is also reported as a potent cancer chemotherapeutic agent (Dimascio *et al.*, 1989; Rao & Agarwal, 1999). A few studies have been reported on the use of tomato in meat and meat products (Candogan, 2002; Yilmaz *et al.*, 2002; Sanchez *et al.*, 2003; Osterlie & Lerfall, 2005; Deda *et al.*, 2007, Kim *et al.*, 2010).

Addition of tomato powder at 2% level in frankfurters decreased the oxidation during 60 days of refrigerated storage (Eyiler & Oztan, 2010). Incorporation of lycopene rich tomato pulp and its extract in beef patties significantly delayed and inhibited the oxidation of lipids and extended the shelf life of beef patties from 4 to 12 days under modified atmosphere packaging (Escalante *et al.*, 2003). Restructured chicken slices added with fruit powders (red grapes, gooseberry and tomato) at 1% and BHT (200 ppm) had significantly ($P<0.01$) lower TBA values than control products except on 1st day where difference in the TBA values of control and tomato powder added products were not significant (Najeeb, 2012).

Kim *et al.* (2010) reported that incorporation of tomato powder at 0.8, 1.2 and 1.5% level significantly ($P<0.05$) reduced TBA value in low fat pork sausages and the lowest value at highest concentration of incorporation could be due to the antioxidant property of lycopene present in tomato powder. Up to 1.5% tomato powder incorporation in low fat pork sausage was found to be well acceptable even up to a period of 30 days of storage in refrigerator (4 ± 1°C). It was observed that addition of tomato paste at 5, 10 and 15% concentrations significantly reduced TBA value in beef patties than control during 9 days of refrigerated storage (Candogan, 2002).

The pH values of fresh pork sausages at 1.2 and 1.5% level of incorporation were significantly ($P<0.05$) lower than 0.8% level of incorporation and control. The reduction in pH of fresh sausages with higher level of tomato powder could be due to its acidic nature (Kim *et al.*, 2010). It was also reported that beef patties formulated with different levels (10 and 15%) of tomato paste had lower pH than the control over the nine days refrigerated storage (Candogan, 2002). Kim *et al.* (2010) further reported that there were no significant differences in moisture and fat contents of pork sausages by the addition of tomato powder upto 1.5% level. However, protein content at 1.5% level of tomato powder incorporation was significantly (P<0.05) higher than the 0.8% treated products and control. This higher protein content might be due to the added tomato powder in the sausages.

Tomato paste at 5% and 10% in patty formulation did not affect moisture, protein, and ash contents of beef patties. But tomato paste at 15% level of incorporation resulted in higher moisture and lower protein content as compared to control patties. Fat content also decreased significantly (P<0.05) with the addition of 10% and 15% tomato paste. Differences in proximate composition might be due to dilution effects resulting from tomato paste at various concentrations in beef patty formulation (Candogan, 2002). FFA values of restructured chicken slices added with red grapes and tomato powders (1%) were insignificant throughout the storage period (Najeeb, 2012).

Tomato paste addition at the level of 10 and 15% increased the juiciness of beef patties. It was reported that tomato paste incorporation in beef patties up to 15% level did not reveal any adverse effect on the flavour characteristics (Candogan, 2002). The appearance, flavor, texture and juiciness scores of restructured chicken slices added with fruit powders (red grapes, gooseberry and tomato) at 1% and BHT (200 ppm) were not affected even up to 20 days of storage under refrigeration and scores remained above 6 indicating good to very good appearance of the products.

The total plate counts of pork sausages at 1.2 and 1.5% level of tomato powder incorporation were significantly lower than control after fifteen days of storage. These low bacterial counts of tomato powder added sausages seemed to be due to lower pH and antimicrobial activity of tomato powder (Kim *et al.*, 2010).

REFERENCES

Agarwal, S. and Rao, A.V. (2000). Tomato lycopene and low density lipoprotein oxidation: A human dietary intervention study. *Lipids*, 33: 981–984.

Ahn, J., Grun, I.U. and Fernando, L.N. (2002). Antioxidant properties of natural plant extracts containing polyphenolic compounds in cooked ground beef. *J. Food Sci.*, 67: 1364–1369.

Anastasiadi, M., Chorianopoulos, N.G., George, E.N. and Haroutounian, S.A. (2009). Antilisterial activities of polyphenol rich extracts of grapes and vinification byproducts. *J. Agric. Food Chem.*, 57: 457–463.

Anilakumar, K.R., Nagaraj, N.S. and Santhanam, K. (2004). Protective effect of amla (*Embilica officinalis*) on oxidative stress and toxicity in rats challenged with dimethyl hydrazine. *Nutri. Res.*, 24: 313–319.

Baker, N.H.A., Sukari, M.A., Rahmani, M., Shariff, A.M., Khalid, K. and Yusuf, U.M. (2007). Chemical constituents from stem barks and roots of *Murraya koenigii* (Rutaceae). *Malaysian J. Anal. Sci.*, 11: 173–176.

Banon, S., Diaz, P., Rodriguez, M., Garrido, M.D. and Price, A. (2007). Ascorbate, green tea and grape seed extracts increase the shelf life of low sulphite beef patties. *Meat Sci.*, 77: 62–633.

Beecher, G.R. (1998). Nutrient content of tomatoes and tomato products. *Proc. Soc. Exp. Biol. Med.*, 218: 98–100.

Bhattacharya, A., Ghosal, S. and Bhattacharya, S.K. (2000). Antioxidant activity of tannoids principles of *Emblica officinalis* (amla) in chronic stress induced changes in rat brain. *Indian J. Expt. Biol.*, 38: 877–880.

Biswas, A.K., Kondaiah, N. and Anjaneyulu, A.S.R. (2006). Effect of spice mix and curry leaf (*Murraya koenigii*) powder on the quality of raw and precooked chicken patties during refrigerated storage. *J. Food Sci. Technol.*, 43: 438–441.

Biswas, A.K., Chatli, M.K. and Sahoo, J. (2012). Antioxidant potential of curry (*Murraya koenigii L.)* and mint (*Mentha spicata*) leaf extracts and their effect on colour and oxidative stability of raw ground pork meat during refrigeration storage. *Food Chem.*, 133: 467–472.

Boles, J.A. and Shand, P.J. (1998). Effects of comminution method and raw binder system in restructured beef. *Meat Sci.*, 49: 297–307.

Bourre, J.M. (2005). Where to find omega-3-fatty acids and how feeding animals with diet enriched in omega-3-fatty acids to increase nutritional value derived products for human: What is actually useful? *J. Nutri. Health Aging*, 9: 232–242.

Brannan, R.G. (2008). Effect of grape seed extract on physicochemical properties of ground salted chicken thigh meat during refrigerated storage at different relative humidity level. *J. Food Sci.*, 73: 121.

Brannan, R.G. and Mah, E. (2007). Grape seed extract inhibits lipid oxidation in muscle from different species during refrigerated and frozen storage and oxidation catalyzed by peroxynitrite and iron/ascorbate in a pyrogallol red model system. *Meat Sci.*, 77: 540–546.

Caceres, A., Cabrera, O., Morales, O., Mollinedo, P. and Mendia, P. (1991). Pharmacological properties of *Moringa oleifera*: Preliminary screening for antimicrobial activity. *J. Ethnopharm.*, 33: 213–216.

Candogan, K. (2002). The effect of tomato paste on some quality characteristics of beef patties during refrigerated storage. *Eur. Food Res. Technol.*, 215: 305–309.

Carpenter, R., Grady, M.N., Callaghan, Y.C., Brien, N.M. and Kerry, J.P. (2007). Evaluation of the antioxidant potential of grape seed and bearberry extracts in raw and cooked pork. *Meat Sci.*, 76: 604–610.

Chakraborty, D.P., Bhattacharyya, P., Roy, S., Bhattacharyya, S.P. and Biswas, A.K. (1978). Structure and synthesis of mukonine: A new carbazole alkaloid from *Murraya koenigii. Phytochem.*, 17: 834–835.

Das, A.K., Rajkumar, V. and Dwivedi, D.K. (2011). Antioxidant effect of curry leaf (*Murraya koenigii*) powder on the quality of ground and cooked goat meat. *J. Int. Food Res.*, 18: 559–565.

Deda, M.S., Bloukas, J.G. and Fista, G.A. (2007). Effect of tomato paste and nitrite level on processing and quality characteristics of frankfurters. *Meat Sci.*, 76: 501–508.

Devatkal, S.K., Thorat, P.R., Manjunatha, M. and Anurag, R.K. (2011). Comparative antioxidant effect of aqueous extracts of curry leaves, fenugreek leaves and butylated hydroxytoluene in raw chicken patties. *J. Food Sci. Technol.*, DOI 10.1007/s 13197-011-0511-0.

Dillard, C.J. and German, J.B. (2000). Phytochemicals, nutraceuticals and human health: A review. *J. Food Sci. Agric.*, 80: 1744–1756.

Dimascio P; Kaiser S and Sies S (1989). Lycopene as the most effective biological carotenoid singlet oxygen quencher. *Arch. Biochem. Biophys.*, 274: 532–538.

Dorman, H.J.D., Kosar, M., Kahlos, K., Holm, Y. and Hiltunen, R. (2003). Antioxidant properties and composition of aqueous extracts from Mentha species, hybrids, varieties and cultivars. *J. Agric. Food Chem.*, 51: 4563–4569.

Dutta, B.K., Rahman, I. and Das, T.K. (1998). Antifungal activity of Indian plant extracts. *Mycoses*, 41: 535–536.

Escalante, A.S., Gaston, T., Djenane, J.A.B. and Pedro, R. (2003). Stabilization of colour and odour of beef patties by using lycopene-rich tomato and pepper as a source of antioxidants. *J. Sci. Food Agric.*, 87: 187–194.

Eyiler, E. and Oztan, A. (2010). Production of frankfurters with tomato powder as a natural additives. *J. Food Sci. Technol.*, 44: 307–311.

FAO, (2010). Animal Production Year Book. Food and Agricultural Organization, Rome.

Fernandez, J.M., Fernandez, L.J., Sayas, B.E. and Perez, J.A. (2004). Lemon Albedo as a new source of dietary fibre: Application to bologna sausage. *Meat Sci.*, 67: 7–13.

Fernandez, J.M., Fernandez, L.J., Sayas, B.E. and Perez, J.A. (2005). Meat products as functional foods: A review. *J. Food Sci.*, 70: 37–43.

Frankel, E.N. (1984). Recent advances in chemistry of rancidity of fats. *In*: Recent advances in chemistry of meat (*Ed.* Baiely, A.J.). *The Royal Society of Chemistry Special Publication*. 47: 87–118.

Frankel, E., Waerhouse, A.L. and Teissedre, P.L. (1995). Principal phenolic chemicals in selected California wines and their antioxidant activity in inhibiting oxidation of human lowdensity lipoprotein. *J. Agric. Food Chem.*, 43: 890–894.

Fukumoto, L.R. and Mazza, G. (2000). Assessing antioxidant and prooxidant activities and phenolic compounds. *J. Agric. Food Chem.*, 48: 3597–3604.

Godbole, S.H. and Pendse, G.S. (1960). Antibacterial property of some plants. *Indian J. Pharmacol.*, 22: 39–42.

GOI. (2010). Basic animal husbandry statistics. Department of Animal Husbandry, Dairying & Fisheries, Ministry of Agriculture, India.

Gray, J.I. (1978). Measurement of lipid oxidation: A review. *J. Am. Oil Chem. Soc.*, 55: 539–546.

Gray, J.I. and Pearson, A.M. (1987). Restructured meat and poultry products. 1st edn. Von Reihold Co. Newyork.

Gupta, K., Barat, G.K., Wagle, D.S. and Chawla, H.K.L. (1989). Nutrient content and antinutritional factors in conventional and non-conventional leafy vegetables. *Food Chem.*, 31: 105–116.

Gupta, P., Nahata, A. and Dixit, V.K. (2011). An update of *Murrya Koenigii* Spreng: A multifunctional ayurvedic herbs. *J. Chinese Integrative Med.*, 9: 824–832.

Habiba, B., Chaker, A.N., Belhadj, H., Sahli1, F., Ramdhani1, M., Laouer, H. and Harzallah, D. (2011). Chemical composition and antibacterial activity of *Mentha pulegium* L. and *Mentha spicata* L. essential oils. *Der. Pharmacia Letter*, 4: 267–275.

Hazra, S., Biswas, S., Bhattacharyya, D., Das, S.K. and Khan, A. (2011). Quality of cooked ground buffalo meat treated with the crude extracts of *Moringa oleifera* (Lam.) leaves. *J. Food Sci. Technol.* DOI 10.1007/s13197-011-0383-3.

Hedrick, B.H., Aberle, E.D., Forrest, J.C., Judge, M.D. and Merkel, R.A. (1994). Principles of Meat Science, 3rd edn. W.H. Freeman & Company.

Ito C; Thoyama Y; Omura M; Kajiura I and Furukawa H (1993). Alkaloidal constituents of *Murraya koenigii*: Isolation and structural elucidation of novel binary carbazole quinones and carbazole alkaloids. *Chem. Pharmacol. Bullet.* 41: 2096–2100.

Jayaprakasha, G.K., Selvi, T. and Sakariah, K.K. (2003). Antibacterial and antioxidant activities of grape (*Vitis vinifera*) seed exstracts. *Food Res. Int.*, 36: 117–122.

Jensen, C.C., Lauridson, A. and Bertelsen, G. (1998). Dietary vitamin E: Quality and storage stability of pork and poultry. *Trends Food Sci. Technol.*, 9: 62–72.

Kannat, S.R., Chander, R. and Sharma, A. (2005). Antioxidant potential of mint (*Mentha spicata* L.) in radiation-processed lamb meat. *Food Chem.*, 100: 451–458.

Kanner, J., Frankel, E., Granit, R. and German, B. (1994). Natural antioxidants in grapes and wines. *J. Agric. Food Chem.*, 42: 64–69.

Kemin Europa, N.V. (2009). The interaction between meat quality, lipid oxidation and antioxidants in animal diets. *Technical Literature*, pp. 1–6.

Kim, I.S., Jin, S.K., Mandal, P.K. and Kang, S.N. (2010). Quality of low-fat pork sausages with tomato powder as colour and functional additive during refrigerated storage. *J. Food Sci. Technol.*, 48: 591–597.

Love, J.D. and Pearson, A.M. (1971). Lipid oxidation in meat and meat products: A review. *J AOCS*, 48(10): 547–549.

Lu, Y.R. and Foo, L.Y. (1999). The polyphenol constituents of grape pomace. *Food Chem.*, 65: 8–12.

Makkar, H.P.S. and Becker, K. (1996). Nutritional value and antinutritional components of whole and ethanol extracted *Moringa oleifera* leaves. *Anim. Feed Sci. Technol.*, 63: 211–228.

Mandal, P.K., Pal, U.K., Das, C.D. and Rao, V.K. (2002). Changes in the quality of restructured cured chicken during refrigerated storage. *Indian J. Poult. Sci.*, 37: 151–154.

Mandal, P.K., Sudheer, K., Pal, U.K., Das, C.D. and Rao, V.K. (2011). Preparation of low fat restructured chicken slice utilizing gizzard. *J. Meat Sci.*, 7: 56–59.

Marinova, E.M. and Yanishlieva, N.V. (1997). Antioxidative activity of extracts from selected species of the family Lamiaceae in sunflower oil. *Food Chem.*, 58: 245–248.

Meyer, A.S., Yi, O.S., Pearson, D.A., Waterhouse, A.L. and Frankel, E.N. (1997). Inhibition of human low density lipoprotein oxidation in relation to composition of phenolic antioxidants in grapes *(Vitis vinifera). J. Agric. Food Chem.*, 45: 1638–1643.

Moroney, M.A., Alacaraz, M.J., Forder, R.A., Carey, F. and Hoult, J.R.S. (1988). Selectivity of neutrophil 5-lipoxygenase and cyclooxygenase inhibitor by an anti-inflammatory flavonoid glycoside and related flavonoids. *J. Pharmacol.*, 40: 784–787.

Najeeb, A.P. (2012). Efficacy of certain leaves and fruits powder as natural preservatives in restructured chicken slices. MVSc thesis, submitted to Pondicherry University.

Naseem, U., Muhammad, K., Muhammad, U.A., Hamid, H.A., Farhat, A.K,, Sahibzada, M.U.K., Saleem, U., Najeeb, U., Javid, H. and Asif, K.M. (2011). Comparison of phytochemical constituents and antimicrobial activities of *Mentha spicata* from four northern districts of Khyber Pakhtunkhwa. *J. Applied Pharm. Sci.*, 1(7): 72–76.

Ningappa, M.B., Dinesha, R. and Srinivas, L. (2008). Antioxidant and free radical scavenging activities of polyphenol-enriched curry leaf (*Murraya koenigii* L.) extracts. *Food Chem.*, 106: 720–728.

Osterlie, M. and Lerfall, J. (2005). Lycopene from tomato products added minced meat and its effect on storage quality and colour. *Food Res. Int.*, 38: 925–929.

O'Byme, D.J., Devaraj, S., Grundy, S.M. and Jialal, I. (2002). Comparison of antioxidant effects of concord grape juice flavonoids and alphatocopherol on markers of oxidative stress in healthy adults. *American J. Clinical Nutri.*, 76: 1367–1374.

Pearson, A.M. and Dutson, T.R. (1987). Restructured meat and poultry products. 1st edn. Von Reihold Co. New York.

Pearson, A.M. and Gillet, T.A. (1996). Processed Meat. 3rd edn. CBS Publishers, New Delhi.

Rababah, T., Hettiarachchy, N.S., Eswaranandam, S., Meullenet, J.F. and Davis, B. (2005). Sensory evaluation of irradiated and nonirradiated poultry breast meat infused with plant extracts. *J. Food Sci.*, 70: 228–235.

Rahman, M.M., Mominul, I.S., Shamima, A.S., Soriful, I.M., Mizanur, R.M. and Alam, M.F. (2009). Antibacterial activity of leaf juice and extracts of *Moringa oleifera* Lam. against some human pathogenic bacteria. *J. Nat. Sci.*, 8: 219.

Rajak, S., Banerjee, S.K., Dinda, A.K., Gupta, Y.K., Gupta, S.K. and Maulik, S.K. (2004). *Emblica officinalis* causes myocardial adaptation and protects against oxidative stress in ischemic reperfusion injury in rats. *Phytother. Res.*, 18: 54–60.

Rani, P. and Khullar, N. (2004). Antimicrobial evaluation of some medicinal plants for their anti-enteric potential against multidrug resistant *Salmonella typhi*. *Phytother. Res.*, 18: 670–673.

Rao, A.V. and Agarwal, S. (1999). Role of lycopene as antioxidant carotenoid in the prevention of chronic diseases: A Review. *Nutr. Res.*, 19: 305–323.

Reddy, V., Urooj, A. and Kumar, A. (2004). Evaluation of antioxidant activity of some plant extracts and their application in biscuits. *Food Chem.*, 90: 317–321.

Renaud, S. and Longeril, M. (1992). Wine, alcohol, platelets and French paradox for coronary heart disease. *Lancet.*, 39: 1523–1526.

Rojas, M.C. and Brewer, M.S. (2007). Effect of natural antioxidants on oxidative stability of cooked, refrigerated beef and pork. *J. Food Sci.*, 72: 282–288.

Rotava, R., Zanella, I., D'Silva, L.P., Manfron, M.P., Ceron, C.S., Alves, S.H., Karkow, A..K and Santos, J.P.A. (2009). Antibacterial, antioxidant and tanning activity of grapes byproduct. *Cienc. Rural.*, 39: 941–944.

Sanchez, E.A., Torrescano, G., Djenane, D., Beltrán, J.A. and Roncales, P. (2003). Stabilization of colour and odour of beef patties using lycopene-rich tomato and peppers as a source of antioxidants. *J. Sci. Food Agric.*, 83: 187–194.

Saeed, S. and Thariq, P. (2007). Antimicrobial activity of *Embilica officinalis* and *Coriander sativum* against gram negative urinary pathogen. *Pakistan J. Pharmacol. Sci.*, 20: 32–35.

Selani, M.M., Packer, V.G., Silva, T.Z., Benato, R.T., Mourao, G.B. and Contreras, C.C.J. (2010). Addition of grape residues extracts and their effect on the oxidative stability and color of cooked chicken meat stored under freezing. International conference on food innovation. 25–29th October, Vol. 3: 44–46.

Shahidi, F., Janita, P.K. and Wanasundara, P.D. (1994). Phenolic antioxidants: Critical reviews. *Food Sci. Nutr.*, 32: 67–103.

Sheard, P.R. and Jolley, P.D. (1988). Restructured and reformed meat products. *Food Technol. Int. Europe*, 14: 129–132.

Shobana, S. and Naidu, K.A. (2000). Antioxidant activity of selected Indian spices. *Int. J. Vitamin Nutri. Res.*, 62: 107–110.

Sreelatha, S. and Padma, P.R. (2009). Antioxidant activity and total phenolics content of *Moringa oleifera* leaves in two stages of maturity. *Plant Foods Human Nutri.*, 64: 303–311.

Sudheer, K., Mandal, P.K., Das, C.D., Pal, U.K., Santoshkumar, H.T. and Rao, V.K. (2011a). Development of restructured chicken block utilizing gizzard and its refrigerated storage stability. *J. Food Sci. Technol.*, 48: 96–1101.

Sudheer, K., Das, C.D., Mandal, P.K., Pal, U.K. and Rao, V.K. (2011b). Effect of frozen storage on physicochemical, microbiological and sensory quality of low fat restructured chicken block incorporated with gizzard. *Int. J. Meat Sci.*, 1: 62–69.

Tachibana, Y., Kikuzaki, H., Lajis, N.H. and Nakatani, N. (2003). Comparison of antioxidative properties of carbazole alkaloids from *Murraya koenigii* leaves. *J. Agric. Food Chem.*, 51: 6461–6467.

Tang, S., Kerry, J.P., Sheehan, D., Buckley, D.J. and Morrissey, P.A. (2001). Antioxidant activity of added tea catchins on lipid oxidation of raw minced red meat, poultry and fish muscle. *Int. J. Food Sci. and Technol.*, 36: 685–692.

Vimela, Y., Vijaya, K. and Uma, K. (2009). Antimicrobial activity of *Embilica officinalis* seeds on human pathogen. *J. Pure Applied Micro.*, 3: 219–222.

Xie, Q.E., Fang, G.D., Jun, Y.G. and Bin, L.H. (2010). Biological activities of polyphenols from grapes *Int. J. Mol. Sci.*, 11: 622–646.

Yilmaz, I., Simsek, O. and Isikli, M. (2002). Fatty acid composition and quality characteristics of low-fat cooked sausage made with beef and chicken meat, tomato juice and sunflower oil. *Meat Sci.*, 62: 253–258.

Zheng, W. and Wang, S.Y. (2001). Antioxidant activity and phenolic compounds in selected herbs. *J. Agric. Food Chem.*, 49: 5165–5170.

15

Processing and Storage of Meat Soup

CHIDANANDAIAH AND M.K. SANYAL

ABSTRACT

Soup is a liquid food made by cooking vegetable, meat etc. together in a stock or water. Soups are popular because of their nutritive values, palatability, easy digestibility, low cost and simple techniques involved in their preparation. Though a large number of soups of different kinds are prepared world over, details of their manufacturing techniques, composition etc. are not properly available in literature. A large number of soups of different varieties are available in the world market. Meat and bone broth have been used extensively in the preparation of soups. Soup made from skimmed bone broth is considered as an important convalescent food. This chapter is covering discussion about types of soup and their chemical composition, methods of preparation of soups, role of additive ingredients, like salt, condiments, spices, monosodium glutamate, nutritive value of soup, packaging of soup. It also discusses about the changes in, physico-chemical, microbiological and in sensory properties during storage and shelf life of soup.

1. INTRODUCTION

Soup is a liquid food made by cooking vegetable, meat etc. together in a stock or water. It normally serves as a pleasant opening to a large meal particularly in cold weather and forms a good vehicle for vegetable, meat, milk, cheese etc. Soups are popular because of their nutritive values, palatability, easy digestibility, low cost and simple techniques involved in their preparation. A large number of soups of different varieties are available in the world market. Chicken meat is widely used for making a variety of soups such as chicken corn soup, yang soup, chicken peanut soup etc. Though a large number of soups of different kinds are prepared world over, details of their manufacturing techniques, composition etc. are not properly available in literature as most of these soups are either patented to safeguard business interests or are made by entrepreneurs who do not find it necessary to publish the results.

2. SOUP AND ITS CHEMICAL COMPOSITION

Soup is a liquid food made by cooking vegetables, meat etc. in water. Normally, it is made as a non-alcoholic beverage and served hot. Instant soups are primarily sold as convenience beverages in multiple packs of single serving sachets. Meat and bone broth are widely used in the preparation of soup. Soup made from skimmed bone broth is usually recommended as a convalescent food. Among various kinds of meat, chicken meat finds extensive use in the preparation of soup for its contribution towards flavour, nutritive value and acceptability of the soup. Two broad categories of soups are available world over *i.e.*, dried soups (including instant soup) and liquid soup in cans or cartons. A number of soup preparations made by utilizing meat, bones etc. are quite popular in some countries of South East Asia and other parts of the world. Yoo *et al.* (1994a,b) reported on the preparation of Komtang (Korean shank bone beef soup) and Doganitang (Korean knee bone beef soup) prepared by using shank bones and knee bones respectively from Hanwoo, Holstein or imported cattle from Australia and New Zealand. Recipes used for preparing three different types of Korean soup such as Kalbitang (beef rib soup), Yukgaejang (hot meat soup) and seolnong tang (meat soup) were standardized by Kye *et al.* (1995) after examining recipes used by food service institutions and restaurants, and cook book recipes. The authors concluded that standard methods for manufacturing soups could benefit the food service industry. Park *et al.* (1993) prepared Sam-gye-tang broth (chicken based soup) and investigated its quality. Bone extract has also been used by Ochi (1981) for making soup.

Kasahara and Nishibori (1995) reported on the preparation of Niboshi soup stock made by using Sardine niboshi (dried fish). A process for manufacture of lettuce and onion soup based on separate preparation of two components (an onion/butter/bouillon/egg component and a lettuce/poultry bouillon/cream component) and mixing them immediately before serving was described by Robuchon (1996). Lentil soup cubes were developed using lentil by Ereifej (1995). Sea weed has also been used for making Kombu soup stock (Hatae *et al.*, 1994). Composition of a few varieties of soup as reported by Hugh and Dorothy (1969) are given in Table 1.

Table 1: Composition of different types of soup.

	Constituents per 100 g of soup					
Soup	***Protein (g)***	***Fat (g)***	***Available carbohydrate as monosaccharide (g)***	***Ca (mg)***	***Fe (mg)***	***Calories***
Bone and vegetable broth, homemade	3.7	4.6	1.1	16.9	0.28	60
Ditto, commercial	4.4	–	0.3	10.2	0.28	–
Hospital soups, mean of 7 samples	2.0	1.3	4.3	33.8	1.39	36

Table 1: (*Contd...*)

Table 1: *(Contd...)*

Soup	*Constituents per 100 g of soup*					
	Protein (g)	*Fat (g)*	*Available carbo-hydrate as mono-saccharide (g)*	*Ca (mg)*	*Fe (mg)*	*Calories*
Lentil soup, home made	5.3	4.8	9.1	38.8	0.10	96
Potato soup, homemade	2.1	4.4	11.1	46.0	0.39	90
Tomato soup, canned	0.9	3.1	9.4	18.5	0.30	67
Vegetable soup, canned	1.9	0.4	8.1	20.5	0.44	42
Chicken noodle soup mix, dry	13.3	6.3	58.9	34.8	2.84	331
Thick pea soup mix, dry	18.1	7.2	60.9	75.6	4.62	366

3. METHODS OF PREPARATION OF SOUP

Bones are normally processed while fresh but chilled stored bones can also be used for making soup. The conventional method for preparing bone soup involves water boiling of bones for as long as six hours which is known as simmering (Keshri & Sharma, 1990). The authors prepared soup utilizing goat and buffalo bones. Bone broth was prepared by simmering (70–75°C for 6 h) or pressure cooking (1.05 kg/cm^2 for 1.5 h) of bones immersed in water. Ratio of bone to water was kept constant initially (simmering 1:2 kg/L and pressure cooking 1:1.5 kg/L). The broth was condensed to 1L for every kg of bones and filtered through a fine mesh screen followed by chilling to solidify the fat layer on the top of the broth which was then skimmed off. Soup was made by diluting the broth and using thickener, salt, spices etc.

A recipe of chicken stock for soup preparation by utilizing chicken meat, water, onion, carrot, pepper, coriander and bay leaf was described by Selvaraj (1973). Chicken soup was prepared by boiling 75–125 g of a mixture consisting of minced chicken meat, stomach, liver and heart, bread crumbs, eggs, onions, spices and seasonings, chicken fat etc. (Niemann, 1970). Lachhiramani (1979) described a procedure for making soup by using chicken shanks. The process involved pressure cooking of 100 shanks in 1.5L of water along with spices. Chizzolini (1982) reported on the preparation of bone broth by pressure cooking for 1h, a mixture of 3L water, 750 g beef, 250 g each of bone, onion, carrots and celery and 45 g of sodium chloride. Shimakov *et al.* (1982) made chicken puree soup using chicken meat and broth with small addition of dried milk, rice flour and vegetables. Shank and knee bones from the carcasses of Hanwoo, Holstein or imported cattle from Australia and New Zealand were removed and separately extracted 3x for 4h each time and 4x for 6 h each time respectively for making Korean soups (Yoo *et al.*, 1994a, b). Park *et al.* (1993) cooked sam-gye-tang broth, a chicken based soup, in a cauldron or in a pressure cooker to assess the effect of cooking time on the mineral content of the broth. Keshri *et al.* (1998) described a procedure for making soup from bony cuts of spent hens and whey. The procedure essentially consists of chopping of bony cuts and mixing them with whey and other ingredients followed by heating to boiling temperature and subsequently simmering at 70°C for 4 h. Chidanandaiah and Sanyal (2000) reported pressure cooking (at 121°C for 30 min) of neck, back and wings and shanks of spent hens at 0, 10, 15 and 20% levels in whey and water. They found that chicken-whey soup prepared at 20% level in whey was scored highest for sensory attributes along with higher nutrient density when compared with control and other levels.

4. ROLE OF ADDITIVE INGREDIENTS

Additive ingredients *viz.*, sodium chloride, spices and condiments, thickener, monosodium glutamate etc. are used in the preparation of nutritious soup. These ingredients improve sensory attributes of the soup along with preservative action. The information regarding additive ingredients is also not properly available in literature. The following are the commonly used ingredients in the preparation of different kinds of soups.

4.1. Common Salt

Sodium chloride is known as common salt. It is an inorganic chemical compound and has been used as seasoning agent, preservative or taste modifier of foods since long. Common salt is one of the most important non-meat ingredients in meat and poultry products. According to Mandigo (1991), common salt contributes to the flavour of a variety of spices and other flavouring. It has diverse functions ranging from simple flavouring or flavour enhancement to providing complex functional properties in various food systems. Sodium chloride exerts a growth repressing action on certain microorganisms in foods. It is a known dehydrating agent and can cause plasmolysis in microorganisms by dehydrating protoplasm. Alteration of osmotic pressure by sodium chloride is also a cause of inhibition of bacterial growth. Certain lactic acid bacteria, yeasts and moulds, either tolerate or adapt to moderate salt solutions. Sodium chloride in solution inhibits spore forming aerobes and anaerobes, proteolytic bacteria and pectolytic organisms to a great extent (Pearson & Gillett, 1997; Desrosier & Desrosier, 1987).

Common salt performs an important flavouring function in meat both in production and enhancement of taste (Gillette, 1985). It is used in meat products to promote functionality, improve binding, flavour and shelf-life (Terrel, 1983; Reddy & Marth, 1991). Generally, type of products and customer preference dictate the level of sodium chloride to be added to a product. It ranges from 0.5% in nugget products to 2.0% in sausages (Cordray & Huffman, 1987) and from 0.75% in canned chicken broth to 1.0% in canned clear chicken noodle soup (Binsted & Devey, 1970). There are no legal restrictions on the quantity of sodium chloride that can be added to meat and poultry products. However, their effect on flavour acts as the limiting factor (Trout & Schmidt, 1987). In Finland, permissible concentration of sodium chloride in soup is 10 g per kg.

But average sodium chloride content was 10 and 7 g/kg in sausage and soup, respectively (Skrokki, 1993). Gillette (1985) noticed that addition of sodium chloride to low-sodium and no-added-salt food products enhanced salty taste and masked off-flavour notes. It also enhanced mouth feel, sweetness and particularly enhanced flavour balance and overall flavour intensity. According to Rosett *et al.* (1996), overall flavours of chicken soup were influenced by added sodium chloride level. Higher level of sodium chloride increased saltiness perception, chicken and overall flavour of the product.

During the last decade, use of sodium chloride in processed meats has been a major health issue because of its association with hypertension. Among the food processing industries, the meat industry is the largest single user of salt. Several efforts are being made to reduce sodium chloride or substitute it with other chemicals leading to the availability of a number of low-sodium meat products in the market. However, finding a substitute for sodium chloride is difficult because of its unique pure salty taste and flavour enhancing properties (Reddy & Marth, 1991).

4.2. Spices and Condiments

Spices, the aromatic substances derived from plants or herbs, are used as seasoning ingredients for meat and meat products. Besides contributing to flavour, spices in some instances provide bacteriostatic as well as antioxidant effects (Pearson & Gillett, 1997). Spices may be used as whole or in crushed or ground form to assure proper control of the flavour complex.

The antioxidant activities of some natural spices and condiments such as cinnamon leaves, red chilli, turmeric, black pepper, cloves, cumin, onion, garlic, ginger etc. have been studied by Lee *et al.* (1986). Frazier and Westhoff (1989) stated that spices and condiments did not have any marked bacteriostatic effect in the concentrations normally used in foods but might help in preventing growth of organisms in food. Essential oils from aromatic plants including many herbs used in foods had bacteriostatic and fungistatic effect (Bourrel *et al.*, 1995). Antifungal effect of Turkish spice extracts was assessed by Boyraz and Ozcan (1997). The authors reported that mycelia growth of all the fungi was completely inhibited by savoury and oregano essential oils and extracts. Garlic could be added to suitable spice at effective levels to prevent mould growth (Abdel-Hafez & El-Said, 1997). However, spices may add high numbers of undesirable kinds of microorganisms to food of which they are ingredients if they are not treated to reduce microbial load.

4.3. Thickeners

Various types of hydrocolloids are used as gelling or thickening agents in food industry like algae extract (agar agar, carrageenan, furcellaran, alginic acid, alginates etc.), plant extracts (pectin, pectic salts etc.), seed meals (guar, carob, tara etc.), plant exudates (Gummi arabicum, tragacanth, karaya etc), cellulose and cellulose derivatives, starch and starch derivatives, xanthan, gellan, gelatin, milk proteins etc. (Niederbaumer, 1994). Tapioca, potato, corn, mung bean, waxy maize, wheat, sweet potato and modified potato starches are used as thickening agents for Chinese cuisine sauces such as fish-smelling, szechwang and curry (Huang *et al.*, 1992). Toasted wheat is commonly used as thickener in sauces, soups and gravies in the form of flour (Lazar *et al.*, 1974). Hydrocolloids, starches and flours used as thickeners, stabilizers, or water binders in high moisture food system such as soups could reduce overall flavour intensity if Na^+ got bound by such ingredients (Rosett *et al.*, 1996). Commonly used hydrocolloid ingredients, such as xanthan and K-carrageenan gums affected taste perception in model systems, particularly in suppressing salt taste (Rosett *et al.*, 1994; 1995).

Mudde (1985) prepared a corn starch product for use as food thickener that formed a uniform viscous dispersion when added to boiling water. Arro and Leon-Sy-de (1982) reported that rice bran could substitute the wheat flour as a filler or thickener in dry cream soup mixes and 30% bran substitution was found to be the most acceptable. Ruales *et al.* (1988) described the use of maize starch in dry soup mixes. Optionally modified maize starch can be used as a thickener in many foods including tomato sauce and cherry pie filling (Wurzburg & Fergason, 1984). Lindemann and Reinhold (1980) replaced 75% of the wheat flour used as a thickener by corn starch. According to Andon (1987), soup containing 4% starch did not provide any fiber, however, reducing the starch content to 2% and adding 1% each of low viscosity guar gum and low viscosity carboxy-methylcellulose (soluble dietary fibers), the fiber content could be increased to 1.03 g/2 oz. without affecting the rheological characteristics.

Corn starch used as a thickener provided body to the low-sodium chicken soup and enhanced saltiness and overall flavours of the soup (Rosett *et al.*, 1996). In potato starch thickened low-sodium soup, saltiness was similar to corn starch, but soup was perceived to have less chicken and overall flavours. Blenford (1994) emphasized on the use of pea starches in sauces, gravies requiring freeze thaw stability, dry soup mixes etc. Singhal and Kulkarni (1991) studied the effect of treated waxy *Amaranthus paniculatus* starch as a thickener in canned tomato soup and recommended it as a useful thickener for foods processed under retort conditions. Yanez *et al.* (1986) noticed that amaranthus starch had low pasting temperature and peak viscosity, did not form a solid gel and had many applications in instant soups, gravies, canned foods etc. The use of pigeon pea (*Cajanus cajan*) flour as thickener in chicken shank-whey soup improve the consistency, flavor, overall palatability and total solid content of soup (Chidanandaiah, 2011).

4.4. Monosodium Glutamate

Monosodium-L-glutamate monohydrate, popularly known as monosodium glutamate (MSG), is a sodium salt of the naturally occurring amino acid, glutamic acid. It is widely used in food processing industry as a flavour enhancer

(Anon, 1987; Solen *et al.*, 1985). In USA, MSG has been accepted by the Food and Drug Administration (FDA) as a 'Generally Recognised as Safe' (GRAS) substance. Vazquez *et al.* (1982) reported that malnourished infants of 2–24 months of age preferred soup seasoned with 0.4% MSG as compared to the same soup without MSG. It has been suggested that glutamates in foods may provide a fifth basic taste in addition to sweet, sour, salty and bitter taste (Anon, 1987; McBride, 1991). There seems to be an optimal concentration of MSG in foods beyond which the palatability of food decreases (Yamaguchi and Takahashi, 1984a). A level of MSG ranging from 0.2 to 0.8% by weight of food is recommended to give the best enhancement of natural food flavour (Anon, 1987; Bellisle *et al.*, 1989). The palatability of MSG varies with the presence of sodium chloride in food system (Yamaguchi & Takahashi, 1984b).

Chi and Chen (1992) observed that the effect of MSG and sodium chloride on the hedonic scores of non-spiced and spiced chicken broths were highly significant ($P<0.005$). An increase in MSG and sodium chloride levels had a quadratic effect ($P<0.005$) on hedonic scores of broth. The authors recommended that the levels of MSG and sodium chloride for maximum hedonic score in non-spiced broth should be 0.33 and 0.83%, respectively while in spiced broth the levels recommended were 0.38 and 0.87%, respectively. Roininen *et al.* (1996) studied the effect of Umami substances (MSG and 5′-Ribonucleotides) on the acceptance of low-salt soups and obtained higher ratings when soups contained Umami in both the low and high salt groups.

Nutritious chicken whey soup/chicken soup/shank-whey soup prepared by using additive ingredients *viz.* spice mix, condiments, thickener, common salt, and monosodium glutamate at 0.2, 1.75, 2.0, 1.0 and 0.1% (w/w) level provides good taste, flavour with improved overall acceptability (Chidanandaiah,1999; Chidanandaiah & Sanyal, 2001; Chidanandaiah *et al.*, 2002).

5. NUTRITIVE VALUE OF SOUP

Bone soup is considered to be nutritious as it provides protein, minerals and variable quantity of lipids exudating from bone marrow (Field *et al.*, 1974). Keshri and Sharma (1990) also reiterated that bone can serve as a good material for soup preparation as it is rich in marrow. The major structural component of bone is calcium phosphate and calcium makes up to 37% of bone ash in poultry bone (Taylor *et al.*, 1960). Dietary calcium has been linked to osteoporosis, hypertension and cancer, the risk of which can be minimized through increase in calcium intake (Marcus, 1982; Newmark *et al.*, 1984; Weaver, 1988). Masood and Chen (1995) reported that though broth or soup from bones is assumed to be rich in calcium but cooking bones only in water does not extract calcium from bone. The authors indicated that the content (mg/100 g) of phosphorus and calcium in bone soup ranged from 7.96–20.5 and 1.74-1.80, respectively. However, FAO (1972) reported that soup made from bone contained 12 mg/100 g of phosphorus and 1 mg/100 g of calcium.

Sam-gye-tang, a chicken based soup when cooked in a pressure cooker showed increase in calcium, potassium, magnesium and zinc, the contents of which increased with increase in cooking time (Park *et al.*, 1993). Maximum contents of iron, sodium and phosphorus were obtained in the soup after 120 min. of pressure cooking. Sharma *et al.* (1990) observed that pressure cooking released greater amount of neutral lipids, but lesser amount of saturated fatty acids from bone marrow into soup when compared with the cooking method of simmering in open kettle. According to Rosen *et al.* (1994), prolonged cooking of bones during soup making increased calcium content of soup in acidic medium but not at neutral pH. During cooking of chick pea soup, stewed lentils, bean sausage/ham dish etc., total available carbohydrate content decreased by 12–25% in starch rich products and 8–18% in sugar rich products (Diaz-Marquina & Villanueva-Suarez, 1987). The average content (% w/v) of soup prepared from spent hen heads (at 20% level) and whey are fat 1.13 ± 0.05, protein 0.89 ± 0.04, lactose 3.85 ± 0.06, ash 0.71 ± 0.04 and total solids 8.98 ± 0.08 (Chidanandaiah & Sanyal, 2001), while shank-whey soup had (% w/v) fat 1.40 ± 0.06, protein 1.09 ± 0.07, lactose 3.80 ± 0.08, ash 0.77 ± 0.05 and total solids 9.25 ± 0.09 (Chidanandaiah, 1999).

6. PACKAGING OF SOUP

Packaging protects the food stuffs from contamination, damage, degradation and spoilage, ensures ease in storage, transport and provides convenience in handling and marketing of the food material. Package must ensure the same quality of the product to the consumer as that of freshly prepared ones. Gnanendra Kumar (1982) emphasized on the suitability of flexible polyethylene (PE) and laminates for packaging fluid or solid food products as these are economical, convenient and have a good functional properties. Aseptic pouches are used for packaging of UHT milk and other liquid foods, made up of environmentally friendly, inexpensive co-extruded laminate of polyester with up to 6 layers affording protection against light, tearing and taste contamination (Anon, 1997).

A microwaveable container made up of microwave transparent material like filled polystyrene, plasticized paper board, polyethylene, polypropylene or polycarbonate was designed for condensed soup as a packaging material (Topolski, 1988). Rectangular container with two transverse pockets at the bottom to provide maximum exposure to cooling medium was developed by Dunden (1987) for frozen soups, soup concentrates, stews etc. which was suitable for large scale food catering operations because of ease in handling. Katsura and Suzuki (1982) reported on the retort-sterilizable laminated structure made of aluminium foil and heat resistant propylene type resin layer for soup, hamburger steak and pilaf. Oriental chicken pea soup packaged in 400-ml cartons made from polyethylene and aluminium foil-coated paper board had advantage of easy use/disposal and occupied 30% less super market shelf space as compared to tinned soups apart from maintaining soup flavour for longer period (Anon., 1986).

Heat sealed multilayered film stock could be used for packaging of chicken soup (Anon., 1984). Retention of ascorbic acid content in tomato soup and of thiamin in chicken soup was much higher in aseptic canning than conventional canning (Green, 1976). Bag in box type packaging materials are quite suitable for syrups (Anon., 1995a). Such a system consists of a bag made up of food compatible hyper polyethylene and outer layer of nylon for filling with the liquid and a box made up of single wall corrugated board containing 30% recycled material. The bag is put inside the box. Spouted 500-ml or 1l tetrarex laminated table top cartons have been used for filling with chilled juices (Anon., 1995b). Warner (1986) used 500-ml tetrapack cartons for packaging of UHT processed soup for improving shelf-life. Achhammer and Spang (1997) discussed the packaging of beverages in plastic bottles. The oxidation rate of fatty acid was lower in laminated paper than film packs during storage of dehydrated soup at 30°C and 80% relative humidity (RH) (Doronin *et al.*, 1984). Physico-chemical and sensory quality of soup maintained up to 6 days in aluminium-polyethylene laminated pouches at refrigeration (4 ± 1°C) storage (Chidanandaiah *et.al.*, 2002).

7. QUALITY CHANGES IN SOUP DURING STORAGE

7.1. Physico-Chemical Changes

Vitamin C concentration in Gazpacho soup decreased increasing severity of pasteurization condition and increasing duration of storage (Jimenez *et al.*, 1988). Thiamin, riboflavin and ascorbic acid contents of various frozen dishes including cauliflower soup, Frankfurt soup etc. decreased during storage at –18°C for 3 months (Mnukova, 1980). Ziombski *et al.* (1971) observed that low moisture content favored fat oxidation in dehydrated pea soup, whereas, high moisture content favored fat hydrolysis during storage at 18–20°C. Low temperature storage (0–4°C), however, decreased fat hydrolysis and to a lesser extent fat oxidation in soup. Though fat was found to be the most susceptible component in gherkin soup to change, no appreciable change in moisture content, titratable acidity of the product was noticed by Anisimov *et al.* (1974a) during storage for 12 months at 22 ± 3°C in aluminium tubes. Gradual deterioration in quality of meat soup (solyanka) during storage was caused mainly due to fat oxidation when large numbers of unsaturated fatty acids were present in the lipids of solyanka (Anisimov *et al.*, 1974b). However, moisture content, pH and titratable acidity did not change appreciably during the entire storage period of 12 months at –18°C.

The lipid peroxidation was indicated by slight increase in TBA values after 50 days of storage at 2°C in corn soup (Miyamoto *et al.*, 1989). Krishnaiah *et al.* (1989) noticed faster decrease in pH of whey beverage with increased storage time both at room temperature and refrigeration temperature. Viscosity of the product, however, showed more increase during storage at room temperature as compared to refrigeration temperature. Pectin and vitamin contents in milk based fruit juice beverage decreased upon storage (Guleria &

Jain, 1979). Fat, protein, ash, total solid content and pH of both chicken-whey soup and chicken soup decreased at each interval of 0, 2, 4, and 6 days during storage at 4 ± 1°C in polyethylene-aluminium pouches. The decrease in pH of chicken-whey soup was to a greater extent (5.95 to 5.52) as compared to that of chicken soup after 6 days of storage which might be due to conversion of lactose into lactic acid. However, thiobarbituric acid value and titratable acidity of both soups were increased at each interval of 0, 2, 4 and 6 days (Chidanandaiah *et al.*, 2002).

7.2. Microbiological Changes

It is necessary to assess the microbiological quality of meat product for determining its impact on product shelf-life and consequently upon human health. Depending on the processing schedule, meat extracts often contain a mixed microflora dominated by Bacillus spp. and the more heat-resistant vegetative microorganisms such as enterococcus. Stability of the product depends on water activity level (a_w), sodium chloride content, ingredients used and other factors. The level of heat-resistant endospores can be a problem when meat extracts are used in canned products such as soups (Varnam & Sutherland, 1995).

Miyamoto *et al.* (1989) studied the viable bacterial count in pouched corn soup. The bacterial counts increased rapidly from 5.5×10^2 cells/ml to 10^7 cells/ml after 10 days at 10°C followed by putrefaction of the soup. The total count of bacteria and mould cells decreased during storage of dried soup, rate being faster at 18–22°C than at 2°C (Egorova, 1982). During 5 months storage period, a slight drop in the geometric average count of the proteolytic bacteria was observed in frozen poultry meat, cream soup, poultry soup with vegetables and poultry giblet soup while the lipolytic organisms showed increasing trend (Brodnanova & Zdichavska, 1976). Total plate count ($\log_{10}$ cfu/ml), enterobactariaceae count and yeast and mould count of both chicken-whey soup and chicken soup increased significantly ($P<0.05$) at each interval of 0, 2, 4 and 6 days. But the counts were within acceptable limit after 6 days of storage (Chidanandaiah *et al.*, 2002).

7.3. Changes in Sensory Properties

During storage of Gazpacho soup, yield stress increased and the rheological properties of soup were least affected by the less severe pasteurization regimes (Jimenez *et al.*, 1988). The soup pasteurized under low temperature-short time regime also showed less change in sensory properties during storage as compared to that pasteurized by using severer conditions. According to Ziombski *et al.* (1971), peroxide content of fat did not have any clear-cut correlation with organoleptic assessment of soup. However, appearance of soup was related to increase in fat acidity. Suyama *et al.* (1988) concluded that degradation product of gelatin influenced the taste of turtle soup. Flechtenmacher (1980) noticed more rapid deterioration in sensory properties of unseasoned potato soup during storage as compared to that in seasoned potato soup. Sensory scores *viz.*, colour and appearance, flavour, consistency, meat flavour intensity and overall palatability of both chicken-whey soup and chicken soup decreased at each interval of 0, 2, 4 and 6 days at refrigerated (4 ± 1°C) storage and both the products were acceptable for consumption on day 6 (Chidanandaiah *et al.*, 2002).

7.4. Shelf Life of Soup

Panda and Ahlawat (1994) reported that the keeping quality of pasteurized kinnow egg beverage was about 4 days under refrigeration temperature. Orangegg drink showed off-flavour and became unpalatable on 6th day during storage at 4°C (Cunningham, 1972). Soup powder (18.46% protein, 5.4% fat) had shelf life of 4 months at ambient temperature in polyethylene bags (Nambair *et al.*, 1988). Spray dried tomato powder was acceptable for less than 9 months at room temperature and 3 months at 37°C while chicken soup cubes kept well for 12 and 6 months at room temperature and 37°C, respectively (Jayakumar *et.al.*, 1976). Non-carbonated beverages when stored at 3–5°C in aluminium and polyethylene laminate did not show any adverse change for 2.5–3 months (Gonzales *et al.*, 1985). Kacharava *et al.* (1973) reported that the shelf-life of hot bottled sochi beverage was 12 months. The shelf life of chicken-whey soup, chicken soup and chicken shank-whey soup was reported 6 days during storage at refrigeration (4 ± 1ºC) temperature in laminated/polyethylene pouches. (Chidanandaiah *et al.*, 2002; Chidanandaiah *et al.*, 2011).

8. RESEARCH NEED

Though a large number of soups of different kinds are prepared world over, details of their manufacturing techniques, composition etc. are not properly available in literature as most of these soups are either patented to safeguard business interests or are made by entrepreneurs who do not find it necessary to publish the results. Utilization of by-products of meat and Dairy industry adds value to the products, generates employment and provides nutritive beverage at low cost needs attention. Meat soup powders in sachet are commercially available at super markets throughout the world with demand. There is increasing trend in consumption of this hot beverage worldwide. In view of discussion mentioned above there is tremendous scope for scientific research needed in this line.

REFERENCES

Achhammer, K.H. and Spang, D. (1997). Packaging in plastic bottles. *Brauwelt*, 137(48): 2246–2207 [cf FSTA (1998) 30: 4Ha599].

Andon, S.A. (1987). Applications of soluble dietary fiber food technology. *Food Tech.*, 41(1): 74–75.

Anisimov, B.N., Blagova, R.N., Gurova, L.A., Kuznetsova, L.I., Pupynina, V.D. and Tikhomirova, T.A. (1974a). Changes in the quality of preserved foods in aluminum tubes during long term storage. *Trudy-Vsesoyuzin-Nauchno-issledovatel'skii-Institut-Konservnoi-i-voshchesushil'noi-promyshelnnonsti.*, 21: 23–27.

Anisimov, B.N., Blagova, R.N., Gurova, L.A., Pavlova, L.S., Tikhomirova, T.A. and Cherkasova, L.P. (1974b). Study of the quality of quick frozen ready-to-serve dishes during storage. *Trudy-Vsesoyuzni-Nauchno-issledovatel'skii-Institut-Konservnoi-i-voshchesushil'noi-promshelnnosti.*, 21: 13–23.

Anon, (1984). Retortable plastic cans. *Food Eng. Intern.*, 56(4): 58–59.

Anon, (1986). Chunky soups in aseptic cartons. *Food Eng. Intern.*, 11: 32.

Anon, (1987). Monosodium glutamate (MSG). A scientific status summary by the Institute of Food Technologists. *Expert Penal on food safety and Nutrition. Food Technol.*, 41(5): 143–154.

Anon, (1995a). A classic move at Coca-Cola. *Canadian Packaging*, 48(2): 20–21 [cf FSTA (1996) 28: 1H146].

Anon, (1995b). Chilled juice factory adds packaging line. *Food Eng. Intern.*, 20(6): 23–24 [cf FSTA (1996) 28: 9H131].

Anon, (1997). World premier for the Elopouch aseptic packaging system. European Diary Magazine, 5: 45–46 [cf FSTA (1998) 30: 2Pe318].

Arro, R.I. and Leon-Sy-de (1982). Utilization of rice bran in dry soup mixes. *UP Home Economics J.*, 10(2): 76–90.

Bellisle, F., Tournier, A. and Louis-Sylvestre, J. (1989). Monosodium glutamate and the acquisition of food preferences in a European context. *Food Quality and Preferences*, 1(3): 103–108

Binsted, R. and Devey, J.D. (1970). Soups for canning. *In*: Soup Manufacture Canning, Dehydration and Quick Freezing, Food Trade Press Ltd, London, pp. 56–57.

Blenford, D. (1994). Pea starches-unique natural ingredients. *International-Food Ingredients*, 6: 27–32.

Bourrel, C., Vilarem, G., Michel, G. and Gaset, A. (1995). Studies on bacteriostatic and fungistatic properties in solid media and analyses of 24 essential oils. *Rivista Italiana EPPOS*, 16: 3–12 [cf FSTA (1996) 28: 4T52].

Boyraz, N. and Ozcan, M. (1997). Antifungal effects of some Turkish spice extracts and essential oils on plant pathogen fungi. *Gida*, 22(6): 457–462 [cf FSTA (1998) 30: Te562].

Brodnanova, O. and Zdichavska, M. (1976). Effect of storage on changes in the count of proteolytic and lipolytic microorganisms in frozen poultry soups. *Hydinarsky-Priemysel.*, 18(4/5): 168–179.

Chi, S.P. and Chen, T.C. (1992). Predicting optimum monosodium glutamate and sodium chloride concentrations in chicken broth as affected by spice addition. *J. Food Proc. and Preserv.*, 16: 313–326.

Chidanandaiah, (1999) Studies on preparation and self life of Chicken-whey soup. M.V. Sc. thesis submitted to the Deemed Univ. IVRI. Izatnagar. India-243122.

Chidanandaiah, and Sanyal, M.K. (2000). Quality of Chicken-whey soup prepared from bony cuts of spent hens and whey. *Indian J. Poultry Sci.*, 35(3): 367–369.

Chidanandaiah, and Sanyal, M.K. (2001). Quality of soup prepared from spent hen heads and whey. *Indian J. Poultry Sci.*, 36(2): 240–242.

Chidanandaiah, Sanyal, M.K. and Keshri, R.C. (2002). Changes in quality of chicken soup and chicken-whey soup in laminated pouches during refrigerated storage. *J. Food Sci. Technol.*, 39(3): 288–291.

Chidanandaiah, Srivastava, A.K., Pathak, V., Biswas, A.K. and Kumar, R.R. (2011). Changes in quality of whey-shank soup thickened with pigeon pea (*Cajanus cajan*) flour during storage at 4 ± 1°C in polyethylene pouches. *Milch Wissenschaft-Milk Science International*, 66(2): 165–168.

Cordray, J. and Huffman, D. (1987). Non-meat ingredients. Part I. *Meat Poult.*, 33(2):18.

Cunningham, F.E. (1972). Shelf life of pasteurized, refrigerated orangegg. *Poultry Sci.*, 51(5): 1776–1777.

Desrosier, N.H. and Desrosier, J.N. (1987). The Technology of Food Presrvation , 4th edn. CBS Publishers and Distributors, Delhi, pp. 340–342, 394–395.

Diaz-Marquina, A. and Villanueva-Suarez, M.J. (1987). Changes in available carbohydrate content during cooking of some common dishes. *Anales-de-Bromatologia*, 39(1): 173–186.

Doronin, A.F., Sultnovich-Yua., Kiseleva, T.V. and Gulyev, V.N. (1984). Changes in fatty acid composition of soup concentrates in relation to conditions of storage. *Knoservnaya-i-Ovoshchesushil'naya-promyshelennost*, 11: 42–44.

Dunden, W.J. (1987). Container for freezing and storage of foodstuffs. United States Patent.

Egorova, Z.E. (1982). Changes in micro flora in dried soups during storage. *Izvestiya-Vyssikh-Uchebnykh-Zavedenii-Pishchevaya-Technologiya*, 6: 51–55.

Ereifej, K.I. (1995). Preparation of acceptable traditional dehydrated lentil soup at small scale. *J. Food Sci. Technol. India*, 32(6): 486–488.

FAO, (1972). Food composition table for use in East-Asia. Food Policy and Nutrition Division. Food and Agriculture Organization of the United Nations, Rome.

Field, R.A., Riley, M.L., Mellow, F.C., Corbridge, M.H. and Kotula, A.W. (1974). Bone composition in cattle, pigs, sheep and poultry. *J. Anim. Sci.*, 39(3): 493.

Flechtenmacher, W. (1980). Sensory evaluation of spay-dried protein concentrates. *Information-fuer die-Fischwirtschaft*, 27(2): 73–75.

Frazier, W.C. and Westhoff, D.C. (1989). Food Microbiology, pp. 160–164. McGraw Hill Book Co., New York.

Gillette, M. (1985). Flavour effects of sodium chloride. *Food Technol.*, 39(6): 47–52, 56.

Gnanendrakumar 1982. Need for suitable packaging of burfi. *Packaging India*, 14(2): 8–10.

Gonzales, O.N., Alejo, J.V., Brillante, J. and Valdecanas, M. (1985). Process for preparing noncarbonated and carbonated coconut water. *Cocomunity Apcc/QS/*03: 31–42.

Green, J. (1976). Aseptic canning. *BNF-Bulleti.*, 18: 376–384.

Guleria, S.P. and Jain, S.C. (1979). The development of milk based fruit juice beverage using buffalo milk. *In*: Proceedings of the first Indian Convention of Food Scientists and Technologist, 8.2:85 [cf FSTA (1979) 11: 12A871].

Hatae, K., Wakida, M., Miyago, E., Sato, Y. and Shimado, A. (1994). Relationship between the amount, time and temperature for extracting kombo components to produce soup stock. *J. Jap. Soc. Food Sci. Technol.*, 41(11): 755–762 [cf FSTA (1996), 28: 2R36].

Hugh Smclair, M. and Dorothy Hollingsworth, F. (1969). Foods taken chiefly for protein. In: *Hutchinson's Food and the Principles of Nutrition*. 12th edn. Edward Arnold Ltd., London.

Jay Kumar, K.S., Goverdhana, T., Bhatia, B.S. and Nath, A. (1976). Studies on the development of some dehydrated instant soup cubes for use in emergency rations. *J. Food Sci. Technol. India*, 13(1): 29–34.

Jimenez, L., Sanchez, I. and Chica, A. (1988). Changes in some characteristic of pasteurized gazpacho during storage. *Sciences-des-Aiments*. 8(4): 523–535.

Jong-Jang-Huang Yen-Pyng-Lin and Shun-Jen-Tsai (1992). Texture of starch-thickened Chinese cuisine sauces. *J. Chinese Agric. Chem. Soc.*, 30(4): 519–531.

Kacharava, I.A., Fan-Yung, A.F., Kaminskaya, F.I., Biryukova, S.N., Itskovskaya, L.S. and Palvichenko, V.I. (1973). Sochi beverage. *Lonservanaya-i-ovoshcheesushil'naya-promyshlennost*. 10: 10–11.

Kasahara, K. and Nishibori, K. (1995). The suppressing effect of garland *Chrysanthemum* on the odour of niboshi soup stock. *Fisheries Sci.* 61(4): 672-74 [cf FSTA (1996) 28: 11R32].

Katsura, T. and Suzuki, K. (1982). Retort-sterilizable laminated structure and process for preparation thereof. United States Patent.

Keshri, R.C. and Sharma, N. (1990). Sensory quality of soup processed with goat and buffalo bones. *Indian J. Meat Sci. Tech.*, 3(1): 54–57.

Keshri, R.C., Sanyal, M.K. and Yadav, P.L. (1998). Chicken-whey soup for human consumption. *Poultry Punch*, 14(4): 41–44.

Krishnaiah, N., Reddy, C.R., Sastry, P.M. and Rama Rao, M. (1989). Studies on the keeping quality of whey beverage. *Asian J. Dairy Res.*, 8(1): 8–4.

Lachhiramani, R.S. (1979). Utilization of chicken shanks in chicken soup. *Poultry Guide*, 16(3): 23–25.

Lazar, M.E., Mossman, A.P. and Wallace, J.M. (1974). Air-fluidized toasting of whole kernel wheat-processing variables and functional properties for food Applications. *J. Food Sci.*, 39(2): 239–243.

Lee, Y.B., Ki, Y.S. and Ashmore, C.R. (1986). Antioxidative property of ginger rhizome and its application to meat products. *J. Food Sci.*, 51: 21.

Lindemann, C. and Reinhold, M. (1980). Industrial manufacture of gravy. I. *Fleisch.*, 34(3): 49–52.

Mandigo, R. (1991). Non-meat ingredients. *Meat Poultry*, 37(3): 10.

Marcus, R (1982). The relationship of dietary calcium to the maintenance of skeletal integrity in man: An interface of endocrinology and nutrition. *Metabolism*, 31: 93–102.

Masood, M. and Chen, T.C. (1995). Gelatin, calcium and phosphorus contents of chicken broth from broiler deboning by-products. *J. Food Proc Preserv.*, 19: 165–173.

McBride, R.L. (1991). MSG and the bliss point factor. *Food Australia*, 43(7): S16–S18.

Miyamoto, T., Tanaka, R., Kishikawa, T., Mori, S., Hatano, S. and Shimoda, M. (1989). Studies on chilled storage of corn soup. Science Bulletin. Faculty of agriculture, *Kyushu University*. 44(1/2): 9–15.

Mnukova, J. (1980). Some traditional changes in frozen foods. *Vyziva-Lidu.*, 25(9): 134–135.

Mudde, J.P. (1985). Hot-water dispersible corn starch having high paste viscosity. United States Patent.

Nambair, V.N., Valsan, H.P., Garg, D.K. and Damle, S.P. (1988). Value added consumer products from non-penaeid prawns (*Acetes indicus*) of Maharashtra coast. *Indian Food Packer*, 42(5): 12–16

Newmark, H.L., Wargovich, M.J. and Bruce, W.R. (1984). Colon cancer and dietary fat, phosphate and calcium: A hypothesis. *J. Nat. Cancer Inst.*, 72:1323–1325.

Niederbaumer, T. (1994). Manufacture, properties and application of gelling and thickening agents. *Zucker-und-suesswarenwirtschaft*, 47(4):134-136; (5): 186-187.

Niemann, E. (1970). Production process for clear chicken soup. West German Patent Application.

Ochi, H. (1981). Production and application of bone extracts. *Food Technol.*, 35(7): 58–59.

Panda, P.C. and Ahlawat, S.S. (1994). Keeping quality of pasteurized kinnow egg liquid. *Poultry Guide*, 31(6): 27–29.

Pearson, A.M. and Gillett, T.A. (1997). Processed Meats, 3rd edn. CBS Publishers and Distributors, New Delhi, pp. 54-55.328-331.

Reddy, A.K. and Marth, E.H. (1991). Reducing the sodium content of foods: A review. *J. Food Prot.*, 54(2): 138–150.

Robuchon, J. (1996). Lettuce and cream of onion soup. French Patent Application. FR 272729341 [cf FSTA (1996) 28: 11G61].

Roininen, K., Lahteenmaki, L. and Tuorila, H. (1996). Effect of umami taste on pleasantness of low salt soups during repeated testing. *Physiol. Behav.*, 60(3): 953–958.

Rosen HN, Salemme H, Zeind AJ, Moses AC Shapiro A and Greenspan SL 1994. Chicken soup revisited: Calcium content of soup increases with duration of cooking. *Calcif-Tissue-Int.*, 54(6): 486–488.

Rosett, T.R., Kendregan, S.L., Ying-Gao-Schmidt, S.J. and Klein, B.P. (1996). Thickening agent effect on sodium binding and other taste qualities of soup systems. *J. Food Sci.*, 61(5): 1099–1104.

Rosett, T.R., Shirle, L.L., Klein and Schmidt, S.J. (1994). Na^+ binding as measured by Na nuclear magnetic resonance spectroscopy influences the perception of saltiness in gum solutions. *J. Food Sci.*, 59: 206–210.

Rosett, T.R., Wu, Z., Schmidt, S.J., Ennis, D.M. and Klein, B.P. (1995). KCl, $CaCl_2$, Na^+ binding and salt taste of gum systems. *J. Food Sci.*, 60: 849–853.

Ruales, J., Polit, P. and Nair, B.M. (1988). Production and nutritional evaluation of some dehydrated food products based on deodorized fish flesh and maize. *Food Sci. Nutr.*, 42F (2): 97–109.

Selvaraj, A. (1973). Studies on the production of canned soups from marine invertibrates. M.Sc. thesis. CFTRI, Mysore (India).

Seung-Hee-Kye, Hyun-Kyung-moon, Cho-Ae-yum and Eun-mipark (1995). Standardization of the recipe for preparation of Korean foods. I. For the focus on soup. *J. Korean Soc Food Sci.*, 11(1): 1–8 [cf FSTA (1996) 28: 11S8].

Sewon, Park.,Suntae, Kim. and Yangja-Yoo (1993). Mineral content in sam-gye-tang broth according to cooker and boiling time. *J. Korean Soc. Food Sci.*, 9(1): 52–56 [cf FSTA (1994) 26: 11G9].

Sharma, N., Keshri, R.C. and Gandemer, G. (1990). Effect of cooking on lipid composition of buffer bone marrow. *J. Food Sci. Technol.*, (India). 27(6): 365–67.

Shimakov, E.G., Korotaeva, M.M., Krainyayo, V.S., Chernova, G.G., Khtevovaya, V.V and Brentsmya, (1982). Rational process and quality of poultry meat products for infant. *Proc. Europ. Meeting Meet Res. Workshop. II* (28):533–55 [cf FSTA (1984) 76:5S816].

Singhal, R.S. and Kulkarni, P.R. (1991). Studies on cross-linked *A. paniculatas* (Rajgeera) starch. *Starch*, 43(1): 15–18.

Skrokki, A. (1993). Sodium chloride concentrations in Finnish sauces and soups. *Nahrung*, 37(5): 494–497 [cf FSTA (1994) 26: 1G7].

Solen, F.S., Latham, M.C., Guirriec, R., Florentino, R., Williamson, D.F. and Auguilar, J. (1985). Fortification of MSG with vitamin A: The Philippine experience. *Food Technol.*, 39(11): 71–77.

Suyama, M., Hirano, T. and Suzuki, T. (1988). Changes in gelatin in hot water extract of snapping turtle during heating. *Bullet Jap Soc Scientific Fisheries* (Nihon-Suisan –Gokki-Shi). 54(4): 635–638.

Taylor, T.G., Moore, J.H. and Hertelendy, F. (1960). Variations in the mineral composition of individual bones of skeleton of the domestic fowl. *Brit. J. Nutr.*, 14: 49.

Terrel, R.N. (1983). Reducing the sodium content of processed meats. *Food Technol.*, 37(7): 66–71.

Topolski, A.S. (1988). Combination microwaveable packaging and service container. European Patent Application.

Trout, G.R. and Schmidt, G.R. (1987). Non protein additives. *In*: Advances in Meat Research-Restructured Meat and Poultry Products, Pearson, A.M., Duston, T.N. (*Eds.*), Van Nostrand Reinhold Co., New York, pp. 307–329.

Varnam, A.H. and Sutherland, J.P. (1995). Dried meats, intermediate moisture meats and meat extracts. *In*: Meat and Meat Products, Technology, Chemistry and Microbiology, Chapman and Halls, London.

Vazquez, M., Pearson, P.B. and Beauchamp, G.K. (1982). Flavour preferences in molnourished Mexican infants. *Physiol. Behav.*, 28: 513.

Warner, E. (1986). What's cooking at Healds? Milk Industry. *U.K.*, 88(3): 15–17 [cf FSTA (1986) 18:11G54].

Weaver, C.M. (1988). Calcium and hypertension. *Cereal Foods World*, 33: 792–794.

Wurzburg, O.B. and Fergason, V.L. (1984). Starch thickener characterized by improved low-temperature stability. United States Patent.

Yamaguchi, S. and Takahashi, C. (1984a). Hedonic functions of monosodium glutamate and four basic taste substances used at various concentration levels in single and complex systems. *Agric Biol. Chem.*, 48: 1077.

Yamaguchi, S. and Takahashi, C. (1984b). Interactions of monosodium glutamate and sodium chloride on saltiness and palatability of a clear soup. *J. Food Sci.*, 49: 82–85.

Yanez, G.A., Messinger, J.K., Walker, C.E. and Rupnow, J.H. (1986). *Amaranthus pochondriacus*: Starch isolation and partial characterization. *Cereal Chem.*, 63(3): 273–276.

Yoo, I.J., Park, B.S. and Yoo, S.H. (1994a). Comparison of physico-chemical characteristics among Hanwoo, Holstein and kneebone soup (doganitang). *Korean J. Anim. Sci.*, 36(5): 515–33 [cf FSTA (1995) 27: 4S78].

Yoo, I.J., Yoo, S.H. and Park, B.S. (1994b). Comparison of physico-chemical characteristics among Hanwoo, Holstein and imported shank bone soup (Komtang). *Korean J. Ani. Sci.*, 36(5): 507–514 [cf FSTA (1995) 27: 4S77].

Ziombski, H., Kwasniewska, I. and Chudzikiewichz, M. (1971). Effect of storage on nutritive value of dehydrated pea soups. III. Chemical changes in the fat and organoleptic changes in the soups. *Roczniki-Panstwowego-Zakladu-Higieny*, 22(2): 169–178.

16

Pesticides and Mycotoxins in Animal Products

A.K. Biswas, P.K. Mandal and K.P. Mini

ABSTRACT

The pesticide residues and mycotoxins are a major concerns of the consumers of animal products. Various conventional methods of analysis indicating most analytical techniques are error proof, sensitive but costlier and some are not compatible with the speed which the products need to be delivered. However, all this techniques come under quantitative methods of analysis. For rapid screening of samples ELIZA, LFA, biosensors, molecular imprinted polymers (MIPS), Microarray, Luminex xMAP^R technology shown to have promising results and meet the present day need. But antibodies or other recognition reagents must be commercially available for methods development. The use of novel labels such as nanoparticles, the development of non-competitive immunoassays for more sensitive, simpler and rapid assays; and the development of formats for multi-analyte detection such as microarrays and microspheres must be further investigated. Relevant metabolites and degradation products for antibody development and hapten design need to be considered as food products are complex mixtures making food analysis a complex issue. The quantitative analysis of food contaminants is essential with respect to safeguarding public health, preventing illicit use of compounds, and facilitating government regulations and surveillance.

1. INTRODUCTION

The office of the United States Trade Representative has estimated that international trade has increased fivefold since signing of General Agreement on Tariff and Trade (GATT) in 1947. Following this agreement, the creation of World Trade Organization (WTO) in 1995 resulted in significantly increased trade in foods of animal origin and live animals between different countries. But chemical contaminants due pesticides, mycotoxins and their metabolites are the key issue of the new pattern of food trades. Pesticides for example, immediately its discovery were extensively used in agricultural field for benefits of society, but now causing serious health hazard problems leaving their residues in foods. "Residues" in the widest sense may be defined as undesirable substances present in meat. These substances are chemical or biological in nature and have always been present in a small amount or can be introduced into the environment by various technological practices, can arise as a result of incorrect storage of foodstuff, can get in to the food chain due to modern agricultural practices and thus introduced into the foods or they may be results of medicines given to the animals or of processing methods.

Surveillance/monitoring on the occurrence of residues in meat and their products were relatively a neglected area until last decade. But with the advancement of technological intervention regarding livestock rearing, disease control and intensive crop production system, the chances of residues in foods of animal origin increased tremendously. This results a potential risk of various life threatening diseases such as cancer, leukemia, reproductive disorder besides disruption of body's immune, endocrine and nervous system (van Wendel de Joode *et al.*, 2001; Horrigan *et al.*, 2002). So, the monitoring and strict implementation of emerging contaminants continues to demand many positive benefits such as improved national nutritional statuses, major economic advantages including job creation, and improved diplomatic relation between the countries concerned (Buzby & Roberts, 1997), but this reliance on testing and other forms of inspection by either exporting or importing country or both.

The Codex Alimentarius Commission (CAC) is the reference point of World Trade Organization for disputes involving food. This committee achieved a great deal of international consensus on food export or import inspection and certification system. So, if any country is willing to export meat and meat products to other country, they should need regular monitoring of food products according to international standards or guidelines set by that country, since it is scientifically impossible for importing country to inspect or test the safety of all foodstuffs. However, the importing country can maintain a limited inspection and sampling program to act as vigilant against accidental or intentional contamination (JECFA, 2002).

Many developed countries in the world have already been tracking this problem by fixing statutory limits and their enforcement through monitoring set ups to ensure safe food supply to consumers. The tolerance limits of pesticides and mycotoxins in animal foods have also been fixed under several international bodies/organizations such as the Codex Alimentarious Commission (CAC), Food and Drug Administration (FDA), the United States Department of Agriculture-Food Safety Inspection Services (USDA-FSIS), the World Health Organization (WHO) and the Food and Agricultural Organization (FAO). The monitoring of toxicants in foods can reveal current status, and thus enabling preventive and control measures to be initiated before their level becomes so serious or widespread that it threatens human health or causes serious economic losses.

Pesticides residues encountered in meat industry are organochlorines (OC), organophosphates (OP), *N*-methylcarbamates, synthetic pyrethroids and their metabolites. Among these, OC groups of pesticides enjoy unique status because of their high insecticidal efficacy. However, in view of their persistence in the environment, susceptibility to biomagnifications and toxicity to higher animals and non-human primates, ecological catastrophy measures have been taken to restrict use of some of the chlorinated pesticides, and others are banned by imposing regulatory Acts/Rules. So, now replacement chemicals available for numerous pest control operations are OP, *N*-methylcarbamates and synthetic pyrethroid group of pesticides (ICMR, 1996). Organophosphates (OP), synthetic pyrethroids (*e.g.*, cypermethrin, deltamethrin/decamethrin, fenvalerate) and *N*-methylcarbamate (carbaryl, propoxur and aldicarb) compounds have been extensively used chemicals for crop protection and animal husbandry practices (Bogialli *et al.*, 2004). They are used at a low concentration compared to OC compounds and attributed low persistency in the environment. Despite their low persistency, both the pesticides exhibit toxicity to humans and animals and their presence in food has raised safety issues (Argauer *et al.*, 1997).

Mycotoxins are secondary metabolites of fungi. The major fungal genera producing mycotoxins include Aspergillus, Fusarium and Penicillium. The most common mycotoxins are aflatoxins, ochratoxin A, fumonisins, deoxynivalenol, T-2 toxin and zearalenone. Regulations for major mycotoxins in commodities and food exist in at least 100 countries, most of which are for aflatoxins; maximum tolerated levels differ greatly among countries.

2. HISTORICAL BACKGROUND

Historically, among all the food contaminants, pesticides have received unique status worldwide because of its risk-benefits relationship. However, the fact, these chemicals are acutely toxic to human, non-human primates and animals, even ingestion of low levels over a long period of time can have adverse effects. Early chemicals insecticides included inorganic materials such as sulfur, arsenic and mercury followed by several naturally occurring plant derived materials such as pyrethrum, nicotine, and rotenone. These materials along with a number of tars and oils and few synthetic organics used for insect control until the time of World War II (Baker & Wilkinson, 1990). DDT was first synthesized in 1874, but its insecticidal properties were discovered in 1939 (IARC, 1991). By 1943, the height of commercial production raised up to 4,00,000 metric tonnes worldwide annually. So, there is little doubt that during the 1940s and 1950s, world-wide use of DDT to control mosquitoes and insect vectors of human and plant diseases resulted in markedly reduced human suffering and enormous benefits to agriculture (Metcalf, 1973).

The introduction of DDT and other chlorinated hydrocarbons had a profound impact on pest control practices. But their adverse effect became increasingly apparent due to stability and persistence in the environment, their accumulation in the tissues of living organism and their lack of selectivity in the target species. Furthermore, these chemicals have the ability to undergo biomagnification through food chain causing serious problems for raptorial birds, carnivorous as well as human beings (McEwen & Stephenson, 1979). For this, the chemical industry began intensive research and development programmes for other synthetic organic pesticides.

By the year 1950, synthetic organic chemicals accounted 60 percent of the total pesticide markets. During the mid 1980s, the Environmental Protection Agency (EPA) estimated that there were 600 active pesticide ingredients marketed some 45,000 to 50,000 formulations. This continued development of new more effective pesticides that meet the more stringent modern standards of acceptability has become increasingly difficult, time consuming and costly. In 1993, among 20 new pesticides only one chemical meet to actual agricultural use. Presently, India alone produced over 90,000 metric tonnes pesticides annually (Srivastava *et al.*, 2005), though some of these compounds were banned and a few of them have restricted use in agricultural field imposed by regulatory authority. However, in contrast to DDT, Lindane, BHC and other organochlorine pesticides, the new synthetic pyrethroids, *N*-methylcarbamates and organophosphate (OP) compounds require only few grams per acre of land for its efficacy. But these pesticides still associated with certain measure of risk and vary considerably depending on the nature of inherent toxicity, intensity, frequency and duration of exposure, which in turn, related to the circumstances under which exposure occurs.

Though, the potentially harmful effects of feeding moldy grain and foods has been known for many years, mycotoxicology, the study of mycotoxins, began in earnest in 1960 with the outbreak of Turkey- X disease in the United Kingdom. This outbreak was linked to peanut meal imported from Brazil. Because of an intensive multidisciplinary research effort a blue-fluorescent toxin was isolated and mycelium of *Aspergillus flavus* was observed. *A. flavus* was soon shown to produce the same toxic compound(s) found in the toxic peanut meal. The toxin was characterized chemically and biologically and was given the trivial name, aflatoxin. Aflatoxin was shown to be very toxic and carcinogenic in some of the test animal species used and that it resulted in a toxic metabolite, aflatoxin M_1, in milk of dairy cows. The discovery of aflatoxin and elucidation of some of its effects led to research on other livestock health and production problems linked with moldy feedstuffs and to the discovery of additional mycotoxins produced by other fungi. In dairy cattle, swine, and poultry, mycotoxin contamination of feeds reduces growth, milk production, egg production, lower reproduction, and immunity. Mycotoxins have also been involved in outbreaks of human diseases.

3. GENERAL DESCRIPTION

3.1. Pesticides

3.1.1. *Organochlorine pesticides*

Organochlorine pesticides are hydrocarbon compounds containing multiple chlorine substitutions. There are four main types of OC pesticides; dichlorodiphenylethanes; cyclodienes; chlorinated benzenes; and cyclohexanes. All share a similar pair of carbon rings, one ring being heavily chlorinated. They are hydrophobic, lipophilic and extremely stable. Once in the environment they are subject to global deposition processes and bioaccumulate in the food chain. Diet is the main source of human exposure, primarily through consumption of animal products where OC pesticides have bioaccumulated. Toxicity appears to be via disruption of neural function and specific disturbances vary by chemistry. Studies support both acute and chronic effects of OC pesticides, potentially via damage to reproductive and neurological functions, carcinogenesis and endocrine disruption.

Dieldin Endosulfan

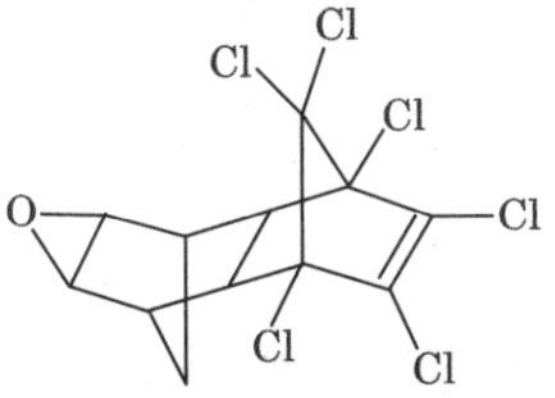

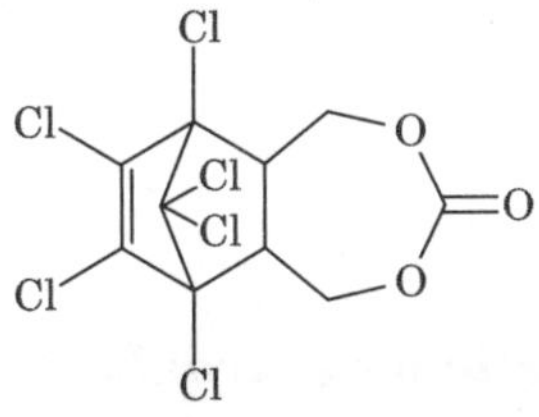

Heptachor 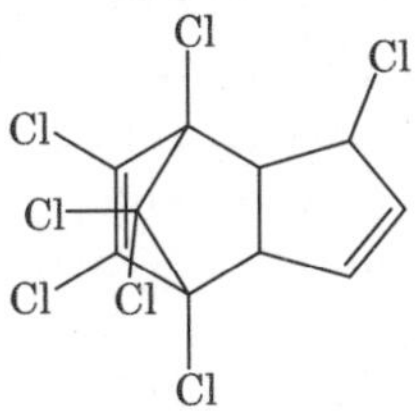 Lindane

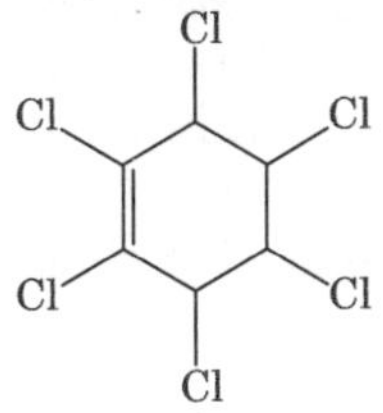

3.1.2. *Organophosphorous*

All OP compounds share a common chemical structure. The organophosphorus insecticides are normally esters, amides, or thiol derivatives of phosphoric or phosphonic acid. The basic components include a phosphorus atom that forms a double bond with either oxygen (P = O) or sulphur (P = S), and three side chains. Group X differs widely in individual OP agents and to a great extent determines the physical and chemical characteristics of the compounds. The R_1 and R_2 side chains are typically alkoxy groups, but may be aliphatic or aromatic hydrocarbons. In phosphoramidates, carbon is linked to phosphorus through an –NH group. Some OP compounds are branched aliphatic, aromatic, or heterocyclic groups linked to phosphorus via a bond of some lability (usually

–O– or –S–) and are referred to as the leaving group. The double-bonded atom is oxygen or sulfur and related compounds are called phosphates or phosphorothioates. The P = O form of a thioate ester may be referred to as the oxon, and this is often incorporated in the trivial name *e.g.*, parathion is the parent P = S compound of paraoxon.

However, there are 3 main commercial groups—(1) phosphates (without a sulfur atom), (2) phosphorothioates (with one sulfur atom), and (3) phosphorodithioate (with 2 sulfur atoms). Since the P = S form is intrinsically more stable, many insecticides are manufactured in this form which can be converted to the biologically active oxon in tissues. Oxidation of phosphorothioates to phosphates is potentially dangerous, since the phosphates are more volatile and are directly toxic agents. The biological action of OP depends on the phosphorylating ability of the OP, which in turn is determined by the electrophilicity of the phosphorus atom. Electrophilicity is mostly determined by the substituent groups in the OP molecule. *In vivo* activity of the OP depends also on the lipid solubility of the OP compound. Highly lipid soluble OP compounds can cross the biological membranes and the blood brain barrier, to reach the neural tissues where the OP can exert its main action.

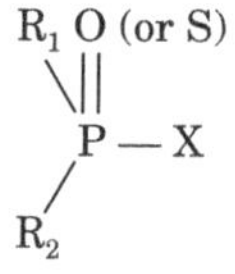

R_1 O (or S)
P — X
R_2

3.1.3. *Synthetic pyrethroids*

The pyrethroids are synthetic insecticides modeled after pyrethrins, which are natural constituents of the flower of certain *Chrysanthemums* (Newmann, 1990). They all are containing a 3-phenoxyphenyl moiety in their stereo chemistry. The pyrethroids, like most other insecticides are neurotoxic to insects. They are used at a low concentrations compared with other agricultural chemicals on a wide variety of crops, including many fruits and vegetables, ornamentals, and animals, and for the control of household, industrial and veterinary pests (Argauer *et al.*, 1997; EMEA, 2004). Because the established tolerance levels for pyrethroid insecticides in meat are low and metabolism in animals is high, residues exceeding established tolerance levels for meat and other commodity might occur only because of accidental or deliberate adulteration.

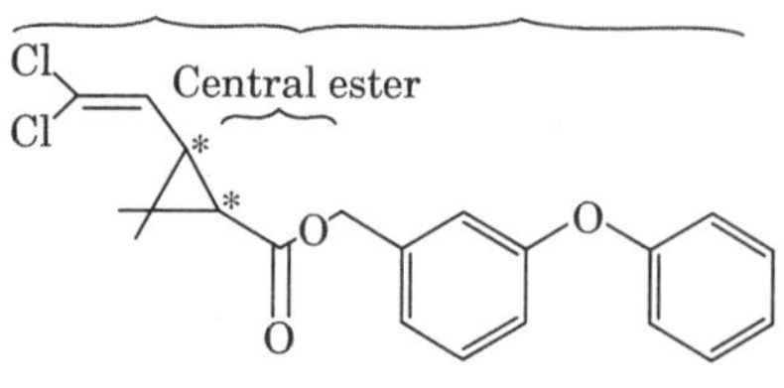

Central ester

Permethrin (I)

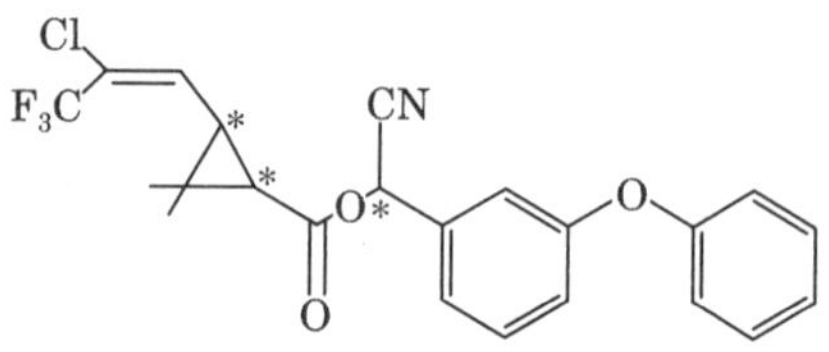

λ-Cyhalothrin (II)

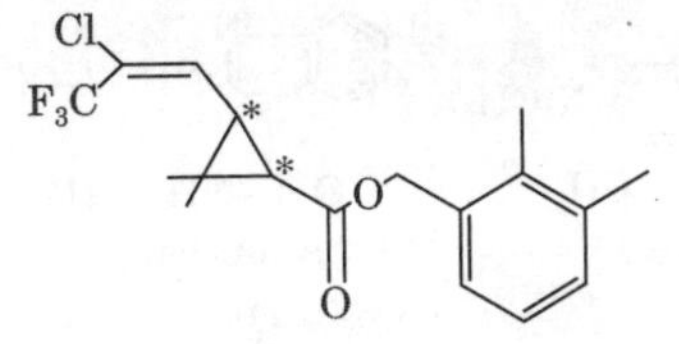

Bifenthrin (I)

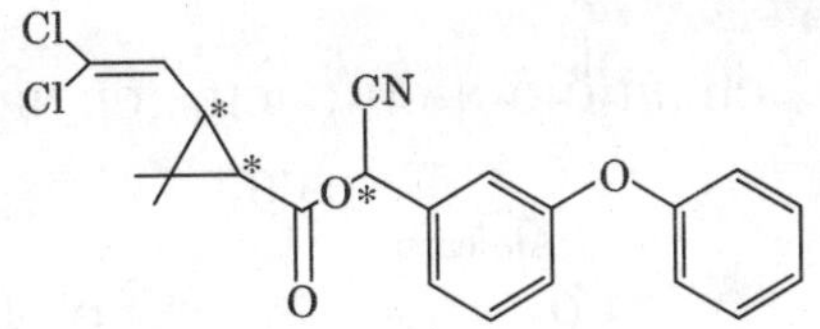

Cypermethrin (II)

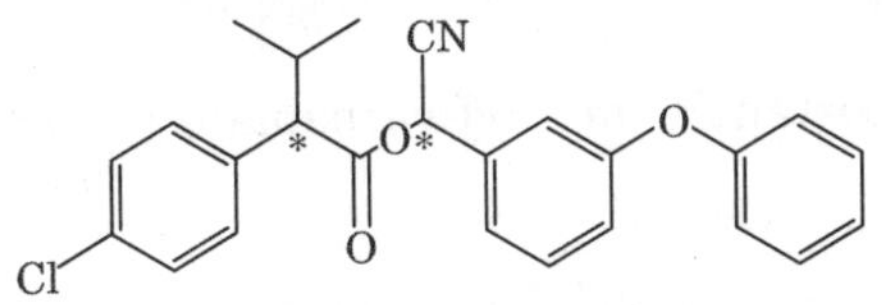

Esfenvalerate (II)

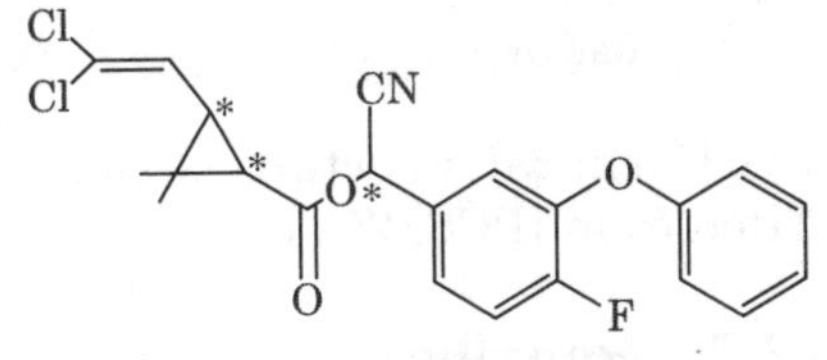

Cyfluthrin (II)

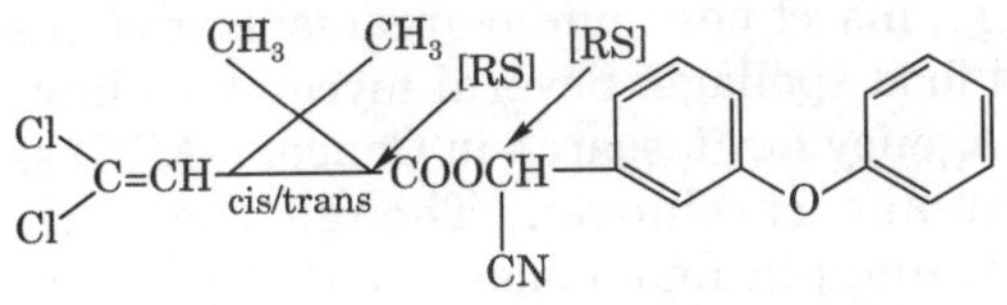

Cypermethrin isomers

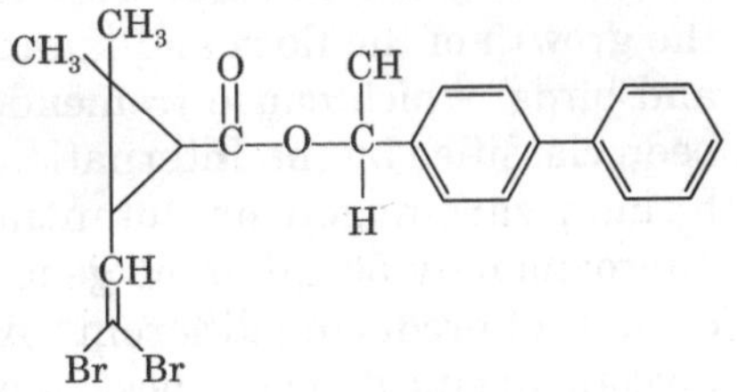

Deltamethrin

Chemical structure of some important synthetic pyrethroid insecticides (Cited from IPCS, 2011b).

3.1.4. *N-Methylcarbamates*

Carbamates are carbamic group containing broad-spectrum insecticides that are widely applied to fruit and vegetable crops including veterinary practices because of their low mammalian toxicity and their effectiveness against insects. A variety of substitutions can occur around the carbamic group, which often will determine both the degree of toxicity and potential use. They act as cholinesterase inhibitors and are usually neurotoxic. Nearly all carbamate pesticides are rapidly metabolized at carbamic side chain. However, research during last few years has shown that carbamates have possible chronic toxic effect. In fact the United States Department of Agriculture has recently changed the health hazard status of carbamate pesticides to a high-risk category (Bushway, 1988). Because of possible chronic toxic effects and wide distribution, there is a need to know if and to what extent this group of pesticides may be present in process foods and in particular food of animal origin.

$CH_3NH-C(=O)-O-N=CH-C-(CH_3)$ with SCH_3 — Aldicarb

$CH_3NH-C(=O)-O-C_6H_3(CH_3)-N(CH_3)_2$ — Aminocarb

Benzimidazole $-NH-C-OCH_3$, $O=C-NH-C_4H_9$ — Aminocarb

$CH_3NH-C(=O)-O-$ (naphthyl) — Carbaryl

$CH_3NH-C(=O)-O-N=C(SCH_3)-C(=O)-N(CH_3)_2$ — Oxamyl

$CH_3NH-C(=O)-O-C_6H_4-OCH(CH_3)_2$ — Propoxur

Chemical structure of some important *N*-methyl carbamate pesticides (Cited from IPCS, 2011a).

3.2. Mycotoxins

In most of the developing countries where the weather conditions are favourable for growth of agricultural crops throughout the year, also ideal for promoting the growth of the flora and fauna (*e.g.*, insect pest, micro-organisms, rodents and birds) which cause tremendous food spoilage. Several mycotoxins have been classified by the International Agency for Research in Cancer (IARC) as human carcinogen or potential human carcinogen. Though there are approximately 50 different genera of fungi in four major taxonomic groups capable of producing different toxins. Epidemiological studies suggested that mycotoxins incidence is most common in country like Mozambique, Thailand and Uganda where mould damage to foods and feeds is frequent. Mycotoxicosis can result from the ingestion of toxins produced by four types of fungi. (1) Macroscopic fungi: The best known being the toxic mushrooms such as the death cap mushroom (*Amanita phalloides*), (2) Parasitic fungi: Which infect and cause disease in plants growing in the field, (3) Storage Fungi: Which under suitable conditions are capable of rapid growth on crops during cultivation, harvesting, drying, transportation and storage, and (4) Advanced deterioration fungi: Normally fungi do not infest intact grains but easily attack damaged ones and require high moisture content. Chemical structure of different mycotoxins are given as under

Trichothecene	R_1	R_2	R_3	R_4	R_5
Type A					
T-2 toxin	OH	$OCOCH_3$	$OCOCH_3$	H	$OCOCH_2CH(CH_3)_2$
HT-2 toxin	OH	OH	$OCOCH_3$	H	$OCOCH_2CH(CH_3)_2$
Neosolaniol {NEOS}	OH	$OCOCH_3$	$OCOCH_3$	H	OH
Diacetoxyscirpenol	OH	$OCOCH_3$	$OCOCH_3$	H	H
Type B					
Deoxynivalenol	OH	H	OH	OH	=O
Nivalenol {NIV}	OH	OH	OH	OH	=O

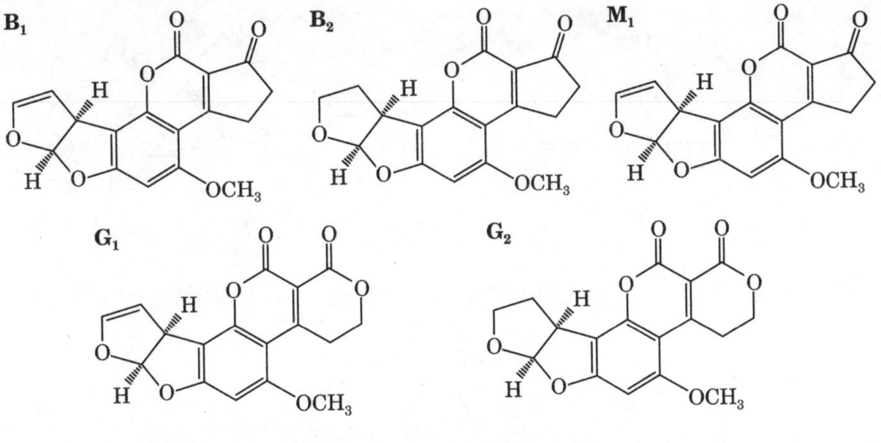

Fig. 1: Structures of aflatoxins {AFs} B_1, B_2, M_1, G_1, G_2.

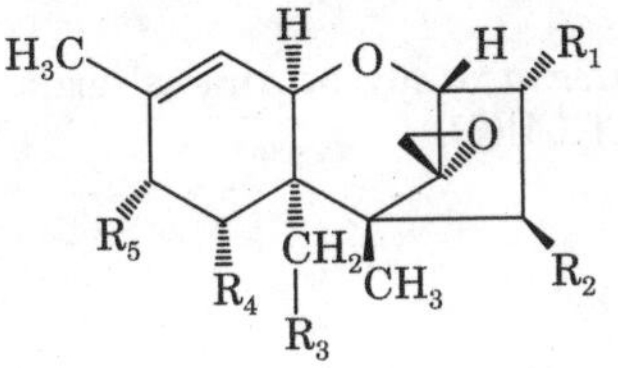

Fig. 2: Structures of some important trichothecenes.

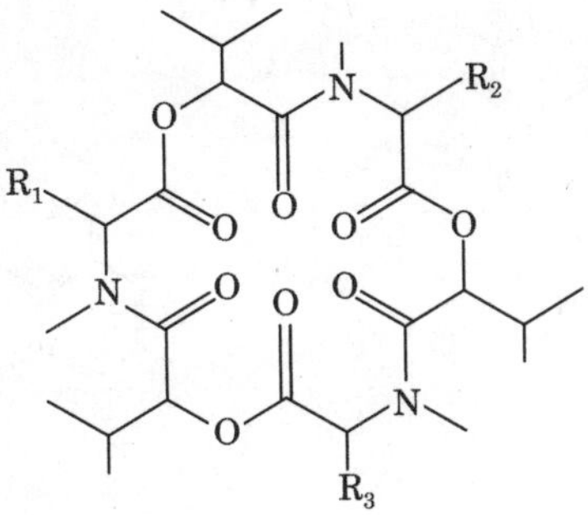

Fig. 3: Structures of enniatins.

Compound	R_1	R_2	R_3
Enniatin A	$CH(CH_3)C_2H_5$	$CH(CH_3)C_2H_5$	$CH(CH_3)C_2H_5$
Enniatin A_1	$CH(CH_3)C_2H_5$	$CH(CH_3)C_2H_5$	$CH(CH_3)_2$
Enniatin B	$CH(CH_3)_2$	$CH(CH_3)_2$	$CH(CH_3)_2$
Enniatin B_1	$CH(CH_3)_2$	$CH(CH_3)_2$	$CH(CH_3)C_2H_5$
Beauvericin	$CH_2C_6H_5$	$CH_2C_6H_5$	$CH_2C_6H_5$

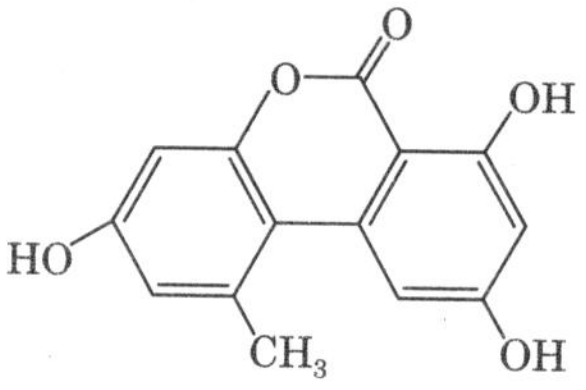

Fig. 4: Structures of some *Alternaria* mycotoxins: alternariol {AOH} (1), altenuene {ALT} (2) and altertoxin I{ATX-I}(3).

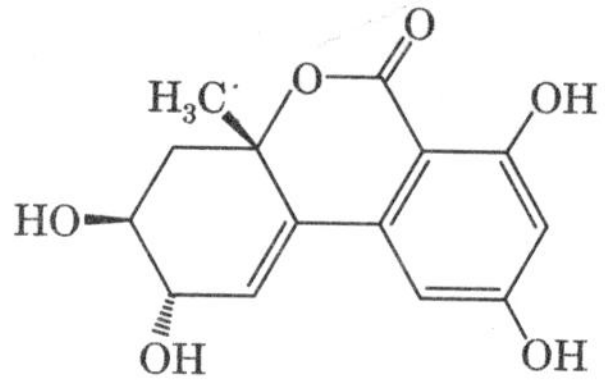

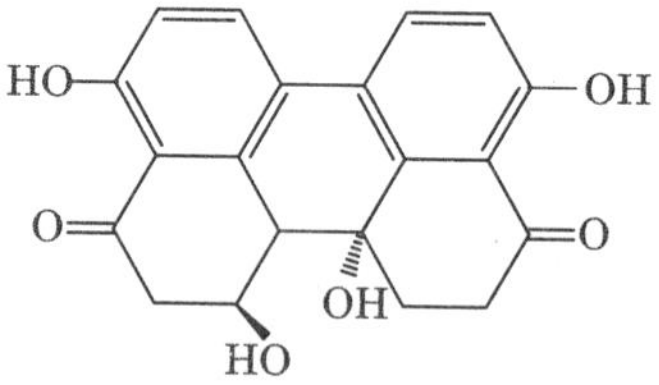

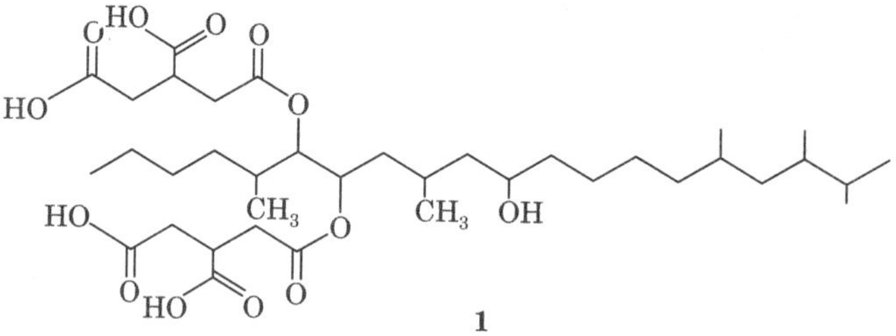

Fig. 5: Structures of some *Fusarium* mycotoxins: FUM B_1 (1), moniliformin (usually as sodium or potassium salt) (2), zearalenone {ZON} (R = O) and zearalenol {ZOL} (R–OH), (3) and possible metabolite zearalanone {ZAN} (4).

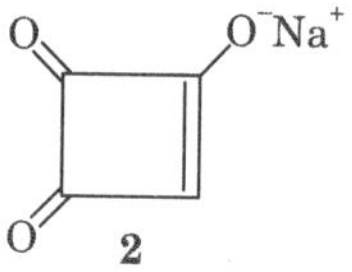

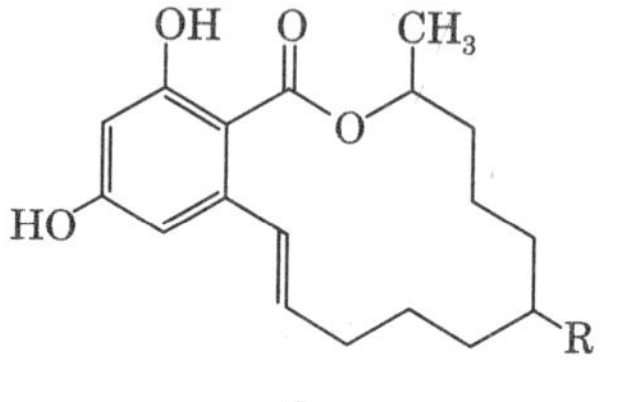

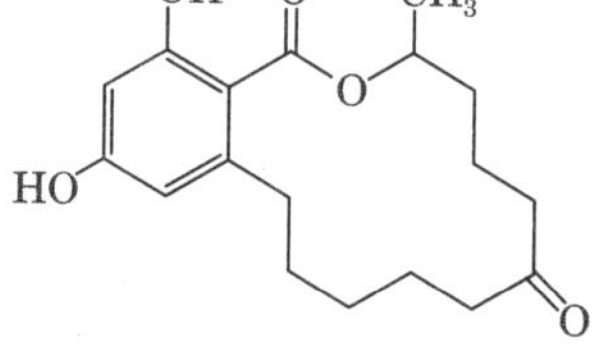

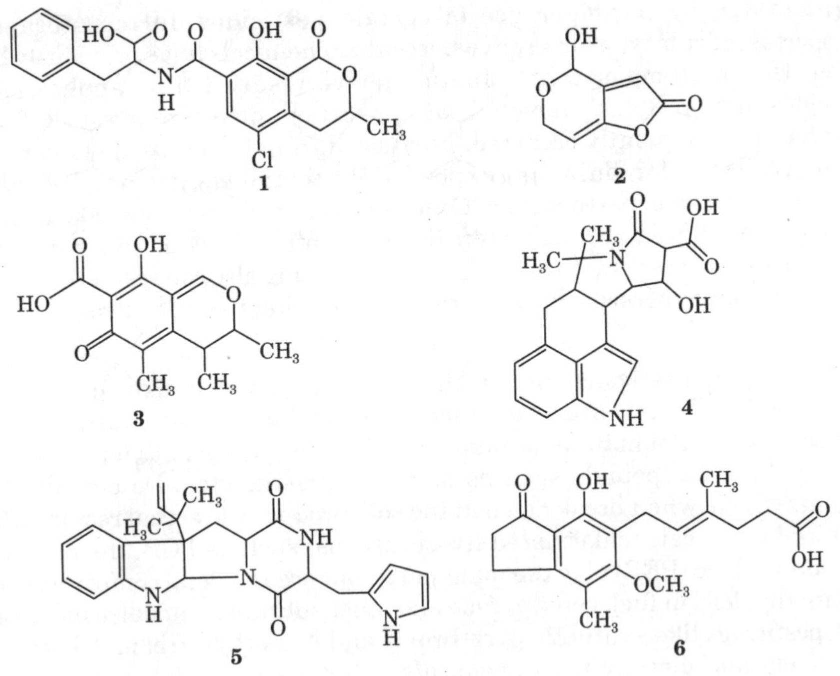

Fig. 6: Structures of some important mycotoxins produced by *Penicillium* and *Aspergillus* species: ochratoxin A {OTA} (1), patulin {PAT} (2), citrinin {CIT} (3), cyclopiazonic acid {CPA} (4), roquefortin C {RQ} (5) and mycophenolic acid {MPA} (6).

4. SOURCES OF EXPOSURE

4.1. Pesticides

Pesticides have unique status of all food contaminants because these compounds are regularly used in agricultural fields to meet worldwide food demands. A number of pesticides are used in modern agriculture for producing food commodities as cheaply and efficiently as possible. These are either synthetic or natural compound could leave residues in edible food products from animal exposed to that compound. It is estimated that, without pesticides, world production of food would be reduced by 30 percent. Despite the obvious benefits, however, occasional misuse of such chemicals has resulted in intoxication of animals and/or accumulation of residues in meat, poultry and meat products. Although violative levels of pesticide are uncommon, nearly 80 percent of all residues in food animals are estimated to occur through the feed. The most common exposures are from accidental inclusion of pesticides in feed formulations as has similar granular appearance to some feed ingredients (Buck, 1973).

In addition, special feed ingredients such as oil and mineral mixture may be contaminated during their manufacturing process. Feed may also be contaminated by improper use of certain pesticides during storage or transportation or by storing grain where other chemicals have previously been stored. However, major contaminations have resulted from application of pesticides in crop fields. During arial application, drip into livestock forage crops has also frequently occurred. Further, during lean season when fodder is not available for animals, exposure of pesticides occurred through ingestion of soil from pasture land. Other sources of exposure include drinking water contaminated through crop field run off pollution, the mixture of misapplication of pesticide to animals or their housing also can result in toxicosis (Heise, 1992). Laypersons with no training for using pesticides may result in residue problem.

Among all pesticides, organochlorine (OC) compounds such as DDT, lindane or hexachlorobenzene, heptachlor, heptachlor epoxide, aldrin/dieldrin are mostly persistent in the environment and cause major health hazard effects. Some of these compounds such as mercury, arsenic etc. also contain toxic elements which, when break down in the soil, produce the actual residues. The industrial chemicals (chlorinated hydrocarbons) such as PCB, the dioxin, or perchlorethylene (PER) play the main parts, but organometal compounds such as tetraethyl lead in fuel, benzopyrene and other substances are also important. Some pesticides like synthetic pyrethroids and *N*-methylcarbamates are also extensively and directly used chemicals as ectoparasiticidal treatment and regular household practices (EMEA, 2004). The reason for this is that they prove to have high and broad insecticide efficacy but low toxicity towards variety of warm-blooded animals.

4.2. Mycotoxins

Mycotoxicosis is confronted most commonly in some African and Southeast Asian countries as fungal growth and toxin formation in stored commodities is much more frequent. It is also in this region that the existing gap between production and demand forces the people and animals to desist from rejecting any material that can be used as food, even when it is infected with fungi and is aesthetically and organoleptically unacceptable. Often crops which have been rejected for export market are sold on the home markets. Any type of processing to make the food safe will enhance its cost and therefore make it unavailable to the people who need it. It should also be realized that as most of the people in the developing regions are under-nourished they may be more susceptible to the toxic effects of mycotoxins. When the mycotoxin problem is viewed against this background, it becomes clear that the bulk of the population in many developing countries is exposed to the toxic effects of mycotoxins. However, the mycotoxin problem is certainly not confined to the tropical and sub-tropical developing countries. It is frequently encountered in more developed regions. Here, rejection of contaminated material is geared not so much to avoid creating food shortages but more to avoid the pollution of food which would increase the health risk of the community.

5. RAPID SCREENING METHODS

5.1. Enzyme-Linked Immunoassay (ELIZA)

Immunoassay techniques are one of the recently developed methods of analysis of pesticide residues mycotoxins in foods. These methods are based on the interactions between antibodies and antigens. Antibodies are highly specific for structurally very different compounds, but can show considerable cross-reactivity for structural analogs, as they recognize only specific chemical groups - the epitope. Pesticides and mycotoxins do not elicit immunological response in test animals as they are small molecules. Hence analogues are synthesized with a spacer arms of 5–6 atoms with a reactive end (like –OH, –SH, –COOH) that can be coupled to carrier proteins. These conjugates are antigenic and test animals like rabbits are immunized with them. Antisera isolated from these rabbits contain antibodies that are specific to pesticides. Using these antibodies enzyme-linked immunosorbent assays (ELISAs) are formulated.

A major concern is the degree of cross reactivity (affinity) the antibodies show for related chemical structures. Thus, antibodies that were developed to one molecule may react with other related chemical structures. Usually this cross reactivity is characterized by a lower affinity for the related structures than for the analyte itself. The merits of immunoassays include their high selectivity and the simplicity, speed and moderate cost of the procedure relative to other methods. The principal disadvantage of the immunochemical approach is the extensive effort and time required to elicit the antibodies in the vertebrate host.

5.2. Lateral Flow Assay (LFA)

Lateral flow assay also known as solid phase immunochromatography has a great demand for developing rapid and reliable analytical techniques to control multiple pesticides in foods, agriculture and environmental samples. Several immunochromatographic assay (ICA) were developed (Suri *et al.,* 2009; Guo *et al.,* 2009 & Wang *et al.,* 2005) and all of them had successful applications in residue analysis. Guo *et al.* (2009) proposed two gold-based ICAs to detect simultaneously carbofuran and triazophos in water samples. In this study, one test strip format employed a bispecific monoclonal antibody against both carbofuran and triazophos. The second test format was based on two different monoclonal antibodies raised specifically against the two pesticides. Gold-monoclonal antibody conjugates were applied on the conjugate pads of the developed test strips to serve as detector reagents. The capture reagents consisted of two test lines (T_1 and T_2) coated with carbofuran-ovalbumin conjugate and triazophos-ovalbumin conjugate, respectively. The visual detection limits of the pesticides by the two ICA formats were from 4–62 μg/ml. The strip showed low cross-reaction to other structurally related pesticide compounds including carbosulfan, carbofuran phenol, isoprocarb, chlorpyrifos, parathion and diazinon.

A single test strip format for one-site verification of active ingredients in commercial insectide formulations of endosulfan and three pyrethryroids (cypermethrin, deltamethrin and fenvalerate) also investigated. Colloidal gold were conjugated with two antisera raised against endosulfan and cypermethrin. The developed test strip consisted of two capture test lines coated separately with endosulfan-BSA conjugate or cypermethrin-BSA conjugate. Finally, the test strip was designed to allow the detection of standardized concentrations of endosulfan (1800 mg/ml), cypermethrin (800 μg/ml), deltamethrin (1000 mg/ml) and fenvalerate (1400 mg/ml). These concentrations were found sufficient to inhibit colour development on the test lines. The test strip developed here is very useful for the farmers, and it also could have potential applications in environmental testing.

For detection of AFT B_1 in foods a LFA was proposed by Xiulan *et al.* (2006). The assay used gold-labeled polyclonal antibodies as detector reagent against AFT B_1, and AFT B_1-BSA as capture reagents. In the first step of the study, the efficacy of the test strip was evaluated through analysis of AFT B_1 in food sample extracts. With this method, the limit of detection of AFT B_1 by visual observation was 2.5 ng/ml, and it ranged 0.05–0.150 ng/ml when using photometric strip reader. The analytical recoveries for AFB_1 in sample extract (rice, corn, and wheat) spiked at 2–50 ng/kg were from 80.79–110.56 percent with mean coefficient of variations ranging 6.27–14.63 percent. In all assay, the analysis of AFT B_1 in one sample was completed within 10 min.

More recently, a novel membrane-based LFA was developed by Tang *et al.* (2009) for the fast screening of aflatoxin B_2 (AFT B_2) in food samples. In this study, the detector reagent consisted of magnetic nanogold microspheres (MnGMs) with nano-Fe_2O_3 particles as core and gold nanoparticles as shell, and bio-functionalized with monoclonal anti-AFT B_2 antibodies. Result showed that the cutoff value of conventional strip with pure gold nanoparticles as detection reagent was 3.1 ng/ml AFT B_2, while the cutoff value with MnGMs as detection reagent was 3 times lower at 0.9 ng/ml AFT B_2. The assay showed sensitivity rate of 100 percent and a specificity rate of 75 percent. This study demonstrates a new approach to develop higher sensitive LFA which makes use of bio-functionalized magnetic nanogold microspheres as detection reagent. The developed method would be convenient for use to detect other toxins and residue contaminants.

Wang *et al.* (2005) developed a colloidal gold-based LFA against Ochratoxin (OTA). The test strip showed a detection limit at 5 ng/ml for OTA, and it can be completed within 10 min. A more sensitive and rapid LFA for the detection OTA in food samples was proposed by Wang and Yu (2008). The detection limit was 1 ng/ml with an assay time of less than 10 min. The handling of pure toxin in traditional test strips may pose a toxicity risk to strip manufacturers, users, and the environment. For this purpose, Lai *et al.* (2009) developed a test strip to detect OTA by using a mimotpe peptide. In this study, two kinds of lateral flow strips (OTA–BSA and mimotope peptide in the nitrocellulose membrane) were evaluated for screening OTA in corn and wheat samples. Results of analysis showed that the two kinds of lateral flow strips had the same detection limit at 10ng/ml, and the assay time was accomplished within 10 min. This study demonstrates the convenient use of nontoxic chemical as an immunochemical reagent which could replace toxic compounds in test strip development to enhance laboratory and environmental safety.

5.3. Biosensor Techniques

Biosensor is analytical device consisting of biologically active component like an enzyme, antibody or binding protein and an electronic or optical system to convert the biochemical event into quantifiable electronic signals that can be processed and measured accurately. Biosensors can be categories in to two main types that are configurated based on signal transduction and biorecognition principles. Signal transducing biosensors can be of electrochemical, optical, piezoelectric, and thermal sensors. Biorecognition biosensors are immunochemical, enzymatic, non-enzymatic receptor, whole-cell, and DNA biosensors. Immunosensors present the advantages of sensitivity and selectivity inherent to the use of immunochemical interactions. Limitations are the troubles derived from the regeneration of the immunosurface, and cross-reactivity.

Conventional techniques such HPLC/MS and GC/MS though considered as 'golden techniques' for the analysis of pesticides, new assays with sensor based techniques are cheaper and faster, and hence called on-site online sensors. Enzymatic sensors, based on the inhibition of a selected enzyme, are the most extended biosensors used for the determination of these compounds. Biosensors based on enzyme inhibition although sensitive, are not selective and cannot, therefore, be used for quantification of either an individual or a class of pesticides. But introduction of recombinant enzyme for biosensor applications can solve the problem (Choi *et al.,* 2001; Andres *et al.,* 1997). The organophosphorous hydrolase (OPH) is able to hydrolyze a number of OP pesticides such as paraoxon and parathion, and chemical warfare agents such as sarin and soman.

Hydrolysis of these OP pesticides generates *p*-nitrophenol, which is an electroactive and chromophoric product. Thus, OPH could be combined with an optical transducer to measure the absorbance of *p*-nitrophenol or with an amperometric transducer to monitor the oxidation or reduction current of this product. In a different approach, biosensors based on immunological assays have been developed with limit of detection of 0.1 μg/l. Mallat *et al.* (2001) applied the "River Analyzer" (RIANA) immunosensor in the determination of pesticides such as atrazine, simazine, isoproturon, 2, 4-D, alachlor, and paraquat in natural waters. Recently, a label-free direct piezoelectric immunosensor built on a flow-through cell was used for the determination of 2, 4-D in water with a limit of detection around 0.2 μg/l. However, there are very limited applications of biosensor detection of pesticide compounds in meat system.

For detection of mycotoxins, a great number of specific sensors have been developed for food and environmental control (Delehanty & Ligler, 2002). Thus, an integrated optical sensor has been reported for the analysis of aflatoxin B in corn. A light-addressable potentiometric immunosensor based on the commercial device for the analysis of saxitoxin and ricin has also been described. An impedance- based immunosensor has been prepared by using an ultrathin platinum film with an immobilized layer of antibodies against the staphylococcal enterotoxin B (Kumar *et al.,* 1994). Various evanescent wave immunosensors have also been reported to be capable of detecting botulin with very low limits of detection. A rapid and sensitive immunosensor for the detection of the *Clostridium botulinum* toxin A has also been developed. This fiber optic-based biosensor utilizes the evanescent wave of a tapered optical fiber where antibodies antitoxin A has been covalently immobilized at the distal end. The toxin could be detected by means of a rhodamine label, within a minute concentrations as low as 5 ng/ml.

In some study it has been reported that the bacterial spores can acts as nano-detector which could have unique ability to sense environmental changes in response to specific "germinant" and transform rapidly into growing vegetative cells. This characteristic can be effectively used as biosensor for tracking microbial and non–microbial contaminants (Kumar *et al.*, 2005; Rotman 2001 & 2003).The present hypothesis is based on the specific spore germination inhibition principle in presence of specific analyte *i.e.*, aflatoxin M_1. In case where analyte is absent in milk system, specific indicator enzyme (s) are produced by active bio-sensing molecules which will act specifically on chromogenic/or fluorogenic substrate resulting in colored reaction/or fluorescence as end product which is measured semi-quantitatively by either visually/or using optical system at specific excitation/emission spectra (Kumar *et al.*, 2010). The end product response is significantly different in case of buffer/ or milk system containing specific analyte *i.e.*, aflatoxin M_1. The developed test assay was validated by analyzing 25 samples of each raw and pasteurized milk procured from different organized/or private own dairies and other reputed brands using AOAC approved RIA and ELISA based system and a significant correlation with ELISA at Codex MRL Limit (0.5 ppb) of Aflatoxin M_1 was established.

Despite the high number of biosensors under development only few are currently enjoying market acceptance. SPR biosensors constitute the most successful type in the commercial instruments for environmental monitoring. The pioneers of SPR-based biosensing were Pharmacia Biosensor AB, now BIACORE AB (Uppsala, Sweden), who launched the original BIAcore system in 1990. The company now has a large range of biosensors, such as BIAcore series 1000, 2000, and 3000 with varying degrees of automation and parameter specifications. The BIACORE 3000 biosensor, which is based on SPR principle, monitors bimolecular interactions as they proceed over time. The advantage of the method is that there is no need for labeling the interactants. Other SPR commercial biosensors are provided by Windsor Scientific, Ltd. (Berks, UK),

Nippon Laser and Electronics Labs (Hokkaido, Japan). Texas Instruments (Dallas, TX) integrates SPR detector Spreeta that can be configured for industrial, environmental, and biological applications. On the other hand, Affinity Sensors (Franklin, MA) manufactures the IAsys line of instruments, which use evanescent wave technology.

5.4. Molecular Imprinted Polymers (MIPS)

Molecular imprinting is a process of template induced formation of specific recognition sites in materials where the template directs the positioning and orientation of the materials structural components by a self-assembling mechanism. A polymer is formed around a molecule acting as a template. After the template is removed the imprints contain functional groups complementary to those of the template. MIPs have potential applications in mycotoxin analysis adopting—(1) solid phase extraction; (2) biosensor devices; and (3) chromatographic matrices for separations (Shephard *et al.,* 2005).

The advantages of MIPs are: (1) MIPs are arguably the most generic and cost effective technique for preparing synthetic receptors; (2) MIPs can be prepared for practically any compound; (3) MIPs have similar affinity to natural biomolecules but often have better specificity; (4) MIPs can work in organic solvents; (5) MIPs are stable at low/high pHs, pressure and temperature; (6) polymers are compatible with microfabrication; (7) polymers are inexpensive.

However, MIPs also have some limitations: (1) template costs may vary considerably from one compound of interest to another; (2) the exact stereo chemical structure of the imprint is not known; (3) the unfavourable adsorption isotherm and slow mass transfer in the polymer matrix are among the major limitations for MIPs as separation media; (4) MIPs also tend to swell when solvent composition is altered leading to irreversible deformation of the imprinted cavity and loss of selectivity.

5.5. Microarray Technology

Microarray technology, essentially a reverse dot blotting technique, is a new technology for the identification of DNA fragments such as PCR products, and may be a useful technique for identification and differentiation of high numbers of microorganisms in parallel. Arrays of genus, species or sub-species specific capture oligonucleotides, representing the microorganisms to be identified, are immobilized on a solid support such as a glass slide. This glass slide may then be probed with labelled DNA from the sample of interest. Hence applications of microarray technology restricted to determination of mycotoxins. Using this technique Fusarium species was identified and differentiated from a range of mycotoxins by designing and testing an oligonucleotide array (Nicolaisen *et al.*, 2005).

The microarray technique has several advantages such as: high throughput analysis; small volume of sample needed and a wide range of microorganisms can be detected. However, it has several limitations in the application for mycotoxin analysis. Many researchers still do not trust data obtained from microarrays because of their high variability (low reproducibility). RNA extraction, amplification, and hybridization are all procedures involved in microarray research, and all have inherent problems leading to statistical errors. The method is designed to detect mycotoxin-producing fungal species, but not mycotoxins.

5.6. Luminex xMAP^R Technology

Luminex's xMAP^R technology is comprised of existing technologies—flow cytometry, microspheres, lasers, digital signal processing and traditional chemistry, and now applying for determination of mycotoxins in foods. The ranges of applications are considerable throughout the drug-discovery and diagnostics fields, as well as in basic research (Vignali, 2000). Microspheres are dyed to create 100 distinct colours. Each microsphere has a 'spectral address' based on red/infrared content. The suspendable microspheres are coated with capture reagents such as antibody or oligonucleotides. Sample is then added to microspheres and the analyte is captured by the microspheres. A fluorescent reporter tag is then added and results are read using a compact microsphere analyzer. The advantages of the technology are: high speed, high throughput, multi-analyte detection, versatility and reproducibility. There is no current method for mycotoxin analysis based on this technology. Future developments in the analysis of mycotoxins should be for multi-mycotoxin analyses, further miniaturization, and portability for on-site mycotoxin testing. Also, the general trend is that methods must be user-friendly and cost-effective.

6. QUANTITATIVE ANALYTICAL METHODS

Consumers' expectations are healthy foods with affordable price, good shelf-life and high safety, while meat inspections require good manufacturing practices, safety, labeling and compliance with the regulations. The creation of quantitative analytical methods and their adoption in residues monitoring may satisfy the regulatory requirements and also helps to improve the safety standards of livestock products. There are voluminous technical literatures on pesticide analyses. Several methods have been used widely and undergone extensive testing and validation for many years. Colorimetric method, thin layer chromatography (TLC), high performance- thin layer chromatography (HP-TLC), gas liquid chromatography (GLC) and high performance liquid chromatography (HPLC) methods though conventionally used for determination pesticide residues have several limitations. All these methods are time consuming, cost affair, complex and required highly toxic chemicals for extracting, purification and quantification of pesticide residues in animal tissues (Biswas *et al.,* 2009). Recently, gas chromatography- mass spectrometry (GC-MS) and liquid chromatography-mass spectrometry-mass spectrometry (LC-

MS-MS) methods were developed for effective and confirmatory determination of pesticide residues with satisfactory results but still they are time consuming. Immunoassays and use of sensor based techniques are comparatively new, but getting wider popularity for their applications in the determination of pesticide residues in food products.

6.1. Types of Analytical Methods

As mentioned previously, there are several approaches to pesticide analysis. These methodological approaches vary in their degree of complexity, in the time, efforts, and analytical instrumentation required to complete them, in the degree of confidence that can be placed in the final results. Procedure may be quantitative-multi or single residue methods, or only semi-quantitative or even quantitative.

6.1.1. *Multiresidue methods*

These methods come closest to meeting regulatory requirements of pesticide residues enforced by several monitoring agencies. They have been designed to detect and measure a multiplicity of residues in ranges of food products. They are sufficiently precise to provide reliable estimates of residue levels for many pesticides at or below the established tolerance levels. These multistep methods typically contain the steps of sample preparation, extraction, clean-up, derivatization and chromatographic separation with on-line detection using highly selective detector and automated quantification. There are several multiresidue methods adopted by FDA and USDA in many years, but yet they are continue to be modified, expanded, and optimized because none of these methods can be detect all residues on all food types.

6.1.2. *Single residue methods*

In contrast to multiresidue methods, single residue methods have been designed to measure a single analyte and often, its principal metabolites and transformation products of toxicological importance. The majority of single residue methods have been developed in support of applications for registration including tolerance setting and/or research into the metabolism and environmental fate of the analyte. The majority of single residue methods are based on the same sequential steps as multiresidue methods; however, each step has been optimized for analytes of interest. Generally they are less time consuming to perform and often provide lower limits of detection than multiresidue methods.

6.1.3. *Semi-quantitative and quantitative methods*

Semi-quantitative and quantitative methods range widely in their abilities to estimate the level of particular pesticide residue in a sample. In general, they are capable of detecting a limited number of somewhat similar pesticide residues. These methods are often referred as screening methods in the sense that they are capable of assaying a large number of samples for the presence of a limited number of pesticide residues in a relatively short time. Additionally they are robust in character because they are less sensitive to small changes in the purity of reagents, quantities of reagents, time factors, reaction temperatures and/or environmental conditions and are not limited to highly controlled laboratory environment. The semi-quantitative methods provide an estimate of the concentration range for a detected residue, while in quantitative methods will detect the pesticide if present above some predetermined level. The principal benefits of semi-quantitative methods are their low cost, relative speed, and simplicity. These methods make use of such techniques as a thin-layer chromatography, enzyme inhibition and immunoassay.

6.2. Sample Preparation and Pre-Concentration

Foods with high moisture-low fat, high moisture-high fat, low moisture-low fat and low moisture-high fat may behave different way during extraction of clean up. Again the polarity, volatility, chemical reactivity, and thermal stability of the analytes will determine the efficacy of recovery from the food matrix, the behaviour on clean up columns, the choice of an appropriate analytical column and the selection of analytical instrument.

6.2.1. *Sample preparation*

Sample immediately after collection or receiving in the laboratory need to be maintained at appropriate temperature and relative humidity conditions to avoid unusual changes take place in that sample in a way that would affect determination of analytes or concentration of the analytes. Meat, egg, milk or any other biological samples are held at refrigeration temperature or even at freezing temperature to avoid decomposition. If samples are frozen, need to be thawed at refrigeration temperature. Usually a composite sample is prepared by cutting, mincing, followed by blending and mixing. These steps have two objectives-to reduce the structural features of the sample and facilitate the subsequent extraction and to produce a homogenate composite sample from which subsamples can be taken. Care should be taken to avoid contamination and/or exposing it to unnecessary heat, which can cause loss of volatile analytes and/or accelerate decomposition.

6.2.2. *Extraction*

As the objective of this stage of analysis to recover as much of analytes as possible by solubilizing them into added solvents, so choice of solvents should be based on the polarity of analytes of concerns. It has been considered that non-polar compounds can be well extracted with polar solvents. However, toxin solvents like hexane, diethyl ether which posing potential environmental pollution could be restricted in sample extraction. On the other hand, acetonitrile has been considered as choice of solvent for the extraction of many pesticides. This solvent was found to have advantages over a number of organic solvents such as methanol, acetone, ethyl acetate, acetonitrile metaphosphoric acid, and methanol/metaphosphoric acid in terms of deproteinization (removal of >99 percent of protein and fat), recovery, chromatographic interference, and emulsion formation (Biswas *et al.*, 2009). Similarly, other researchers also reported that acetonitrile is the best solvent for extraction because it is easy to evaporate and the lipid/oil interference could readily be removed from the extract solution by introducing *n*-hexane (Biswas *et al.*, 2010).

The only disadvantage of this solvent was its binding with highly polar compounds from the sample matrices, which ultimately come first in the reverse phase chromatography. Anhydrous sodium salts like sodium sulphate or potassium sulphate can be added to absorbed water, or water may be sometimes intentionally added so that the crude extract can be purified with a subsequent partitioning step with a second water-immiscible solvent. The solvent is separated from insoluble solids by filtration. The organic phase is dried over sodium sulfate, reduced in volume to 3–5 ml, and diluted with 5 ml of appropriate solvent and reconcentrated. The dilution-reconstitution steps are repeated to ensure complete removal of those extraction solvents that can disrupt the operation of the detector. The resulting concentrate can be used for analysis by gas chromatography without further purification unless an electron capture detection is to be used.

6.2.3. *Cleanup*

The extracted sample is partially purified before separation in chromatogram. Though this step is the most time consuming, labor intensive, and error prone step, a number of techniques have been used successfully. In this stage of analysis, the analytes are separated from coextractives that can interfere with the detection of analytes. This step starts with partitioning in which water-acetone or water-acetonitrile extract is partitioned with a relatively non-polar solvent. The organic phase is dried and reduced in volume. In some methods the residues are sometimes purified in packed column either using adsorption chromatography or size exclusion chromatography. The choice of packaging material is both analyte and matrix dependant; however Florisil, silica gel, alumina and carbon are most commonly used in packing of column for adsorption chromatography. The activity of the adsorbent is standardized by heating in an oven overnight, and then deactivated by equilibrating with water. The compounds are eluted either by isocratic elution or a gradient elution with the increase of solvent polarity. The size exclusion chromatography represents an attractive alternative because it represents a different chromatographic mode to adsorption, it provides excellent recoveries usually more than 80%, and it can be automated readily.

6.2.4. *Derivatization*

This represents structural modification of analytes by chemical reaction which improves thermal stability and/or volatility of the analytes, modifies chromatographic behaviour, increase the selectivity of detection, and/ or increase the sensitivity of detection. Most often derivatization is used to overcome limitations of sensitivity. Typically, substitution or addition reactions are used to introduce a chromophore, flurophoreor other functional group into the analytes to augment the detector response to the resulting products. Typically the derivatization reaction is performed precolumn, but post column reactions can also be used provided that the reaction can be performed *in situ* and automated. The requirement is that the identity of the original analytes not be compromised by the reaction. The advantages of derivatization must outweigh the disadvantages of increased sample handling, analysis time, and typically, a decrease in precision.

6.3. Chromatographic Separation and Detection

6.3.1. *Thin-layer chromatography* (*TLC*)

Thin-layer chromatography (TLC) is a low-cost, rapid analytical technique, yielding qualitative or semi-quantitative estimations by visual inspection, but with densitometric measurements also reliable quantitative results (Lin *et al.*, 1998; Krska *et al.*, 2008). These methods were mostly developed before the great development and affordability of HPLC and LC-MS instruments and some were established as the official AOAC methods. Their many advantages include suitability for crude extract analysis, a wide choice of stationary and mobile phases, as well as an array of spraying agents used for the detection.

A TLC method with previous immunoaffinity clean-up and densitometric quantification has been developed for aflatoxins in food samples. Advantages of the method include fast sample processing and limits of quantification lower than established by regulatory organs. Over pressured-layer chromatography, two-dimensional TLC and high performance TLC (HPTLC) are other possibilities for efficient separation and determination of pesticides and aflatoxins. But TC method is now rarely used for screening or quantification of pesticides. While aflatoxins are naturally fluorescent compounds and their detection in TLC is not problematic, trichothecenes require post-development visualization with spraying reagents: *p*-anisaldehyde, 4-*p*-nitrobenzylpyridine, and aluminum chloride for deoxynivalenol (DON).

Recent methods include quantitative determination of DON and T-2 toxin in cereals by high performance TLC. Fumonisins have also been determined in corn using reverse-phase TLC plate, spraying with fluorescamine in borate buffer/acetonitrile mixture, and densitometric detection. In contrast to older methods, this application has had acceptable limits of detection for routine analysis. In another method, fumonisin B1 has been analyzed by TLC after clean-up on ion-exchange column. The spraying reagent was *p*-anisaldehyde, detection was by laser-scanning densitometry. Zearalenone emits fluorescent light upon excitation. Quantitative methods are based on TLC with densitometry or HPTLC. Moniliformin, a low molecular weight *Fusarium* mycotoxin, can also be determined by TLC and visualized by 3-methyl-2-benzothiazolinone hydrazone (MBTH) or 2, 4-dinitrophenylhydrazine with low limits of detection.

TLC determination of ochratoxin A was compared to different HPLC methods. Recoveries for the TLC method were lower than average of HPLC methods. There hasn't been much development in recent years for the TLC methods for patulin. Generally, the analyte has to be visualized by MBTH or ammonia fumes. Separation from the commonly present contaminant in apple juice, 5 hydroxymethylfurfural, was often problematic. The situation with citrinin is similar, as the TLC methods used before the expansion of HPLC-based protocols were generally not sensitive enough. Citrinin was visualized by exposure to aluminum chloride, pyridine or acetic anhydride. Silica plates had to be impregnated with acid to achieve symmetrical spots. Other chromatographic techniques besides HPLC, GC and TLC are a rarity in mycotoxin analysis.

6.3.2. *Capillary electrophoretic (CE) methods*

Even great separation power and versatility of capillary electrophoretic (CE) techniques, they have never gained such popularity as HPLC might be in better detection limits and greater user-friendliness of HPLC methods. Achieving low enough detection limits might present a serious problem in CE, therefore the mycotoxins for which CE methods have been developed are mostly those for which fluorescence detection (FL). Ochratoxin A was determined by CE laser induced fluorescence (LIF) in human serum, food samples after immunoaffinity clean-up, and in wine after combined extraction with SPE and supported liquid membrane. Capillary zone electrophoresis (CZE) coupled to LIF was applied to OTA analysis in house dust and human serum. Cyclodextrins were applied to enhance native zearalenone fluorescence in CE-LIF analysis, and for the same purpose in CE-LIF determination of T-2 toxin, which was previously derivatized with pyrene-1-carbonyl cyanide.

Another detector commonly used in CE is UV-spectrometric. Methods have been developed for determination of patulin in apple juice, employing micellar electrokinetic capillary chromatography (MEKC) (Murillo *et al.*, 2008) or microemulsion electrokinetic chromatography (MEEKC) (Murillo-Arbizu *et al.*, 2008) and UV detection at 276 nm. Limits of detection were in the μg/L range, which is satisfactory for routine analysis. In contrast to the CE-FL methods for ochratoxin A mentioned above, UV detection of OTA after CE separation has given quantification limits that were too high for the analysis of real samples.

6.3.3. *Gas chromatography*

Gas Chromatography (GC) has been used routinely for determination of pesticide residues in foods. For this purified extract dissolve in volatile solvent is added to head of a glass column containing an adsorbing material, which is maintained at elevated temperature inside an oven. Components of sample mixture are volatilized at the normal operating temperatures of the instrument, and they partitioned in between a stationary phase and a mobile phase (a carrier gas). Since components of the sample can only advanced through the column when dissolve in the mobile phase, separations are achieved because different components have different relative affinities for the two phases. Typically for routine multi-residue analysis, the column is maintained under isothermal conditions; however, a temperature programme that increases the column temperature with time can also be used.

The detection of compounds is accomplished with the use of sensitive and highly selective detectors. In general, flame photometric, electron capture, electrolytic conductivity and thermoionic detectors are widely used for detection of pesticide residues. A flame photometric detector (FPD) is operated with hydrogen-rich flame under conditions that favour the formation of S_2 or HPO decomposition products when S- and P- containing pesticides are combusted. By adjusting optical filter between the flame and photomultiplier tube, operating conditions can be chosen so as to detect either of these species with a high degree of selectivity. The response of this detector to sulfur-containing analytes is not linear, hence calibration requires several standard concentrations to accurately define calibration curve.

Electron capture detection is characterized by only moderate selectivity but can be made to be very sensitive. Relative to other detectors, sample extracts must be purified more extensively when using this detector. When run in constant current mode, the detector becomes saturated, easily resulting in a narrow linear range. The linear range is increased appreciably by operating the detector in a pulse mode. Halogen and aromatic solvent must be avoided, and when possible, alkenes are used. However, in certain applications, acetone must be used to avoid loss of certain Ops that can become adsorbed on the linear surfaces of the syringe. In electrolytic conductivity detector analyte is converted to an ionic species, interfering ions are scrubbed from the effluent, and the ionic analyte-transformation product is detected within an electrolytic conductivity cell. When operated in the nitrogen mode, analyte is mixed with H_2 gas and hydrogenated over a nickel catalyst at 850°C in a quartz tube. Acidic hydrogenation products are removed from the effluents by passage over a Sr $(OH)_2$ trap, and the NH_3 from the analyte passes to conductivity cell. Detection limits can be as little as 0.1 ng.

The thermionic emission detector is a modified flame ionization detector in which a nonvolatile ceramic bead is used to suppress the ionization of hydrocarbons as they pass through the low temperature fuel-poor hydrogen plasma. The bead is heated to 600–800°C and controlled independently of the plasma temperature. The detector responds to both nitrogen and phosphorous at low He flow rates to the plasma. A highly selective response to phosphorus containing analytes is achieved with higher flow rates of H_2 and inverting the polarity between the plasma tip and the collector.

6.3.4. *Gas chromatography-mass spectrometry*

GC-Mass spectrometry has been used for the analysis of many pesticide residues such as organochlorines, triazine herbicides, pyrethroids and even *N*-methylcarbamates. MS offers enormous possibilities as a high selective means of detection and quantitation of pesticides residues. This technique offers unparallel performance in terms of selectivity, in confirmatory power, and universality of analytes that can be detected.

Dedicated systems for chromatography (mass selective and ion trap types) continue to decrease in cost and to improve in performances. These devices provide maximum response when only a few masses are monitored selectively (selected ion monitoring SIM, as opposed to recording complete mass spectra). For multi-residue monitoring, there is a requirement that the mass detector be capable of rapidly switching between preselected masses as the chromatogram develops. The GC is equipped with a HP-5MS or equivalent capillary column 30 meter long, 0.25 mm i.d. × 0.25 mm film thickness. The injector temperature is maintained above 230°C. Oven temperature is monitored initially at low temperature and then gradually increased up to 300°C. The detector is kept at selective ion monitoring mode with EM voltage: Abs 3000 volt (max). The limit of quantitation is in the range of 0.017 to 0.3 ppb.

Modern gas chromatography combines superior separation on the capillary columns with a variety of general and specific detectors. The obvious disadvantage when compared to LC is the fact that only thermally stable and volatile analytes can be analyzed, although this problem can partially be solved by derivatization. However, as mycotoxins are generally speaking rather large and polar molecules, they are better amenable to LC and GC methods are not common.

6.3.5. *High performance liquid chromatography*

The application of high performance liquid chromatography to pesticide multiresidue analysis has largely been restricted to those analytes that do not posses either the volatility or the thermal stability required for GC. These include *N*-methylcarbamates and some synthetic pyrethroid pesticides that are decomposed at normal GC operating temperatures while in HPLC samples are separated, normally at ambient temperatures, by partitioning between stationary phase (column) and a liquid mobile phase. The samples are generally injected into a loop maintaining a constant temperature, and then trigger for transformation to the head of the analytical column, typically a 4 mm (i.d.) × 15–25 cm stainless steel tube packed with different particle size of silica. Smaller the particle size of the column, faster will be the mobile phase flow rate without affecting the resolving power of the column.

The optimal chromatography could be achieved by changing mobile phase flow rate and introducing a reliable quadruple pumping system to deliver solvent in isocratic or gradient elution profile against back pressure of 1200 to 3000 psi. The column is considered as heart of the HPLC and that are coming in the market with different packing materials for variety of separation mode. Reversed phase (RP) HPLC is most popular followed by normal phase. In RP-HPLC modified silica is employed as stationary phase in which the accessible surface silanol functional groups have been substituted with C_{18}, C_8 or even C_2 normal alkyl chains. The treated packing material preferentially absorbs nonpolar analytes. A very polar mobile phase, typically a mixture of water with either methanol or acetonitrile forces the solutes to interact with nonpolar stationary phase. Hence, from a sample mixture a most polar compound elute first, and the order of elution is reversed relative to the order observed for untreated silica. The pesticide residue of concern is detected using either photo-diode-array (PDA), UV or Fluorescence detector.

The UV detector operated at fixed wavelength which have advantages of sensitivity and low cost while PDA detector operated at variable wavelengths between 200–700 nm, and are less sensitive because a small fraction of the radiation from their polychromatic light sources is actually used. Fluorescence detectors exploit the fact that the fluorescence emitted by an analyte is at longer wavelength of light required to induce this phenomenon. Typically a filter monochromator is used to select the wavelengths of radiation incident on the sample, and a second and separate filter or monochromator is used to select the wavelengths to monitor for the light that is re-emitted from the analyte. Fluorophores can readily be added to reactive functional groups of those analytes that are not naturally fluorescent. This process of derivatization, which increases the number of analytes and/or the response to the analytes, can be performed either pre-or post column.

The detector is responsible for quantification of sample as the analog signal from this device must be related to the quantity of analyte via a separate process of calibration. A recording device provides a record of the changes in detector signal with time. The time required to reach the maximum of a particular peak is referred to as its retention time, and the peak height or peak area is used to determine the quantity of analyte that must have caused that displacement. Components of the mixture are identified, tentatively, based solely on their retention times as the retention time is identical to the retention time of an authentic standard that has been chromatographed under identical conditions. The retention time for an analyte is expressed relative to the retention time for an internal standard that has been added intentionally to the sample.

However, external standard method is most prevalently followed for quantification, in which a series of standard solutions is prepared by dissolving known quantities of authentic standard pesticide in suitable solvent. The standard solutions are separately chromatographed to develop the response of detector for the quantity of analyte injected into the instrument. A relationship between response and quantity of analyte is then established by regression analysis.

Fluorescence detection is by its nature highly specific and sensitive. Several well-established, reliable, robust and sensitive LC methods with fluorescence detection exist especially for the determination of mycotoxins with natural fluorescence, *e.g.*, aflatoxins, ochratoxin-A, citrinin (Zollner & Mayer-Helm, 2006) and many of them have been established as the official AOAC methods. For the others, pre- or postcolumn derivatization with suitable reagents has been widely applied, but has been recently superseded by relative simplicity, but comparative specificity of LC-MS methods after the lower cost has made these instruments available to most laboratories. In spite of its specificity for the fluorescing compounds, they have to be well separated on the chromatographic column to enable a reliable quantification. Usually, a reverse-phase stationary phase is used. Base deactivation additionally improves peak shape for polar mycotoxins with carboxylic groups: citrinin, OTA, some fumonisins. Ion-pair HPLC has also been applied for the polar toxins, using tetrabutylammonium hydroxide. Mobile phase should be composed of acidic aqueous phase (acetic acid, trifluoroacetic acid, acidic buffer) to prevent ionization of carboxylic groups, and gradient with methanol or acetonitrile is preferred (Wilkes & Sutherland, 1998).

6.3.6. *LC-MS-MS methods*

Liquid chromatography-mass spectrometry (LC/MS) quadrupole instruments have become the most widely used analytical tools for the analysis of pesticides. However, due to some limitations, the technical progress in LC-MS is rapid on the mass front, especially with the development of tandem MS, which improves traditional MS methods with two stages of mass analysis-one to pre-select an ion and the second to analyze fragments induced. More recently, quadrupole ion trap (QIT) and hybrid quadrupole/time of flight (TOF) instrument have joined the previous ones to increase the possibilities of identification.

An idealized quadrupole mass spectrometer consists of four parallel hyperbolic rods. Each rod has an applied direct current (DC) voltage and a superimposed radiofrequency (rf) potential. One pair of diagonally opposite rods has positive DC value, and the other has a negative one. The rf phase is opposite for the positive and negative electrodes. The ions are sufficiently accelerated to pass through the rods, but not so much that they fly through the rods without any mass separation. Once they reach the applied field, the ion oscillate according to their mass size, the ion remain in the quadrupole for hundreds of microseconds.

The QIT analyzer operates just like any other quadrupole instrument with the exception that it has only three electrodes: one ring electrode between two hyperbolic endcap electrodes, which form a three dimensional trap. The oscillating potential difference established between the ring and endcap electrodes form a substantially quadrupolar field. The ions pass in and out of the traps by holes in the endcaps.

The TOF analyzers are the simplest and the fastest mass spectrometers. All of the sample ions are accelerated with the same amount of energy, and travel toward the detector over the same length of an evacuated flight tube (typically 1–2 meters).

LC-MS methods are relatively uncommon for determination of pesticides and mycotoxins; however, they are often used for determination of thermolabile pesticide compounds. LC-MS methods are avoiding for determination of mycotoxins in particular aflatoxins because of well-established, reliable and sensitive LC methods with fluorescence detectors measuring the natural fluorescence of aflatoxins. Most newly developed LC-MS methods enable analysis of a whole group of similar mycotoxins. The reasons are the development of efficient electrospray (ESI) and atmospheric pressure chemical ionization (APCI) interfaces for LC-MS coupling, the developments in the field of mass analyzers, and the considerably greater simplicity of use and affordability of tandem mass spectrometers. Probably the greatest merit goes to the ESI and APCI. Before their development, the LC-MS analyses of mycotoxins were performed using thermospray and fast-atom bombardment interfaces, but with significant difficulties.

Modern LC-MS instruments with ESI or APCI enable ionization in both positive and negative mode, as well as switching between them in the same chromatographic run, which means the best possible detection conditions for all analytes. For A-trichothecenes, a method using LC-APCI (+)-MS and deuterated T-2 toxin as an internal standard (IS) was developed (Razzazi-Fazeli *et al.*, 2002). Electrospray with MS/MS was applied to the semi-quantitative analysis of non-macrocyclic and macrocyclic trichothecenes in samples from indoor environments. Six B-trichothecenes and metabolites were determined by LC-APCI (–)-MS with dexamethasone as an internal standard. Trichothecenes from groups A and B were analyzed using APCI and tandem MS without IS (Berthiller *et al.*, 2005) or by ESI and MS/MS with deepoxydeoxynivalenol as IS (Klotzel *et al.*, 2005) Several LC-MS methods have been published for the determination of estrogenic *Fusarium* mycotoxins-RALs with a single mass analyzer or tandem MS (Maragou *et al.*, 2008; Hartmann *et al.*, 2007). The ionization was possible both by APCI and ESI.

7. RESIDUES AND HEALTH RISK

7.1. Pesticides

Pesticides were first agricultural chemicals to come to the attention of an environmentally conscious public in early 1960s. Most pesticides are acutely toxic to human and animals, even ingestion of low levels over a long time can have adverse effects. However, in recent year, violative levels pesticides are relatively uncommon. But despite this, pesticide remains an important public health consideration because of their widespread use in meat and poultry production, their persistence in the environment, and their varying toxicity. Pesticides via food are the source of several mysterious diseases unheard in the past; the very horror of these would drag the patient fast to the graveyard. The catalogue of such diseases and disorders include cancer, epilepsy, liver and kidney dysfunction, somatic cell growth, depression and neuritis. The United Nations has estimated that about 2 million poisoning and 10,000 deaths occur each year from pesticides, with about three-fourths of this occurring in developing countries (Quiyano *et al.*, 1993). The acute or malicious consumption involving higher dose result in death, whereas, chronic insidious intake lead to elevated cancer risk and disruption of body's reproductive, immune, endocrine and nervous system (van Wendel de Joode *et al.*, 2001; Horrigan *et al.*, 2002). The association between various classes of pesticide relating to various forms of cancer is shown in Table 2.

The carcinogenicity of organochlorine (OC) pesticides has been studied in a number of laboratory animals including non-human primates. Lifetime or limited period of time treatment of mice with DDT induced liver tumors including malignant metastasizing hepatoblastomas. DDT also increased incidences of lung tumors, and lymphoma in mice, liver tumors in rats and adrenal adenomas in hamster. However DDT did not induce DNA damage in bacteria or cultured rodent and human cells. It induced chromosomal aberration in mouse but not in rat bone marrow cells. DDT and its metabolites inhibited gap-junctional intracellular communication.

Other areas of concerns are childhood brain cancer and cancers of nervous system. Studies suggest an increase in risk in brain cancer, leukemia, Wilm's tumors, Ewing's sarcoma and germ cell tumors associated with paternal occupational exposure to pesticides prior to and during pregnancy. However, maternal occupational exposure during pregnancy was less frequent but was also associated leukemia, Wilm's tumors and germ cell tumors (Daniels *et al.*, 1997). Pesticides can suppress immune system, as epidemiological evidence shows an association between pesticide exposure and increase incidence of human disease, particularly those disease to which immuno-compromised individuals are especially prone (Repetto and Baliga, 1996). The endocrine disruptor includes atrazine and alachlor. However, there is strong need of sufficient data for complete health hazard evaluation data in this context.

Many organophosphates are potent nerve agents, functioning by inhibiting the action of acetylcholinesterase (AChE) in nerve cells. They are one of the most common causes of poisoning worldwide, and are frequently intentionally used in suicides in agricultural areas. Organophosphorus pesticides can be absorbed by all routes, including inhalation, ingestion, and dermal absorption. Their toxicity is not limited to the acute phase, however, and chronic effects have long been noted. Neurotransmitters such as acetylcholine are profoundly important in the brain's development, and many OPs have neurotoxic effects on developing organisms, even from low levels of exposure. Other organophosphates are not toxic, yet their main metabolites, such as their oxons are.

Repeated or prolonged exposure to organophosphates may result in the same effects as acute exposure including the delayed symptoms. Other effects reported in workers repeatedly exposed include impaired memory and concentration, disorientation, severe depressions, irritability, confusion, headache, speech difficulties, delayed reaction times, nightmares, sleepwalking and drowsiness or insomnia. An influenza-like condition with headache, nausea, weakness, loss of appetite, and malaise has also been reported.

Organophosphate and carbamate insecticides inhibit the activity of cholinesterase, resulting in a buildup of acetylcholine in the body. An increase in acetylcholine results in the uncontrolled flow of nerve transmissions between nerve cells. The nervous system becomes "poisoned"; the accumulation of acetylcholine causes the continual transmission of impulses across the synapses. The effects of organophosphate or carbamate poisoning can result in both systemic and topical symptoms. Direct exposure of the eye, for example, can cause topical symptoms such as constriction of the pupils, blurry vision, an eyebrow headache, and severe irritation and reddening of the eyes. Symptoms and signs of systemic poisonings are almost entirely due to the accumulation of acetylcholine at the nerve endings. The onset of symptoms depends on the route of entry and the severity of the poisoning. Gastric symptoms such as stomach cramps, nausea, vomiting, and diarrhea appear early if the material has been ingested. Similarly, salivation, headache, dizziness, and excessive secretions that cause breathing difficulties are initial symptoms if the material has been inhaled. Involvement of the respiratory muscles can result in respiratory failure. Stomach, intestinal, and respiratory symptoms usually appear at the same time if the pesticide is absorbed through the skin. In children, the first symptom of poisoning may be a convulsion.

In advanced poisonings, the victim is pale, sweating, and frothing at the mouth. The pupils are constricted and unresponsive to light. Other symptoms include changes in heart rate, muscle weakness, mental confusion, convulsions, and/or coma. The victim may die if not treated.

Pesticides like synthetic pyrethroids and *N*-methycarbamates received importance in recent years because of their acute mammalian toxicity (Ozhan *et al.*, 2003; EMEA, 2004). The acute toxicity of cypermethrin and deltamethrin are very variable and depends on factors such as age, sex and strain of the animal, nature of the solvent vehicle and the environmental conditions. The toxic signs of these chemicals characterized by salivation, increase startle response, ataxia, splayed gait, tremors and convulsions. Myelin and axon degeneration observed at lethal dose levels. Though cypermethrin affect central nervous system, it is not mutagenic in *in vitro* assay for gene mutation in bacteria and in V79 Chinese hamster cells, in a cytogenetic assay and in a sister chromatid exchange (SCE) assay in human lymphocytes. But these compounds shown to have a potential skin sensitizing and immune modulatory effects besides transient paraesthesia on the face other exposed area of the body.

Carbaryl is an inhibitor of acetylcholinesterase and can produce adverse effects in humans through skin contact, inhalation, or ingestion. Several studies indicate that carbaryl can react with nitrite under certain conditions to give rise to N-nitrosocarbaryl (Mirvish, 1975), which has been shown to be carcinogenic and mutagenic in laboratory test systems. Carbaryl degrades readily to 1-naphthol, which is several times more toxic than its precursor (EHS, 2003).

7.2. Mycotoxins

It is difficult to assess the overall extent of the problem since only a few of the known mycotoxins have definitely been implicated in outbreak of mycotoxicosis. These toxins are pathologically classified as hepatotoxin, nephrotoxins vomitoxin, and neuro-musculotoxins, some of which are potentially carcinogenic and mutagenic. Aflatoxin is the most potent hepatotoxin and mutagen among mycotoxins like Ochratoxin-A, patulin, zearalenone, trichothecene, citrinine, penicillic acid, T-2 toxin, sterigmatocystin, PR toxin, Kozic acid, deoxynivalenol, diacetoxyscirphenol, sporidesmin, ergot alkaloids, fuminosin, alternariol and tenuazonic acid, there is only circumstancial evidence linking a few of them with a particular mycotoxicoses, the remainder being only produced in laboratory culture. The incidence and levels of contamination reported for aflatoxin, Ochratoxin-A and patulin were of some concern, whereas few findings were reported for trichothecenes, zearalenone, citrinine and penicillic acid. The problem of mycotoxin contamination of corn outweighed that of any other commodity. The temperature ranges under which certain fungi grow and produce toxin vary from *A. flavus* producing aflatoxin at 32°C to 38°C. *Fusarium tricinctum* producing trichothecene toxins at 8°C or lower with *Penicillium* sp. in a wide range of 4°C to 35°C.

Mycotoxins exert their effects through three primary mechanisms: (1) A reduction in amount of nutrients available for use by the animal. This occurs in a multifactorial process. First, an alteration in nutrient content of feed may occur during the molding process. Mold growth can reduce the content of nutrients such as vitamins and amino acids such as lysine in feedstuffs. The energy value of feeds is usually reduced by mold growth. Second, some mycotoxins reduce feed intake which lowers nutrient intake: Third, a mycotoxin-produced irritation to the digestive tract can reduce nutrient absorption, and forth, certain mycotoxins interfere with normal nutrient metabolism such as the inhibition of protein synthesis by T-2 toxin, (2) Effects on the endocrine and exocrine systems. An example is the effect of zearalenone on reproductive performance due to its estrogenic effects. Zearalenone's estrogenic effect results from the affinity of zearalenone and its derivatives to bind with the animal's estrogen receptors, and (3) Suppression of the immune system. The effects of mycotoxins on immunity have been reviewed. Trichothecenes such as DON and T-2 toxins reduce immunity by inhibiting protein synthesis and thus cell proliferation. Some mycotoxins are cytotoxic to lymphocytes *in vitro*. Corticosteriods produced in response to stress also reduces immune function.

8. REGULATIONS

8.1. Pesticides

In 1979, a US General Accounting Office (GAO) report identified 143 pesticides and drugs as likely to leave residues in raw meat and poultry. Of these, 42 are known to cause or are suspected to causing cancer, 20 are suspected teratogens, and 6 are suspected mutagens (Shull & Cheeke 1983). In the ensuing 32 years these number have probably risen

The regulations of pesticide used by each country throughout the world have a tendency towards uniform approach. In the United States, pesticide use is regulated under FIFRA (US, Federal Insecticide, Fungicide and Rodenticide Act) on the basis of risk-benefit standard (Driver *et al.*, 1996). This balancing takes into account "the economic, social and environmental cost as well as potential benefits of the use of any pesticide in relation to its efficacy, inherent toxicity to mammals, wild-life and plants". A given pesticide may have many different uses, but required individual approval by the U.S. Environmental Protection Agency (EPA) under FIFRA. In fact, in the United States, pesticide regulation is under auspices of three government agencies, the Environmental Protection Agency (EPA), the Food and Drug Administration and the United States Department of Agriculture-Food Safety Inspection Services (USDA-FSIS). The FDA and USDA-FSIS are responsible for monitoring pesticide residues in foods based on level set by the EPA. The FSIS is responsible for meat and poultry products and FDA covers all other types of raw commodities and processed food products.

However, regulation of pesticides in European Union is done on the basis of their toxicological properties. The council directive of the European Economic Community 91/414/EEC specified criteria for plant protection products in relation to acceptable daily intake (ADI) for man, acceptable operator exposure level (AOEL), acute reference dose, and maximum admissible concentration in water (Anadon *et al.*, 2001). The international organizations such as Codex Alimentarius Commission (CAC) have also set their standards using the term maximum residue limits (MRLs) and acceptable daily intake (ADI) levels of various pesticide residues in meat and meat products.

In India, regulations of contaminants are governed under different Acts/ Rules, that is, Insecticides Acts, 1968; Environmental (Protection) Act, 1986; Hazardous (Management and Handling) Rules, 1989; Water (Prevention and Control of Pollution) Act, 1974; Air (Prevention and Control of Pollution) Act,

1981; Prevention of Food Adulteration Acts, 1954; The Factories acts, 1948; Bureau of Indian Standards Act, 1986 and Drug and Cosmetic Act, 1940. The Ministry of Agriculture (Govt. of India) regulates the manufacture, sale, import, export and use of pesticides through the Insecticides Act. Central Insecticides Boards (CIB) constituted under the section 4 of the Act advises central and state governments on technical matters. The registration committee (RC) constituted under section 5 of the Act approves the use of pesticides and new formulations to control the pest problem associated with various crops or ectoparasiticides measure onto the animals. The monitoring of residue level in food comes under the purview of Ministry of Health and Family Welfare (Govt. of India). The Central Committee on Food Standards (CCFS), a statutory committee under the PFA Act, 1954. Ministry of Health and Family Welfare, is responsible for laying down standards for various food items (Dikshit & Parmar, 2005).

8.2. Mycotoxins

For regulation of mycotoxins currently the Joint Expert Committee on Food Additives (JECFA), a scientific advisory body of FAO and WHO, has recently evaluated their hazards. The evaluation of toxicological data carried out by JECFA normally results in the estimation of a Provisional Tolerable Weekly Intake (PTWI) or a Provisional Tolerable Daily Intake (PTDI). This hazard assessment approach does not apply for toxins where carcinogenicity is the basis for concern as is, for example, the case with the aflatoxins. Assuming that a no-effect concentration limit cannot be established for genotoxic compounds, any small dose will have a proportionally small probability of inducing an effect. Imposing the absence of any amount of genotoxic mycotoxins would then be appropriate, if these toxins were not natural contaminants that can never completely be eliminated without outlawing the contaminated food or feed. In these cases, JECFA does not allocate a PTWI or PTDI. The number of countries regulating aflatoxins has significantly increased over the years. However, some countries have many regulations specifying different tolerated levels for individual foods and feeds, while others have set only one tolerated level for instance for “all foods” or for “all feeds”. Therefore simplifications were made.

9. HARMONIZATION OF REGULATION

9.1. Pesticides

International Organization such as Codex Alimentarius Commission have taken initiation of harmonization of chemical residues in food through establishment of statutory limitations *viz.*, maximum residue limits (MRLs), acceptable daily intake (ADI) levels, acute reference dose (ARfD), No Observed Adverse Effect Levels (NOAEL) etc. However, all these statutory limitations are important principals for risk assessment of residues in foods including meat. For statutory limits of residues, the terms tolerances or maximum residue limits (MRLs) are frequently used by regulatory agencies. But both the term tolerances and MRLs are synonymous; former is used in United States and other countries, while later is used in Canada and the European Union. The term MRL may be defined as the maximum concentration of marker residue (*e.g.*, parent compound, metabolites etc.), expressed in parts per million (ppm) or parts per billion (ppb) on fresh weight basis, that is legally permitted or recognized as acceptable in or on food. The MRLs are established on the basis of package of toxicology and residue data and these are referred to as safety file and the residue file.

The toxicology data are used not only to characterize the biological properties of the molecules, but also to identify a suitable 'no observed effect levels' (NOELs), which in turn is used to calculate an acceptable daily intake (ADI), the quantity, which if consumed over a human lifetime will have no adverse effect on consumer health. This in turns expressed on a body weight basis and can be considered the safety standard for that compound (Brynes *et al.*, 1996). The MRL is then elaborated from this ADI along with knowledge of the depletion kinetics of the contaminants in the animal or its milk and with reference to standard intake values for particular types of food. The identification of NOELs, the calculation of ADI values and the establishment of MRLs is a complex scientific process involving toxicology, pharmacology, microbiology, residue kinetics and analytical chemistry but the MRL is established at a magnitude which ensures that the ADI will not be exceeded by the consumer when eating food of animal origin (Hartzell, 1996). Another term total residue levels (TRL) refers to the safe concentration of total residues that corresponds to the MRL.

The term acceptable operator exposure level (AOEL) and acute reference dose (ARfD) also practiced most commonly by European Union (Anadon *et al.*, 2001). The AOEL is defined as the maximum amount of active substance to which the operator may be exposed without any adverse health effect and is based on the highest level at which no adverse effect is observed in tests in the most sensitive relevance animal species. The AOEL is expressed as mg of chemical/kg body weight of the operator. On the other hand, the ARfD of a chemical is an estimate of the amount of substance in food or drinking water, expressed on the body weight basis, that can be ingested over a short period of time, usually during 1 meal or 1 day, without appreciable health risk to the consumer, based on all the known fact at the time of evaluation, this calculated reference dose is compared to an estimate or measurement of exposure in the risk assessment process.

9.2. Mycotoxins

European Union has harmonized EU regulations for aflatoxin B_1 in various feeds since 1976 including official protocols for sampling and analysis. In 1998, the first EU-harmonized regulations for mycotoxins in food came into force and has gradually expanded to various mycotoxins in different foodstuffs. In 2004 and subsequent years, a significant further expansion of EU-harmonized mycotoxin regulations may be expected for foods and feeds. For foods, this concerns patulin, aflatoxin B_1, aflatoxin M_1, ochratoxin A and DON in infant formulae and follow-up formulae; ochratoxin A in coffee, wine, beer, spices, grape juice, cocoa and cocoa products; several *Fusarium*-produced mycotoxins, *i.e.*, trichothecenes (T-2 and HT-2 toxins, in addition to DON), fumonisins and zearalenone in cereal-based foodstuffs. In addition, new limits will probably be established for feeds in the years to come for several mycotoxins including ergot alkaloids, DON, zearalenone and ochratoxin A, and relevant scientific opinions of the EFSA panel on contaminants in the food chain are currently in preparation. In the mycotoxin area, CCFAC established standards for total aflatoxins in unprocessed peanuts, aflatoxin M_1 in milk and patulin in apple juice in 2003.

A draft standard has been developed for ochratoxin A in wheat, barley, rice and derived products, and proposed standards for DON in cereals are currently under discussion. The CCFAC has, apart from its goal to develop standards (Maximum Limits) where necessary, also decided to devote much attention to developing Codes of Practice in which principles and advice about practical measures to control mycotoxins during cultivation, storage and processing are assembled. Examples of these include the codes of practice developed for: (*i*) the reduction of aflatoxin B_1 in raw materials and supplemental feedstuffs for milk producing animals; (*ii*) prevention and reduction of patulin contamination in apple juice and apple juice ingredients in other beverages, and (*iii*) the prevention and reduction of mycotoxin contamination in cereals including annexes on ochratoxin A, zearalenone, fumonisins and trichothecenes. Association of Southeast Asian Nations (ASEAN) comprise Brunei Darussalam, Cambodia, Indonesia, the Lao People's Democratic Republic, Malaysia, Myanmar, the Philippines, Singapore, Thailand and Viet Nam. Most of these countries have specific regulations for mycotoxins. Whereas, harmonized regulations are obviously not (yet) established by ASEAN, an ASEAN Task Force on Codex Alimentarius has taken a common position to support the 0.5 mg/kg level for aflatoxin M_1 in milk.

REFERENCES

Anadon, A., Martinez-Larrangaga, M.R. and Matrinez, M.A. (2001). Scientific basis and regulatory aspect for the toxicology of plant protection products in the European Union. *Vet. Human Toxicol.*, 43: 297–301.

Andres, R.T. and Narayanaswamy, R. (1997). Fibre-optic pesticide biosensor based on covalently immobilized acetylcholinesterase and thymol blue. *Talanta*, 44: 1335–1352.

Argauer, R.J., Eller, K.J., Pfeil Richard, M. and Brown Richard, T. (1997). Determining ten synthetic pyrethroids in lettuce and ground meat by using ion-trap mass spectrometry and electron capture gas chromatography. *J. Agric. Food Chem.*, 45: 180–184.

Berthiller, F., Schuhmacher, R., Buttinger, G. and Krska, R. (2005). Rapid simultaneous determination of major type A- and B-trichothecenes as well as zearalenone in maize by high performance liquid chromatography-tandem mass spectrometry. *J. Chromatogr A.*, 1062: 209–216.

Biswas, A.K., Kondaiah, N., Anjaneyulu, A.S.R. and Rao, G.S. (2009). Simple simultaneous determination of cypermethrin and deltamethrin residues in buffalo meat using HPLC equipped with photodiode-array detector. *Fleischwirt Int.*, 24: 82–86.

Biswas, A.K., Kondaiah, N., Anjaneyulu, A.S.R., Rao, G.S. and Singh, R.P. (2010). Optimization of liquid chromatographic methods for the monitoring of carbaryl Residues in buffalo meat. *Analytical Methods*, 2: 393–396.

Bogialli, S., Curini, R., Carcia, A.D., Nazzari, M. and Tamburro, D. (2004). A simple and rapid assay for analyzing residues of carbamate insecticides in vegetable and fruits: Hot water extraction followed by liquid chromatography-mass spectrometry. *J. Agric Food Chem.*, 52: 665–671.

Brynes, S.D., Sundlof, S.F., Vilim, A., Lambert, G., Yong, M.S. and Fitzpatrick, S.Z. (1996). International harmonization of maximum residue levels for veterinary drugs via dietary intake estimates. *Proceedings of the Euroresidue III Conference*, Veldhoven, The Netherlands, 6–8 May.

Buck, B.W. (1973). Colour identification of granular insecticide for mutation. *In*: Diechmann, W.B. (*Ed.*). Pesticide and environment a continuing controversy. Intercontinental book export, New York.pp565-568.

Bushway, R.J. (1988). High-performance liquid chromatographic determination of carbaryl in fruit juices. *J. Chromatogr.*, 457: 437–444.

Buzby, J.C. and Roberts, T. (1997). Economic cost and trade impact of microbial food borne illness. *World Health Statistics Quarterly*, 50: 57–66.

Choi, J.W., Kim, Y.K., Lee, I.H., Min, J. and Lee, W.H. (2001). Determination of binary pesticide mixtures by an acetylcholinesterase–choline oxidase biosensor. *Biosens. Bioelectron.*, 16: 937–943.

Daniels, J.L., Olshan, A.F. and Savitz, D.A. (1997). Pesticide and childhood cancers. *Environmental Health Perspectives*, 105: 1068–1077.

Dikshit, A.K. and Parmar, B.S. (2005). Pesticide residue and food safety. *In*: National Seminar on Food Quality and Safety Standards of agricultural Raw Produce.apri26-27.New Delhi, pp. 74–77.

Driver, J.H., Wilkinson, C.F. and Johason, E.L. (1996). Pesticide and human health. *In*: Kolluru, R.V., Bartell, S.M. and Stricoff, R.S. (*Eds.*). First edn. Risk assessment and management and book for environmental health and safety professionals. McGraw-Hill, Inc., New York, pp. 7.1–7.35.

EHS, (2003). 1-Naphthol, Material Safety Data Sheet. Environmental Health and Safety Available at: www.jtbaker.com/msds/englishhtml/N0840.htm/

EMEA. (2004). Committee for medicinal products for veterinary use. European Medicines Agency *Available at:* http://www.emea.in.int/

Guo, Y.R., Liu, S.Y., Gui, W.J. and Zhu, G.N. (2009). Gold immunochromatographic assay for simultaneous detection of carbofuran and triazophos in water samples. *Anal Biochem.*, 389: 32–39.

Hartmann, N., Erbs, M., Wettstein, F.E., Schwarzenbach, R.P. and Bucheli, T.D. (2007). Quantification of estrogenic mycotoxins at the ng/L level in aqueous environmental samples using deuterated internal standards. *J. Chromatogr A.*, 1138: 132–140.

Hartzell, G.E. (1996). Overview of combustion toxicology. *Toxicol.*, 31: 7–23.

Heise, S. (1992). Pesticide residues in food as a potential health hazards in the third world. *In*: *Proceeding of 3rd Congress on food borne infection and intoxication held at Berlin*, June16, 1992, Vol.II. pp. 706–707.

Horrigan, L., Robert, S.L. and Walker, P. (2002). How sustainable agriculture can address the environmental and human health harms of industrial agriculture. *Environ Health Pers.*, 110(5): 445–456.

IARC. (1991). DDT and associated compounds. *International agency for Research on Cancer Monography Evaluation of Carcinogen Risk Humans*, 53: 179–249.

ICMR. (1996). Surveillance, prevention and control of food contaminants. *Proceedings of National Symposium*, ICMR, New Delhi.

JECFA. (2002). Future trends in veterinary public health. *Fifty-eight report of the Joint FAO/WHO Expert Committee on Food Additives*. Report of a WHO study group. WHO technical Report Series. No. 907.

International Programme on Chemical Safety. (2011a). Environmental health criteria for carbamate pesticides: A general introduction. Access on 09.12. 2011. Website address: http://www.inchem.org/documents/ehc/ehc/ehc64.htm

International Programme on Chemical Safety. (2011b). Data sheet on pesticides no. 58. Access on 09.12. 2011. Website address: http://www.inchem.org/documents/pds/pds/pest58_e.htm

Klötzel, M., Gutsche, B., Lauber, U. and Humpf, H.U. (2005). Determination of 12 type A and B trichothecenes in cereals by liquid chromatography-electrospray ionization tandem mass spectrometry. *J Agric Food Chem.*, 53: 8904–8910.

Krska, R., Schubert-Ullrich, P., Molinelli, A., Sulyok, M., Macdonald, S. and Crews, C. (2008). Mycotoxin analysis: An update. *Food Addit Contam.*, 25: 152–163.

Kumar, P., Colston, J.T., Chambers, J.P., Rael, E.D. and Valdes, J.J. (1994). Detection of botulinum toxin using an evanescent wave immunosensor. *Biosens Bioelectron.*, 9: 57–63.

Kumar, N., Sawant, S., Malik, R.K. and Patil, G.R. (2005). Development of analytical process for detection of antibiotic residues in milk using bacterial spores as biosensor (Patent Reg # IPR/4.9.1.4/05074/1479/del/2006).

Kumar, N., Singh, N., Singh, V.K., Bhand, S. and Malik, R.K. (2010). Development of spore inhibition based–enzyme substrate assay (SIB-ESA) for monitoring aflatoxin M1 in milk (Patent Regd. # IPR /4/14.4/10045).

Lai, W., Fung, D.Y.C., Xu, Y., Liu, R. and Xiong, Y. (2009). Development of a colloidal gold strip for rapid detection of ochratoxin A with mimotope peptide. *Food Control*, 20: 791–795.

Lin, L., Zhang, J., Wang, P., Wang, Y. and Chen, J. (1998). Thin-layer chromatography of mycotoxins and comparison with other chromatographic methods. *J. Chromatogr. A.*, 815: 3–20.

Liu, B.H., Tsao, Z.J., Wang, J.J. and Yu, F.Y. (2008). Development of a monoclonal antibody against ochratoxin A and its application in enzyme-linked immunosorbent assay and gold nanoparticle immunochromatographic strip. *Anal Chem.*, 80: 7029–7035.

Mallat, E., Barzen, C., Abuknesha, R., Gauglitz, G. and Barcelo, D. (2001). Fast determination of paraquat residues in water by an optical immunosensor and validation using capillary electrophoresis-ultraviolet detection. *Anal. Chim. Acta*, 427: 165–171.

Maragou, N.C., Rosenberg, E., Thomaidis, N.S. and Koupparis, M.A. (2008). Direct determination of the estrogenic compounds 8-prenylnaringenin, zearalenone, alpha- and beta-zearalenol in beer by liquid chromatography-mass spectrometry. *J. Chromatogr A.*, 1202: 47–57.

McEwen, F.L. and Stephenson, G.R. (1979). The use and significance of pesticides in the environment, Wiley, New York.

Metcalf, R.L. (1979). A century of DDT. *J. Agric. Food Chem.*, 21: 511–519.

Mirvish, S.S. (1975). Formation of N-nitroso compounds: Chemistry kinetics, and *in-vivo* occurrence. *Toxicol. Appl. Pharm.*, 31: 325–351.

Murillo, M., González-Peñas, E. and Amézqueta, S. (2008). Determination of patulin in commercial apple juice by micellar electrokinetic chromatography. *Food Chem. Toxicol.*, 46: 57–64.

Murillo-Arbizu, M., Gonzalez-Peñas, E., Hansen, S.H., Amezqueta, S. and Ostergaard, J. (2008). Development and validation of a microemulsion electrokinetic chromatography method for patulin quantification in commercial apple juice. *Food Chem. Toxicol.*, 46: 2251–2257.

Newmann, K. (1990). *Synthetic Pyrethroid Insecticides: Structure and Properties*. Vol 4. Springer-Verlag, Berlin, Germany.

Nicolaisen, M., Justesen, A.F., Thrane, U., Skouboe, P. and Holmstrom, K. (2005). An oligonucleotide microarray for the identification and differentiation of trichothecene producing and non-producing Fusarium species occurring on cereal grain. *J. Microbiol. Methods*, 62: 57–69.

Ozhan, G., Topur, S. and Alpertunga, B. (2003). A simple method for determination of carbaryl and 1-napthol in fruit juices by high performance liquid chromatography-diode-array detection. *J Food Prot.*, 66(8): 1510–1513.

Quiyano, R., Panganiban, L. and Cortex-Maraba, N. (1993). Time to blow whistle, danger of toxic chemicals. *World Health*, 46(5): 26–27.

Razzazi-Fazeli, E., Rabus, B., Cecon, B. and Böhm, J. (2002). Simultaneous quantification of Atrichothecene mycotoxins in grains using liquid chromatography atmospheric pressure chemical ionisation mass spectrometry. *J. Chromatogr A.*, 968: 129–142.

Repetto, R. and Baliga, S. (1996). Pesticides and immune system: The public health risk. *World Resource Institute*, Washington.

Rotman, B. (2001). Using living spores for real-time biosensing. *Gen. Eng. News*, 21: 65.

Rotman, B. and Cote, M.A. (2003). Application of a real-time biosensor to detect bacteria in platelet concentrates. *Biochem. Biophys. Res. Comm.*, 300: 197–200.

Shephard, G.S., van der Westhuizen, L., Gatyeni, P.M., Katerere, D.R. and Marasas, W.F.O. (2005). Do fumonisin mycotoxins occur in wheat? *J. Agric. Food Chem.*, 53: 9293–9296.

Shull, L.R. and Cheeke, P.R. (1983). Effect of synthetic and natural toxicants as livestock. *J. Ani. Sci.*, 57(2): 330–354.

Srivastava, A.K., Kotwal, S.K. and Chidanandaiah, (2005). Public health aspects of chemical residues in foods of livestock origin. In: 3rd Annual Conference and National Symposium on New Approaches in Food Safety & Quality Control with Special reference to Emerging food borne diseases and intoxications. February 9–10, Ludhiana, India. pp. 77–89.

Suri, C.R., Boro, R., Nangia, Y., Gandhi, S., Sharma, P., Wangoo, N., Rajesh, K. and Shekhawat, G.S. (2009). Immunoanalytical techniques for analyzing pesticides in the environment. *Trends Anal. Chem.*, 28: 29–39.

Tang, D., Sauceda, J.C., Lin, Z., Basova, S.O.E., Goryacheva, I., Biselli, S., Lin, J., Niessner, R. and Knopp, D. (2009). Magnetic nanogold microspheres-based lateral-flow immunodipstick for rapid detection of aflatoxin B_2 in food. *Biosens. Bioelectron.*, 25: 514–518.

van Wendelde Joode, B., Wesselling, C., Kromhout, H., Moge, P., Garcia, M. and Mergler, D. (2001). Chronic nervous system effects of long-term occupational exposure to DDT. *Lancet*, 357: 1014–1016.

Vignali, D.A. (2000). Multiplexed practile-based flow cytometric assays. *J. Immunol. Methods*, 243: 243–255.

Wang, S., Zhang, C., Wang, J. and Zhang, Y. (2005). Development of colloidal gold-based flow-through and lateral-flow immunoassays for the rapid detection of the insecticide carbaryl. *Anal. Chim. Acta*, 546: 161–166.

Wilkes, J.G. and Sutherland, J.B. (1998). Sample preparation and high-resolution separation of mycotoxins possessing carboxyl groups. *J. Chromatogr B.*, 717: 135–156.

Xiulan, S., Xiaolian, Z., Jian, T., Xiaohong, G., Jun, Z. and Chu, F.S. (2006). Development of immunochromatographic assay for detection of aflatoxin B_1 in food. *Food Control*, 17: 256–262.

Zöllner, P. and Mayer-Helm, B. (2006). Trace mycotoxin analysis in complex biological and food matrices by liquid chromatography-atmospheric pressure ionisation mass spectrometry. *J. Chromatogr. A*, 1136: 123–169.

17

Functional Dairy Beverages

SUBROTA HATI, SURAJIT MANDAL AND SHILPA VIJ

ABSTRACT

Fermented foods are prepared using starter cultures, mostly lactic acid bacteria that initiate rapid acidification of milk. The lactic acid bacteria produce antimicrobial substances, sugar polymers, sweeteners, aromatic compounds, vitamins, or useful enzymes, or that have probiotic properties. Recently, new starter cultures of lactic acid bacteria with industrially important functionality are being developed which offer one or more organoleptic, technological, nutritional or health advantages. Today, a number of dairy-based beverages are marketed all over the world. The dairy-based beverages market is still a niche market compared with the sales of yogurt and plain milk. Dairy beverages containing probiotics and / or prebiotics dominate the functional dairy beverages market. Apart from milk based beverages, whey or soya based functional beverages are also gaining popularity. Functional dairy beverages may offer several health, marketing, and technological advantages. However, fundamental and applied research is still needed to optimally implement functional starter cultures in the existing production technology and to obtain quantitative data. Viability of probiotic bacteria in fermented dairy products, the physical, chemical and organoleptic properties of such products should be validated in vitro and in vivo condition or through human trials.

1. INTRODUCTION

Fermented foods are of great significance since they provide and preserve vast quantities of nutritious foods in a wide diversity of flavors, aromas and textures, which enrich the human diet. The fermentation is used as a method of value addition and conversion of raw materials by microorganisms and enzymes into various types of products with distinct nutritional and sensory properties. Over 3500 traditional, fermented foods exist worldwide. Fermented foods have been with us since humans arrived on earth and of these fermented milks have long been an important component of nutrition and diet. Fermented milk products have been reported to have therapeutic properties like anti-cholesterolemic, and anti-carcinogenic properties beyond their basic nutritive value. They, contributing to a variety in our gustative desire, have been recognized to provide important nutrients and considered superior over nonfermented dairy products in terms of nutritional attributes as the microflora present produce simple compounds like lactic acid, amino acids and free fatty acids that are easily assimilable. Originally fermented milks were developed as a means of Lactic acid bacteria (LAB) have been widely used as starter culture for the manufacturing of various fermented dairy products such as dahi, lassi and whey beverages etc. LAB and their food products are thought to confer a variety of important nutritional and therapeutic benefits and have many documented health promoting or probiotic effects in human such as inhibition of pathogenic organism, antimutagenic and reduction of blood cholesterol. Those LAB with scientifically supported health claims define as probiotic and have an increasingly high market potential preserving nutrients.

2. FERMENTED MILK PRODUCTS

Fermented milks are manufactured throughout the world and approximately 400 generic names are applied to traditional and industrialized products but in actual essence the list may only include a few varieties.

2.1. Lactic Fermentated Products

(*a*) Mesophilic type, *e.g.*, cultured buttermilk, filmjolk, tatmjolk and langofil;

(*b*) Thermophilic type, *e.g.*, yoghurt, Bulgarian buttermilk, zabadi, dahi and

(*c*) Therapeutic or probiotic type, *e.g.*, acidophilus milk, Yakult, ABT, Onka, Vifit; products within this group constitute by far the largest number known worldwide;

Yeast—lactic fermentations (kefir, koumiss, acidophilus yeast milk); and

Mould—lactic fermentations (villi).

Certain, closely related products are manufactured from fermented milks by de-wheying; examples include labneh, skyr, ymer and shrikhand. According to a study by global market analyst Euromonitor international global sales of dairy products reached 211.5 billion. The manufacture of cultured dairy products represents the second most important fermentation industry (after the production of alcoholic drinks). A dynamic category, fermented dairy drinks were reported to grow at six times the rate of total dairy growth between 1998 and 2003 in value terms. Also, probiotic drinking yoghurt was the fastest growing dairy product sector between 1998 and 2003, followed by soy milk, (spoonable) probiotic yoghurt, flavoured milk drinks with juice and fermented dairy drinks. The increasing demand from consumers for dairy products with 'functional' properties is a key factor driving value sales growth in developed markets. This led to the promotion of added-value products such as probiotic and other functional yoghurts, reduced-fat and enriched milk products and fermented dairy drinks and organic cheese. There are several principal reasons for the success of fermented dairy products, which relate to nutrition and health, versatility and marketing.

The consumption of milk drinks and fermented products has been recently reviewed by the International Dairy Federation, shown briefly in Table 1. It is quite clear from the data that the consumption of fermented milks has generally increased around the globe over a period from 2001 to 2004.

Table 1: Broad classification of fermented milk.

Name	***Country of origin***	***Microflora***
Acidophilus milk	Australia	*L. acidophilus*
Yoghurt (Bio-ghurt)	Middle Asia, Balkans	*S. salivarius* ssp. *thermophilus, L. delbruckii* ssp. *bulgaricus*, Micrococcus and other lactic acid bacteria, cocci, yeasts, molds
Kefir	Caucasus	*L.lactis* ssp. *lactis, Leuconostoc* spp. *L. delbruckii* ssp. *caucasius, Saccharomyces kefir, Torula kefir, micrococci,* spore forming bacilli
Kumiss	Asiatic steppes	*L. delbruckii ssp. bulgaricus, L. acidophilus, Torulla kumiss, Saccharomyces lactis,* micrococci, spore forming bacilli
Dahi (dadhi)	India, Persia	*L. lactis* ssp. *lactis, S. salivarius* ssp. *thermophilus, L. delbruckii* ssp. *bulgaricus*, lactose fermenting yeasts, mixed cultures
Srikhand (chakka)	India	*S. salivarius* ssp. *thermophilus, L. delbruckii* ssp. *bulgaricus*
Lassi	India	*L. lactis* ssp. *lactis, S. salivarius ssp. thermophilus, L. delbruckii* ssp. *bulgaricus*
Cultured buttermilk	Scandinavian and European countries	*L. lactis* ssp. *lactis, L. lactis* ssp. *diacetylactis, Leuconostoc dextranicum* ssp. *citrovorum*
Leben, Labneh	Lebanon, Arab countries	*L. lactis* ssp. *lactis, S. salivarius* ssp. *thermophilus, L. delbruckii* ssp. *bulgaricus,* Lactose fermenting yeasts

The occurrence of various bioactive peptides in fermented milks, *e.g.*, yoghurt, sour milk and "Dahi", has been reported in many studies. ACE-inhibitory, immunomodulatory and opioid peptides, *e.g.*, have been found in yoghurt and in milk fermented with a probiotic *Lb. casei* ssp. *rhamnosus* strain. Also, ACE-inhibitory peptides have been detected in yoghurt made from ovine milk and in kefir made from caprine milk.

Table 2: Commercial dairy products/ingredients functional claims based on bioactive peptides.

Brand name	*Type of product*	*Claimed functional bioactive peptides*	*Health/function claims*
Calpis	Sour milk	Val-Pro-Pro, Ile-Pro-Pro, derived from casein and k-casein	Reduction of blood pressure
Evolus	Calcium enriched fermented milk	Val-Pro-Pro, Ile-Pro-Pro, derived from casein and k-casein	Reduction of blood pressure
BioZate	Hydrolysed whey protein isolate	b-Lactoglobulin fragments	Reduction of blood pressure
BioPURE-GMP	Whey protein isolate	k-casein f (106–169) (Glycomacropeptide)	Prevention of dental carries, influence the clotting of blood, protection against viruses and bacteria
PRODIET F200/ Lactium	Flavoured milk, drink, confec-tionery, capsules	as1-casein f(91-100) (Tyr-Leu-Gly-Tyr-Leu-Glu-Gln-Leu-Leu-Arg)	Reduction of stress effects
Festivo	Fermented low-fat hard cheese	as1-casein f (1–6), (1–7), (1–9)	No health claim
Cysteine peptide	Ingredient/ hydrolysate	Milk protein derived peptide	Aids to raise energy level and sleep
C12	Ingredient/ hydrolysate	Casein derived peptide	Reduction of blood pressure
Capolac	Ingredient	Casein phosphopeptide	Helps mineral absorption
PeptoPro	Ingredient/ hydrolysate	Casein derived peptide	Improves athletic performance and muscle recovery
Recaldent	Chewing gum	Calcium casein peptone-calcium phosphate	Anticarcinogenic

3. WHEY-BASED BEVERAGES

(*a*) Lactic fermented acido whey soft drink.

(*b*) Alcoholic fermented wine and beer.

(*c*) Low alcoholic whey beverages fortification with grape juice.

In the present scenario of consumption of fermented whey drinks such as Molke in West Germany, Rivella in Switzerland etc and these products are showing increasing trends in most of the countries around the world. Keeping in view increased demand of soft drinks and juices these days in India, there is a tremendous scope and need to exploit commercial production of these fermented whey beverages since it is the best proposition to convert largest by-product (whey) of dairy industries into value added product by simple and indigenous processes.

3.1. Benefits of Whey Based Fermented Drinks

(*i*) Whey is an excellent growth medium for Lactic Acid Bacteria to ferment lactose in whey to form lactic acid.

(*ii*) Whey is a genuine thirst quencher unlike most of the soft drinks.

(*iii*) Whey as a drink can replace much of the lost organic and inorganic salts to the extracellular fluid.

(*iv*) Whey is rapidly adsorbed due to absence of fat emulsion.

(*v*) Whey has been used to treat various ailments such as arthritis, liver complaints and dyspepsia.

(*vi*) It also possesses almost all the electrolytes of Oral Rehydration Solution (ORS) which is invariably used to control dehydration.

(*vii*) On fermentation with LAB, it becomes a suitable drink for lactose-intolerant people.

(*viii*) Fermentation of whey with LAB also masks the effect of curdy flavor of whey.

(*ix*) At industrial scale, large volumes of whey can be used directly from paneer/cheese vats, thus eliminating transportation and disposal problems.

(*x*) Conversion of whey into beverages involves very simple processes.

(*xi*) Utilization of whey generates additional revenue to the dairy plant.

Above all, its utilization also solves the problems of environmental pollution.

4. PROBIOTIC DAIRY BEVERAGES

A probiotic is defined as a 'living organism which when administered in certain numbers exerts health benefits in the host' (FAO, 2001). Owing to this property, incorporation of probiotic micro-organisms in dairy foods has increased rapidly during the last two decades. Consumption of probiotic bacteria via food products is an ideal way to reestablish the balance of intestinal microbiota. These include alleviation of lactose intolerance symptoms, lowering cholesterol, curing antibotic-associated diarrhoea, prevention of intestinal tract infections, prevention of colon cancer, control of rotavirus, prevention of ulcers related to *Helicobacter pylori*, improvement of immune system, irritable bowel syndrome and antihypertensive effects. In order to produce therapeutic benefits, a suggested range for the minimum level for probiotic bacteria in probiotic milk is from 10^6 to 10^7 colony-forming units (cfu) DmL (IDF 1992). In recent years, probiotic beverages based on fruit juice, cereal products and daily dose dairy drinks have also become popular commercially. Today, a wide range of probiotic products is available for consumers in the market.

Table 3: Various fermented dairy products available in market.

Product	***Starter organisms***
Acidophilus milk	*L. acidophilus*
Sweet Acidophilus milk	*L. acidophilus*
Acidophilin	*L. acidophilus, L. lactis* ssp. *lactis,* kefir yeast
Diphilus milk	*L. acidophilus, B. bifidum*
Biomild	*L. acidophilus, B. bifidum*
A/B milk	*L. acidophilus, B. bifidum*
Bifigurt	B. longum (DSM 2054)
Acidophilus Buttermilk	*L. acidophilus, L. lactis* ssp. *lactis, L. lactis* ssp. *cremoris, L. lactis* ssp. *lactis biovar. diacetylactis*
Bifidus milk	*B. bifidum, B. longum*
Yakult	*L. casei Shirota*
Yakult miru miru	*L. casei, B. bifidum*
A-38 fermented milk	*L. acidophilus*, Mesophilic cultures
Actimel	*L. casei* Immunitas

5. DAIRY BEVERAGES WITH ADDED BIOACTIVE COMPONENTS

The a-linoleic acid (C18:3 n-3, ALA), eicosapentaenoic acid (C20:5 n-3, EPA) and docosahexanoic acid (C22:6 n-3, DHA) are the principal members of the Omega-3 fatty acids which can be added in milk products for improving their functional value. Europe's first commercial Omega-3 enriched milk was introduced by Dawn Dairy (Ireland) and the desired amount of Omega-3 was added at the final stage of bottling by means of a dosing system developed by Tetra Pak. Parmalat, an Italian dairy company, launched Omega-3 enriched milk 'Omega-3 Plus' (containing 80 mg Omega-3 per litre of milk) in 1998. Consumption of Omega- 3 enriched semi-skimmed milk at a dose of 500 mL per day. Fortification of Omega-3 enriched milk with vitamins A, C and E is a common application. In order to attract consumers, chocolate flavoured milk fortified with Omega-3 has been introduced into the functional foods market by Neilson Dairy (Dairy Oh, Canada) and Parmalat (Beatrice, Canada). Two potent ACE-inhibitory peptides, Valine-Proline-Proline (VPP) and Isoleucine-Proline-Proline (IPP), derived from caseins during milk fermentation with *Lactobacillus helveticus* and *Saccharomyces cerevisiae*, are responsible for the anti-hypertensive activity shown by Calpis_ sour milk (Calpis Co. Ltd, Tokyo, Japan). Other examples of commercial dairy-based beverages with added bioactive peptides are Evolus_and Evolus. The former product is manufactured by incorporating two tripeptides, Val-Pro-Pro and Ile-Pro-Pro, and is claimed to reduce blood pressure upon regular consumption. The latter product contains the same tripeptides added to evolus plus plant sterols which help to reduce blood cholesterol levels.

Melatonin is a naturally occuring hormone found in animals and in some other living organisms, including algae. Melatonin controls the body's day and night rhythm. A Finnish company, Ingman Dairy, introduced the world's first ever high-melatonin premium milk (under the Night Time brand) in 1999. The same company produces high-melatonin milk based on the standardized milking system at night. A similar product has been launched in Japan, and the UK has two milk brands inspired by the Finnish example.

Conjugated linoleic acid is found almost exclusively in animal products, with a natural level of approximately 6 mgD g of fat. Normal daily intake of CLA in the diet is 150–400 mgD g. addition of linoleic acid at a level of 0.1% increased cis9-trans11-CLA content of nonfat yogurt significantly without affecting the sensory properties of the final product. CLA level in fermented milk made with the standard yogurt culture (0.57 mgDg lipid).

6. OTHER FERMENTED DAIRY PRODUCTS

Cultured dairy products are an excellent medium to generate an array of products that fit into the current consumer demand for health-driven foods. Owing to expanding market share and size of dairy companies, there has been a reduction of clearly structured markets *i.e.*, merging of dairy products and fruit beverage markets with introduction of juiceuticals that include hybrid products like fruit based cultured milk beverages. In India *lassi* made out of *dahi* is a widely consumed fermented milk beverage, popular in all parts of the country and has a great potential in the domestic as well as overseas markets. However, problems like short shelf life, post acidification, whey syneresis, hinder the market saleability of *lassi*. Accordingly, studies were conducted for the production of fruit *lassi* with extended shelf life using biopreservatives. Khurana (2006) developed a process for making fruit *lassi*, using buffalo milk. Standardized buffalo milk was heated to 90°C for 10 minutes followed by cooling to 30°C and inoculated with 1.5 per cent culture and incubated at 300C for 10–12 hrs. The curd obtained was then mixed with sugar syrup and fruit pulp. Homogenization was done followed by packaging, cooling and storage at 4°C. For further stabilization and improvement of consistency of each type of fruit *lassi* exopolysaccharide producing (EPS+) cultures were used along with pectin at different levels. An enhancement in the rheological and overall sensory characteristics of all the 3 types (banana, mango and pineapple) of fruit *lassi* was observed with increase in proportion of EPS+ culture up to a certain level. Nisin and MicroGARD™ were used to extend the shelf life of fruit *lassi* up to 30 and 50 days, respectively as compared to control mango *lassi* which had a shelf life of 15 days at 4 ± 1°C. The technology developed for manufacture of fruit *lassi* with extended shelf life appeared to have considerable potential to facilitate commercial manufacture and marketing of this popular fermented milk beverage. Inclusion of fruits and artificial sweeteners in *lassi* would not only enhance nutrition, help diabetic/obese people, aid product diversification but also help in curtailing the post harvest losses in fruits. Such a product would not only serve as a low calorie-quick meal snack but also offer stiff competition to expensive soft drinks in the beverage market.

7. CEREAL BASED FERMENTED DAIRY BEVERAGES

Cereal grains constitute a major source of dietary nutrients all over the world. Addition of cereals into milk not only enriches its mineral value but also supplements fiber. Fermentation further enhances the nutritive value, palatability and functionality of cereals by reducing the antinutritional factors. Development of technologies for the manufacture of cereal based fermented milk beverages will lead to the utilization of underutilized cereals like sorghum, pearl millet, finger millet etc. *Raabdi*, prepared by cooking cereal (maize) flour with buttermilk, is a traditional popular beverage of Haryana, Punjab and Rajasthan. The technology of producing these traditional fermented foods from cereals and milk remains a household art. Traditional process of *raabdi* preparation yields a product with limited shelf life (one to two days) with unpredictable sensory quality. A technology has been developed at National Dairy Research Institute, Karnal for manufacture of *raabdi*-like fermented beverage using cereals like Pearl millet/Sorghum/wheat possessing health attributes. Development of *raabdi*-like fermented milk beverages was based on the selection of milk solids source; selection of levels of milk solids and cereal solids, stage of addition of cereal solids addition; fermentation conditions and stabilization of developed product in terms of preventing sedimentation and wheying-off in the product during storage. Cereal solids were incorporated in milk at two stages *i.e.*, before fermentation with *dahi*/yoghurt culture or after fermentation. The fermented mass will be blended into beverage by the addition of water, salt and spices. The cereal based beverage thus developed was having very good sensory qualities and it was accepted well by the consumers but the only limitation of the beverage was its limited shelf life. Attempts were made to increase the shelf life of the product in which, this beverage was made in ready to reconstitute form Thus, it not only increased the shelf life of the beverage but at the same time it aided to the convenience in use. In other study to increase the shelf life of the liquid beverage, various preservatives viz. biological, chemical and thermal treatments were tried and the shelf life increased to about a month at refrigerated temperature. Gupta *et al.* (2007) prepared *raabdi* from buttermilk and mothbean by two methods. Buttermilk was mixed with mothbean flour to get homogeneous paste and this was diluted with plain water and boiled for 7 min with constant stirring while salt was added during stirring. In the first method the mix was fermented before cooking and in the second method the mix was fermented after cooking. They reported that 4 h fermented and cooked and 12 h cooked and fermented *raabadi* had better acceptability. Kindumu is a fermented milk beverage popular in central African region and is prepared by sun drying the mixture of fermented milk and germinated/non-germinated sorghum flour. Grewal and Chauhan (1993) prepared Soy *raabdi* by blending autoclaved dehulled soybean slurry and the curd mass (obtained from reconstituted skim milk). The blend was added with water and fermented at 30°C for 12 h. Soy *raabdi* appeared to be a low cost wholesome food which was free from beany flavour and rich in protein, fat and dietary essential minerals. Mugocha *et al.* (2000) developed a composite finger millet and milk based fermented beverage. Various parameters including level of finger millet gruel in skim milk, type of starter culture, incubation temperature etc, were optimized to develop the finger millet- dairy beverage.

8. FUTURE RESEARCH

Today, a number of dairy-based beverages are marketed all over the world. The dairy-based beverages market is still a niche market compared with the sales of yogurt and plain milk. Dairy beverages containing probiotics and/or prebiotics dominate the functional dairy beverages market. Apart from milk based beverages, whey- or soya-based functional beverages are also gaining popularity. Functional dairy beverages may offer several health, marketing, and technological advantages. However, fundamental and applied research is still needed to optimally utilize existing production technology and to obtain quantitative data. Viability of probiotic bacteria in fermented dairy products, the physical, chemical and organoleptic properties of such products should be validated *in vitro* and *in vivo* condition or through human trials.

REFERENCES

Adwan, L. (2003). Fermented dairy drinks under pressure (online). Euromonitor international archive; 2003 Jul 25 (cited 2003 Jul 25). Available from: http://www.euromonitor.com/article.asp?id=1371

Aneja, R.P., Mathur, B.N., Chandan, R.C. and Banerjee, A.K. (2002). Heat–acid coagulated products. *In*: Technology of Indian milk products. Delhi, India.

Chandan, R.C. (1999). Enhancing market value of milk by Adding Cultures. *J. Dairy Sci.*, 82: 2245–2256.

Grewal, R.B. and Chauhan, B.M. (1993). Preparation, organoleptic acceptability and nutritional value of soy rabadi - an indigenous fermented food of India. *Int. J. Trop-Agric.*, 11: 172–177.

Gupta, V., Sharma, A. and Nagar, R. (2007). Preparation, acceptability and nutritive value of *rabadi*- A fermented mothbean food. *J. Food Sci. Tech.*, 44 (6): 600–601.

IDF, (2005). The World Dairy Situation. Bulletin of IDF. 399: 81.

Jindal, A.R., Shore, M., Shukla, F.C. and Singh, B (2004). Studies on the use of chhana and paneer whey in the preparation of puras (pancakes). *Int. J. Dairy Technol.*, 57: 221–225.

Khamrui, K. and Rajorhia, G.S. (1998). Making profits from whey. *Indian Dairyman*, 50: 13–18.

Khurana, H.K. (2006). Development of technology for extended shelf life fruit *lassi*. Ph.D. Thesis, National Dairy Research Institute, Karnal, India.

Kosikowski, F.V. and Mistry, V.V. (2002). *Cheese and Fermented Milk Foods*, Volume I, Origins and Principles, third edition, F. V. Kosikowski, L.L.C., Westport, Connecticut, 1997, Chapter 26. Ling, K. Charles. *Marketing Operations of Dairy Cooperatives*, USDA Rural Business-Cooperative Service, RBS Research Report 201.

LeBlanc, A.M., Matar, C., LeBlanc, N. and Perdigón, G. (2005). Effects of milk fermented by *Lactobacillus helveticus* R389 on a murine breast cancer model. *Breast Cancer Research*, 7: 477–486.

Mugocha, P.T., Taylor, J.R.N. and Bester, B.H. (2000). Fermentation of a composite finger millet-dairy beverage. *W.J. Microbiol. Biotech.*, 16: 341–344.

Narva, M., Halleen, J., Väänänen, K. and Korpela, R. (2004). Effects of *Lactobacillus helveticus In vitro. Life Sci.*, 75: 1727–1734.

Richelsen, B., Kristensen, K. and Pedersen, S.B. (1996). Longterm (6 months) effect of a new fermented milk product on the level of plasma lipoproteins: A placebo-controlled and double-blind study. *Eur. J. Clin. Nutr.*, 50: 811–815.

Rudrello, F. (2004). Health trends shape innovation for dairy products (online). Euromonitor international archive; http://www.euromonitor.com/article.asp?id =4011 (2004 Oct 5.)

Siso, M.I.G. (1996). The biotechnological utilization of cheese whey: A Review: *Bioresource Technology*. 57: 1–11.

Smithers, G.W. (2008). Whey and whey proteins-from 'gutter –to-gold'. *International Dairy Journal*, 18: 695–704.

Spreer, E. (1998). Whey and whey utilization. In Milk and Dairy Technology, Chapter 10. Marcel Dikker INC, New York.

Webb, B.H. and Whittier, E.O. (1948). The Utilization of Whey: A Review. *J. Dairy Sci.*, 31: 39–64.

18

Modern Packaging of Meat and Meat Products

ASHIM K. BISWAS AND PRABHAT K. MANDAL

ABSTRACT

Changes in consumer preferences have led to innovations and developments in new packaging technologies for meat and meat products. Active packaging is useful for extending the shelf life of fresh, cooked and other meat products. Forms of active packaging relevant to muscle foods include; oxygen scavengers, carbon dioxide scavengers and emitters, drip absorbent sheets and antimicrobial packaging. Recognition of the benefits of active packaging technologies by the food industry, development of economically viable packaging systems and increased consumer acceptance opens new frontiers for active packaging technology. Commercially, there is widespread use of oxygen scavengers in pre-packed cooked sliced meat products. Antimicrobial packaging is gaining interest from researchers and industry due to its potential for providing quality and safety benefits. Future research in the area of microbial active packaging should focus on naturally derived antimicrobial agents, biopreservatives and biodegradable packaging technologies. The possibility of utilising additional active packaging technologies, as currently applied to other foodstuffs, for safe and effective storage of meat and meat products also merits investigation.

1. INTRODUCTION

Modern food packaging is believed to have originated in the 19th century with the invention of canning by Nicholas Appert in 1809. In the beginning of 20th century, 3 piece tin-plated steel cans were used for food distribution. Many packaging innovations occurred during World War I and World War II.

Packaging of fresh meat is carried out to avoid contamination, delay spoilage, permit some enzymatic activity to improve tenderness, reduce weight loss, and to maintain desirable bright red colour at retail outlets. Processed meat products, however, affected by certain factors like dehydration, lipid oxidation, discoloration and loss of aroma. Many meat packaging systems currently exist, each with different attributes and applications. These systems range from overwrap packaging for short term chilled storage and/or retail display, to a diversity of specified modified atmosphere packaging (MAP) systems for longer-term chilled storage and/or display, to vacuum packaging, bulk-gas flushing or MAP systems using 100% carbon dioxide for long term chilled storage. Due to the diversity of product characteristics and basic meat packaging demands and applications, any packaging technologies offering to deliver more product and quality control in an economic and diverse manner would be favourably welcomed (Sceter, 2010).

In 20th century, innovations include active packaging and intelligent or smart packaging. Retort pouch packaging, stand-up flexible pouches, self-sealing flexible pouches, co-extrusion films and bottles emerged as rigid and semi-rigid packaging. Multilayer barrier plastic cans, microwave suseptors, dispensing colours, gas barrier bags for prime cuts of meat, modified atmosphere packaging, rotogravure printed full-panel shrink film labels, and dual ovenable trays are examples of innovations for the convenience attributes that have propelled food packaging in 21st century. More recently, innovations are related to nanotechnology whose future may lie in improving barrier and structural/ mechanical properties of packaging materials and development of sensing technology (Brody, 2010).

2. PACKAGING OF FRESH MEAT

Fresh meat packaging is only minimally permeable to moisture and so surface desiccation is prevented (Faustman & Cassens, 1990), while gas permeability varies with the application. A single layer plastic film does not have all the properties required for a food packaging applications. So lamination or co extrusion is used to create layers of plastic with the desired properties. Heat sealing and barrier properties are often improved by application of coatings to the surfaces of plastic films. Therefore, most meat packaging films are of multilayer construction which incorporates a variety of polymer resins. However, to provide extended storage, low gas and moisture vapour transmission rates are essential, since packaging effects on product storage life have been reported to be mostly related to the gas and moisture vapour permeability of the packaging materials. Films used for retail packaging must have a high oxygen permeability to prevent oxidative browning. However, only films with oxygen transmission rates of no more than 10 ml/m^2/24 h/bar at 23°C and 0% RH can be used for preservative packaging. Storage life can be increased 10–15% by using films with oxygen permeabilities less than 2 mL m^2/24 h/bar at 23°C and 0% RH.

All plastics are permeable to gases to some degree. Film permeability varies with the partial pressures inside and outside the package, the type of plastic, the thickness of the plastic layers, and the temperature of the plastic. Some plastics are also markedly influenced by relative humidity and contamination with nutrients, such as fat. Oxygen permeable films, such as cellophane, cellulose, and polyethylene permit microbiological growth at maximal rates until dehydration or mould growth interferes. Aerobic, high permeability, gas flushed, and tray overwrapped packages have relatively short storage lives, when compared to anaerobic packages or packages with low permeability (Cole, 1986). Bacterial growth in impermeable packages is dependent upon the atmosphere in the package and is characterized by longer lag phases. Entrapped oxygen and oxygen ingress produce oxidation and pronounced surface discoloration. Oxygen ingress into a package can only be prevented by using completely gas impermeable packaging materials, with oxygen impermeable heat seals on all parameters. Sealing of packages is a critical step, but several sealants can be used, and usually represent the thickest layer in laminates. Surlyn is a widely used ionomer, since it has a wide heat sealing range, a low heat requirement for sealing, and maintains its strength while hot. The use of aluminium foil in a laminate was required to prevent discoloration during storage of meat at 4°C.

Prolonged storage in PE bags or PVC film wrap should be avoided to prevent microbial growth. Storage of beef quarters or wholesale cuts in PVC film for more than 8 days increased both microbiological counts and the incidence of off odours. However, PE bags and PVC film wrap reduced shrinkage and improved muscle colour and the subcutaneous fat appearance of beef quarters and wholesale cuts during shipment. Beef quarters or cuts can be wrapped in PVC film and be shipped for up to 6 days in air, but shipments over 6 days in duration should be in modified atmospheres (Marriott *et al.*, 1977). Polyethylene wrapped beef samples had lower weight losses and thiobarbituric acid (TBA) values than PE wrapped, vacuum packaged, and modified atmosphere packaged samples. However, PE wrapped and vacuum packaged samples had higher reflectance values. Polyethylene wrapped cow beef samples, however, had higher bacterial counts than vacuum packaged and modified atmosphere samples, packed in carbon dioxide, even after 3 months of frozen storage.

2.1. Shrink Packaging

Plastic shrink films are used for wrapping large and uneven cuts of fresh meat. It is a technique in which heat shrinkable polymer film is shrunk around the meat product by application of heat to achieve a skin-tight and compact pack. The packaging film should have structural strength. It should be a good water vapour barrier and be capable of withstanding storage temperature of about –45°C. The advantages of plastic shrink film include neat appearance, ease in handling and a contour fit. Hot tunnels are used to effect a tight wrap. Heat shrinkable PVC, PP, irradiated PE and PVDC are used to shrink wrap fresh meat.

2.2. Skin Packaging

Skin packaging offers advantages for presentation as well as packaging design variety. The process allows the packaging film to conform exactly to the profile of the product. This gives rise to many opportunities for enhanced product presentation as well as further improving the integrity of the pack itself. In a skin pack, the product becomes the die for the thermoform packaging operation. The semi-rigid bottom web may or may not be thermoformed. The top web is heated in an evacuating chamber until it is near its melting point, at which it drapes over the product and forms a skin around all the contours. Upon sealing and cooking, it retains its new shape, ensuring intimate contact with the product, irrespective of surface irregularities. Skin packs are prepared with an oxygen barrier plastic film. There are four categories of preservative packaging that can be used with raw muscle foods. These are vacuum packs (VP), high oxygen modified atmosphere packs (high O_2 MAP), low oxygen modified atmosphere packs (low O_2 MAP), and controlled atmosphere packs (CAP).

2.3. Gas Packaging

Gas packaging is another technique to extend the shelf- life of fresh meat. In this case, an atmosphere composition surrounding the product which is different from that of normal air, usually consisting of reduced O_2, N_2 and CO_2 concentration. Each gas has specific functions in the extension of the refrigerated shelf-life of meat, however, in most cases combination of gases are used. Atmosphere modification can be done in one of two ways: flushing the package with a gas having the required composition when the product is packed or relying on naturally occurring processes that tend to modify the composition over time. The first case is usually referred to as active packaging and is commonly used for meat products, while the second is called passive packaging and it may be used for fresh and minimally processed fruits and vegetables.

Passive modified atmosphere packaging is also designated as equilibrium atmosphere packaging. Due to product respiration and gas transfer through the package, the CO_2 level increases and the O_2 level decreases with time after the product has been packed. After a period of time, equilibrium is achieved and the amount of CO_2 generated in the product respiration equals the amount being transferred out of the package and the amount of O_2 consumed equals the amount being transferred into the package. When the packaging system is well designed, the equilibrium composition is very close to the optimum required for maximizing the product shelf-life. Passive packaging may be used in practice when the equilibrium composition is achieved within a short period of time. Beef, buffalo meat and mutton-they need fairly high O_2 content to maintain the bright red colour.

Pork-due to high fat content (unsaturated fat) require less O_2 and some N_2 for glow. Chicken in comparison to beef, buffalo meat and pork needs little O_2 and reduced amount of CO_2 which can flavour the meat. A mixture of 75% N_2 + 15% CO_2 and 10% O_2 has been reported to maintain the colour of red meat up to 15 days at 4°C. It has been reported that a mixture of 80% N_2 and 20%

CO_2 keeps the fresh meat well up to 15 days at 4°C. Such a combination of gases has been found to be ideal in obtaining attractive meat colour and restricting the bacterial growth in providing reasonable shelf-life of meat.

Advantages: There is attractive colour of meat.

Disadvantages

(*i*) The modified atmospheric packaged meat has relatively shorter shelf-life than vacuum packaged meat.

(*ii*) Require more space during storage.

(*iii*) Generally not suitable for frozen storage of meat.

2.4. Vacuum Packaging

Vacuum packaging is an alternative method of packaging of fresh meat for longer shelf- life up to 2–3 wks. But because of purple red colour of meat, it is not popular at retail level. Bigger cuts (primal cuts) of meat are vacuum packaged and sold of institutional packs (*e.g.*, hotel, hostel, restaurant, catering establishment etc.). Vacuum packaging can be done two ways—(*i*) conventional vacuum packaging using of non-heat shrinkable film; and (*ii*) vacuum shrink packaging using of heat shrink film/laminates. In suitable vacuum packaging, film should have a good barrier to water vapour and gases with high mechanical strength. Some vacuum packaging films are polyester/PE laminate, polyamide (nylon)/PE laminate, PVC/PVDC copolymer (cryovac), PVC film, polyamide film etc. The growth of aerobic bacteria is inhibited due to absence of oxygen. However, growth of facultative anaerobic bacteria is not inhibited. Pseudomonas and Achromobacter are common genera associated with spoilage of meat. The vacuum packaged meat products should have shelf-life up to 2–3 wks at 0–2°C and 85–90% R.H (Table 1).

2.5. Master Packaging

The master pack system consists of four to eight air permeable overwrapped packages that are placed in a large pouch (master pack) that is impermeable to oxygen and moisture. The master pack is evacuated to remove air and then back flushed with the desired gas mixture. Typical gas mixtures range from 100% to 80% carbon dioxide with nitrogen used as the remaining gas. The presence of CO_2 inhibits the growth of bacteria and therefore extends shelf-life. When a retailer needs product to stock the shelf, he/she opens the master pack and removes the packages from the bag with the CO_2 gas blend. The product blooms in approximately 30 minutes from the atmospheric O_2 it is exposed to after removal from the bag and forms the desirable, bright, cherry-red colour. The expected shelf-life of products packaged in a master pack system is ten to 14 days with a retail display-life of two to seven days (Buffo & Holley, 2005).

Table 1: Shelf life of meat and meat products under different packaging.

Food/Treatment	*Packaging materials/methods*	*Shelf life*
Sliced marinated beef beef loin High-pressure treatment (600 MPa, 6 min, 31°C)	Bottom: PS-EVAL based (RSC03X60); Vacuum-packed	120 days at 4°C
Beef and pork	Air: PE Vacuum: PA/PE	10–11 days at 4°C 11 days at 4°C
Beef steaks	Expanded PS-tray + PE(80)/PA(20). MAP 1: 70% O_2/20% CO_2/10% N_2 MAP 2: 60% O_2/40% CO_2	22–28 days at 1°C
Pork chops, treated: 500 ppm ascorbic acid, 250 ppm citric acid	Vacuum: PVC+PA/PE MAP: 80% N_2/20% CO_2	12–14 days at 4°C
Sausages, heat-treated	Vacuum-packed	6–11 weeks at 7°C
Turkey Bologna In-package pasteurization	Biocide film: nisin (7%, w/w) wheat gluten protein and glycerol, vacuum pack	8 weeks at 4°C
Precooked chicken meat product	PE-LD/PA/PE-LD, air PE-LD/PA/PE-LD MAP: 30% CO_2/70% N_2	8 days at 4°C 12 days at 4°C
Beef loin steaks	Top: PET/PE, Bottom: PA/PE Top: PET/PE; Bottom: PA/PE, Vacuum MAP: 0.4% CO/60% CO_2/40% N_2 MAP: 70% O_2/30% CO_2	14 days at 4°C 27 days at 8°C 10 days at 4°C 7 days at 8°C
Ground beef	PE-tray wrapped in Cryovac BDF 550 shrinking film. MAP: 0.4% CO/60% CO_2/40% N_2 MAP: 70% O_2/30% CO_2	11 days at 4°C 6 days at 8°C 8 days at 4°C 4 days at 8°C
Pork chops	MAP: 0.4% CO/60% CO_2/40% N_2 MAP: 70% O_2/30% CO_2 MAP: 60% CO_2/40% N_2 + 7O2-absorber (Ageless[R])	21 days at 4°C 14 days at 8°C 14 days at 4°C 7 days at 8 °C 17 days at 4°C 10 days at 8°C
Pork	PA(20)/PE-E(70.5), vacuum	4 weeks at 4°C
Beef	PA(20)/PE-E(70.5), vacuum	7 weeks at 4°C
Cooked ham and Frankfurters	PA/PE (120)	4 weeks at 4°C
Lumbar vertebrae	Tray: PP/EVAL Overwrap: barrier film MAP 1: 80% N_2/20% CO_2 MAP 2:0.4% CO/30% CO_2 /69.6% N_2	2 weeks at 4°C 6 weeks at 4°C
Beef ribs	MAP: 80% O_2 /20% CO_2	1 weeks at 4°C

Table 1: *(Contd...)*

Table: *(Contd...)*

Food/Treatment	***Packaging materials/methods***	***Shelf life***
Ostrich steaks	PA(20)/PE(100), vacuum	9 days at 4°C
	Tray: PS-E/barrier film:	
	Air	8 days at 2°C
	Vacuum	12 days at 2°C
	MAP 1: 80% CO_2/20% N_2	12 days at 2°C
Ostrich steaks	MAP 2: 30% CO_2/69.8% Ar/0.2% CO	12 days at 2°C
Ostrich meat, under lighting (1700 ± 100 lux)	PE-LLD/EVAL	
	Vacuum: polyolefin/PVDC	< 6 days at 4°C
	MAP 1: 80% N_2/20% O_2	< 6 days at 4°C
	MAP 2: 80% O_2/20% CO_2	< 3 days at 4°C
Pork sausages	Tray: PS-E, overwrapp:	
	(*a*) PE (*b*) PE/PA	8 days at 2°C
	MAP 1: 0% O_2/20% CO_2/80% N_2	16 days at 2°C
	MAP 2: 0% O_2/20% CO_2/80% N_2+O_2-Absorber	> 20 days at 2°C
	MAP 3: 20% O_2/20% CO_2/60% N_2	
	MAP 4: 40% O_2/20% CO_2/40% N_2	
	MAP 5: 60% O_2/20% CO_2/20% N_2	
	MAP 6: 80% O_2/20% CO_2/0% N_2	12 days at 2°C
	MAP 7: 0% O_2/20% CO_2/80% N_2	
	Vacuum	> 20 days at 2°C
Chicken wings	Vacuum-packaged and cooked (sous vide)	7 weeks at 2°C

References: Garriga *et al.*, 2004; Samelis *et al.*, 2000; Mancini *et al.*, 2005; Capita *et al.*, 2006; Mantis *et al.*, 2005; Blixt and Botch, 2002; Patsias *et al.*, 2006; Sorheim *et al.*, 1999; McCormick *et al.*, 2005; Mayr *et al.*, 2003; Huang *et al.*, 2005; Fernandez-Lopez *et. al.*, 2008; Martinez *et al.*, 2006; Seydim *et al.*, 2006; Wang *et al.*, 2004

3. MODERN PACKAGING TECHNIQUES

3.1. Active Packaging

Active packaging allows packages to interact with food and the environment and play a dynamic role in food preservation (Lopez-Rubio *et al.*, 2006). Active packaging refers to the incorporation of additives into packaging systems with the aim of maintaining or extending meat product quality and shelf-life. Developments in active packaging have led to advances in many areas, including delayed oxidation and controlled respiration rate, microbial growth, and moisture migration. Other active packaging systems include carbon dioxide absorbers and emitters, odour absorbers, ethylene removers, and aroma emitters. While purge and moisture control and oxygen removal have been prominent in active packaging, purge control is the most successful commercially. For example, drip absorbing pad used in the poultry industry.

3.1.1. *Oxygen scavengers*

The presence of oxygen in package can trigger or accelerate oxidative reactions that result in food deterioration: oxygen facilitates the growth of aerobic microbes or molds. Oxidative reactions result in adverse qualities such as off-odours, off-flavours, undesirable colour changes, and reduced nutritional quality. Oxygen scavengers are easily oxidizable substances included in the packaging system to remove the oxygen by means of a chemical reaction. The use of scavengers led to faster reduction and to lower levels of residual oxygen, as compared to nitrogen flushing. Ferrous oxide is the most commonly used scavenger (Kerry *et al.*, 2006). Others include ascorbic acid, sulfites, catechol, some nylons, photosensitive dyes, unsaturated hydrocarbons, ligands, and enzymes such as glucose oxidase. They are available in the market in various forms (sachets in headspace, labels, or direct incorporation into package material and/or clousers) and names. Ageless^{R} is an iron and ascorbic acid containing oxygen scavenger. The main types of absorbents are types Z, S, FX, E and G, although other types are available in markets. Types Z, S and FX are based on iron and are single function. Type E contains, besides iron, $Ca(OH)_2$, which is responsible for carbon dioxide scavenging. Type G is based on ascorbic acid and generates an equal volume of carbon dioxide to the oxygen volume scavenged. The FreshPax^{TM} has packets configuration with absorbing capacities ranging from 10 to 2 000 cc of oxygen. Four main types are commonly available: type B, D, R and M.

Various studies have been conducted with oxygen scavengers to prevent permanent discolouration of red meats in atmospheres. Although some success with the atmospheres, the general utility of such an approach has some doubt because the muscle tissue itself acts as a very efficient oxygen scavenger (Day, 2008; Kerry *et al.,* 2006). Gill and McGinnis (1995) performed an oxygen absorption kinetics study with a commercial oxygen scavenger (FreshPax^{TM} 200R) and reported that discoloration could be prevented in ground beef if large numbers of scavengers were used in each pack to bring residual oxygen to <10 ppm within 2 h at a storage temperature of –1.5°C. The inclusion of oxygen scavengers (Ageless^{R} S/S200) in master packs flushed with 50% CO_2: 50% N_2 significantly improved the colour stability of *M. Longissimus dorsi* and *M. Psoas major*, relative to controls. Tewari *et al.* (2001) examined the effect of two commercial oxygen scavengers (Ageless^{R} FX-100 and FreshPax^{TM} R-2000), in conjunction with CAP, on the discoloration of *M. Psoas major* in master packs filled with nitrogen and stored at 1 ± 0.5°C. Steaks packaged without oxygen scavengers had more discoloration and significantly higher proportions of metmyoglobin when compared to steaks packaged with oxygen scavengers. Prevention of metmyoglobin formation was influenced by the number but not the type of oxygen scavenger employed. Payne *et al.* (1998) examined the effect of vacuum CAP with carbon dioxide, packs flushed with carbon dioxide, packs flushed with carbon dioxide and containing Ageless^{TM} (Z50) oxygen scavengers and packages containing oxygen scavengers alone on the drip loss, microbial and sensorial properties of *M. Longissimus lumborum* stored for up to 20 weeks at –1.5°C. Beef in packs flushed with carbon dioxide and flushed containing the oxygen scavenger had lower drip loss than the standard CAP system. The packages flushed with carbon dioxide and those containing the oxygen scavenger alone have the best results depending on the storage shelf-life required. In addition to fresh beef oxygen scavenging technology has also been applied to pork and pork products, where, Martinez *et al.* (2006) reported that fresh pork sausages stored in 20% CO_2: 80% N_2 plus an oxygen scavenger (Ageless^R FX-40) for up to 20 days at 2 ± 1°C had reduced psychrotrophic plate counts and an extended shelf-life in terms of colour and lipid stability. Oxygen scavenging labels are widely used commercially as oxygen scavengers in pre-packed cooked meat products. Emco Packaging Systems, specialists in active and intelligent packaging, are a UK manufacturer and distributor for self-adhesive oxygen absorbing labels. Emco supply labels for use in pre-packed sliced cooked meats, especially hams, to meat processors in Ireland, throughout the UK and in Europe. While labels used in sliced cooked meat packages scavenge between 10 and 20 cm^3 of oxygen, Emco have recently launched larger oxygen scavenging labels onto the market, which scavenge between 100 and 200 cm^3 oxygen, for use in larger capacity packaging applications. An alternative to sachets involves the incorporation of the oxygen scavenger into the packaging structure itself.

3.1.2. *Carbon dioxide absorbers and emitters*

Carbon dioxide may be added for beneficial effects like to suppress microbial growth in certain products such as fresh meat and poultry. Carbon dioxide is also used to reduce the respiration rate of fresh produce and to overcome package collapse or partial vacuum caused by oxygen scavengers, and thus increasing the product shelf-life. Carbon dioxide is available in various forms, such as moisture activated bicarbonate chemicals in sachets and absorbent pads. Since the permeability of carbon dioxide is 3/5 times higher than that of oxygen in most plastic films, it must be continuously produced to maintain the desired concentration within the package. High carbon dioxide levels (10–80%) are desirable for foods such as meat and poultry in order to inhibit surface microbial growth and extend shelf life. When a package is flushed with a mixture of gases including carbon dioxide, the carbon dioxide dissolves in the product creating a partial vacuum. In such cases, the simultaneous release of carbon dioxide from inserted sachets, which consume oxygen, is desirable. Such systems are based on either ferrous carbonate or a mixture of ascorbic acid and sodium bicarbonate.

Examples of commercially available dual action combined carbon dioxide generators/oxygen scavengers are Ageless^R, Fresh Max^R and Fresh Pax^{TM} (Table 2). Carbon dioxide emitting sachets or labels can also be used alone. The Verifrais^{TM} package, manufactured by SARL Codimer has been used to extend the shelf life of fresh meats. This innovative package consists of a standard MAP tray but has a perforated false bottom under which, a porous sachet containing sodium bicarbonate/ascorbate is positioned. When juice exudates from the packaged meat drips onto the sachet, carbon dioxide is emitted, thus replacing any carbon dioxide absorbed by the meat and preventing package collapse. The inhibition of spoilage bacteria utilizing active packaging technology may reduce bacterial competition and thus permit growth and toxin production by non-proteolytic *C. botulinum* or the growth of other pathogenic bacteria (Sivertsvik, 2003). Lovenklev *et al.* (2004) reported that while a high concentration of carbon dioxide decreased the growth rate of nonproteolytic *C. botulinum* type B, the expression and production of toxin was greatly increased which means the risk of botulism may also be increased, instead of reduced if used in MAP systems.

Table 2: Properties and types of FreshMax^R and FreshPax^{TM} oxygen scavengers.

Type	*Ideal applications*	a_w	*Absorption speed*	*Capacities*
FreshMax ^{R} B	Moist or semi-moist foods	> 0.65	1–3 days	10cc, 20cc
FreshMax^R M	Moist or semi-moist foods where a CO_2 flush is used	> 0.65	2–5 days (30% to 60% of capacity); 4–7 days (entire capacity)	10cc, 20cc
FreshPax^{TM} B	Moist or semi-moist foods	> 0.70	1–3 days	10–2000cc
FreshPax^{TM} D	Dehydrated or dry foods	> 0.70	0.5–4 days	50–2000cc
FreshPax^{TM} R	Refrigerated or where rapid deoxygenation is required	All	8 hr–2 days	50–2000cc
FreshPax^{TM} M	Moist or semi-moist foods where a CO_2 flush is used	> 0.65	1–6 days	20–1000cc

3.1.3. *Moisture absorbers*

Control of moisture is important for food preservation. Depending on the food product in question, it might be required to keep the moisture out of the package or in the package. In most cases, the packaging material itself is responsible for the control of moisture transfer between the internal and external environment, providing an adequate barrier. There are situations however, where a greater control is needed to avoid the build-up of liquid water inside the package, therefore requiring liquid water control or humidity buffering. Soaking up moisture by using various absorbers is very effective at maintaining food quality and extending shelf life by inhibiting microbial growth and moisture related degradation of texture and flavour. In addition to moisture absorber sachets for humidity control in packaged dried foods, several companies manufacture moisture drip absorbent pads, sheets and blankets or liquid water control in high a_w foods such as meats, fish, poultry, fruit and vegetables. Basically they consist of two layers of a microporous non-woven plastic film, such as PE or PP, between which is placed a superabsorbent polymer that is capable of absorbing up to 500 times its own weight with water. Typical superabsorbent polymers include polyacrylate salts, carboxymethyl cellulose (CMC) and starch copolymers, which have a very strong affinity for water (Reynolds, 2007). Moisture drip absorber pads are commonly placed under packaged fresh meats, fish and poultry to absorb unsightly tissue drip exudate. Commercial moisture absorber sheets, blankets and trays include Toppan Sheet™, Thermarite™, Luquasorb™ and Fresh-R-Pax™.

3.1.4. *Antimicrobial packaging*

Antimicrobial packaging is a promising form of active packaging especially for meat products. Since microbial contamination of meat products occurs primarily at the surface, due to post-processing handling, attempts have been made to improve safety and to delay spoilage by the use of antibacterial sprays or dips. Limitations of such antibacterial include neutralization of compounds on contact with the meat surface or diffusion of compounds from the surface into the meat mass. Incorporation of bactericidal agents into meat formulations may result in partial inactivation of the active compounds by meat constituents and therefore exert a limited effect on surface microflora (Quintavalla & Vicini, 2002). Antimicrobial food packaging materials have to extend the lag phase and reduce the growth phase of microorganisms in order to extend shelf life and to maintain product quality and safety. Many classes of antimicrobials listed range from acid anhydride, alcohol, bacteriocins, chelators, enzymes, organic acids and polysaccharides but a very few packages have commercial successes except in Japan where Silver substituted zeolite is the most common antimicrobial agent incorporated into plastics. Ag-ions inhibit a range of metabolic enzymes and have strong antimicrobial activity (Vermeiren *et al.*, 1999). Coating of films with antimicrobial agents can result in effective antimicrobial activity. Natrajan and Sheldon (2000) carried out a study to evaluate the potential use of packaging materials as delivery vehicles for carrying and transferring nisin containing formulations onto the surfaces of fresh poultry products. The efficacy of nisin coated (100 μg/mL) polymeric films of varying hydrophobic ties (PVC, LLDPE and PA) in inhibiting *Salmonella typhimurium* on fresh broiler drumstick skin was evaluated. It was concluded that packaging films coated with nisin were effective in reducing *S. typhimurium* on the surface of fresh broiler skin and drumsticks. Ha *et al.* (2001) examined the effect of grape fruit seed extract (GFSE), a natural antimicrobial agent, incorporated (0.5% or 1% concentration) by co-extrusion or a solution-coating process in multilayered polyethylene (PE) films, on the microbial status and quality (colour (L, a, b), thiobarbituric acid reactive substances (TBARS) and pH of fresh minced beef. The antimicrobial activity of the fabricated multilayer films was also evaluated using an agar plate diffusion method. It was reported that coating the PE film with GFSE with the aid of a polyamide binder resulted in greater antimicrobial activity compared to GFSE incorporation by co-extrusion (Kerry *et al.*, 2006). The use of naturally derived antimicrobial agents is important as they represent a lower perceived risk to the consumer. Skandamis and Nychas (2002) studied the combined effect of volatiles of oregano essential oil and modified atmosphere conditions (40% CO_2: 30% O_2: 30% N_2, 100% CO_2, 80% CO_2, vacuum packaged and aerobic storage) on the sensory, microbiological and physicochemical attributes of fresh beef stored at 5 and 15°C. Filter paper containing absorbed essential oil was placed in the packages but not in direct contact with the beef samples. The shelf life of beef samples followed the order: aerobic storage < vacuum packaged < 40% CO_2: 30% O_2: 30% N_2 < 80% CO_2: 20% air < 100% CO_2. Longer shelf life was observed in samples supplemented with the volatile compounds of oregano essential oil (Kerry *et al.*, 2006).

3.1.5. *Ethanol emitters*

The use of ethanol as an anti-microbial agent, particularly for surface sterilization and disinfection, is well known. It acts against vegetative cells of microorganisms in high concentrations, and it also has a preserving action in low concentrations. Ethanol creates an anti-mold atmosphere, particularly effective in extending the shelf-life of high water activity baked products. The use of ethanol generating sachets or strips avoids the ethanol spraying directly onto the product surface prior to packaging. The ethanol is absorbed or encapsulated in a carrier material enclosed in sachets of selective permeability to ethanol. These allow for the ethanol vapor to be released into the packaging headspace. The packaging material should have high barrier characteristics to ethanol, preferably below $2g/m^2/day$ according to suppliers' recommendations, to allow for ethanol accumulation in the headspace. The level of ethanol in the packaging headspace depends obviously on the sachet size and on product water activity. The most commercially used system is the Ethicap. Ethicap consists of silicon dioxide powder (35%) containing adsorbed alcohol (35%) and water (10%). The sachet is made of a laminate of paper/ethyl vinyl acetate, which presents one of the lowest barriers to ethanol vapor of the commonly used materials.

3.1.6. *Flavour and odour absorbers*

In packaged products, flavour and odour absorbers take in unwanted gaseous molecules such as volatile package ingredients, chemical metabolites of foods, microbial metabolites, respiration products, or off-flavours in raw foods. During storage various sulfurous compounds and amines produced biochemically from protein degradation, aldehydes and ketones produced from lipid oxidation or anaerobic glycolysis, and that are removed by use of flavour and odour absorbing systems employing mass transfer mechanisms. They are typically available as films, sachets, tapes/labels, and trays. Flavour and odour absorbers are usually placed inside packages or combined with other flavour permeable materials. In mass transfer phenomenon, porous/fine particulate materials absorb flavour molecules due to their large surface areas, though have low absorption rates, overall absorption is very significant. Porous materials-such as zeolites, clays, molecular sieves, active carbon, maltodextrin, and cyclodextrin have been found to increase flavor. Besides absorption rate, the maximum absorption, which is also significant factor, can be controlled by the thickness of films, with thicker films absorbing more flavours. In another aspect, polar materials or corona treatment of film surface and/or high barrier materials can be used to prevent non-polar flavor absorption. Alkali chemicals such as ammonia and amines can be absorbed and neutralized by packaging materials containing acidic ingredients. This acid-alkali reaction mechanism between volatile chemicals and polymer ingredients has been utilized to eliminate various alkali chemicals and aldehydes.

3.2. Intelligent Packaging

Intelligent packaging systems are those that monitor the condition of packaged foods to give information regarding the quality of the packaged food during transport and storage. The potential of biosensor technologies like freshness indicators and time-temperature indicators (TTI) and radio frequency identification (RFID) are evaluated for potential use in meat and meat products. These smart devices may be incorporated in package materials or attached to the inside or outside of the package. Recognition of the benefits of active and intelligent packaging technologies by the food industry, development of economically viable packaging systems and increased consumer acceptance is necessary for commercial realization of these packaging technologies (Kerry *et al.,* 2006).

3.2.1. *Freshness indicators*

The information provided by intelligent packaging systems on the quality of meat products may be either indirect or direct (Kerry *et al.,* 2006). Freshness indicators provide direct product quality information resulting from microbial growth or chemical changes within a food product. Microbiological quality may be determined through reactions between indicators included within the package and microbial growth metabolites. The chemical detection of spoilage of foods and the chemical changes in meat during storage provide the basis for which freshness indicators may be developed based on target metabolites associated with microbiologically induced deterioration. Using the marker concept in this manner may result in the more widespread commercial development of freshness indicators for meat products in the not too distant future. A number of validation studies have been undertaken in order to establish the usefulness of TTIs in food products (Welt *et al.,* 2003). Yoon *et al.* (1994) showed a positive correlation between oxidative stability and TTI colour change using a phospholipid based TTI in frozen pork. Smolander *et al.* (2004) and Vainionpaa *et al.* (2004) determined the applicability of TTIs for monitoring the quality of MA packed broiler cuts at different temperatures and in comparison with several standard analytical methods, respectively.

The TTI's presently on the market have working mechanisms based on different principles: biological, chemical and physical. For the first type, the change in biological activity, such as microorganisms, spores or enzymes is the basic working principle. The others are based on a purely chemical or physical response towards time and temperature, such as an acid-base reaction, melting, polymerization, etc. The devices must activate easily, they must have an accurate response to temperature or to the cumulative effect of time and temperature, and the response must be quick enough and irreversible. In order to be used as food quality and safety monitors, the rate of change in the device must correlate well with the rate of deterioration of the food product. The rate of device changes must also correlate with the temperature variation over time during transportation and distribution.

TTI's are devices that show an irreversible change in a physical characteristic, usually color or shape, in response to temperature history. The TTI's are expected to mimic the change of a certain quality parameter of the food product undergoing the same exposure to temperature. Some are designed to monitor the evolution of temperature with time along the distribution chain and others are designed to be used in the consumer packages. The devices can have a threshold indicator response to temperature that is independent of a threshold, which implies that they start monitoring immediately after activation—these devices are called full-history devices.

In commercial application Fresh-Check^{TM} LifeLines (self-adhesive labels) is used as full history indicator whose working mechanism is based on the color change of a polymer formulated from diacetylene monomers. It consists of a small circle of polymer surrounded by a printed ring for color reference. The polymer, which starts lightly colored, gradually deepens in color to reflect the cumulative exposure to temperature. The polymer changes color at a rate proportional to the rate of food quality loss: the higher the temperature, the more rapidly the polymer changes in color. Consumers may be advised on the package not to consume the product if the color of the center is darker than the reference ring. This device is self-activated, *i.e.*, it is produced in an active form. Therefore, it is supplied and requires storage under deep-frozen conditions, typically –20°C.

Vitsab^{TM} Indicator is another full history indicator but this based on an enzymatic reaction. The device consists of a bubble-like dot containing two compartments: one for the enzyme solution, lipase plus a pH indicating dye compound and the other for the substrate, consisting primarily of triglycerides. The dot is activated at the beginning of the monitoring period by application of pressure on the plastic bubble, which breaks the seal between compartments. The ingredients are mixed and as the reaction precede a pH change results in a color change. The dot, initially green in color, becomes progressively yellow as product approaches the end of shelf-life. The reaction is irreversible and will proceed faster as temperature is increased and slower as temperature is reduced. The integrator, in the format of adhesive labels, is available in two basic configurations: single or triple-dot. Single-dot tags are used for transit temperature monitoring of cartons and pallets of product and for consumer packages as well. Triple-dot tags are typically used in the wholesale distribution chain and incorporate three graded responses in a single label. The sequential development of color is appropriate for "signposts" in the management of the shelf-life of the product.

Partial-history devices will not respond unless a certain temperature threshold has been exceeded. Abuse indicators show exposure above a reference temperature and are not intended to show the history of exposure above a certain critical temperature. It is based on a special substance having a selected melting point and blue dye. The heart of the indicator is a porous wick layer over a reservoir pad containing the blue dye. A film strip separates the wick from the reservoir that is removed at the activation stage. At this point, the porous wick, white in color, is shown in the window. Upon exposure to a temperature exceeding the critical temperature, the substance melts and begins to diffuse through the porous wick, causing a blue coloring to appear. There are available indicators with different critical temperatures from 15°C to 26°C. The consumer label is a partial history integrator that changes color when exposed to higher than recommended storage temperature and will also change as the product reaches the end of its shelf-life. The working principle is based, as above, on the melting and diffusion of a dye.

Presently, although several prototypes are being developed, commercial biosensors for intelligent packaging are not available. For example, a biosensor/barcode called Food Sentinel System is being developed to detect pathogens in food packages. A specific-pathogen antibody is attached to the barcode's membrane-forming part; the presence of contaminating bacteria will cause the formation of a localized dark bar, rendering the barcode unreadable upon scanning. A diagnostic system called Toxin Guard is being developed that incorporates antibodies into plastic packaging films to detect pathogens. When the antibodies encounter a target pathogen, to alert the consumer, inspector, or retailer the packaging material displays a clear visual signal. This system is planned for detecting gross contamination, because it is not sensitive enough for detecting very low levels of pathogens that can cause disease (Ayala & Park, 2000).

REFERENCES

Ayala, C.E. and Park, D.L. (2000). New bar code will help monitor food safety. *The Magazine of the Louisiana Agricultural Experiment Station,* 43(2): 8.

Blixt, Y. and Borch, E. (2002). Comparison of shelf life of vacuum-packed pork and beef. *Meat Sci.*, 60: 371–378.

Brody, A.L. (2010). Intelligent packaging: A powerful tool for food. *Food Technol.*, 5: 79–80.

Buffo, R.A. and Holley, R.A. (2005). Centralized packaging systems for meats. *In*: Han, J.H. *Ed*, *Innovations in food packaging*, Elsevier Academic Press, Amsterdam, pp. 227–236.

Capita, R., Diaz-Rodriguez, N., Prieto, M. and Alonso-Calleja, C. (2006). Effects of temperature, oxygen exclusion, and storage on the microbial loads and pH of packed ostrich steaks. *Meat Sci.*, 73: 498–502.

Cole, A.B.Jr. (1986). Retail packaging systems for fresh red meat cuts. Reciprocal Meat Conference. *Proc. Am. Meat Sci. Assoc.*, 39: 106–111.

Day, B.P.F. (2008). Active packaging of food, smart packaging technologies for fast moving consumer goods, edited by Joseph, Kerry. and Paul, Butler. John Wiley & Sons, Ltd.

Faustman, C. and Cassens, R.G. (1990). The biochemical basis for discoloration in freshmeat: A review. *J. Muscle Foods*, 1: 217–243.

Fernandez-Lapez, J., Sayas-Barbera, E., Munoz, T., Sendra, E., Navarro, C. and Perez-Alvarez, J.A. (2008). Effect of packaging conditions on shelf-life of ostrich steaks. *Meat Sci.*, 78 (1–2): 143–152.

Garriga, M., Grebol, N., Aymerich, M.T., Monfort, J.M. and Hugas, M. (2004). Microbial inactivation after high pressure processing at 600 MPa in commercial meat products over its shelf life. *Innovative Food Science and Emerging Technologies*, 5(4): 451–457.

Gill, C.O. and McGinnis, J.C. (1995). The use of oxygen scavengers to prevent the transient discoloration of ground beef packaged under controlled, oxygen depleted atmospheres. *Meat Sci.*, 41: 19–27.

Ha, J.U., Kim, Y.M. and Lee, D.S. (2001). Multilayered antimicrobial polyethylene films applied to the packaging of ground beef. *Packaging Technology and Science*, 15: 55–62.

Huang, N., Ho, C. and McMillen, K.W. (2005). Retail shelf life of pork dipped in organic acid before modified atmosphere or vacuum packaging, *J. Food Sci.*, 70: M382–M387.

Kerry, J.P., O'Grady, M.N. and Hogan, S.A. (2006). Past, current and potential utilization of active and intelligent packaging systems for meat and muscle based products: A review. *Meat Sci.*, 74: 113–30.

López-Rubio, A., Gavara, R. and Lagarón, J.M. (2006). Bioactive packaging: Turning foods into healthier foods through biomaterials. *Trends in Food Science and Technology*, 17: 567–575.

Lovenklev, M., Artin, I., Hagberg, O., Borch, E., Holst, E. and Radstrom, P. (2004). Quantitative interaction effects of carbon dioxide, sodium chloride, and sodium nitrite on neurotoxin gene expression in nonproteolytic *Clostridium botulinum* type B. *Appl. Env. Microbiol.*, 70: 2928–2934.

Mancini, R.A., Hunt, M.C., Hachmeister, K.A., Kropf, D.H. and Johnson, D.E. (2005). Exclusion of oxygen from modified atmosphere packages limits beef rib and lumbar vertebrae marrow discoloration during display and storage. *Meat Sci.*, 69: 493–500.

Mantis, F.N., Tsachev, I., Sabatakou, O., Burriel, A.R., Vacalopoulos, A. and Ramantanis, S.B. (2005). Safety and shelf-life of widely distributed vacuum packed, heat treated sausages. *Bulgarian Journal of Veterinary Medicine*, 8(4): 245–254.

Marriott, N.G., Smith, G.C., Hoke, K.E., Carpenter, Z.L. and West, R.L. (1977). Long-distance transoceanic shipments of fresh beef. *J. Food Sci.*, 42 (2): 316–320.

Martinez, L., Djenane, D., Cilla, I., Beltran, J.A. and Roncales, P. (2006). Effect of varying oxygen concentrations on the shelf-life of fresh pork sausages packaged in modified atmosphere. *Food Chem.*, 94: 219–225.

Mayr, D., Margesin, R., Klingsbichel, E., Hartungen, E., Jenewein, D., Schinner, F. and Mark, T.D. (2003). Rapid detection of meat spoilage by measuring volatile organic compounds by using proton transfer reaction mass spectrometry. *Appl. Env. Microbiol.*, 69: 4697–4705.

McCormick, K.E., Han, I.Y., Acton, J.C., Sheldon, B.W. and Dawson, P.L. (2005). In-package pasteurization combined with biocide-impregnated films to inhibit *Listeria monocytogenes* and *Salmonella* Typhimurium in turkey bologna. *J. Food Sci.*, 70: M52–M56.

Natrajan, N. and Sheldon, B.W. (2000). Efficacy of nisincoated polymer films to inactivate Salmonella typhimurium on fresh broiler skin. *J. Food Prot.*, 63: 1189–1196.

Patsias, A., Chouliara, I., Badeka, A., Savvaidis, I.N. and Kontominas, M.K. (2006). Shelf-life of a chilled precooked chicken product stored in air and under modified atmospheres: Microbiological, chemical and sensory attributes. *Food Microbiol.*, 23: 423–429.

Payne, S.R., Durham, C.J., Scott, S.M. and Devine, C.E. (1998). The effects of non-vacuum packaging systems on drip loss from chilled beef. *Meat Sci.*, 49: 277–287.

Quintavalla, S. and Vicini, L. (2002). Antimicrobial food packaging in meat industry. *Meat Sci.*, 62: 373–380.

Reynolds, G. (2007). Superabsorbent soaks up packaging problems. Available at: http://www.foodproductiondailyusa.com/., Accessed: 02.05.2010.

Samelis, J., Bedie, G.K., Sofos, J.N., Belk, K.E., Scanga, J.A. and Smith, G.C. (2002). Control of *Listeria monocytogenes* with combined antimicrobials after postprocess contamination and extended storage of frankfurters at 4C in vacuum packages. *J. Food Prot.*, 65: 299–307.

Scetar, M., Kurek, M. and Galic, K. (2010). Trends in meat and meat products packaging—a review. *Croat J. Food Sci. Technol.*, 2(1): 32–48.

Seydim, A.C., Acton, J.C., Hall, M.A. and Dawson, P.L. (2006). Effects of packaging atmospheres on shelf-life quality of ground ostrich meat. *Meat Sci.*, 73: 503–510.

Sivertsvik, M. (2003). Active packaging in practice: Fish. *In*: Ahvenainen, R. (*Ed.*), Novel food packaging techniques, Cambridge, UK: Woodhead Publishing Ltd., pp. 384–400.

Skandamis, P.N. and Nychas, G.J.E. (2002). Preservation of fresh meat with active and modified atmosphere packaging conditions, *Int. J. Food Microbiol.*, 79: 35–45.

Smolander, M., Alakomi, H.L., Ritvanen, T., Vainionpaa, J. and Ahvenainen, R. (2004). Monitoring of the quality of modified atmosphere packaged broiler cuts stored in different temperature conditions. A. Timetemperature indicators as quality-indicating tools, *Food Control*, 15: 217–229.

Sorheim, O., Nissen, H. and Nesbakken, T. (1999). The storage life of beef and pork packaged in an atmosphere with low carbon monoxide and high carbon dioxide. *Meat Sci.*, 52(2): 157–164.

Tewari, G., Jayas, D.S., Jeremiah, L.E. and Holley, R.A. (2001). Prevention of transient discoloration of beef. *J. Food Sci.*, 66: 506–510.

Vainionpaa, J., Smolander, M., Alakomi, H.L., Ritvanen, T., Rajamaki, T., Rokka, M. and Ahvenainen, R. (2004). Comparison of different analytical methods in the monitoring of the quality of modified atmosphere packaged broiler chicken cuts using principal component analysis. *J. Food Engineer*, 65: 273–280.

Vermeiren, L., Devlieghere, F., Van Beest, M., de Kruijf, N. and Debevere, J. (1999). Developments in the active packaging of foods. *Trends Food Sci. Technol.*, 10: 77–86.

Wang, S.H., Chang, M.H. and Chen, T.C. (2004). Shelf-life and microbiological profiler of chicken wing products following Sous vide treatment. *Int. J. Poult. Sci.*, 3: 326–332.

Welt, B.A., Sage, D.S. and Berger, K.L. (2003). Performance specification of time-temperature integrators designed to protect against botulism in refrigerated fresh foods. *J. Food Sci.*, 68: 2–9.

Yoon, S.H., Lee, C.H., Kim, D.Y., Kim, J.W. and Park, K.H. (1994). Time-temperature indicator using a phospholipids-phospholipase system and application to storage of frozen pork. *J. Food Sci.*, 59: 490–493.

19

Low-Fat Meat Products

S.K. Das and P.K. Mandal

ABSTRACT

Meat fat acts as a reservoir for flavor compounds and contributes to the texture of product. There are diverse possible strategies for developing low fat meat and meat products. Reducing the fat content in products leads to a firmer, rubbery, less juicy product with dark color and more cost. Technological problems like reduction in particle binding, reduced cook yields, soft and mushy interiors, rubbery skin formation, excessive purge and shortened shelf life are also associated with reduction in fat levels. This paper describes role of fat in diet, rationale for reducing fat in meat products, fat replacer and their classification, strategies for fat reduction and low fat meat products, effect of fat reduction on product quality, procedure of reducing fat content, selection of additives, protein, carbohydrate and fat based fat replacer and super critical fluid extraction.

1. INTRODUCTION

The increasing concerns for health led the efforts made by the food industries to develop new foods with positive health benefits. Obesity, cardiovascular disease, stroke, and cancer have been implicated with high fat intake in excess of body needs. The contribution of fat to obesity is due to the well known fact that fat is energy dense providing 2.25 times more energy per unit as carbohydrate and protein. International medical institutions have been recommending for the last 40 years that dietary fat intake should be controlled both as quantity of calories and type of fatty acids. Nutritional guidelines suggest that dietary fat should provide between 15 and 30 percent of total calories and that saturated fats should be limited between 0 and 10 percent of caloric intake.

Meat is an excellent source of valuable nutrients. Undoubtedly, a major source of food proteins with high biological value in many countries is meat. Meat is also an excellent source of some essential fatty acids, vitamins and minerals, and all these components have specific function to our body. One of the most important commodities of total expenditure for food in most of the European and American countries is meat and meat products. It accounts for about 30% of expenditure in Italy, Spain, Ireland and Turkey, 25% in the UK, The Netherlands and Greece and 35% in Denmark, France and Belgium (Anonymous, 2002; Tuley, 1996). Although meat and meat products are highly nutritious, fat content in these products, of late are receiving bad publicity. Since the last few decades, negative campaigns about the muscle foods and their possible health hazards made the consumers think about the food which not only taste good, but also attractive, safe and healthy. The increasing concern of the consumers for health has led the meat scientists and workers to develop new products with positive health benefits.

The relationships between high cholesterol level and low polyunsaturated/ saturated fatty acids (PUFA/SFA) ratio and the rise in coronary heart diseases has resulted in focusing on low fat food products including some meat products (Giese, 1992). Researchers have been working on strategies to reduce animal fat usage in meat products. Thus, there is an expanding new category of products that acquire added nutritional value as a result of a significant reduction in their fat and calorie contents, and in their saturated fat and cholesterol levels. In order to be labeled as low fat, a meat product must contain no more than 10% fat (USDA, 1991). Fat content is essential to meat product characteristics such as flavour, juiciness and texture and therefore fat content cannot be reduced by using less fat or direct replacement with another type of fat. Fat reduction in finely comminuted meats results in products of less desirable texture due to significant changes in hardness. Fat replacers or substitutes are ingredients that contribute a minimum of calories to formulated meat products and do not dramatically alter flavour, juiciness, mouth feel, viscosity or other organoleptic and processing properties (Keeton, 1994).

2. FUNCTIONS OF FAT IN DIET

It is very difficult to replace the fat from a diet since it has many important functions. It is one of the major nutrients, and is important for maintaining good health. Fats carry the fat-soluble vitamins A, D, E, and K and other phytochemicals and aid their absorption in the intestine. They are the only source of the essential fatty acids linoleic and linolenic acids. They are also an important source of calories for adults and infants that have high-energy needs (Fat provides 9 calories per gram, compared with 4 calories per gram for protein and carbohydrates). Fat is a calorie-dense nutrient that provides flavour, texture, juiciness and water entrapment in meat product. As a food component, fat contributes key sensory and physiological benefits. Fat contributes to flavor (the combined perception of mouth-feel, taste and aroma/ odor), creaminess, appearance, palatability, texture and lubricity of foods and increases the feeling of satiety during meals. Fat can also carry lipophilic flavor compounds, act as a precursor for flavor development (*e.g.*, by lipolysis or frying), and stabilize flavour. From a physiological standpoint fat is a source of fat soluble vitamins, essential fatty acids, precursors for prostaglandins and is a carrier for lipophilic drugs (Akoh, 1998). Finally, since lipid digestion is slow, fats produce a satiated feeling. If reformulated products do not produce a feeling of satiety, food intake may be increased—another task which has to be considered when formulating fat reduced products. The key roles of fat in a diet are depicted in Fig. 1.

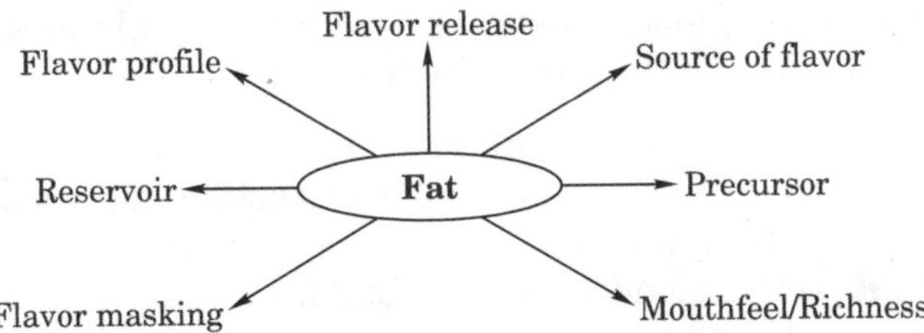

Fig. 1: Functions of fat (*Source*: Pearson & Gillett, 1999).

Fat content of a food influences its eating qualities, binding properties, rheological and textural properties. Removal or lowering of fat in food products poses challenges in the form of various technological problems such as rubbery and dry texture, reduced product yield, soft mushy interiors, excessive purge and changes in flavour and mouth-feel (Keeton, 1994). During cooking also, fat in food has multiple functions. Its heat transfer properties enable rapid heating and attainment of very high temperatures. High temperature in frying and deep-fat frying creates many browning (maillard reaction) taste compounds and absorb many flavour compounds that have positive sensory attributes. In meat, fat carries flavour and contributes to juiciness and tenderness, key to the difference in taste of various kinds of meat and poultry (Grossklaus, 1996; Jones, 1996). Functionally, fats affect the melting point, viscosity and body, crystallinity, and spreadability of many foods (Drewnowski, 1998; Sandrou & Arvanitoyannis, 2000). Fats are responsible for the aroma and texture of many foods, and thereby affecting the overall palatability of the diet. In baked products, fat inhibits the tough gluten strands, softens the crumb, imparts tenderness, and delays staling. Crispiness in cookies is due to fat in combination with some of the other ingredients. In flaky products such as croissants and pastries, fat's ability to pool in layer and coat gluten strands is crucial (Omayma & Yussef, 2007).

3. RATIONALE FOR LOW-FAT MEAT PRODUCTS

Over the past few decades, there has been a trend to reduce or modify the consumption of fat and saturated fatty acid content of our diet, especially of animal origin. Traditionally, processed meat contains a high percentage of fat. For example, frankfurters and bologna contain over 20% fat in Australia and as much as 50% fat in fresh pork sausage is allowed in USA. Link of high fat and saturated fatty acid intake to some chronic diseases like high blood cholesterol, stroke, high blood pressure, coronary heart disease, obesity, diabetes and some types of cancer has been established (Grossklaus, 1996; AHA, 1996; USDHHS, 1988; ADA, 2005). Over the past decade, obesity rate have increased by more than 60% among adults, with approximately 59 million adults being obese. Among youth ages 6 to 19 years, approximately 9 million are considered to be overweight (CDCP, 2000). Overconsumption of fat, an energy-dense substance, providing 2.25 times of more energy per unit as carbohydrate and protein, is accountable for these hazards. With the obesity levels rising across the globe, consumers are increasingly looking for low-fat and low-calorie versions of their favorite foods.

More than twenty governments and health organizations from many parts of the world have formulated recommendations asking for a reduction in total fat and saturated fat from the diet. The American Dietary Guidelines (USDA & USDHHS, 1995) recommended to limit total fat intake to not more than 30% of daily energy intake, with saturated fats not more than 10% and monounsaturated and polyunsaturated fats accounting for at least two-thirds of daily energy intake. The American Heart Association, National Cholesterol Education Program and several other professional health organizations throughout the world have recommended lowering the daily intake of dietary fat to an average of 30% of total calories, consuming less than 300 mg of cholesterol per day, and limiting saturated fat to less than 10% of total calories as a means of reducing the risk of developing coronary heart disease. Consumers are advised to choose lean cuts of red meat, use low-fat cooking methods, eat moderate amount of meat (approximately 170 g/day, cooked), and remove visible fat from the meat cuts (Keeton, 1994). As a result reduction of fat in food products is a growing area of interest to food manufacturers especially in the meat sector. Consumption of meat products is therefore under pressure and in order to maintain their market share of convenience foods, the meat industry must change the image and quality of traditional meat products by developing low-fat versions.

The demand for low fat meat products has greatly increased in recent years as high fat intake, especially saturated fats, is associated with increased risk of chronic diseases of the circulatory system and some types of cancers. Since fat acts as a reservoir for flavour compounds and contributes to the texture of product, reducing the fat content in meat products may alter product quality. Long chain polyunsaturated n-3 fatty acids (PUFA) have been implicated as critical nutrients for human health. In order to improve healthy lipid profile, reformulation is generally based on the partial replacement of the animal fat normally present in the product with another fat whose characteristics are more in line with health recommendations and if possible cholesterol free. A variety of non-meat fats of plants and marine origin have been added to different meat products as partial substitutes for meat fats (Cengiz & Gokoglu, 2007).

4. CLASSIFICATION OF FAT REPLACERS BASED ON FUNCTIONALITY

Fat replacers are the substances that can be used to provide some or all of the functional properties of fat in a complex food system but yields fewer calories than fat. Fat replacers are either fat substitutes or fat mimetics (Fig. 2).

4.1. Fat Substitutes

They are macromolecules that physically and chemically resemble triglycerides (conventional fats and oils) and which can theoretically replace the fat in foods on a one-to-one, gram-for-gram basis. Often referred to as lipid- or fat-based fat replacers, fat substitutes are either chemically synthesized or derived from conventional fats and oils by enzymatic modification. Many fat substitutes are stable at cooking and frying temperatures. Olestra, sucrose polyesters, sucrose fatty acid esters, trehalose, raffinose, stachylose polyesters and sorbestrin, alkyl glycoside polyesters, salatrim etc. are some of the examples of fat substitutes or lipid-based fat replacers.

4.2. Fat Mimetics

They are substances that imitate organoleptic or physical properties of triglycerides but cannot replace fat on a one-to-one, gram-for-gram basis. Fat mimetics, often called protein- or carbohydrate- based fat replacers, are common food constituents, *e.g.*, starch and cellulose, but may be chemically or physically modified to mimic the function of fat. The caloric value of fat mimetics ranges from 0–4 kcal/g. Fat mimetics generally are not suitable for frying because they bind excessive water and denature or caramelize at high temperatures. Many fat mimetics, however, are suitable for baking and retorting. Fat mimetics are less flavorful than fats that they are intended to replace. They carry water soluble flavors but not lipid soluble flavor compounds (Akoh, 1998).

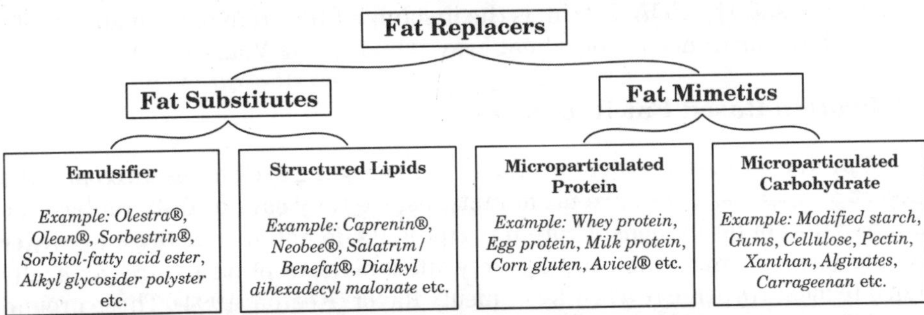

Fig. 2: Classification of fat replacers based on functionality.

Fat analogs are compounds with many of the characteristics of fat but have an altered digestibility and altered nutritional value. *Fat extenders* optimize functionality of fat, thus allowing a decrease in the usual amount of fat in the products. *Fat barriers* are different systems which reduce the fat content of a food by retarding fat absorption into the product during frying.

5. CLASSIFICATION OF FAT REPLACER BASED ON INGREDIENT

Fat replacers can also be classified as: 1. Starch based, 2. Fiber based, 3. Protein based, 4. Gums, gels and thickeners, 5. Emulsifiers, 6. Bulking agents, 7. Low-calorie fats, 8. Fat extenders, 9. Synthetic fat-substitutes and 10. Combination system (Omayma and Youssef, 2007). This classification is based partially on chemical composition and partially on functionality and includes combination system (*i.e.*, blends) (Jones, 1996). Some of the major categories are described here briefly.

5.1. Carbohydrate Based Fat Replacer

These fat replacers achieve fat mimicking mainly by stabilizing the added water in gel like matrix resulting in lubricity and moisture release similar to high fat products (Glicksman 1991). Pre formed carbohydrate gels such as Konjac flour can also mimic particle definition. Malto dextrins and dextrins are found to contribute significantly to mouth feel, body and viscosity (Lucca and Tepper, 1994). Modified starches are found to be of great use for reasonable cost, wide availability and conventional handling procedures. Fiber based fat replacers are also basically coming under this category. Oat fiber/bran, rice bran, wheat bran etc. are widely used as fibrous fat replacer.

5.2. Modified Starch

They are mainly used as texture modifier and bodying agent used along with emulsifier, protein, gums and other modified food starches. Commercially available modified starches are mainly Oatrim and Z-trim. Oatrim, the water soluble products of oat flour are to be used as fat replacer and as a texturizing agent, produces 1–4 kcal/gram. Oatrim is classified as GRAS (generally regarded as safe) by FDA. Z-trim is the insoluble fiber from oat, soyabean, pea and rice hulls or from corn or wheat bran (Omayma & Youssef, 2007).

5.3. Protein Based Fat Replacers

Use of proteins in low fat meat products is an ongoing area of research and the subject of reviews. These are fat mimetics since they cannot fully replace the functionality of fat in foods. Protein ingredients do not appear to tie up water soluble flavor components and hydrolyzates of these proteins are generally added to meats to achieve a brothy or meaty flavor (Keeton, 1994). These protein based fat replacers have been used successfully in the manufacture of comminuted meat products and are reported to have good potential as extenders in several meat products due to their high nutritive value and wide range of functional properties like solubility, viscosity and water binding capacity. Protein based fat replacers were widely studied and reported; whey protein concentrates in sausages (Laroia *et al.*, 1995, Serdaroglu & Sapanci-Ozsumer, 2003), use of wheat flour and caseinates in chicken nuggets (Reddy and Rao, 1997), sodium caseinate, milk power and whey powder in turkey rolls, milk proteins in chicken sausages (Rao *et al.,* 1999), soy flour in buffalo meat burgers (Modi *et al.*, 2003) milk co-precipitates in the buffalo meat (Sunil kumar *et al.*, 2003).

5.4. Micro-Particulated Proteins

These are reduced-calorie (1–2 calorie/gram) ingredient made from mainly whey protein, milk protein or egg-protein. Micro particulated proteins can mimic certain properties of fat. The process of micro particulation involves simultaneous application of heat and high shear to proteins. While processing, these particles form micelle with uniform heating and size distribution (Setser & Racette, 1992). The microparticles function in the mouth as small "balls" that imitates the creaminess of fat. Also these particles coat the taste buds and mouth with fat-mimicking effect (*e.g.*, Simplesse®). This coating action besides allowing flavors to reach receptors slowly and gradually also masks up some of the off flavors that seem more pronounced in low fat high water added systems (Lucca & Tepper1994). Simplesse (NutraSweet Kel Co), Trail blazer (Kraft general foods), Finess (Reach associates), Dairy light (Ault foods) are some of the examples for microparticulated protein based fat replacers.

5.5. Non-Protein Additives

Microcrystalline cellulose, carboxy-methyl cellulose, methylated cellulose derivatives, and other gums fall into this category (mainly fiber based). Most of them provide viscosity or thickening and in some cases gel formation, contribute no calories and have potential health benefits as dietary fibers. Some gums can provide a slippery, creamy and lubricacious mouth-feel like that of fat and hence used as fat replacer in wide range of food (Glicksman, 1991). Carboxy methyl cellulose like water soluble celluloses can be used as bulking agents because of their high water binding capacity (Sester & Racette, 1992). Methylated cellulose derivatives have reversible thermal gelation property and even though not true fat mimetics, but are useful as fat barriers in fried foods as they reduce oil absorption into batter. Gums are also hydro colloids that provide viscosity or thickening and in some cases gel formation. These gums contribute no calories and have potential health benefits as dietary fibers. There are a number of gums available to the food industry (Dziezak, 1991).

Carrageenan, a sea weed gum also used widely as fat replacer in meat product. Carrageenan is being used with good results in ground beef patties (Keeton, 1998), Pork nuggets (Berry, 1994), Pork sausages (Lyons *et al.*, 1999). Other gums such as xanthan gum and locust bean are frequently included in reduced fat formulations to help stabilize the system. Alginates and pectins are also used as fat replacers in meat product. Usage levels of microcrystalline cellulose can range from 0.1 to 10%, but standard levels are from 0.4 to 3.0%.

Animal proteins like milk proteins including non fat drymilk, calcium reduced non fat dry milk, dried whey, whey protein concentrate, butter milk proteins, the caseinates and skim milk co-precipitates contain appreciable amounts of protein. The dairy products are widely utilized in processed meat and poultry products such as whey protein concentrate in sausages (Laroia *et al.*, 1995; Lyons *et al.*, 1999; Serdaroglu & Deniz, 2003), milk co-precipitates in pork patties (Manish and Sharma, 2002). Other animal proteins utilized in meat and poultry include mechanically deboned meat and poultry which are both widely used in processed products (Pearson & Gillet, 1995).

Slendid is a non-amidated low methyl pectin derived from citrus peel, which is being promoted as a fat replacer for wide range of products in recent time (Thestrup, 1993). Microbiological gums such as xanthan gum from xanthonas, capestris and gellan gum from culturing of *Pseudomonas elodea* can also can be used as potential partial fat replacers. Konjac gels can also be used to stimulate the sensory properties and appearance of hardened fat and connective tissue in the mouth (Tye, 1991). Konjac gels are being used in low fat bologna (Chin *et al.*, 2000), low fat cured lamb sausages (Osburn & Keeton, 2004).

5.6. Lipid Based Fat Replacers

These are either triglycerols or having similar configuration to tryglycerols, intended to replace the functionality of natural fats on a one-to-one weight basis.

5.6.1. *Emulsifiers*

These are partial esters of fatty acids with chain lengths from C12 to C22 and polyvalent alcohols like glycerol, sorbitol/sorbitan, and sucrose or organic acids like acetic acid. Emulsifiers possess many of the properties of fats and oils as exemplified in the fatty consistency, lubricity, and texture and cohesiveness of fat, they form films or spread on surface, build or increase viscosity through hydration, and soften or weaken structures created by polysaccharides or proteins (Flack, 1996). Olestra (also known as synthetic fat replacer), sobestrin etc. are the example of emulsifiers.

5.6.2. *Structured lipids*

These are triacylglycerols produced either by chemical or by enzymatic inter-esterification or via genetic engineering. They have lower melting points, high solubility in water and are resistant to oxidation and metabolized directly in the liver. Examples of this class are salatrim and caprenin.

5.6.3. *Synthetic fat replacers*

Olestra, source polyester is being used widely. It has an appearance, flavor, heat stability, flash point and shelf life similar to those of its constituent fats

(Anon, 1990). Olestra would probably work well as a zero-calorie frying medium for fried or pre browned meat items (Aylward, 1996). It has similar functionality to conventional fats and oils in comminuted meat products. Other experimental synthetic fat substitutes include tri-alloxy-tricarballylate, frito-lay, esterified propoxylated glycerols etc.

5.7. Super Critical Fluid Extraction

This technique is based upon altering the proportion of fluid with high pressure and temperature to produce a super-critical phase, so that the fluid gets an unique extraction properties. This procedure was found effective for fat extraction from meat products, lipid and cholesterol from ground beef (Chao *et al.*, 1991) and cholesterol from dried egg-yolk (Froning *et al.*, 1990). However, this process needs further research to know about its full mechanism.

6. APPROACHES FOR PRODUCING LOW FAT MEAT PRODUCTS

Among the various strategies, selection of meat and trimming of excess fat from carcass and cuts, modification of carcass composition, reformulation of meat products with reduced fat content along with modification of the fatty acid profile, reduction of cholesterol and calories provide positive results and can greatly reduce the fat content. Trimming of excess fat from carcass and cuts even though greatly reduces the fat content of over fat carcasses and cuts, this is short term solution for providing low fat products. Replacement of fat with added water, addition of carbohydrate or protein based fat mimetics and use of synthetic fat substitutes is promising technologies for reducing fat. This is usually accomplished by adding substances that bind water which overcomes the dryness and rubberiness and improves juiciness. The additives that bind water can be classified as protein additives, non- protein additives and other procedures like super critical fluid extraction.

There are different types of additives like fat replacers, fat mimetics, fat substitutes and analogs and fat barrier compounds. Fat replacers are the ingredients that contribute fewer or no calories to the formulated foods without altering flavor, mouth feel, viscosity ororganoleptic properties (Keeton, 1994). Fat mimetics are defined as a partial replacement for fat by mimicking or imitating a particular function, but not all functions of fat in a food. Fat mimetics can replace the mouth feel of fats that cannot substitute for fat on an equivalent weight basis. Fat substitutes and analogs are molecules whose physical and thermal properties resemble fat but have either fewer calories than fat or no calories. Fat barriers are different systems which reduce the fat content of a food by retarding fat absorption into the product during frying.

Fat may be replaced in food products by traditional techniques such as substituting water or air for fat, using lean meat in frozen meat product, skim milk instead of whole milk in frozen desserts, and baking instead of frying for manufacturing or preparing snack foods. Fat may also be replaced in foods by reformulating the foods with lipid, protein, or carbohydrate based ingredients, individually or in combination (Akoh, 1998). Many non-meat ingredients of plant and animal origin are commonly employed as fat replacers. They include water (Ahmed *et al.*, 1990), starches, hydrocolloids (Bloukas *et al.*, 1997), soya products like soya protein concentrates/isolates, whey proteins and dietary fibers (Akoh, 1998; Mendoza *et al.*, 2001).

7. MANUFACTURE OF LOW FAT MEAT PRODUCTS

The manufacture of low fat products generally follows two basic approaches: the use of leaner raw materials and/or the reduction of fat and calorie contents by adding water and other ingredients that contribute few or no calories. These approaches can be supplemented by the use of a number of technological procedures that help to offset undesirable side effects produced as a result of changes to the product's composition and nature. The aim is to reduce fat levels and/or modify fat characteristics and yet produce a product with acceptable levels of functionality, safety, sensory properties and stability (Yun *et al.,* 2009). There are a number of procedures for the development of such products, which may be followed on their own or in combination, and are based on the following approaches.

7.1. By Selection of Raw Material

The composition of raw material can be adjusted by conditioning of the carcass composition using breeding and/or feeding strategies and by reducing fat levels in meat using a number of physical and/or chemical techniques. The level of carcass fat has been considerably reduced over the past 20 years. Further, research has been also directed towards inducing changes in lipid components by modifying the feeding regimes of monogastric animals, so as to reduce the levels of saturated fatty acids and increase the levels of oleic acid present in muscle and adipose tissue. The procedures to reduce fat levels in meat range from fat trimming to physicochemical techniques like cryoconcentration, centrifugation, and supercritical fluid extraction technology.

7.2. Use of Ingredients to Enhance Water Holding Capacity

The substances used for this purpose must contribute to particular characteristics without a major calorie input. Combination of fat replacers that mimic the mouth feel and textural characteristics of fat offer potential for development of low fat meat products (Keeton, 1994). The major ingredients and/or additives used to reduce fat levels include added water, non meat proteins (soy, surimi, dairy proteins, gluten, albumin, etc.), carbohydrates (gums or hydrocolloids, starches and maltodextrins and cellulose derivatives) and others (functional mixtures, vegetable oils and synthetic products).

(*a*) *Water*: In low-fat products the fat that is removed must be replaced to a greater or lesser extent by water. The level of substitution will depend on the type of product being formulated and the legal regulations governing this.

(b) *Non meat protein*: Both animal and vegetable proteins have been used in the development of meat products to increase yield (water and fat binding properties) and reduce formulation costs or enhance specific functional properties (water holding ability, emulsifying properties). Some of these properties can be useful for reducing fat content. However, incorporation of higher levels of soy protein flours and concentrates into meat products leads to decrease in cooking loss, caloric density, meat aroma, shear force and increase in tenderness and juiciness (Keeton, 1994). Other vegetable proteins such as wheat and maize derivatives or oat flours have also been used (Sen *et al.*, 2008). Dairy proteins are of particular interest in terms of functionality, especially with regard to emulsifying and gel forming properties. Non-fat dried milk, caseinates and whey proteins have been used up to 3.5% to offset the effects of fat reduction. Blood plasma proteins have been noted as potentially suitable for reducing fat levels, but have not yet been used very much in low fat products. Egg protein is said to be of considerable interest for the development of low fat products. Egg white has been found to affect the texture, but not the binding properties of sausage formulated with different fat levels (Jimenez *et al.*, 1996).

(c) *Carbohydrates*: The carbohydrates used in the formulation of low fat products have been basically gums or hydrocolloids of varying origin. They are generally used to improve cooking yield, enhance water holding ability, reduce formulation costs, modify texture and improve freezing stability. Carrageenans (iota and kappa), sodium alginate, gum arabic, guar gum, xanthan gum and other gums have been added at concentrations in the range 0.1–1% to low fat ground meat and emulsions (Suman and Sharma, 2003). Konjac flour (a hydrocolloid capable of forming a strong, elastic hydrophilic gel with a very low calorie content) and pectins, either alone or in combination with other ingredients, can produce sensory properties similar to those provided by fats. Starches and maltodextrins are glucose polymers that are found in maize, oats, rice, tapioca, potato etc. Starches are of interest in emulsified systems, as they can absorb or bind water that is not bonded to the matrix, thus enabling the emulsion to retain more water than it otherwise could. Maltodextrins can give rise to a gel which, when hydrated, displays some characteristics that are similar to those of fat. Cellulose derivatives, such as carboxymethylcellulose, methylcellulose, hydroxyl propylmethyl cellulose or micro crystallized cellulose and a number of dietary fibres have been tested in the formulation of low fat products to induce modifications to the binding properties, texture and flavor (Desmond *et al.*, 1998; Grigelmo Miguel *et al.*, 1999, Cofrades *et al.*, 2000).

7.3. Use of Technology to Change the Composition of Final Product

This includes use of technological treatments such as pre-blending, physical manipulation or high pressure during processing and/or preparation to modify properties that determine the stability and texture of meat products. Pre-emulsification of part of the fat with a non-meat protein (before addition to the meat batter) improves the system's fat binding ability. This leaves more meat protein available to act in gel formation and enhance water holding ability (Bishop *et al.,* 1993; Jimenez, 2000).

8. PROBLEMS ASSOCIATED WITH LOW FAT MEAT PRODUCTS

However, fat generally is an important constituent of meat products and affects product characteristics such as juiciness, texture, handling and heat transfer. To simply reduce fat or replace it with another type is therefore not an easy option. In principle, full-fat product attributes are the benchmark for reduced-fat products—a challenge for the meat industry as taste is the most decisive factor for the success of processed meat products.

Acceptability of food products with fat substitutes are higher only when they taste good; are reasonably priced; have nutritional merits; and elicit a widespread perception of healthiness. Fat is a calorie-dense nutrient that provides flavour, texture, juiciness and water entrapment in processed meats. Meat fat acts as a reservoir for flavor compounds and contributes to the texture of product. Fat plays an important role in formation of a stable meat emulsion. When the amount of fat in the formulation is reduced, this affects the structure and appearance of raw meat emulsion. In addition, there are no longer enough fat particles to immobilize the meat juices and added water (by adsorption), thus salt addition in the formulation becomes more important. The optimum amount of salt is necessary to obtain maximum extraction of salt soluble meat proteins and thus increase water holding capacity as much as possible in order to make up for the reduced water adsorption for fat (Tume & Yang, 1993).

Development of low-fat products while assuring the palatability demanded by the consumers is not an easy task. With excessive fat reduction, these products become bland and dry and the texture can be hard, rubbery or mealy. Reformulation using fat substitutes has caused a particle binding, darkened product colour and shortened product shelf-life because of dry, swollen packages. Reducing the fat content in meat products leads to a firmer, rubbery, less juicy product with dark color and more cost (Trout *et al.*, 1992; Cavestany *et al.*, 1994; Keeton, 1994; Paneras *et al.*, 1996; Desmond *et al.,* 1998 & Kirchan *et al.*, 2000). Other problems that occur with low-fat, meat products are reduction in particle binding, reduced cooking yields, soft and mushy interiors, rubbery skin formation, excessive purge in vacuum packages shortened shelf life and changes in mouth-feel of the product after conventional and microwave reheating. In making a product with more added water and less fat, it's likely that the level of spices need to be adjusted. Fat absorbs and masks some of the spicy flavours. So, as the fat levels are decreased, the spicy flavour in the formulation may become much more pronounced and therefore, need to be adjusted.

The first approach taken by the meat processing industry in overcoming these difficulties was to use other non-meat ingredients such as water, milk proteins, starches and carrageenan to substitute for fat. Replacement of fat with added water, addition of carbohydrate or protein based fat mimetics and uses of synthetic fat substitutes are the promising technologies for reducing fat. This is usually accomplished by adding substances that bind water which overcomes the dryness and rubberyness and improves juiciness (Naga Mallika *et al.*, 2009). These are supposed to mimic the functional properties of fat in the cooked product. While many have proven successful in performing some of these functions, no substitute has yet been found that imparts all the characteristics of fat to the product without any undesirable effects such as unnatural tastes.

9. FUTURE RESEARCH NEEDS

Improving the leanness of meat may be possible by breeding and genetics, gene mapping of meat producing species, transgenic animal production. Altering the leanness may be tried by feeding and management techniques, use of repartitioning agents, use of protein and non-protein additives and their mechanism of action and super critical fluid extraction. Further research is needed to understand the mechanism of sensation of juiciness. By understanding the factors that contribute to juiciness, the development of low fat products can be exploited. Recent studies are trying to incorporate some plant starches like sorghum flour, finger millet flour (Das *et al.*, 2011) etc in meat product to fetch a synergistic effect of fat replacement vis-à-vis cost efficiency to prepare the comminuted meat products. Replacement of fat in meat and meat products is an intricate job and sensory attributes of fat substituted meat products are more complicated to achieve up to the mark of all meat products. We still do not know enough about all the functions of fat in a complex food-matrix in order to mimic its properties adequately. No one fat replacer is perfect for all the diversified applications. Problems with flavor, juiciness, texture, mouth-feel and other organoleptic properties have not been solved adequately by current fat replacers. There is particular lack of information about the microparticulated protein, emulsifiers and synthetic fat replacers, which need a further research. To select most effective and cost effective fat replacer, we need to develop a better understanding of the specific mechanism of action of various fat replacers as well as synergistic and additive effects. Entrapment of water on processing, meat protein gelation and other important functionalities of different fat replacers should be studied separately.

REFERENCES

AHA (1996). Dietary guidelines for healthy Americans. *American Heart Association. Circulation*, 94: 1795–1800.

Ahmed, P.O., Miller, M.F., Lyon, C.E., Vaughters, H.M. and Reagan, J.O. (1990). Physical and sensory characteristics of low-fat fresh pork sausages processed with various levels of added water. *Journal of Food Science*, 55: 625–628.

Akoh, C.C. (1998). Fat replacers. *Food Technology*, 52: 47–53.

ADA, (2005). Position of the American Dietetic Association: Fat replacers. *J. Am. Diet. Assoc.*, 105: 266–275.

Anonymous, (2002). Meat Industry Statistics. Retrieved April 01, 2002 from the World Wide Web: http://www.mhr-viandes.com/en/docu/them/7stat.htm.

Bishop, D.J., Olson, D.G. and Knipe, C.L. (1993). Pre-emulsified corn oil, pork fat, or added moisture affect quality of reduced fat bologna quality. *J. Food Sci.*, 58: 484–487.

Bloukas, J.G., Paneras, E.D. and Papadima, S. (1997). Effect of carrageenan on processing and quality characteristics of low-fat frankfurters. *Journal of Muscle Foods*, 8; 68–78.

Centers for Disease Control and Prevention (CDCP), Chronic Disease Prevention, (2000). Fact sheet: Actual causes of Death in United States. Available at: http://www.cdc.gov/nccdphp/factsheet/death_cause2000.html, Accessed May, 6, 2006.

Cengiz, E. and Gokoglu, N. (2007). Effects of fat reduction and fat replacer addition on some quality characteristics of frankfurter-type sausages. *Int. J. Food Sci. Tech.*, 42: 366–372.

Chao, R.R., Mulvaney, S.J., Bailey, M.E. and Fernando, L.N. (1991). Supercritical CO_2 conditions affecting extraction of lipid and cholesterol from ground beef. *J. Food. Sci.*, 56: 183–187.

Cofrades, S., Guerra, M., Carballo, J., Fernandez-Martin, F. and Jimenez Colmenero, F. (2000). Plasma protein and soyafibre content effect on bologna sausages properties as influenced by fat level. *J. Food Sci.*, 65: 281–287.

Das Sudip, Kumar., Pazhaniandi, P.P., Tanwar, V.K., Biswas, S. and Khan, A. (2011). Effect of sorghum flour and finger millet flour as fat replacers for preparing low-cost chicken – Investigation on some physical and sensory properties of chicken patties. *Fleischwirtschaft International*, 26 (3): 64–68.

Desmond, E., Troy, D.J. and Bucley, J. (1998). Comparative studies on non meat ingredients used in the manufacture of low fat burgers. *J. Muscle Foods*, 9: 221–224.

Drewnowski, A. (1998). Energy density, palatability and satiety implications for weight control. *Nutr. Rev.*, 56: 347–353.

Giese, J. (1992). Developing low fat meat products. *Food Technol.*, 46: 100–108.

Flack, E. (1996). The role of emulsifiers in low-fat products. *In*: Roller, S. and Jones, S.A. (*Eds*.). Handbook of fat replacers. CRC press, Boca Roton, New York, pp. 213–234.

Froning, G.W., Wehling, R.L. and Cuppett, S.L. (1990). Extraction of cholesterol and other lipids from dried egg yolk using supercritical carbon dioxide. *J. Food Sci.*, 55: 95–98.

Glicksman, M. (1991). Hydrocolloids and the search for the oily grails. *Food Technol.*, 45(10): 94–103.

Grigelmo Miguel, N., Abadfa Seros, M.I. and Martin Belloso, O.A. (1999). Characterisation of low fat high dietary fibre frankfurters. *Meat Sci.*, 52: 247–256.

Grossklaus, R. (1996). Fat Replacers: Requirement from a nutritional physiological point of view. *Fett/Lipid*, 98: 136–141.

Jimenez, C.F., Barreto, G., Fernhdez, P. and Carballo, J. (1996). Frozen storage of bologna sausages as a function of fat content and of levels of added starch and egg white. *Meat Sci.*, 42: 325–332.

Jimenez, C. (2000). Relevant factors in strategies for fat reduction in meat products. *Tr. Food Sci. Technol.*, 11: 56–66.

Jones, J.M. (1996). Fat substitutes: Nutritional promises or potential disaster. Chemistry snd Industry, July issue, pp. 494–498.

Keeton, J.T. (1994). Low-fat meat products—technological problems with processing. *Meat Science*, 36: 261–276.

Laroia, S., Magoli, E.L. and Hnasen, P.M.T. (1995). in IFT annual meeting, pp. 205.Chicago, IL.

Lucca, P.A. and Tepper, B.J. (1994). Fat replacers and the functionality of fat in foods. *Trends in Food Sci. Technol.*, 5: 12–19.

Mendoza, E., Garcia, M.L., Casa, C. and Selgas, M.D. (2001). Inulin as fat substitute in low-fat, dry fermented sausages. *Meat Science*, 57: 387–393.

Modi, V.K., Mahendrakar, N.S., Rao, D.N. and Sachindra, N.M. (2003). Quality of buffalo meat burger containing vacuum flours as binders. *Meat Sci.*, 66: 143–149.

Naga Mallika, E., Prabhakar, K. and Reddy, P.M. (2009). Low-fat meat products—an overview. *Veterinary World*, 2(9): 364–366.

Omayma, E.S. and Youssef, M.M. (2007). Fat replacers and their applications in food products: A review. *Alex. J. Fd. Sci. and Technol.*, 4(1): 29–44.

Pearson, A.M. and Gillett, T.A. (1999). Effects of Fat on Flavor In Processed Meats. 3rd ed., pp: 356–358. Aspen Publication®, Aspen Publisher, Inc. Gaithersburg, Maryland, (ISBN: 08342-1304-4).

Rao, K.H., Anjaneyulu, A.S.R., Singh, R.R.B. and Yadav, P.L. (1999). Quality and storage stability of smoked chicken sausages incorporated with milk proteins. *Ind. J. Poult. Sci.*, 34(3): 373–377.

Reddy, P. and Rao, S. (1997). Influence of binders and refrigerated storage on certain quality characteristics of chicken and duck meat patties. *J. Food Sci. Tech.*, 34: 446–449.

Sandrou, D.K. and Arvanitoyannis, I.S. (2000). Low-fat calorie foods. Current state and perspective. *Crit. Rev. Food Sci. Nutr.*, 40: 427–447.

Sen, A.R., Naveena, B.M. and Muthukumar, M. (2008). Effect of wheat fibre on sensory, physico-chemical traits and storage stability of low fat mutton nuggets. *Fleischwirtschaft Int.*, 24: 71–74.

Serdaroglu, M. and Sapanci-ozumer, M. (2003). Electronic journal of Polish Agricultural Universities Series: *Food Sci. and Technol.*, 6: 1–9.

Setser, C.S. and Racette, W.L. (1992). Macromolecule replacers in food products. *Critical Reviews in Science and Nutrition*, 32(3): 275–297.

Sharma, B.D., Biswas, A.K. and kumar, Sunil. (2004). Influence of milk co-precipitates on the quality of restructured buffalo meat blocks. *Asian-australian J. Animal Sci.*, 17(4): 564–568.

Suman, S.P. and Sharma, B.D. (2003). Effect of grind size and fat levels on the physico-chemical and sensory characteristics of low-fat ground buffalo meat patties. *Meat Sci.*, 65: 973–976.

Tuley, L. (1996). The search is on for low fat meat. *Int. Food Manufact.*, 1: 29–32.

Tume, R.K. and Yang, A. (1993). The changing perspective of fat—adaptive to the market. Meat, 93', CSIRO: The Australian Meat Industry Research Conference, *Session No.*, 4: 1–6.

USDA, (1991). Labelling of low fat and reduced fat/salt products. *In*: Production and processing of healthy meat, poultry and fish products. Advances in meat research, Volume 11. Pearson, A.M. and Dutson, T.R. (*Eds.*), Blakie Academic and Professional London, pp. 48–52.

USDA & USDHHS, (1995). Nutrition and your Health; Dietary guidelines for Americans. 4th ed., Home and Garden Bulletin, No. 232., U.S. Dept. Agriculture and U.S. Dept. Health and Human Services, Washington, D.C.

USDHHS, (1988). The surgeon general's report on nutrition and health. Publ. No. 88–50210. U.S. Govt. Print Office, Washington D.C.

Yun, S.C., Ji, H.C., Doo, J.H., Hack, Y.K., Mi, A.L., Hyun, W.K., Jong, Y.J. and Cheon, J.K. (2009). Characteristics of low fat meat emulsion systems with pork fat replaced by vegetable oils and rice bran fibre. *Meat Sci.*, 82: 266–271.

20

High Pressure Processing of Meat

A.K. Biswas, Gauri Jairath and P.K. Mandal

ABSTRACT

The ability of HHP processing to deliver a fresh clean ingredient statement, fresher flavour, and better nutrition provide a unique point of difference for producers. HHP processing is a paradigm-shifting technology for meat industry. HHP processing caused protein denaturation, aggregation, or gelation which can result in meat becoming either tenderized or toughened, depending on meat protein system, the temperature, the pressure, and its denaturation. When processing conditions are carefully controlled, HHP could prove to be a useful tool to improve meat and seafood safety and quality. Regulations recognize HHP as a post-packaging step in the control of food-borne pathogens in meat products, and combined HP-thermal treatment is effective in sterilizing foods with limited impact on their nutritional and sensory qualities. Texturizing effects over proteins has also been used to enhance the characteristics of already existing products and the development of new formulates. The development of high-efficiency HHP machines has reduced the processing costs to acceptable levels. Again, HHP as a low-temperature treatment is an environmentally friendly and waste-free technology.

1. INTRODUCTION

In recent years the consumers have become more health conscious in their food choices but have less time to prepare healthful meals. As a result the market demand for "minimally processed" or "lightly processed" foods has rapidly increased. Consumers increasingly demand foods which retain their natural flavor, colour and texture and contain fewer additives such as preservatives. High pressure technology is becoming of increasing importance in the production of minimally processed foods for which shelf-life can be extended primarily through inactivation of bacteria and enzymes with minimum loss of nutritional value and sensory quality. However, few studies were published until 1970s because of technical difficulties and cost associated with high hydrostatic pressure (HHP) processing units and packaging materials. A successful implementation of high pressure technology in the meat industry depends on a qualitative and quantitative description of the meat improvement attributes and the deteriorative processes.

The application of high pressure for food processing started one century ago. This technique was first time used by Royer (1895) to kill bacteria and subsequently in 1899 by Hite. Hite demonstrated that high-pressure processing (HPP) could coagulate egg albumen thereby prolonging the shelf-life of it. In 1924, the potential of HPP in commercial fruit and vegetable processing was recognized. Serious attempts to use this technology for food processing on a commercial scale was initiated in Japan by Japanese Ministry of Agriculture and Fisheries in collaboration with several food industries in 1989. Presently, with the growing demands of consumers for natural, fresh, safe, and "minimally processed" food products, HHP is currently of great interest in food research, primarily as an alternative to thermal processing for preservation of meat and meat products.

Many developed countries like United States; Italy already got marketplaces for selling of HHP treated meat and meat products. Sliced ham and turkey, pork and poultry cuts, thick sliced ham, chicken and turkey products, whole and sliced Serrano ham, salami and Chorizos are available in Spain. Natural, minimally processed cooked sliced meat, roasted chicken, sliced chicken and turkey, chicken sausages, sliced turkey and strips of chicken in modified atmosphere packaging (MAP) and Prosciutto are sold in United States. Nitrite free HHP treated cooked pork products are sold in Japan, while Parma ham, salami and pancetta are sold in Italy (Sun & Holley, 2010).

High pressure treatment, up to 1000 MPa, can affect protein conformation and lead to its denaturation, aggregation or gelation. So, HHP treatment can be used to develop new variety of products which may exhibit distinctly different texture than obtained with thermal gelation properties. Many research findings indicated that HHP treatment can induce changes in muscle structure through changes in myofibriller proteins. Suzuki *et al.* (1991) investigated the effect of HHP (100–300 MPa, 2–4°C, 5 min) on isolated myofibrils and whole meat. They reported that at a lower pressure (100 MPa), changes in sarcomere structure occurred in both myofibrils and whole meat system, however, at 200 MPa, z-disks disappeared and dense material aggregated on both sides of the lost M-lines but only in whole meats.

2. PRINCIPLES OF HIGH PRESSURE PROCESSING

Various physical and chemical changes result from the application of high pressure treatment. Generally, physical compression during pressure treatment results in a volume reduction and increase in the temperature and energy (Heremans, 2003). In predicting the effect of HHP on foods, it is necessary to consider the net combined pressure temperature effect of the process. In HHP processing, food is subjected to very high pressure (up to 8.4 kg/cm^2) to kill bacteria present in the raw food. High pressure works like heat to kill bacteria, but the food remains fresh and rich. For this, pre-packaged raw product is placed in a pressure chamber and subjected to very high pressures for specific time (< 10 minutes).

There are three general principles of the application of HHP in food processing: Le Chatelier's principle, principle of microscopic ordering and Pascal's isostatic rule (Zhang & Mittal, 2008). The Le Chatelier's principle states the phenomena of phase transition; chemical changes etc which are accompanied by a decrease in volume by enhanced pressure and vice-versa. The principle of microscopic ordering explains at constant temperature, an increase in pressure increases the degree of ordering of a given substance.

Therefore, pressure and temperature exert antagonistic forces on molecular structure and chemical reactions (Balny & Masson, 1993). The isostatic principle indicates that pressure is instantaneously and uniformly transmitted throughout a sample, whether in direct contact with the pressurized medium or insulated from it in a flexible container. Thus, the pressurization process is independent of sample volume, product size, and geometry, in contrast to thermal processing. Processing by high hydrostatic pressure is usually carried out in a low compressibility liquid such as water. Pressure influences most biochemical reactions occurring in foods since they often involve a change in volume. Pressure may inhibit the availability of energy by affecting energy producing enzymatic reactions.

The effect of hydrostatic pressure in food product is dependent upon the amount of pressure applied, duration of compression, depressurization rate, temperature of treatment, product pH, water activity (a_w), salt concentration etc. There are also other salient features which are unique for HPP of foods. As compared with thermal processing, HPP treatment consumes less energy; energy required for pressurization at 400MPa is equivalent to heating the same material to 300°C. Currently, the main areas of interest for the application of HPP to food processing include microorganism inactivation or activation; modification of biomacromolecules, *e.g.*, protein denaturation, starch gelation and enzyme modification; and quality retention and improvement (flavor, color, and functionality). These could be used advantageously in several segments of food industry including sea food and meat industry.

3. EFFECT OF HHP PROCESSING ON MEAT QUALITY

3.1. Effect on Texture/Tenderness

The textural changes in meat of different food animals follow same principle of decrease in volume of proteins which can be determined by the constitutive volume of the atoms, by the volume of the internal cavities, and by the salvation of peptide bonds and amino acid side chains (Masson, 1992). The volume of a treated protein decreases because of the compression of the internal cavities. High-pressure effects on proteins are primarily related to the rupture of electrostatic and hydrophobic interactions within protein molecules, and to the subsequent re-formation of intra- and inter-molecular bonds within or between protein molecules. However, upon constant HHP pressure and variation in heat treatment the secondary, tertiary, and quaternary structures of protein behave differently and ultimately affect texture of meat and meat products (Galazka & Ledward, 1998). Pressure denaturation of proteins seems to occur because of the destabilization of non-covalent interactions in tertiary structure. A small degree of unfolding exposes hydrophobic regions of the protein. The treated proteins retain much of their secondary structure and this is assumed to be the cause of protein aggregation. The unfolding of the protein during high-pressure treatment is followed by formation of hydrophobic and disulfide bonded aggregates after pressure release.

Muscle tissues exhibited severe contraction (either reversible or irreversible) when HHP applied immediately after slaughtering or on pre-rigor meat, however intensity of contraction depends on pressure level. A 35–50% reduction of muscle fiber length is most common with disruption of muscle cell structure, particularly when pressure levels exceed above 100 MPa (Kennic *et al.*, 1980). The extent of structural damage of meat proteins is greater when the period between animal slaughter and HHP processing is shorter than that needed for development of muscle stiffness through rigor onset. HHP treatment also caused Z-disk disruption of the myofibril yielding what were termed “contraction bands”. It has also been observed that a high pressure treatment lead to disappearance of H-zones and M-lines of muscle *L. dorsi* from beef.

In the year 1973, Macfarlane first introduced high pressure treatment for tenderization of pre-rigor meat. According to his observation, beef and sheep muscles when exposed to high pressure treatment at 103 MPa exhibited very firm and compact structure raw meat, but that meat was more tender and had higher moisture content than non-pressurized meat cut from carcasses pre-rigor (hot-boned), post-rigor, or after cooking. Later, MacFarlane *et al.* (1980) when worked with high pressure treatment, they reported that strip of muscle maintained under tension showed instantaneous and severe contraction, but the muscle strip became extended. From this, they assumed that extension was due to pressure disruption of myofibrillar proteins which ultimately tenderized the meats. As a whole, pressure induced contraction has direct correlation with changes in tenderness of pre-rigor meat.

Pressure application above 100 MPa for 1 min or more at low temperature (<30°C) did not show any beneficial effects on post-rigor beef tenderness. However, extensive modifications in sarcomere structure were noticed with no contraction bands. The most noticeable changes reported due to high pressure treatment (100 MPa, 60 min, 25°C) were the absence of M-lines in the central region of the A-bands and a loss of I-band filament integrity adjacent to the Z-disks in ovine *Semimembranosus* muscle. The Z-disks remain nearly unaltered. In addition, the initial bridges present between thick filaments were no longer observed in transverse sections made in the M-line region. Therefore, it was proposed that a translocation of thin filaments onto the thick filaments occurred, which was used to explain why the H-zone in the centre of the sarcomeres was no longer visible.

The pressure-induced destruction of the Z-line was possibly due to the release of α-actinin (Iwasaki *et al.,* 2006). They again reported that upon HHP processing at 200 MPa, there were disappearance of the M-line, dissociation of the thin and thick filaments, and destruction of the Z-line (chicken myofibril) in the presence of 0.2 M NaCl. The depolymerization of thin filaments led to high apparent elasticity of a heat-induced myofibriller gel after 200 MPa HHP treatments while pressurized at 300 MPa, the decreased apparent elasticity of a myofibriller gel observed was thought to have been caused by pressure-induced shortening of myosin filaments. This may have been due to depolymerization of thin filaments, which in turn inhibited head-to-head interaction among myosin filaments. Nonetheless, the rheological properties of heat-induced gels were improved by HHP treatment.

3.2. Effect on Gelation Properties

HHP induced gelation properties of myofibriller proteins such as myosin, actomyosin and actin have been well reviewed (Macfarlane, 1985). HHP processing has been shown to depolymerize the actin and actomyosin and thereby helps in solubilization of predominant cytoskeletal protein *i.e.*, titin or connection. Actin was very sensitive to HHP processing and underwent depolymerization at 100 MPa, but dependent upon temperature, the nature and the concentration of the salt, as well as the pH of the meat. Pressure-induced gelation of meat proteins also depends upon the protein system, when myosin in the solution, with increased pressure and treatment time, changes began to occur at 140 MPa.

The myosin molecules undergoes head-to-head interaction to form oligomers (Yamamoto *et al.,* 2002) while tail area of the myosin monomers remain unchanged as their original helical structure. At HHP processing 300 to 400 MPa, the myosin and actin both were denatured along with many of the sarcoplasmic proteins. In general, HHP treatment increases the solubility of myofibrillar protein and causes depolymerization, therefore improving gelation of fresh meat and results in desirable tenderization; however, pressure should be applied at relatively low levels to avoid toughness. The effects of temperature, salt concentration and fat level can also influence HHP effects on meat gelation properties.

3.3. Effect on Proteolytic Enzymes

Several biochemical events take place in post mortem muscle. During anaerobic glycolysis, glycogen reserves are depleted, ATP gradually hydrolyses as a consequence of a pH fall due to ion pump failure, and actin and myosin bind to form an irreversible acto-myosin complex, with the onset of rigor. As rigor begins, muscle elasticity decreases and the tissue reach its maximum toughness at the completion of rigor. Post-rigor tenderization occurs within hours and differences can be observed depending on the muscle fiber type, muscle type, individual animal and species, the main determinant of ultimate tenderness is the extent of proteolysis of key target proteins within muscle fibers.

Moreover, enzymes are involved in flavour and taste development in dry-cured meat products. Myofibrillar proteins, such as actin, myosin, tropomyosin, troponin-T, nebulin and titin, along with cytoskeletal desmin and costamere vinculin, are subjected to cleavage by endogenous proteolytic enzymes after death. Little or no changes are reported in connective tissue as a consequence of pressure treatment. The best studied are the calpains/calpastatin system and the lysosomal cathepsins. Caspases, a family of cysteine aspartate specific proteases, along with the proteasome complex, are also involved in muscle tenderization, although their role is still controversial.

HHP readily inactivated calpain enzymes at 300 MPa, probably due to structural modification and dissociation of calpain subunits. Evidence suggests that inhibition of calpain activity by high pressure prevents the degradation of cytoskeletal proteins such as desmin, reducing the water-holding capacity of muscle and increasing the drip loss. The ability of muscle to retain water is strictly related to post-mortem events such as pH decline, proteolysis and protein oxidation.

Cathepsins are a group of enzymes including both exo- and endopeptidases. Cathepsins known to express in muscle tissue include six cysteine peptidases (cathepsins B, L, H, S, F and K) and one aspartic peptidase (cathepsin D). They are located in lysosomes and are mostly active at acidic pH. After death, lysosomal enzymes become accessible to muscle structural proteins. This is due to the progressive disruption of lysosomes through membrane breakdown, which occurs by the decrease in pH at high post-mortem temperature or by the failure of ion pumps in lysosomal membranes during rigor development following the depletion of ATP stores.

The application of high pressure to fresh meat enhances the endogenous cathepsin proteolytic activity in meat, probably due to the release of proteases from lysosomes to the cytoplasm and by denaturation of the tissue protein (Ohmori *et al.,* 1991). Kurth (1986) reported the maintenance of cathepsin B activity under pressure (150 MPa) and even an increase in some pressure heat combinations. It has also been found that upon HHP in bovine muscle (100–500 MPa, 5 min) activity of cathepsins B and L were increased and inactivation of aminopeptidase B and cathepsin H, an amino-endopeptidase take place. However, Campus *et al.* (2008) observed reduced activity of cathepsins B and L at pressure of 400 MPa, along with the loss of activity of other enzymes.

3.4. Effect on Meat Colour

The colour of meat depends on state and concentration of myoglobin in meat, although other proteins such as haemoglobin and cytochrome C may also play a role in beef, lamb, pork and poultry; in addition, the optical (scattering) properties of the meat surface can influence colour evaluation. Myoglobin contains a prosthetic group, the haem ring, with a centrally located iron atom. The ligand present (oxygen, carbon dioxide and carbon monoxide) and the valence of iron dictate the colour of muscle (Mancini & Hunt, 2005). The colour of dry-cured products is mainly due to the presence nitrosylmyoglobin and metmyoglobin. It has been observed that high pressure treatment of myoglobin caused partial denaturation, but the process was reversible (Defaye *et al.*, 1995). It is also known that the effect of high pressure treatment on myoglobin solutions depends on the temperature at which the pressure treatment occurs. Carlez *et al.* (1995) reported that meat discolouration in pressure-processed meat was due to the following: (1) a whitening effect due to myoglobin denaturation and/or to haem displacement or release and (2) oxidation of the ferrous myoglobin to ferric myoglobin above 400 MPa.

However, colour stability in beef muscles could be improved with the application of pressure of 80–100 MPa for 20 min, as measured by the rate of formation of metmyoglobin post-slaughter. This effect precludes when pressure treatment applied at 7–20 days post-slaughter. The HHP processing of cured meat is unlikely to cause problems in this respect. Studies on dry-cured ham reported an increase in lightness and a decrease in redness when the ham was pressurized. Serra *et al.* (2007) only reported higher lightness in several muscles of dry-cured ham, while Marcos *et al.* (2007) found no effect on colour after pressure treatment at 400 MPa of low acid fermented sausages. Some reports also indicated changes during the storage of pressurized dry-cured loin under vacuum at refrigerated temperatures.

3.5. Effect on Lipid Oxidation

HHP seems to promote lipid oxidation in meat products. Studies have reported a generally more rapid increase in the values of thiobarbituric acid reactive substances (TBARS) in pressurized minced meat, deboned turkey meat and chicken breast muscle with respect to control samples (Tuboly *et al.*, 2003). Catalysis of lipid oxidation seems to be related to level of the applied pressure and release of non-haem iron and membrane damage. Cheah and Ledward (1997) reported that a HHP in between 300 and 400 MPa is adequate to induce lipid catalysis in pork meat. They also reported that denatured forms of proteins play an important role in catalysing lipid oxidation. However, contradictory results also reported and according to Orlien *et al.* (2000) lipid oxidation at HHP was not related to the release of non-haem iron or catalytic activity of metmyoglobin but could be linked to membrane damage.

A very few studies have reported the effect of pressure treatment on TBARS in dry-cured products. Andres *et al.* (2006) reported significantly higher values of TBARS in dry cured ham treated at 400 MPa and stored for 39 days in a modified atmosphere with 5% residual oxygen, indicating a decrease in oxidative stability during storage. Marcos *et al.* (2007) found no differences in

TBARS values in pressurized fermented sausages at 400 MPa, indicating that most of the lipid oxidation had already occurred during ripening.

3.6. Effect on Microbiological Quality of Meat

High pressure processing (HPP) involves submerging the packaged food in a liquid within an enclosed vessel. The pressure is generated either by pumping more liquid into the pressure vessel or by reducing the volume of the pressure chamber. HPP kills microorganisms by interrupting their cellular function without the need for heat. Studies show that the process extends product shelf life by inactivating spoilage organisms. When appropriately used, HPP does not alter the texture, appearance or flavour of foods (Hugas *et al.*, 2002). Pressures of 101 MPa to 1013 MPa have been explored as potential food safety treatments for meat. The effects of extreme pressure on microorganisms are not fully understood, but substantial reductions (> 5 log cycles) in numbers of *Pseudomonas fluorescens*, *Citrobacter freundii* and *Listeria innocua* in ground beef have been demonstrated (Carlez *et al.*, 1993), and high pressure treatment slowed the development of spoilage organisms during subsequent storage of ground beef (Carlez *et al.*, 1994). Microbial reductions are enhanced when high pressure treatment is combined with mild heating or chilling, but colour changes were observed after 10 minutes of treatment.

The use of pulsed high pressure can be more effective than continuous single application, so treatment times can be reduced (Hayakawa *et al.*, 1994). Hayman *et al.* (2004) reported that ready-to-eat (RTE) meats pressure treated at 600 MPa at 20°C for 180 s, there was no deterioration in sensory quality, no difference in consumer acceptability; a 4 log reduction in *Listeria monocytogenes* in inoculated product and the refrigerated shelf life was extended. Hugas *et al.* (2002) reported that HPP treatment (600 MPa for 10 minutes at 30°C) could extend the shelf-life of food including cooked ham, dry cured ham and marinated beef loins.

The resistance of microorganisms to pressure is extremely variable. The extent of inactivation depends on several parameters, such as type of microorganism, the pressure level, the process temperature and time and the composition of the dispersion medium. Vegetative cells in early growth phase are more sensitive than cells in stationary phase. Gram positive bacteria are more resistant than gram negative bacteria. Vegetative cells and also spores of yeasts and molds are inactivated at 200–300 MPa, whereas, bacterial spores are very resistant and can even survive pressures above 1000 MPa.

Some viruses are very resistant to pressure. DNA replication and transcription and protein synthesis are inhibited at much lower pressure levels in *E. coli*. The effect of HPP is essentially due to change in permeability of cell membrane and associated leakage intercellular material. Further, membrane bound ATPase is denatured and displaced. The decline of pH due to high hydrostatic pressure also leads to loss of viability of microorganisms. Venugopal *et al.* (2001) reported that *S. typhimurium* and *L. monocytogenes* could be inactivated by a factor of 7 and 2 log cycles, respectively, by exposure to 200 MPa pressure for 10 min. A combination of gamma-radiation (2–4 Gy) and 400 MPa could reduce bacterial spore load more effectively than the individual treatment *per se*.

The effects of HHP on bacterial spores were studied over 100 years ago. The major barrier for the effective use of HHP for sterilization of products is the strong resistance of spores to pressurization. To cope with this critical drawback of HHP, processing methods and combination with other treatments and agents have been investigated. The most important of spore forming bacteria for food preservation are *Bacillus* and *Clostridium*. Many studies on the effect of HHP on spores were carried out on spores of *Bacillus* and *Clostridium*. The physical and chemical characteristics of the environment can strongly affect the survival of spore formers. Spore germination and growth are inhibited in foods with an equilibrium pH below 4.5. Thus, most studies have focused on spore inactivation in low-acid foods.

4. EFFECT OF HHP ON VIABILITY OF *T. SPIRALIS* OF PORK

The U.S. Department of Agriculture considers all fresh pork products to be potentially infected with trichinae. Thus all ready-to-eat products must be treated to assure that trichinae are inactivated by processes prescribed in the Code of Federal Regulations-heating, refrigerating and curing. HHP technique can be used as an effective alternative to that of heat processing for inactivation of trichinae. Hydrodynamic pressure has been shown to be effective in causing instantaneous tenderization in pork, beef, and lamb meat (Solomon *et al.*, 1997). This instantaneous tenderization results from myofibriller fragmentation in the region adjacent to the Z-lines of myofibrils, causing increased myofibriller spaces with longitudinal gaps or splits in the myofibril lattice. As a result of this filamentous damage within the muscle proteins, we felt that damage might also occur with tissue dwelling parasites such as *T. spiralis*.

It has been observed that treatment of infected loin muscle by the hydrodyne process resulted in statistically fewer numbers of worms recovered than from untreated loin muscle. These results suggested that the hydrodyne process was partially effective in killing *T. spiralis*; however, the remaining larvae retained their ability to infect a subsequent host, thereby showing that the process, as performed here, was not sufficient to eliminate the risk of trichinae infection from infected pork.

Higher hydrodynamic forces of magnitude up to 140 MPa may be necessary to achieve successful inactivation or destruction of trichinae. Increased hydrodynamic pressure has been shown to increase the level of tenderization without adversely affecting other qualities of the meat. It is possible that the application of higher pressures will have a greater effect on the destruction of trichinae and could make products safe to eat.

HPP foods differed significantly from their thermally treated counterparts, whereas, heat treatments makes large changes in colour, density, flavour and elasticity. Such changes are minimal in pressure treated foods. On the other hand, significant improvements in transparency, smoothness, adhesiveness and gloss of the product could be achieved by pressure processing. While both thermal and pressure treatments cause denaturation and coagulation of proteins, gelatinization of starch, chemical changes such as Mallard reaction and loss of vitamins and minerals during pressure treatment. While both heat and pressure inactivate vegetative cells of microorganisms as well as insects and their larvae, pressure processing does not inactivate spores of microorganisms.

5. APPLICATION IN RAPID FREEZING AND THAWING OPERATIONS

Conventional freezing techniques are known to have detrimental effects. The optimum freezing rate for food products or living cells is highly system dependant. Slow freezing generally results in larger ice-crystal formation, which may cause extensive mechanical damage, accelerated enzyme and microbiological activities as well as potentially increased oxidation rates, resulting from increasing substrate concentration and the insolubility of oxygen in ice. Ultra-rapid freezing may cause lethal intracellular ice crystallization or mechanical cracking. Pressure assisted freezing occurs as an impact of pressure induced freezing point depression which enables water still in the liquid phase at high pressures.

The behavior of water under pressure offers potential applications in food processing, such as pressure assisted freezing (pressure-shift freezing) and pressure assisted thawing (pressure-shift thawing), non-frozen storage under pressure at sub-zero temperature and formation of different ice polymorphs. Pressure assisted freezing and pressure assisted thawing have been examined on a variety of foods. For example, pressure assisted thawing of tuna and carp meat resulted in better quality of the products. Pressure assisted freezing and thawing are particularly beneficial for preventing quality loss during freezing and thawing of fishery products since such frozen products from the major components of international sea food trade.

The use of moderate hydrostatic pressure in combination with sub-zero temperatures has been proposed as a means of storing food products without the formation of ice, avoiding damage due to freezing and microbial spoilage. Further significant energy could be saved using pressure assisted rather than frozen storage. Product deterioration due to freezing and thawing effects is also avoided.

6. APPLICATION IN SEA FOODS WITH NOVEL TEXTURE

The possibilities for modification of proteins through pressure-induced changes in non-covalent interactions of polypeptide moieties offer potentials for development of fish products having novel textural attributes. Although, at present, most of these applications of high pressure have been carried out only at laboratory scale, these could be scaled up on a commercial scale. For example, the reddish fresh mackerel meat could be changed to opaque, as if it has been grilled or boiled. Similarly, the transparent nature of cod meat also can be converted into opaque by HHP at 608 MPa for 15 min.

Surimi is a washed fish meat, which forms a gel with excellent texture when heated. Surimi when mixed with small amount of salt and subjected to heating can form gel. The gelation property has been used commercially to develop high value items such as lobster tails, shrimp etc from meat of several low cost fish. Using ingredients like egg white and suitable moulds, it is possible to prepare fish and shellfish analogues. Instead of heat, HHP can be used to induce gelation of surimi. Pressure induced surimi gels from marine species were smother and more elastic than those produced by heat and were considered to be organoleptically superior. Excellent gel could be produced from pollack, sardine, skipjack and tuna meat at 400 MPa and from squid at 600 MPa. High pressure is very effective in producing highly appealing kamaboko, a Japanese product made from surimi. It has been however, generally observed that instead of surimi, if other fish are used for HHP treatment, they should be of low fat content, since HHP can enhance oxidation of lipids present in the meat.

Effect of high pressure on quality of some meat food products

Food items	*Pressure (MPa)*	*Quality*
Beef	200–350	Pinker meat
Sausages	300	Changes in functionality, increase in lightness
Beef	400 (10 min)	No changes in flavor. When surface is slightly baked, the Product tasted like steaks
Ham	400–500	HP helps reduce salt content
Fish Paste	400–500	HP helps reduce salt content
Oyster	300–400 (10 min)	No change in flavor
Shrimp	400 (10 min)	Meat coagulation as boiled shrimp. No change in colour
Surimi	300–500	Gelation of protein without heat treatment

7. EFFECT OF HHP PROCESSING

7.1. Effect on Meat Quality

High pressure can modify the structure and function of many proteins. It is reported that isolated myosin from both meat and fish will be denatured by pressure and develop a gel-like structure. These structural changes will affect the texture of the muscle, and since the effect of pressure on protein structure is different from the effect caused by thermal denaturation, the textures of heat- and pressure-treated proteins may also be very different. In a heat-treated system, of the weak bonds maintaining the secondary, tertiary, and quaternary structures, it is the hydrogen bonds that are most labile. In a pressure-treated system, it is the hydrophobic and electrostatic interactions which are most vulnerable, hydrogen bonds being mostly unaffected by pressure.

The pressure-induced denaturation of protein was caused by a decrease in protein volume, while heat-induced denaturation of protein was caused by the violent movement of molecules that result in destruction of hydrogen and covalent bonds. In the study by Iwasaki *et al.* (2006) evaluated pressure and heat-induced chicken myofibrillar gels. Heat-induced gels were formed from strands, which consisted of bundles of myofibrils, whereas gels from pressurized treatment at 200 MPa before heating consisted of a fine, 3-dimensional network of strands. The latter treatment also increased the apparent elasticities of both chicken myofibrillar gels and pork patties. Similarly Lee *et al.*, (2007) reported that both HHP and heat caused denaturation of bovine *semitendinosus* myofibrillar proteins; however, the mechanism of HHP-induced denaturation was different from the denaturation induced by heat.

7.2. Effect on Microbiological Quality

Many studies have been made on the effect of HHP on microorganisms in foods. A Japanese study on pressure-pasteurization using various microorganisms has shown that the decimal reduction time of 2–11 min at 300 MPa and 20°C. Pressure processing of pork homogenate at neutral pH and 400 Mpa caused 6 log cycle inactivation of several microorganisms including *E. coli*, *Campylobacter jejuni*, *S. typhimurium*, *Y. enterocolitica*, *Sacharomyces cerevisae*. *Bacillus cereus* spores were not inactivated.

When the pressure level in each volume element is the same, the heat of compression is homogeneously distributed and can be utilized to improve microbial inactivation and to maximize process efficiency. Instantaneous adiabatic heating can help to reach the sterilizing end-temperature quickly and can result in a new approach to food sterilization with a significant improvement in food quality. Additional heat flow across the system must be taken into account as a transient temperature field, after which thermal equilibration will occur during pressure holding time between the warmest point in the center of the product and the metal high pressure vessel. It is noteworthy that after pressure release, the product will return to its initial temperature or even below, an aspect of high interest for processing of high quality foods. Over-processing and fouling at the inner surface of the packaging may inhibit increasing the temperature gradient. Industrial scale systems for sterilization of cans therefore are limited to heat recovery rates in the range of 50%. The autoclave is heated from 80 to 140°C, and after 130 minutes to reach an F-value of 2.4 min in the center of the can, the autoclave is cooled.

7.3. Effect on Bacterial Spores

A combined application of high pressure and heat can be utilized to achieve a similar inactivation of spores of *Clostridium*. The temperature of the high pressure vessel will be maintained at 105°C. After 105 minutes, when a core temperature above 80°C is obtained, a pressure of 800 MPa will be applied for 10 minutes. Adiabatic heat of compression will cause a temperature increase to 110°C, after which decompression will cause a temperature drop to approximately 87°C and the can will be cooled. From the temperature-time-profile, it is obvious that the thermal load of the product will be significantly reduced due to shorter residence times, in particular within the boundary regions of the can.

The HHP processing at 981 MPa for 40 min and at 588 MPa for 120 min at 5 to 10°C did not inactivate the spores of *B. stearothermophilus* IAM12043, *B. subtilis* IAM12118, and *B. licheniformis* IAM1166. Spores of *C. sporogenes* were resistant to HPP, as 600 MPa for 30 min at 20°C caused no significant inactivation. No reduction of *C. botulinum* type E spores was observed up to 827 MPa at temperatures below 35°C. Spores of *B. cereus* and *B. subtilis* were not significantly affected by processing at 400 MPa and 10°C for 20 min. Thus bacterial spores are difficult to inactivate by HHP alone while inactivation of spores by combined high pressure and temperature is considered for the production of shelf stable foods. A 4 log reduction of *B. coagulans* spores was observed when the temperature was increased from 25°C to 70°C at 400 MPa.

Combining heat and pressure simultaneously or sequentially proved more effective in inactivating *C. sporogenes* spores than pressure alone. At 827 MPa for 5 min, the log reduction of *C. botulinum* type E spores increased as temperatures increased from 35 to 55°C. A 10 min processing at 827 MPa and 40°C resulted in a 5 log reduction of the spores. Pressurization of *B. subtilis* 168 spores at 70°C, pH 6.0 or 7.0 for 15 min at 404 MPa resulted in a 5 log reduction as compared to a 0.5 log reduction for the same pressure at 25°C. *B. stearothermophilus* spores suspended in egg matrix achieved 4 log reductions in 5 min at 700 MPa and 105°C in comparison to 1.5 log reduction by thermal treatment at 121°C for 15 min. Spore inactivation by high pressure and temperature achieved 3 log reduction in 100 s at 700 MPa and 105°C, but higher holding time up to 5 min had comparatively limited effect.

The use of HPP along with ultrasonic energy, pulsed electric fields (PEF) and irradiation could inactivate spores, with reduced pressure, treatment time and equipment cost. It has been reported HHP treatment at 50 MPa and ultra sonication at 117 μm of ultrasonic amplitude for 12 min inactivated approximately 99% of the *B. subtilis* spores. The inactivation effect of HHP was significantly improved when the amplitude of ultrasonic waves was increased. B*acillus* spores germinated by HHP were not inactivated by pulsed electrical field (PEF) up to 40°C. Use of both HHP at 6800 atm. up to 60 min and irradiation dose of 3 kGy for 20 min was sufficient to eliminate *C. sporogenes* spores in chicken breast. The irradiation sensitized the spores to pressure, and mild dose of irradiation and high pressure was more effective in inactivating *Clostridium* spores than application of either process alone. The combination treatments were synergistic in their sporicidal effect, and could decrease the intensity of one or both treatments while retaining similar inactivation. It was postulated that irradiation had disruptive effect on peptidoglycan in the spore cortex, leading to partial rehydration of the spore core. This made the spores more sensitive to both radiation and HHP.

8. EQUIPMENT FOR HHP

Initial equipment used for food processing was those used for ceramic processing for electronic components, which were modified with additional sensors. The machines presently available are batch type having limited volume. It may be mentioned that in HHP, a 50 liter equipment is comparable to 200 liter heating vessel and a time of 15 min for HHP is equivalent to one hr in heating cooling operations. Conventionally, food items packaged in pouches made from conventionally plastic materials is charged into a high pressure vessel filled with normal tap water. The vessel is closed and the pressure therein is increased to the necessary value and then decompressed to ambient pressure again.

The treated package is dried which is ready for distribution. The former is used for packaged foods and the latter for liquid items. Some of the company manufacturing equipments for high pressure food processing include ABB Autoclave Systems Inc, Columbus, Ohio, USA and Mark Freeman, Stansted Fluid Power Ltd., Essex, CM 24, 8HT, UK. The Quintus research high pressure food system developed by ABB Autoclave systems Inc., S-721 66, Sweden consists of pre-stressed, wire-bound isostatic pressure containment system, intensifier-based electrohydraulic pumping system, water circulation system and control module. The equipment can withstand 900 MPa (130, 500 psi) and a temperature of 80°C. A semi-continuous system of 1–4 tonne per hr treatment is also available.

9. REGULATORY ASPECTS OF HHP PROCESSING

Regulatory agencies such as the US FDA, USDA/FSIS, Health Canada, and others require processes that are reliable, traceable, and effective. Commercially proven, HHP processing technology from Avure meets or exceeds the demands of consumers, processors, and regulatory agencies by giving enterprising processors the ability to create clean label products and extend market reach while at the same time lowering safety risks. The science and technology behind Avure's HHP systems are well proven—exposing food to the pressures generated by HHP (600 MPa) inactivates food borne pathogens while leaving the natural organoleptic qualities of food intact. HHP enables consumers to get the natural and safe RTE meat products they demand, right on their local grocer's shelf. When RTE meat products, including cold cuts and hams, are subjected to the very high pressures of HHP for a few minutes or less, the levels of spoilage organisms are reduced significantly, and pathogens are inactivated. HHP treated meats retain their original sensory qualities such as texture, color and nutritional content throughout their shelf-life. This quality extension reduces product spoilage and returns.

10. LIMITATION

The greatest limitations for the application of HHP technology for meat texture enhancement have been the substantial capital equipment cost and need for its application to hot-boned (pre-rigor) meat for consistent benefit. An additional problem is that current batch sizes are a restriction for operations with high line speeds. It would be valuable if conditions could be established that would optimize both antimicrobial effects of HHP treatment and its beneficial texture effects simultaneously. In addition, it appears that undesirable meat color changes (browning) when used in conjunction with thermal treatment are a limitation. Although pressurization of frozen meat is being studied to avoid color changes, the use of HHP treatment of frozen hot-boned meat to prevent shortening caused by thaw-rigor and to enhance overall tenderness would appear to be an area worthy of study.

REFERENCES

Andres, A.I., Adamsen, C.E., Moller, J.K.S., Ruiz, J. and Skibsted, L.H. (2006). High pressure treatment of dry-cured Iberian ham. Effect on colour and oxidative stability during chill storage packed in modified atmosphere. *Eur. Food Res. Technol.*, 222: 486–491.

Balny, C. and Masson, P. (1993). Effects of high pressure on proteins. *Food Rev. Int.*, 9: 611–628.

Campus, M., Flores, M., Martinez, A. and Toldra, F. (2008). Effect of high pressure treatment on colour, microbial and chemical characteristics of dry cured loin. *Meat Sci.*, 80: 1174–1181.

Carlez, A., Rosec, J.P., Richard, N. and Cheftel, J.C. (1993). High pressure inactivation of *Citrobacter freundii, Pseudomonas jluorescens* and *Listeria innocua* in inoculated minced beef muscle. *Lebensm Wiss- Technol.*, 26: 357–363.

Carlez, A., Rose, J.P., Richard, N. and Cheftel, J.C. (1994). Bacterial growth during chilled storage of pressure treated minced meat. *Lebenson Wiss. Technol.*, 27: 48–54.

Carlez, A., Veciana-Nogues, T. and Cheftel, J.C. (1995). Changes in colour and myoglobin of minced beef meat due to high pressure processing. *Lebenson Wiss. Technol.*, 28: 528–538.

Cheah, P.B. and Ledward, D.A. (1997). Catalytic mechanism of lipid oxidation following high pressure treatment in pork fat and meat. *J. Food Sci.*, 62: 1135–1139.

Defaye, A., Ledward, D.A., MacDougall, D.A. and Tester, R.F. (1995). Renaturation of metmyoglobin subjected to high isostatic pressure. *Food Chem.*, 52: 19–22.

Galazka, V.B. and Ledward, D.A. (1998). High pressure effects on biopolymers. *In*: Hill, S.E., Ledward, D.A. and Mitchell, J.R. *Eds*. Functional properties of food macromolecules. Gaithersburg, Md.: Aspen Publishers Inc. pp. 278–301.

Hayakawa, I., Kanno, T., Yoshiyama, K. and Fujio, Y. (1994). Oscillatory compared with continuous high pressure sterilization on *Bacillus steatothermophilus* spores. *J. Food Sci.*, 59: 164–167.

Hayman, M., Baxter, I., Oriordan, P.J. and Stewart, C.M. (2004). Effects of high-pressure processing on the safety, quality, and shelf life of ready-to-eat meats. *J. Food Prot.*, 67: 1709–1718.

Heremans, K. (2003). The effect of high pressure on biomaterials. *In*: Ultra High Pressure Treatments of Foods. *Ed.* Marc, E.G., Hendrickx, and Dietrich, K. pp. 23–52. Kluwer Academic/Plenum Publishers, New York.

Hugas, M., Garriga, M. and Monfort, J.M. (2002). New mild technologies in meat processing: High pressure as a model technology. *Meat Sci.*, 62: 359–371.

Iwasaki, T., Noshiroya, K., Saitoh, N., Okano, K. and Yamamoto, K. (2006). Studies of the effect of hydrostatic pressure pretreatment on thermal gelation of chicken myofibrils and pork meat patty. *Food Chem.*, 95: 474–83.

Lee, E.J., Kim, Y.J., Lee, N.H., Hong, S.I. and Yamamoto, K. (2007). Differences in properties of myofibrillar proteins from bovine semitendinosus muscle after hydrostatic pressure or heat treatment. *J. Sci. Food Agric.*, 87: 40–6.

Kennick, W.H., Elgasim, E.A., Holmes, Z.A. and Meyer, P.F. (1980). The effect of pressurization of pre-rigor muscle on post-rigor meat characteristics. *Meat Sci.*, 40: 33–40.

Kurth, L.B. (1986). Effect of pressure-heat treatment on cathepsin B_1 activity. *J. Food Sci.*, 51: 663–664.

Macfarlane, J.J. (1985). High pressure technology and meat quality. In: Lawrie RA, editor. Developments in meat science. Vol. 3. London : Elsevier Applied Science. pp. 155–84.

MacFarlane, J.J., McKenzie, I.J., Turner, R.H. and Jones, P.N. (1980). Pressure treatment of meat: Effects on thermal transitions and shear values. *Meat Sci.*, 5: 307–17.

Mancini, R.A. and Hunt, M.C. (2005). Current research in meat color. *Meat Sci.*, 71: 100–121.

Marcos, B., Aymerich, T., Guardia, M.D. and Garriga, M. (2007). Assessment of high hydrostatic pressure and starter culture on the quality properties of low-acid fermented sausages. *Meat Sci.*, 76: 46–53.

Masson, P. (1992). Pressure denaturation of proteins. *In*: Balny, C., Hayasht, R., Heremans, K. and Masson, P. *Ed.* High pressure and biotechnology (Colloque INSERM Vol. 224). London: John Libbey Eurotext. pp. 89–99.

Ohmori, T., Shigehisa, T., Taji, S. and Hayashi, R. (1991). Effect of high pressure on the protease activities in meat. *Agric. Biol. Chem.*, 55: 357–361.

Orlien, V., Hansen, E. and Skibsted, L.H. (2000). Lipid oxidation in high-pressure processed chicken breast muscle during chill storage: Critical working pressure in relation to oxidation mechanism. *Eur. Food Res. Technol.*, 211: 99–104.

Serra, X., Sarraga, C., Grebol, N., Guardia, M.D., Guerrero, L., Gou, P., Masoliver, P., Gassiot, M., Monfort, J.M. and Arnau, J. (2007). High pressure applied to frozen ham at different process stages 1. Effect on the final physicochemical parameters and on the antioxidant and proteolytic enzyme activities of dry-cured ham. *Meat Sci.*, 75: 12–20.

Solomon, M.B., Long, J.B. and Eastridge, J.S. (1997). The hydrodyne: A new process to improve beef tenderness. *J Anim. Sci.*, 75: 1534–1537.

Sun, X.D. and Holley, R.A. (2010). High hydrostatic pressure effects on the texture of meat and meat products. *J Food Sci.*, 75: R17–R21.

Suzuki, A., Suzuki, N., Ikeuchi, Y. and Saito, M. (1991). Effects of high pressure treatment on the ultrastructure and solubilization of isolated myofibrils. *Agric Biol. Chem.*, 55: 2467–73.

Tuboly, E., Lebovics, V.K., Gaal, O., Meszaros, L. and Farkas, J. (2003). Microbiological and lipid oxidation studies on mechanically deboned turkey meat treated by high hydrostatic pressure. *J. Food Eng.*, 56: 241–244.

Venugopalan, V., Kamat, A.S. and Bongorwar, D.R. (2001). Processing of foods using high hydrostatic pressure. *Indian Food Industry*, 20: 65–69.

Yamamoto, K., Yoshida, T. and Iwasaki, T. (2002). Hydrostatic pressure-induced solubilization and gelation of chicken myofibrils. *Prog. Biotechnol.*, 19: 461–8.

Zhang, H. and Mittal, G. (2008). Effects of high-pressure processing (HPP) on bacterial spores: An overview. *Food Rev. Int.*, 24: 330–351.

21

Functional Meat and Meat Products

SUNIL KUMAR, Z.F. BHAT, PAVAN KUMAR AND P.K. MANDAL

ABSTRACT

Evidence of health benefits associated with the presence of specific components in meat and meat products is progressively gaining credibility. Recently, much attention has been paid to develop meat and meat products with physiological functions to promote health and prevent the risk of diseases. This chapter focuses on strategies to improve the functional value of meat and meat products. Functional value can be improved by genetic manipulation and by adding functional compounds including conjugated linoleic acid, ω3 fatty acids, vitamins and antioxidants and minerals in animal feeds to improve carcass composition and meat quality. Functional ingredients such as vegetable proteins, whey proteins, dietary fibers, garlic, sage, minerals, herbs and spices, probiotics and prebiotics can be directly incorporated into meat products during processing to improve their functional value. Reduction of specific unhealthy compounds like fat and cholesterol, sodium, allergens is another strategy. Functional compounds, especially antihypertentive peptides, opioid peptides, antioxidative peptides can also be generated from meat and meat products during processing such as fermentation, curing and aging. The use of these ingredients in meat products offers processors the opportunity to improve the nutritional and health qualities of their products. Future research needs for functional meat and meat products are also discussed. Meat and meat products thus offer exciting opportunities in the area of functional foods.

1. INTRODUCTION

Meat and meat products have both positive and negative nutritional attributes. Although muscle foods are major sources of many compounds like iron, zinc, conjugated linoleic acid and B vitamins but they are also associated with nutrients and nutritional profiles that are often considered negative including high levels of saturated fatty acids, cholesterol, sodium and high fat and caloric contents. However, meat and its derivatives may also be considered functional foods to the extent that they contain numerous compounds thought to be functional. Furthermore, meat and its associated products can be modified by adding ingredients considered beneficial for health or by eliminating or reducing components that are considered harmful.

Some of these negative nutrients in meats can be minimized by certain strategies like selection of lean meat cuts, removal of adipose fat, dietary manipulation to alter fatty acid composition and proper portion control to decrease fat consumption and caloric intake. Moreover, by the addition of potentially health promoting nutrients, the nutritional profile of meat products could be further improved. In this way, a series of meat based foods can be obtained which, without altering their base, are considered "healthy". This idea of using meat and meat products for health purposes rather than for nutrition opens up a whole new field for the meat industry. In addition to traditional presentations, meat industry can explore various possibilities, including the control of the composition of raw and processed materials via addition of a variety of functional ingredients.

This chapter focuses on strategies to improve the functional value of meat and meat products that can be realized by adding functional compounds such as conjugated linoleic acid, ω3 fatty acids, vitamin E and selenium in animal diets or by directly incorporating the functional ingredients such as vegetable proteins, dietary fibers, herbs and spices, and lactic acid bacteria into meat products during processing. Furthermore, various compounds, especially peptides, can also be generated from the products during processing such as fermentation, curing and aging, and enzymatic hydrolysis to improve their functional value for consumers.

2. FUNCTIONAL FOODS

Scientific progress in understanding the relationship between nutrition and health has an increasingly profound impact on consumer's approach to nutrition. It is a practical and new approach to achieve optimal health status by promoting the state of well-being and possibly reducing the risk of disease by producing a healthier food supply as a preventive health care strategy. This concept has moved the scientific attention from the primary role of food as the source of energy and body-forming substances to the more subtle action of biologically active food components on human health. Thus an increasing amount of attention has been focused on the emergence of a value added food category founded on the specific health and preventive properties provided by certain foods. The category has been described by a wide and creative variety of nomenclatures, including functional foods, nutraceuticals, designer foods, nutritional foods, medical foods, longevity foods, hypernutritional foods, super foods, pharma foods, prescriptive foods, phyto-foods, therapeutic foods, fitness foods, foodiceuticals, and numerous other terms.

2.1. Definition of Functional Foods

Functional foods have been defined as foods that may provide health benefits beyond basic nutrition and are similar in appearance to conventional foods that are intended to be consumed as part of a normal diet, but have been modified to subserve physiological roles beyond the provision of simple nutrient requirements (Bech-Larsen & Grunert, 2003). Functional foods contain significant levels of biologically active components that provide specific health benefits beyond the traditional nutrients they contain (Drozen & Harrison, 1998). A food marketed as functional contains added, technologically developed ingredients with a specific health benefit (Niva, 2007). However, Hasler's opinion (1998) is that all foods are functional, as they provide taste, aroma, or nutritive value.

2.2. Functional Meat and Meat Products

In the industrialized world, animal products contribute more than 60 percent of total lipids, 70 percent of saturated fats and 100 percent of the cholesterol of the diet. Consumer's preference for the animal products will probably continue. Thus it would be of national strategic importance in the fight against certain diseases (like heart disease) to design or modify animal products in such a way that dietary risks are minimized. Furthermore, using food for health purposes rather than for nutrition opens up a whole new field for the meat industry that can explore various possibilities including the control of the composition of raw and processed materials by reformulation of fatty acid content or by inclusion of dietary fiber, antioxidants, probiotics, etc. (Jiménez-Colmenero *et al.,* 2001; Kovács *et al.,* 2007; Ricondo & Ayo, 2007; Scollan, 2007).

Furthermore, various compounds, especially peptides, can also be generated from the products during processing such as fermentation, curing and aging, and enzymatic hydrolysis to improve their functional value for consumers. According to Chizzolini *et al.* (1999), selection of breeds and genetic lines within breeds, changes in animal feeding practices including some feed additives (like probiotics) and intervention in animal metabolism (like anabolic implants) are the main tools used to achieve a reduction in carcass fat content, although many such practices are not authorized in many countries. Thus many strategies could be undertaken to change the nutritional profile of meat in order to improve its functional value and change its negative image.

3. IMPROVING MEAT QUALITY THROUGH PRODUCTION STRATEGIES

3.1. Through Genetics

Genetics play a crucial role in the development of farm animals and thus on the quality and nutritional profile of meat obtained from them. There are many breeding strategies that have been proposed in recent years to reduce the fat content of the carcass by reducing the backfat thickness and intramuscular fat with an increment in lean percentage of the carcass. Some of the research studies have been directed towards the production of more tender meat (Johnson *et al.*, 1990; Wheeler *et al.*, 1990). Several strategies for increasing the ratio of lean to fat in farm animals have been proposed which may be based on the repartitioning effects of somatotropin, specific β-adrenergic agonists, some growth promoting agents, and gene manipulation techniques (Solomon *et al.*, 2002). Other strategies may be based on different breeding schemes, but most entail a backcross or a three-way or four-way cross. It has been noted that leaner carcasses are generally associated with a more glycolytic muscle fibre type and hence more rapid postmortem metabolism, pale meat colour and lower water holding capacity (De Semet *et al.*, 2002). Some changes in lipids have been further reported to be dependent on sire genetic type (Armero *et al.*, 2002).

The meat from an aged animal may have a stronger flavor and colour (due to greater generation of volatile compounds and myoglobin respectively) but it correlates directly with increased intramuscular fat content and connective tissue (Armero *et al.*, 1999a) which is a consequence of the action of muscle proteolytic and lypolytic enzymes, which are affected by age. Muscles from heavy pigs have been reported to have a higher peptidase to proteinase ratio and higher lipase activity than muscles from younger pigs (Rosell & Toldra, 1998). Meat from barrows has more fat, marbling and a thicker layer of subcutaneous fat than meat from gilts as sex has an influence on the fat content of the carcass (Armero *et al.*, 1999b).

3.2. Through Dietary Modifications

Since last many decades there has been a good deal of research on altering the fatty acid composition of the meat to attain the levels recommended by the nutritionists. A ratio of above 0.4 between polyunsaturated and saturated fatty acids being the initial target and later attempts were tried to manipulate the PUFA composition in the direction of a lower ω6: ω3 ratio to reduce the adverse effect of arachidonic acid and its eicosanoid products (Simopoulos *et al.*, 2000). This ω6: ω3 ratio constitutes a risk factor in coronary heart diseases and cancers (Enser *et al.*, 1998) and a ratio of less than 4 is highly recommended (Wood *et al.*, 2003). Meat and meat based products far exceed these target values, but there are strategies to improve them like by changing the type of feed fed to the farm animals.

However, manipulation of fats to improve the content of PUFA may cause certain problems during processing, such as softer fats due to its low melting point; oxidation and generation of off-flavours, and colour deterioration are likely to arise in absence of antioxidants (Toldra & Flores, 2004). Further, it may be mentioned that the amount of feed also affects the fat content in meat. An excess feed may increase the amount of intramuscular fat whileas feed deprivation may reduce the fat especially in glycolytic muscles by inducing lypolysis (Fernandez *et al.*, 1995).

3.2.1. *Dietary supplementation of conjugated linoleic acid*

CLA is naturally found in meat (Chin *et al.*, 1992) and was originally identified as an anti-cancer component from ground beef extract (Pariza and Hargraves, 1985; Ha *et al.*, 1987). During last few decades interests in conjugated linoleic acid (CLA) have increased as a result of its potential effects on animal production and human health-related benefits (Khanal, 2004; Roy & Antolic, 2009). It has been associated with many physiological and biological properties including antioxidant and antiobesity (Park *et al.*, 1997; Smedman & Vessby, 2001), anticarcinogenic (Munday *et al.*, 1999), antiatherosclerotic (Gavino *et al.*, 2000), antidiabetogenic (Wahle *et al.*, 2004), protection of immune system (Corino *et al.*, 2002), and contribution to bone formation (Roy & Antolic, 2009) and body composition (Smedman & Vessby, 2001).

Numerous amount of research has been done on the dietary effects of CLA in improving the animal performance, meat quality, and meat products with high amounts of CLA. The results of the dietary CLA trials on the growth, body composition and meat quality reported are inconsistent and could be explained by different animal species, different breeds, age, duration and levels of CLA, husbandry conditions and the composition of feed. No significant effects of dietary CLA (0, 0.5, 1.0, and 1.5%) on feed efficiency and body weight gain in broiler chickens was observed by Szymczyk *et al.* (2001) however, Du and Ahn (2002) reported that feeding broilers with diet containing CLA (0.25, 0.5, or 1%) for 3 weeks before slaughter improved the body composition through reducing fat deposition and backfat thickness although, no significant effects on bodyweight and body composition were observed. However, Park *et al.* (1997) were the first to report the reduction of the body fat by 60% in rats on addition of CLA (0.5% on the weight basis) in diet.

Further, Du and Ahn (2002) reported a reduction of 16% and 14% body fat in broilers by feeding a feed containing 2% and 3% CLA respectively for 5 weeks. Ostrowska *et al.* (1999) observed a reduction in fat deposition in pigs and a linear increase in the ratio of lean to fat with the dietary CLA. Along with the decrease of fat deposition, many other workers reported an increase in the protein and ash content by the dietary CLA (Szymczyk *et al.*, 2001; Terpstra *et al.*, 2002). Szymczyk *et al.* (2001) further reported that dietary CLA not only reduced fat deposition but also altered the fatty acid composition of tissue lipids.

The proportion of saturated fatty acids increased significantly, while that of monounsaturated and polyunsaturated fatty acids decreased significantly. Feeding 1% CLA oil to genetically lean pigs decreased the proportion of unsaturated fatty acid significantly and increased saturated fatty acids in both belly fat and longissimus muscle (Eggert *et al.*, 2001). Similar effects of dietary CLA on the modification of fatty acid in pig tissues were also reported by others (Joo *et al.*, 2002; Wiegand *et al.*, 2002).

Usually, greater concentration of CLA are present in ruminant meat than that from non-ruminants and it can be naturally synthesized in the rumen of ruminant animals by bacteria *Butyrivibrio fibrisolvens* via the Δ-9-desaturase of trans 11 octadecanoic acid pathway (Pollard *et al.*, 1980). Thus feeding the diets rich in polyunsaturated fatty acids, it is possible to increase the content of CLA in meat from ruminant animals (Lawson *et al.*, 2001). Further, it has been reported that the total CLA content in intramuscular fat from steers fed with pasture was greater than that fed with concentrates (Shantha *et al.*, 1997; Rule *et al.*, 2002; Realini *et al.*, 2004). Other polyunsaturated fatty acids-rich ingredients as dietary supplements also increased the CLA content in muscle lipids.

Safflower oil supplementation significantly increased the level of CLA isomers in safflower oil-fed sheeps (Ivan *et al.*, 2001; Kott *et al.*, 2003; Boles *et al.*, 2005). By gastric bacterial biohydrogenation in pigs, CLA can be produced with very limited amount resulting in low amount of CLA in pork (Dugan *et al.*, 2004). However, pork is an ideal candidate for CLA enrichment by feeding chemically synthesized CLA because CLA cannot be further saturated and can be deposited in tissues with relatively high efficiency (Dugan *et al.*, 2004).

However, many workers have reported the ill effects of dietary CLA on the sensory attributes of meat and meat products. A harder, drier and darker cooked meat than that of control broiler meat was reported by Du and Ahn (2002) after adding 2% and 3% CLA in diet. An improved hardness and decreased juiciness in chicken breast rolls was observed with increased dietary levels of CLA (Du *et al.*, 2003). However, a decreased lipid oxidation in raw chicken meat along with decreased content of haxanal and pentanal in cooked chicken meat and improved color stability in cooked chicken and pork was observed (Du *et al.*, 2000; Joo *et al.*, 2002). Furthermore, four weeks of CLA feeding resulted in lower purge loss associated with increased intramuscular fat in pig. Thiobarbituric acid-reactive substance (TBARS) value in CLA-added group was lower than that of control in pork loin. Dietary addition of 5% CLA resulted in lower lightness and yellowness after 7 days of refrigerated storage (Joo *et al.*, 2002).

3.2.2. *Dietary supplementation of omega-3 (ω3) fatty acids*

Due to the fact that unsaturated fatty acids are subjected to biohydrogenation in the rumen, the fatty acid composition of meat from ruminants is generally more saturated. The lack of a rumen means that the fatty acid composition of muscle foods from animals such as pigs, poultry and fish can be altered by diet. Many papers have been published on increasing the unsaturated fatty acids composition of pigs and poultry (Bou *et al.*, 2009). This practice already occurs in specialty products such as Iberian hams which are high in oleic acid due to consumption of acorns (Narvaez-Rivas *et al.*, 2008) and aquaculture salmon which are fed fish oils high in omega-3 fatty acids (Blanchet *et al.*, 2005).

However, these practices are limited by the fact that increasing levels of unsaturated fatty acids decreases the oxidative stability of the meat product. Increasing unsaturated fatty acids is especially a problem in muscle foods since they are high in prooxidative metals, generally low in endogenous antioxidants and subjected to processing operations that greatly increase oxidative stress (*e.g.*, cooking to produce warmed-over flavor). Thus, alteration of dietary fatty acids to change muscle composition is most easily accomplished with oleic acid since it is at least 10 times more oxidatively stable than polyunsaturated fatty acids such as linolenic (McClements & Decker, 2008). However, if dietary manipulation is performed to increase polyunsaturated fatty acids, then antioxidant technologies must also be employed to minimize oxidative deterioration. Inhibition of oxidative deterioration could be accomplished by increasing muscle antioxidants by diet or food ingredients, decreasing storage temperature and/or oxygen exclusion by vacuum packaging.

Dietary supplementation of fat and oils is an efficient method to increase the content of ω3 PUFA in animal muscles. All forms of ω3 PUFA content could be significantly increased by feeding diets supplemented with fish oil for 38 days in broiler chickens (Lopez-Ferrer *et al.*, 2001). Vegetable oils including linseed oil and rapeseed oil when supplemented in diets could also increase ω3 fatty acid content in the form of linolenic acid, which could be used to synthesize long chain ω3 PUFA (Lopez-Ferrer *et al.*, 2001). Feeding pigs with a diet containing 2% rapeseed oil plus 1% fish oil increased the content of ω3 PUFA in the longissimus muscle, backfat and sausage (Leskanich *et al.*, 1997).

Lambs on pasture usually present more ω-3 fatty acids than when they are fed concentrates (Jakobsen, 1999). When lambs are fed with linseed, the amount of n3 fatty acids increases and same findings with EPA and DHA have been reported in lambs fed a mixture of fish oil and marine algae (Elmore *et al.*, 2005). Feeding with palm oil supplements enriches the meat lipids with saturated fatty acids, especially palmitic acid (Castro *et al.*, 2005). It has been reported that the composition of fatty acid is affected by genotype (Salvatori *et al.*, 2004).

The fatty acid composition of ruminants is difficult to change because unsaturated fatty acids are biohydrogenated in the rumen. Attempts have been made to use protected dietary fats that are unavailable to ruminant bacterial enzymes but are then released in the small intestine where they are absorbed and incorporated into muscle or milk. However, protected lipids have been largely unsuccessful due to their higher cost especially in relation to the low level of fatty acid change they produce in tissue. There are also many references of improved fatty acid compositions in grass fed beef. These references relate to the increased levels of omega-3 fatty acids (especially α-linolenic acid) in beef on pasture. However, the nutritional importance of increased α-linolenic acid concentration is not clear since α-linolenic acid is not as bioactive as longer chain omega-3 fatty acids such as EPA and DHA (Harris, 2007) and since the total increase in omega-3 fatty acids in an 85 g portion of grass feed beef is generally less than 50 mg (Scollan *et al.*, 2006) which is less than 10% of current recommendation of 500 mg omega-3 fatty acids per day (Harris, 2007).

3.2.3. *Dietary supplementation of vitamins and antioxidants*

Vitamin E supplementation in animal diet and meat products can improve the quality of fresh meat and meat products by limiting protein and lipid oxidation. Concentration and time of supplementation are the factors that determine the deposition of vitamin E in the poultry and pig muscles (Mercier *et al.*, 1998). Vitamin E supplementation can improve meat color and reduce lipid oxidation in pork, beef and lamb (Chan *et al.*, 1996; Lanari *et al.*, 1995; Guidera *et al.*, 1997). Vitamin E is possibly involved in fresh meat quality by regulating the conversion of muscle to meat by inhibiting protein oxidation. Rowe *et al.* (2004) reported that dietary vitamin E caused faster degradation of troponin-T at 2 days postmortem in beef steaks through decreasing the levels of protein oxidation. A lower shear force in beef steaks from longissimus dorsi after 14 day of postmortem storage was reported by Carnagey *et al.* (2008) after feeding a diet supplemented with 1000 IU vitamin E for 104 days before slaughter.

However, 1000 IU dietary vitamin E in combination with injection of calcium chloride improved proteolysis and the rate of tenderization resulting in decreased shear force in beef steaks (Harris *et al.*, 2001). Feeding 500 mg vitamin E/kg diet in Landrace pigs reduced drip loss by 45% and 54%, respectively, in longissimus thoracis of Halothane positive and Halothane negative pigs; 1000 mg vitamin E/kg diet significantly decreased the occurrence of PSE carcass in PSE-prone Landrace x Large White Halothane positive pigs (Cheah *et al.*, 1995).

Although meat contains several natural antioxidant enzymes like superoxide dismutase and glutathione peroxidase, their activity is weakened during postmortem storage. Many antioxidants have therefore been assayed as an effective alternative for protection against oxidation. Vitamin E or α-tocopherol acetate is an effective antioxidant and is commonly added to feed today. Many other endogenous antioxidants like the histine containing dipeptides carnosine and anserine are there but their content depents on the anatomical location and species. These dipeptides are more abundant in light muscles than in dark muscles (Aristoy & Todra, 1998). Carnosine content is higher in beef and pork and anserine content is higher in poultry (Toldra & Reig, 2004).

3.2.4. *Dietary supplementation of minerals*

Meat is a rich source of iron, and also of trace elements like selenium, magnesium and zinc. Red meats (beef and lamb) contain higher levels of iron than pork and poultry. The iron is present in the haem part of myoglobin *i.e.*, the meat pigment and is highly bioavailable for humans. Oxidative muscles contain higher amounts of iron in comparison to glycolytic muscles. Iron, phosphorus, copper and zinc contents are little affected by dietary levels; however, the levels of selenium in meat are highly dependent on dietary intakes (Lynch & Kerry, 2000).

Supplementation with sodium selenite or selenium rich yeast makes for high selenium content in meat. Kim and Mahan (2001) reported that dietary supplementation of 5% or less organic and inorganic selenium increased selenium levels significantly in blood and tissues including kidney, liver, pancreas, spleen, heart and muscle in growing–finishing pigs. Mahan and Parret (1996) reported that feeding of growing–finishing swine with 0.5 ppm of inorganic and organic selenium increased the selenium content in loin by 66% and 218%, respectively. Selenium-enriched pork "Selen Pork" was produced in Korea by feeding yeastbound selenium to the pigs and sold as a functional food containing approximately 10 times the selenium content of traditional pork and was leaner and juicier with a noticeably redder color (Fisinin *et al.*, 2009). "Selen Chicken" has been developed as a premium chicken brand with high content of selenium and α- tocopherol in Korea by 24 weeks of selenium-enriched yeast and selenium-enriched alga chlorella feeding in laying hens (Skrivan *et al.*, 2008).

Beef is a major source of dietary selenium for humans and the concentration of selenium in beef varies dramatically among countries and regions. Dietary supplementation of 5% selenium-enriched yeast for 112 days in beef cattle increased the content of selenium in psoas major and longissimus muscle from 0.26 ppm to 0.63 and 0.66 ppm (Juniper *et al.*, 2008a). Dietary supplementation of 5% selenium-enriched yeast in lamb increased the selenium contents in psoas major and longissimus muscle from 0.29 and 0.30 ppm in control group to 7.02 and 7.82 ppm in selenium enriched (Juniper *et al.*, 2008b). These increased selenium contents in meat products can be an excellent way to improve selenium status for people living in selenium-deficient areas.

Different magnesium salts, such as magnesium aspartate, magnesium aspartate hydrochloride, and magnesium fumarate have been used as dietary supplements to improve pork meat quality although cheaper sources like magnesium sulphate and magnesium chloride can also be effective (D'Souza *et al.*, 1999).

4. FUNCTIONAL MEAT PRODUCTS THROUGH PROCESSSING

4.1. Reduction of Specific Unhealthy Compounds

Like any other food, meat and meat products may contain a number of components that can have adverse effects on health if more than certain amounts are consumed. Some of these components are either present naturally in meat or are part of various ingredients added during their preparation. Processing and storage of meat products involves a number of steps that can modify the concentration and bioavailability of some of their components. These steps may cause the density of some components to increase while other components may show a decline in density (due to cooking).

Meat and meat products undergo major chemical changes during storage, processing or cooking. Some components may be health promoting and may have positive effects like antioxidants, antimicrobial agents or probiotics (Knorr, 1998) whereas others may have potentially adverse effects on human health like nitrosamines, polycyclic aromatic hydrocarbons, heterocyclic amines, biogenic amines or lipid oxidation products (Goldberg *et al.*, 2004; Ruiz-Capillas & Jimenez Colmenero, 2004; Vitaglione & Fagliono, 2004). However, by applying various strategies we can either prevent or atleast reduce the concentration of these compounds within the established and acceptable limits. Any practical approach/strategy to promoting potential functional meat products therefore must reduce the formation of these compounds within the acceptable limits.

4.1.1. *Fat and cholesterol*

Due to the recommendations regarding the qualitative and quantitative aspects of fat included in population intake goals for preventing diet-related chronic diseases (WHO, 2003), many consumers are currently limiting the amount of fat and calories in their diet. Therefore, these recommendations must be key consideration before designing a new formulation for a meat product. However, fat is the major determinant of the sensory attributes of meat products (Pearson & Gillett, 1999), and thus acceptable low fat meat products cannot be effectively achieved by simply reducing the fat content in the product formulation.

Reducing the fat content in products leads to a firmer, rubbery, less juicy product with dark color and more cost (Keeton, 1994; Paneras *et al.*, 1996; Desmond *et al.*, 1998; Kirchner *et al.*, 2000). Other technological problems like reduction in particle binding, reduced cook yields, soft and mushy interiors, rubbery skin formation, excessive purge and shortened shelf life are also associated with reduction in fat levels. Various aspects of interest relating to product design and fat reducing strategy have been reported. These aspects concerning nutritional, sensory, technological, safety, appreciation, legal and cost procedures are important to an improved new product design.

Among the various strategies selection of meat and trimming of excess fat from carcass and cuts, modification of carcass composition, reformulation of meat products with reduced fat content along with modification of the fatty acid profile, reduction of cholesterol and calories provide positive results and can greatly reduce the fat content. Trimming of excess fat from carcass and cuts even though greatly reduces the fat content of over fat carcasses and cuts; this is short term solution for providing low fat products. The fat reduction technology is generally governed by two basic criteria *i.e.*, the use of leaner meat raw materials and the reduction of fat and energy density through the addition of water and other ingredients.

However, when fat content falls and water content increases, attention must be paid to the water holding ability, for which purpose protein or carbohydrate based fat replacers/mimetics and use of synthetic fat substitutes and other procedures like super critical fluid extraction are promising fat reducing technologies. The feasibility of developing low-fat products depends on a number of factors such as desired fat level, the nature of the product to be reformulated and the kind of the processing required.

Cholesterol content can vary considerably in meat products. There are two main approaches to cholesterol reduction among processing strategies, one entailing alteration of the composition of meat products, and another entailing direct removal of cholesterol by supercritical carbondioxide extraction (Clarke, 1997). A number of different meat products have been reformulated, their original composition has been altered by reducing animal fat and/or partially replacing it with vegetable oils and adding various plant based proteins or other fat replacers. This dilution method has made significant cholesterol reduction possible in the meat products (Jimenez-Colmenero, 2005).

4.1.2. *Sodium*

A significant sector of the population in Western societies is suffering of cardiovascular diseases being hypertension the most extended. Part of them experiences an increase in blood pressure when sodium is present in the diet. Jiménez-Colmenero *et al.* (2001) reported that approximately 20% to 30% of common salt intake comes from meat and meat derivatives, taking eating habits into account. As such public health and regulatory authorities have recommended a reduced dietary intake of sodium chloride but the intake still exceeds the nutritional recommendations in many countries (Ruusunen, 2003a). Hypertension has become an important health concern because it may increase the incidence of symptoms associated with coronary heart disease and stroke (Jacobson & Liebman, 1981).

The concern of reducing sodium in the food supply is important to the processed meat industry. Sodium chloride is essential for processed meat products as it has numerous functions in the emulsion making. It solubilizes meat proteins and assists in binding meat, moisture and fat in the formation of desirable gel texture. It activates proteins to increase hydration and water binding capacity. It increases the viscosity of meat batters, facilitating the incorporation of fat to form stable batters. It increases the pH of meat systems and is essential for flavour. It exerts preservative effects by its ability to lower water activity by drawing water from the tissues. It decreases fluid loss in vacuum packaged product, which has been thermally processed.

Studies on reduction or replacement of sodium chloride in processed meats suggest that replacement of sodium chloride with other chloride salts may not be totally desirable for flavour and that reducing levels of sodium chloride may affect shelf stability of such products (Olson & Terrell, 1981). Furthermore, reduction of sodium chloride is important from economic point of view as it decreases the product yield. Many attempts have been made to reduce the sodium content of the meat products by replacing sodium chloride with other chloride salts like calcium, lithium, magnesium and potassium (Hand *et al.*, 1982b). However, salt reduction considerably affected the textural changes, flavour differences, decreased moisture retention (thus yields), shortened shelf life (microbial spoilage), safety (pathogenic microorganisms) and even product appearance. Studies indicate that salt reduction has an adverse effect on water and fat binding in meat products with increasing cooking loss and weakening the texture (Ruusunen, 2003a & b).

The feasibility of reducing the sodium chloride content of processed meat products can be done by various approaches *i.e.*, by reducing the NaCl content only, replacing part or all of the NaCl with other chloride salts, replacing part of NaCl with non-chloride salt compounds or altering processing techniques with any of the above. Several studies have been done on the effects of replacing NaCl by other chlorides in the 1980s (Hand *et al.*, 1982b; Marsh, 1983; Terrell, 1983; Terrell *et al.*, 1983). NaCl and other chlorides were re-evaluated for their influences on flavour, water binding factors and microbial inhibition (Marsden, 1980; Maurer, 1983; Terrell *et al.*, 1983; Smith, 1988).

Partial replacement of NaCl with KCl, up to about 50% replacement, was generally successful without the major negative effects of KCl on flavour. Increasing the KCl proportion was accompanied by increased bitterness and off-flavour (Olson & Terrell, 1981; Marsh, 1983). Hand *et al.* (1982b) and Terrell (1983) found that LiCl replacement of NaCl came closest to producing the same flavour as NaCl with little off taste. LiCl is not approved as food ingredient and is considered to be toxic and as such has not been applied in commercial practice. Thus, 50:50 blends of NaCl and KCl are often sold and used as a salt replacer/substitute in several applications.

Food acids, particularly lactic, acetic or citric acid, enhances or intensifies saltiness impact of the salt, or contributes to the saltiness itself. Sodium and potassium lactate were used with the corresponding reduction in the NaCl level (Anders *et al.*, 1989; Askar *et al.*, 1994). Hofmann and Marggrander (1989) demonstrated that a particular collagen hydrolysate enhanced saltiness in certain products and might have potential for salt reduction. Natura (1993) has developed a non-salt substitute, called eXalt, composed of lysine and succinic acid which has a salty flavour besides some antimicrobial and antioxidant properties. It may substitute for some of the functions of salt in meat products. Phosphates, gums or starches can be used to maintain the water binding function lost due to salt deletion.

Ruusunen *et al.* (2003a) studied the preparation of low-salt phosphate-free frankfurters and observed that below 1.5% salt content, phosphate-free frankfurters require additional non-meat ingredients. Ruusunen *et al.* (2003b) studied the quality characteristics of low-salt bologna-type sausage prepared with sodium citrate, carboxymethyl cellulose, and carrageenan and with less than 1.4% NaCl, the use of these ingredients decreased frying loss and increased saltiness. Gimeno *et al.* (2001) prepared dry-fermented sausages using calcium ascorbate as a partial substitute for NaCl and seems to be a viable way of decreasing sodium in dry-fermented sausages although, substitution caused higher acidification.

4.1.3. *Allergens*

There are certain ingredients used during the production of meat products that contain some substances that cause allergic reactions in some consumers. Such ingredients, like egg, vegetables, milk proteins, etc., can be excluded from the formulations to produce "allergen-free" meat products like gluten-free meat products or lactose-free meat products. Such type of meat products have been included in the FOSHU category and approved in the Japan (Arihari, 2004).

5. PROMOTING THE SPECIFIC HEALTHY COMPOUNDS

5.1. Addition of Vegetable Proteins during Processing

5.1.1. *Soya protein based meat products*

Soy proteins (flours, concentrates, and isolates) are more commonly used in processed meat products for their functional properties and relatively low cost compared with lean meat (Chin *et al.*, 1999). Soy proteins have been incorporated in these products for their water-binding and fat-binding ability, enhancement of emulsion stability, and increased yields (Chin *et al.*, 2000). Soy oil also contains approximately 0.2 g plant sterols per 100 g that are associated with lowering plasma LDL cholesterol at intakes of 2 to 3 g/d (Ferrari & Torres, 2003; Sadler, 2004). Soy proteins' benefits relate to the reduction of cholesterol levels and menopause symptoms and the reduction of the risk for several chronic diseases, *i.e.*, cancer, heart disease, and osteoporosis (Jiménez-Colmenero *et al.*, 2001; Halsted, 2003). Many studies have found that adding soy protein to the diet or replacing animal protein in the diet with soy lowers blood cholesterol (Caroll, 1991; Sirtori *et al.*, 1993). The cholesterol-lowering effect of soy has been attributed to isoflavones, a class of phytochemicals found in soybeans (Potter, 1998).

Soy protein binds, emulsifies, and hydrates meat products. It makes meat products juicer, which improves flavor, color, texture, shelf life, sliceability, and yield. It can be used to lower the percentage of fat in ham, sausage, luncheon meats, and hot dogs. Up to 30% hydrated soy protein is permitted in commercially made hamburger patties provided the final product meets specified water-to-protein requirements (Lusas & Riaz, 1995). Soy concentrates have the ability to bind and hold natural flavors and moisture, resulting in products that stay moist and flavorful even after reconstitution in conventional or microwave ovens. In coarsely ground meat, such as pizza topping, beef patties, and Mexican fillings, the concentrate helps retain fat and prevent grease pooling, as well as retain moisture through freeze/thaw and second cooking.

The addition of soy concentrates in meat applications does not change current manufacturing processes, and soy concentrates can be used in virtually any meat system. A traditional application for soy protein concentrates in the meat industry is in cooked-sausage manufacturing. Soy concentrates can be added in ground meat, whole muscle meats and poultry emulsion meats, dry sausages, seafood, and vegetarian foods (McCue, 1994). Some of the approved products are sausage, meat balls, and hamburger patties containing soy protein isolate, which may carry the claim that these meat products lower serum cholesterol (Burke, 1998).

The textured concentrates impart a meaty texture and also perform well in both raw and precooked meat patties. Textured soy-protein concentrates have high natural-protein content, neutral taste, and structure-forming characteristics, which contribute to an improvement in the nutritional profile of standard food products and offer food manufactures new alternatives in product development. The nutritional properties of soy protein products, especially soy protein concentrates (absence of cholesterol, reduced content of fat and calories) and their physicochemical attributes (water absorption, emulsifying capacity) are important advantages for their use in many food applications. Textured soy protein ingredients have been advocated as seafood extenders. Many processors of seafood items, including surimi products, have become interested in soy concentrates because of their water-binding and holding capacity.

Soy protein isolates are very hydrophilic and thus can be incorporated into meat products to reduce cooking loss. In Argentina sausage "Chorizo", addition of 2.5% soy protein isolate decreased drip loss during 14 d refrigerated storage without introducing any changes in flavor, aroma, juiciness characteristics, oxidation and microbiological stability (Porcella *et al.*, 2001). In frankfurters and fish frankfurter analogs, incorporated soy protein hydrolysates reduced bacterial counts and extended their shelf-life stored at 25°C without influencing the flavor and texture properties of the products (Vallejo-Cordoba *et al.*, 1987). However, soy flour produced some beany flavor and soy protein concentrates and isolates provided some undesirable palatability in soy-added meat products (Rakosky, 1970; Smith *et al.*, 1973).

To overcome these disadvantages, dried soy tofu powder was added in frankfurters and pork sausage patties. Incorporation of tofu powder resulted in lower fat and higher protein and moisture content, but did not affect sensory parameters in lean pork sausages. Lean frankfurters added with tofu powder had lower moisture content, but their texture and overall acceptability was better than control (Ho *et al.*, 1997). Chin *et al.* (1999) concluded that 4.4% SPI (soy protein isolates) gives a softer texture to low-fat bologna without altering other chemical parameters. Feng *et al.* (2002) studied the effect of thermally/ enzymatically obtained soy protein isolates (2%) on pork frankfurters and observed that heat obtained soy protein improved hardness whereas enzyme-hydrolyzed soy proteins reduced hardness, cohesiveness, and breaking strength.

5.1.2. *Whey proteins based meat products*

In low-fat meat products, whey proteins show excellent nutritional and functional properties (Perez-Gago & Krochta, 2001). It can replace 100% of ice in the formula of frankfurter-type sausages when used in liquid form (Yetim *et al.*, 2001). Yetim *et al.* (2001) used whey proteins in frankfurter-type sausages and reported an improved emulsion stability, better color properties, and lower chewiness and elasticity, but with higher brittleness and hardness. Pre-heated whey protein isolates formed gel at low protein concentrations and low temperature in the presence of added salt (Hongsprabhas & Barbut, 1997).

Hongsprabhas and Barbut (1999) used pre-heated whey protein in poultry raw and cooked meat batter, it resulted in increased water holding capacity, improved rheological properties, and reduced cooking loss. Whey proteins can also be incorporated into films and coatings for various meat products. Shon and Chin (2008) used whey protein coating in sausages and reported that the growth of aerobic bacteria and *Listeria monocytogenes* were inhibited and moisture loss was decreased by 31.3%. Whey protein coatings reduced the TBARS and peroxide value by 31.3% and 27.1%, respectively, in low-fat pork sausages stored under refrigerated conditions for 8 week.

5.1.3. *Wheat proteins based meat products*

Due to their ability to form viscoelastic mass of gluten through the interaction with water, wheat proteins could be added to the formulations of various meat products (Pritchard & Brock, 1994). Gluten produced from wheat flour can be used as a binder or extender in sausage products (Janssen *et al.*, 1994). Chymotrypsin-hydrolyzed wheat gluten resulted in improved thermal gelation and emulsifying properties of myofibrillar protein isolates and lower microbial transglutaminase activity (Xiong *et al.*, 2008).

Hardness of the smoked sausages made with mechanically separated poultry meat increased but springiness decreased when wheat proteins at 3% and 6% were added (Li *et al.*, 1998). Gnanasambandam and Zayas (1992) added wheat protein flour to the frankfurters and observed that 3.5% of wheat protein flour increased water holding capacity and decreased cooking loss whereas the textural and sensory properties including viscosity, adhesiveness and batter stability were also improved.

5.2. Addition of Fiber During Processing

Studies have proven that dietary fibers have the potential to reduce blood low density lipoprotein cholesterol (Brown *et al.*, 1999), risk of diabetes mellitus type 2 (Willet *et al.*, 2002), coronary heart disease (Bazzano *et al.*, 2003), blood pressure (Streppel *et al.*, 2005), obesity (Liu *et al.*, 2003) and colorectal cancer (Schatzkin *et al.*, 2007). Fiber is suitable for addition to meat products and has previously been used in cooked meat products to increase the cooking yield due to its water-binding and fat-binding properties and to improve texture (Cofrades *et al.*, 2000).

Fiber from various types of sources like vegetables, legumes, cereal byproducts, fruits etc have been studied alone or combined with other ingredients for formulations of reduced-fat-high-fiber meat products, largely ground and restructured products (Desmond *et al.*, 1998; Mansour & Khalil, 1999), and meat emulsions (Chang & Carpenter, 1997; Grigelmo-Miguel *et al.*, 1999). Thus addition of dietary fiber is one of the options to provide meat products with low-fat and high fibers.

Fruits are one of the important sources of fiber which can also be obtained as by-products of plant food processing. Verma *et al.* (2010) studied the effect of apple pulp on the quality characteristics of low fat chicken nuggets and reported a significantly lower ($P < 0.05$) emulsion stability and cooking yield. Incorporation of apple pulp significantly increased ($P < 0.01$) dietary fibre content, redness, yellowness and chroma index of the product. Textural properties of the products significantly decreased ($P < 0.01$) with addition of apple pulp. Addition of 17% and 29% of peach dietary fiber suspensions to frankfurters increased viscosity and decreased pH without influencing cooking loss, protein and collagen contents, and sensory evaluation of the sausages (Grigelmo-Miguel *et al.*, 1999).

High level (3%) of fruit (peach, apple and orange) fibers cause increased hardness and cohesiveness and decreased sensory and textural properties in low-fat and dry fermented sausages. Addition of 1% and 2% of orange fiber to Spain dry fermented sausages decrease the residual of nitrite and increase the amounts of micrococcus during fermentation (Garcia *et al.*, 2002). Dietary fibers result in changes in pH, water activity and nitrite residue during the dry-curing (Fernandez-Lopez *et al.*, 2008).

Citrus byproducts (lemon albedo and orange fiber powder) have been added, at different concentrations, to cooked and dry-cured sausages with excellent results (Aleson-Carbonell *et al.*, 2003; 2004; Fernández-Ginés *et al.*, 2003, 2004; Fernández-López *et al.*, 2004). Fernández-Ginés *et al.* (2004) added lemon albedo at different concentration (2.5% to 10%) to cooked sausages whereas Aleson-Carbonell *et al.* (2003, 2004) added lemon albedo at different concentration to dry-cured sausages. The addition of lemon albedo to both sausages had healthy effects due to the presence of active biocompounds, which induced a decrease in residual nitrite levels.

Sausages with 2.5% to 7.5% lemon albedo added had sensory properties similar to conventional sausages. Fernández-Ginés *et al.* (2003) studied the effects of orange fiber powder at different concentrations (0.5% to 2%) on cooked sausages (bolognas). The results showed that the addition improved the nutritional value, decreased the residual nitrite level, and delayed the oxidation process as determined by TBA values and the red color.

Taludker and Sharma (2009) developed dietary fiber rich chicken meat patties by incorporating wheat and oat bran to chicken meat at 5, 10 and 15% levels. Incorporation of bran significantly increased the water holding capacity (WHC), emulsion stability (ES), cooking yield, firmness, TDF, USFA and reduction in sensory attributes, moisture, protein, fat and cholesterol content. Oat bran showed better effect on WHC and ES than wheat bran. Yilmaz (2004) used rye bran as a fat substitute in the production of meatballs. The addition of rye bran to meatballs at the levels assayed (5% to 20%) improved their nutritional value and health benefits.

The total trans-fatty acid content was lower and the ratio of total unsaturated fatty acids to total saturated fatty acids was higher in the samples with added rye bran. The same samples were lighter and yellower than the control samples. Steenblock *et al.* (2001) added oat to determine its effects on the quality characteristics of light bologna and fat-free frankfurters. Different types of oat fiber were used; high absorption (HA) or bleached oat (BL) fiber at levels up to 3%. The results indicated that the addition of both types of oat fiber produced greater yields and a lighter red color. Purge was reduced with oat fiber at 3%. Product hardness increased for bologna. It has been reported that oat bran and oat fiber provide the flavor, texture, and mouthfeel of fat in ground beef and pork sausages (García *et al.*, 2002).

Nuts provide high levels of protein and also contain dietary fiber and various bioactive compounds such as plant sterols, which have cholesterol lowering properties (Halsted, 2003, Sadler, 2004). The addition of walnuts to restructured beef steak was studied by Jimenez-Colmenero *et al.* (2003). The results showed that the addition of walnuts affects the cooking properties, color, texture, and sensory attributes, making the product softer and providing it with better water-binding properties.

Anderson and Berry (2000) studied the effects of inner pea fiber on lower-fat beef patties (10% and 14%), in which it improved tenderness and cooking yield without having negative effects on juiciness and flavor. Cardoso *et al.* (2008) studied the addition of dietary fiber obtained from inner pea and chicory root on low fat fish sausages and reported an improved gel strength and hardness without influencing textural and color parameters of the sausages. Archer *et al.* (2004) reported that a breakfast sausage product added with lupin-kernel fiber was rated more satiating than full-fat sausages and the total fat intake with lupin-kernel fiber-added breakfast sausage was 18 g lower and that with inulin-added one was 26 g lower than control.

Cáceres *et al.* (2004) studied the effect of short-chain fructooligosaccharide (FOS), a generic name for all nondigestible oligosaccharides composed mainly of fructose, on cooked sausages. The addition did not affect the pH, aw or weight losses because the presence of soluble dietary fiber (SDF) leads to a compact gel structure and therefore prevents proteins from retaining the water. The energy values decreased from 279 kcal/100 g in the conventional control to 187 kcal/100 g in the reduced-fat sausages with 12% added fiber at 12% SDF. The hardness of the samples with SDF was lower, and the overall acceptability in the sensory analysis was higher in samples with 12% SDF. Mendoza *et al.* (2001) prepared low-fat, dry-fermented sausages with a fat content close to 50 and 25% of the original amount and supplemented with 7.5 and 12.5% of inulin, another SDF which can be used as a fat substitute. The results indicate that inulin impacts a softer texture and a tenderness, springiness, and adhesiveness very similar to that of conventional sausages.

A low-calorie product (30% of the original) can be obtained with approximately 10% inulin. Todd *et al.* (1989) added pure cellulose (Solka-Floc), microcrystalline cellulose (Avicel), and a soluble gum (Nutriloid Fiberplus) to ground pork at 3.5% and 7.0% based on total dietary fiber content of each ingredient. The two cellulose products at 7.0% exhibited more hardness, whereas the gum products at 3.5% and 7.0% showed less springiness (elasticity). Cooking losses declined as fiber concentration increased from 3.5% to 7.0% for the cellulose products, but increased for the gum products. The Avicel products at 3.5% and 7.0% and Solka-Floc product at 7.0% exhibited significantly less change in diameter (patty shrinkage).

5.3. Addition of Antioxidants During Processing

Dietary antioxidants have been suggested to have the capability to prevent the oxidative damage in the body and reduce the risk of several diseases like CVD, some cancers, Alzheimer's disease, and impairment of vision (Holm, 2003). The antioxidants also reduce the susceptibility of foods to oxidation and reduce the production of harmful products. Production of functional meat products with high content of PUFAs are generally associated with increased susceptibility to lipid peroxidation. Despite of their high cholesterol lowering response than MUFAs, there has been caution in recommending the high PUFA diets because of potentially adverse health effects of their lipoperoxidation products (Williams, 2000).

Thus increase in the degree of unsaturation during meat processing should always be accompanied with an increase in concentration of some suitable antioxidants. The prime most antioxidants with a proven antioxidant effect in meat and meat products are tocopherols, vitamin C, carotenoids, flavonoids, and phenolic compounds. Several meat products containing vitamin E have been produced (Jimenez Colmenero, 2005; Jensen *et al.*, 1998). Various plant obtained ingredients rich in vitamin E have been added to the meat products during processing, including walnut in restructured steak (Serrano *et al.*, 2005), rice bran oil in roast beef (Kim *et al.*, 2000) and wheat germ in frankfurters (Gnanasambandam & Zayaz, 1992).

Ascorbic acid has been used in several meat products either in isolated form or in the form of some incorporated ingredients like in beef patties (Sanchez-Escalante *et al.*, 2003), dry cured sausages (Fernandez Lopez *et al.*, 2004) and cooked pork (O'Connell *et al.*, 2002). Some of the carotenoid rich vegetables are used as ingredients in some processed meat products as non-meat ingredients. Several carotenoids have been used as exogenous added antioxidants in meat products.

Lycopene has been added in beef patties (Desmond & Troy, 2004). Beef patties, restructured beef steak, frankfurters, and liver or meat loaves have been prepared with plant derived ingredients such as tomato pulp/juice that is lycopene rich (Sanchez-Escalante *et al.*, 2001; Yilmaz *et al.*, 2002), carrot and sweet potato rich in provitamin A (Devatkal *et al.*, 2004; Saleh & Ahmed, 1998) or spinach rich in lutein and zeaxanthin (Pizzocaro *et al.*, 1998). Phytates have been incorporated in meat products in the form of sodium phytate (commercial preparation) in restructured beef (Lee *et al.*, 1998) or as components of any of a number of phytate rich plants added to different meat products, such as rice fiber in beef roast (Kim *et al.*, 2000).

5.4. Addition of Herbs and Spices During Processing

Another important functional modification is addition of natural extracts with antioxidant properties during processing. Lipid oxidation is one of the causes for the deterioration of meat and derivatives causing undesirable changes in flavor, texture, colour, and nutritional value (Gil *et al.*, 2001; Kanner, 1994). Lipid oxidation can be effectively retarded by the use of synthetic antioxidants, such as butylated hydroxytoluene (BHT), butylated hydroxyanisole (BHA) and tertiary-butylhydroquinone, which were widely used in the meat industry but concerns over safety and toxicity pressed the food industry to find natural antioxidants (Coronado *et al.*, 2002). Use of synthetic antioxidants has recently come into dispute due to a suspected carcinogenic potential (Chen *et al.*, 1992) and the general rejection of synthetic food additives by consumers.

Natural antioxidants are extracted from plants such as rosemary, sage, tea, soybean, citrus peel, sesame seed, etc. as alternatives to the synthetic antioxidants because of their equivalent or greater efficiency (Tang *et al.*, 2001b). Some natural ingredients including herbs and spices have been studied especially in Asian countries as potential antioxidants in meat and meat products (McCarthy *et al.*, 2001). Herbs and spices contain many phytochemicals which are potential sources of natural antioxidants including phenolic diterpenes, flavonoids, tannins and phenolic acids (Dawidowicz *et al.*, 2006) which can improve flavor, retard lipid oxidation-induced food deteriorations, inhibit the growth of microorganisms, and play roles in decreasing the risk of some diseases (Achinewhu *et al.*, 1995; Tanabe *et al.*, 2002). Clove is reported to have the strongest antioxidant capacity among the spices followed by rose petals, cinnamon, nutmeg and other spices (Al-Jalay *et al.*, 1987).

5.4.1. *Rosemary extracts*

Rosemary extracts exhibit a potent antioxidant activity and are widely used in the food industry. A number of authors have reported the effectiveness of rosemary for lowering lipid oxidation in various foods (Chang *et al.*, 1977; Barbut *et al.*, 1985). The leaves of rosemary contain a large number of phenolic compounds with antioxidant activities (Gil *et al.*, 2001; Fernández-López *et al.*, 2003). Rosemary extract has been proposed as a natural means to stabilize meat oxidation. Guntensperger *et al.* (1998) tested the effectiveness of this extract both prior to and after cooking of ground pork meat.

Mirshekar *et al.* (2009) evaluated the effect of rosemary extract on the color, subsequent lipid oxidation (TBARS) and rancidity development during frozen storage of chicken thigh meat. The results showed that rosemary extract (extract at 1000 ppm) had significant effect on lipid oxidation as measured by TBARS value during frozen storage at –20°C for 120 days. Lipid oxidation only occurred to a limited extent and was insufficient to cause rancid flavor development. The results also demonstrated that rosemary was one of the two most effective antioxidants in stabilization of a* value.

Rosemary extracts added to pork sausages at 2500 ppm level was equal to or more effective than BHA/BHT in delaying TBARS values in raw and precooked sausage during refrigerated and frozen storage (Sebranek *et al.*, 2004). Rosemary extracts improved the color and freshness of pork sausages (Sebranek *et al.*, 2004). Addition of water-soluble rosemary extract in cooked turkey products was effective in retarding lipid oxidation and preventing color loss evidenced by decreased L value and increased a* value during refrigerated storage (Yu *et al.*, 2002). Rosemary extracts resulted in better color retention evidenced by decreased metmyoglobin concentration and increased oxymyoglobin values during 8 days storage in irradiated minced beef (Formanek *et al.*, 2003). Coronado *et al.* (2002) prepared wiener sausages with rosemary extract and observed no lipid oxidation in the product during long-term frozen storage.

5.4.2. *Green tea*

Green tea leaves are rich in phenolic compounds like epicatechin, epicatechin gallate, epigallocatechin, epigallocatechin gallate, teaflavin gallate, teaflavin monogallate A and B and teaflavin digallate (Kuroda & Hara, 1999; Wanasundara & Shahidi, 1996). These tea compounds promote health by preventing lipid oxidation and providing antibacterial, anticarcinogenic and antiviral ability (Katiyar & Mukhtar 1996; Yang *et al.*, 2000). Tea catechins were reported to reduce the formation of peroxides even more effectively than α-tocopherol and BHA in porcine lard and chicken fat (Chen, *et al.*, 1998).

Tea polyphenols could inhibit the formation of mutagens, which was known to be associated with the breast and colon cancer, during cooking of ground beef hamburger style meat (Weisburger *et al.*, 2002). Tang *et al.* (2001a) studied addition of tea catechins to cooked red meat and poultry and observed a significantly controlled lipid oxidation in cooked muscle patties. Jo *et al.* (2003) studied the functional properties of raw and cooked pork patties with added irradiated green tea leaf extract and reported to have beneficial biochemical properties. Tang *et al.* (2001b) reported that addition of tea catechins at 300 ppm level significantly reduced the TBARS values of beef, duck, ostrich, pork and chicken during 10 day refrigerated storage. At the same concentration, tea catechins provided two to four times more antioxidative ability than α-tocopherol depending on meats from different animal species.

Bozkurt (2006) reported that addition of green tea had no significant effects on pH, color and overall sensory quality to sausages however; it decreased the formation of TBARS and the concentration of putrescine and tyramine in a dry fermented turkey sausage. In pork sausages, green tea powder could partly substitute nitrite, and resulted in lower TBARS value and decreased volatile basic nitrogen contents compared to samples prepared with nitrite alone (Choi *et al.*, 2003).

5.4.3. *Clove*

Clove (*Eugenia caryophyllus*) is reported to have antibacterial (Everting and Deibel, 1992; Suresh *et al.*, 1992), antimycotic (Karapinar, 1990; Illicim *et al.*, 1998) and yeast inhibitory (Farag *et al.*, 1989) activity. Such properties of clove may be attributed to its 2-methoxy-4-(2-propenyl) phenol content (Beuchat & Golden, 1989; Suresh *et al.*, 1992). Clove also possesses good antioxidant activity (Gulcin *et al.*, 2004). Kumar and Tanwar (2011) explored the possibilities of utilization of clove powder as phytopreservative in chicken nuggets prepared from spent hen meat under refrigerated storage at 4 ± 1°C.

Addition of clove powder in standard formulation caused a slight change in emulsion stability, cooking yield, water activity and composition of chicken nuggets. The findings showed antimicrobial effect of clove powder as microbial load in optimized preparation was found to be significantly lower throughout the storage period when compared with control preparation.

Shan *et al.* (2009) studied the inhibitory effects of clove extract on *Listeria monocytogenes, Staphylococcus aureus* and *Salmonella enterica* in raw pork at room temperature (~20°C). The results showed that clove was effective against the bacteria and increased the stability of raw pork against lipid oxidation. Menon and Garg (2001) studied the antilisteric activity of clove oil in meat and cheese at both 30°C and 7°C. At concentrations of 0·5% and 1% clove oil restricted the growth of *Listeria monocytogenes* in the food items at both temperatures. The inhibitory activity of clove oil was more pronounced at a *concentration* of 1%. The results revealed the potential of clove oil as a natural preservative in meat and cheese.

Clove oil at 1% and 2% level inhibited the growth of *L. monocytogenes* during storage at 5°C and 15°C in ready-to-eat chicken frankfurters (Mytle *et al.*, 2006). Clove oil was also effective in inhibiting other food borne pathogens including *C. jejuni, S. enteritidis, Escherichia coli* and *Staphylococcus aureus* (Smith-Palmer *et al.*, 1998). Naveena *et al.* (2006) studied the addition of clove oil in combination with lactic acid and vitamin C in buffalo meat during retail display and reported that it decreased lipid oxidation, maintained high color a* value, and improved the sensory color.

5.4.4. *Garlic*

There are a lot of reports concerning the antibacterial and antioxidant effects of garlic on meat products (Kourounakis & Rekka, 1991; Lin *et al.*, 1991). The extract of garlic is effective in reducing the growth of many pathogens including *S. aureus, S. albus, S. typhi, E. coli, L. monocytogenes, A. niger, Acari parasitus,*

Pseudomonas aeruginosa, and *Proteus morganni* (Kumar & Berwal, 1998; Maidment *et al.*, 1999). Allicin is the main ingredient of garlic that has antimicrobial activity against both gram-positive and gram-negative bacteria and is enzymatically produced from its precursor aliin via the intermediate product of allylsulfenic acid (Ellmore & Feldberg, 1994). Aqueous garlic extract inhibited the growth of microbial contaminants including facultative aerobic, mesophilic, and faecal coliforms on the surface of refrigerated poultry carcasses (Oliveira *et al.*, 2005).

In emulsified sausage during cold storage, addition of 1% and 3% of garlic juice could lead to decreased peroxide value, TBARS, residual nitrite and total microbiological counts than those of control (Park & Kim, 2009). Kim *et al.* (2010) studied the effects of garlic juice (GJ) on lipid oxidation and meat quality of fresh semimembranosus pork during cold storage and reported significantly lower Hunter lightness and yellowness values than control. The TBARS values of GJ treated samples were significantly lower whereas tenderness and flavor attributes were significantly higher than control. Lilian *et al.* (2008) reported the effect of sage and garlic on lipid oxidation in high-pressure processed chicken meat.

5.4.5. *Sage*

Dried leaf of a mint family with many antioxidant compounds, sage contains carnosol, carnosic acid, rosmadial, rosmanol, epirosmanol, and methyl carnosate (Cuvelier *et al.*, 1994) and is commonly used in pork and pizza sausages. Fasseas *et al.* (2008) reported a decrease in the TBARS values in raw and cooked pork sample by 75% and 86%, respectively, while those of raw and cooked beef decreased by 57% and 62% compared with control by using sage oil at 3% level. Karpinska-Tymoszczyk (2007) reported a better taste in cooked meatballs with a decrease in water activity, pH, mesophilic and coliforms counts by using sage extract alone or in combination with sodium isoascorbate. Mariutti *et al.* (2008) reported that use of sage protected minced chicken breast from lipid oxidation during subsequent chilled storage for 2 weeks in high-pressure processed chicken meat.

5.4.6. *Oregano*

The essential oil obtained from oregano, a traditional Mediterranean spice, contains more than 30 compounds with carvacrol and thymol being responsible for its major antioxidant capacity (Vekiari *et al.*, 1993). Fasseas *et al.* (2008) reported lowered levels of oxidation after 12 days of refrigerated storage of pork and beef added with 3% oregano oil. By reducing the growth of microorganisms during refrigerated storage, oregano oil could extend the shelf-life of fresh chicken breast meat however, it can introduce very strong unfavorable flavor to food products resulting in low sensory quality if used at 1% level (Burt, 2004; Chouliara *et al.*, 2007). Under modified atmosphere conditions it can delay the growth of microorganisms and decrease the final counts of spoilage microorganisms at 0.05%, 0.5% and 1% levels (Skandamis & Nychas, 2001). Shan *et al.* (2009) reported the inhibitory effects of cinnamon stick, oregano, clove, pomegranate peel and grape seed extracts on *Listeria monocytogenes*, *Staphylococcus aureus*, *Salmonella enterica* and on lipid oxidation in raw pork at room temperature (~20°C). During storage the colour parameters of the extract-treated pork samples changed slightly, in comparison with significant changes in the control.

5.5. Addition of Probiotics and Prebiotics

A culture of living microorganisms which are mainly lactic acid bacteria or bifidobacteria is known as probiotic whereas prebiotic is a non digestible food ingredient that beneficially affects the host by selectively stimulating the growth and/or activity of one or limited number of bacteria in the colon and thus improves the health of the host (Arihara, 2004). Probiotics have been reported to have numerous functions including modulation of intestinal flora; prevention of diarrhea; improvement of constipation; prevention and treatment of food allergies; reduction of cancer risk; lowering plasma cholesterol level; and lowering faecal enzyme activities (Agrawal, 2005; Arihara, 2006; Stanton *et al.*, 2003).

Dry sausages are processed by fermentation without heat treatments and thus probiotic bacteria are mainly used in them. Lactic acid bacteria can contribute to flavor generation due to lactic and acetic acids, and the volatiles resulted from carbohydrate fermentation (Molly *et al.*, 1996). Germany and Japan are technically the first two countries to incorporate probiotic lactic acid bacteria into meat products (Arihara, 2006). Most of the scientific literature supports the idea that probiotic lactic acid bacteria would not cause significant differences in overall sensory properties of the product (Muthukumarasamy & Holley, 2006; Pidcock *et al.*, 2002).

Prebiotics utilized in a number of meat products include dietary fibers and oligosaccharides. Consumption of either soluble or insoluble fiber results in distinct physiological effects. Dietary fiber from oats, sugar beet, soy, apples, peas, citrus, inulin and other foods have been incorporated in the formulation of several meat products including ground meat and sausages.

5.6. Addition of Minerals

A number of meat products enriched with minerals have been developed like magnesium, calcium, phosphorus or iron to cater the requirement of specific groups. As dietary calcium has been associated with many conditions like osteoporosis, hypertension and cancer (Weaver & Liebman, 2002), some meat products have been developed with additional calcium (Arihara, 2004; Harris, 2000) and fluorated salt. They are intended to help tooth and bone development and aimed at children. Content of a number of minerals could be increased in the processed meat products by the addition of non-meat ingredients. Addition of wall nut has been reported to increase the magnesium and manganese content of restructured beef steaks (Serrano *et al.*, 2005).

5.7. Addition of Plant Sterols and Stanols

Plant sterols are natural constituents of plants and are generally present in relatively low concentrations in diet (Pennington, 2002; Ashwell, 2002). Stanols are the saturated derivatives of sterols and both resemble cholesterol. They have the ability to reduce total and low density lipoprotein (LDL) cholesterol in humans, inhibit the absorption of dietary cholesterol, and reabsorb the endogenous cholesterol from the digestive tract (Holm, 2003). Use of plant sterol and stanol esters to reduce the risk of cardiovascular diseases has been approved by FDA in the United States. Some vegetable oils and other plant ingredients are said to contain phytosterols which are/maybe used in the meat processing. Phytosterols may be isolated from them to add the pure phytosterols to the meat products or vegetable oils may be used as such during the course of meat processing. Some plant sterol enriched meat products like frankfurters, sausages, meat balls, etc, containing a mixture of plant strerols or their esters and mineral salts have been already developed (Scientific Committee on Food, 2003; Leino, 2001).

6. FUNCTIONAL INGREDIENTS DURING PROCESSING

6.1. Curing

Due to the availability of new effective preservation technologies, curing is nowadays mainly utilized to provide aroma and flavor to the meat products whereas originally it was used as a method to preserve meats (Flores, 1997). The process of 'curing' has different meaning in different countries and products. It (curing) means a long ripening (aging) process of the products in Mediterranean regions and China. Spanish Iberian and Serrano hams, Italian Parma and San Daniele hams, French Bayonne ham, and Chinese Jinhua ham are typical cured meat products in which curing process can extent up to 2–3 years without smoking and addition of nitrite. However, 'curing' has a more general meaning in Northern Europe and America and includes the meat products with added nitrite or nitrate which are usually smoked and cooked before consumption (Flores, 1997).

Many biochemical changes such as proteolysis, lipolysis and oxidation can occur in meat products especially in dry-cured meat products during this processing with the degradation of ribonucleotides being the prime most in the typical aromatic volatile compounds development. The degradation of major myofibrillar proteins; the generation of polypeptides as substrates for peptidases to produce small peptides; and the production of free amino acids are generally the three main steps of proteolysis during curing (Toldrá, 2006). Numerous muscle endogenous proteases are possibly involved in meat protein hydrolysis including calpains, cathepsin, dipeptidyl peptidases, and aminopeptidases with cathepsins and calpains being the most important for muscle proteolysis (Luccia *et al.*, 2005).

Latest and effective techniques have been used by many researchers like SDS-polyacrylamide electrophoresis (Larrea *et al.*, 2006), Free Solution Conjugate Electrophoresis (FSCE) and Reversed Phase-High Performance Liquid chromatography (RP-HPLC) (Rodriguez-Nuñez *et al.*, 1995), and two-dimensional gel electrophoresis (2-DGE) (Luccia *et al.*, 2005) to detect the protein changes during curing and map peptides. Production of many small peptides and free amino acids has been reported during long-term ripening of meat products especially in dry-cured products by these techniques. Alanine, leucine, valine, arginine, lysine, glutamic and aspartic acids are the main free amino acids generated from curing and their levels depend on aminopeptidase activity and the type of meat products (Toldrá *et al.*, 2000).

These compounds directly attribute to flavor characteristics (Spanier *et al.*, 1997) and taste properties (Koutsidis *et al.*, 2007) of meat products and also serve as water-soluble flavor precursors which can further react with reducing sugars to form Maillard reaction products and Strecker degradation products contributing to meat flavor (Imafidon & Spanier, 1994). Cysteine has been reported to play very important role for meat flavor formation among many flavor precursors. Each free amino acid can provide special taste properties: glycine and alanine are associated with sweet taste, hydrophobic amino acids contribute to bitter taste, and sodium salt of glutamic and aspartic acids can enhance taste (Nishimura & Kato, 1988; Rodriguez-Nuñez *et al.*, 1995).

Lipolysis and auto-oxidation are responsible for the changes in lipids during the curing process (Toldrá, 1998; Coutron-Gambotti & Gandemer, 1999). Phospholipases and lipases release free fatty acids from degradation of phospholipids and triglycerides which could undergo oxidation to form peroxides which could further react with peptides, amino acids leading to secondary oxidation products to form aroma compounds (Toldrá, 2006; Zhou & Zhao, 2007). Neutral lipase (hormone sensitive lipases), basic lipases (lipoprotein lipases) and acid lipase (Coutron-Gambotti & Gandemer, 1999) are the three lipase systems that are involved in the break down of triglycerides.

Whereas, phospholipases A1 (responsible for the hydrolysis of fatty acids in sn1 of the glycerol backbone of phospholipids), phospholipases A2 (responsible for the hydrolysis of fatty acids in sn2 of the glycerol backbone of phospholipids), and lysophospholipases (hydrolyse the remaining fatty acids) are the three main groups of phospholipases that are involved in the break down of phospholipids (Coutron-Gambotti & Gandemer, 1999). Action of these enzymes result in increase and accumulation of free fatty acids in meat products and provide substrates for further oxidation that is a crucial reaction to develop typical flavor of meat products, especially for many kinds of dry-cured meat products with long-term ripening process (Chizzolini *et al.*, 1998).

Oxidation is an important process for the development of typical flavor of cured products although it is recognized as the main cause of deterioration of meat quality during storage and processing. Autooxidation is the main oxidation occurring during meat processing (Gandemer, 1999) involving initiation, propagation and termination steps (Frankel, 1984). During meat products processing, the phospholipids which contain greater proportion of polyunsaturated fatty acids undergo rapid auto-oxidation compared to triglycerides (Chizzolini *et al.*, 1998) and are thus more important source for a number of volatiles such as alkanes, aldehydes, alcohols, esters and carboxylic acids (Toldrá, 1998). Volatiles with low odor threshold like aldehydes and several unsaturated ketones and furan derivatives such as C3–C10 aldehydes, C5 and C8 unsaturated ketones and pentyl or pentenyl furans (Bolzoni *et al.*, 1996; Ruiz *et al.*, 1999) play important roles for meat flavor perception development and produce oily, deep-fried, green, tallowy, metallic, cucumber, mushroom and fruity odor notes in meat products (Toldrá, 1998).

Ribonucleotides like 5′-Ribonucleotides, adenosine monophosphate (AMP), inosine monophosphate (IMP) and Guanosine monophosphate (GMP) are important sources of umami compounds in meat flavor development (Durnford & Shahidi 1998; Spurvey *et al.*, 1998) that can also enhance flavor properties, such as meaty, brothy, mouth-filling, dry and astringent qualities and suppress sulfurous perception (Kuninaka, 1981) in addition to their umami taste characteristics. Sweet amino acids such as glycine, alanine, and serine are reported to intensify umami taste of Inosine monophosphate (Kawai *et al.*, 1999). Glutamate is the major free amino acid found in the final product obtained from the curing of meat products (Córdoba *et al.*, 1994) and because of the taste synergism of inosinate with glutamate, it is an important factor in the taste of meats (Kato & Nishimura, 1987).

6.2. Fermentation

Fermentation is one of the oldest and most economical methods for producing and preserving foods. Many more traditional fermented foods originating from different countries are available (Soni & Sandhu, 1999; Sahlin, 1999; Nout & Sarkar, 1999). Dry fermented sausages are known in China since 1000 years ago (Wani and Sharma, 1999). *German salami* (smoked), *Italian salami*, *Lebanon bologna* (sausage), *Longaniza* (sausage), and *Teewurst* are typical fermented meat products of Europe. While *paak* made of fish and cereal by lactic acid fermentation and *pin dang* and *tarama* made of fermented roe are typical fermented fish products of the Far Eastern countries. India is given the credit of first evidence of fermented meat product where they produced a meat product using ghee (clarified butter) by the process of fermentation (Hamm *et al.*, 2008). Fermented meat products account for 20–40% of the total processed meat of the European Union countries which are the major producers of fermented meat products (Hamm *et al.*, 2008) with fermented sausages playing a major role among their meat products (Campbell-Platt & Cook, 1995; Lucke, 1998).

Fermentation plays a major role in meat industry as an ancient method of extending shelf-life of meat products. Production of bacteriocins and/or low-molecular mass peptides or proteins could aid in the safety of non-cooked fermented meat items. These antimicrobial compounds have the additional advantage of being readily degraded after digestion to common nutrients. Selections of specific strains of bacteria that can also produce antimicrobial compounds appear to be a useful additional hurdle in producing safer fermented uncooked products. Lactic acid bacteria has been reported to produce the peptides with antibacterial properties known as bacteriocins which can reduce or inhibit the growth of other Gram-positive bacteria (Cintas *et al.*, 1995; Cleveland *et al.*, 2001; Diep & Nes, 2002), and thus they can be used to control the growth of food borne pathogens such as *L. monocytogenes* in food products (Ennahar *et al.*, 1999).

Cintas *et al.* (1995) prepared the Spanish dry fermented sausages and reported that *Pediococcus acidilactici* isolated had a strong inhibitory effect against members of gram-positive genera. Hugas *et al.* (1995) and De Martinis and Franco (1998) reported that starter cultures containing *Lactobacillus sakei* reduced the growth of *Listeria* in fermented sausages. Also, *Lactobacillus curvatus* and *L. plantarum* in sausage starter cultures have shown antilesterial effect (Campanini *et al.*, 1993; Dicks *et al.*, 2004). Lactic acid bacterium isolated from Italian salami has been reported to show antilisterial effect (Teixeira de Carvalho *et al.*, 2006). Furthermore, nine strains of *Lactobacilus casei* and three strains of *L. plantarum* isolated from dry fermented sausages have been reported to show antagonistic activity against the indicator species tested (Vignolo *et al.*, 1993).

6.2.1. *Chemical changes during fermentation*

Selected strains of homofermentative *Lactobacilli* (Lactic acid bacteria, LAB) and/or *Pediococci*, and Gram-positive catalase positive cocci (GCC), nonpathogenic, coagulase-negative staphylococci and/or kocuriae are the basic starter cultures used in meat industry. The quality and safety of the fermented products is primarily due to the rapid production of lactic acid in those products (Campbell-Platt & Cook, 1995; Hugas & Monfort, 1997; Lucke, 1998). Despite of some problems associated with fermented meat products like detrimental effect caused by the growth of other unwanted bacteria, the growth of spoilage causing *Clostridium bacillus* and other mesophillic bacteria when the lactic acid production is low (Ray, 2004), or undesirable over acidity caused by some LAB (Coventry & Hickey, 1991; Hugas & Monfort, 1997; Garriga *et al.*, 1996), meat industry is interested in fermented meat products due to improvement in functional qualities such as sensory characteristics and nutritional aspects of the products (Jimenez-Colmenero *et al.*, 2001). Furthermore due to the increased demand for functional foods during last few decades, meat industry is looking for "functional starter cultures" which can improve sensory, nutritional quality, health and microbial safety of meat products (De Vuyst, 2000; De Vuyst *et al.*, 2003).

Numerous reactions take place during the fermentation process that remarkably change the original characteristics of raw materials resulting in products with improved functionality. Fermented sausages with characteristic aroma (Flores *et al.*, 2004; Schmidt & Berger, 1998), dry fermented sausages with improved texture (Ordonez *et al.*, 1999), semi-dry fermented sausages with improved texture and flavor are some of the examples. The production of aromatic substances is the most prominent among those changes that determines the sensory characteristics of the end product (Rantsiou & Luca, 2008). Since sausages are subjected to mild heating and may enhance the survival of probiotic bacteria in the digestive system, they can be potential candidates for probiotics (Arihara, 2006; De Vuyst *et al.*, 2008). Erkkila and Petaja (2000) reported that strains of *Lactobacillus sakei* and *Pediococcus acidilactici* have the best survival capacities under acidic conditions and high levels of bile salt out of eight meat starter cultures evaluated. However, as per Erkkila *et al.* (2001) the use of probiotics in dry fermented meat products is not common. In order to increase the survival ability of probiotics during the meat fermentation, microencapsulation has been suggested as a promising method (Audet *et al.*, 1988; Sheu & Marshall, 1993).

No significant difference was observed in the sensory attributes of a fermented dried sausage containing either unencapsulated or microencapsulated probiotic bacterium of *L. reuteri* (Muthukumarasamy & Holley, 2006). Rebucci *et al.* (2007) evaluated potential use of lactobacillus strains (*L. casei, L. paracasei paracasei, Lactobacillus rhamnosus* and *L. sakei sakei*) isolated from a traditional Italian dry fermented as probiotics. They found that *L. casei* and *L. rhamnosus* had an antibacterial activity against *E. coli* and *Salmonella enterica* ssp. *enterica* (serovar *Typhimurium*). However, development of fermented meat products with probiotics seems challenging since the viability of bacteria is affected by high content of curing salt and low pH due to acidification and low water activity due to drying (De Vuyst *et al.*, 2008).

Acidification (carbohydrate catabolism), solubilization and gelation of myofibrilla and sarcoplasmic proteins, degradation of proteins and lipids, reduction of nitrate into nitrite, formation of nitrosomyoglobin and dehydration are some of the changes (Hamm *et al.*, 2008) that are mainly caused by endogenous and microbial enzymatic activities (Molly *et al.*, 1997) which eventually result in functional characteristics of the fermented products. Lactic acids and low molecular weights flavor compounds such as peptides and free amino acids, aldehydes, organic acids and amines from proteolysis of meat are mainly responsible for the taste of fermented meat products (Naes *et al.*, 1995) whereas, the aroma of a meat product is mainly determined by lipid oxidation products, free fatty acids, and volatile compounds (Ordonez *et al.*, 1999; Claeys *et al.*, 2004). Lactic acid and acetic acid are produced from carbohydrates during fermentation process (Molly *et al.*, 1997) with lactic acid being the major flavor compound in the fermented meat products whereas acetic acid plays an important role in fully dried meat products (Mateo & Zumalacárregui, 1996).

The desirable lactate to acetate ratio is in the range of 7:1 to 20:1 (Erkkila *et al.*, 2001; Hamm *et al.*, 2008).

One of the key factors involved in the improvement of functional value of meat products during the fermentation process is the degradation of proteins. Johansson *et al.* (1994) evaluated the profiles of sarcoplasmic proteins in the fermented sausages and reported that sarcoplasmic proteins with molecular weights 20 and 30 kDa disappeared at the end of the fermentation period of 7 days. Diaz *et al.* (1997) prepared the fermented sausages and also reported that proteins with molecular weights of 40, 44, 84 and 100 kDa completely disappeared in the sausages during fermentation at 22°C for 24 h and ripening for 26 days, while polypeptides with molecular weights of 8, 10, 11, 16, 38 and 49 kDa appeared over the same time period. Verplaetse *et al.* (1989) reported similar observations in myofibrillar proteins of fermented (22°C for 3 days) and dried sausages (15°C for 18 days).

Molly *et al.* (1997) reported 75% and 57% degradation of myosin and actin, respectively, in dried (at 15°C for 18 days) and fermented (24°C for 3 days) sausages. Hughes *et al.* (2002) evaluated the semi-dried fermented sausages and concluded that the initial degradation of sarcoplasmic proteins was due to indigenous proteinases but the degradation of myofibrillar proteins was due to both indigenous and bacterial enzymes. Proteolysis of meat by endogenous enzymes such as cathepsin D-like enzymes producing peptides during the fermentation process has also been reported (Hierro *et al.*, 1999; Molly *et al.*, 1997). Free amino acids produced during the fermentation play a major role in the production of volatile compounds, which is important for the production of characteristic flavors of dry sausages. Aldehydes, alcohols and acids produced from the degradation of free amino acids have low threshold values (Montel *et al.*, 1996).

Free fatty acids produced by lipolysis get easily oxidized to produce alcohols, aldehydes, ketones, esters and lactones (Viallon *et al.*, 1996; Chizzolini *et al.*, 1998) and have a significant effect on the development of characteristic flavor (Samelis *et al.*, 1993; Galgano *et al.*, 2003) ultimately affecting the sensory qualities of the fermented meat products. Production of the above-mentioned compounds by the oxidation of free fatty acids is mainly attributed to bacteria during the fermentation process (Molly *et al.*, 1997; Lizaso *et al.*, 1999). Ansorena *et al.* (2001) found that short chain fatty acids (Cb6) are mainly responsible for strong cheesy odor.

Thus it may be concluded that the biochemical changes occurring during fermentation play an important role in enhancing the functional value of meat products. However, the production of these compounds which affect the flavor and aroma during fermentation is a very complex procedure that varies with raw materials (meat, spice and starter culture) and technology (salting, fermentation, ripening drying, fermentation and drying procedures) used for the production of fermented meat products.

7. BIOACTIVE PEPTIDES FROM MEAT

Due to increasing concerns about health in recent years, much attention has been paid to the physiological functions of foods, including anticarcinogenicity, antimutagenicity, antioxidative activity and antiaging activity. Efforts have been made by the food industries in many countries to develop new foods with physiological functions. Several attractive meat based bioactive substances have been studied (Arihara, 2004) and in addition to these compounds, protein derived peptides are another group of promising functional components of meat. Although the activities of these peptides in the sequence of proteins are latent, they are released by proteolytic enzymes (*i.e.*, muscle, microbial and digestive proteinases) or during food processing. Thus, meat proteins have possible bioactivities beyond a nutritional source of amino acids alone.

Bioactive peptides, which can exert different physiological effects, have been generated *in-vivo* in various living organisms or *in-vitro* by enzymatic hydrolysis of various proteins (Smacchi & Gobbetti, 2000). Many kinds of bioactive peptides with different activities have been reported like antioxidant (Elias *et al.*, 2008), anticancer (Song *et al.*, 2000), antihypertensive (Arihara, 2004), antimicrobial (Minervini *et al.*, 2003), opioid (Leppala, 2001), mineral binding (Jiang & Mine, 2000), immunomodulatory (Nelson *et al.*, 2007), cholesterol-lowering (Jeong *et al.*, 2007) and anti-diabetic (Jianyun *et al.*, 2008). People are showing a growing interest in potential uses of bioactive molecules in food and health care sectors (McCann *et al.*, 2005). Meat has been used as a valuable protein source for the production of bioactive peptides with special emphasis on the production of ACE inhibitory bioactive peptides.

7.1. Antihypertentive Peptides

People are showing a growing interest in potential uses of bioactive molecules in food and health care sectors (McCann *et al.*, 2005). Meat has been used as a valuable protein source for the production of bioactive peptides with special emphasis on the production of ACE inhibitory bioactive peptides. Food-based strategies to control high blood pressure would be an excellent intervention to improve health and wellness and decrease health care costs. Naturally occurring peptides with ACE inhibitory activity have been identified from the proteolytic degradation products of food proteins that can be of animal (*e.g.*, dairy, fish, meat and eggs), plant (*e.g.*, soy, rice and garlic) or microbial (*e.g.*, yeast) origin (FitzGerald & Meisel, 2000; Hong *et al.*, 2008). Bioactive peptides have been identified from the hydrolysis of skeletal muscle proteins including myosin, tropomyosin, troponin, actin and collagen (Vercruysse *et al.*, 2005).

A large number of studies have shown that peptides originating from food proteins have ACE inhibitory activity *in vitro* and can decrease blood pressure in laboratory animal studies (FitzGerald & Meisel, 2000; Hong *et al.*, 2008). However, evidence also exists for the ability of these peptides to decrease blood pressure in human populations. Arihara (2004) studied eight different enzymatic hydrolyzates of porcine skeletal muscle proteins by using exogenous enzymes for their ACE inhibitory activity and reported that the thermolysin digest had the most potent inhibitory activity and the two identified ACE inhibitory peptides, Met-Asn-Pro-Pro-Lys and Ile-Thr-Thr-Asn-Pro, had significant blood pressure-reducing effect in spontaneous hypertensive rats (Nakashima *et al.*, 2002). Saiga *et al.* (2003) evaluated chicken breast meat extract treated with an Aspergillus protease and gastric proteases (trypsin, chymotrypsin, and intestinal juice) and observed ACE inhibitory effect in both the extract and hydrolysate of the extract whereas three ACE inhibitory peptides having common sequence of Gly-X-X-Gly-X-X-Gly-X-X had been identified.

The strongest ACE inhibitory activity was observed with Gly-Phe-Hyp-Gly-Thr-Hyp-Gly-Leu-Hyp-Gly-Phe peptide. Fu-Yuan *et al.* (2008) reported the highest ACE inhibitory activity by hydrolysate obtained by alkalase enzyme while evaluating the hydrolysates of chicken leg bones. Jang and Lee (2005) and Kazunori *et al.* (2003) reported that a peptide with Val-Leu-Ala-Gln-Tyr-Lys sequence and RMLGQTPT amino acid sequence, respectively from hydrolysates of sarcoplasmic protein extracts of beef and pepsin hydrolysate of porcine skeletal troponin C, respectively showed a very strong ACE inhibitory ability. Kim *et al.* (2001) isolated two peptides with amino acid sequence of Gly-Pro-Leu and Gly-Pro-Val with high ACE inhibitory activity while sequentially digesting the bovine skin gelatin with Alcalase, Pronase E and collagenase.

7.2. Opioid Peptides

There is no documented proof on the generation of opioid peptides from the muscle proteins however, possible opioid sequences, such as Tyr-X-Phe or Tyr-X1-X2-Phe are found in the sequences of muscle proteins. Therefore, it should be possible to find opioid peptides in meat proteins by proteolytic treatment.

Studies on hemoglobin peptic hydrolyzate have revealed the presence of biologically active peptides with affinity for opioid receptors (Nyberg *et al.*, 1997; Zhao *et al.*, 1997). These peptides were named as hemorphins and were first time isolated from enzymatically treated bovine blood and later were found in brain, plasma, and cerebrospinal fluid. Meat products which are rich in blood like blood sausages could be explored in this field. Poljak *et al.* (2004) have reported that these peptides could be candidates for the pharmacological prevention of Alzheimer's disease whereas another study has demonstrated that valorphin, a fragment of hemorphins, suppressed the proliferation of tumor cells (Blishchenko *et al.*, 2002).

7.3. Antioxidative Peptides

Carnosine and anserine are the two endogenous antioxidative dipeptides found in skeletal muscle (Lynch & Kerry, 2000). These peptides have been reported to play many physiological roles, such as prevention of oxidative stress related diseases, and are known to be the most abundant antioxidants in meats (Hipkiss & Brownson, 2000).

Several antioxidant peptides have been reported to be generated from meat proteins by enzymatic digestion. The hydrolyzates obtained from porcine myofibrillar proteins by protease treatment has been reported to exhibit high antioxidant activity in a linolenic acid peroxidation system induced by Fe^{2+} (Tanabe & Nishimura, 2003). Hattori *et al.* (1998) reported the antioxidant activities of pepsin-solubilized elastin and acid solubilized elastin. Arihara *et al.* (2005) studied the antioxidant activities of enzymatic hydrolyzates of porcine skeletal muscle actomyosin using a hypoxanthine xanthine oxidase system as the source of superoxide anion. The three peptides isolated (Asp-Leu-Tyr-Ala, Ser-Leu-Tyr-Ala, and Val-Trp) not only showed *in-vitro* antioxidant activity but also exhibited *in-vivo* physiological activity.

Sakanaka *et al.* (2005) observed a strong antioxidant activity of casein calcium peptides while evaluating the ground beef homogenates incorporated with casein calcium peptides obtained by using microbial enzyme hydrolysis. In another investigation the effect of hydrolyzed potato proteins on the oxidation of isolated myofibril proteins in induced (iron-catalyzed and metmyoglobin) oxidizing systems was evaluated and it was observed that the hydrolyzed potato proteins reduced the oxidation of myofibril proteins in all physicochemical conditions tested (Wang & Xiong, 2008). Rossini *et al.* (2009) observed that casein peptides produced using flavourzyme had greater antioxidant capacity than alcalse-derived ones and were effective in inhibiting lipid peroxidation of ground beef homogenates and mechanically deboned poultry meat.

Zhang and Zhou (2010) observed significant reduction in lipid peroxidation upon incorporation of three fractions of soy bean hydrolysates into ground beef obtained from neutral protease treatment. All these results point towards the potential of antioxidative peptides derived from different food ingredients in the development of functional meat products. Moreover, due to potential health hazards related to synthetic antioxidants, the use and application of artificial antioxidant has become challenging (Becker, 1993; Mendis *et al.*, 2005). Furthermore, consumers are becoming more aware as evident from the significantly increased utilization of natural antioxidants recently (Shahidi *et al.*, 2006). Thus utilization of bioactive antioxidant peptides in meat products makes them functional food by avoiding the potential health risk associated with artificial anti-oxidants.

7.4. Peptides with Organoleptic Properties

Peptides also contribute to the organoleptic properties of foods (Gonzalez de Llano & Polo Sanchez, 2003; Pihlanto & Korhonen, 2003). Foods that involve protein hydrolysis processes during their preparation such as fermentation and aging show the generation of flavor peptides. Savory flavor peptides are known to be generated from food proteins. An octapeptide with delicious taste was isolated from beef treated with papain (Yamazaki & Maekawa, 1980) which was later called as 'beef meaty peptide' or 'savory taste enhancing peptide' (Hau *et al.*, 1997). Umami-taste enhancing peptides were found in chicken protein hydrolysate (Maehashi *et al.*, 1999). Okumura *et al.* (2004) reported sourness-suppressing peptides generated in cooked pork loins.

Natural products could be made by the addition of protein hydrolysates to enhance the flavor of meat products that plays an important role in replacing synthetic flavor enhancers. Formation of bitter tastes has been identified as a problem associated with food hydrolysates. However, hydrolysates of meat, fish and gelatin are less bitter than those from other food sources (Johanna, 2007). Thus meat proteins have a high potential to produce bioactive peptides and be used as functional ingredients for meat products. Incorporation of these bioactive peptides in meat products in order to enhance the functional value of meat products may not be practical at this point, but meat products with bioactive peptides could open door for a new market since demands for functional foods, especially natural functional foods, is increasing rapidly (Arihara, 2006).

8. RESEACH NEEDS

The functional food components in meat products serve to illustrate the value of these products in the human diet. Further research is needed to understand their interactions with meat products constituents and thus to improve their safety in potential industrial applications. There is need for more research in this fascinating area regarding the marketing of these new generation meat products and to assess long-term effects of their consumption and ultimately to convince customers of the benefits of eating these products.

REFERENCES

Achinewhu, S.C., Ogbonna, C.C. and Hart, A.D. (1995). Chemical composition of indigenous wild herbs, spices, fruits, nuts and leafy vegetables used as food. *Plant Food Hum. Nutr.*, 48: 341–348.

Agrawal, R. (2005). Probiotics: An emerging food supplement with healthy benefits. *Food Biotechnol.*, 19: 227–246.

Aleson-Carbonell, L., Fernandez-Lopez, J., Sayas-Barbera, E., Sendra, E. and Perez-Alvarez, J.A. (2003). Utilization of lemon albedo in dry-cured sausages. *J. Food Sci.*, 68: 1826–30.

Aleson-Carbonell, L., Fernández-López, J., Sendra, E., Sayas-Barberá, E. and Pérez-Alvarez, J.A. (2004). Quality characteristics of a non-fermented dry-cured sausage formulated with lemon albedo. *J. Sci. Food Agric.*, 84: 2077–84.

Al-Jalay, B., Blank, G., McConnel, B. and Al-Khayat, M. (1987). Antioxidant activity of selected spices used in fermented meat sausage. *J. Food Prot.*, 50: 25–27.

Anders, R.J., Cerveny, J.G. and MIlkowski, A.L. (1989). Method for delaying *Clostridium botulinum* growth in fish and poultry. *US Patent*, 4: 888, 191.

Anderson, E.T. and Berry, B.W. (2000). Sensory, Shear and cooking properties of lower fat beef patties made with inner pea fiber. *J. Food Sci.*, 65(5): 805–10.

Ansorena, D., Gimeno, O., Astiasaran, I. and Bello, J. (2001). Analysis of volatile compounds by GS-MS of a dry fermented sausage: Chorizo de Pamplona. *Food Res Int.*, 34: 67–75.

Archer, B.J., Johnson, S.K., Devereux, H.M. and Baxter, A.L. (2004). Effect of fat replacement by inulin or lupin-kernel fibre on sausage patty acceptability, postmeal perceptions of satiety and food intake in men. *Brit. J. Nutr.*, 91: 591–599.

Arihara, K. (2004). Functional foods. *In*: Encyclopedia of Meat Sciences, Vol. 1, Jensen, W., Devine, C. and Dikemann, M. (*Eds.*), Elesevier Science, London, pp. 492–499.

Arihara, K. (2006). Strategies for designing novel functional meat products. *Meat Sci.*, 74: 219–229.

Arihara, K., Ishikawa, S., Itoh, R., Ikimoto, M., Nakshima, Y. and Kanai, S. (2005). Antihypertensive peptides generated from pork proteins. Japan Patent (Submitted to government).

Aristoy, M.C. and Toldra, F. (1998). Concentration of free amino acids and dipeptides in porcine skeletal muscles with different oxidative patterns. *Meat Sci.*, 50: 327–332.

Armero, E., Barbosa, J.A., Toldra, F., Baselga, M. and Pla, M. (1999b). Effect of terminal sire and sex on pork muscle cathepsin (b, B+L, H) cysteine proteinase inhibitors and lypolytic enzyme activities. *Meat Sci.*, 51: 185–189.

Armero, E., Flores, M., Toldra, F., Barbosa, J.A., Olivet, J., Pla, M. and Baselga, M. (1999[a]). Effects of pig sire types and sex on carcass traits, meat quality and sensory quality of dry cured ham. *J. Sci. Food Agric.*, 79: 1147–1154.

Armero, E., Navarro, J.L., Nadal, M.I., Baselga, M. and Toldra, F. (2002). Lipid composition of pork muscle as affected by sire genetic type. *J. Food Biochem.*, 26: 91–102.

Ashwell, M. (2002). Concepts of functional foods (International Life Science Institute Europe Concise Monograph Series). Brussels, Belgium: International Life Science Institute.

Askar, A., El-Samahy, S.K., Sheheta, H.A. and Tawfek, M. (1994). Pasterma and beef bouillon. The effect of substituting KCl and K lactate for sodium chloride *Fleishwirschaft*, 1: 50.

Audet, P., Paquin, C. and Lacroix, C. (1988). Immobilized growing lactic acid bacteriawith ê-carrageenan–locust bean gumgel. *Appl. Microbiol. Biot.*, 29: 11–18.

Barbut, S., Josephson, D.B. and Maurer, A.J. (1985). Antioxidant properties of rosemary oleoresin in turkey sausage. *J Food Sci.*, 50: 1356–1359.

Bazzano, L.A., He, J., Ogden, L.G., Loria, C.M. and Whelton, P.K. (2003). Dietary fiber intake and reduced risk of coronary heart disease in US men and women: The National Health and Nutrition Examination Survey I Epidemiologic Follow-up Study. *Arch. Intern. Med.*, 163(16): 1897–1904.

Bech-Larsen, T. and Grunert, K.G. (2003). The perceived healthiness of functional foods-A conjoint study of Danish, Finnish and American consumers' perception of functional foods. *Appetite.*, 40: 9–14.

Becker, G.L. (1993). Preserving food and health: Antioxidants make functional, nutritious preservatives. *Food Processing* (Chicago), 12: 54–56.

Beuchat, L.R. and Golden, D.A. (1989). Antimicrobials occurring naturally in foods. *Food Technol.*, 43: 134–142.

Blanchet, C., Lucas, M., Julien, P., Morin, R., Gingras, S. and Dewailly, E. (2005). Fatty acid composition of wild and farmed Atlantic salmon (Salmo salar) and rainbow trout (Oncorhynchus mykiss). *Lipids*, 40(5): 529–531.

Blishchenko, E., Sazonova, O., Surovoy, A., Khaidukov, S., Sheikine, Y. and Sokolov, D. (2002). Antiproliferative action of valorphin in cell cultures. *J Pept Sci.*, 8: 438–452.

Boles, J.A., Kott, R.W., Hatfield, P.G., Bergman, J.W. and Flynn, C.R. (2005). Supplemental safflower oil affects the fatty acid profile including conjugated linoleic acid of lamb. *J Anim Sci.*, 83: 2175–2181.

Bolzoni, L., Barbieri, G. and Virgili, R. (1996). Changes in volatile compounds of Parma hard during maturation. *Meat Sci.*, 43: 301–310.

Bou, R., Codony, R., Tres, A., Decker, E.A. and Guardiola, F. (2009). Dietary strategies to improve nutritional value, oxidative stability, and sensory properties of poultry products. *Crit. Rev. Food Sci.*, 49(9): 800–822.

Bozkurt, H. (2006). Utilization of natural antioxidants: Green tea extract and *Thymbra spicata* oil in Turkish dry-fermented sausage. *Meat Sci.*, 73: 442–450.

Brown, L., Rosner, B., Willet, W.W. and Sacks, F.M. (1999). Cholesterol-lowering effects of dietary fiber: A meta-analysis. *Am. J. Clin. Nutr.*, 69: 30–42.

Burke, C. (1998). Functional foods' rising sun. *Prepared Foods,* 167: 31.

Burt, S. (2004). Essential oils: Their antibacterial properties and potential applications in foods—A review. *Int. J. Food Microbiol.*, 94: 223–253.

Cáceres, E., García, M.L., Toro, J. and Selgas, M.D. (2004). The effect of fructooligosacharides on the sensory characteristics of cooked sausages. *Meat Sci.*, 68: 87–96.

Campanini, M., Pedrazzoni, I., Barbuti, S. and Baldini, P. (1993). Behaviour of *Listeria monocytogenes* during the maturation of naturally and artificially contaminated salami: Effect of lactic acid bacteria starter cultures. *Int. J. Food Microbiol.*, 20: 169–175.

Campbell-Platt, G. and Cook, P.E. (1995). Fermented meats. London, Blackie Academic and Professional.

Cardoso, C., Mendes, R. and Nunes, M.L. (2008). Development of a healthy low-fat fish sausage containing dietary fibre. *Int. J. Food Sci. Technol.*, 43: 276–283.

Carnagey, K.M., Huff-Lonergan, E.J., Trenkle, A., Wertz-Lutz, A.E., Horst, R.L. and Beitz, D.C. (2008). Use of 25-hydroxyvitamin D3 and vitamin E to improve tenderness of beef form *longissimus dorsi* of heifers. *J. Anim. Sci.*, 86: 1649–1657.

Carroll, K.K. (1991). Review of clinical studies on cholesterol-lowering response to soy protein. *J. Am. Diet. Asso.*, 91: 820.

Castro, T., Manso, T., Mantecon, A.R., Guirao, J. and Jameno, V. (2005). Fatty acid composition and carcass characteristics of growing lamb fed diets containing palm oil supplements. *Meat Sci.*, 69: 757–764.

Chan, W.K.M., Hakkarainen, K., Faustman, C., Schaefer, D.M., Scheller, K.K. and Liu, Q. (1996). Dietary vitamin E effect on color stability and sensory assessment of spoilage in three beef muscles. *Meat Sci.*, 42: 387–399.

Chang, H. and Carpenter, J.A. (1997). Optimizing quality of frankfurters containing oat bran and added water. *J. Food Sci.*, 62(194–197): 202.

Chang, S.S., Matijasevic, B.O., Hsieh, O.L. and Huang, C.L. (1977). Natural antioxidants from rosemary and sage. *J Food Sci.*, 42: 1102–1106.

Cheah, K.S., Cheah, A.M. and Krausgrill, D.I. (1995). Effect of dietary supplementation of vitamin E on pig meat quality. *Meat Sci.*, 39: 255–264.

Chen, C., Pearson, A.M. and Gray, J.I. (1992). Effects of synthetic antioxidants (BHA, BHT and PG) on the mutagenicity of IQ-like compounds. *Food Chem.*, 43: 177–183.

Chen, Z.Y., Wang, L.Y., Chan, P.T., Zhang, Z.S., Chung, H.Y. and Liang, B. (1998). Antioxidative activity of green tea catechin extract compared with that of rosemary extract. *J. Am. Oil Chem. Soc.*, 75: 1141–1145.

Chin, K.B., Keeton, J.T., Longnecker, M.T. and Lamkey, J.W. (1999). Utilization of soy protein isolate and konjac blends in a low-fat bologna (model system). *Meat Sci.*, 53: 45–57.

Chin, K.B., Keeton, J.T., Miller, R.K., Longnecker, M.T. and Lamkey, J.W. (2000). Evaluation of konjac blends and soy protein isolate as fat replacements in low-fat bologna. *J Food Sci.*, 65(5): 756–63.

Chin, S.F., Liu, W., Storkson, J.M., Ha, Y.L. and Pariza, M.W. (1992). Dietary sources of conjugated dienoic isomers of linoleic acid, a newly recognized class of anticarcinogens. *J. Food Compos Anal.*, 5: 185–197.

Chizzolini, R., Novelli, E. and Zanardi, E. (1998). Oxidation in traditional Mediterranean meat products. *Meat Sci.*, 49: S87–S99.

Chizzolini, R., Zanardi, E., Dorigoni, V. and Ghidini, S. (1999). Calorific value and cholesterol content of normal and low fat meat and meat products. *Trends Food Sci Tech.*, 10: 119–128.

Choi, S.H., Kwon, H.C., An, D.J., Park, J.R. and Oh, D.H. (2003). Nitrite contents and storage properties of sausage added with green tea powder. *Korean J. Food Sci. Anim. Res.*, 23: 299–308.

Chouliara, E., Karatapanis, A., Savvaidis, I.N. and Kontominas, M.G. (2007). Combined effect of oregano essential oil and modified atmosphere packaging on shelf-life extension of fresh chicken breast meat stored at 4°C. *Food Microbiol.*, 24: 607–617.

Cintas, M.L., Rodriguez, J.M., Fereandez, M.F., Sletten, K., Nes, I.F. and Hernandez, P.E. (1995). Isolation and characterization of Pediocin l50, a new bacteriocin from *Pediococcus acidilactici* with a broad inhibitory spectrum. *Applied Environmental Microbiol.*, pp. 2643–2648.

Claeys, E., De Smet, S., Balcaen, A., Raes, K. and Demeyer, D. (2004). Quantification of fresh meat peptides by SDS-PAGE in relation to ageing time and taste intensity. *Meat Sci.*, 67: 281–288.

Clarke, A.D. (1997). Reducing cholesterol levels in meat, poultry and fish products. *In*: Production and processing healthy, meat and fish products, Pearson, A.M. and Dutson, T.R. (*Eds.*), London, Blackie Academic and Professional, pp. 101–117.

Cleveland, J., Montville, T.J., Nes, I.F. and Chikindas, M.L. (2001). Bacteriocins: Safe, natural antimicrobials for food preservation. *Int. J. Food Microbiol.*, 71: 1–20.

Cofrades, S., Guerra, M.A., Carballo, J., Fernández-Martín, F. and Jiménez-Colmenero, F. (2000). Plasma protein and soy fiber content effect on bologna sausage properties as influenced by fat level. *J. Food Sci.*, 65: 281–7.

Córdoba, J.J., Rojas, T.A., González, C.G. and Barroso, J.V. (1994). Evolution of free amino acids and amines during ripening of Iberian cured ham. *J. Agr. Food Chem.*, 42: 2296–2301.

Corino, C., Bontempo, V. and Sciannimanico, D. (2002). Effects of dietary conjugated linoleic acid on some specific immune parameters and acute phase protein in weaned piglets. *Canadian J. Anim. Sci.*, 82: 115–117.

Coronado, S.A., Trout, G.T., Dunshea, F.R. and Shah, N.P. (2002). Antioxidant effects of rosemary extract and whey powder on the oxidative stability of wiener sausages during 10 months frozen storage. *Meat Sci.*, 62: 217–24.

Coutron-Gambotti, C. and Gandemer, G. (1999). Lipolysis and oxidation in subcutaneous adipose tissue during dry-cured ham processing. *Food Chem.*, 64: 95–101.

Coventry, J. and Hickey, M.W. (1991). Growth-characteristics of meat starter cultures. *Meat Sci.*, 30: 41–48.

Cuvelier, M., Berset, C. and Richard, H. (1994). Antioxidant constituents in sage (*Salvia officinalis*). *J. Agr. Food Chem.*, 42: 665–669.

D'Souza, D.N., Warner, R.D., Dunshea, F.R. and Leury, B.J. (1999). Comparison of different dietary magnesium supplements on pork quality. *Meat Sci.*, 51: 221–225.

Dawidowicz, A.L., Wianowska, D. and Baraniak, B. (2006). The antioxidant properties of alcoholic extracts from *Sambucus nigra* L. (antioxidant properties of extracts). *Lebensmittel-Wissenschaft und Technologic*, 39: 308–315.

De Martinis, E.C.P. and Franco, B.D.G.M. (1998). Inhibition of *Listeria monocytogenes* in a pork product by a *Lactobacillus sake* strain. *Int. J. Food Microbiol.*, 42: 119–126.

De Smet, S., Claeys, E. and Demeyer, D. (2002). Muscle enzymes in relation to meat quality and muscularity. *In*: Research advances in the quality of meat and meat products. Toldra, F. (*Ed*,) Trivandrum, India, Research Signpost, pp. 123–148.

De Vuyst, L. (2000). Technology aspects related to the application of functional starter cultures. *Food Technol. Biotech.*, 38: 105–112.

De Vuyst, L, Falony, G. and Leroy, F. (2008). Review: Probiotics in fermented sausages. *Meat Sci.*, 80: 75–78.

De Vuyst, L., Foulquié Moreno, M.R. and Revets, H. (2003). Screening for enterocins and detection of hemolysin and vancomycin resistance in enterococci of different origins. *Int. J. Food Microbiol.*, 84: 299–318.

Desmond, E. and Troy, D. (2004). Nutrient claims on packaging. *In*: Encyclopedia of meat Science, Vol II, Jensen, W.K., Devine, C. and Dikeman, M. (*Eds.*), Oxford UK, Elsevier Academic, pp. 903–910.

Desmond, E., Troy, D.J. and Bucley, J. (1998). Comparative studies on non-meat ingredients used in the manufacture of low-fat burgers. *J. Muscle Foods*, 9: 221–4.

Devatkal, S., Mendiratta, S.K. and Kondaiah, N. (2004). Quality characteristics of loaves from buffalo meat, liver and vegetables. *Meat Sci.*, 67: 377–383.

Devendra, Kumar. and Tanwar, V.K. (2011). Utilization of clove powder as phytopreservative for chicken nuggets preparation. *J. Stored Products and Postharvest Research,* 2(1): 11–14.

Diaz, O., Fernandez, M., Garcia de Fernando, G.D., de la Hoz, L. and Ordonez, J.A. (1997). Proteolysis in dry fermented sausages: The effect of selected exogenous proteases. *Meat Sci.*, 46: 115–128.

Dicks, L.M.T., Mellet, F.D. and Hoffman, L.C. (2004). Use of bacteriocin producing starter cultures of *Lactobacillus plantarum* and curvatus in production of ostrich meat salami. *Meat Sci.*, 66: 703–708.

Diep, D.B. and Nes, I.F. (2002). Ribosomally synthesized antibacterial peptides in Grampositive bacteria. *Current Drug Targets*, 3: 107–122.

Drozen, M. and Harrison, T (1998). Structure/function claims for functional foods and nutraceuticals. *Nutraceuticals World*, 1: 18.

Du, M. and Ahn, D.U. (2002). Effect of dietary conjugated linoleic acid on the growth rate of live birds and on the abdominal fat content and quality of broiler meat. *Poult Sci.*, 81: 428–433.

Du, M., Ahn, D.U., Nam, K.C. and Sell, J.L. (2000). Influence of dietary conjugated linoleic acid on volatile profiles color and lipid oxidation of irradiated raw chicken meat. *Meat Sci.*, 56: 387–395.

Du, M., Nam, K.C., Hur, S.J., Ismail, H., Kim, Y.H. and Ahn, D.U. (2003). Quality characteristics of irradiated chicken breast rolls from broilers fed different levels of conjugated linoleic acid. *Meat Sci.*, 63: 249–255.

Dugan, M.E.R., Aalhus, J.L. and Kramer, J.K.G. (2004). Conjugated linoleic acid pork research. *Am. J. Clin. Nutr.*, 79: 1212–1216.

Durnford, E. and Shahidi, F. (1998). Flavour of fish meat. *In:* Shahidi, F. (*Ed.*), Flavor of meat, meat products and seafoods (2nd Ed), London: Blackie Academic and Professional. pp. 131–158.

Eggert, J.M., Belury, M.A., Kempa-Steczko, A., Mills, S.E. and Schinckel, A.P. (2001). Effects of conjugated linoleic acid on the belly firmness and fatty acid composition of genetically lean pigs. *J. Anim. Sci.*, 79: 2866–2872.

Elias, R.J., Kellerby, S.S. and Decker, E.A. (2008). Antioxidant activity of proteins and peptides. *Crit. Rev. Food Sci. Nutr.*, 48: 430–441.

Ellmore, G.S. and Feldberg, R.S. (1994). Allin lyase localization in bundle sheaths of the garlic glove (*Allium sativum*). *Am. J. Bot.*, 81: 89–94.

Elmore, J.S., Cooper, S.L., Enser, M., Mottram, D.S., Sinclair, L.A., Wilkinson, R.G. and Wood, J.D. (2005). Dietary manipulation of fatty acid composition in lamb meat and its effect on the volatile aroma compounds of grilled lamb. *Meat Sci.*, 69: 233–242.

Ennahar, S., Sonomoto, K. and Ishizaki, A. (1999). Class IIa bacteriocins from lactic acid bacteria: Antibacterial activity and food preservation. *J. Biosci. Bioeng.*, 87: 705–716.

Enser, M., Hallet, K.G., Hawett, B., Fursey, G.A., Wood, J.D. and Harrington, G. (1998). Fatty acid content and composition of UK beef and lamb muscle in relation to production system and implications for human nutrition. *Meat Sci.*, 49: 329–341.

Erkkila, S. and Petäjä, E. (2000). Screening of commercialmeat starter cultures at lowpHand in the presence of bile salts for potential probiotic use. *Meat Sci.*, 55: 297–300.

Erkkila, S., Petaja, E., Eerola, S., Lilleberg, L., Mattila-Sandholm, T. and Suihko, M.L. (2001). Flavor profiles of dry sausages fermented by selected novel meat starter cultures. *Meat Sci.*, 58: 111–116.

Everting, W.T. and Deibel, K.E. (1992). Sensitivity of *Listeria monocytogens* to spices at two temperatures. *J. Food Safety*, 12: 129–131.

Farag, R.S., Daw, Z.Y., Hewedi, F.M. and El-Baroty, G.S.A. (1989). Antibacterial activity of some Egyptian spices essential oils. *J Food Protect,* 52: 665–667.

Fasseas, M.K., Mountzouris, K.C., Tarantilis, P.A., Polissiou, M. and Zervas, G. (2008). Antioxidant activity in meat treated with oregano and sage essential oils. *Food Chem.*, 106: 1188–1194.

Feng, J., Xiong, Y.L. and Mikel, W.B. (2002). Textural properties of pork frankfurters containing thermally/enzymatically modified soy proteins. *J. Food Sci.*, 68(1): 1220–4.

Fernandez Lopez, J., Fernendez Gines, J.M., Aleson-Carbonell, L., Sendra, E., Sayas Barbera, E. and Perez Alvarez, J.A. (2004). Applications of functional citrus byproducts to meat products. *Trends Food Sci. Technol.*, 15: 176–185.

Fernandez, X., Mourot, J., Mounier, A. and Ecolan, P. (1995). Effect of muscle type and food deprivation for 24 hrs on the composition of the lipid fraction in muscles of large White pigs. *Meat Sci.*, 41: 335–343.

Fernandez-Gines, J.M., Fernandez-Lopez, J., Sayas-Barbera, E., Sendra, E. and PerezAlvarez, J.A. (2003). Effect of storage conditions on quality characteristics of bologna sausages made with citrus fiber. *J. Food Sci.*, 68: 710–714.

Fernández-Ginés, J.M., Fernández-López, J., Sayas-Barberá, E., Sendra, E. and PérezÁlvarez, J.A. (2004). Lemon albedo as a new source of dietary fiber: application to bologna sausages. *Meat Sci.*, 67: 7–13.

Fernández-López, J., Sayas-Barberá, M.E., Navarro, C., Marín, F. and Pérez-Alvarez, J.A. (2003). Evaluation of the antioxidant potential of hyssop (*Hyssopus officinalis* L.) and rosemary (*Rosmarinus officinalis* L.) extract in cooked pork meat. *J Food Sci.*, 68: 660–4.

Fernandez-Lopez, J., Sendra, E., Sayas-Barbera, E., Navarro, C. and Perez-Alvarez, J.A. (2008). Physico-chemical and microbiological profiles of "salchichon" (Spanish dry-fermented sausage) enriched with orange fiber. *Meat Sci.*, 80: 410–417.

Ferrari, C.K.B. and Torres, E.A.F.S. (2003). Biochemical pharmacology of functional foods and prevention of chronic diseases of aging. *Biomed. Pharmacother*, 57: 251–60.

Fisinin, V.I., Papazyan, T.T. and Surai, P.F. (2009). Producing selenium-enriched eggs and meat to improve the selenium status of the general population. *Crit. Rev. Biotechnol.*, 29: 18–28.

FitzGerald, R.J. and Meisel, H. (2000). Milk protein-derived peptide inhibitors of angiotensin-I-converting enzyme. *British J. Nutr.*, 84(1): S33–S37.

Flores, J. (1997). Mediterranean vs Northern European meat products. Processing technologies and main differences. *Food Chemistry*, 59: 505–510.

Flores, M., Dura, M., Marco, A. and Toldr, F. (2004). Effect of *Debaryomyces* spp. on aroma formation and sensory quality of dry-fermented sausages. *Meat Sci.*, 68: 439–446.

Formanek, Z., Lynch, A., Galvin, K., Farkas, J. and Kerry, J.P. (2003). Combined effects of irradiation and the use of natural antioxidants on the shelf-life stability of overwrapped minced beef. *Meat Sci.*, 63: 433–440.

Frankel, E.N. (1984). Lipid oxidation: Mechanisms, products and biological significance. *J. Am. Oil Chem. Soc.*, 61: 1908–1917.

Fu-yuan, C., Yu-tse, L., Tien-chun, W., Liang-chuan, L. and Sakata, R. (2008). The development of angiotensin I-converting enzyme inhibitor derived from chicken bone protein. *Anim Sci. J.*, 79: 122–128.

Galgano, F., Favati, F., Schirone, M., Martuscelli, M. and Crudele, M.A. (2003). Influence of indigenous starter cultures on the free fatty acids content during ripening in artisan sausages produced in the Basilicata region. *Food Technol. Biotech.*, 41: 253–258.

Gandemer, G. (1999). Lipids and meat quality: Lipolysis, oxidation, maillard reaction and flavour. *Sciences des Aliments*. 19: 439–458.

Garcia, M.L., Dominguez, R., Galvez, M.D., Casas, C. and Selgas, M.D. (2002). Utilization of cereal and fruit fibres in low fat dry fermented sausages. *Meat Sci.*, 60: 227–236.

Garriga, M., Hugas, M., Gou, P., Aymerich, M.T., Arnau, J. and Monfort, J.M. (1996). Technological and sensorial evaluation of *Lactobacillus* strains as starter cultures in fermented sausages. *Int. J. Food Microbiol.*, 32: 173–183.

Gavino, V.C., Gavino, G., Leblanc, M.J. and Tuchweber, B. (2000). An isomeric mixture of conjugated linoleic acids but not pure cis-9, trans-11-octadecadienoic acid affects body weight gain and plasma lipids in hamsters. *J. Nutr.*, 130: 27–29.

Gil, M.D., Bañón, S.J., Cayuela, J.M., Laencina, J. and Garrido, M.D. (2001). Utilización de extractos de plantas como antioxidantes naturales en carne y productos cárnicos: Revisión. *Eurocarne*, 101: 1–10.

Gimeno, O., Astiasarán, I. and Bello, J. (2001). Calcium ascorbate as a potential partial substitute for NaCl in dry fermented sausages: Effect on color, texture and hygienic quality at different concentrations. *Meat Sci.*, 57: 23–9.

Gnanasambandam, R. and Zayas, J.F. (1992). Functionality of wheat germ protein in comminuted meat products as compared with corn germ and soy proteins. *J. Food Sci.*, 57: 829–833.

Goldberg, T., Cai, W., Peppa, M., Dardaine, V., Baliga, B.S., Uribarri, J. and Vassara, H. (2004). Advanced glycoxidation end products in commonly consumed foods. *J. Am. Diet. Assoc.*, 104: 1287–91.

Gonzalez de Llano, D. and Sanchez, C.P. (2003). Peptides: *In* Encyclopedia of food science and nutrition. 2nd edition, Caballero, B., Trugo, L.C. and Finglas, P.M. (*Eds.*), London: Acedemic Press, pp. 4468–4473.

Grigelmo-Miguel, N., Abadias-Seros, M.I. and Martin-Belloso, O. (1999). Characterisation of low-fat high-dietary fibre frankfurters. *Meat Sci.*, 52: 247–256.

Guidera, J., Kerry, J.P., Buckley, D.J., Lynch, P.B. and Morrissey, P.A. (1997). The effect of dietary vitamin E supplementation on the quality of fresh and frozen lamb meat. *Meat Sci.*, 45: 33–43.

Gulcin, I., At, G., Beydemir, U., Elmasta, M. and Lu, O.R.K. (2004). Comparison of antioxidant activity of clove (*Eugenia caryophylata* Thunb) buds and lavender (*Lavandula stoechas* L.). *Food Chem.*, 87(3): 393–400.

Guntensperger, B., Hammerli-Meier, D.E. and Escher, F.E. (1998). Rosemary extract and precooking effects on lipid oxidation in heat-sterilized meat. *J. Food Sci.*, 63(6): 955–957.

Ha, Y.L., Grimm, N.K. and Pariza, M.W. (1987). Anticarcinogens from fried ground beef: Heat-altered derivatives of linoleic acid. *Carcinogenesis*, 8(12): 1881–1887.

Halsted, C.H. (2003). Dietary supplements and functional foods: 2 sides of a coin? *Am J Clin Nutr.*, 77: 1001–7.

Hamm, W.P., Haller, D. and Ganzle, M.G. (2008). Fermented meat products. *In*: Edward, R. Farnworth (*Ed.*), Handbook of fermented functional foods, 2nd Ed. USA: CR Press.

Hand, L.W., Terrel, R.N. and Smith, G.C. (1982b). Effects of complete or partial replacement of sodium chloride on processing and sensory properties of hams. *J. Food Sci.*, 47: 1776.

Harris, C. (2000). Meat products are perfect as functional foods. Meat Processing January/February: 19. NAS (National Academy of Sciences), Dietary Reference Intakes for Thiamin, Riboflavin, Niacin, Vitamin B6, Folate, Vitamin B12, Pantothenic Acid, Biotin, and Choline, National Academy Press, Washington, D.C.

Harris, S.E., Huff-Lonergan, E., Lonergan, S.M., Jones, W.R. and Rankins, D. (2001). Antioxidant status affects color stability and tenderness of calcium chloride injected beef. *J Anim Sci.*, 79: 666–677.

Harris, W.S. (2007). International recommendations for consumption of long-chain omega-3 fatty acids. *Journal of Cardiovascular Medicine*, 8(1): S50"S52 (Hagerstown, Md).

Hasler, C.M. (1998). Functional foods: Their role in disease prevention and health promotion. *Food Technol.*, 52(2): 57–62.

Hattori. M/, Yamaji-Tsukamoto. K., Kumagai, H., Feng, Y. and Takahashi, K. (1998). Antioxidant activity of soluble elastin peptides. *J. Agr. Food Chem.*, 42: 2167–2170.

Hau, J., Cazes, D. and Fay, L.B. (1997). Comprehensive study of the 'beefy meaty peptide'. *J. Agr. Food Chem.*, 45: 1351–1355.

Hierro, E., de la Hoz, L. and Ordonez, J.A. (1999). Contribution of the microbial and meat endogenous enzymes to the free amino acid and amine contents of dry fermented sausages. *J. Agr. Food Chem.*, 47: 1156–1161.

Hipkiss, A.R. and Brownson, C.A. (2000). A possible new role for the antiageing peptide carnosine. *Cell Mol. Life Sci.*, 57: 747–753.

Ho, K.G., Wilson, L.A. and Sebranek, J.G. (1997). Dried soy tofu powder effects on frankfurters and pork sausage patties. *J. Food Sci.*, 62: 434–437.

Hoffman, K. and Marggrander, K. (1989). Reducing common salt content of meat products by using collagen hydrolysates. *Fleishwirschaft Int.*, 3: 34.

Holm, F. (2003). New functional foods ingredients cardiovascular health. Project No. QLK1-CT-2000-00040, SMEs No. 5, Retrived October 2004 from www.nutrition.org.uk.

Hong, F., Ming, L., Yi, S., Zhanxia, L., Yongquan, W. and Chi, L. (2008). Theantihypertensive effect of peptides: A novel alternative to drugs? *Peptides* 29(6): 1062–1071.

Hongsprabhas, P. and Barbut, S. (1997). Effect of gelation temperature on Ca^{2+}-induced gelation of whey protein isolate. *Food Sci. Technol.*, 30: 45–49.

Hongsprabhas, P. and Barbut, S. (1999). Effect of pre-heated whey protein level and salt on texture development of poultry meat batters. *Food Res. Int.*, 32: 145–149.

Hugas, M. and Monfort, J.M. (1997). Bacterial starter cultures for meat fermentation. *Food Chem.*, 59: 547–554.

Hugas, M., Garriga, M., Aymerich, M.T. and Monfort, J.M. (1995). Inhibition of Listeria in dry fermented sausages by the bateriocinogenic *Lactobacillus sakei* CTC494. *J Appl Bact.*, 79: 322–330.

Hughes, M.C., Kerry, J.P., Arendt, E.K., Kenneally, P.M. and McSweeney, P.L.H. (2002). Characterization of proteolysis during the ripening of semi-dry fermented sausages. *Meat Sci.*, 62: 205–216.

Illicim, A., Digrak, M. and Bagei, E. (1998). The investigation of antimicrobial effect of some plant extracts. *Turkish J. Biology*, 22: 119–125.

Imafidon, G.I. and Spanier, A.M. (1994). Unraveling the secret of meat flavor. *Trends Food Sci. Tech.*, 5: 315–321.

Ivan, M., Mir, P.S., Koenig, K.M., Rode, L.M., Neill, L. and Entz, T. (2001). Effects of dietary sunflower seed oil on rumen protozoa population and tissue concentration of conjugated linoleic acid in sheep. *Small Ruminant Res.*, 41: 215–227.

Jacobson, M.F. and Liebman, B.F. (1981). Statement to subcommittee on investigations and oversight, Committee on science and Technology U.S. House of Representatives, Washington, DC (April 14).

Jakobsen, K. (1999). Dietary modifications of animal fat: Status and future perspectives. *Lipids*, 101: 475–483

Jang, A. and Lee, M. (2005). Purification and identification of angiotensin converting enzyme inhibitory peptides from beef hydrolysates. *Meat Sci.*, 69: 653–661.

Janssen, F.W., de-Baaij, J.A. and Hagele, G.H. (1994). Heat-treated meat-products: Detection of modified gluten by SDS-electrophoresis, western-blotting and immunochemical staining. *Fleischwirtschaft*. 74(168–170): 176–178.

Jensen, C., Lauridsen, C. and Bertelsen, G. (1998). Dietary vitamin E: Quality and storage stability of pork and poultry. *Trends Food Sci. Tech.*, 9: 62–72.

Jeong, J.B., Jeong, H.J., Park, J.H., Lee, S.H., Lee, J.R. and Lee, H.K. (2007). Cancerpreventive peptide lunasin from *Solanum nigrum* L. inhibits acetylation of core histones H3 and H4 and phosphorylation of retinoblastoma protein (Rb). *J. Agr. Food Chem.*, 55: 10707–10713.

Jiang, B. and Mine, Y. (2000). Preparation of novel functional oligophosphopeptides from hen egg yolk phosvitin. *J. Agr. Food Chem.*, 48: 990–994.

Jianyun, C., Hu, C., Ren, F. and Peng, C. (2008). Enzyme hydrolysis of silk fibroin and the antidiabetic activity of the hydrolysates. *International J. Food Eng.*, 4: 13.

Jimenez-Colmenero (2005). *In*: Derivados carnicos funcionales: Estrategias y perspectivas, Sanchez Muniz, F., Jimenez Colmenero, F. and Olmidilla, B. (*Eds.*), Madrid, Spain, pp. 65–76.

Jimenez-Colmenero, F., Carballo, J. and Cofrades, S. (2001). Healthier meat and meat products: Their role as functional foods. *Meat Sci.*, 59: 5–13.

Jiménez-Colmenero, F., Serrano, A., Ayo, J., Solas, M.T., Cofrades, S. and Carballo, J. (2003). Physicochemical and sensory characteristics of restructured beef steak with added walnuts. *Meat Sci.*, 65: 1391–7.

Jo, C., Ho Son, J., Bae Son, C. and Woo Byun, M. (2003). Functional properties of raw and cooked pork patties with added irradiated, freeze-dried green tea leaf extract powder during storage at 4°C. *Meat Sci.*, 64: 13–7.

Johanna, M. (2007). Metalloproteases. *In*: Julio, Poaina. and MacCble Andrew, P (*Eds.*), Industrial enzymes, structure, function and applications. The Netherlands: Springer Publisher.

Johansson, G., Berdague, J.L., Larsson, M., Tran, N. and Borch, E. (1994). Lipolysis, proteolysis and formation of volatile components during ripening of a fermented sausage with *Pediococcus pentosaceous* and *Staphylococcus xylosus* as starter cultures. *Meat Sci.*, 38: 203–218.

Johnson, M.H., Calkins, C.R., Huffman, R.D., Johnson, D. and Hargrove, D.D. (1990). Differences in cathepsin B+L and calcoium dependent protease activities among breed type and their relationship to beef tenderness. *J Anim Sci.*, 68: 2371–2378.

Joo, S.T., Lee, J.I., Ha, Y.L. and Park, G.B. (2002). Effects of dietary conjugated linoleic acid on fatty acid composition, lipid oxidation, color, and water-holding capacity of pork loin. *J. Anim. Sci.*, 80: 108–112.

Juniper, D.T., Phipps, R.H., Ramos-Morales, E. and Bertin, G. (2008a). Effect of dietary supplementation with selenium-enriched yeast or sodium selenite on selenium tissue distribution and meat quality in beef cattle. *J Animal Sci.*, 86: 3100–3109.

Juniper, D.T., Phipps, R.H., Ramos-Morales, E. and Bertin, G. (2008b). Effect of high dose selenium enriched yeast diets on the distribution of total selenium and selenium species within lamb tissues. *Livestock Science*, 122: 63–67.

Kanner, J. (1994). Oxidative processes in meat and meat products: Quality implications. *Meat Sci.*, 36: 169–174.

Karapinar, M. (1990). Inhibitory effect of anethole and euginol on growth and toxin production of *Aspergillus parasitius. Int. J. Food Microbiol.*, 10: 193–200.

Karpinska-Tymoszczyk, M. (2007). Effects of sage extract (*Salvia officinalis* L.) and a mixture of sage extract and sodium isoascorbate on the quality and shelf life of vacuum-packaged turkey meat balls. *J. Muscle Foods*, 18: 420–434.

Katiyar, S.K. and Mukhtar, H. (1996). Tea in chemoprevention of cancer: Epidemiologic and experimental studies. *Int. J. Oncol.*, 8: 221–238.

Kato, H. and Nishimura, T. (1987). Umami: A basic taste. *In*: Kawamura, Y. and Kare, M. (*Eds.*), New York, NY: Marcel Dekker.

Kawai, M., Okiyama, A. and Ueda, Y. (1999). Taste interaction between sweet L-a-Amino acids and IMP. *Taste Smell Research*, 6: 69–694.

Kazunori, K., Tomatsu, M., Fuchu, H., Sugiyama, M., Kawahara, S. and Yamauchi, K. (2003). Purification and characterization of an angiotensin I-converting enzyme inhibitory peptide derived from porcine troponin C. *Anim Sci. J.*, 74: 53–58.

Keeton, J.T. (1994). Low fat meat products. Technological problems with processing. *Meat Sci.*, 36: 261–276.

Khanal, R.C. (2004). Potential health benefits of conjugated linoleic acid (CLA): A review. *Asian-Austral J. Anim.*, 17: 1315–1328.

Kim, J.S., Godber, J.S. and Prinaywiwatkul, W. (2000). Restructured beef roasts containing rice bran oil and fiber influences cholesterol oxidation and nutritional profile. *J. Muscle Foods*, 11(2): 111–127.

Kim, S.K., Byun, H.G., Park, P.J. and Shahidi, F. (2001). Angiotensin I converting enzyme inhibitory peptides purified from bovine skin gelatin hydrolysate. *J. Agr. Food Chem.*, 49: 2992–2997.

Kim, Y., Jin, S., Park, W., Kim, B., Joo, S. and Yang, H. (2010). The effect of garlic or onion marinade on the lipid oxidation and meat quality of pork during cold storage. *J. Food Qual.*, 33: 171–185.

Kim, Y.Y. and Mahan, D.C. (2001). Comparative effects of high dietary levels of organic and inorganic selenium on selenium toxicity of growing–finishing pigs. *J. Anim. Sci.*, 79: 942–948.

Kirchner, M.J., Beasleyi, L.C., Harris, K.B. and Savell, J.W. (2000). Evaluating the cooking and chemical characteristics of low-fat ground beef patties. *J. Food Comp. Anal.*, 3: 253–254.

Knorr, D. (1998). Technology aspects related to microorganisms in functional foods. *Trends Food Sci. Tech.*, 9: 295–306.

Kott, R.W., Hatfield, P.G., Bergman, J.W., Flynn, C.R., Van Wagoner, H. and Boles, J. (2003). Feedlot performance, carcass composition, and muscle and fat CLA concentrations of lambs fed diets supplemented with safflower seeds. *Small Ruminant Res.*, 49: 11–17.

Kourounakis, P.N. and Rekka, E. (1991). The activity of garlic powder and alliin as antioxidants and catchers of free radicals. *Medizinische Welt*, 0(7A): 44.

Koutsidis, G., Koutsidis, J.S., Elmore, M.J., Oruna-Concha, M.M., Campo, J.D.W. and Mottram, D.S. (2007). Water-soluble precursors of beef flavor: II. Effect of postmortem conditioning. *Meat Sci.*, 79: 270–277.

Kova´ cs, A., Zsarno´ czay, G. and Gasparik Reichardt, J. (2007). Development of new functional meat products. In Proceedings of the fourth international FFNet meeting on functional foods.

Kumar, M. and Berwal, J.S. (1998). Sensitivity of food pathogens to garlic (*Allium sativum*). *J. Appl. Microbiol.*, 84: 213–215.

Kuninaka, A. (1981). Taste and flavor enhancers. *In*: Teranishi, R., Flath, R.A. and Sugisawa, H. (*Eds.*), Flavour research recent advances, New York and Basel: Marcel Dekker, pp. 305–353.

Kuroda, Y. and Hara, Y. (1999). Antimutagenic and anticarcinogenic activity of tea polyphenols. *Mutat Res.*, 436: 69–97.

Lanari, M.C., Schaefer, D.M. and Scheller, K.K. (1995). Dietary vitamin E supplementation and discoloration of pork bone and muscle following modified atmosphere packaging. *Meat Sci.*, 41: 237–250.

Larrea, V., Hernando, I., Quiles, A. and Pérez-Munuera, I. (2006). Changes in proteins during Teruel dry-cured ham processing. *Meat Sci.*, 74: 586–593.

Lawson, R.E., Moss, A.R. and Givens, D.I. (2001). The role of dairy product in supplying conjugated linoleic acid to man's diet: A review. *Nutr. Res. Rev.*, 14: 153–172.

Lee, B.J., Hendricks, D.G. and Cornforth, D.P. (1998). Effcet of sodium phytate, sodium pyrophosphate and sodium tripolyphosphate on physico-chemical characteristics of restructured beef. *Meat Sci.*, 50: 273–283.

Leino, M. (2001). Health trends in meat production. Paper presented at IFAJ 45th Annual Congress. Savvonlina, Finland. Retrieved Oct 5 2004 from www.maataloustoimittajat.fi/ifaj-2001/speeches/monday//leino_en.htm.

Leppala AP 2001. Bioactive peptides derived from bovine whey proteins: Opioid and ACEI peptides. *Trends Food Sci Tech.*, 11:347-356.

Leskanich, C.O., Matthews, K.R., Warkup, C.C., Noble, R.C. and Hazzledine, M. (1997). The effect of dietary oil containing (n"3) fatty acids on the fatty acid, physicochemical, and organoleptic characteristics of pig meat and fat. *J. Anim. Sci.*, 75: 673–683.

Li, R., Carpenter, J.A. and Cheney, R. (1998). Sensory and instrumental properties of smoked sausage made with mechanically separated poultry (MSP) meat what protein. *J Food Sci.*, 63: 1–7.

Lilian, R., Mariutti, B., Vibeke, Orlien., Neura, Bragagnolo. and Skibsted Leif, H. (2008). Effect of sage and garlic on lipid oxidation in high-pressure processed chicken meat. *Eur. Food Res. Technol.*, 227(2): 337–344.

Lin, X., Li, I., Zhang, Q. and Mei, X. (1991). The preventive effect of garlic against toxicity of dimethyluitrosamine in rats fed with aminopyrine and nitrite. *ACTA Nutrimenta Sinica*, 13(2): 126–132.

Liu, S., Willett, W.C., Manson, J.E., Hu, F.B., Rosner, B. and Colditz, G. (2003). Relation between changes in intakes of dietary fiber and grain products and changes in weight and development of obesity among middle-aged women. *Am. J. Clin. Nutr.*, 78(5): 920–927.

Lizaso, G., Chasco, J. and Beriain, M.J. (1999). Microbiological and biochemical changes during ripening of salchichon, a Spanish dry cured sausage. *Food Microbiol.*, 16: 219–228.

Lopez-Ferrer, S., Baucells, M.D., Barroeta, A.C. and Grashorn, M.A. (2001). n-3 Enrichment of chicken meat. 1. Use of very long-chain fatty acids in chicken diets and their influence on meat quality: Fish oil. *Poult Sci.*, 80: 741–752.

Luccia, A.D.I., Picariello, G., Cacace, G., Scaloni, A., Faccia, M. and Liuzzi, V. (2005). Proteomic analysis of water soluble and myofibrillar protein changes occurring in dry-cured hams. *Meat Sci.*, 69: 479–491.

Lucke, F.K. (1998). Fermented sausages. *In*: Wood, B.J.B. (*Ed*,), Microbiology of fermented foods, London: Blackie Academic and Professional, pp. 444–483.

Lusas, E.W. and Riaz, M.N. (1995). Soy protein products: Processing and Use. *J. Nutr.*, 125: 573S.

Lynch, P.B. and Kerry, J.P. (2000). Utilizing diet to incorporate bioactive compounds and improve the nutritional quality of muscle foods. *In*: Antioxidants in muscle foods, Decker, E., Faustman, C. and Lopez-Bote, C.J. (*Eds.*), New York: Willey, pp. 455–480.

Maehashi, K., Matsuzaki, M., Yamamoto, Y. and Udaki, S. (1999). Isolation of peptides from an enzymatic hydrolyzate of food proteins and characterization of their taste properties. *Biosci. Biotech. Bioch.*, 63: 555–559.

Mahan, D.C. and Parret, N.A. (1996). Evaluating the efficatcy of selenium-enriched yeast and sodium slenite on tissue selenium retention and serum glutathione peroxidase activity in grower and finisher swine. *J. Anim. Sci.*, 74: 2967–2974.

Maidment, D.C.F., Dembny, Z. and Harding, C. (1999). A study into the antibiotic effect of garlic *Allium sativum* on *Escherichia* coli and *Staphylococcus albus*. *Nutr. Food Sci.*, 4: 170–172.

Mansour, E.H. and Khalil, A.H. (1999). Characteristics of low-fat beef burgers as influenced by various types of wheat fibers. *J. Sci. Food Agr.*, 79: 493–8.

Mariutti, L.R.B., Orlien, V., Bragagnolo, N. and Skibsted, L.H. (2008). Effect of sage and garlic on lipid oxidation in high-pressure processed chicken meat. *Eur. Food Res. Technol.*, 227: 337–344.

Marsden, J.L. (1980). The importance of sodium in porcessed meats, *In*: Proceedings of the meat industry Research conference. Amer. Meat Inst. Fdn. Arlington, VA, pp. 77–87.

Marsh, A.C. (1983). Processes and formulations that affect the sodium content of foods. *Food Technol.*, 37(7): 45.

Mateo, J. and Zumalacárregui, J.M. (1996). Volatile compounds in chorizo and their changes during ripening. *Meat Sci.*, 44: 255–273.

Maurer, A.J. (1983). Reducing sodium usage in poultry muscle foods. *Food Technol.*, 37(7): 60.

McCann, K.B., Shiellb, B.J., Michalskibm, W.P., Leec, A., Wanc, J. and Roginskia, H. (2005). Isolation and characterization of antibacterial peptides derived from the f (164–207) region of bovine aS2-casein. *Int Dairy J.*, 15: 133–143.

McCarthy, T.L., Kerry, J.P., Kerry, J.F., Lynch, P.B. and Buckley, D.J. (2001). Assessment of the antioxidant potential of natural food and plant extracts in fresh and previously frozen pork patties. *Meat Sci.*, 57: 177–184.

McClements, D.J. and Decker, E.A. (2008). Lipids. *In*: Damodaran, S., Parkin, K.L. and Fennema, O.R. (*Eds.*), Food chemistry, Boca Raton, FL: CRC Press, pp. 155–216.

McCue, N. (1994). Soy transforms burger into vegetarians' dream. *Prepared Foods.* 163: 147.

Mendis, E., Rajapakse, N. and Kim, S. (2005). Antioxidant properties of a radicalscavenging peptide purified from enzymatically prepared fish skin gelatin hydrolysate. *J. Agr. Food Chem.*, 53: 581–587.

Mendoza, E., García, M.L., Casas, C. and Selgas, M.D. (2001). Inulin as fat substitute in low fat, dry fermented sausages. *Meat Sci.*, 57: 387–93.

Menon, K.V. and Garg, S.R. (2001). Inhibitory effect of clove oil on *Listeria monocytogenes* in meat and cheese. *Food Microbiol.*, 18(6): 647–650.

Mercier, Y., Gatellier, P., Viau, M., Remignon, H. and Renerre, M. (1998). Effect of dietary fat and vitamin E on colour stability and on lipid and protein oxidation in turkey meat during storage. *Meat Sci.*, 48: 301–318.

Minervini, F.F.A., Rizzello, C.G., Fox, P.F., Monnet, V. and Gobbetti, M. (2003). Angiotensin I-converting-enzyme-inhibitory and antibacterial peptides from *Lactobacillus helveticus* PR4 proteinase-hydrolyzed caseins of milk from six species. *Appl. Environ. Microb.*, 69: 5297–5305.

Mirshekar, R., Dastar, B. and Shabanpour, B. (2009). Effect of rosemary, Echinacea, green tea extracts and ascorbic acid on broiler meat quality. *Pakistan J. Biol. Sci.*, 12: 1069–1074.

Molly, K., Demeyer, D., Civera, T. and Verplaetse, A. (1996). Lipolysis in a Belgian sausage: Relative importance of endogenous and bacterial enzymes. *Meat Sci.*, 43: 235–244.

Molly, K., Demeyer, D., Johansson, G., Raemaekers, M., Ghistelinck, M. and Geenenc, I. (1997). The importance of meat enzymes in ripening and flavor generation in dry fermented sausages. *Food Chem.*, 59: 539–545.

Montel, M.C., Reitz, J., Talon, R., Berdague, R. and Rousset-Akrim, J.L. (1996). Biochemical activities of *Micrococcaceae* and their effects on the aromatic profiles and odours of a dry sausage model. *Food Microbiol.*, 13: 489–499.

Munday, J.S., Thompson, K.G. and James, K.A.C. (1999). Dietary conjugated linoleic acids promote fatty streak formation in the C57BL/6 mouse atherosclerosis model. *Brit. J. Nutr.*, 81: 251–255.

Muthukumarasamy, P. and Holley, R.A. (2006). Microbiological and sensory quality of dry fermented sausages containing alginate-microencapsulated *Lactobacillus reuteri*. *Int. J. Food Microbiol.*, 111: 164–169.

Mytle, N., Anderson, G.L., Doyle, M.P. and Smith, M.A. (2006). Antimicrobial activity of clove (*Syzgium aromaticum*) oil in inhibiting *Listeria monocytogenes* on chicken frankfurters. *Food Control.*, 17: 102–107.

Naes, H., Holck, A.L., Axelsson, L., Anderson, H.J. and Blom, H. (1995). Accelerated ripening of dry fermented sausage by addition of a *Lactobacillus proteinase*. *Int. J. Food Sci. Technol.*, 29: 651–659.

Nakashima, Y., Arihara, K., Sasaki, A., Ishikawa, S. and Itoh, M. (2002). Antihypertensive activities of peptides derived from porcine skeletal muscle myosin in spontaneously hypertensive rats. *J. Food Sci.*, 67: 434–437.

Narvaez-Rivas, M., Vicario, I.M., Constante, E.G. and Leon-Camacho, M. (2008). Changes in the fatty acid and triacylglycerol profiles in the subcutaneous fat of Iberian ham during the dry-curing process. *J Agr Food Chem.*, 56(16): 7131–7137.

Natura, (1993). Metal free and low metal salt substitutes containing lysine. *US Patent*, 5: 229, 161.

Naveena, B.M., Muthukumar, M., Sen, A.R., Babji, Y. and Murthy, T.R.K. (2006). Improvement of shelf-life of buffalo meat using lactic acid, clove oil and vitamin C during retail display. *Meat Sci.*, 74: 409–415.

Nelson, R., Katayama, S., Mine, Y. and Duarte, J.M.C. (2007). Immunomodulating effects of egg yolk low lipid peptic digests in a murine model. *Food Agric. Immunol.*, 18: 1–15.

Nishimura, T. and Kato, H. (1988). Taste of free amino acids and peptides. *Food Res Int.*, 4: 175–194.

Niva, M. (200)7. All foods affect health': Understandings of functional foods and healthy eating among health-oriented Finns. *Appetite*, 48: 384–393.

Nout, M.J.R. and Sarkar, P.K. (1999). Lactic acid food fermentation in tropical climates. *Antonie Van Leeuwenhoek*, 76: 395–401.

Nyberg, F., Sanderson, K. and Glamsta, E.l. (1997). The hemorphins: A new class of opioid peptides derived from the blood protein hemoglobin. *Biopolymers*, 43:1 47–156.

O'Connell, M., Kerry, J.P., Fannen, S. and Gilroy, D. (2002). Assesment of honey as a functional ingredient in cooked meat pork. *In*: Proceedings of the 48th International Congress of Meat Science and Technology. Vol. II, pp. 1016–1017, Rome, Italy, ICOMST.

Okumura, T., Yamada, R. and Nishimura, T. (2004). Sourness-supressing peptides in cooked pork loins. *Biosci. Biotech. Bioch.*, 68: 1657–1662.

Oliveira, K.A.M., Santos-Mendonca, R.C., Gomide, L.A.M. and Vanetti, M.C.D. (2005). Aqueous garlic extract and microbiological quality of refrigerated poultry meat. *J. Food Process Pres.*, 29: 98–108.

Olson, D.G. and Terrell, (1981). Sensory properties of processed meats using various sodium salt substitutes. In proceedings of the meat industry Research conference. Amer. Meat Inst. Fdn. Arlington, VA, pp. 59–66.

Ordonez, J.A., Hierro, E.M., Bruna, J.M. and de la Hoz, L. (1999). Changes in the components of dry-fermented sausages during ripening. *Critical Rev. Food Sci. Nutr.*, 39: 329–367.

Ostrowska, E., Muralitharan, M., Cross, R.F., Bauman, D.E. and Dunshea, F.R. (1999). Dietary conjugated linoleic acids increase lean tissue and decrease fat deposition in growing pigs. *J. Nutr.*, 129: 2037–2042.

Paneras, E.O., Boulkas, J.G. and Papadina, S.N. (1996). Effect of meat source and fat level on processing and quality characteristics of frankfurters. *Lebensm Wiss. Technol.*, 29: 507–514.

Pariza, M.W. and Hargraves, W.A. (1985). A beef-derived mutagenesis modulator inhibits initiation of mouse epidermal tumors by 7, 12-dimethylbenz[a]anthracene. *Carcinogenesis*, 6(4): 591–593.

Park, W.Y. and Kim, Y.J. (2009). Effect of garlic and onion juice addition on the lipid oxidation, total plate counts and residual nitrite contents of emulsified sausage during cold storage. *Korean J. Food Sci. Anim. Resour.*, 29: 612–618.

Park, Y., Albright, K.J., Liu, W., Storkson, J.M., Cook, M.E. and Pariza, M.W. (1997). Effect of conjugated linoleic acid on body composition in mice. *Lipids*, 32: 853–858.

Pearson, A.M. and Gillett, T.A. (1999). Effects of fat on flavor in processed meats. 3rd ed., pp: 356–358. Aspen Publisher, Inc. Gaithersburg, Maryland, (ISBN: 08342-1304-4).

Pennington, J.A.T. (2002). Food composition databases for bioactive food components. *J. Food Comp. Anal.*, 15: 419–434.

Perez-Gago, M.B. and Krochta, J.M. (2001). Denaturation time and temperature effects on solubility, tensile properties and oxygen permeability of whey protein edible films. *J. Food Sci.*, 66: 705–710.

Pidcock, K., Heard, G.M. and Henriksson, A. (2002). Application of nontraditional meat starter cultures in production of Hungarian salami. *Int. J. Food Microbiol.*, 76: 75–81.

Pihlanto, A. and Korhonen, H. (2003). Bioactive peptides and proteins. *Adv. Food Nutr. Res.*, 47: 175–276.

Pizzocaro, F., Senesi, E., Veronese, P. and Gasparoli, A. (1998). Mechanically deboned poultry meat hamburgers: II. Protective and antioxidant effect of the carrot and spinach tissues during frozen storage. *Industrie Alimentari.*, 37: 710–720.

Poljak, A., McLean, C.A., Sachdev, P., Brodaty, H. and Smythe, G.A. (2004). Quantification of hemorphins in Alzheimer's disease brains. *J. Neurosci. Res.*, 75: 704–714.

Pollard, M.R., Gunstone, F.D., James, A.T. and Morris, L.J. (1980). Desaturation of positional and geometric isomers of monoenoic fatty acids by microsomal preparations from rat liver. *Lipids*, 15: 306–314.

Porcella, M.I., Sanchez, G., Vaudagna, S.R., Zanelli, M.L., Descalzo, A.M. and Meichtri, L.H. (2001). Soy protein isolate added to vacuum-packaged chorizos: Effect on drip loss, quality characteristics and stability during refrigerated storage. *Meat Sci.*, 57: 437–443.

Potter, S.M. (1998). Soy protein and cardiovascular disease: The impact of bioactive component in soy. *Nutr Rev.*, 56: 231.

Pritchard, P.E. and Brock, C.J. (1994). The glutenin fraction of wheat protein: The importance of genetic background on its quantity and quality. *J. Sci. Food Agr.*, 65: 401–406.

Rakosky, J.J. (1970). Soy products for the meat industry. *J. Agr. Food Chem.*, 18: 1005–1009.

Rantsiou, K. and Luca, C. (2008). Fermented meat products food. *In:* Cocolin, L. and Ercolini, D. (*Eds.*), Microbiology and food safety-Molecular techniques in the microbial ecology of fermented foods. New York: Springer.

Ray, B. (2004). Fundamentals of food microbiology. USA: CRS Press.

Realini, C.E., Duckett, S.K., Brito, G.W., Dalla Rizza, M. and De Mattos, D. (2004). Effect of pasture vs. concentrate feeding with or without antioxidants on carcass characteristics, fatty acid composition, and quality of Uruguayan beef. *Meat Sci.*, 66: 567–577.

Rebucci, R., Sangalli, L., Fava, M., Bersani, C., Cantoni, C. and Baldi, A. (2007). Evaluation of functional aspects in *Lactobacillus* strains isolated from dry fermented sausages. *J. Food Quality*, 30: 187–201.

Rodriguez-Nuñez, E., Aristoy, M.C. and Toldrá, F. (1995). Peptide generation in the processing of dry-cured ham. *Food Chem.*, 53: 187–190.

Rosell, C.M. and Toldra, F. (1998). Comparison of muscle proteolytic and lipolytic enzyme levels in raw hams from Iberian and white pigs. *J. Sci. Food Agr.*, 76: 117–122.

Rossini, K., Noren, C.P.Z., Cladera-Olivera, F. and Brandelli, A. (2009). Casein peptides with inhibitory activity on lipid oxidation in beef homogenates and mechanically deboned poultry meat. *Food Sci. Technol-Leb.*, 42: 862–867.

Rowe, L.J., Maddock, K.R., Lonergan, S.M. and Huff-Lonergan, E. (2004). Oxidation environments decrease tenderization of beef steaks through inactivation of μ-calpain. *J. Anim. Sci.*, 82: 3254–3266.

Roy, B.D. and Antolic, A. (2009). Conjugated linoleic acid (CLA) and bone health: A review. *Current Topics in Nutraceutical Research*, 7: 27–36.

Ruiz, J., Garcia, C., del Carmen, D.M., Cava, R., Tejeda, J.F. and Ventanas, J. (1999). Drycured Iberian ham non-volatile components as affected by the length of the curing process. *Food Res Int.*, 32: 643–651.

Ruiz-Capillas, C. and Jimenez Colmenero, F. (2004). Biogenic amines in meat and meat products. *Crit. Rev. Food Sci. Nutr.*, 44: 489–599.

Rule, D.C., Broughton, K.S., Shellito, S.M. and Maiorano, G. (2002). Comparison of muscle fatty acid profiles and cholesterol concentrations of bison, beef cattle, elk, and chicken. *J. Anim. Sci.*, 80: 1202–1211.

Ruusunen, M., Vainionpaa, J., Poulanne, E., Lyly, M., Lähteenmäki, L., Niemistö, M. and Ahvenainen, R. (2003a). Physical and sensory properties of low-salt phosphate free frankfurters composed with various ingredients. *Meat Sci.*, 63: 9–16.

Ruusunen, M., Vainionpaa, J., Poulanne, E., Lyly, M., Lähteenmäki, L., Niemistö, M. and Ahvenainen, R. (2003b). Effect of sodium citrate, carboxymethyl cellulose and carrageenan levels on quality characteristics of low-salt and low-fat bologna type sausages. *Meat Sci.*, 64: 371–81.

Sadler, M.J. (2004). Meat alternatives—market developments and health benefits. *Trends Food Sci. Technol.*, 15: 250–60.

Sahlin, P. (1999). Fermentation as a method of food processing. Licentiate thesis, Lund Institute of Technology, Lund University, Lund, Sweden.

Saiga, A., Okumura, T., Makihara, T., Katsuta, S., Shimizu, T. and Yamada, R. (2003). Angiotensin I-converting enzymes inhibitory peptides in a hydrolyzed chicken breast muscle extract. *J. Agr. Food Chem.*, 51: 174–1745.

Sakanaka, S., Tachibana, Y., Ishihara, N. and Juneja, L.R. (2005). Antioxidant properties of casein calcium peptides and their effects on lipid oxidation in beef homogenates. *J. Agr. Food Chem.*, 53: 464–648.

Saleh, N.T. and Ahmed, Z.S. (1998). Impact of natural sources rich in provitamin A on the cooking characteristics, colour, texture and sensory attributes of beef patties. *Meat Sci.*, 50: 285–293.

Salvatori, G., Pantaleo, L., Di Cesare, C., Mairano, G., Filletti, F. and Oriani, G. (2004). Fatty acid composition and cholesterol content of muscles as related to genotype and vitamin E treatment in crossbred lambs. *Meat Sci.*, 67: 45–55.

Samelis, J., Aggelis, G. and Metaxopoulos, J. (1993). Lipolytic and microbial changes during the natural fermentation and ripening of Greek dry sausages. *Meat Sci.*, 35: 371–385.

Sanchez-Escalante, A., Dnejane, D., Torrescano, G., Beltran, J.A. and Roncales, P. (2001). The effcts of ascorbic acid, taurine, carnosine and rosemary powder on colour and lipid stability of beef patties packed in modified atmosphere. *Meat Sci.*, 58: 421–429.

Sanchez-Escalante, A., Torrescano, G., Djenane, D., Beltran, J.A. and Roncales, P. (2003). Combined effect of modified atmosphere packaging and addition of lycopene rich romato pulp, oregano and ascorbic acid and their mixtures on the stability of beef patties. *Food Sci. Technol. Int.*, 9: 77–84.

Schatzkin, A., Mouw, T., Park, Y., Subar, A.F., Kipnis, V., Hollenbeck, A., Leitzmann, M.F. and Thompson, F.E. (2007). Dietary fiber and whole-grain consumption in relation to colorectal cancer in the NIH-AARP Diet and Health Study. *Am. J. Clin. Nutr.*, 85(5): 1353–1360.

Schmidt, S. and Berger, G. (1998). Aroma compounds in fermented sausages of different origins. *Lebensm Wiss Technol.*, 31: 559–567.

Scientific Committee on Food (2003). Opinion of the scientific committee on food on an application from multibene for approval of plant sterol enriched foods (Scientific Committee on Food SFC/CS/NF/DOS/24ADD2Final). Brussels, Belgium: Health and consumer protection Directorate–General, European Commission.

Scollan, N. (2007). Enhancing the content of beneficial fatty acids in beef and improving meat quality for the consumer. Functional Food News. http://www.functionalfoodnet.eu/images/site/assets/1-HealthyBeef %20Nigel.pdf.

Scollan, N., Hocquette, J., Nuernberg, K., Dannenberger, D., Richardson, I. and Moloney, A. (2006). Innovations in beef production systems that enhance the nutritional and health value of beef lipids and their relationship with meat quality. *Meat Sci.*, 74(1): 17–33.

Sebranek, J.G., Sewalt, V.J.H., Robbins, K.L. and Houser, T.A. (2004). Comparison of a natural rosemary extract and BHA/BHT for relative antioxidant effectiveness in pork sausage. *Meat Sci.*, 69: 289–296.

Serrano, A., Cofrades, S., Ruiz Cappillas, C., Olmidilla-Alonso, B., Herrero-Barbudo, C. and Jimenez Calmenero, F. (2005). Nutritional profile of resructured beef steak with added walnuts. *Meat Sci.*, 70: 647–654.

Shahidi, F., Liyana-Pathirana, C.M. and Wall, D.S. (2006). Antioxidant activity of white and black sesame seeds and their hull fractions. *Food Chem.*, 99: 478–483.

Shan, B., Cai, Yi.Z., Brooks, J.D. and Corke, H. (2009). Antibacterial and antioxidant effects of five spice and herb extracts as natural preservatives of raw pork. *J. Sci. Food Agr.*, 89: 1879–1885.

Shantha, N.C., Moody, W.G. and Tabeidi, Z. (1997). A research note: Conjugated linoleic acid concentration in semimembranosus muscle of grass- and grain-fed and zeranol-implanted beef cattle. *J. Muscle Foods*, 8: 105–110.

Sheu, T.Y. and Marshall, R.T. (1993). Micro-encapsulation of lactobacilli in calcium alginate gels. *J Food Sci.*, 54: 557–561.

Shon, J. and Chin, K.B. (2008). Effect of whey protein coating on quality attributes of lowfat, aerobically packaged sausage during refrigerated storage. *J. Food Sci.*, 73: 469–475.

Simopoulos, A.P., Leaf, A. and Salem, N. (2000). Workshop on the essentiality of and recommended dietary intakes for omega 3 and omega 6 fatty acids. *Food Rev. Int.*, 16: 113–117.

Sirtori, C.R., Evan, R. and Lovati, M.R. (1993). Soybean protein diet and plasma cholesterol: From therapy to molecular mechanisms. *Ann. NY Acad. Sci.*, 676: 188.

Skandamis, P.N. and Nychas, G.J.E. (2001). Effect of oregano essential oil on microbiological and physico-chemical attributes of minced meat stored in air and modified atmospheres. *J. Appl. Microbiol.*, 91: 1011–1022.

Skrivan, M., Marounek, M., Dlouha, G. and Sevcikova, S. (2008). Dietary selenium increases vitamin E contents of egg yolk and chicken meat. *Brit. Poult. Sci.*, 49: 482–486.

Smacchi, E. and Gobbetti, M. (2000). Bioactive peptides in dairy products: Synthesis and interaction with proteolytic enzymes. *Food Microbiol.*, 17: 129–141.

Smedman, A. and Vessby, B. (2001). Conjugated linoleic acid supplementation in humans-Metabolic effects. *J. Nutr.*, 36: 773–781.

Smith, D.M. (1988). Meat Proteins: Functional property in comminuted meat products. *Food Technol.*, 42(4): 116.

Smith, G.C., Hynunil, J., Carpenter, Z.L., Mattil, K.F. and Cater, C.M. (1973). Efficacy of protein additives as emulsion stabilizers in frankfurters. *J Food Sci.*, 38: 849–855.

Smith-Palmer, A., Steward, J. and Fyfe, L. (1998). Antimicrobial properties of plant essential oil and essences against five important food-borne pathogens. *Lett. Appl. Microbiol.*, 26: 118–122.

Solomon, M.B., Pursel, V.G. and Mitchell, A.D. (2002). Biotechnology for meat quality enhancement. *In*: Research advances in the quality of meat and meat products. Toldra, F. (*Ed*), Trivandrum, India, Research Signpost, pp. 17–31.

Song, E.K., Kim, H.H., Kim, J.Y., Kang, Y.I., Woo, H.J. and Lee, H.J. (2000). Anticancer activity of hydrophobic peptides from soy proteins. *J. Bio. Factors*, 12: 4151–4155.

Soni, S.K. and Sandhu, D.K. (1999). Fermented cereal foods. *In*: Biotechnology: Food Fermentations, Vol. I, Joshi, V.K. and Panday, A. (*Eds*.), Educational Publishers and Distributors, New Delhi, pp. 1–24.

Spanier, A.M., Spanier, M., Flores, K.W. and McMillin, B.T.D. (1997). The effect of postmortem aging on meat flavor quality in Brangus beef: Correlation of treatments, sensory, instrumental and chemical descriptors. *Food Chem*., 59: 531–538.

Spurvey, S., Pan, B.S. and Shahidi, F. (1998). Flavour of shellfish. *In*: Shahidi, F. (*Ed*.), Flavor of meat, meat products and seafoods, 2nd Edn, London: Blackie Academic and Professional, pp. 159–196.

Stanton, C., Murphy, J., McGrath, E. and Devery, R. (2003). Animal feeding strategies for conjugated linoleic acid enrichment of milk. *In*: Sébédio, J.L., Christie, W.W. and Adlof, R. *Eds*., Advances in conjugated linoleic acid research 2, AOAC Press, Champaign, IL, USA, pp. 123–145.

Steenblock, R.L., Sebranek, J.G., Olson, D.G. and Love, J.A. (2001). The effects of oat fiber on the properties of light bologna and fat-free frankfurters. *J. Food Sci.*, 66(9): 1409–15.

Streppel, M.T., Arends, L.R., Veer, P.V., Grobbee, D.E. and Geleijnse, J.M. (2005). Dietary fiber and blood pressure. A meta-analysis of randomized placebo-controlled trials. *Arch. Int. Med.*, 165: 150–156.

Suresh, P., Ingel, V.K. and Vijayalakshmi, N. (1992). Antibacterial activity of eugenol in comperision with other antibiotics. *J. Food Sci. Technol.*, 29(4): 254–257.

Szymczyk, B., Pisulewski, P.M., Szczurek, W. and Hanczakowski, P. (2001). Effects of conjugated linoleic acid on growth performance, feed conversion and subsequent carcass quality in broiler chickens. *Brit. J. Nutr.*, 85: 465–473.

Taludkar, S. and Sharma, D.P. (2009). Development of dietary fiber rich chicken meat patties using wheat and oat bran. *J. Food Sci. Technol.*, 47(2): 224–229.

Tanabe, H., Yoshida, M. and Tomita, N. (2002). Comparison of the antioxidant activities of 22 commonly used culinary herbs and spices on the lipid oxidation of pork meat. *Anim. Sci. J.*, 73: 389–393.

Tang, S., Kerry, J.P., Sheehan, D. and Buckley, D.J. (2001a). A comparative study of tea catechins and α-tocopherol as antioxidants in cooked beef and chicken meat. *Eur. Food Res. Technol.*, 213: 286–289.

Tang, S., Kerry, J.P., Sheehan, D., Joe, D., Bukley and Morrissey, P.A. (2001b). Antioxidative effect of added tea catechins on susceptibility of cooked red meat, poultry and fish patties to lipid oxidation. *Food Res. Int.*, 34: 651–7.

Teixeira de Carvalho, A.A., Aparecida de Paulaa, R., Mantovani, H.C. and Alencar de Moraes, C. (2006). Inhibition of *Listeria monocytogenes* by a lactic acid bacterium isolated from Italian salami. *Food Microbiol.*, 23: 213–219.

Terpstra, A.H.M., Beynen, A.C., Everts, H., Kocsis, S., Katan, M.B. and Zock, P.L. (2002). The decrease in body fat inmice fed conjugated linoleic acid is due to increases in energy expenditure and energy loss in the excreta. *J. Nutr.*, 132: 940"945.

Terrel, R.N., Qeintanilla, M., Vanderzant, C. and Gardener, F.A. (1983). Effects of reduction or replacement of sodium chloride on growth of *Micrococcus, Moraxella* and *Lactobacillus* inoculated ground pork. *J Food Sci.*, 48: 122.

Terrell, R.N. (1983). Reducing the sodium content of processed meats. *Food Tecnol.*, 37(7): 66

Todd, S.L., Cunningham, F.E., Claus, J.R. and Schwenke, J.R. (1989). Effect of dietary fiber on the texture and cooking characteristics of restructured pork. *J Food Sci.*, 54(5): 1190–1192.

Toldrá, F. (1998). Proteolysis and lipolysis in flavour development of dry-cured meat products. *Meat Sci.*, 49: S101–S110.

Toldrá, F. (2006). The role of muscle enzymes in dry-cured meat products with different drying conditions. *Trends Food Sci. Technol.*, 17: 164–168.

Toldra, F. and Reig, M. (2004). Analysis of meat in food. *In*: Handbook of Food Analysis. Ed. Nollet, L.M.L. 1941-1959. New Yolk, Mercel Dekker.

Toldrá, F., Aristoy, M.C. and Flores, M. (2000). Contribution of muscle aminopeptidases to flavor development in dry-cured ham. *Food Res. Int.*, 33: 181–185.

Vallejo-Cordoba. B., Nakai. S., Powrie, W.D. and Beveridge, T. (1987). Extended shelf life of frankfurters and fish frankfurter-analogs with added soy protein hydrolysates. *J Food Sci.*, 52: 1133–1136.

Vekiari, S.A., Oreopoulou, V., Tzia, C. and Thomopoulos, C.D. (1993). Oregano flavonoids as lipid antioxidants. *J. Am. Oil Chem. Soc.*, 70: 483–487.

Vercruysse, L., Van Camp, J. and Smagghe, G. (2005). ACE inhibitory peptides derived from enzymatic hydrolysates of animal muscle protein: A review. *J. Agr. Food Chem.*, 53(21): 8106–8115.

Verma, A.K., Sharma, B.D. and Banerjee, R. (2010). Effect of sodium chloride replacement and apple pulp inclusion on the physico-chemical, textural and sensory properties of low fat chicken nuggets. *Food Sci. Technol-Leb.*, 43(4): 715–719.

Verplaetse, A., De Bosschere, M. and Demeyer, D. (1989). Proteolysis during dry sausage ripening. *In*: Proceedings 35th international congress on meat science and technology. Copenhaegen, Denmark.

Viallon, C., Berdague, J.L., Montel, M.C., Talon, R., Martin, J.F. and Kondjoyan, N. (1996). The effect of stage of ripening and packaging on volatile content and flavour of dry sausage. *Food Res. Int.*, 29: 667–674.

Vignolo, G.M., Suriani, F., de Ruiz Holgado, A.P. and Oliver, G. (1993). Antibacterial activity of *Lactobacillus* strains isolated from dry fermented sausages. *J. Appl. Microbiol.*, 75: 344–349.

Vitaglione, P. and Fagliono, V. (2004). Use of antioxidants to minimize the human health risk associated to mutagenic carcinogenic heterocyclic amines in food. *J. Chromatogr.*, 8002: 189–199.

Wahle, K.W.J., Heys, S.D. and Rotondo, D. (2004). Conjugated linoleic acids: Are they beneficial or detrimental to healthy? *Prog. Lipid Res.*, 43: 553–587.

Wanasundara, U.N. and Shahidi, F. (1996). Stabilization of seal blubber and menhaden oil with green tea catechins. *J. Am .Oil. Chem. Soc.*, 73: 1183–1190.

Wang, L.L. and Xiong, Y.I. (2008). Inhibition of oxidant-induced biochemical changes of pork myofibrillar protein by hydrolyzed potato protein. *J. Food Sci.*, 73: C482–C487.

Wani, S.A. and Sharma, B.D. (1999). Fermented meat products. *In*: Biotechnology-Food Fermentations, Vol. I, Joshi, V.K. and Panday, A. (*Eds.*), Educational Publishers and Distributors, New Delhi, pp. 971–1002.

Weaver, C.M. and Liebman, M. (2002). Biomarkers of bone health appropriate for evaluating functional foods designed to reduce the risk of osteoporosis. *Brit. J. Nutr.*, 88: S225–S232.

Weisburger, J.H., Veliath, E., Larios, E., Pittman, B., Zang, E. and Hara, Y. (2002). Tea polyphenols inhibit the formation of mutagens during the cooking of meat. *Mut. Res-Gen. Tox. En.*, 516: 19–22.

Wheeler, T.L., Savell, J.W., Cross, H.R., Lunt, D.K. and Smith, S.B. (1990). Mechanisms associated with the variations in the tenderness of meat from Brahman and Hereford cattle. *J. Anim. Sci.*, 68: 4206–4220.

Wiegand, B.R., Sparks, J.C., Parrish Jr, F.C. and Zimmerman, D.R. (2002). Duration of feeding conjugated linoleic acid influences growth performance, carcass traits, and meat quality of finishing barrows. *J Anim Sci.*, 80: 637–643.

Willet, W., Manson, J. and Liu, S. (2002). Glycemic index, glycemic load, and risk of type 2 diabetes. *Am J. Clin. Nutr.*, 76: 274S–280.

Williams, C.M. (2000). Dietary fatty acids and human health. *Ann. Zootech.*, 49: 165–180.

Wood, J.D., Richardson, R.I., Nute, G.R., Fisher, A.V., Campo, M.M., Kasapidou, E., Sheard, P.R. and Enser, M. (2003). Effects of fatty acids on meat quality-A review. *Meat Sci.*, 66: 21–32.

World Health Organization (2003). Diet, nutrition and the prevention of chronic diseases (WHO technical report series 916), Geneva, Switzerland.

Xiong, Y.L., Agyare, K.K. and Addo, K. (2008). Hydrolyzed wheat gluten suppresses transglutaminase mediated gelation but improves emulsification of pork myofibrillar protein. *Meat Sci.*, 80: 535–544.

Yamazaki, Y. and Maekawa, K. (1980). Synthesis of a peptide with delicious taste. *Agr. Biol. Chem.*, 44: 93–97.

Yang, C.S., Chung, J.Y., Yang, G.Y., Chhabra, S.K. and Lee, M.J. (2000). Tea and tea polyphenols in cancer prevention. *J. Nutr.*, 130: S472–S478.

Yetim, H., Muller, W.D., Dogan, M. and Klettner, P.G. (2001). Using fluid whey in comminuted meat products: Effects on textural properties of frankfuter-type sausages. *J. Muscle Foods*, 17: 354–366.

Yilmaz, I. (2004). Effects of rye bran addition on fatty acid composition and quality characteristics of meat balls. *Meat Sci.*, 67:245-249.

Yilmaz, I., Simsek, O. and Isikli, M. (2002). Fatty acid composition and quality characteristics of low-fat cooked sausages made with beef and chicken meat, tomato juice and sunflower oil. *Meat Sci.*, 62: 253–8.

Yu, L., Scanlin, L., Wilson, J. and Schmidt, G. (2002). Rosemary extracts as inhibitors of lipid oxidation and color change in cooked turkey products during refrigerated storage. *J Food Sci.*, 67: 582–585.

Zhang, L. and Zhou, J.L.K. (2010). Chelating and radical scavenging activities of soy protein hydrolysates prepared from microbial proteases and their effect on meat lipid peroxidation. *Bioresource Technol.*, 101: 2084–2089.

Zhao, Q., Girreau, I., Sannier, F. and Piot, J.M. (1997). Opioid peptides derived from hemoglobin: Hemorphins. *Biopolymers*, 43: 75–98.

Zhou, G.H. and Zhao, G.M. (2007). Biochemical changes during processing of traditional *Jinhua ham. Meat Sci.*, 77: 114–120.

22

Dairy Ingredients in Meat Products

Prabhat Kumar Mandal

ABSTRACT

Comminuted meat products are complex meat systems in which water absorption, emulsion formation and gelation influence stability and texture of the cooked product. Dairy ingredients are known for their sensory and nutritional value. They have important applications in food formulation because of the versatile functionality of their proteins. Milk proteins can be used to enhance the functional properties of meat emulsions particularly conformation and structural properties. Milk proteins including whey protein concentrate (WPC), non-fat dry milk, calcium–reduced NFDM, dried whey, butter milk, caseinates and skim milk co-precipitates contain appreciable amounts of protein. These dairy ingredients are widely utilized in processed meat and poultry products. This chapter discusses about the following aspects: Role of dairy ingredients in meat products, function of whey proteins in meat products, use of WPC in meat products like sausages, nuggets, patties; Use of WPC in low fat meat products like low fat sausages, low fat nuggets. Use of caseinates, NFDM in meat products, NFDM in low fat kabab; Use of dairy ingredients in restructured and reformed meats, effect of dairy ingredients on colour of meat products; Miscellaneous dairy ingredients skim milk powder, co-precipitates, dry whole milk in meat products, curd in enrobed meat products, benefits of using dairy ingredients in meats etc.

1. INTRODUCTION

Although milk products are primarily known for their sensory and nutritional value, they have important applications in food formulation because of the versatile functionality of their proteins. Milk proteins can be used to enhance the functional properties of meat emulsions. In particular their conformation and structural properties are important in terms of functionality (Ensor *et al.*, 1987; Baardseth *et al.*, 1992; Zorba *et al.*, 1998; Yetim *et al.*, 2001; Srinivasan *et al.*, 2002). Casein proteins contain many proline amino acids and few disulphide bonds. They cannot be easily denatured after being heated at low temperatures. In addition, they have high electrical potential which increases solubility, thereby causing an increase in emulsion stability (ES). Conversely whey proteins contain few proline amino acids and many disulphide bonds; they can contribute to gel properties which help to retain water and oil in the products. They can also absorb a higher amount of water than caseinates (Smith, 1988; Ozdemir *et al.*, 1994; Isýk and Gokalp, 1999). Milk proteins can act both as emulsifiers and as water and fat binders in foods (Sebranek, 1996). Whey products have also been used as binders and extenders (Ellekjær *et al.*, 1996; Hughes *et al.*, 1997, 1998).

2. ROLE OF DAIRY INGREDIENTS IN MEAT PRODUCTS

Comminuted meat products are complex meat systems in which water absorption, gelation and emulsion formation influence stability and texture of cooked product (Ellekjaer *et al.*, 1996). Use of non meat ingredients in meat products decreases costs and increases products quality and yield. The use of dairy ingredients to improve the functionality of meat proteins and quality of processed meats is well established. Milk proteins can act as emulsifiers in comminuted meat products. They can also be used as water and fat binders in the products. In particular nonfat dry milk (NFDM) and whey powder (WP) have been used for improving the technological properties of comminuted meat products (Ensor *et al.*, 1987; Baardseth *et al.*, 1992). The functionality of milk proteins in meat emulsions has been estimated by using model systems to determine emulsion properties (Parks & Carpenter, 1987; Zorba *et al.*, 1995; Zorba *et al.*, 2004). A number of milk protein derivatives are widely used, which include non fat dry milk, calcium reduced NFDM, dried whey, whey protein concentrate, buttermilk solids, the caseinate and skim milk co precipitates.

Dairy ingredients have been used as fillers and binders in comminuted meat products to improve functional and sensory quality of meat products. Skim milk powder is widely used as neutral filler with good water binding effect in comminuted meat products. Use of lactose may cause discolouration due to Mailard reaction. Extensive work has been done on the use of whey protein concentrate in processed meats. It can be used as functional ingredients to improve the technological and sensory quality of meat products.

Table 1: Composition of some milk protein derivatives utilized in processed meats.

	Composition in % Dry Basis				
Products	***Protein***	***Moisture***	***Fat***	***Lactose***	***Ash***
Non fat dry milk (NFDM)	35.9	3.0	0.8	52.3	8
Calcium reduced NFDM	36-39	3	0.8	52.3	4–7
Dried whey	12	4.5	1.1	73.5	8
Whey protein concentrates	20–60	2	2–9	18–60	3–18
Dried buttermilk	34.4	2.8	5.3	50	7.6
Caseinates (Na^+, K^+, $Ca\ 2^+$)	92	4	0.8	–	1.5
Skim milk co_precipitates	83	4	1.5	1	10

Source: Hugunin and Ewing (1977)

3. USE OF WHEY PROTEIN CONCENTRATE IN MEAT PRODUCTS

These are byproducts of cheese production. The principal protein in whey is lactalbumen as most of casein is removed in cheese production. Dried whey has only 12% protein, some modified whey protein has higher protein and lower in ash and lactose, modified whey protein include whey protein concentrate, reduced lactose whey and reduced mineral whey, these not only increase protein but also serves as binders and thickeners in various sausages (Pearson & Tauber, 1984).

Whey protein concentrate (WPC) represents a protein source with considerable potential for blending with various food products to improve their nutritional value and functional properties. WPC is a substance obtained by the removal of sufficient non-protein components from whey so that the finished dry product contains not less than 25% proteins (Renner & Abd EL-Salam, 1991). WPC are obtained by ultra-filtration of acid or sweet whey and are carefully spray dried to maintain protein functionality. WPC are available with protein levels ranging from 35–75%. Having excellent nutritional and functional properties WPC finds numerous applications in food and dairy industries (Mann, 1998). Morr and Foegeding (1990) used WPC in various foods as functional ingredients supplying flavour, texture, colour and aeration properties. Whey protein possesses very good functional properties such as solubility, foaming, emulsifying, gelling and water binding (Patel & Kilara, 1990; Jayaprakasha, 1992; De Wit & Kessel, 1996). WPCs have minimal lactose and lipid contents and are highly acceptable as functional protein sources for food industry. WPCs are competitive with other proteins such as caseinate, soy and egg protein products for numerous applications in different food systems (Morr, 1982). Whey proteins are unique among proteins used in food applications, as in their native form, they are soluble at low strength over the entire pH range required in food applications.

3.1. Function of Whey Proteins in Meat Products

Whey proteins can be used for partial replacement of meat proteins, partial or total replacement of non-meat binders/fillers, in processed meats. This can improve emulsion stability, water binding, slicing characteristics, cook yields, flavor and reduce costs (Keaton, 1999).Whey proteins can bind a high amount of water through physical and chemical means. This aids in improving firmness, texture and retention of moisture during processing and cooking. When properly formulated whey ingredients can be used as low cost solid replacers and fat replacers. Whey proteins add positive textural attributes: chew, bite, firmness, and smoothness, non-coarse, non-grainy, non-gritty mouth feel improvements. Whey proteins are very efficient emulsifiers of fat and oil. They form stable emulsions easily and can be used to replace emulsifiers, meat proteins, and other non-meat proteins. Additionally, the bound fat in whey protein products is relatively high in phospholipids adding to the emulsification capacity of any given whey ingredient (Keaton, 1999).

Whey products have a sweet flavor, to virtually no perceivable flavor of their own. The amount and type of whey ingredients used and typical seasoning and spices used in processed meats very rarely impact finished product flavor. When using whey products to increase finished product yields, slight increases in spices and seasonings may be required to maintain the flavor of original meat product (Keaton, 1999). The price-value relationship of whey products, in particular whey proteins, is such that there are few equivalent sources of whey. Other valuable nutrients are calcium and a variety of health-enhancing and prebiotic components. The indirect impact on the nutrient content of mixes such as in reduced or low-fat products is also a valuable contribution of whey products. An important aspect of whey product performance in processed meats is the ability to control mix ingredient costs and improve cook yields. By selecting the right whey product, significant savings can be achieved (Keaton, 1999). Whey proteins showed excellent nutritional and functional properties in low-fat meat products (Perez-Gago & Krochta, 2001).

3.2. Use of Whey Proteins in Processed Meats

Whey and whey products are used successfully in processed meats (ground meats, emulsion products, coarse ground products, whole muscle products) to improve flavor, texture, emulsification, water binding, cooking yield and finished product functional performance (Keaton, 1999). Sweet whey, whey protein concentrates (34–80% protein), and whey protein isolates (90% protein) are commonly used in processed meats. They are mainly used for technological and nutritional quality improvement in processed meats.

WPC has been reported to have higher emulsifying capacity than non fat dry milk (Melachouris, 1984) and denatured whey protein has higher water binding capacity (Short *et al.*, 1978). Smith and Allan (1994) reported that WPC formed heat induced gels which could hold more water.

3.2.1. *WPC in sausages*

Ensor *et al.* (1978) found that addition of WPC in emulsion type sausages increased fat binding properties, product firmness, enhances flavour, texture and juiciness of the product. Pre-heated whey protein isolates formed gel at low protein concentrations and low temperature in the presence of added salt (Hongsprabhas & Barbut, 1997). When pre-heated whey protein was used in poultry raw and cooked meat batter, it resulted in increased water holding capacity, improved rheological properties, and reduced cooking loss (Hongsprabhas & Barbut, 1999). Pawar *et al.* (2000) used WPC in chevon patties prepared from Osmanabadi goat with good results. When liquid whey was used in frankfurter-type sausages, it could replace 100% of ice in frankfurter formula (Yetim *et al.*, 2001). Whey proteins improved emulsion stability, provided better color properties, and resulted in lower chewiness and elasticity, but caused higher brittleness and hardness in frankfurter-type sausages (Yetim *et al.*, 2001).

In a study the effects of fat level (5, 10 and 20%) and soy protein (SP), whey powder (WP) or wheat gluten (WG) on binding properties, color, textural and sensory characteristics of cooked beef sausages were evaluated (Serdaroðlu & Özsümer, 2003). Fat reduction decreased emulsion stability and water holding capacity (WHC) and resulted in higher cooking losses. Addition of SP, WP and WG increased WHC and emulsion stability (ES). The most effective additive on WHC and ES was SP. Reduced fat products were darker compared with 20% fat controls, adding fat replacers increased inside yellowness value. Fat level had no effect on shear force values, WP and WG reduced shear force values regardless of the fat level. Fat replaces resulted in no detrimental effects on sensory characteristics in low fat sausages (Serdaroðlu & Özsümer, 2003). Decreasing fat content significantly alters WHC, emulsion stability and cooking loss. Decreasing fat level had no effect on flavour scores and shear force values. Fat replacers (soy protein, whey powder and wheat gluten) were successfully used to improve binding properties and had no detrimental effects on sensory characteristics. Neither fat level nor fat replacers had an effect on external colour parameters.

Whey proteins can be incorporated into films and coatings for meat products. During 8 weeks of refrigerated storage, whey protein coatings reduced the TBARS and peroxide value by 31.3% and 27.1%, respectively, in low-fat pork sausages. The growth of aerobic bacteria and *Listeria monocytogenes* were inhibited and moisture loss was decreased by 31.3% in sausages with whey protein coating (Shon & Chin, 2008).

3.2.2. *WPC in nuggets*

Food grade polyphosphate and/or milk proteins have been used in meat product formulation to improve the yield. Anjaneyulu *et al*. (1997) reported the combined effect of whey protein concentrate (WPC) and pyrophosphate (PP) on the physicochemical and sensory attributes of restructured buffalo meat nuggets. Incorporation of WPC alone slightly improved the emulsion stability, yield and sensory attributes of nuggets. Addition of WPC + PP or PP alone significantly increased the emulsion stability, yield and all the sensory attributes of products. The study indicates that use of WPC and PP in meat product formulations has substantial beneficial effect rather than addition of WPC alone (Anjaneyulu *et al*., 1997).

Devaki *et al*., (2006) investigated the optimum level of inclusion of whey protein concentrate and egg white for preparation of chicken nuggets with 80 percent deboned spent hen meat and 20% chicken fat was carried out with three levels of whey protein concentrate (1, 2 & 3%). They observed that nuggets with 1 and 2% of whey protein concentrate had no significant difference in cooking yield and sensory characteristics, whereas, incorporation of whey protein concentrate at 3% level improved emulsion stability, cooking yield and also markedly improved the sensory characteristics of nuggets. Hence, 3% level of whey protein concentrate was selected as optimum. Based on the findings it was concluded that nuggets with 3% whey protein concentrate was superior for preparing chicken nuggets with 80% deboned spent hen meat and 20% fat (Devaki *et al.*, 2006).

3.2.3. *WPC in patties*

Quality of chicken patties formulated with the proportionate replacement of lean broiler meat with various levels (0, 3, 6 & 9%) of whey protein concentrate (WPC) was assessed on the basis of different physicochemical characteristics and sensory evaluation (Sakunde *et al.*, 2006). pH and fat contents of the chicken patties increased with the increase in the level of WPC. Cooking yield, moisture and protein contents were also improved while cooking loss was reduced with added levels of WPC over control. Incorporation of WPC up to 3% improved flavour, juiciness, texture and overall palatability of the product (Sakunde *et al.*, 2006).

3.3. WPC in Low fat Meat Products

Fat is a major determinant of the sensory characteristics of a product, and therefore the reduction of fat levels is not merely a matter of using less fat in the formulation. The feasibility of fat-reduced products will depend on a number of factors that determine to what degree, from a technological point of view, fat can be reduced; these factors include the desired fat level, the nature of the product to be reformulated and the kind of processing required.

The manufacture of low-fat products generally follows two basic approaches: the use of leaner raw materials (which raises the cost) and/or the reduction of fat and calorie contents by adding water and other ingredients that contribute few or no calories. These approaches can be supplemented by the use of a number of technological procedures that help offset undesirable side effects produced as a result of changes to the product's composition and nature. The aim is to reduce fat levels and/or modify fat characteristics and yet produce a product with acceptable levels of functionality, safety, sensory properties and stability. Dairy proteins are of particular interest in terms of functionality, especially with regard to emulsifying and gel forming properties. Varying proportions (generally up to 3.5%) of such derivatives as non-fat dried milk, caseinates and whey proteins have been used to offset the effects of fat reduction.

3.3.1. *WPC in low fat sausages*

In finely ground meat emulsions (frankfurters and the like) the limit is –10% (with the substitution of fat by protein and water); in dry fermented sausage it is 20–30% (depending on the degree of drying); and in spreadable dry sausage it is –20% (percentages refer to the minimum fat content in the finished product in all cases).

3.3.2. *WPC in low fat chicken nuggets*

Cholan (2008) prepared low fat chicken nuggets (7.5% added fat) with three different levels of whey protein concentrate (1, 2 and 4%). Incorporation of WPC at different levels did influence the emulsion stability but not organoleptic attributes of low fat chicken nuggets (Table 2). Cooking yield with 1% level of incorporation was higher. Significant difference (P< 0.05) was noticed with emulsion stability and the cooking yield of low fat chicken nuggets incorporated with 1% WPC was higher compared to other two products. Cooking yield and sensory attributes of 1% WPC incorporated chicken nuggets recorded higher values than the products incorporated with 2 and 4% WPC. It was also observed that the low fat nuggets incorporated with 2 and 4% WPC had a rough grainy visual appeal, was less cohesive and dry and hence were scored lower by the panelist for most of the sensory attributes and also had lower yield.

Table 2: Effect of whey protein concentrate on the quality of low fat nuggets.

	Low fat chicken nuggets incorporated with WPC		
Parameters	***1(%)***	***2(%)***	***4(%)***
Emulsion stability (%)	84.68 ± 0.26[c]	85.57 ± 0.23[b]	88.55 ± 0.25[a]
Cooking Yield (%)	94.68	92.86	93.32
pH	6.16 ± 0.10	6.15 ± 0.00	6.10 ± 0.00
Appearance	6.50 ± 0.17	6.50 ± 0.14	6.45 ± 0.17
Flavour	6.54 ± 0.17	6.41 ± 0.15	6.14 ± 0.15
Juiciness	6.54 ± 0.02	6.37 ± 0.17	6.37 ± 0.16
Texture	6.20 ± 0.21	6.16 ± 0.18	6.41 ± 0.16
Overall palatability	6.50 ± 0.17	6.41 ± 0.13	6.70 ± 0.15

(Cholan, 2008)

4. CASEINATES IN MEAT PRODUCTS

Caseinates are produced from pasteurized skim milk either by treating with rennet, by direct acidification or by culturing with lactic acid bacteria, with casein being precipitated at pH 4.3–4.5 the ash content is lower at this pH than at higher pH (Pearson & Tauber, 1984). Curd is easily separated, washed and dried. The casein curd is neutralized with either calcium, sodium or potassium bases at pH 6.8–7.5 to produce corresponding caseinnates. Caseinates are widely used as protein extenders. Sodium and potassium caseinates are generally used because of their higher solubility than calcium. They impart some water and fat binding properties but do not have fat emulsification properties at normal pH of meat.

Various caseinates have also been used in emulsion type of meat products. Sodium caseinate containing 90% protein is an ideal emulsifier and contributes to water retention and texture of meat products (Baardseth *et al.*, 1992). Chevon loaf formulation and processing was standardized utilizing meat from old and unproductive goats (Sharma *et al.*, 1995). The possibility of further improving the loaf quality was explored by incorporating sodium caseinate at 1% and 2% levels in the pre-standardized formulation. There was significant improvement (P< 0.05) in the emulsion stability and cooking yield at both levels of sodium caseinate, although shear force value of chevon loaf was not significantly affected (P< 0.05). Incorporation of sodium caseinate at 1% level showed significant (P< 0.05) improvement in flavour, juiciness, texture and overall palatability of chevon loaf as compared to control. However, further improvement was not observed in these sensory attributes, except for texture at 2% level of sodium caseinate. These findings indicated that meat from old and unproductive goats could be utilized for the production of highly palatable loaf by incorporating 1% sodium caseinate in the formulation (Sharma *et al.*, 1995).

Chempaka *et al.* (1996) used sodium caseinate (SCA) with soy protein isolate (SPI) in different ratios into bologna. Formulations with 5:1 and 5:5 SPI:SCA had lower liquid loss resulting in higher yields. Firmer texture was exhibited by formulations 1:1, 5:1 and 10:1 SPI:SCA but formulation with 1:1 SPI:SCA showed better gelation. Formulation with 5:1 SPI:SCA was more acceptable for texture, chicken taste and overall acceptance while formulation with 1:1 SPI:SCA was more acceptable for the chewiness, juiciness and saltiness attributes.

5. NON FAT DRY MILK IN MEAT PRODUCTS

NFDM (36% protein, 0.8% fat) is prepared by using pasteurized centrifugally separated skim milk and concentrating it to 45–50% solids followed by drying either spray or roller. Although former has higher quality mostly NFDM by roller is used in meat industry. In sausages NFDM improves water and fat binding, texture and flavour in emulsion type sausages, cheaper than meat protein (Pearson & Tauber, 1984).

The disadvantage of using NFDM is high lactose and ash content. The effects of different concentrations (0.00, 0.25, 0.50%) of either non-fat dry milk (NFDM) or whey powder (WP) on emulsifying capacity (EC) and emulsion stability (ES) of beef, chicken and turkey meats were studied by using a computer model system. The effects of meats types and additives (NFDM and WP) on EC and ES were found to be statistically significant (P < 0.01). Of the meats, chicken had the highest and beef the lowest values of EC and ES. When using different concentrations of WP and NFDM, 0.25% WP and 0.50% NFDM had the highest values of EC. The effect of different concentrations (0.25 and 0.50%) of WP was insignificant on ES; however, the effects of different concentrations (0.0, 0.25 and 0.50%) of NFDM were significant on ES.

NFDM in low fat kabab: Effects of microbial transglutaminase (MTG) and its combinations with sodium caseinate (SC) or non-fat dry milk (NFDM) on quality of salt-free, low fat turkey kabab were investigated. The purpose of this study was to prevent quality deteriorations (*e.g.*, cooking loss, textural problems) caused by a lack of salt in the product. The results of this study indicated that the use of MTG can minimize textural quality loss caused by a lack of salt in turkey kabab manufacture. When the enzyme was combined with SC, or NFDM, its effect on texture was enhanced and improved cooking yield, moisture and sensory properties compared to the corresponding control manufactured without the enzyme, salt, SC and NFDM. However, there was no significant effect on pH and color as a result, the possible quality problems which may occur in salt free, low fat turkey can be overcome by the use of combinations of MTG with SC or NFDM.

Hung and Zayas (1992) studied the effects of nonfat dry milk (NFDM), whey protein concentrate (WPC), and sodium caseinate (SC) on quality characteristics of comminuted meat products. Water holding capacity (WHC) was lowest and cooking loss was highest for the control formulation, whereas, formulations extended with milk proteins were higher in WHC and lower in cooking losses. Shear force and firmness increased as extenders were added, except WPC. All frankfurters with extenders were firmer than the control, except those extended with WPC. Frankfurters with SC had a stronger atypical aroma than the control. Meaty aroma score of WPC added product was comparable to that of control.

6. DAIRY INGREDIENTS IN RESTRUCTURED MEAT

Reformed and restructured meats are two important categories of processed meat products. In reformed meat products the intact meat pieces are bound together while restructured meat products are processed using minced meat. Sodium chloride and alkaline phosphates together with mechanical and heat treatments are used to bind meat pieces together. In the process the proteins in muscle become soluble, bind large amounts of water by forming gel during cooling which is made stable by heating. Transglutaminase (TG) is a food-grade commercially available enzyme which helps in the formation of a stable protein matrix (gel). This property can be used to improve the water-binding capacity of dairy proteins in restructured meat products.

Reformed meats: The combined use of sodium caseinate and TG into reformed meat systems resulted in binding of the individual meat pieces together after storage at 5°C overnight and the resulting product was found to be stable to heat during cooking ensuring its stability and sliceability (O'Kennedy *et al.*, 1999). It was also reported that increasing the whey protein (WPC-75) concentration decreased the binding strength of the meat. Optimum binding strength was obtained at 84:16 (casein:whey) in raw meat and 80:20 in cooked at a TG concentration of 500 EU/kg meat.

Restructured meats: O'Kennedy *et al.* (1999) used a model meat system to assess the ability of non meat proteins in the presence of TG but in the absence of salt, to improve the water binding of restructured meats following cooking. Sodium caseinate proved to be best of the non-meat proteins assessed, but the ratio between the caseinate and TG concentrations were required to be adjusted. At a sodium caseinate concentration of 3.3% of the total meat and a TG concentration of 1500 EU/kg, cook losses were reduced.

7. EFFECT OF DAIRY INGREDIENTS ON COLOUR OF MEAT PRODUCTS

Dobson and Cornforth (1992) worked on the ability of nonfat dry milk (NFDM) and whey protein concentrate (WPC) to reduce pink color development of stored turkey samples. Rongey and Bratzler (1966) reported that NFDM was able to lighten the color of bologna. It was also observed that NFDM could reduce red color development in the samples. Calcium caseinate was found to have a whitening effect on chicken meat (van den Hoven, 1987). However, WPC was reported to increase the redness to a level that was "visibly pink" upon cutting. Dobson and Cornforth (1992) postulated that NFDM's ability to reduce the pink color in turkey could be related to an increase of the oxidation-reduction potential leading to a reduction in the formation of denatured hemochromes.

Slesinski *et al.* (2000) evaluated different dairy ingredients for their ability to reduce the spontaneous and unexplainable occurrence of a pink color in cooked, uncured ground turkey meat samples. Sodium nitrite and nicotinamide were added to induce pink color formation. Nonfat dry milk (NFDM) and the whey protein concentrates (WPC) reduced redness values in samples containing 10 ppm sodium nitrite. All of the dairy proteins tested also reduced redness values in nicotinamide-treated samples. In samples prepared without nicotinamide or nitrite, only WPC reduced redness values, while the other proteins tested had no effect or increased redness. It was concluded that NFDM or specific WPC proteins could be used to reduce the pink color defect and increase yield (Slesinski *et al.*, 2000).

Pink discoloration in turkey rolls was prevented by incorporating 3% nonfat dry milk (NFDM) in the formulation (Dobson & Cornforth, 1992). Coarsely ground, cooked turkey rolls formulated with NFDM also had higher levels of undenatured myoglobin than did controls. Nonfat dry milk may inhibit formation of reduced heme-denatured globin pigments (globin hemochromes) in cooked meat by inhibiting myoglobin denaturation or promoting heme oxidation. The mechanism of inhibition of pink discoloration by NFDM remains to be determined (Dobson & Cornforth, 1992).

8. MISCELLANEOUS DAIRY INGREDIENTS IN MEAT PRODUCTS

8.1. Skim Milk Powder/Co-Precipitates

They are produced by co-precipitation of both casein and whey proteins mainly lactalbumen. Co-precipitation involves heating and adjusting pH by acid, alone or in combination of hexametaphosphate or calcium chloride addition. Co precipitaion produces greater yield of proteins, enchances functional value and also contain low lactose levels compared to other milk derived proteins (Pearson & Tauber, 1984). Composition varies depending on preparation process. Contain less lactose when compared to other milk derived protein extender used as protein extender. Improves fat and water binding capacity and acts as thickening agent.

Baardseth *et al.* (1994) investigated the effects of dairy ingredients (1, 3, 5%) sodium caseinate, skim-milk powder, whey protein or whey powder on sensory properties and instrumental texture and color of sausages. Sausages were formulated with 2 or 4% potato starch and cooked to a core temperature of 76 or 82°C. Principal component analysis (PCA) revealed three dominating factors for sensory properties; the first related to dairy ingredients and starch concentrations, the two others to type of dairy ingredients.

8.2. Dry Whole Milk and Dry Butter Milk

Both are produced by drying and have similar composition to NFDM, except the higher fat content. Dried buttermilk contains casein and lactalbumen and prepared from either sweet or sour cream. Dried buttermilk has ability to emulsify fat and as binder and thickening agent (Pearson & Tauber, 1984). The use of dry whole milk, skimmed milk, caseinate, regular and modified whey, at 2% level (w/w) and with 2% additional protein level was studied in a chicken breast meat system with 51% water addition. At the 2% (w/w) level, all dairy proteins significantly reduced cooking loss compared with the control, with caseinate showing the best results. When compared on an equal protein level (2% total protein), the best performing ingredients were the whole milk and modified whey. A similar observation was made in their effect on the products' hardness and fracturability. A cost analysis revealed that modified whey provided the most economical ingredient even when used in quantities three times greater than that of caseinate. Microscopy results showed the formation of larger fine-protein-matrix regions in the treatments that provided higher fracturability values (Barbut & Choy, 2007).

Sardarodlu and Deniz (2004) prepared turkey rolls with 3% each of sodium caseinate, milk powder, whey powder separately and a mixture of all the three @1%. Addition of these dairy ingredients increased water holding capacity and emulsion stability. Rolls added with dairy ingredients had lower cooking losses than control. Mixture of sodium caseinate, milk powder, whey powder gave better sensory scores. They recommended that mixture of dairy ingredients may have the potential of improving technological and sensory quality of emulsion type meat products.

8.3. Calcium Reduced NFDM

Removal of 20–70% from NFDM and substituting sodium resulted with a pH 7.5, which is undesirable in meat products. The sodium is replaced by hydrogen ions which brings back pH to 6–7. These processes produce NFDM with superior solubility characteristics than ordinary NFDM, but otherwise have similar properties (Pearson & Tauber, 1984). Ensor *et al.* (2006) investigated the effects of various nonmeat binders on the yield and textural properties of knockwurst, an emulsion-type sausage product. Three whey protein concentrate levels, calcium-reduced nonfat dry milk and soy protein isolate and an all-meat control were evaluated. Whey protein concentrate proved to be a viable binder alternative for specific emulsion-type meat products by providing similar stability, textural and sensory attributes in comparison to equal levels of soy protein isolate and calcium-reduced nonfat dry milk.

8.4. Curd in Enrobed Meat Products

Enrobing with different cereal flours are used to add value to a product for improving texture, flavour, weight and volume and reducing water loss during frying which in turn increases juiciness. Biswas *et al.* (2006) investigated the acceptability of deep fat fried chicken meat chunk using model systems based on Bengal gram flour(BGF), corn flour (CF) and refined rice flour (RRF) after overnight marination in curd (200 g/kg) along with 2% added salt of each. Crispiness was higher in RRF based coated meat chunks; however, coating containing CF had better adhesive qualities than those containing other cereals. Batter consistency and pick up was higher in BGF based chicken meat chunks followed by RRF > CF. Marination before enrobing reduced times required for doneness. Temperature of heating was negatively correlated with the oil absorption. From the experiment it has also been observed that addition of whole egg liquid instead of combination of WEL and water offer the promising results in terms of adhesiveness, coating thickness, crispiness, color and attractive appearance (Biswas *et al.*, 2006).

9. BENEFITS OF USING DAIRY INGREDIENTS IN MEATS

9.1. Benefits of Whey Protein Concentrates

- Enhances sensory attributes of meat products such as bologna, reduced-fat sausage and roast beef
- Delivers exceptional nutritional value to meat products, including amino acids that are readily digestible and completely bioavailable
- Provides an excellent source of calcium, an essential nutrient not readily available in other regularly consumed foods
- Contributes to a food's healthful image and clean label, and ultimately, to the sale of the food
- WPC, properly processed, provides outstanding fat distribution in meat products
- Whey proteins adsorb on the surface of fat globules, providing a physical separation and preventing the formation of large fat globule

- In sausage and processed meat, whey ingredients assist in the adhering of pieces of meat to each other after the meats have been chopped and blended with other ingredients
- Exceptional water-binding capacity increases the yield value of various meat products such as ham and surimi
- Improved water retention affects the texture of meat products, with moistness being directly related to consumers' perception of freshness

9.2. Benefits of Using Concentrated and Dry Milk Ingredients

- Enhance stable emulsifications
- Form rigid, heat-induced irreversible gels that hold water and fat, and provide structural support to meat products
- Provide water-binding properties, which may reduce the cost of food and improve sensory perception
- Enhance the flavor of meat products
- The milk fat present in the milk ingredient acts as a flavor carrier for fat-soluble ingredients, spices and herbs

9.3. Benefits of Using Lactose

- Enhances the color intensity and stability of many meat products
- Brightens and preserves color
- In sausage and meat products, the addition of lactose helps control browning, especially important during heat treatments like frying or microwave heating
- Masks salty and bitter flavors, such as the characteristically strong taste of liver in liver-based spreads or pâtés
- Injecting brine solutions containing lactose can improve product and structure formation
- Provides an effective source of carbohydrate for the starter culture in preparation of fermented sausages

REFERENCES

Askin, O.O. and Kilic, B. (2009). Effect of microbial transglutaminase, sodium caseinate and non-fat dry milk on quality of salt-free, low fat Turkey doner kebab. *Food Sci. Technol.*, 42: 1590–1596.

Anjaneyulu, A.S.R., Lakshmanan, V., Rao, K.H. and Murthy, T.R.K. (1997). Effect of whey protein concentrate and polyphosphate on the quality of restructured buffalo meat nuggets, *In*: Proceedings of "Emerging trends in Indian food industry in the 21st century-Indian convention of food scientists and technologies." September 25–26, Mumbai. p. 9.

Baardseth, P., Naes, T., Mielnik, J., Skrede, G., Holland, S. and Eide, O. (1992). Dairy ingredients effects on sausage sensory properties studied by principal component analysis. *J. Food Sci.*, 57: 822–828.

Barbut, S. and Choy, V. (2007) Use of dairy proteins in lean poultry meat batters – a comparative study. *Int. J. Food Sci. Technol.*, 42: 453–458.

Biswas, A.K., Kumar, R.R., Chidanandaiah and Pathak, V. (2006). Effect of enrobing on the acceptability of curd marinated chicken meat chunks, *In*: Proceedings of "National symposium on prospects and challenges in Indian meat industry-IMASCON-II." July 27–29, Chennai. p. 179.

Cholan, P (2008). Effect of fat replacers on the quality and storage stability of low fat chicken nuggets. M.V.Sc Thesis submitted to Pondicherry University, Puducherry.

Chempaka, S., Yusof, M. and Babji, A.S. (1996). Effect of non-meat proteins, soy protein isolate and sodium caseinate, on the textural properties of chicken bologna. *J. Food Int.*, 47: 323–329.

Devaki, A., Venkataramanujam, V. and Dushyanthan, K. (2006). Effect of whey protein concentrate and egg white on physicochemical, textural and sensory properties of chicken nuggets from spent hen meat, In Proceedings of "National symposium on prospects and challenges in Indian meat industry-IMASCON-II". July 27–29, Chennai, p. 200.

De wit, J.N. and Kessel, J. (1996). Effects of ionic strength on the solubility of whey proteins products. A colloids chemical approach. *Food Hydrocolloids*, 10: 143–149.

Dobson, B.N. and Cornforth, D.P. (1992). Nonfat dry milk inhibits pink discoloration in Turkey rolls. *Poult. Sci.*, 71(11): 1943–1946.

Ensor, S.A., Mandigo, R.W., Calkins, C.R. and Quint, L.N. (1978). Comparative evaluation of whey protein concentrate, soy protein isolate and calcium-reduced non-fat dry milk as binders in an emulsion-type sausage. *J. Food Sci.*, 52: 1155–1158.

Ellekjær, M.R., Naes, T. and Baardseth, P. (1996). Milk proteins affect yield and sensory quality of cooked sausage. *J. Food Sci.*, 61(3): 660–666.

Hughes, E., Cofrades, S. and Troy, D.J. (1997). Effect of fat level, oat fiber and carrageen on frankfurters formulated with 5, 12 and 30% fat. *Meat Sci.*, 45(3): 273–281.

Hughes, E., Mullen, A.M. and Troy, D.J. (1998). Effects of fat level tapioca starch and whey protein on frankfurters formulated with 5% and 12% fat. *Meat Sci.*, 48(12): 169–180.

Hugunin, A.G. and Ewing, N.L. (1977). Dairy based ingredients for food products. Dairy Research Inc. Rosemont, IL.

Hung, S.C. and Zayas, J.F. (1992). Functionality of milk proteins and corn germ protein flour in comminuted meat products. *J. Food Quality*, 15: 139–152.

Isýk, F. and Gokalp, H.Y. (1999). Peyniraltý suyu ve peyniraltý suyu tozunun emulsiyon ozellikleri ve gýda sanayiinde bu amaclą kullanýmlarý II (in Turkish). Standard (Kasým), 38: 61–72.

Jayaprakash, H.M. (1992). Membrane processing applications for production of whey powder and whey protein concentrates. Ph D Thesis, NDRI, Deemed University, Karnal, India.

Keaton, J. (1999). Whey protein and lactose products in processed meats, Application Monograph-Meat, U.S. Dairy Export Council,

Lyons, H.P., Kerry, J.F., Moriscy, P.A. and Buckley, D.J. (1999). The influence of added whey protein/carrageenan gels and tapioca starch on the textural properties of low- fat sausages. *Meat Sci.*, 51: 43–52

Mann, E.J. (1998). Various dairy products. *Dairy Industries Inter.*, 63: 16–17.

Melachouris, N. (1984). Critical aspects in the development of whey protein concentrates. *J. Dairy Sci.*, 67: 2693–2702.

Morr, C.V. (1982). Development in Dairy Chemistry. *In*: Fox, P.F. (*Ed.*) Applied Science Publications, London and New York, p. 375.

O'Kennedy, B.T., Neville, D.P. and Kelly, P.M. (1999). Use of novel dairy ingredients in processed meats, Project Report: DPRC No. 15, Dairy Products Research Centre, ISBN: 1 901138 55 0, DPRC No. 15, © Teagasc, October 2000.

Ozdemir, S., Zorba, O. and Gokalp, H.Y. (1994). Yagsýz sut tozu, yagsýz sut ve peyniraltý suyunun emulsiyon ozellikleri (in Turkish). *Doga Tr. Agriculture and Forestry*, 18: 507–513.

Parks, L.L. and Carpenter, J.A. (1987). Functionality of six non-meat proteins in meat emulsion systems. *J. Food Sci.*, 52: 271–274.

Patel, M.T. and Kilara, A. (1990). Studies on WPC-I. Compositional and thermal properties. *J. Dairy Sci.*, 73: 1439–1445.

Pawar, D.D., Malik, S.V.S., Bhilegoanker, K.N. and Barbhudda, S.B. (2000). Effects of nisin and its combination with sodium chloride on the survival of *Listeria monocytogenes* added to raw buffalo meat mince. *Meat Sci.*, 56: 215–219.

Pearson, A.M. and Tauber, F.W. (1984). Processed Meats, 2nd Edn. AVI Pub., Westport Cnnectcut pp. 264–266.

Renner, E. and Abd El-Salam (1991). Application of ultra filtration in the dairy industry. Elsevier Applied Science, London.

Sebranek, J.G. (1996) Frankfurters with lean finely textured tissue as affected by ingredients. *J. Food Sci.*, 61: 1275–1280.

Short, J.L., Mathews, M.E., Marshal, K.R. and Tuttiet, P.T. (1978). Process economics in a seasoning industry. NZ. *J. Dairy Sci. Technol.*, 13: 178–182.

Smith, D.M. and Allan, J.R. (1994). Gel properties of WPC as influenced by ionized calcium. *J. Food Sci.*, 59: 1115–1122.

Sakunde, D.T., Ambadkar, R.K. and Zanjad, P.N. (2006). Effect of addition of various levels of whey protein concentrate on the quality of chicken patties, in proceedings of "National symposium on prospects and challenges in Indian meat industry-IMASCON-II." July 27–29, Chennai, p. 184.

Sharma, B.D., Rao, V.K., Sanyal, M.K. and Yadav, P.L. (1995). Effect of sodium caseinate incorporation on the quality of chevon loaf, In proceedings of "Prospects of production, processing and marketing of goat meat." January 10–11, Makdoom p. 28.

Serdaroðlu, M. and Özsümer, M.S. (2003). Effects of soy protein, whey powder and wheat gluten on quality characteristics of cooked beef sausages formulated with 5, 10 and 20% fat. Electronic Journal of Polish Agricultural Universities, Food Science and Technology, Volume 6, Issue 2. Available Online http://www.ejpau.media.pl

Slesinski, A.J., Claus, J.R., Anderson-Cook, C.M., Eigel, W.E., Graham, P.P., Lenz, G.E. and Noble, R.B. (2000). Ability of various dairy proteins to reduce pink color development in cooked ground turkey breast. *J. Food Sci.*, 65: 417–420.

Sardarodlu, M. and Deniz, E.E. (2004). Chemical composition and quality characteristics of emulsion type turkey rolls formulated with dairy ingredients. *J. Food Technol.*, 2: 109–113.

Smith, D.M. (1988). Meat proteins: Functional properties in comminuted meat products. *Food Technol.*, 42: 116–121.

Srinivasan, M., Singh, H. and Munro, P.A. (2002). Formation and stability of sodium caseinate emulsion: Influence of retorting (121°C for 15 min) before or after emulsification. *Food Hydrocolloids*, 16: 153–160.

Yetim, H., Muller, W.D. and Eber, M. (2001). Using fluid whey in comminuted meat products: Effects on technological, chemical and sensory properties of frankfurter type sausages. *Food Research Inter.*, 34: 97–101.

Zorba, O., Yetim, H., Ozdemir, S. and Gokalp, H.Y. (1995). The possibility of using fluid whey in comminuted meat products: Capacity and viscosity of the model emulsions prepared with whey and muscle proteins. *Z. Lebensm. Unters. Forsch.*, 200: 425–427.

Zorba, O., Ozdemir, S. and Gokalp, H.Y. (1998). Stability of model emulsions prepared using whey and muscle proteins. *Nahrung*, 42: 16–18.

Zorba, O., Kurt, S. and Genccelep, H. (2005). The effects of different levels of skim milk powder and whey powder on apparent yield stress and density of different meat emulsions. *Food Hydrocolloids*, 19: 149–155.

23

Nanotechnology for Animal Products

A.K. Biswas, Gauri Jairath and P.K. Mandal

ABSTRACT

The use of nanotechnology in the food industry is here and if projections are correct, it will continue to grow rapidly over the next few years. Its potential for providing safer and more nutritious foods is important; however, there are still many uncertainties about the technology and its applications. We must make sure that the technology is safe and properly regulated to ensure maximum food safety and personal health protection. In the long run nanotechnology will change the fabrication of whole packaging. Nanotechnology can make products cost-effective. Production will be carried out by self-replicating nano-devices using small assent of material, energy, low capacity, less labour and land. Thus production is more efficient. The potential of nano-technology in meat industry cannot be fully appreciated yet because of lack of sufficient knowledge. If nano-technology continues to advance at its current pace, we could expect that soon we will be able to create unlimited amount of meat by synthesis at the atomic level, which would eradicate hunger.

1. INTRODUCTION

In today's competitive market, technology is essential to keep leadership in the food and food processing industry. Consumers demand fresh authentic, convenient and flavourful food products. The future belongs to new products and new processes, with the goal of enhancing the performance of the product, prolonging the product shelf life and freshness, and improving the safety and quality of food. Nanotechnology is an enabling technology that has the potential to revolutionise the food industry. Nanotechnology, shortened to "nanotech", is the study of the controlling of matter on an atomic and molecular scale. Generally nanotechnology deals with structures of the size 100 nanometers or smaller in at least one dimension, and involves developing materials or devices within that size.

Nanotechnology is very diverse, ranging from extensions of conventional device with dimensions on the nanoscale to investigating whether we can directly control matter on the atomic scale, completely new approaches based upon molecular self-assembly, from developing new nanomaterials. So nanotechnology is the engineering of functional systems at the molecular scale. This covers both current work and concepts that are more advanced. Nanotechnology can be applied to develop nanoscale materials, controlled delivery systems, contaminant detection and to create nanodevices for molecular and cellular biology. In its original sense, 'nanotechnology' refers to the projected ability to construct items from the bottom up, using techniques and tools being developed today to make complete, high performance products. In today's competitive market technology is essential to keep leadership in the food and food processing industry.

Nanotechnology involves creating and manipulating organic and inorganic matter at the nanoscale. It promises to provide the means for designing nanomaterials; materials with tailor-made physical, chemical and biological properties controlled by defined molecular structures and dynamics. The present molecular biology techniques of genetic modification of crops are already forms of what has been termed nanotechnology. Nanotechnology can provide for the future development of far more precise and effective methods of, and other forms of, manipulation of food polymers and polymeric assemblages to provide tailor-made improvements to food quality and food safety. Nanotechnology promises not only the creation of novel and precisely defined material properties, it also promises that these materials will have self-assembling, self-healing and maintaining properties.

2. HISTORICAL BACK GROUND

Nanotechnology has huge historical background. The core concepts were introduced in early fifties. The concept of nano-technology was first given by Noble Laureate Physicist Richard P. Feynman in Southern California in 1959. In 1959 Feynman observed that the principles of physics do not deny the possibility of manipulating things atom by atom. He suggested using small machines to make even tinier machines, and so on down to the atomic level itself. Although there were conflicts about the early theories, in 1974 Norio Taniguchi a Japanese scientist first introduced the term "nano-technology" to describe semiconductor processes such as thin film deposition and ion beam milling exhibiting characteristic control on the order of a nanometre. His definition still stands as the basic statement today: "Nano-technology' mainly consists of the processing of separation, consolidation, and deformation of materials by one atom or one molecule'.' In 1977, Drexler proposed the official concept of nanotechnology at MIT and Eric Drexler popularized the term 'nanotech' in 1980's.

In 1981 first technical paper on molecular nanotechnology was written which was an inspiration for building a machine using atomic precision. Scanning Tunnelling Microscopy (STM) was the first machine invented using nanotechnology in its construction. STM allows scientists to see and manipulate individual atoms. In 1985 first book on nanotechnology was published and given first Feynman prize in early nineties. In 1988 nanotechnology became a part of course for the first time at university of Stanford. International business machine (IBM) spelled in the form of individual atom to represent the very idea was presented in 1989 and the concept was declared original and practical in national conference of engineering in 1989.

In 1991 the bottom up approach was announced by MITI Japan and the institute also committed $200 million at carbon nanotube discovery during the routine projects of nanotech. In 1992 the textbook for students and testimonials published also supported the concept very well. Engineering students started thinking atomically with the publication of the textbook. In 1995 first military application think tank was conceptualized by a panel of military engineers. About 196 grand prize of US$ 250,000 was announced for the most genius machine developed by using nanotechnology. NASA begins working with its computational project on nanotech. First robot using nanotech systems was designed in 1997 which took the concept of nanotechnology a step ahead. After the appearance of first report US government started officially supporting the nanotech development teams and implemented the concept even on regional basis in the continent of United States of America.

With the start of new century the nanotechnology provide the world the high advancements in the field of biology, chemistry, technology, medicine, literature, electronics and many other. The very recent development of nanotechnology is the bimolecular nanotechnology which holds great importance in the fundamental research in bio physics and computational improvements. In future the nanotechnology will play great role in discovering the new components and to enhance the existing technologies and this technology will last longer as it's the latest one in the field of sciences. According to the U.S. Department of Agriculture (USDA), by 2015 the global impact of products in which nanotechnology plays a key role will be approximately $1 trillion annually.

3. NANOTECHNOLOGY AND NANOPARTICALS

Nanotechnology is defined as "a science that involves the design and application of structures, devices and systems on an extremely small scale, called the nanoscale—that is, billionths of a metre, or about 1-millionth the size of a pinhead". The word "NANO" is derived from the Latin word which mean Dwarf. The term 'nanotechnology' does not describe a singular technology, but rather encompasses a range of technologies that operate at the scale of the building blocks of biological and manufactured materials—the nanoscale.

Nanotechnology has been provisionally defined as relating to materials, systems and processes which operate at a scale of 100 nanometres (nm) or less. In layman's terms, nanotechnology is the science behind the intentional creation, manipulation, and characterization of extremely small particles and macro molecules. The chemical, physical, and biological properties of materials differ in fundamental and valuable ways from those of individual atoms, molecules, or bulk matter. Research and development activities in nanotechnology are directed toward understanding and creating improved materials, devices, and systems that exploit these new properties.

The smallest particle ever found as the result of molecular processing and nanotechnology are known as nano particles. These are the particles with extra ordinary properties of deliberating new components and transmitting energies at nano metre scale. The foremost feature of these particles is that they have the ability to reassemble themselves again and again which is very useful for finding out new components from them. At present they have applications in the field including, chemistry, biological sciences, garments, wear and tear, electronics, computer and fiction sciences, mathematics, mechanical engineering, artificial intelligence, fertilization, purification and many more. They were originated with the concept of nanotechnology.

4. APPLICATIONS OF NANOTECHNOLOGY IN FOOD

Today food materials are often considered not only a source of nutrients but also as having to contribute to the health of consumers. Most of the nanoparticles used traditionally belong to the group of colloids *i.e.*, emulsions, micelles, mono and bi-layers. The colloidal particles were attracted to each other through Van der Waals forces, which give them stability. Again stearic stabilization is achieved by adsorbing polymers and surfactants on the surface. Nanoparticles could further be stabilized by coating them with molecules that can form chemical bonds. For food applications, nanotechnology can be applied by either 'bottom up' or 'top down' approaches. The top-down approach is achieved basically by means of a physical processing of the food materials, such as grinding and milling. For example, dry-milling technology can be used to obtain wheat flour of fine size that has a high water-binding capacity.

This technology has been used to improve antioxidant activity in green tea powder. As the powder size of green tea is reduced to 1000 nm by dry milling, the high ratio of nutrient digestion and absorption resulted in an increase in the activity of an oxygen-eliminating enzyme. By contrast, self assembly and self organization are concepts derived from biology that have inspired a bottom-up food nanotechnology. The organization of casein micelles or starch and the folding of globular proteins and protein aggregates are examples of self-assembly structures that create stable entities. Self organization on the nanometer scale can be achieved by setting a balance between the different non-covalent forces.

Owing to the greater surface area of nanoparticles per mass unit, they are expected to be more biologically active than larger sized particles of the same chemical composition. This offers several perspectives for food applications. Nanoparticles can, for instance, be used as bioactive compounds in functional foods. Bioactive compounds that can be found naturally in certain foods have physiological benefits and might help to reduce the risk of certain diseases, including cancer. By reducing particle size, nanotechnology can contribute to improve the properties of bioactive compounds, such as delivery properties, solubility, prolonged residence time in the gastrointestinal tract and efficient absorption through cells. Omega 3 and omega 6 fatty acids, probiotics, prebiotics, vitamins and minerals have found their applications in food nanotechnology as bioactive compounds.

Several other novel applications of nanotechnologies have become apparent, including the use of nanoparticles, such as micelles, liposomes, nanoemulsions, bio polymeric nanoparticles and cubosomes, as well as the development of nanosensors, which are aimed at ensuring food safety. However, despite the increased marketing efforts in the nanotechnology sector, research into nanotechnology of food and food-related products is only just beginning to develop.

Some examples of the use of nanotechnology in food products are cooking oils that contain nutraceuticals within nanocapsules, nanoencapsulated flavour enhancers and nanoparticles that have the ability to selectively bind and remove chemicals from food. For meat, a great revolution has been brought by this technology. Nano-encapsulated active ingredients including vitamins and fatty acids are now sold commercially for use in processing and preservation of meats and other foods. Nanoparticles are added intentionally in many foods to improve flow properties, colour stability during processing, or to increase shelf life. For instance, aluminium-silicates are commonly used as anti-caking agents in granular or powdered processed foods, while anatase titanium dioxide is a common food whitener and brightener additive, used in confectionery, some cheeses and sauces. The encapsulation system has numerous benefits:

(*i*) *Ease of handling*: The system can be utilised to transform volatile liquids such as flavours into a powder, which are in many cases easier to handle.

(*ii*) *Enhanced stability*: The system can be utilised to isolate active ingredients as well as flavours that may interact with the other food ingredients. This provides long-term product shelf life.

(*iii*) *Protection against oxidation*: The micro spheres have very low surface oil (less than 0.5%) at very high payloads (30–40%) compared to conventional spray-dried particles utilising materials such as gum arabic or starch.

(*iv*) *Retention of volatile ingredients*: The moisture-sensitive matrix provides excellent retention of highly volatile ingredients, such as flavours, over an extended period of time to reduce the flavour loss during the product shelf life.

(*v*) *Taste masking*: Unwanted taste can be masked by preventing interaction between the active molecule and the oral mucosal surface. The nanospheres are hydrophobic and can prevent bitter ingredients encapsulated within their structure from going into solution and interacting directly with taste receptors.

(vi) *Moisture triggered controlled release*: As discussed above, the micro spheres dissolve in the presence of water or saliva to release the active ingredients or flavours, thereby providing a high impact flavour "burst".

(vii) *pH triggered controlled release*: Ingredients can be encapsulated in the microspheres to enhance their stability during the product shelf life and to release them when needed or upon food consumption. This pH triggered release was initially designed to deliver drugs to different regions of the gastrointestinal tract.

(viii) *Heat-triggered release*: The hydrophobic nanospheres are temperature sensitive and can be utilised to release active ingredients and flavours at a certain temperature.

(ix) *Consecutive delivery of multiple active ingredients*: Two or more ingredients that would react with each other if put together can be separated and provided consecutively by placing one in the nanosphere and the other in the microsphere. An example is encapsulation of folic acid and iron that work synergistically. Other examples would be the delivery of one flavour after another, or the delivery of a flavour or sensation (in the microsphere) to indicate that the active ingredient (in the nanospheres) has been delivered.

(x) *Change in flavour character*: Encapsulation of a flavour in the nanospheres that is different from the flavour encapsulated in the microsphere can provide a perceivable change in the organoleptic perception in response to moisture during the use of the product.

(xi) *Long-lasting organoleptic perception*: As a result of the bio adhesive properties of the nanospheres and their residence in the oral cavity, flavour perception and mouth-feel can be extended over a longer period of time.

(xii) *Enhanced bioavailability and efficacy*: As a result of their hydrophobic/lipophilic nature, the nanospheres can enhance the bioavailability of various active ingredients, such as vitamins, nutrients and other biologically active agents encapsulated within their structure.

Table 1: Range of sizes of nanomaterials in the food sector.

	***Diameter or length* (*nm*)**
DNA	12
Glucose	21–75
Liposome	30–10000
LDH	40–300
Amylopectin	44–200
Casein micelle	60–100
PLA nanosphere	100–300
Zein	200
Cubosomes	500
Nanosensors	<1000

There are potential applications of nanotechnology in the food and food-packaging industries. Applications in food packaging are considered highly promising because they can improve the safety and quality of food products. This includes intelligent packaging, which is reactive to the environment and active packaging. Extensive use of nanoadditives to the food itself is less likely in the near future owing to safety concerns. Applications of nanoparticles in food are illustrated in following Table 2.

Category	***Examples of the different applications***	***Reference(s)***
Food processing	Interactive foods and beverages give desired flavours and colours by the addition of nanocapsules which burst at different microwave frequencies	ETC Group, 2005
	Initiative has explored the applications of nano-technology in water purification and treatment focusing on the areas such as membranes and membrane processes, biofouling and disinfection, and contaminants removal	INNI, 2006
	Development of nano-scale formulations of different traditional herbal plants by reducing the herbs to nanoscale powder or emulsion	ElAmin, 2007
	Micronization of ganoderma spore to ultrafine powder by top-down approach, resulting in the rupture of cell walls and release of potential active ingredients	Liu, Wang, & Yuan, 2005
	Frying oil refining catalytic device inhibits thermal polymerization of frying oil and reduce off-odors	OilFresh, 2005
	Micrometres long stiff hollow nanotubes made of milk protein by self-assembly have potential to be used as novel ingredients for viscosifying, gelation, nanoencapsulation, and controlled release purposes	Graveland-Bikkera *et al.*, 2006
	Adding nanocomposites or nanoparticles (*e.g.*, silver, titanium dioxide, silicon dioxide, and nano-clay) into packaging materials to ensure better protection of foods by modifying the permeation behaviour of foils, deodorizing, increasing barrier properties, blocking ultraviolet light, improving mechanical and heat-resistance properties, and developing antimicrobial and antifungal surfaces	ElAmin, 2005, Fletcher, 2006, IFST, 2006 & Roach, 2006
Packaging	Nylon nanocomposites providing barriers to oxygen and carbon dioxide flow have been used in food packaging (*i.e.*, multi-layer PET bottles for beer and other alcoholic beverages) to keep freshness and block out smells	Sherman, 2005
Nutraceutical delivery	Nanotechnology renders hydrophilic substances fat soluble and lipophilic ones water soluble, allowing nanoparticles of some functional ingredients (*e.g.*, carotenoids, phytosterols, and antioxidants) to be dispersed in water or fruit drinks to improve their bioavailability	Chen *et al.*, 2006 & IFST, 2006

Table 2: (*Contd...*)

Table 2: (*Contd...*)

Category	*Examples of the different applications*	*Reference(s)*
	Synthetic nanoparticles of lycopene are developed and accepted as GRAS-affirmed by the FDA for use in food in the USA	–
	Minute micelles (nanocapsules) are used as carriers for essential oils, flavour, antioxidant, coenzyme Q10, and vitamins, minerals, and photochemical to improve their bioavailability	ElAmin, 2005, Fletcher, 2006; Pitman & Halliday, 2006
	Encapsulating the nanoparticles of active ingredients (*e.g.*, polyphenols, minerals, and micronutrients) to protect them from oxidation and getting to the taste receptor site, thus to reduce their undesirable off-tastes in the finished application	Heller, 2006
	Food industry application of liposomal nanovesicles for the encapsulation and delivery of nutrients and functional ingredients such as proteins, enzymes, flavours, and antimicrobial compounds	Wen *et al.,* 2006
Safety and sensing	Whey protein nanospheres (40 nm), which are internalized by cells and degraded therein to release the nutraceutical compounds, can be used as carriers for oral administration of nutraceutical agents to improve their bioavailability	Chen *et al.*, 2006
	Protein-coated nanocantilever, naturally vibrating at a specific frequency, is a new class of ultra-small silicon sensors for the quick detection of viruses, bacteria and other pathogens. When contaminants land on the devices, the slight mass changes can cause the nanocantilever to vibrate at a different frequency and be quickly detected	ElAmin, 2006
	Development of synthetic tree-shaped DNA being tagged with color-coded probes, as a nanobarcode device, enables the identification of food pathogens	Li *et al.*, 2004
Silver nanoparticles have	A miniature portable microbiodetector was developed using different nanowires, specific pathogen antibodies, and fluorescent antibodies for the simultaneous detection of toxins, pathogens, and chemicals in foodstuffs	Roach, 2006

5. APPLICATION OF NANOPARTICLES IN FOOD PACKAGING

5.1. Improvement of Shelf Life of Meat by Nano-Packaging

Nanotechnology has the capability to transform the nature of food packaging materials in future. Certain nanoscale innovations could bring amazing improvements to food packaging in the forms of detection of pathogens, smart and active packaging, and barrier and mechanical properties. Packaging that incorporates nano materials can be "smart", which means that it can respond to environmental conditions or repair itself or alert a consumer to contamination and/or the presence of pathogens. It offers several extraordinary benefits to improve food packages like advancements in fundamental characteristics of food packaging materials such as antimicrobial properties, barrier properties, strength, and stability to heat and cold. Packaging of meat is improved by following ways.

5.1.1. *Incorporation of nanoparticles in plastic films*

Polymer nanocomposites incorporating metal or metal oxide nanoparticles are utilized mainly for their antimicrobial action, abrasion resistance, UV absorption, and strength. Some nanomaterials have been used to develop active packaging that can absorb oxygen and therefore keep food fresh. Other nanomaterials have been incorporated as UV absorbers to prevent UV degradation in plastics such as PS, PE and PVC.

5.1.2. *Silicate nanoparticles*

Silicate nanoparticles are incorporated in plastic film used for packaging to provide a barrier to gasses. Durethan, from Bayer Polymers, is a nanocomposite film enriched with an enormous number of silicate nanoparticles which reduce entry of oxygen and other gases and the exit of moisture, thus preventing food from spoiling.

5.1.3. *Zinc oxide nanoparticles*

These particles are incorporated into plastic packaging to block UV rays and provide anti-bacterial protection, while improving the strength and stability of the plastic film.

5.1.4. *Nanosilver*

Nanosilver is a biocidal agent intended to maintain the hygienic conditions of the surface by preventing or reducing microbial growth. It acts as surface biocide and not intended to have preservative effect on food.

5.1.5. *Titanium dioxide*

These nanoparticles are incorporated in transparent plastic to protect the meat from UV degradation.

5.2. Application of Nanosensors in Intelligent Packaging

Nanotechnology has also enabled the development of nanosensors that can be applied as labels or coatings to add an intelligent function to food packaging in terms of ensuring the integrity of the package through detection of leaks, indications of time-temperature variations (freeze-thaw-refreezing), or microbial safety. The use of nanosensors in plastic packaging can detect gases given off by food when it spoils and the packaging itself changes colour to alert as food gone bad. Nanosensors could also detect bacteria and other contaminates such as salmonella on the surface of food at a packaging plant. The use of sensor allows frequent testing of products at a much lower cost than is incurred by sending samples to a lab for analysis. This point-of-packaging testing, if conducted properly, has the potential to dramatically reduce the chance of contaminated food reaching grocery store shelves.

Examples include an indicator that turns from transparent to blue, informing the consumer that air has entered the modified atmosphere of the packaged materials. For this type of application, nanotechnology-derived printable inks have been developed. One example is an oxygen detecting ink containing light-sensitive (TiO_2) nanoparticles, which only detect oxygen when they are "switched on" with UV light. Other conductive inks for ink jet printing based on copper nanoparticles have also been developed (Park *et al.*, 2007). Food safety also requires confirmation of the authenticity of products. This is where application of nanobarcodes incorporated into printing inks or coatings has shown the potential for use in tracing the authenticity of the packaged product (Han *et al.*, 2001).

Food quality indicators have also been developed that provide visual indication to the consumer when a packaged foodstuff starts to deteriorate. An example of such food quality indicators is a label based on detection of hydrogen sulphide, which is designed for use on fresh poultry products. The indicator is based on a reaction between hydrogen sulphide and a nanolayer of silver (Smolander *et al.*, 2004). The Nanosilver layer is opaque light brown, but when meat starts to deteriorate silver sulphide is formed and the layer becomes transparent, indicating that the food may be unsafe to consume.

Other materials developed for potential food packaging applications are based on nanostructured silicon with nanopores. The potential applications include detection of pathogens in food and variations of temperature during food storage. Another relevant development is aimed at providing a basis for intelligent preservative packaging technology that will release a preservative only when a packaged food begins to spoil.

5.3. Smart Packaging

Today, meat packaging and monitoring are a major focus of meat industry-related nanotech R&D. Packaging that incorporates nano materials can be "smart", which means that it can respond to environmental conditions or repair itself or alert a consumer to contamination and/or the presence of pathogens. According to industry analysts, the current US market for "active, controlled and smart" packaging for foods and beverages is an estimated $54 billion by 2008.

5.4. Controlled Transmission of Gasses through Packaging Material

Through nanotechnology we can increase or decrease gas transmission rate to suit the packaging requirements of the product. Nanotechnology enables designers to alter the structure of the packaging materials on the molecular scale. With different nanostructure, plastics can gain various gas and water vapour permeability to fit the requirements of preserving fruit, vegetable, beverage, wine and other foods. By adding nanoparticles, people can also produce bottles and packages with more light- and fire-resistance, stronger mechanical and thermal performance and less gas absorption. Such nano tweaking can increase the shelf life of meats and preserve flavour and colour. SINTEF-Nanotech to create small particles in the film and improve transportation of some gases through the plastic film to pump out dirty air such as carbon dioxide has been developed. Dispersing a filler material into nano particles, creates multiple, parallel layer that force gases to flow through the polymer in a "tortuous path", forming complex barriers to gases and water vapour.

5.5. Other Utility of Nano-packaging

5.5.1. *Carbon nanotubes*

They are cylinders with nanoscale diameters that can be used in food packaging to improve its mechanical properties. It has been recently discovered that they might exhibit powerful antimicrobial effects. *E. coli* bacteria died on immediate direct contact with aggregates of carbon nanotubes.

5.5.2. *Radio frequency identification (RFID) tags*

This could be incorporated into food packages in the future, do not require line-of-sight for reading like bar-codes and enable registration of hundreds of tags in a second. Retail chains such as Wal-Mart are testing this technology.

5.5.3. *Nanowheels*

This adds to the improvement in food packaging. These molecules are incorporated into plastics in order to enhance their barrier and mechanical properties.

5.5.4. *Nanovesicles*

Capable of detecting *E. coli* 0157:H7, *Listeria monocytogens* and *Salmonella* spp. Liposome nanovesicles have been devised to detect peanut allergen proteins.

5.5.5. *DNA biochips*

They have been under development for quite some time now, to detect pathogens. A carbon nanotube functions as the transmitter while a single strand of DNA function as the sensor.

5.5.6. *Electronic tongue nanosensors*

It is developed to detect substances in parts per trillion which could be used to trigger a color changes in food packages to alert the consumer to food that has been spoiled.

5.5.7. *Antimicrobial nanoemulsions*

Nano-emulsions have been developed for use in the decontamination of food packaging equipment and in the packaging of food. Baker and co-workers have used high-shear mixing of a lipid oil discontinuous phase with an aqueous continuous phase to develop nano-emulsions, which is an effective solution for disinfecting sensitive equipment safely (Lerner, 2000). Nano-emulsions consist of oil droplets of 400–800 µm in diameter that are able to fuse with and subsequently disrupt the membrane of a variety of different pathogens, such as bacteria, spores, enveloped viruses, and fungal spores.

6. ADVANTAGES OF NANOTECHNOLOGY IN FOOD SECTOR

Nanoparticles may be able to detect bacteria, extend food shelf life, add health benefits, or improve flavour. Nonetheless, nanotechnology offers some exciting potential benefits for the quality and safety of our foods.

(*i*) *Contamination sensor*: Flash a light to reveal the presence of *E. coli* bacteria.

(*i*) *Antimicrobial packaging*: Edible films made with cinnamon or oregano oil, or nano particles of zinc, calcium other materials that kill bacteria.

(*iii*) *Improved food storage*: Nano-enhanced barrier keeps oxygen-sensitive foods fresher.

(*iv*) *Enhanced nutrient delivery*: Nano-encapsulating improves solubility of vitamins, antioxidants, healthy omega oils and other 'nutraceuticals'.

(*v*) *Green packaging*: Nano-fibres made from lobster shells or organic corn is both antimicrobial and biodegradable.

(*vi*) *Pesticide reduction*: A cloth saturated with nano fibers slowly releases pesticides, eliminating need for additional spraying and reducing chemical leakage into the water supply.

(*vii*) *Tracking, tracing; brand protection*: Nanobarcodes can be created to tag individual products and trace outbreaks.

(*viii*) *Texture*: Food spreadability and stability improve with nano-sized crystals and lipids for better low-fat foods.

(ix) *Flavour*: Trick the tongue with bitter blockers or sweet and salty enhancers.

(x) *Bacteria identification and elimination*: Nano carbohydrate particles bind with bacteria so they can be detected and eliminated.

(xi) *Thermal stability*: Nanotechnology helps to maintain the thermal stability of the product and protect the product against extreme temperature variations.

(xii) *Improved packaging:* It provides high mechanical strength, good optical density and recyclability to the packaging.

7. HAZARDS ASSOCIATED WITH NANOTECHNOLOGY

Today, hundreds of consumer products incorporating nanomaterials are on the market. Many of these contain nano-silver which has potential environmental and human health risks. Silver is known as a potent antibacterial agent, but in recent years the use of silver as a germ killer has experienced a dramatic revival. When produced at the nano level, silver become exponentially more potent. Nano-silver can be found lining cups, bowls and cutting boards, soaps and menstrual pads, children's toys, infant products and health supplements. Scientific evidence suggests silver, and in particular nano-silver, is toxic and may be detrimental to human health. Bacterial resistance to antibiotics is an ever increasing problem globally, and to use nano-silver indiscriminately is not only unnecessary, but may further increase bacterial resistance to a dangerous level. Nanoparticles may pose threat to liver cells, lung damage, can also damage DNA and increase cancer risk

8. REGULATIONS

Increasing use of nanotechnology is one of the major concerns to impose regulations for public health safety. Different country is having their own regulations. In some countries (USA), some existing laws such as notably the Toxic Substances Control Act; the Occupational Safety and Health Act; the Food, Drug and Cosmetic Act, and the major environmental laws may at least provide some legal basis for regulating nanotechnology. To date, there is no international regulation of nanotechnology or nanoproducts. Only a few government agencies or organizations from different countries have established standard and regulation to define and regulate the use of nanotechnology. Codex Alimentarius Commission (CAC), an intergovernmental agency established jointly by FAO and WHO, is promoting coordination of all food standards work undertaken by international governmental and non-governmental organizations. In order to prevent regulatory gap, CAC should harmonise international food safety regulation relating to nanoparticles and other nanotechnologies in food and agriculture. The existing regulations of nanotechnology in several representative countries such as the USA, UK, Japan, and mainland China are discussed as follows:

US FDA states that it regulates "products, not technologies". So, this may affect the stage at which the FDA becomes engaged in the regulation of nanotechnology. FDA only regulates certain categories of products and anticipates many nanotechnology products. FDA has traditionally regulated many products with particulate materials in same size range as the cells and molecules nano in size; hence, the existing requirements of FDA as well as the existing battery of pharmacotoxicity tests may be adequate for most nanotechnology products. Accordingly, FDA states that particle size is not the issue. However, a close coordination in knowledge and policy among different government agencies is important for harmonisation of regulation on the use of nanotechnology. FDA is collaborating with the National Institutes of Health; National Institute of Environmental Health Sciences (NIH/NIEHS) on nanotoxicity studies, and is a direct contributor to the evaluations of the toxicity of materials. In 2006, FDA has formed an internal FDA Nanotechnology Task Force for determining regulatory approaches that encourage the continued development of innovative, safe and effective FDA regulated products that use nanotechnology materials.

In UK, the Royal Society and the Royal Academy of Engineering (RS&RAE) have carried out independent study on nanotechnologies and reviewed the benefits and problems associated with those technologies. Health and Safety Executive (HSE) has also participated in the negotiation, agreement, and enforcement of the regulations of manufactured nanoparticles. It has been suggested that the regulations available from the HSE dealing with the harmful chemicals, fire risk, and explosion risk of materials, such as the Approved Code of Practice (ACOP) under the Control of Substances Hazardous to Health Regulations (COSHH), guidance on COSHH risk assessments, ACOP from the Dangerous Substances and Explosive Atmosphere Regulations (DSEAR), and guidance of precautions against the explosion risk of combustible dusts, might be applied for regulating the use of nanomaterials.

Furthermore, it was recommended that chemicals in the form of nanoparticles or nanotubes could be treated as new substances under the chemical safety regulations such as Notification of New Substances (NONS) act and also the forthcoming Registration, Evaluation, and Authorization Chemicals (REACH) regulation from the European Commission. The ideal goal of the REACH system being developed is to ensure that new chemicals be denied a market in the absence of convincing safety data. Nevertheless, whether naturally occurring substances and chemical elements at nanoscale be considered as "new chemicals" or not are important to sustain the safety of these nanomaterials under REACH.

According to the European Inventory of Existing Commercial Chemical Substances (EINECS) under the NONS regulation, nanoscale materials could be regarded as "existing" or "new" substances in relation to their ways of production. Under such a circumstance, the nanoscale titanium dioxide, a "top-down" material, is regarded as an "existing" chemical and is not subject to notification under NONS. On the other hand, carbon fullerene, a "bottom-up" nanomaterial, is regarded as distinct and separate allotrope of carbon, which is not an EINECS entry. Thus, carbon fullerenes and carbon nanotubes would be regarded as "new" chemicals.

In Japan, the development of nanotechnology industry are highly supported by the Japanese government agencies including the Ministry of Education, Culture, Sports, Science, and Technology (MEXT), the Ministry of Economic, Trade, and Industries (METI), the Ministry of Health, Labour, and Welfare (MHLW), and the Ministry of the Environment (MOE). MEXT is responsible for research and development, and is building industry-academia-government cooperative research platforms to promote research and development in nanotechnology and materials science. METI works for the standardization of testing methods for safety evaluation of nanoparticles and MHLW develops the evaluation methods for health impacts of nanomaterials. Therefore, MEXT is responsible to perform researches on basic science for public acceptance of nanotechnology while METI, MHLW, and MOE are working on the regulatory science. Furthermore, the Nanotechnology Researchers Network Center (Nanonet), established by National Institute for Materials Science (NIMS), provides informational support for the development of nanotechnology.

In mainland China, the National Center for Nanoscience and Technology (NCNST) was co-founded by Chinese Academy of Sciences (CAS) and Ministry of Education in March 2003. The NCNST consists of several divisions such as (1) nanoprocessing and nanodevices laboratory; (2) nanomaterials and nanostructures laboratory; (3) nanomedicine and nanobiotechnology laboratory; (4) nanostructure characterization and test laboratory; and (5) coordination laboratory. It has established a public technological platform and engaged basic and applied researches in nanoscience.

In 2005, the Commission on Nanotechnology Standardization affiliated with the NCNST was established and responsible for developing national standards including terminology, protocol, and safety requirements for particle size measurement, nanoprocessing, nanomaterials and nanodevices, nanobiotechnology, and nanomedicines. Based upon the existing China national standard, the commission takes part in guiding and governing the assessment and accreditation of nanoproducts, enabling the nanotechnology industries to improve product quality, reduce health risks, sustain safety, and foster new products development.

In India, it should be noted that the safety researches of nanotechnology and the development of nanoregulation among most countries are still mainly focusing on the non-food nanomaterials and aspects. As the application of nanotechnology in food industry is growing rapidly, it is foreseeable that the safety researches, development of regulatory standards, and public awareness and acceptance of the use of nanotechnology in the food industry will also be coming to the forefront.

9. CERTIFICATION SYSTEM OF NANOPRODUCTS

In 2004, the Industrial Technology Research Institute was commissioned by the Industrial Development Bureau, Ministry of Economic Affairs of Taiwan to implement a certification system for nanotechnology-based consumer products, and hence to validate and issue the world's first certificate "Nano Mark" for the verified nanoproducts. Under this government certificate system, certificate is issued on the basis of promoting industrialization of nanotechnology, encouraging the development of prominent and high standard nanoproducts, protecting the customers, and upgrading the consumers' acceptability of nanoproducts. With the assistance of an Advisory Committee, an on-site evaluation and inspection of the manufacturing plant as well as an assessment on the criteria including particle size, functional properties, processing, quality control, and relevant safety reports and proofs of the products are required.

Products being considered for the certificate of "Nano Mark" should at least conform to the following two basic requirements: (*i*) one of the dimensions of the major functional ingredient or component in the product is within the length scale of approximately 1–100 nm range and (*ii*) the nanoproduct possesses novel properties or improved functionality because of its small size. Until January 2007, there were three categories of nanotechnology products certified and awarded with the "Nano Mark", *e.g.*, antimicrobial photocatalytic fluorescent lamp, antimicrobial photocatalytic tile, and photocatalytic coating for deodorization effects. As yet the main active ingredient in a total of 42 certified nanoproducts among these three categories is nanoscale photocatalytic titanium dioxide. Up to now, all of the certified nanoproducts belong to non-food category. The scale limit (1–100 nm) of the certificate system may possibly confine the product development of nanofoods to a certain extent.

Besides the governmental product certificate system, a non-governmental association "Nano Manufacture and Inspection Union" was established in 2006 by different domestic industries of Taiwan. Unlike the nanoproduct certificate system, the union is a platform to develop analytical techniques, measurement standard, inspection protocol, and standard operation procedure, and also provide services such as nanomaterials processing, analysis, product inspection, information exchange, and technical consultation.

Recently, the union is scheming out an issue to develop an efficient process and standard inspection protocol for the production of fine pearl powder (100–500 nm in size) for which its larger form is a common edible material in the traditional Chinese medicine. The assessment criteria for the submicrosized pearl powder include particle size, heavy metal, nutritional labelling, amino acids profile, functional property, and place of origin. The union regards the micronized pearl powder as a nanoproduct. It is obvious that there is a discrepancy between the government and private union in the recognition of size limit for a nanoproduct. Perhaps, this instance can be used as a reference for considering what the practical particle size range of the micro- or nanosized biological or food products should be.

10. FUTURE USES OF NANOTECHNOLOGY

Nanotechnology is already found in minute particles used in everyday products such as sunscreens. Many sunscreens today contain zinc oxide or titanium oxide nanoparticles that allow the creams to be absorbed into the skin, leaving no whitish residue behind. Nanoparticles are also used in scratch-resistant glass, wrinkle-resistant clothes, and stronger but lighter tennis rackets. In Food, nanotechnology will be used to preserve food for much longer time periods by forming a grid-lock "net" over the food and/or bag that will take air an extremely long time to get through. Nanotechnology will also allow companies and scientists to put any time of vitamin into the food that they want simply by injecting preloaded nanites into the food. In fact, some say that nanotechnology will one day allow us to entirely change the food and drinks that we consume at any time we want. If nano-technology continues to advance at its current pace, we could expect that soon we will be able to create unlimited amount of meat by synthesis at the atomic level, which would eradicate hunger (Carmen *et al.*, 2003).

REFERENCES

Carmen, I.M., Chithra, P.P., Qingrong, H., Paul, T., Sean, L. and Jozef, L.K. (2003). Nanotechnology: A new frontier in food science. *Food Technol.*, 57(12): 24–29.

Chen, L.Y., Remondetto, G.E. and Subirade, M. (2006). Food protein-based materials as nutraceutical delivery systems. *Trends Food Sci. Technol.*, 17: 272–283.

ETC Group. (2005). Nanotech's 'Second Nature' Patents: Implications for the Global South. Canada: June.

ElAmin, A. (2005). Nanotechnology targets new food packaging products. Available at: http://www.foodproductiondailyusa.com/news/ng.asp?id=63147

ElAmin, A. (2006). Nanocantilevers studied for quick pathogen detection. Available at: http://www.foodproductiondaily-usa.com/news/ng.asp?id=70159

ElAmin, A. (2007). The evolution of the nanotech revolution. Available at: www.foodproductiondaily.com/Processing/The-evolution-of-the-nanotech-revolution

Fletcher, A. (2006). Nanotech food conference targets future opportunities. Available at: http://www.foodnavigator.com/news/ng.asp?id=67113

Graveland-Bikker, J.F. and de Kruif, C.G. (2006). Unique milk protein-based nanotubes: Food and nanotechnology meet. *Trends Food Sci. Technol.*, 17: 196–203.

Han, M., Gao, X., Su, J.Z. and Nie, S. (2001). Quantum-dot tagged microbeads for multiplexed optical coding of biomolecules. *Nature Biotechnol.*, 19: 631–635.

Heller, L. (2006). Flavor firm uses nanotechnology for new ingredient solutions. Available at: www.confectionerynews.com/news/ng.asp?id=69008.

IFST. (2006). Nanotechnology. Institute of Food Science and Technology. Available at: www.ifst.org/uploadedfiles/cms/store/ATTACHMENTS/Nanotechnology.pdf

INNI. (2006). Researchers in Israel and U.S. select top four nanotech projects. Israel National Nanotechnology Initiative. Available at: www.nanoisrael.org/download/nanowater1/USIL%20NanoWater%20Workshop%20Press%20Release%20JUL-06.pdf

Lerner, E.J. (2000). Nano is now at Michigan. Medicine at Michigan, Summer issue, pp.14-21. Available at: www.medicineatmichigan.org.

Li, Y., Tseng, Y.D., Kwon, S.Y., Despaux, L., Bunch, J.S. and Mceuen, P.L. (2004). Controlled assembly of dendrimer-like DNA. *Nature Materials*, 3: 38–42.

Liu, X., Wang, J.H. and Yuan, J.P. (2005). Pharmacological and anti-tumor activities of ganoderma spores processed by top-down approaches. *Journal of Nanoscience and Nanotechnology*, 5: 2001–2013.

OilFresh. (2005). OilFresh™ 1000: Frying oil reforming catalytic device. Available at: www.oilfresh.com/of1000.html.

Park, J., Bauer, S., von der Mark, K. and Schmuki, P. (2007). Nanosize and vitality: TiO_2 nanotube diameter directs cell fate. *Nano Lett.*, 7: 1686–1691.

Pitman, S. and Halliday, J. (2006). Inventory groups nanotech supplement launches. Available at: http://www.nutraingredients-usa.com/news/ng.asp?id=66526

Roach, S. (2006). Instant, portable, simultaneous pathogen inspection. Available at: http://www.foodproductiondaily-usa.com/news/ng.asp?id=69938

Sherman, L.M. (2005). Chasing Nanocomposites. Available at: www.ptonline.com/articles/200411fa2.html

Smolander, M., Hurme, E., Koivisto, M. and Kivinen, S. (2004). *PCT International Patent Application* WO 2004/102185 A1, 2004.

Wen, H.W., DeCory, T.R., Borejsza-Wysocki, W. and Durst, R.A. (2006). Development of neutravidin tagged-liposomal nanovesicles as universal detection reagents in Bioassay. *Talanta*, 68: 1264–1272.

24

Natural Preservation of Meat

K.S. Bhuvana, P.K. Mandal and K.P. Mini

ABSTRACT

Microbial growth and oxidative rancidity are the major cause of quality deterioration in meat and meat products. Biopreservatives can be useful in extending shelf life of foods, by reducing spoilage bacteria and protecting from oxidative rancidity. Use of herbs, plants, fruits and vegetables extracts or powders possessing antioxidant and antimicrobial properties have made this possible. This chapter discusses the use of some natural antioxidants and antimicrobials like clove, Ceratonia siliqua essential oil, pomegranate extract, mint and curry leaves, celery powder, leafy green vegetable extracts, rosemary extract, citrus extracts, propolis etc. that can be used in meat. Future studies may be focused on determining the applicability of such systems or their combination to meat and meat products with the objective of replacing or reducing nitrite and other synthetic / chemical preservatives.

1. INTRODUCTION

Microbial growth and oxidative rancidity are the major problems causing quality deterioration in meat and meat products. Bio-preservatives can extend the shelf life of foods by reducing or eliminating spoilage bacteria and protecting overall quality of food products through inhibition of oxidative rancidity. In the present days where use of chemical preservatives are frowned upon by the health conscious population, use of herbs, plants, fruits, and vegetables extracts or powders as preservatives seems to be the most feasible solution. World over, spices and herbs have been traditionally used to provide distinct flavours to food and beverages. Several of them are recognized for their antimicrobial and antioxidant properties, that depend on various factors, including type, composition and concentration of these preservatives, treatment and storage conditions.

Natural antioxidants from herbs, spices and vegetables may have more consumer appeal as these plants are accepted as healthy and safe for consumption. The recent revival of consumer interest in the therapeutic properties of herbs has led to an increase in consumer demand for herbal based food products. The Western society appears to be experiencing a trend of 'green' consumerism desiring fewer synthetic food additives and products with a smaller impact on the environment. Essential oils and other components of herbs and spices are widely used in the food preparation and are classified as GRAS (generally recognized as safe). The use of such naturally occurring plant products may provide an additional barrier to the growth of food borne pathogens. It has been evident that these spices and herbs may play more than one role in foods to which they are added. In addition to imparting flavour, certain spices prolong the storage life of foods by bacteriostatic or bacteriocidal activity, while others prevent rancidity by their antioxidant activity.

It is envisaged that the use of phytochemicals from plants in meat based products will be an interesting area for the meat industry to explore as these functional extracts with anti-oxidative and antimicrobial properties are able to replace the current usage of synthetic food additives in the development and improvement of healthy food products. Though spices and herbs and their oil fractions are used in foods as flavoring compounds, their antimicrobial activities are said to be too low to play a significant role as preservatives at the amounts acceptable by consumers (Shelef, 1983).

The importance of interactions of various antimicrobial agents to maintain safe and stable products has been widely recognized (Roberts, 1989; Kabara, 1993). Often, combinations of antimicrobial substances reduce effective concentrations of each antimicrobial component required to induce antimicrobial activity (Raccach, 1984). Kabara (1993) proposed a new approach to food preservation by formulation of 'preservative systems' that utilize the multiple effects of several natural chemicals simultaneously to create an environment hostile to a broad spectrum of organisms. The next interesting thing to do with the scientific knowledge obtained would be to introduce calculated safe dosage of these plant extracts in the form of paste, marinades and seasonings in meat based food system.

2. BIOPRESERVATIVES FROM HERBS AND SPICES

In view of present national and international barriers regarding the use of chemical food additives in food processing and preservation, search for biological and plant origin food additives have notably increased for the production of high-quality, ready-to cook or ready-to-eat products in recent years. Consumers' interest in meat products formulated with natural antioxidants has motivated researchers to evaluate the effectiveness of naturally occurring compounds in plant leaves and herbs for functional purposes. Utilization of natural preservatives offers adequate microbiological safety, replaces the toxic effect of chemical preservatives and contributes health promoting effect hence making the food functional. However, crude extracts of many spices and herbs at higher level impart an objectionable colour, odour and taste to food (Draughon, 2004). The use of spices with other food ingredients, such as sodium chloride, sugar and organic acid might provide a synergistic effect in controlling microbial growth.

Extensively studied sources of natural preservatives are fruits and vegetables, seeds, cereals, berries, wine, tea, onion bulbs, olive oil and aromatic plants. Attempts are also made to identify and evaluate agricultural by-products, ethnic and traditional products, herbal teas, cold pressed seed oils, exudates resins, hydrolysis products, not evaluated fruits and edible leaves and other raw materials rich in antioxidant phenols that have nutritional importance and/or the potential for applications in the promotion of health and prevention against damages caused by free radicals.

Essential oil, a complex mixture of compounds, has received a great interest for their potential as alternative medicine and natural food preservatives (Alim *et al.*, 2009). Commercial essential oil (EO) has been classified as GRAS at low concentration for food application based on the 21 Code of Federal Regulations (CFR, 2003; Burt, 2004). The antifungal and antibacterial effects of EO on different microorganisms have been reported in several studies (Chanthaphon *et al.*, 2008).

Essential oils *in vitro* studies have demonstrated antibacterial activity against *Listeria monocytogenes*, *Salmonella typhimurium*, *Escherichia coli* O157:H7, *Shigella dysentriae*, *Bacillus cereus* and *Staphylococcus aureus*. Gram-negative organisms are slightly less susceptible than Gram-positive bacteria. A number of essential oil components have been identified as effective antibacterials, *e.g.*, carvacrol, thymol, eugenol, perillaldehyde, cinnamaldehyde and cinnamic acid. However, higher concentrations are needed to achieve the effect in foods such as fresh meat, meat products, fish, milk, dairy products etc. Essential oils comprise a large number of components and it is likely that their mode of action involves several targets in the cell. The hydrophobicity of essential oils enables them to partition in the lipids of the cell membrane, rendering them permeable and leading to leakage of cell contents. The action of essential oils is improved by physical conditions like low pH, low temperature and low oxygen levels. Synergism has been observed between carvacrol and its precursor *p*-cymene and between cinnamaldehyde and eugenol. There is a synergy observed between essential oil components and mild preservation methods. Some components are legally registered flavourings in the EU and the USA. Undesirable organoleptic effects can be limited by careful selection of essential oils according to the type of food.

2.1. Clove Oil

Clove is a naturally occurring spice, which has been shown to possess antioxidant and antimicrobial activity in food products. The oil of clove is used in dentistry as a local antiseptic. The terpenoids in clove are thought to exert inhibitory action against microorganisms by membrane disruption (Lambert *et al.*, 2001). Its inhibitory action against various gram positive and gram negative bacteria in meat and meat products is also recorded (Mannie, 1999).

Eugenol, the essential oil of clove, is a 2-methoxy phenolic derivative, which has been reported to possess 90% of the antioxidant activity of butylated hydroxy anisole (Cort, 1974). Al-Jalay *et al.* (1987) reported the highest antioxidant index for clove followed by rose petals, cinnamon, nutmeg and other spices in pastourma sausages. They further reported that clove was the strongest antioxidant in an oil/water emulsion. However, to have an inhibitory effect on microorganisms higher levels of cloves are required in the formulation (Mannie, 1999).

2.2. *Ceratonia siliqua* Essential Oil

Ceratonia siliqua is a food and medicinal plant widely distributed in Tunisia. Chemical composition, cytotoxicity and antimicrobial activity of *Ceratonia siliqua* essential oil (CsEO), with preservative effects against *Listeria* inoculated in minced beef have been reported. A study described the phytochemical role and the protective effects of CsEO. Twenty different components were identified in the CsEO, among them, the major components were: lonadecane, heneicosane, naphthalene, 1, 2-benzenedicarboxylic acid dibutyl-ester, heptadecane, hexadecanoic acid, octadecanoic acid, 1, 2-benzenedicarboxylic acid, phenyl ethyl tiglate, eicosene, farnesol 3, camphor, nerolidol and n-eicosane. The antimicrobial activity of CsEO was evaluated against a panel of 13 bacteria and 8 fungal strains including *Aspergillus* species such as *A. niger, A. flavus, A. fumigatus* and *A. nidulans* which are responsible for spoilage of many foods. Results have shown that CsEO exhibited moderate to strong antimicrobial activity against the tested species.

The inhibitory effect of CsEO was evaluated *in vivo* against the food borne pathogen *Listeria monocytogenes*, experimentally inoculated in minced beef meat (2×10^2 CFU/g of meat) amended with different concentrations of the CsEO and stored at 7°C for 10 days. The antibacterial activity of CsEO in minced beef meat was clearly evident and its presence led to a strong inhibitory effect against the pathogens at 7°C. The *in vivo* application of CsEO was found to be effective in controlling *L. monocytogenes* in minced beef. This essential oil could be used for the preservation of meats against *L. monocytogenes* and for increasing their shelf life in foods. On the other hand, the cytotoxic effects of the essential oil against two tumoral human cell lines HeLa and MCF-7 were examined by MTT assay. The results suggest that the EO of *C. siliqua* possesses antimicrobial and cytotoxic properties, and is therefore a potential source of active ingredient for food and pharmaceutical industry. The bioprotective effect exhibited by the CsEO could therefore be used as natural preservative (Hsouna *et al.*, 2011).

2.3. Celery Powder

One of the most recent and novel developments in the use of celery powder and similar natural nitrite sources for meat products is not for development of cured meat properties, but rather for development of fresh meat color in low-oxygen package environments. A recent patent granted to Tyson Foods, Inc. (Baublitis & Sawyer, 2011) describes the use of celery powder for surface application on fresh meat that is packaged in vacuum or in low-oxygen modified atmosphere. The result is development of nitrosylmyoglobin on the product surface, which is a bright cherry red color that is visually indistinguishable from the oxymyoglobin color of fresh meat in an aerobic environment. The vacuum or low-oxygen package environment achieves a much longer shelf life for these products than for products exposed to an aerobic environment.

2.4. Mint and Curry Leaves

Curry and mint leaf extracts have a substantial amount of phenolic compounds and significant superoxide anionic scavenging activity and demonstrated to have good antioxidant effects. Both these extracts significantly reduced auto-oxidation, and salt promoted lipid oxidation in raw ground pork meat during refrigerated storage. Among the two different extracts, the curry leaf extract was more effective in reducing malonaldehyde formation. Curry leaf and mint leaf extracts could be successfully added to raw ground pork meat to function as natural antioxidants with added health benefits and increasing consumer appeal (Biswas *et al.*, 2012). Restructured chicken slices prepared with leaf powders (drumstick, mint and curry leaves) showed significantly lower microbial counts compared to both control and BHT (200 ppm) added products (Najeeb, 2012).

3. BIO-PRESERVATIVES FROM FRUITS AND LEAVES

Plant extracts have been traditionally used in folk medicine and also as additives to extend the shelf life of food (Bagamboula *et al.*, 2004). Most of their properties are due to secondary metabolites such as the volatile constituents.

3.1. Pomegranate Extract (P.E.)

Pomegranate fruit contains many different kinds of polyphenolic antioxidants and commercial pomegranate juice has been shown to possess antioxidant activity three times higher than those of red wine and green tea (Gil *et al.*, 2000). The soluble polyphenolic content of pomegranate juice includes punicalagins, anthocyanins, catechins, ellagic tannins, and gallic and ellagic acids (Dornfeld *et al.*, 2000; Gil *et al.*, 2000). A linear relationship between total phenolic content, antioxidant capacity and antibacterial activity against several microorganisms has been shown (Shan *et al.*, 2007). Antimicrobial activities of phenolic compounds may involve multiple modes of action. The type of the microorganism and its cell wall structure are thought to play an important role (Shan *et al.*, 2007). Phenolic compounds can denature enzymes (Furneri *et al.*, 2002) but they can also bind to substrates such as minerals, vitamins and carbohydrates making them unavailable for microorganisms

(Stern *et al.*, 1996; Shahidi & Naczk, 2004). Furthermore, phenols can be absorbed to the cell wall, resulting in disruption of the membrane structure and function (Hugo & Bloomeld, 1971).

The antibacterial properties of *Punica granatum* have been reported extensively for several pathogens such as *Escherichia coli, Staphylococcus aureus,* including several methicillin resistant strains, *Vibrio cholerae* and *Bacillus subtilis* (Guevara *et a1.*, 1994; Machado *et al.*, 2002; Braga *et al.*, 2005; Melendez & Capriles, 2006). Ellagic and gallic acids are two of the well known phenolic constituents of pomegranate fruit (Nasr *et al.*, 1996; Poyrazoglu *et al.*, 2002; Chidambaramurthy *et al.*, 2002; Zhou *et al.*, 2008; Martin *et al.*, 2009) and both have shown antibacterial properties (Naz *et al.*, 2007; Zambuchini *et al.*, 2008). Anti-listerial activity of the two pure compounds gallic and ellagic acid was tested and was found that the application of pure components rather than the complex compound like PE might be more practical in sense of consistency and characteristics of the compound.

3.2. Citrus Fruit Extracts

Citrus essential oils are the most widely used essential oils in the world (Bousdia *et al.*, 2009). They are extracted from the peel of citrus fruits (Dharmawan *et al.*, 2009). The phenolics that occur in citrus oil include the flavonoids (flavanones, flavones, and flavonols), anthocyanins and coumarins among others (Berhow *et al.*, 1996). Flavonoids have shown antimicrobial activity in several studies (Tim & Lamb, 2005). Citrus essential oils have been found to inhibit Gram-positive and Gram-negative bacteria as well as yeasts, moulds and food poisoning bacteria (Deans & Ritchie, 1987) both in the form of liquid and vapour.

3.3. Leafy Green Vegetable (LGV) Extracts

LGV are widely consumed in East Asia and are prepared using various cooking methods. These vegetables have been eaten for centuries and are classied as GRAS. An abundance of research has shown that fresh LGV contain important functional components, such as b-carotene, ascorbic acid, riboflavin, and folic acid, as well as minerals (Grusak & DellaPenna, 1999). In particular, these foods contain a large amount of polyphenols (*e.g.*, phenolic acids, flavonoids, and aromatic compounds), the most abundant phytochemicals in the human diet. LGV are also known for their characteristic color, flavor, and therapeutic value (Faller & Fialho, 2009; Gupta *et al.*, 2005). Their bioactive substances and phytonutrients have a wide range of biological functions, including antioxidant and antimicrobial activities (Burt, 2004; Gutierrez *et al.*, 2008). Some epidemiological evidence has shown that consumption of fruits and vegetables with high levels of natural antioxidants helps to prevent chronic diseases, such as cardiovascular diseases and cancer (Alia *et al.*, 2003). Consequently, LGV have received substantial attention from researchers in recent years as a potential source of natural antimicrobial and antioxidant agents.

As opposed to synthetic compounds, natural preservatives obtained from edible plants are rich in phenolic compounds, as mentioned above, and can enhance overall quality through decreases in lipid oxidation and microbial growth. However, many LGV are only available seasonally, and the scientific literature contains little information regarding their nutrient contents and bioactivities.

3.4. Rosemary Extract

Rosemary is composed of carnosol, carnisic acid, rosmaric acid, urosolic acid, betulinic acid, a-pinene, camphor and bornyl acetate. Rosemary extract also possesses antioxidant and antimicrobial properties. The di-terpenes, carnosol and carnosic acid, have been shown to account for the antioxidant properties of rosemary (Offord, 2004; Riznar *et al.*, 2006). The phenolic compounds in rosemary extract are believed to enhance antimicrobial properties by affecting the function of the bacterial cellular membrane, the synthesis of DNA, RNA, protein and lipids, and the function of the mitochondrion (Raccach, 1984; Belantine *et al.*, 2006). Although rosemary extract was not identified as a good antimicrobial, its antimicrobial and antioxidant properties were improved when used in combination with sulpher dioxide (Mathenjwa *et al.*, 2012).

4. BIOPRESERVATIVES FROM ANIMAL AND MICROBIAL ORIGIN

4.1. Chitosan

Chitosan, a novel alternative preservative with GRAS status in the US, is a deacetylated form of chitin derived from the shell of crabs and shrimps and the cell wall of fungi (Roller *et al.*, 2002; Soultos *et al.*, 2008; Aldemir & Bostan, 2009). In meat, its combination with various plant preservatives has proved to be useful in prolonging the shelf life. Boerewors, a South African fresh sausage conventionally preserved with 450 mg/kg sulphur dioxide (SO), combinations of Chi+SO and Ros+Chi (Ros; 260 mg/kg) and chitosan (Chi;10 g/kg) showed comparable antimicrobial and antioxidant properties to the conventionally preserved sausages. Kanatt *et al.* (2008) studied the effectiveness of combinations of chitosan and mint for meat and meat products. They reported both antimicrobial as well as antioxidant activities of the combination. Antioxidant properties were attributed to the mint extract while antimicrobial activities were attributed to chitosan. Chitosan-mint mixtures (0.05%) inhibited growth of *S. typhimurium, Pseudomonas, E. coli, B. cereus* and *S. aureus* during storage at 0–3°C for 28 days.

4.2. Propolis

Propolis is a resinous natural product, produced by bees (*Apis mellifera*) from vegetable secretions. Bees use propolis mainly to cover the hive interior and the breeding cells and also to repair cracks and fissures. These uses are significant, because propolis avoids hive colonization with diseases. Propolis is increasingly recognized by their antibacterial and antifungal properties. Its chemical composition is complex and varies according to its botanical and phyto-geographical origin (Greenaway *et al.*, 1991; Bonvehi *et al.*, 1994; Bankova & Marcucci, 2000). It contains a wide variety of substances such as polyphenols, quinones, coumarins, steroids, amino acids and inorganic compounds. Most propolis components are phenolic in nature, mainly flavonoids. It is known that simple phenols, phenolic acids and polyphenols are active antimicrobial agents (Cowan, 1999). Similarly, gallic acid derivatives have shown inhibitory effects against Gram-positive and Gram-negative bacteria (Kayser & Kolodziej, 1997).

Flavonoids are synthesized by plants as a response to microbial infections and are recognized to have effective antimicrobial effects against a wide range of microorganisms (Recio *et al.*, 1989). It was demonstrated that the three –OH group substitution in ring B, and a third –OH group in ring C constitute the necessary structures to *E. coli* antibacterial activity of flavonoids, as is the case of myricetine, present in propolis (Mori *et al.*, 1989; Pimia *et al.*, 2001; Farre *et al.,* 2004). The bactericidal or bacteriostatic effects depend on the concentration of propolis extract, and are influenced by the extraction method (Fuentes & Rojas, 1990). Propolis was tested as food preserver, due to its recognized bactericidal and bacteriostatic properties. Propolis' antioxidants, antibacterial and antifungal properties make it useful in food technology.

4.3. Lactic Acid Bacteria (LAB)

LABs are used as natural or selected starters in food fermentations in which they bring about acidification due to production of lactic and acetic acids flavours. Protection of food from spoilage and pathogenic microorganisms by LAB is through producing organic acids, hydrogen peroxide, diacethyl (Messens & De Vugst, 2002), antifungal compounds such as fatty acids (Corsetti *et al.*, 1998) or phenullactic acid and/or bacteriocins (De Vugst & Vandamme, 1994). LAB play an important role in food fermentation as the products obtained with their aid are characterized by hygienic safety, storage stability and attractive sensory properties.

The preservative ability of LAB in foods is attributed to their competition for nutrients and to the production of antimicrobial metabolites including organic acids, hydrogen peroxide, enzymes and bacteriocins. Bacteriocins are a heterogeneous group of peptides and proteins ribosomally produced that kill or inhibit the growth of close related micro organisms such as food pathogens (*Listeria monocytogenes* and *Clostridium*) and other meat spoilage bacteria such as *B. thermosphacta*. The use of bacteriocins in biopreservation is a growing area of research. Hence, numerous bacteriocins have been isolated over the past three decades. Furthermore, LAB contribute to the flavour of the final product through the formation of noticeable acidic and vinegary tastes. Production of organic acids is the determining factor on which the shelf life and the safety of the final product depend (Olaoye *et al.*, 2008; Olaoye & Onilude, 2009).

5. USE OF NATURAL ANTIOXIDANTS IN MEAT

Including natural antioxidants in the diet has beneficial effects on human health because they protect the biologically important cellular components, such as DNA, proteins, and membrane lipids, from reactive oxygen species (ROS) attacks (Su *et al.*, 2007). Locally resourced natural antioxidants may facilitate towards the formulations of safer alternatives of antioxidants in food systems. The natural antioxidants present in fruits and vegetables could scavenge free radicals and provide oxidative stability to many food items including high fat meat products (Yogesh *et al.*, 2012). However, screening trial of locally grown and consumed herbs and vegetables is of priority in their selection for further use.

5.1. Thuja and Peach

Thuja (*Thuja occidentalis*), cones extract (TCE) and peach (*Prunus persia*) seeds extract (PSE) contain high levels of bioactive phenolics, flavonoids and other free radical scavengers that can help to control lipid oxidation. A study showed that TCE and PSE had effective antioxidant activity in raw ground chicken meat during refrigerated storage because use of these extracts inhibited the formation of lipid peroxide and thiobarbituric acid reactive substances in ground chicken meat. These extracts also positively affected water holding capacity and cooking loss (Yogesh & Ali, 2012).

5.2. Banana and Sapodilla

In another study, antioxidant properties of banana (*Musa paradisiaca*) and Sapodilla/Chikoo (*Manilkara zapota*) peel extracts in chicken patties were evaluated. Banana and sapodilla peel extracts significantly reduced the lipid oxidation in chicken meat during refrigerated storage and antioxidant effect was comparable to that of BHT. Therefore, extracts of these fruits' by-products could be successfully added to meat to function as an antioxidant (Devatkal *et al.*, 2011).

5.3. Curry leaf, Fenugreek and Tomato

Curry leaf and fenugreek extracts showed significant level of total phenolics and antioxidant properties reducing the lipid oxidation in chicken meat during refrigerated storage. Further, it was observed that antioxidant effect of these extracts was higher than that of BHT (Devatkal *et al.*, 2011). Tomato was utilized as powder up to 1.5% in low fat pork sausages and was found to be well acceptable up to 30 days at refrigerated storage. It showed a lower lipid oxidation, lightness, pH, hardness values and higher redness, yellowness, chroma, hue angle and water holding capacity values compared to control during refrigerated storage. It could also be a source of lycopene which acts as a functional additive in human diet (Kim *et al.*, 2011).

5.4. Citrus Extracts

A study by Mexis *et al.* (2012) reported that the addition of citrus extracts had a small preservative effect on fresh ground chicken meat with shelf life extension by 2 days, while the use of the oxygen absorber increased product shelf life by 3 days. The strongest preservative effect was achieved by combining the oxygen absorber with the citrus extract (0.1 ml/100 g) with shelf life extension of 5 days as compared to control samples, which indicates the additive action between the two.

Vitamin C has been used as a steak surface treatment either alone or in combination with other antioxidants to stabilize meat colour during display (Mitsumoto *et al.*, 1991). Various natural antioxidants have also been used along with ascorbic acid for preventing food oxidation (Shahidi & Wanasundara, 1992). Depending on conditions, ascorbic acid can act as a prooxidant or an antioxidant (Elliott, 1999).

5.5. Miscellaneous

The antioxidant effects of oregano essential oil (Martos V *et al.*, 2010), grape seed oil (Ayerdi *et al.*, 2009), citrus fruit by-products (Devatkal & Naveena, 2010) and curry leaf powder (Biswas *et al.*, 2006) have been investigated for their use in meat products. The antioxidant effects of grape seed, oregano extract, and rosemary in frozen vacuum packaged beef and pork was evaluated by Rojas and Brewer (2008). More recently Devatkal and Naveena (2010) evaluated the antioxidant effect of kinnow and pomegranate fruit by-product powders on the lipid stability and colour of raw ground goat meat. The available reports indicate widespread use of modern preservative techniques with a renewed interest in the antimicrobial activities of spice in the last decade.

6. USE OF NATURAL ANTIMICROBIAL AGENTS IN MEAT

Addition of antimicrobial ingredients coupled with refrigeration of food materials was found to provide more protection to the stored food samples. A novel naturally occurring antimicrobial may not serve as an alternative to the use of nitrite in meat products because of the broad spectrum of activity of nitrite and a combination of antimicrobials is required (Sofos, 2008). Ntzimani *et al.* (2010) added EDTA, lysozyme, rosemary and oregano oil to chicken meat stored in vacuum packages at 4°C. They showed that combinations were effective against the growth of Gram-positive bacteria, Gram-negative bacteria and to a lesser degree, yeasts. Combinations of nisin with oregano oil were tested in minced sheep meat during refrigerated storage (Govaris *et al.*, 2010). Growth of *Salmonella enteritidis* was inhibited after the addition of 500–1000 IU/g of nisin and 0.6–0.9% oregano oil.

6.1. Lactic Acid

Lactic acid, a GRAS antimicrobial agent is commonly used in meat and meat products for decontamination. However, discolouration continues to be a major problem associated with organic acid decontamination of meat cuts especially at higher levels (Kotula & Thelappurate, 1994), and improving the colour stability of meat is a matter of concern to retailers.

6.2. Rosemary

Extracts of rosemary (*Rosmarinus officinalis* L.) have been used in preserving meat by way of its addition in the meat products as well as by dietary supplementation to the animal. The latter has shown to improve the functional and nutritional qualities of meat, along with tea catechins, it improved color and lipid stability in beef patties (Grady *et al.*, 2006). Distillate from rosemary leaves fed to pregnant ewes produced lambs with better scores for meat and fat color (Nieto *et al.*, 2010).

6.3. Leafy Green Vegetables

The antimicrobial activity of leafy green vegetable extracts against microbial growth in meat was found to be most effective on coliforms. During storage, a reduction was seen in not only the rate of increase, but also in the coliform counts in the treated meat samples. The addition of natural extracts did not significantly affect lactic acid bacterial (LAB) counts, yeast and mold counts showed a similar tendency with total viable count (Kim *et al.*, 2012).

6.4. Propolis

Substances, which are identified in propolis, generally are typical constituents of food and/or food additives, and are recognized as GRAS substances (Burdock, 1998). It is an interesting alternative to be considered in new applications of food technology. It has been proposed as a chemical preservative in meat products (Han & Park, 1995) and as germicide and insecticide for food packaging (Mizuno, 1989a, 1989b). Donadieu (1979) reported an extension of frozen storage life of fish by 2–3 times. It is shown that it also successfully inhibits the *E. coli* development *in vitro*, at safe levels for human consumption and, consequently, they could be useful as ground fresh beef natural preserver or as unspecific antibacterial food preserver (Tosi *et al.*, 2007).

6.5. Pomegranate

The use of pomegranate extract in meat and meat products was directed towards its action against *Listeria monocytogenes*, an organism of an important public health concern. The potential of pomegranate extract as an anti-listerial agent in RTE products was evaluated using meat paté (Hayrapetyan *et al.*, 2012). The results indicated that the pomegranate extract has a potential to be used as a natural preservative in meat products. Further research could investigate the extract's effect in combination with other hurdles currently in use, which may enhance its effect. Maintenance of cold chain will be important especially taking into account that the inhibitory effect of the extract was temperature dependent. The PE may serve as a good bio-preservative, working both as an antioxidant and antimicrobial compound (Hayrapetyan *et al.*, 2012).

Recent developments in controlling the growth of microorganisms in other products by the use of delivery systems for antimicrobials should be taken into account. Studies have demonstrated that problems of incompatibilities of antimicrobials with food matrix properties can be overcome by delivering naturally occurring antimicrobials in nano and micro scalar carrier systems such as emulsion, liposomes, microemulsions, hydrogel particles and others. Future studies may thus be focused on determining the applicability of such systems to meat and meat products with the objective of replacing or reducing nitrite (Weiss *et al.*, 2010).

7. USE OF ESSENTIAL OILS IN MEAT PRODUCTS

It has been shown that essential oils from plants are useful for reduction of pathogens associated with meat products. However, some authors have recorded low antimicrobial effects against pathogens in contaminated meat products (Grosso *et al.*, 2008). High-fat-content of the food materials has effects on the application of EOs (Burt, 2004; Lis-Balchin *et al.*, 2003). It may be because of the lipid solubility of EOs compared to aqueous parts of food. Combination of 1% cloves and oregano in broth culture showed inhibitory effect against *L. monocytogenes*, however, the same concentration was not effective in meat slurry (Balchin *et al.*, 2003).

Recent studies regarding certain oils such as eugenol and coriander, clove, oregano and thyme oils showed high effect against *L. monocytogenes, Aeromonas hydrophila* and spoilage ora in meat products. However, mustard, cilantro, mint and sage oils were less effective or ineffective (Burt, 2004). Survival and growth of both susceptible and antibiotic-resistant *Campylobacter* strains have been inhibited effectively on agar plates and in contaminated ground beef by application of roselle (*Hibiscus sabdariffa* L.) (Yano *et al.*, 2006; Yin & Chao, 2008). EOs' activity in meat products can be reduced by high levels of fat. Encapsulated rosemary EOs showed better antimicrobial effect compared to standard rosemary EOs against *L. monocytogenes* in pork liver sausage; neither the direct effect of the level used nor manner and effect of encapsulation are reported (Carraminana *et al.*, 2008).

7.1. Oregano, Savory and Sage

The activity of oregano EOs against *Clostridium botulinum* spores in combination with low levels of sodium nitrite enhanced the delay of growth of bacteria more than the sodium nitrite alone, depending on the number of inoculated spores

(Burt, 2004). EO extracted from the aerial parts of cultivated *Salvia officinalis* L. was effective against *Salmonella enteritidis* (Hayouni *et al.*, 2008). Winter savory (*Satureja montana*) EOs in combination with other preservation methods such as reduced temperature, pulsed light, high pressure, pulsed electric and magnetic fields, irradiation, or packaging under a modied atmosphere can be utilized as an economical natural antibacterial substance to control growth of food borne bacteria and improve quality of minced pork (Carraminana *et al.*, 2008). Chinese cinnamon and winter savory EOs were used successfully to increase radio sensitivity in ground beef (Turgis *et al.*, 2008). Combinations of EOs and nisin showed enhanced antimicrobial activities against *L. monocytogenes* (Solomakos *et al*, 2008). Rancidity of heat-treated turkey-meat products was inhibited by aqueous extract of rosemary, sage and thyme (Mielnik *et al.*, 2008). Treatments (homogenization) of meat with oregano and sage EO significantly reduced oxidation, while the heat treatment and storage time significantly affected the antioxidant activity of the meat (Fasseas *et al.*, 2007).

7.2. Clove and Rosemary

Although numerous studies have been done *in vitro* to evaluate the antioxidant and antimicrobial activity of clove, only a few studies have been done with food products. However, use of clove oil along with lactic acid may give synergistic antioxidant/antimicrobial effect and inclusion of Vitamin C also might stabilize the colour. Improvement of shelf-life of buffalo meat steaks using lactic acid, clove oil and vitamin C during retail display at 4 ± 1°C was shown with either lactic acid + clove or lactic acid + clove + vitamin C. Use of these combinations appeared to be advantageous over lactic acid alone (Naveena *et al.*, 2006).

Individual extracts of clove, rosemary, cassia bark and liquorice demonstrated strong antimicrobial activity; but the mixture of rosemary and liquorice extracts was the best inhibitor against all four types of microbes (*L. monocytogenes, Escherichia coli, Pseudomonas fluorescens* and *Lactobacillus sake*) in modified atmosphere-packaged fresh pork and vacuum-packaged ham slices stored at 4°C (Zhang *et al.*, 2009). *Santalum album, Cinnamomum cassia,* and *Artemisia capillaris* were the most effective antimicrobials in raw sheep meat (Luo *et al.*, 2007). There are potential bio-preservative capabilities for application of clove and tea-tree oils to control *Escherichia coli* O157:H7 on blanched spinach and minced cooked beef (Moriera *et al.*, 2007).

7.3. Use of Essential Oils in Packaging

Milk protein-based edible films containing oregano, pimento or oregano/pimento were effective against Escherichia coli O157:H7 or *Pseudomonas* sp. on meat surfaces (Oussalah *et al.*, 2004; Melgar *et al.*, 2008a; 2008b). Chlorophyll-containing films were tested against *Staphylococcus aureus* and *Listeria monocytogenes*, showing that it was possible to reduce growth of these micro-organisms inoculated in cooked frankfurters by covering the frankfurters with sodium magnesium chlorophyll in gelatin films and coatings (López *et al.*, 2008).

Marjoram (*Origanum majorana* L.) EO was effective against several species of bacteria in fresh sausage (Busatta *et al.*, 2008). Chlorophyllins are semi-synthetic porphyrins obtained from chlorophyll. Chlorophyllin-gelatin films and coating applications successfully reduced *Staphylococcus aureus* and *L. monocytogenes* in cooked frankfurter (López-Carballo *et al.*, 2008).

8. FUTURE RESEARCH NEED

The recent revival of consumer interest in the therapeutic properties of herbs has led to an increase in consumer demand for herbal based food products. Owing to their antioxidant and antimicrobial properties, essential oils and other components of herbs and spices are widely used as food bio-preservatives and therefore can be useful in extending shelf life of foods. Further research could investigate the extracts' effect in combination with other hurdles currently in use, which may enhance its effect. Also, various plants, herbs present in nature with potential preservative actions in food systems are yet to be discovered. The dearth of systematic research and availability of relevant toxicology data may be the only limiting factor in their utilization. Future studies may be focused on extensive work determining the applicability of such systems to meat and meat products with the objective of replacing or reducing nitrite and other synthetic/chemical preservatives.

REFERENCES

Aldemir, T. and Bostan, K. (2009). Effects of chitosan on the microbiological quality of ready to cook meatball. *Journal of the Faculty of Veterinary Medicine Istanbul University*, 35: 13–21.

Alia, M., Horcajo, C., Bravo, L. and Goya, L. (2003). Effect of grape antioxidant dietary ber on the total antioxidant capacity and the activity of liver antioxidant enzymes in rats. *Nutr. Res.*, 23(9): 1251–1267.

Alim, A., Goze, I., Goze, H.M., Tepe, B. and Serkedjieva, J. (2009). *In vitro* antimicrobial and antiviral activities of the essential oil and various extracts of *Salvia cedronella*, Boiss. *J. Med. Plants Res.*, 3: 413–419.

Bagamboula, C.F., Uyttendaele, M. and Debevere, J. (2004). Inhibitory effects of thyme and basil essential oils, carvacrol, thyme, estragol, linalool and p-cymene towards *Shigella zonnei* and *S. exneri*. *Food Microbiol.*, 21: 33–42.

Bankova, V. and Marcucci, M.C. (2000). Standardization of propolis: Present status and perspectives. *Bee World*, 8(4): 182–188.

Belantine, C.W., Crandall, P.G., O'Bryan, C.A., Duong, D.Q. and Pohlman, F.W. (2006). The pre- and post-grinding application of rosemary and its effects on lipid oxidation and colour during storage of ground beef. *Meat Sci.*, 73: 413–421.

Berhow, M., Tisserat, B., Kanes, K. and Vandercook, C. (1996). Survey of phenolic compounds produced in citrus. U.S. Department of Agriculture, Agricultural Research Service. Technical Bulletin No. 1856.

Biswas, A.K., Kondaiah, N. and Anjaneyulu, A.S.R. (2006). Effect of spice mix and curry (*Murraya koenigii*) leaf powder on the quality of raw meat and precooked chicken patties during refrigeration storage. *J. Food Sci. Technol.*, 43(4): 438–441.

Biswas, A.K., Chatli, M.K. and Sahoo, J. (2012). Antioxidant potential of curry (*Murraya koenigii* L.) and mint (*Mentha spicata*) leaf extracts and their effect on colour and oxidative stability of raw ground pork meat during refrigeration storage, *Food Chem.*, 133: 467–472.

Bonvehi J.S., Coll F.V., Jorda R.E. (1994). The composition, active components and bacteriostatic activity of propolis in dietetics. *J. Am. Oil Chem.Soc.*, 71: 529–532.

Bousdia, N., Vian, M., Ferhat, M., Meklati, B. and Chemat, F. (2009). A new process for extraction of essential oil from Citrus peels: Microwave hydrodiffusion and gravity. *J. Food Eng.*, 90(3): 409–413.

Braga, L.C., Shupp, J.W., Cummings, C., Jett, M., Takahashi, J.A. and Carmo, L.S. (2005). Pomegranate extract inhibits *Staphylococcus aureus* growth and subsequent enterotoxin production. *J. Ethnopharmacol.*, 96(1–2): 335–339.

Burdock, G. (1998). Review of the biological properties and toxicity of bee propolis. *Food Chemical Toxicol.*, 36: 347–363.

Burt, S. (2004). Essential oils: Their antibacterial properties and potential applications in foods a review. *Int. J. Food Microbiol.*, 94: 223–253.

Busatta, C., Vidal, R.S., Popiolski, A.S., Mossi, A.J., Dariva, C. and Rodrigues, M.R.A. (2008). Application of *Origanum majorana* L. essential oil as an antimicrobial agent in sausage. *Food Microbiol.*, 25(1): 207–211.

Carraminana, J.J., Rota, C., Burillo, J. and Herrera, A. (2008). Antibacterial efciency of spanish *Satureja montana* essential oil against *Listeria monocytogenes* among natural ora in minced pork. *J. Food Prot.*, 71(3): 502–508.

Chanthaphon, S., Chanthachum, S. and Hongpattarakere, T. (2008). Antimicrobial activities of essential oils and crude extracts from tropical Citrus spp. against food-related microorganisms. Songklanakarin. *J. Sci. Tech.*, 30: 125–131.

Chidamabaramurthy, K.N., Jayaprakasha, G.K. and Singh, R.P. (2002). Studies on antioxidant activity of pomegranate peel extracts *in vivo* models. *J. Agric Food Chem.*, 50(17): 4791–4795.

Corsetti, A., Gobbetti, M. and Smacchi, E. (1996). Antibacterial activity of sour dough lactic acid bacteria: Isolation of a bacteriocin like inhibitory substance from *Lactobacillus sanfrancisco* C57. *Food Microbiol.*, 13: 447–456.

De Vugst, L. and Vandamme, E.J. (1994). Bacteriocins of lactic acid bacteria, microbiol, Genet Appl. London: Blackie Acad and professional. ISBND-75 140174-9.

Cort, W.M. (1974). Hemoglobin peroxidation test screens antioxidants. *Food Technol.*, 28: 60–66.

Cowan, M.M. (1999). Plant products as antimicrobial agents. *Clinical Microbiological Review*, 12(4): 564–582.

Deans, S.G. and Ritchie, G. (1987). Antibacterial properties of plant essential oils. *Int. J. Food Microbiol.*, 5(2): 165–180.

Devatkal, S.K. and Naveena, B.M. (2010). Effect of salt and pomegranate fruit byproduct powders on colour and oxidative stability of raw ground goat meat during refrigerated storage. *Meat Sci.*, 85: 306–311.

Devatkal, S.K., Kamboj, R. and Paul, D. (2011). Comparative antioxidant effect of BHT and water extracts of banana and sapodilla peels in raw poultry meat. *J. Food Sci. and Technol.*, DOI 10.1007/s13197-011-0508-8.

Dharmawan, J., Kasapis, S., Sriramula, P., Lear, M. and Curran, P. (2009). Evaluation of aroma-active compounds in Pontianak orange peel oil (*Citrus nobilis* Lour. var. *microcarpa* Hassk.) by gas chromatography eolfactometry, aroma reconstitution, and omission test. *J. Agric. Food Chem.*, 57(1): 239–244.

Donadieu, Y. (1979). La propolis. Paris: Editions Maloine.

Dornfeld, A.M., Rosenblat, L., Volkova, M., Kaplan, N.M. and Coleman, R. (2000). Pomegranate juice consumption reduces oxidative stress, atherogenic modications to LDL, and platelet aggregation: Studies in humans and in atherosclerotic apolipoprotein Eedecient mice. *American J. Clin. Nutr.*, 71(5): 1062–1076.

Draughon, F.A. (2004). Use of botanicals as biopreservatives in foods. *Food Technol.*, 58: 20–28.

Elliott, J.G. (1999). Application of antioxidant vitamins in foods and beverages. *Food Technol.*, 53: 46–48.

Faller, A.L.K. andd Fialho, E. (2009). The antioxidant capacity and polyphenol content of organic and conventional retail vegetables after domestic cooking. *Food Research Int.*, 42(1): 210–215.

Farre, R., Frasquet, I. and Sanchez, A. (2004). El propolis y la salud. *Archiv Pharmaceutica* 45(1): 21–43.

Fasseas, M.K., Mountzouris, K.C., Tarantilis, P.A., Polissiou, M. and Zervas, G. (2007). Antioxidant activity in meat treated with oregano and sage essential oils. *Food Chem.*, 106(3): 1188–1194.

Fuentes, O.A.M. and Hernandez, R.N. (1990). Antimicrobial action of alcoholic extracts of propolis. *Revista Cubana de Farmacia*, 24(1): 34–44.

Furneri, P.M., Marino, A., Saija, A., Uccella, N. and Bisignano, G. (2002). *In vitro* antimycoplasmal activity of oleuropein. *Int J Antimicrobial Agents*, 20(4): 293–296.

Gil, M.I., Tomas-Barberan, F.A., Hess-Pierce, B., Holcroft, D.M. and Kader, A.A. (2000). Antioxidant activity of pomegranate juice and its relationship with phenolic composition and processing. *J. Agric. Food Chem.*, 48(10): 4581–4589.

Govaris, A., Solomakos, N., Pexara, A. and Chatsopouloun, P.S. (2010). The antimicrobial effect of oregano essential oil, nisin and their combination against *Salmonella enteritidis* in minced sheep meat during refridgerated storage. *Int. J. Food Microbiol.*, 137: 175–180.

Grady, M.N.O., Maher, M., Troy, D.J., Moloney, A.P. and Kerry, J.P. (2006). An assessment of dietary supplementation with tea catechins and rosemary extract on the quality of fresh beef. *Meat Sci.*, 73: 132–143.

Greenaway, W., May, J., Scaysbrook, T. and Whatley, F.R. (1991). Identication by gas chromatography–mass spectrometry of 150 compounds in propolis. *Zeitschrift fuer Naturforschung.*, C4: 11–121.

Grosso, C., Ferraro, V., Figueiredo, A.C., Barroso, J.G., Coelho, J.A. and Palavra, A.M. (2008). Supercritical carbon dioxide extraction of volatile oil from Italian coriander seeds. *Food Chem.*, 111(1): 197–203.

Grusak, M.A. and DellaPenna, D. (1999). Improving the nutrient composition of plants to enhance human nutrition and health. *Annual Review Plant Biol.*, 50: 133–161.

Guevara, J.M., Chumpitaz, J. and Valencia, E. (1994). The *in vitro* action of plants on. *Vibrio cholerae*. Revista de gastroenterologia del Peru: Organo ocial de la Sociedad de Gastroenterologia del Peru 14(1): 27–31.

Gupta, S., Jyothi Lakshmi, A., Manjunath, M.N. and Prakash, J. (2005). Analysis of nutrient and antinutrient content of underutilized green leafy vegetables. *LWT Food Sci. and Technol.*, 38(6): 339–345.

Gutierrez, J., Barry-Ryan, C. and Bourke, P. (2008). The antimicrobial efficacy of plant essential oil combinations and interactions with food ingredients. *Int. J. Food Microbiol.*, 124(1): 91–97.

Han, S.K. and Park, H.K (1995). A study on the preservation of meat products by natural propolis: Eect of EEP on protein change of meat products. *Korean J Anim. Sci.*, 37: 551–557.

Hayouni, E.A., Chraief, I., Abedrabba, M., Bouix, M., Leveau, J.Y. and Mohammed, H. (2008). Tunisian *Salvia officinalis* L. and *Schinus molle* L. essential oils: Their chemical compositions and their preservative effects against Salmonella inoculated in minced beef meat. *Int J Food Microbiol.*, 125(3): 242–251.

Hayrapetyan, H., Hazeleger, W.C. and Beumer, R.R. (2012). Inhibition of *Listeria monocytogenes* by pomegranate (*Punica granatum*) peel extract in meat paté at different temperatures. *Food Control*, 23: 66–72.

Hsouna, A.B., Trigui, M., Mansour, R.B., Jarraya, R.M. and SamirJaoua, M.D. (2011). Chemical composition, cytotoxicity effect and antimicrobial activity of *Ceratonia siliqua* essential oil with preservative effects against Listeria inoculated in minced beef meat. *Int. J. Food Microbiol.*, 148: 66–72.

Hugo, W.B. and Bloomeld, S.F. (1971). Studies on the mode of action of the phenolic antibacterial agent fentichlor against *Staphylococcus aureus* and *Escherichia coli*. 3. The effect of fentichlor on the metabolic activities of *Staphylococcus aureus* and *Escherichia* coli. *J. Applied Bacteriol.*, 34(3): 579–591.

Kabara, J.J. (1993). Medium-chain fatty acids and esters. *In*: Davidson, P.M. and Branen, A.L. (*Eds.*), Antimicrobials in Foods. Marcel Dekker, New York, pp. 307–342.

Kanatt, S.R., Ramesh, C. and Sharma, A. (2008). Chitosan and mint mixtures: A new preservative for meat and meat products. *Food Chem.*, 107: 845–852.

Kayser, O. and Kolodziej, H. (1997). Antibacterial activity of extracts and constituents of *Pelargonium sidoides* and *Pelargonium reniforme*. *Planta Medica*, 63(6): 508–510.

Kim, S.J., Jin, S.K., Mandal, P.K. and Kang, S.N. (2011). Quality of low-fat pork sausages with tomato powder as colour and functional additive during refrigerated storage. *J Food Sci. Technol.*, 48(5): 591–597.

Kim, S.J., Cho, A.R. and Han, J. (2013). Antioxidant and antimicrobial activities of leafy green vegetable extracts and their applications to meat product preservation, *Food Control*, 29: 112–120.

Kotula, K.L. and Thelappurate, R. (1994). Microbiological and sensory attributes of retail cuts of beef treated with acetic acid and lactic acid solutions. *J. Food Prot.*, 57: 665–670.

Lambert, R.J.W., Skandamis, P.N., Coote, P.J. and Nychas, G.J.E. (2001). A study of minimum inhibitory concentration and mode of action of oregano essential oil, thymol and carvacrol. *J. Applied Microbiol.*, 91: 453–462.

Lis-Balchin, M., Steyrl, H. and Krenn, E. (2003). The comparative effect of novel Pelargonium essential oils and their corresponding hydrosols as antimicrobial agents in a model food system. *Phytotherapy Res.*, 17: 60–65.

Lopez, M.D., Jordan, M.J. and Pascual-Villalobos, M.J. (2008). Toxic compounds in essential oils of coriander, caraway and basil active against stored rice pests. *J. Stored Products Res.*, 44(3): 273–278.

Luo, H., Lin, S., Ren, F., Wu, L., Chen, L. and Sun, Y. (2007). Antioxidant and antimicrobial capacity of Chinese medicinal herb extracts in raw sheep meat. *J. Food Prot.*, 70: 1440–1445.

Machado, T.D., Leal, I.C.R., Amaral, A.C.F., Santos, D., da Silva, K.R.N.M.G. and Kuster, R.M. (2002). Antimicrobial ellagitannin of *Punica granatum* fruits. *J. Brazilian Chem. Society*, 13(5): 606–610.

Mannie, E. (1999). Natural antimicrobial ingredients for meat products. *Meat Processing*, 12: 80.

Martin, K.R., Krueger, C.G., Rodriquez, G., Dreherd, M. and Reed, J.D. (2009). Development of a novel pomegranate standard and new method for the quantitative measurement of pomegranate polyphenols. *J. Sci. Food Agric.*, 89(1): 157–162.

Mathenjwa, S.A., Hugo, C.J., Bothma, C. and Hugo, A. (2012). Effect of alternative preservatives on the microbial quality, lipid stability and sensory evaluation of boerewors, *Meat Sci.*, 91: 165–172.

Melendez, P.A. and Capriles, V.A. (2006). Antibacterial properties of tropical plants from Puerto Rico. *Phytomedicine*, 13(4): 272–276.

Menssens, W. and De Vugst, L. (2002). Inhibitory substances produced by Lactobacilli isolated from sourdougts a rev. *Intl. J. Food Microbiol.*, 72: 31–43.

Mexis, S.F., Chouliara, E. and Kontominas, M.G. (2012). Shelf life extension of ground chicken meat using an oxygen absorber and a citrus extract. *LWT-Food Sci. Technol.*, 49: 21–27.

Mielnik, M.B., Sem, S., Egelandsdal, B. and Skrede, G. (2008). By-products from herbs essential oil production as ingredient in marinade for Turkey thighs. LWT- *Food Sci. Technol.*, 41(1): 93–100.

Mitsumoto, M., Cassens, R.G., Schaefer, D.M. and Scheller, K.K. (1991). Pigment stability improvement in beef steaks by ascorbic acid application. *J Food Sci.*, 56: 857–858.

Mizuno, M. (1989a). Propolis or its extract containing resin compositions. Japanese Patent No. JP Ol 245 058 [89 245 058].

Mizuno, M. (1989b). Food packaging materials containing propolis as a preservative. Japanese Patent No. JP Ol 243 974 [89 243 974].

Mori, A., Nishino, C., Enoki, N. and Tawata, S. (1989). Antibacterial activity and mode of action of plant avonoids against *Proteus vulgaris* and *Staphylococcus aureus. Phytochemistry*, 26(8): 2231–2234.

Mosqueda-Melgar, J., Raybaudi-Massilia, R.M. and Martin-Belloso, O. (2008a). Inactivation of *Salmonella enterica* Ser enteritidis in tomato juice by combining of high-intensity pulsed electric elds with natural antimicrobials. *J Food Sci.*, 73(2): 47–53.

Mosqueda-Melgar, J., Raybaudi-Massilia, R.M. and Martin-Belloso, O. (2008b). Combination of high-intensity pulsed electric fields with natural antimicrobials to inactivate pathogenic microorganisms and extend the shelflife of melon and watermelon juices. *Food Microbiol.*, 25(3): 479–491.

Najeeb, A.P. (2012). Efficacy of certain leaves and fruits powder as natural preservatives in restructured chicken slices. MVSc thesis, submitted to Pondicherry University.

Nasr, C.B., Ayed, N. and Metche, M. (1996). Quantitative determination of the polyphenolic content of pomegranate peel. *Zeitschrift fuer Lebensmitte-luntersuchung und-Forschung*, 203(4): 374–378.

Naveena, B.M., Muthukumar, M., Sen, A.R., Babji, Y. and Murthy, T.R.K. (2006). Improvement of shelf-life of bualo meat using lactic acid, clove oil and vitamin C during retail display. *Meat Sci.*, 74: 409–415.

Naz, S., Siddiqi, R., Ahmad, S., Rasool, S.A. and Sayeed, S.A. (2007). Antibacterial activity directed isolation of ompounds from *Punica granatum*. *J Food Sci.*, 72(9): 341–345.

Nieto, G., Díaz, P., Bañón, S. and Garrido, M.D. (2010). Dietary administration of ewe diets with a distillate from rosemary leaves (*Rosmarinus officinalis* L.): Influence on lamb meat quality. *Meat Sci.*, 84: 23–29.

Ntzimani, A.G., Giatrakou, V.I. and Savvaidis, I.N. (2010). Combined natural antimicrobial treatmenst (EDTA, lysozyme, rosemary and oregano oil) on semicooked coated chicken meat stored in vacuum packages at 4C: Microbiological and sensory evaluation. *Innovative Food Sci. Emerging Technol.*, 11: 187–196.

Offord, E. (2004). Rosemary. *In:* Packer, I., Ong, C.N. and Halliwell, B. (*Eds.*), Herbal and traditional medicine: Molecular aspects of health (pp. 457–470). New York: CRC Press.

Olaoye, O.A., Onilude, A.A. and Dodd, C.E.R. (2008). Identification of *Pediococcus* spp. From beef and evaluation of their lactic acid production in varying concentrations of different carbon sources. *Advances in Natural and Applied Sciences*, 2(3): 197–207.

Oussalah, M., Caillet, S., Salmiea, S., Saucier, L. and Lacroix, M. (2004). Antimicrobial and antioxidant effects of milk protein-based lm containing essential oils for the preservation of whole beef muscle. *J. Agric Food Chem.*, 52(18): 5598–5605.

Pimia, P.R., Nohynek, L., Meier, C., Kahkonen, M., Heinonen, M. and Hopia, A. (2001). Antimicrobial properties of phenolic compounds from berries. *J Applied Microbiol.*, 90(4): 494–507.

Raccach, M. (1984). The antimicrobial activity of phenolic anti- oxidants in foods: A review. *J Food Safety*, 6: 141–170.

Recio, M.C., Ri os, L.L. and Villar, A. (1989). A review of some antimicrobial compounds isolated from medicinal plants reported in the literature 1978–1988. *Phytotherapy Research*, 3(4): 117–125.

Rižnar, K., Celan, Š., Knez, Ž., Škerget, M., Bauman, D. and Glaser, R. (2006). Antioxidant and antimicrobial activity of rosemary extract in chicken frankfurters. *J. of Food Sci.*, 71: C425–C429.

Roberts, T.A. (1989). Combinations of antimicrobials and processing methods. *Food Technol.*, 43: 156–163.

Rojas, M.C. and Brewer, M.S. (2008). Effect of natural antioxidants on the oxidative stability of frozen vacuum packaged beef and pork. *J. Food Quality*, 31: 173–185.

Roller, S., Sagoo, S., Board, R., O'Mahony, T., Caplice, E. and Fitzgerald, G. (2002). Novel combinations of chitosan, carnocin and sulphite for the preservation of chilled pork sausages. *Meat Sci.*, 62: 165–177.

Sayago-Ayerdi, S.G., Brenes, A. and Goni, I. (2009). Effect of grape antioxidant dietary ber on the lipid oxidation of raw and cooked chicken hamburgers. *LWT-Food Sci. Technol.*, 42: 971–976.

Shahidi, F. and Wanasundara, P.K.J.P.D. (1992). Phenolic antioxidants. *Critical Reviews Food Sci. Nutr.*, 32: 67–103.

Shahidi, F. and Naczk, M. (2004). Phenolics in food and nutraceuticals. Boca Raton. FL: CRC Press. pp. 352–355.

Shan, B., Yi, Z.C., Brooks, J.D. and Corke, H. (2007). The *in vitro* antibacterial activity of dietary spice and medicinal herb extracts. *Int. J. of Food Microbiol.*, 117(1): 112–119.

Shelef, L.A. (1983). Antimicrobial effects of spices. *J. Food Safety*, 6: 29–44.

Sofos, J.N. (2008). Challenges to meat safety in the 21st century. *Meat Sci.*, pp. 783–13.

Solomakos, N., Govaris, A., Koidis, P. and Botsoglou, N. (2008). The antimicrobial effect of thyme essential oil, nisin, and their combination against *Listeria monocytogenes* in minced beef during refrigerated storage. *Food Microbiol.*, 25(1): 120–127.

Soultos, N., Tzikas, Z., Abrahim, A., Georgantelis, D. and Ambrosiadis, I. (2008). Chitosan effects on quality properties of Greek style fresh pork sausages. *Meat Sci.*, 80: 1150–1156.

Stern, J.L., Hagerman, A.E., Steinberg, P.D. and Mason, P.K. (1996). Phlorotanninprotein interactions. *Chemical Ecology*, 22(10): 1887–1899.

Su, L., Yin, J.J., Charles, D., Zhou, K., Moore, J. and Yu, L. (2007). Total phenolic contents, chelating capacities, and radical-scavenging properties of black peppercorn, nutmeg, rosehip, cinnamon and oregano leaf. *Food Chem.*, 100: 990–997.

Tajkarimi, M.M., Ibrahim, S.A. and Cliver, D.O. (2010). Antimicrobial herb and spice compounds in food (review). *Food Control*, 21: 1199–1218.

Tim Cushnie, T.P. and Lamb, A. (2005). Antimicrobial activity of flavonoids. *Int. J. Antimicrobial Agents*, 26: 343–356.

Tosi, E.A., Edmundo, Re., Ortega, M.E. and Cazzoli, A.F. (2007). Food preservative based on propolis: Bacteriostatic activity of propolis polyphenols and flavonoids upon *Escherichia coli*. *Food Chem.*, 104: 1025–1029.

Turgis, M., Borsa, J., Millette, M., Salimieri, S. and Lacroix, M. (2008). Effect of selected plant essential oils or their constituents and modied atmosphere packaging on the radio sensitivity of *Escherichia* coli O157: H7 and *Salmonella typhi* in ground beef. *J. Food Protection*, 71(3): 516–521.

Viuda-Martos, M., Ruiz-Navajas, Y., Fernandez-Lopez, J. and Perez-Alvarez, J.A. (2010). Effect of orange dietary ber, oregano oil and packaging conditions on shelf-life of bologna sausages. *Food Control*, 21: 436–443.

Weiss, J., Gibis, M., Schuh, V. and Salminen, H. (2010). Advances in ingredient and processing systems for meat and meat products. *Meat Sci.*, 86: 196–213.

Yano, Y., Satomi, M. and Oikawa, H. (2006). Antimicrobial effect of spices and herbs on *Vibrio parahaemolyticus*. *Int. J. Food Microbiol.*, 111(1): 6–11.

Yin, M.C. and Chao, C.Y. (2008). Anti-campylobacter, anti-aerobic, and anti-oxidative effects of roselle calyx extract and protocatechuic acid in ground beef. *Int. J. Food Microbiol.*, 127(1–2): 73–77.

Yogesh, K. and Jamshed, Ali. (2012). Antioxidant potential of thuja (*Thuja occidentalis*) cones and peach (*Prunus persia*) seeds in raw chicken ground meat during refrigerated (4 ± 1°C) storage. *J. Food Sci. Technol.*, DOI 10.1007/s13197-012-0672-5.

Zambuchini, B., Fiorini, D., Verdenelli, M.C., Orpianesi, C. and Ballini, R. (2008). Inhibition of microbiological activity during sole (*Solea solea* L.) chilled storage by applying ellagic and ascorbic acids. *J. Food Sci. Technol.*, 41(9): 1733–1738.

Zhang, H., Kong, B., Xiong, Y.L. and Sun, X. (2009). Antimicrobial activities of spice extracts against pathogenic and spoilage bacteria in modied atmosphere packaged fresh pork and vacuum packaged ham slices stored at 4 ± 1°C. *Meat Sci.*, 81: 686–692.

Zhou, B., Wu, Z., Li, X., Zhang, J. and Hu, X. (2008). Analysis of ellagic acid in pomegranate rinds by capillary electrophoresis and high-performance liquid chromatography. *Phytochemical Analysis*, 19(1): 86–89.

25

Detection of Drug Residues in Animal Products

A.K. Biswas, P.K. Mandal and K.P. Mini

ABSTRACT

In the area of banned substances there is a continued trend towards the development of assays to detect ultra trace levels of residues of illegal substances. These methods generally involve more intensive preparation of samples to allow detection of residues to ng / kg levels. It has also been seen that there is a growing trend to pack more and more residues into methods particularly for veterinary drugs. Some research groups are now applying less specific sample preparation approaches including "dilute-and-shoot" or protein precipitation combined with ultra-filtration. A major problem with some of these multi-class methods is that there are often compromises observed in LODs, chromatography and quantitation due to poor linearity. As such many of these methods can be categorised as satisfactory for screening purposes, particularly when analysis is carried out by LC-MS-MS. So, one can expect that routine application of screening assays for licensed veterinary drugs will be reduced. The analysis of antibiotic residues using inhibitory assays will only have application in an industry environment. Regulatory agencies will have to reconsider the approach to the design of the national residue surveillance programs.

1. INTRODUCTION

Food safety is getting considerable attention because of its major impact on public health. In the recent year high profile food-related crises such as bovine spongiform encephalopathy, dioxin contamination outbreaks, and epidemics caused by food borne pathogens have shaken consumer confidence in the ability of the food industry and the authorities to guarantee food safety. To respond to consumers concerns, many developed countries already introduced food safety policy. The EU introduced an integrated "from farm to fork" strategy, which aim to provide the same high level of health protection to consumers all over Europe. Animals produced for food may be exposed to a wide range of veterinary drugs which are used for therapeutic purposes. In addition to their therapeutic value, these compounds are extensively used to enhance growth and increase feed efficiency. Further, impetuous of modern livestock farming system based on the rearing of a large number of animals in a close confined factory like conditions that are quite favourable for the incidence of many food brone pathogens making them more useful of antimicrobial agents (Miller, 1993).

It is well documented that the use of antibiotics can enhance growth rates, improve feed efficiency, and generally improve animal health, but these compounds pose a potential threat to human and animal health through the development of antibiotic-resistant bacteria. For decades, antibiotics have been used abundantly worldwide in animal production, so antibiotic resistance may spread to other microbial populations, as reflected by the emergence of infectious diseases that have become resistant to standard antimicrobial treatments. The access to systematic, updated, and reliable data on the consumption of veterinary drugs is not easy, especially in developing countries. The orders of magnitude of the consumption of antibacterial agents are very high. According to data published by the European Federation of Animal Health, in 1999 about 3,900 tons of antibiotics were used in the European Union (EU) in veterinary medicine and 800 tons as growth promoters.

More recently, a study estimates the use of antibiotics in veterinary medicine in 25 countries of the EU at 5,393 tons. The most commonly used antimicrobials in food animals are β-lactams (penicillin), tetracycline (oxytetracycline, chlortetracycline and tetracycline), aminoglycosides (streptomycin) and sulfonamides (sulfadimidine, sulfamethoxazole, sulfadoxine, sulfadimethoxine etc.) singly or combination with trimethoprim). Among these, tetracyclines and sulfonamides are extensively used because of low cost, easy availability and broad-spectrum antimicrobial activity. The problem again worsens when layman use drugs without knowing correct dose levels or withdrawal period before slaughter.

In France, in 2005, 760 tons were used in human medicine and 1,320 in veterinary medicine, corresponding to 199 and 84 mg per kg of live weight in human and animal populations, respectively. Tetracyclines were the most prescribed antibiotics in veterinary medicine, whereas beta-lactams predominated in human medicine. In comparison, about 12,650 tons of antimicrobials were used in 2007 in the USA in veterinary medicine, 40% of which were tetracyclines. About 13% of the total amount of antimicrobials was used as growth promoters. The administration of these antibacterial to animals by any route leads to leave residues in foods of animal origin such as meat, milk, eggs and other livestock products.

Other veterinary drugs like coccidiostats, hormones, β-agonists and tranquillizers also getting public attention due to its potential toxicity as they are regularly use in animal production. To reach the required level of protection, adequate monitoring data have to be available, to enable ensure risk evaluation and subsequent action. In other words, sophisticated and robust analytical methods have to be developed for a wide variety of, primarily organic, micro contaminants.

2. HISTORICAL BACKGROUND

The use of veterinary drugs as antimicrobials were started much later, though the concept of antibiotics that substances produced by one living organism may kill or inhibit growth of another, has been recognized for centuries. About 2500 years ago, mouldy curd of soybean was used by Chinese to treat boils, carbuncles and similar infection. There are historical references in the medical literature describing the beneficial effects of plants and their extract in infections. Most of these were probably sources of antibiotic forming moulds or bacteria. Pasteur and Jourbert first speculated on the clinical potential of microorganisms as therapeutic agents in 1877. Since then the use of these compounds has increased to such an extent that many feel that today's intensive agricultural production techniques would be impossible without them.

The modern chemotherapy of infections started with the introduction of sulphanilamide in 1936. Manipulation of the chemical structure of sulfanilamide molecule made it possible to synthesize sulfonamides with variable efficacy. Sulfonamides were the first effective and safe antimicrobial agents to be used clinically for the treatment of bacterial infections in man and animals (Mandell & Sande, 1985). The discovery of penicillin in 1928 by Alexander Fleming leads to antibiotic revolution. Benzylpenicillin and procaine benzylpenicillin are still widely used. However, penicillin and streptomycin came into the market much later possibly during World War II. Tetracyclines and sulfonamides were extensively used in the last few decades for the treatment of variety of bacterial infections (JECFA, 1999). But the basic action of antibiotics is binding a specific site on or within microorganism, leading to interference with the vital life processes (Nightingale *et al.*, 1994). Presently, more than 300 feed additives antimicrobials, anticoccidial and hormonal compounds are used in animal agriculture in the world with a worth value of 40.5 billion US $ (Lee *et al.*, 2001). That indicate all food animals produced in future will have received chemotherapeutic or prophylactic agents.

3. GENERAL DESCRIPTION

The β-lactam antibiotics comprise several classes of compounds, including the cephalosporins and penicillins, which are administered to food producing animals both therapeutically and prophylactically. Both classes contain a side-chain attached to the 7-amino-cephalosporanic acid or 6-aminopenicillanic acid nuclei, respectively. The aminoglycosides produced by (*Streptomyces* spp. and *Micromonospora* spp.) are broad-spectrum antibiotics, which consist of an aminocyclitol ring connected to two or more amino sugars via a glycoside link.

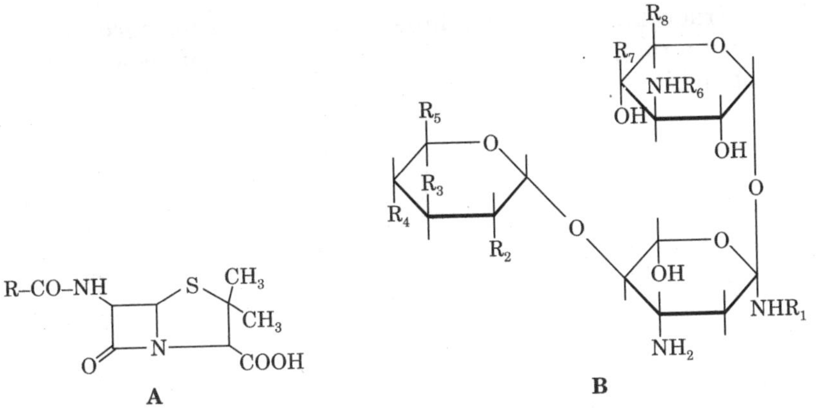

A

B

Fig. 1: Typical structure of β-lactam (A) and aminoglycosides (B).

Chloramphenicol [D-(–)-threo. 2-(dichloroacetamido)-1-(*p*-nitrophenyl)-1, 3-propanediol] (CAP) is a broad-spectrum antibiotic active against a variety of pathogens, including aerobic and anaerobic Gram-negative and Gram-positive bacteria, rickettsia, mycoplasma and chlamydia. Although previously widely used in veterinary and human medicine there have been reports of a plastic anaemia in humans arising from its use.

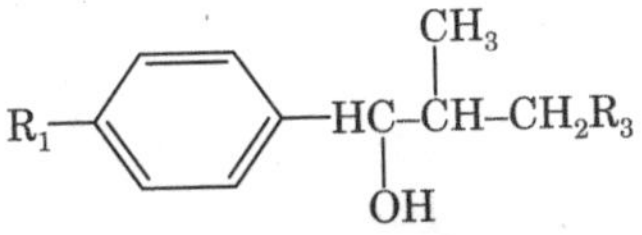

Fig. 2: Typical structure of Chloramphenicol.

The sulfonamides are bacteriostatic, and act by competing with *p*-aminobenzoic acid in the enzymatic synthesis of dihydrofolic acid. This leads to a decreased availability of the reduced folates that are essential for purine synthesis, methionine synthesis, etc. Residues in food are of concern because of their potential carcinogenic character and the possibility of the development of antibiotic resistance in humans. However they are used as veterinary drugs for prophylactic and therapeutic purposes. Furthermore, sulfonamides act as growth promoting substances. The European Union and US regulatory bodies have set an MRL of 100 μg/kg in meat and 100 μg/kg in milk. Each level corresponds to the sum of all parent sulfonamides present in meat and milk.

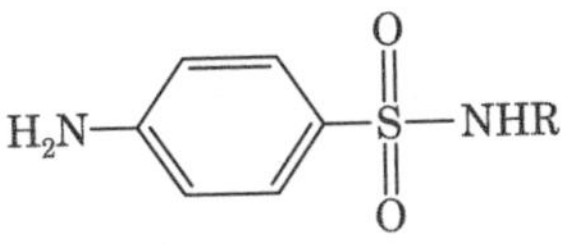

Fig. 3: Typical structure of sulfonamides.

The tetracyclines (TCs) are broad-spectrum antibacterial compounds active against a range of Gram-positive and Gram-negative organisms. They inhibit protein biosynthesis through their binding to the 30S ribosome.

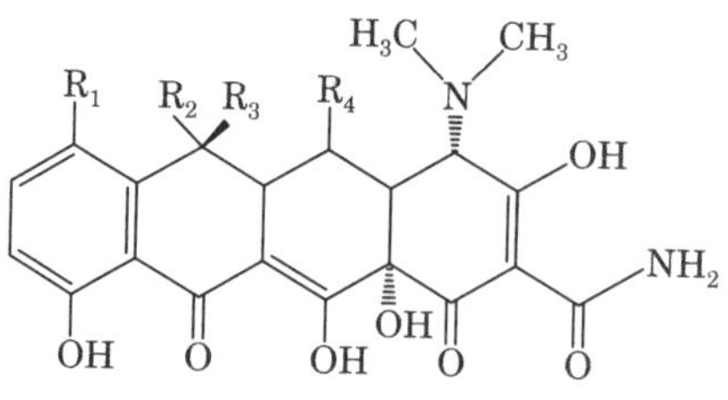

Fig. 4: Typical structure tetracyclines.

The macrolides are a group of compounds that contain a 12-, 14-, or 16-membered macrocyclic lactone ring to which several amino and or neutral sugars are attached. They are widely used in veterinary medicine to treat respiratory diseases or as feed additives to promote growth. The EU has set MRLs for some macrolides in tissues, milk and eggs.

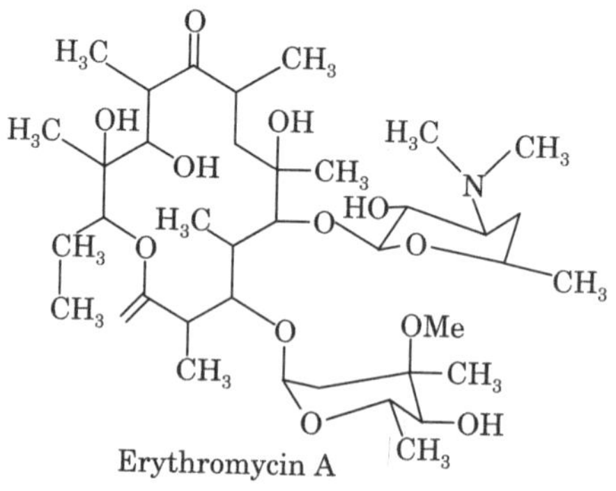

Erythromycin A

Fig. 5: Structure of a macrolide, adopted from Stolker and Brinkman (2005).

Quinolones are broad-spectrum synthetic antimicrobial agents used in the treatment of livestock and in aquaculture. The older quinolones, *e.g.*, nalidixic acid have limited activity against Gram-negative organisms, while the more recently introduced fluoroquinolones, *e.g.*, enrofloxacin, have a very wide spectrum of activity. In general quinolone carboxylic acids are amphoteric with poor water solubility between pH 6 and 8.

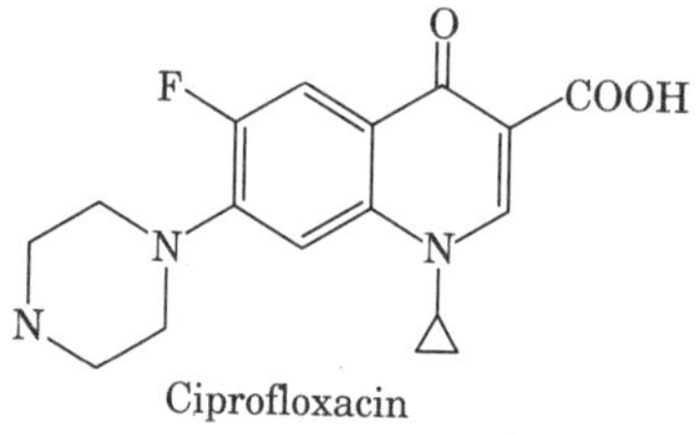

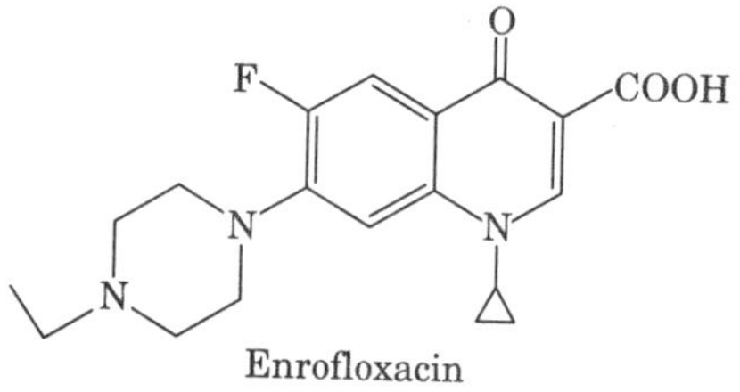

Ciprofloxacin

Enrofloxacin

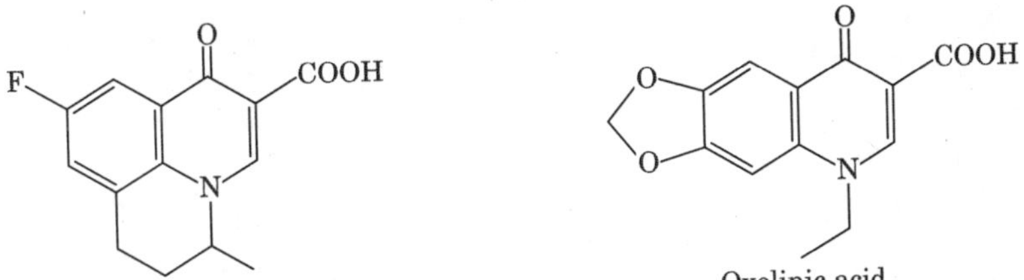

Flumequine

Oxolinic acid

Fig. 6: Typical structure of some quinolone antibiotics.

The nitrofurans are synthetic chemotherapeutic agents, and have a broad-spectrum of activity. They act by inhibiting a number of microbial enzymes involved in carbohydrate metabolism. Hence their use is prohibited in livestock production within the EU.

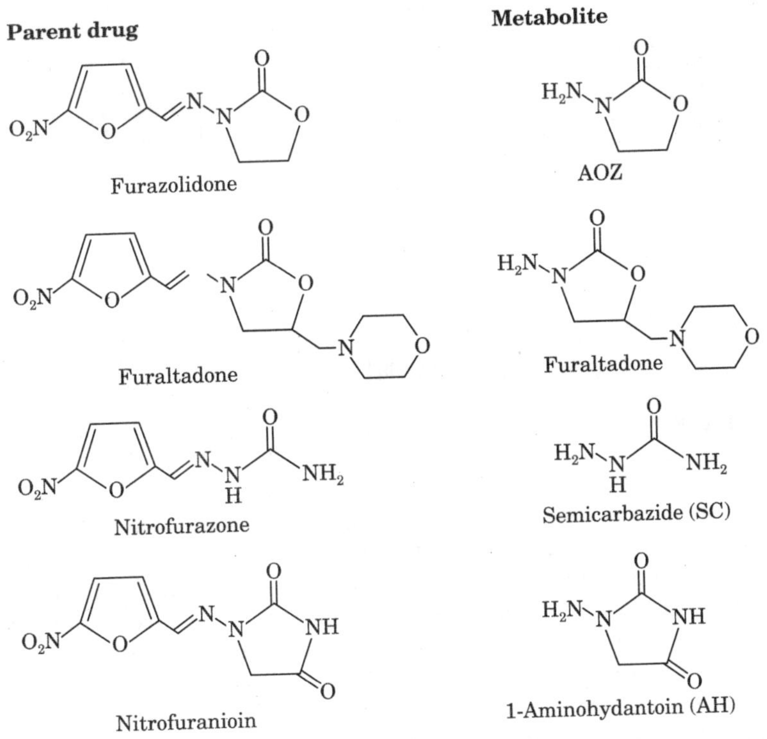

Fig. 7: Typical structure of some nitrofurans, adopted from Leitner *et al.* (2001).

Nitroimidazoles like dimetridazole (DMZ) and ronidazole (RNZ) are compounds used for the prevention of coccidiosis in poultry, and in addition DMZ acts against swine dysentery. Due to the carcinogenic effect of this group, these substances were banned in the EU.

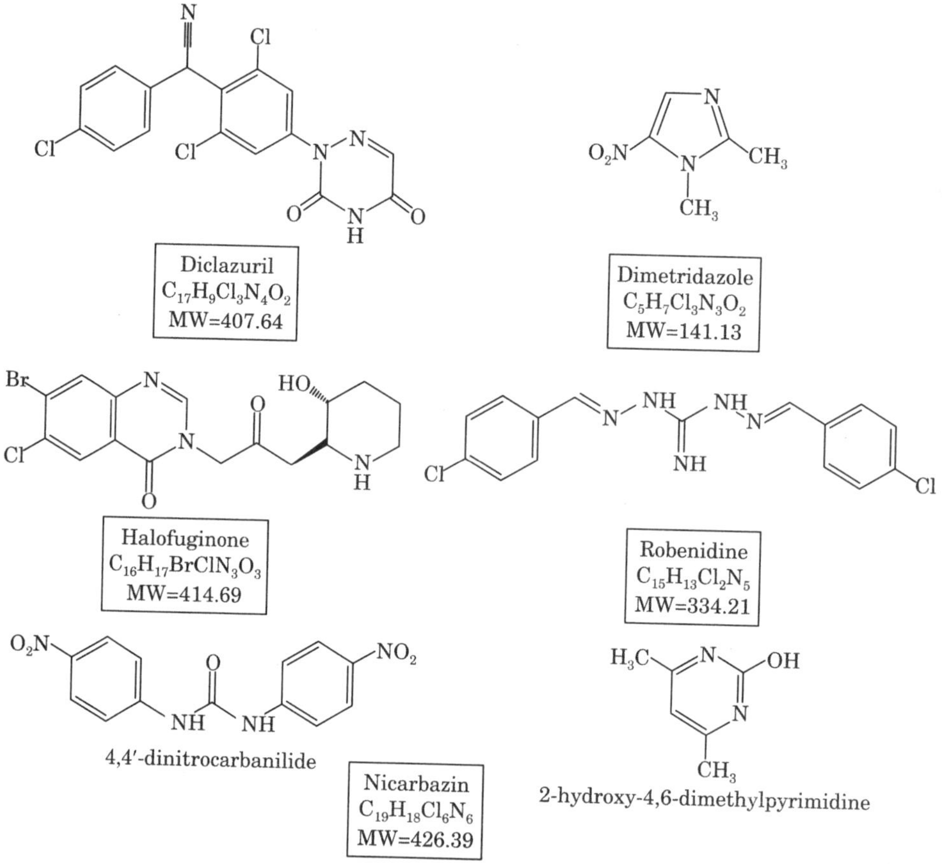

Diclazuril
$C_{17}H_9Cl_3N_4O_2$
MW=407.64

Dimetridazole
$C_5H_7Cl_3N_3O_2$
MW=141.13

Halofuginone
$C_{16}H_{17}BrClN_3O_3$
MW=414.69

Robenidine
$C_{15}H_{13}Cl_2N_5$
MW=334.21

4,4′-dinitrocarbanilide

Nicarbazin
$C_{19}H_{18}Cl_6N_6$
MW=426.39

2-hydroxy-4,6-dimethylpyrimidine

Fig. 8: Structure coccidiostate adoped from (Mortier *et al.*, 2003).

Anthelmintics are drugs acting primarily against intestinal worms, although many are active also against lungworms and liver fluke. The most frequently used anthelmintic compounds are levamisole, several compounds from the benzimidazole group (albendazole, cambendazole, fenbendazole, oxfendazole and thiabendazole) and ivermectin. They are widely used in veterinary medicine against endoparasites in food producing animals. EU regulations require an analytical procedure to provide control at the MRL. The MRL is defined, for example, for fenbendazole as the sum of the extractable residues that may be oxidized to oxfendazole sulfone (that amounts to 50 μg/kg in muscle, fat and kidney and 500 μg/kg in liver).

Flunixin is a non-steroidal anti-inflammatory agent and it is used for the treatment of pulmonary diseases and mastitis. Its MRL value was set at 40 μg per liter in milk. Tranquillizers, such as acepromazine, azaperone, chlorpromazine, propionylpromazine, xylazine and the beta-blocker carazolol are frequently used in veterinary medicine to avoid stress in pigs, during transportation to the abattoir.

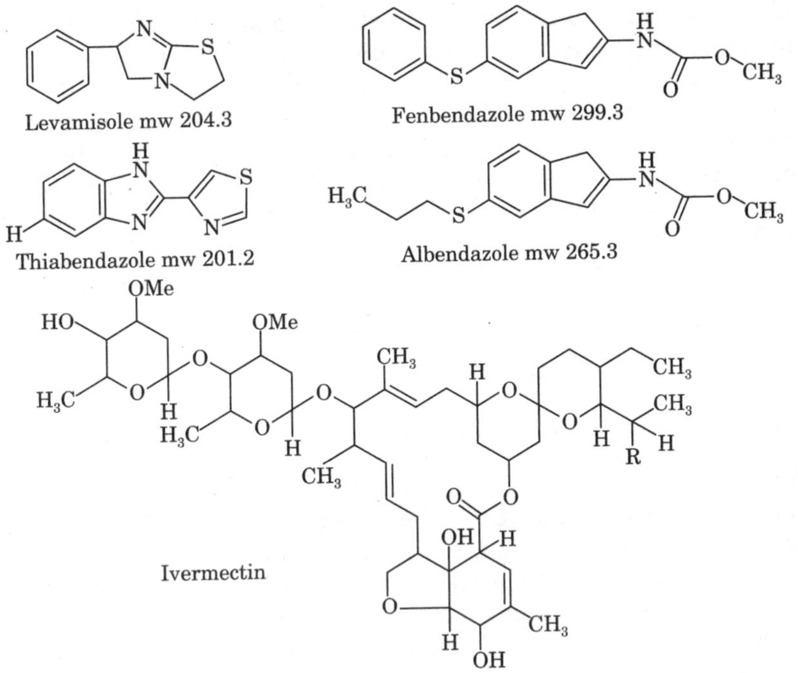

Levamisole mw 204.3

Fenbendazole mw 299.3

Thiabendazole mw 201.2

Albendazole mw 265.3

Ivermectin

Fig. 9: Structure of anthelmintics, adopted from Stolker and Brinkman (2005).

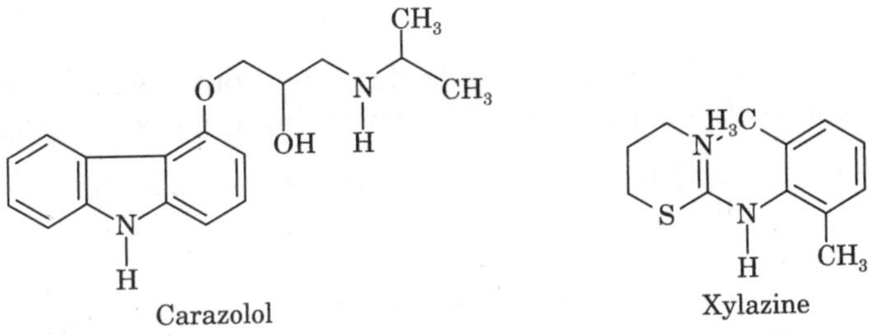

Carazolol

Xylazine

Fig. 10: Chemical structure of some important Tranquillizers.

4. SOURCE OF EXPOSURE

Residues of veterinary medicinal products are most common in foods of animal origin as are directly exposed to the animals. The range of veterinary medicinal products used in regular animal husbandry practices is extremely wide, ranging from teat dips to hormones. Approximately 42% of all veterinary pharmaceuticals used world-wide are used as feed additives, 19% are used as anti-infective, 13% as parasiticides, 11% are used as biologicals and 15% represent other pharmaceuticals (Miller, 1993). The most commonly used antimicrobials in food animals include β-lactams (benzyl-penicillin/procaine benzylpenicillin), tetracyclines (oxytetracycline, chlortetracycline and tetracycline), aminoglycosides (streptomycin and gentamycin), macrolides (erythromycin) and sulfonamides (sulfadimidine).

All the antimicrobials are administered to animals either by injections (intramuscularly, intravenous, subcutaneous) or orally in the feed and water, topically on the skin, and by intra-mammary and intra-uterine infusions (Mitchell *et al.*, 1998). Injectables were responsible for 46% of the violative residues in meat followed by oral administration at 20% (feed, water and bolus) and intra-mammary infusions at 7% (Booth, 1982). Several other factors have contributed to the residues problem such as poor treatment records or failure to identity the animals and result from the use of a drug in same manner that is inconsistent with the labeling. This occurs primarily through not observing label of withdrawal times as well as "extra-label" use of drug (Paige, 1994).

5. ANALYTICAL METHODS

5.1. Sample Extraction Techniques

Residues are typically extracted from samples using simple solvent extraction or liquid–liquid extraction (LLE). The extraction technique adopted may depend on the nature of the samples *i.e.*, liquid or solid and the physico-chemical properties of the residues (polarity and pKa). Simple extraction with aqueous buffer is advantageous for highly polar residues because they reduce non-polar matrix components (lipids) and extracts can be enriched on reversed phase SPE, while eliminating time-consuming evaporation steps. A disadvantage is that strongly protein-bound residues are not fully extracted and polar matrix components are co-extracted. In general, the majority of methods employ more efficient organic solvents as extracting agents. Acetone (ACN) is the preferred extraction solvent as it gives good yields of residues but low levels of matrix co-extractives and is effective at denaturing proteins and inactivating enzymes.

Methanol (MeOH) and ethanol are also widely used solvents but result in the extraction of additional matrix components. Liquid-liquid extraction (LLE) was the most widely applied extraction procedure in residue analysis due to its high selectivity compared to simple solvent extraction. LLE applications can also include polar ionisable compounds, which can be extracted by non-polar organic solvents using the ion-pair technique: transforming positively charged substances into non-polar neutral compounds in the presence of organic anions, or vice versa. Examples of the successful application of ion-pair extraction are β-agonists, aminoglycosides and oxytetracycline. A number of instrumental-based extraction procedures have been developed to isolate residues from food, including microwave, supercritical fluid and pressurised liquid extraction systems.

5.2. Sample Purification

5.2.1. *Solid-phase extraction*

SPE is the most important sample purification technique in residue analysis and has gradually replaced LLE and LLP. However, these are being replaced more and more by polymeric sorbents that offer advantages for analytes that are difficult to purify on conventional phases (Fontanals *et al.*, 2005). However, continued disadvantages of SPE include difficulties in achieving consistent flow and plugging of cartridges, which can lead to difficulties in automation. Disk extraction format overcomes these difficulties but have yet to find widespread application in residue analysis. The use of disks should increase in the future as they offer not only better performance but also faster extraction speeds.

Another form of SPE, dispersive-SPE (DSPE) is a clean-up technique that involves mixing sorbent with a sample that has been pre-extracted with an appropriate solvent. The appropriate sorbent adsorbs matrix co-extractives onto its surface, leaving analytes of interest in the solvent. $MgSO_4$ is added to provide additional clean-up by removing residual H_2O and some other compounds via chelation. Afterwards, the mixture is centrifuged and the resulting supernatant can be analysed directly or can be subjected to a concentration and/or solvent exchange step. It is an extremely fast, simple and inexpensive process that provides high recovery and reproducibility for many LC- and GC-amenable analytes (Lehotay *et al.*, 2005).

The principle of the process is the removal of matrix compounds, while leaving the analytes of interest in the solvent. The physico-chemical properties of the analytes and matrix compounds determine the choice of sorbent. In pesticide analysis, primary secondary amine (PSA) is the most common sorbent used. PSA is effective at retaining fatty acids and other organic acids present in food. For food of animal origin, which has higher lipid content, C_{18} or a combination of PSA/C_{18} is more effective because C_{18} removes lipophilic compounds. In recent research, it has been found that the combination of PSA/C_{18} to provide better clean-up than PSA or C_{18} alone for 38 anthelmintics in liver and milk.

However, PSA/C_{18} gave a lower recovery for some analytes, compared with C_{18} which gave sufficient clean-up and good recovery for all analytes and was therefore chosen as the preferred sorbent. Graphitised carbon black (GCB) has been reported to be a highly effective sorbent for sample clean-up. However, GCB also removes structurally planar analytes and is therefore not useful in many applications. DSPE does not provide the same degree of clean-up as SPE. However, it does provide good recovery and reproducibility, coupled with practical and cost advantages.

5.2.2. *Immunoaffinity chromatography*

Immunoaffinity chromatography (IAC) is particularly advantageous when low detection levels in the µg per kg to ng per kg are required for banned substances, particularly when using less selective HPLC based detection systems. However, it is difficult to see the practical application of this technique in the isolation of residues of licensed veterinary drug from food where there is now a trend to move towards multi-class residue methods with detection by LC–MS/MS.

Despite this, there are some very good multi-residue applications of IAC in residue analysis (Li *et al.*, 2008 and Zhao *et al.*, 2009).

5.2.3. *Molecularly imprinted polymers*

Molecularly imprinted polymers (MIPs) are engineered cross linked polymers that exhibit high affinity and selectivity towards a target compound or class of structurally related compounds. MIPs can be tailored to selectively extract analytes present in complex matrices such as blood, urine, tissue or feed. These materials have demonstrated binding to trace levels of target analytes, and display high selectivity in the presence of other compounds that have similar physico-chemical properties, as well as being extremely stable. A drawback of the technique is the potential leaching or template remaining in the MIP, and also binding site heterogeneity, leading to a range of binding affinities for the target analyte. Aqueous samples, such as milk or urine, generally require purification prior to MIP clean-up. However, there are applications reported that apply the aqueous sample directly to MIP without undergoing an initial extraction step (Ellwanger *et al.*, 2001 and Mohamed *et al.*, 2007).

5.2.4. *Ultra-filtration*

The development of multi-residue assays using LC–MS/MS detection has resulted in the alternative purification systems in the field of residue analysis such as ultra-filtration (UF). In residue analysis of food, UF is primarily used to separate analytes of interest from macromolecules, such as proteins, peptides, lipids and sugars, which may interfere with analysis, particularly affecting ionisation in mass spectrometry. In residue analysis, molecular weight cut-off devices or spin filters coupled to micro-centrifuge tubes are the most commonly used formats. Alternative formats are also available such as 96-well plate, but require dedicated vacuum manifolds and pumps. However, all residue applications use centrifugal devices.

5.2.5. *Size exclusion or gel permeation chromatography*

This mode of purification is widely used in the area of pesticide residue analysis. These materials usually have the appearance of a gel, resulting in the generic name gel permeation chromatography (GPC). Purification is achieved through molecular sieving, which occurs through pores on the surface of a solid sorbent. Size exclusion is not very specific and lacks resolving power, with frequent overlap between similar sized molecules, independent of their chemical structure or properties. However, it can be successfully used to separate low molecular weight drugs from larger interferences.

5.2.6. *Automated off-line/on-line SPE*

Automated sample preparation can be carried out on-line or off-line. Automated sample preparation offers the ability to perform sample clean-up, concentration and analyte separation in a closed system. This reduces the sample preparation time and the whole sample becomes available for analysis, while sensitivity and limits of detection are improved accordingly. It also removes some of the human element from a procedure, thereby improving precision and reproducibility. Furthermore, automated sample preparation reduces cost by using less solvent and fewer personnel. However, there is an increase in initial capital expenditure. The off-line is more common than on-line systems because it can be applied to traditional SPE clean-up. A disadvantage is that extracts require concentration and have to be transferred manually to the analytical system. A particular advantage over some on-line SPE systems is that memory or sample carry over effects are eliminated through single-use SPE cartridges.

A number of applications have been developed using this platform including anabolic steroids in urine, quinolones in milk and tissues; stilbenes in animal tissues, halofuginone in chicken liver and eggs, carbadox and olaquindox in porcine liver. On-line SPE offers better control of the sample preparation process and improved sensitivity through more selective isolation of target residues. A disadvantage of this approach is that some substances can carryover between injections and result in a memory effect.

5.2.7. *Turbulent flow chromatography*

Turbulent flow chromatography (TFC) is a high-throughput sample preparation technique that utilizes high flow rates (4–6 ml) and an analytical column containing sorbent particles with large pore sizes (30–60 µm). Due to the large pore size, there is only moderate back-pressure on the column, which serves as both extraction and analytical column. At the higher flow rate, solvent does not exhibit laminar flow but exhibits turbulent flow instead. This leads to the formation of eddies which promote cross-channel mass transfer and diffusion of the analytes into the particle pores. Samples are applied to the column using aqueous mobile phase. Small molecules diffuse more extensively than macromolecules and are driven into the pores of the sorbent. Due to the high flow rate, the larger molecules are flushed to waste and do not have an opportunity to diffuse into the particle pores.

A simple HPLC pumps and switching valves can be used to carry out TFC, although specialist equipment, termed high turbulent liquid chromatography (HTLC), is also available. The columns used for TFC contain common HPLC sorbents but of larger particle sizes. The chromatographic efficiency of TFC is similar to that of laminar flow but at much lower flow rates. TFC is also effective at separating residues that are bound to sample proteins. The use of TFC eliminates time-consuming sample clean-up in the laboratory and results in a much shorter analysis time, higher productivity and reduced solvent consumption without sacrificing sensitivity or reproducibility.

5.2.8. *Ninety-six-well SPE technology*

This technique allows the simultaneous extraction of up to 96 samples and reduces the sample preparation time drastically. It also reduces handling errors and limits labour input. This technique can also be automated to improve precision and accuracy. Rubies *et al.* (2007) developed a 96-well Oasis HLB SPE method for isolating nine quinolone residues from bovine muscle prior to LC–MS/MS. But, the sample throughput of the method was limited to 24 test samples per day.

5.2.9. *Dialysis*

Dialysis in combination with trace enrichment and LC is a relatively simple on-line sample preparation technique. Although it is not very selective, the dialysis cell is easy to construct and the technique is very efficient at removing macromolecules that may interfere with the subsequent separation and detection process. However, few applications have been reported in the literature on the use of dialysis in residue analysis. The majority of these applications have been applied to the isolation of antibiotic residues from meat, fish and egg.

5.3. Detection Techniques

Currently there are six types of analytical methods commonly used for the detection of antimicrobial residues in foods. These include microbial growth inhibition assays, microbial receptor assays, enzymatic colorimetric assays, receptor binding assays, immunoassays and chromatographic methods. Among all these methods, chromatographic methods are considered as confirmatory methods, while other five are designated as screening methods as use rapid screening of samples. Screening methods can detect the presence of an analyte or group of analytes at the level of interest, and usually provide semi-quantitative or quantitative results. The ideal characteristics for screening methods are low rate of false compliant samples, high throughput, ease of use, short analysis time, good selectivity, and low cost.

However, it is difficult for a method to meet all the requirements. For example, microbiological assays are inexpensive and have high throughput, but selectivity is not good and the time required to obtain results can be quite long. ELISA enables simultaneous processing of a large number of samples in a short time, but is difficult to implement for multiresidue screening, because it gives a wide range of sensitivities for the different compounds, rather than a generic response for a family of compounds. Biosensors are also used as screening methods for antibiotic analysis (Pellegrini *et al.*, 2004), as is liquid chromatography (LC) with UV, fluorimetric, or electrochemical detection, but the technique that is gaining more acceptance for screening purposes is LC coupled to mass spectrometry (MS), as it enables analysis of a large number of analytes in a single run (Weber *et al.*, 2004; Suarez *et al.*, 2009 and Adrian *et al.*, 2009).

The key point, as in all screening methods, is that sample treatment should be as simple as possible, without compromising the operation of the mass spectrometer. Karsourd *et al.* (1998) evaluated a number of common bacterial inhibition test for screening antimicrobial residues in tissues. They found that screening for tetracycline was excellent with German three-plate test, the European Union four-plate test and new Dutch kidney test instead of swab test on premises (STOP), calf antibiotic and sulfa test (CAST) and the fast antibiotic screen test (FAST).

5.3.1. *Screening methods*

5.3.1.1. *Microbial growth inhibition methods*

Screening assays mainly relied on the inhibition of growth of micro-organisms (*Geobacillus stearothermophilus* var. *calidolactis*, *Bacillus subtilis*, *Bacillus megaterium*, *Sarcina lutea*, *Escherichia coli*, *Bacillus cereus* var. *mycoides* or *Streptococcus thermophilus*) by antibiotic residues present in the test sample. These traditional assay procedures are based on either diffusion process or on turbidity. The growth of an indicator organism, in a transparent liquid culture, can be followed by monitoring the increase turbidity with time. In diffusion dependant assays, the material to be assayed diffuses through an agar based nutrient medium that has been uniformly seeded with spores of susceptible organism.

Upon incubation, a zone of inhibition of germination and growth develops, indicating the presence of inhibitors. Both the turbidity and diffusion-based techniques can be carried out manually, and many of the steps may be automated. Two procedures, swab test on premises (STOP) and calf antibiotics and sulfa test (CAST), are typical. Cotton swabs that have been used to sample the suspect tissue are placed in contact with gelled growth media that has been amended with *Bacillus subtilis* (ATCC 6633) spores and incubated at 32°C for 16–18 hr (STOP) or with *Bacillus megaterium* (ATCC 9885) and incubated at 44°C (CAST). In positive cases, these screens are followed by a thin-layer bioautography assay.

In disc assay procedures the test sample is impregnated on filter paper disks (7 mm – 1.27 cm diameter) and placed on solidified agar that has been seeded with *Bacillus subtilis* (ATCC 6633) spores, *Bacillus stearothermophilus* (ATCC 10149) spores or *Bacillus megaterium* (ATCC 9885) spores have been used extensively to screen milk samples for residues of antibiotics.

5.3.1.2. *Enzymatic colourimetric assays*

Penzym test is a qualitative enzymatic colorimetric method for a rapid determination of β-lactam antibiotics in milk. The test principle is based on establishing the level of inactivation of the DD-carboxypeptidase enzyme by β-lactam antibiotics. These residues bind specifically with the enzyme and inactivate it, thus interfering with the bacterial cell wall formation. In the course of the Penzym test, lyophilized enzyme *Streptomyces* DD-carboxypeptidase is enclosed in an ampoule into which a sample of milk is introduced (50 μl). Then the incubation is observed for 5 min at 47°C. β-lactam antibiotics present in the sample create a stable complex with DD-carboxypeptidase in the sample. The degree of inactivation of the enzyme depends on the amount of antibiotics present in the sample. After a reagent tablet containing synthetic D-alanin oligopeptid and d-amino acid-oxidase is added, another incubation process follows for 8 min at 47°C.

If the sample does not contain β-lactam antibiotics residues, after the specific substrate is added, the enzyme DD-carboxypeptidase hydrolyses the tripeptide (Ac-l-Lys-d-Ala-d-Ala) onto the dipeptide (Ac-l-Lys-d-Ala) with a simultaneous release of D-alanin (D-Ala). The amount of the d-Ala released depends on the amount of the active DD-carboxypeptidase enzyme. Free D-Ala is further oxidized by D-amino-acid-oxidase onto pyruvic acid giving rise to hydrogen peroxide. The end products of the substrate and enzyme reaction (pyruvic acid and hydrogen peroxide) are measured using a redox colour indicator and the comparison of the final colour with the colour chart provided with the kit. Hydrogen peroxide is used to oxidize the organic redox indicator that will change into a pink-orange colour compound, indicating a negative result. If the colour is peach-coloured, the sample is close to the detection limit. If a yellow or yellow-orange colour is observed (test-positive outcome), the sample is suspected of containing an antibiotic residue (Mitchell *et al.*, 1998).

5.3.1.3. *Receptor binding assays*

The Delvotest-P-Ampule kit is based on similar screening procedure in which rapid growth and acid production of bacillus stearothermophilus causes whole of the agar medium to turn yellow in the absence of antibiotic inhibitors. However, the Delvo-X PRESS βL test is a qualitative, competitive, receptor-enzyme assay. Although analogous in principle to immunochemical assays, the Delvo-X-PRESS test does not use antibodies to bind specifically β-lactams and therefore cannot be classified as an immunochemical test. The test involves a reagent substance (a tracer) containing a conjugate—a specific receptor protein isolated from the bacteria of *Bacillus stearothermophilus* species conjugated with an enzyme (horseradish peroxidase). One of the reaction compounds is non-specifically adsorbed onto the surface of a solid-the test tubes contain an active specific â-lactam layer. The Delvo-X-PRESS kit uses a special workstation.

The workstation integrates a heater, a manual, a shaker and a reader unit. Via the display, the workstation gives the user step by step instructions. In the first step of the test, the measured volume of milk (0.2 ml) and the conjugate (enzyme bound onto the receptor protein), are mixed and incubated in a test tube (3 min at 64°C). The conjugate binds to the free β-lactam antibiotics which are present in the sample. Only the free conjugate (residual conjugate which did not bind to the β-lactam antibiotics present in the sample) is capable to bind onto the β-lactam layer of the test tube. The more free β-lactam antibiotics the reagent mixture contains, the fewer labeled receptors bind to the immobilized β-lactam antibiotics adsorbed onto the surface of the test tube and *vice versa*. After the removal of the conjugate-free β-lactam antibiotics complex with repeated rinsing of the test tube, an enzymatic substrate (colour developer) is added. It serves for the detection of the residual conjugate binding onto the β-lactam layer. The addition of the enzymatic substrate results in the formation of blue colour which is inversely proportional to the concentration of the β-lactam antibiotics in the sample. The colour is measured photometrically at 660 nm.

The same procedure is applied also in the test tube which contains, instead of the milk sample, 0.2 ml of penicillin G standard solution (concentration 5 ppb). The positivity of the sample is assessed photometrically by comparing optical densities (OD) of the sample and the standard. The samples yielding OD higher than 0 are classified as positive, while those yielding negative OD are assessed as negative (Mitchell *et al.*, 1998; Angelidis *et al.*, 1999; Botsoglou & Fletouris, 2001). In total, these screening assays are efficient monitoring procedures in that they are simple and rapid. However, they are non-specific and respond only to biologically active residues that inhibit the growth of the indicator organism.

5.3.1.4. *IDEXX SNAP test*

SNAP test is an enzyme-linked, receptor binding assay in which β-lactams are captured by a binding protein on a solid support adsorbent matrix housed in a moulded plastic unit. SNAP residues test consists of three components: SNAP device, pipette, sample tube. Using this test, penicillin can be detected in the amount of 4 ppb, ampicillin or amoxicillin in the amounts of 10 ppb, cephapirin 8 ppb, and ceftiofur 50 ppb. The SNAP test utilises a β-lactam receptor protein conjugated to an enzyme. The assay procedure includes three simple steps with a total assay time of about 10 minutes for a sample. In the first step of test, calibrated amounts of milk and conjugate are mixed and incubated in a test tube, placed in a heating block (5 min, 45 ± 5°C). The enzymeconjugate binds with β-lactams present in the milk sample. The mixture is then transferred to the sample well of the SNAP device (plastic unit containing sample and control spots on filter paper strip) where the sample is allowed to migrate on a filter paper strip until it passes to the test spot.

Test spots are coated with β-lactam antibiotic. Any free receptor will be captured at this spot, whereas the receptor protein that interacts with free β-lactams in the sample will not. The substrate is released and reacts with the enzyme attached to the captured receptor protein and a colour develops at the test spot. The results are read either visually or instrumentally (using reflectance) to provide the numerical interpretation of the visual result. The samples are declared positive or negative on the basis of the comparison of the intensity of the colour development between the sample and control spots on the SNAP test. If the colour of the test spot is weaker than that of the control spot, the result is interpreted as positive (Bell *et al.*, 1995 & Mitchell *et al.*, 1998).

5.1.3.5. *Beta star test*

The test involves a specific β-lactam receptor linked to gold particles. It is a dipstick test that detects penicillins and cephalosporins. The milk sample (0.2 ml) is added to a vial containing the test reagents (receptor protein linked to gold particles), mixed and incubated at 47.5°C in the incubator for 3 minutes. During incubation, the receptor will react with the free β lactams contained in the sample. After 3 min of incubation, the dipstick is added and incubation is continued (2 min at 47.5°C). The mixture is transferred to a strip of immunochromatography paper where it migrates towards the test field. With milk samples free of β-lactam residues, the receptor protein will be captured by a biomolecule immobilized at the test field of the chromatography paper.

Since the receptor protein is linked to gold particles, the captured protein-gold complex will appear as a pink-coloured band. With the sample where the receptor protein has interacted with free β-lactam molecules, the receptor protein will not be captured at the test field and no band will occur. The colour intensity of the test band is visually compared with that of the reference band: if the colour intensity of the test band is weaker than that of the reference band, the sample is classified as positive (Gustavsson & Sternesjö, 2004).

5.3.1.6. *Microbial receptor assays (CHARM I and II tests)*

CHARM I and II tests are qualitative microbial receptor assays. The CHARM I test developed for β-lactams in milk was the first rapid test recognised by the with a test time of 15 minutes. In 1984–1985, the extended version the CHARM I test has been referred to as CHARM II test (Botsoglou & Fletouris, 2001). The Charm II is a rapid radioisotopic assay procedure that can detect a numbers of antibiotics namely β-lactams, tetracyclines, macrolids, aminoglycosides, novobiocin, sulfonamides and chloramphenicols. The assay is based on a competition between labeled drug and residues in the milk sample for a limited number of specific binding sites on the surfaces of bacteria that are added to the samples.

The procedure involves the addition of the radiolabeled tracer antibiotic and the binding microorganism to the milk sample, a short incubation, centrifugation, fat removal, resuspension of microbial plug in a scintillation fluid, and counting. The greater the concentration of antibiotic residue, the less radiolabeled tracer will become bound to the microorganism. Two antibiotic groups can be assayed in each tube by using a combination of ^{14}C and ^{3}H tracer antibiotics. The various antibiotic types can be assayed in approximately 12 minutes, and with limits of detection that are very much lower than the more traditional diffusion assays.

5.3.1.7. *Immunoassays*

Immunochemical methods are based on the reaction of an antigen with an antibody—a reaction of antigenic determinants with the antibody linking site. The methods make use of monoclonal or polyclonal antibodies that are labeled with a radio-isotope, enzyme, fluorescence or chemical scintillation agent (Mitchell *et al.*, 1998; Roeder & Roeder 2000) to improve sensitivity. Nonisotopic immunoassays such as ELISA (enzyme linked immunosorbent assay), FPIA (fluorescence polarisation immunoassay), PCIA (particle-concentration immunoassay), PCFIA (particle- concentration fluorescence immunoassay), and monoclonal-based immunoassays will, in all likelihood, play an increasingly important role in antibiotics screening immunoassay determinations (Roeder & Roeder, 2000).

The most frequently used immunochemical method for rapid diagnostics of veterinary drug residues is enzyme immunoanalysis (EIA). As an enzyme label, horse radish peroxidase, alkaline phosphatase, glucose oxidase, pyruvate dehydrogenase and recombinant β-galactosidase are used. These enzymes catalyse the reactions that cause the substrates degradation and form coloured products that can be read spectrophotometrically or visually (Botsoglou & Fletouris, 2001). LacTec test (β-lactam) is based on the immunobinding principles of the enzyme-linked immunosorbent assay system (Botsoglou & Fletouris, 2001).

Among other methods is used *e.g.* the Fluorophos Beta Screen E.U. test—qualitative enzyme immunoassay based on fluorescence detection for the determination of six β-lactam antibiotics—amoxicillin, ampicillin, cloxacillin, dicloxacillin, oxacillin, penicillin G in raw milk. The method is based on competitive fluorescent immunoassays in glass capillary tubes. Most of immunoassays are normally performed in a laboratory and therefore it takes quite a long time before the outcome of the test is known. A number of immunochemical tests are commercially available in a kit format for many drugs (card format, one step strip test). A promising approach in the rapid antibiotics detection is the use of a dipstick format (lateral flow assay).

5.3.1.8. *Lateral flow assay* (*LFA*)

A lateral flow test comprises five different basic components: sample filter, conjugate pad (gold pad) is a fibre glass-like pad which can either be sprayed by, or bathed in gold-, latex- or carbon-conjugated solutions; nitrocellulose membrane are made to display a maze structure with various pore sizes; reservoir—absorption pad; test line and control lines. The sample is applied to the sample filter, housed inside the device casing. The sample runs through the sample filter and conjugate pad. This conjugate pad contains a labeled antibody, specific for the analysis under test. Antibiotics, if present, will form a complex with the conjugate and migrate further slowly to the membrane in the test zone. This zone contains an immobilised antibody, specific for the analysis, but preferably not competing with the conjugated antibody for the same or adjacent epitopes. The test line will thus capture the migrating analyte conjugate complex. The intensity of the test line correlates well with the amount of analyte in the sample (Van Herwijnen, 2006). Immunoassay can be sensitive, class specific, accurate, and can provide a means for a rapid screening of samples for antibiotics.

Lateral flow assay also known as solid state immunochromatography has a great demand for developing rapid and reliable analytical techniques to control multiple drug residues in foods. Several immunochromatographic assay (ICA) were developed and all of them had successful applications in residue analysis. Wang *et al.* (2006) proposed a simple and rapid extraction method for the detection of four sulphonamides in eggs and chicken muscles by LFA. Results from these studies proved the reliability of LFA technology for analyzing sulphonamides in food producing animals. Nevertheless, there are still some limitations to find a novel test strip that could detect simultaneously all sulphonamide compounds.

In the EU, the use of *steroid* and β-*agonist* to enhance animal growth is prohibited according to the council directive 96/22/EC. Monitoring anabolic steroids in meat-producing animals is therefore a challenging task. An LFA test for screening 19-nortestosterone (19-NT) in swine urine was reported by Liqiang *et al.* (2007). The test based on competitive assay format employed a colloidal gold-polyclonal antibody (pAb) conjugate against 19-NT (detector reagent) and 19-NT-ovalbumin conjugate (capture reagent) immobilized on a nitrocellulose membrane. The detection limit of the test strip to 19-NT was 200 ng/ml within 10 min.

However, the developed LFA cannot be applied for specific detection of 19-nortestosterone, and it would be more convenient to develop a monoclonal antibody against 19-nortestosterone. Geertruida *et al.* (2008) proposed a test strip against progesterone (PA). Again Tieming *et al.* (2006) proposed a colloidal gold-based LFAfor the detection of medroxyprogesterone acetate (MPA) residue. The developed LFA test was not very specific to medoxyprogesterone because of the high cross-reactivities values found with other structurally related compounds including Megestrol acetate (48%), Melengestrol acetate (31%), Chlormadinone acetate (43%), Estradiol (21%), 17â-estradiol 3-benzoate (16%), Progesterone (18%) and Pregnenolone (8%).

For Salbutamol (SALB), a commercial β-agonist a LFA was described by Khamta *et al.* (2009). The test strip was constructed by using SALB-BSA conjugate, goat anti rabbit IgG, and anti-SALB-colloidal gold conjugate (SALB-CGC) as test line, a control line, and a detector, respectively. The sensitivity of the assay was estimated to be at 80 ng per ml of SALB in PBS. In the cross-reactivity study, clenbuterol and chloramphenicol were used to probe probable interference with the test strip. With clenbuterol, the test line intensity slightly faded when assaying with standard solution at concentration 100 ng/ml. In contrast, chloramphenicol showed no competition with anti-SABL-CGC conjugate. Similarly, an LFA for the detection of β-adrenergic agonist Clenbuterol (CL) residues was proposed by Zhang *et al.* (2006). Result analyses of clenbuterol by ICA test strip were closely aligned with GC-MS method.

The quantitative analysis was introduced to the test strip format assay by scanning the relative optical density of the test capture lines, and the format of competitive lateral flow analysis agreed with the mathematical model proposed by Qian *et al.* (2004). The test strip could be used as suitable tool for Clenbuterol residues detection at ppb level in swine urine. Lai *et al.* (2007) also presented an ICA against CL. The assay allowed a detection limit at 3 ng per ml within 5 min. Finally, a validation study was carried out by analysing 210 urines samples, and the test strip compared to GC-MS gave a false positive rate of 4.4%, a false negative rate of 0% while the overall relative accuracy obtained was 96.7%. The test strip also showed no cross-reaction with tulobuterol, terbutaline, salbutamol, ritodrine and fenoterol at concentration 100 ng per ml. The CL test strip can be stored at room temperature storage for 12 months.

More recently, a test strip for simultaneous detection of clenbuterol (CL) and ractopamine (RAC) was developed by Zhang *et al.* (2009). The multi-analyte test of CL and RAC was based on competitive principle using colloidal gold-labeled anti-CL and anti-RAC polyclonal antibody and two capture test lines coated with BSA-CL and BSA-RAC. In the experiment protocol of sensitivity tests, a total of 20 blank urine samples obtained from different animals were analyzed in triplicate for both CL and RAC by using the CL and RAC simultaneous test strips with a scanner. The results from visual evaluation of the lateral-flow tests of spiked swine urine samples showed that the cut-off values of CL and RAC were 1.0 and 1.0 ng/ml, respectively. The limit of detection was 0.34 ng per ml for CL and 0.55 ng per ml for RAC in swine urine samples. With the cross-reactivity study, the anti-CL polyclonal antibody exhibited cross-reations with salbutamol (2.24%) and terbutaline (0.56%).

The anti-RAC polyclonal antibody showed cross-reactivity with isoxsuprine (0.284%) and ritodrine (1.42%). Interestingly, low cross-reactivity (<0.071%) was observed with other β-agonists including Bamethane, Cimbuterol, Cimetro, Fenoterol, Isoproterenol, Salmeterol, Zilpaterol and Ractopamine. The developed test strips method was reliable for the CL and RAC detection in swine urine samples. Of particular importance was the fact that anti-CL pAb did not cross-react with RAC and anti-RAC pAb did not cross-react with CL. The method could offer advantages of ease of sample preparation and high throughput. These works demonstrate the potential application of ICA technique to screen β-agonists in food producing animals.

The use of *chloramphenicol* (CAP) in food producing animals is banned because of its associated serious side effects in human reported elsewhere. Li *et al.* (2007) proposed a colloidal gold-based ICA to detect CAP residues in aquaculture tissues. The results indicated good correlation (r^2= 0.97) between the data obtained from both the two methods. The detection limit of CAP in aquaculture tissues was 10 ng per g. In this study, the LFA method provided only a preliminary, semi-quantitative result, however it could be used to judge whether a contaminated fish sample contains CAP at concentration exceeding the detection limit. The *aminoglycosides* are broad-spectrum antibiotics that have bactericidal activity against some gram-positive and many gram-negative organisms. Their clinical use has been limited by the EU because of the side effects of nephrotoxicity and ototoxicity.

A colloidal gold-based ICA for rapid detection of kanamycin and trobramycin in swine tissues was described by Chen *et al.* (2008). The test showed a visual detection limit at 5 ng per ml in PBS, 50 μg per kg in meat or liver, and 100 μg per kg in kidney. The detection time of the assay took only 5–10 min. For suitability, the results from this study have to be validated by using a confirmatory method. *Fluroquinolonones* (FQs) are also used as antimicrobial drugs in animal farmers. A colloidal gold-based LFA test strip for the detection of euroflaxin in chicken tissues was described by Zhao *et al.* (2008). The method showed a detection limit at 0.138 μg per kg of euroflaxin as measured in a strip scanner. Recoveries of the assay with tissue samples spiked at 10, 20 and 30 μg/kg ranged 85.3–96% with an acceptable coefficient of variations at 4.5–7.91%. Parallel analysis of muscle samples from animal experiments showed satisfactory results between the test strip and LC-MS method.

The test had the advantages of simplicity and speed of performance (requires only 5 min), and it could be a useful screening method for quantitative, semiquantitative, or qualitative detection of euroflaxin residues in chicken muscles. More recently, Zhu *et al.* (2008) presented a novel LFA to detect several FQs. The test strip used a monoclonal antibody with broad specificity to twelve FQs including norfloxacin, pefloxacin, enrofloxacin, ciprofloxacin, norfloxacin, flumequine, pefloxacin, ofloxacin, lomefloxacin, enoxacin, danofloxacin, amifloxacin, oxolinic acid and marbofloxacin. The detection limits of FQs in spiked chicken muscle and chicken liver samples were 25 ng per ml for norfloxacin and pefloxacin. For the other FQs, the lowest detection limit was 50 ng per ml. The whole process involving sample preparation and detection can be finished within 10 min. The developed ICA method can be potentially used as a screening tool for the determination of FQ residues in a large amount of samples on site. However, a confirmatory method is required to validate the obtained data.

5.3.1.9. *Biosensor*

Biosensor achievements have revolutionized the detection method and provide us with simple to use device, cost-effective, rapid and appropriate detection method that give immediate and accurate results. The results are comparable to or better than the conventional analytical systems in terms of performance *i.e.*, reliability, sensitivity, selectivity, specificity and robustness. Itcan identify the contaminants much faster, more efficient and can give effective real time monitoring of drug residues and most importantly ensuring customer safety

(Mandal *et al.*, 2011). Variety of biosensing methods has been developed but most of the biosensors developed are aimed at determining them in biological or food samples. The SPR technique has been developed and demonstrated for on-line/at-line detection of veterinary drug residues in milk, porcine bile and bovine urine, including a commercial handling robot. This sensor operates in real time and it may detect up to 8 different veterinary drugs simultaneously with a throughput of up to 600 samples per day.

Surface plasmon resonance (SPR) techniques are having a major impact on the development of new optical biosensors. Sulfamethazine has been determined by Akkoyun *et al.* (2000) with an optical immunosensor in animal urine. Hansen and Sorensen (2000) presented three different reporters gene systems from *V. fischeri*, *E. coli*, and *Aequorea victoria* all combined with a tetracycline inducible promoter in the development of three corresponding whole-cell biosensors. They respond to low levels of tetracyclines by producing galactosidase, light or green fluorescent protein, respectively. In the field of food monitoring, different biosensors were able to determine penicillin G (Setford *et al.*, 1999) or tetracyclines (Hansen & Sorensen, 2000), both in milk.

Bacterial spores appears to have great potential for their application as biosensor as they have the ability to sense environmental changes and to respond using explosive molecular mechanisms that transform dormant spores into rapid growing cells. There are a great number of bacterial species which produce spores for example; genus Bacillus (widely dispersed in soil, plant matter, and air) may be readily grown in the laboratory to form spores: *B. cereus*, *B. licheniformis*, *B. megaterium*, *B. sphaericus*, *B. stearothermophilus*, *B. subtilis*, and *B. thuringiensis*. For the development of bacterial spore as a biosensor, it is a prerequisite to have a complete or descriptive knowledge regarding their germinant which by their action on the dormant spores convert them into vegetative cells.

The germination process of a whole population of spore may be completed in a very short duration of time (15–30 min) followed by a sequence of metabolic reactions and synthesis of enzymes resulting in outgrowth of vegetative cells. After germination de novo acetyl esterase is released from the core of the spore which act upon DAF and its hydrolysis results in fluorescence and the signal can be captured using optical device to quantify the presence of target analyte (Rotman, 2001). An analytical process based on the resistance mechanism of some β-lactamase generating *Bacillus* spp. Some spore forming bacteria such as *B. cereus* and *B. licheniformis* produce β-lactamase enzyme due to induction by β-lactam antibiotics and the enzyme production is proportional to the concentration of inducer present in milk.

A real time microbial assay based on β-lactamase enzyme using starch iodine as color indicator has been developed. The microbial assay is working on principle of non competitive enzyme action on inducer (β-Lactam) resulting in indirect reduction of starch iodine mixture through penicilloic acid. A comparison of the intensity of the test reaction with that of a control was taken as criteria to determines whether the sample is positive or negative (Kumar *et al.*, 2009). The assay can detect specifically β-lactam groups in spiked milk within 15–20 min at regulatory codex limits with negligible sensitivity towards non β-lactam groups. The presence of Inhibitors other than antibiotic residues in milk did not interfere with the working principle of microbial assays.

5.3.1.10. *High performance thin-layer chromatography (HPTLC)*

HPTLC allows the qualitative and quantitative detection of multi-residues in meat but its use has rapidly decreased due to the expansion of other techniques like HPLC. Reported uses of HPTLC applied to meat include the detection of residues like clenbuterol and other agonists (Degroodt *et al.*, 1989) nitroimidazol (Gaugain & Abjean, 1996) and sulphonamides (Van Poucke *et al.*, 1991) and thyreostatic drugs (De Wasch *et al.*, 1998). The plates are sprayed with an appropriate chromogenic reagent or viewed under UV light for visualisation of compounds. Detection by fluorescence is also applied. Quantitation is achieved by measuring the relative intensity of the spot vs that of the internal standard by scanning densitometry. Modern HPTLC has been automatised at a high level.

5.3.1.11. *High performance liquid chromatography (HPLC)*

HPLC methodology is now widely practiced for detection of veterinary drug residues in livestock products because of column availability, good performance, variety of available detectors and possibility of automation. Recent developments like the ultra performance liquid chromatography systems or types of columns with improved packagings in terms of smaller size, geometry and inertness are also very valuable. HPLC is a separative technique where the choice of the detection system is very important for selectivity and sensitivity. Some analytes require chemical modifications to render chromophore, fluorescent or UV-absorbing compounds. Typical detections of multi-residues in meat samples are relatively simple and rapid, requiring a preliminary clean-up through solid-phase extraction followed by filtration before injection into a reverse-phase HPLC with diode array detection.

This procedure has been applied to meat for detection of antibiotics like quinolones (Kirbir *et al.*, 2003), sulphonamides (Prasad *et al.*, 2005; Biswas *et al.*, 2007a), β-lactams and macrolides and tetracyclines (Biswas *et al.*, 2007b), veterinary drugs (Reig *et al.*, 2005) , anabolic steroids (Decock *et al.*, 2001) and corticosteroids like dexamethasone (Reig *et al.*, 2006). In some cases, the compounds can be further identified through diode array or fluorescence detection. Ten quinolone residues (ciprofloxacin, danofloxacin, difloxacin, enrofloxacin, flumequine, marbofloxacin, nalidixic acid, norfloxacin, oxolinic acid, sarafloxacin) in meat have been screened and confirmed with HPLC and fluorescence detection.

5.3.2. *Confirmatory methods*

LC–MS–MS is an instrumental technique that has been established for confirmation of residues in foods as per statutory guidelines. LC–MS procedures enable the determination of compounds with high molecular weights as well as polar, non-volatile and thermolabile analytes without derivatization. However, for complex matrices, such as liver, kidney, muscle or urine used for veterinary drugs, an appropriate sample preparation is necessary in order to avoid matrix effects (*e.g.*, ion suppression) in the ionization processes. For the quantitative determination of β-lactam antibiotics in milk and tissues Rabbolini *et al.* (1998) using an ion trap. McLaughlin *et al.* (1994) developed a multiresidue method for the confirmation of aminoglycosides in bovine kidney by LC–positive electrospray triple quadrupole MS–MS. The interest in LC–MS–MS as a confirmatory method for chloramphenicol has been limited due to the availability of many excellent GC–MS procedures.

Recently Gantverg *et al.* (2003) suggested that LC–MS–MS offered superior sensitivity and selectivity to GC–MS. Ito *et al.* (2000) employed a parallel HPLC and LC–ESI–MS–MS approach for the respective quantification and confirmation of 10 sulfonamides in bovine, swine, and chicken in liver and kidney. For sample preparation, ethyl acetate extract was reconstituted in 50% ethyl acetate-*n*-hexane after evaporation to dryness. Oka *et al.* (1997) developed a method for the analysis of TC, oxytetracycline (OTC), chlortetracycline (CTC) and doxycycline (DC) based on C_{18} SPE clean up followed by electrospray triple quadrupole analysis in the positive mode. Pleasance *et al.* (1992) was reported to use of LC–MS–MS method for determination of erythromycin A in salmon tissues. A simple extraction with acetonitrile was followed by centrifugation and filtration. The filtrate was washed with hexane prior to basification and extraction with dichloromethane. LC–MS and LC–MS–MS of the extract reconstituted in methanol gave detection limits of <0.01 and 0.05 μg/kg, respectively.

The determination of residues of the fluoroquinolone, danofloxacin in cattle and chicken liver, using positive mode electrospray triple quadrupole MS–MS, was reported by Schneider *et al.* (1993). The multiresidue determination of four quinolones namely, flumequine, oxolinic acid, nalidixic acid and piromidic acid, in catfish tissue, was described by Doerge and Bajic (1995). Nitrofurans are known to be unstable and are rapidly metabolized *in vivo*. Typically furazolidone yields the intact side-chain product, 3-amino-2-oxazolidinone (AOZ), which occurs as a free or bound residue in tissue. Horne *et al.* (1996) reported a procedure for the determination of protein bound nitrofurans metabolites. In general bound residues are released under mild acidic conditions and are subsequently derivatized with 2-nitrobenzaldehyde. Recently, HPLC–UV methods were used for the routine screening analysis and LC–MS–MS with APCI or ESI techniques for the confirmatory analysis of nitroimidazole residues in different edible tissues (Sams *et al.*, 1998).

In the past the determination of benzimidazole residues by HPLC with UV detection or gas chromatography after derivatization was frequently used. For confirmatory analysis several methods, based on gas chromatography and mass spectrometry (GC–MS), after silylation (Wilson *et al.*, 1991), methylation and benzylation (Marti *et al.*, 1990) were published. Gas chromatography with high-resolution mass spectrometry (HR-MS) for the identification of thiabendazole, albendazole and fenbendazole residues in muscle tissue was also reported. However, methods based on GC are not suitable for quantification due to thermal decomposition in the GC or due to non-reproducible ionization in the source of the mass spectrometer. The mebendazole together with its hydrolyzed and reduced metabolites in sheep liver and muscle tissues could be analyzed using LC–MS–MS with positive electrospray ionization (ESI+). This quantitative method was optimized and validated according to the new EU requirements with incurred liver and muscle tissue after oral treatment of a sheep. The validation limit was set for sheep muscle tissue at 10 μg/kg.

Several authors (Govaert *et al.*, 1998 & Delahaut *et al.*, 2003) have published the detection and quantification of methods for tranquilizers and the beta-blocker carazolol. Govaert *et al.* (1998) developed a multiresidue method to confirm the presence of the above mentioned compounds using LC–MS–MS (ESI+) in MRM mode. Many methods for measuring anabolic compounds have been developed using GC–MS, since this method provides good sensitivity and is sufficiently specific for use as a confirmatory technique. However, GC–MS required the derivatization of the steroids using silylation, acylation or oxime/silylation reactions (Casademont *et al.*, 1996), depending on the individual properties of the steroid. Confirmatory method for growth promoting agents by LC–MS was also developed. The application of the new EU criteria was demonstrated for the analysis of 16 β-hydroxy stanozolol, 17 α/β-trenbolone, melengestrol acetate, methylboldenone and dexamethasone in bovine urine.

6. PUBLIC HEALTH RISK

Varieties of antimicrobial drugs including antibiotics are used to control the bacterial infections and parasitic infestation. All drugs have side effects when they are exposed to human and animals with higher dose or prolonged time than recommended. These types of toxicity are elucidated by small amount of residual drugs in foods. Sub-therapeutic use of antimicrobial drugs induces resistance of microorganisms that cause difficulties in treatment of animal diseases, and consequently incurs abuse and tissue residue of drugs (Lee *et al.*, 2001). The overuse of antibiotics (tetracyclines, sulfonamides, amino glycosides, β-lactam derivatives) in animal production or their residues in food system pose potential allergic reactions in sensitized individuals, but sub therapeutic and therapeutic levels may perturb human gut micro flora (Paige *et al.*, 1997). The National Research Council and Institute of Medicine have noted a link between the use of antibiotics in food animals and the development of bacterial resistance to these drugs causing human diseases (NRCIM, 1999).

Similarly, scientist at Centers for Disease Control and Prevention (CDC) began tracking a new type of *Salmonella* called Newport 9+ which is resistant to nine antibiotics including ceftriaxone, one of the few drugs that killed most bacteria and the drug of choice for children whose *Salmonella* enter the blood stream (Schmidt, 2002). Nonetheless, toxicity of these antimicrobial chemicals includes aplasia of bone marrow, the emergence of resistant bacteria within animals and the transfer of antibiotic resistance gene to human pathogens. The appearance of resistance among pathogenic organisms such as *Salmonella* DT-104 and *Campylobacter* is more concern (Smith *et al.*, 1999 & Glynn *et al.*, 1998). *E. coli* is an opportunist pathogen capable of infecting people via the food chain and causing enteric infections in young children and travelers as well as range of other infections.

Enteric infections with *Salmonella*, *E. coli* or *Campylobacter* rarely warrant antibiotic treatment and so one might argue that the problem is not nearly as important as methicillin resistant *Staphylococcus aureus* (MRSA) or other major human resistance problems. However, treatment failure has been reported (Smith *et al.*, 1999; Fey *et al.*, 2000) and the livestock industries cannot ignore the problem. Enterococci have only been thought of as foodborne organisms since discovery of vancomycin resistant enterococci (VRE) in pigs and poultry and there is some dispute that their spread occurs from animals to people (Barton, 2000).

The tetracyclines (CTC, OTC and TC) are broad-spectrum antimicrobials that are widely used in human and veterinary medicine to treat a variety of bacterial infections. The tetracyclines are incompletely absorbed from the gastrointestinal tract; they reach readily high concentrations in the intestine, producing perturbations of the intestinal microflora within 48 hours of daily treatment. Experience with tetracyclines in human medicine indicates that therapeutic levels of tetracyclines can perturb the intestinal microflora by inducing emergence of resistant strains and altering the metabolic activity of the microflora, its resistance to colonization by pathogenic, opportunistic or resistant microorganisms barrier effect and its ecological balance, without any identified deleterious effect (JECFA, 1999). Immunodepression and phototoxicity may also occur in animals and human beings besides superinfections related to tetracyclines.

Treatment with oxytetracycline during the second month of pregnancy presents a teratogenic risk to the foetus (Czeizel *et al.*, 1998; Czeizel & Rockenbauer, 2000). As an undesirable side effect, OTC not only discolours the primary and permanent teeth but also causes hypoplasia in developing teeth when administered to the infants, mothers during last two trimesters of pregnancy and children under 12 years of age. However, unwanted risk highest when OTC is given to neonates and babies prior to the first dentition (Tanase *et al.*, 1998).

However, sulfonamides (sulfadimidine and sulfamethoxazole) can induce thyroid, adenoma and hyperplasia in laboratory animals. It has also potential carcinogenic character. To reduce bacteria resistance to sulfonamides to get synergistic effect, pyrimethamines such as trimethoprim and oriprim are used in combination. Other antibiotics like chloramphenicol, β-lactams and aminoglycosides, nitrofurans, anthelmintics, growth promoters and coccidiostats are highly carcinogenic, and they are banned or restricted from use in animal production systems.

7. REGULATION

The USDA-FSIS and the FDA are responsible for monitoring meat and poultry products for animal drug residues. The USDA-FSIS conducts the National Residue Programme (NRP) to prevent animals containing violative amounts of drug residues from market samples through extensive on-site sampling technique. The samples are also sent to FSIS field laboratories for further testing. The FDA Center for Veterinary Medicine (CVM) is responsible for approving new animal drugs, setting tolerances, and enforcement action depends on results of FSIS findings. Other international organizations include the European Agency for the evaluation of medicinal products (EMEA), Office International des Epizootics (OIE) and Consultation Mondiale de I' Industrie de la Sante Animale (COMISA).

Many countries have specialist groups involved such as FDA in USA, the Bureau of Veterinary Drugs in Canada and the Veterinary Products Committee of Ministry of Agriculture, Fisheries and Foods in the United Kingdom. In 1961, the FAO Conference decided to set up the Codex Alimentarius Commission (CAC) and in 1963 the World Health Assembly adopted the Statutes of the CAC and approved the establishment of the Joint WHO/FAO Food Standards Programme. At its 16th Session, the CAC agreed to create the Codex Committee on Residues of Veterinary Drugs (CCRVD) and, finally, this Committee designated the Joint FAO/WHO Expert Committee on Food Additives (JECFA) as the independent scientific body responsible for the assessment of veterinary drugs in relation to MRLs. As result of the work of these committees, several documents concerning the monitoring of residues of veterinary drugs have been issued in recent years.

In 1993, the "Codex Guidelines for the Establishment of a Regulatory Programme for Control of Veterinary Drug Residues in Food" was published. The document is structured in three parts. The first part contains instructions for sampling of commodities such as meat and poultry products, fish, milk, and honey. The second part deals with general considerations on the analytical methods used in determining compliance with MRLs. The third part specifies the performance characteristics of these methods. This document is currently under revision for updating the procedure for validation of the analytical methods used to assess MRLs compliance has received priority attention from the CAC.

REFERENCES

Adrian, J., Pasche, S., Pinacho, D.G., Font, H., Dieserens, J.M., Sa´nchez-Baeza, F., Granier, B., Voirin, G. and Marco, P. (2009). Wavelength-interrogated optical biosensor for multi-analyte screening of sulfonamide, fluoroquinolone, β-lactam and tetracycline antibiotics in milk. *Trends Anal. Chem.*, 28: 769–777.

Akkoyun, A., Kohen, V.F. and Bilitewski, U. (2000). Detection of sulphamethazine with an optical biosensor and anti-idiotypic antibodies. *Sens. Actuators B*, 70: 12–18.

Allan, R.J. and Watson, T.R. (1982). Identification of biliary metabolites of Mebendazole in the rat. *Eur. J. Drug Metab. Pharmakokinet*, 7: 131–136.

Angelidis, A.S., Farver, B.T. and Cullor, J.S. (1999). Evaluation of the Delvo-X-Press assay for detecting antibiotic residues in milk samples from individual cows. *J. Food Prot.*, 62: 1183–1190.

Barton, M.D. (2000). Antibiotic use in animal feed and its impact on human health. *Nutr. Res. Dev.*, 13: 1–22.

Bell, C., Rhoades, J.R., Neaves, P. and Scannella, D. (1995). An evaluation of the IDEXX SNAP test for the detection of beta-lactam antibiotics in ex-farm raw milks. *Netherlands Milk and Dairy Journal*, 49: 15–25.

Biswas, A.K., Rao, G.S., Kondaiah, N., Anjaneyulu, A.S.R. and Malik, J.K. (2007a). Simple multi-residue method for monitoring of trimethoprim and sulfonamide residues in buffalo meat by HPLC. *J. Agric. Food Chem.*, 55(22): 8845–8850.

Biswas, A.K., Rao, G.S., Kondaiah, N., Anjaneyulu, A.S.R., Mendiratta, S.K., Prasad, R. and Malik, J.K. (2007b). A simple multi-residue method for determination of oxytetracycline, tetracycline and chlortetracycline residues in export buffalo meat by HPLC- Photo-diode-array detector. *J. Food and Drug Analysis*, 15(2): 278–284.

Biswas, A.K., Kondaiah, N., Anjaneyulu, A.S.R. and Rao, G.S. (2009). Simple simultaneous determination of cypermethrin and deltamethrin residues in buffalo meat using HPLC equipped with photodiode-array detector. *Fleischwirt- Int.*, 24(1): 82–86.

Botsoglou, N.A. and Fletouris, D.J. (2001). Drug Residues in Foods, Pharmacology, Food,Safety, and Analysis. Marcel Dekker, New York.

Casademont, G., Perez, B. and. Garcia Regueiro, J.A. (1996). Simultaneous determination, in calf urine, of twelve anabolic agents as heptafluorobutyryl derivatives by capillary gas chromatography-mass spectrometry. *J. Chromatogr B.*, 686: 189–198.

Chen, Z.W., Zhanhui, W., Shusheng, T., Yan, Z. and Xilong, X. (2008). Rapid enzyme-linked immunosorbent assay and colloidal gold immunoassay for kanamycin and tobramycin in swine tissues. *J. Agric. Food Chem.*, 56: 2944–2952.

Czeizel, A.E. and Rockenbauer, M. (2000). A population-based case-control teratologic study of oral OTC treatment during pregnancy. *Eur. J. Obs. Gynaecol. Repro. Biol.*, 88: 27–33.

Czeizel, A.E., Rockenbauer, M. and Olsen, J. (1998). Use of antibiotics during pregnancy. *Eur. J. Obs. Gynaecol. Repro. Biol.*, 18: 1–8.

Delahaut, P., Levaux, C., Eloy, P. and Dubois, M. (2003). Validation of a method for detecting and quantifying tranquillisers and a β-blocker in pig tissues by liquid chromatography—tandem mass spectrometry. *Anal. Chim. Acta*, 483: 335–340.

De Ruyck, H., Daeselerie, E., De Ridder, H. and Van Renterghem, R. (2003). Liquid chromatographic-electrospray tandem mass spectrometric method for the determination of mebendazole and its hydrolysed and reduced metabolites in sheep muscle. *Anal Chim. Acta*, 483: 111–123.

De Cock, K.J.S., Delbeke, F.T., Van Eenoo, P., Desmet, N., Roels, K. and De Backer, P. (2001). Detection and determination of anabolic steroids in nitrtional supplements. *J. Pharm Biom. Analysis*, 25: 843–852.

Degroodt, J.M., Bukanski, B.W., Beernaert, H. and Courtheyn, D. (1989). Clenbuterol residue analysis by HPLC-HPTLC in urine and animal tissues. *Zeitschrift fur Lebensmitiel Untersuchung and Forschung*, 189: 128–131.

Doerge, D.R. and Bajic, S. (1995). Multiresidue determination of quinolone antibiotics using liquid chromatography coupled to atmospheric-pressure chemical ionization mass spectrometry and tandem mass spectrometry. *Rapid Commun. Mass Spectrom.*, 9: 1012.

De Wasch, K., De Brabender, H.F., Van Ginkel, L.A., Spaan, A., Sterk, S.S. and Meiring, H.D. (1998). Confirmation of residues of thyreostatic drugs in thyroid glands by multiple mass spectrometry after thin-layer chromatography. *J. Chromatogr A*, 819: 99–111.

Ellwanger, A., Berggren, C., Bayoudh, S., Crecenzi, C., Karlsson, L., Owens, P.K., Ensing, K., Cormack, P., Sherrington, D. and Sellergren, B. (2001). *Analyst*, 126: 784.

Fey, P.D., Safranek, T.J., Rupp, M.E., Dunne, E.F., Ribot, E., Iwen, P.C., Bradfors, P.A., Angulo, F.J. and Hinrichs, H. (2000). Ceftriaxone-resistant salmonella infection acquired by a child from cattle. *New England J. Medicine*, 342: 1242–1249.

Fontanals, N., Marcé, R.M. and Borrull, F. (2005). New hydrophilic materials for solid-phase extraction. *Trends Anal. Chem.*, 24: 394.

Gantverg, A., Shishani, I. and Hoffman, M.K. (2003). Determination of chloramphenicol in animal tissues and urine: Liquid chromatography-tandem mass spectrometry versus gas chromatography-mass spectrometry. *Anal Chim. Acta*, 483: 125–135.

Gaugain, M. and Abjean, J.P. (1996). High-performance thin-layer chromatographic method for the fluorescence detection of three nitroimidazole residues in pork and poultry tissue. *J Chromatogr A*, 737: 343–346.

Geertruida, A., Posthuma, T., Jakob, K. and Amerogen, A.V. (2008). Development of a competitive lateral flow immunoassay for progesterone: Influence of coating conjugates and buffer components *Anal Bioanal. Chem.*, 392: 1215–1223.

Glynn, M.K., Bopp, C., Dewitt, W., Dabney, P., Mokhtar, M. and Angulo, F.J. (1998). Emergence of multidrug resistant *Salmonella enterica* serotype *typhimurium* DT104 infection in the United States. *New England J Medicine*, 19: 1333–1338.

Govaert, Y., Batjoens, P., Tsilikas, K., Degroodt, J.M. and Srebrnik, S. (1998). Multi-residue analysis of tranquillizers in meat: Confirmatory assays using mass spectrometry. *Analyst*, 123: 2507–2512.

Gustavsson, E. and Sternesjo, A. (2004). Biosensor analysis of β-lactams in milk: Comparison with microbiological, immunological, and receptor-based screening methods. *J. AOAC Int.*, 87: 614–620.

Haasnoot, W., Cazemier, G., Koets, M. and Amerongen, A.V. (2003). Single biosensor immunoassay for the detection of five aminoglycosides in reconstituted skimmed milk. *Anal. Chim. Acta*, 488: 53–60.

Hansen, L.H. and Sorensen, S.J. (2000). Detection and quantification of tetracyclines by whole cell biosensors. *FEMS Microbiol. Lett.*, 190: 273–278.

Horne, E., Cadogan, A., O'Keefe, M. and Hoogenboom, L.A.P. (1996). Analysis of protein-bound metabolites of furazolidone and furaltadone in pig liver by high-performance liquid chromatography and liquid chromatography-mass spectrometry. *Analyst*, 121: 1463–1468.

Ito, Y., Ikai, Y., Oka, H., Matsumoto, H., Miyazaki, Y., Takeba, K. and Nagase, H. (2001). Application of ion-exchange cartridge clean-up in food analysis IV.

Confirmatory assay of benzylpenicillin, phenoxymethylpenicillin, oxacillin, cloxacillin, nafcillin and dicloxacillin, in bovine tissues by liquid chromatography-electrospray ionization tandem mass spectrometry. *J. Chromatogr A*, 911: 217–223.

JECFA. (1999). Evaluation of certain veterinary drug residue in food. *Fiftieth report of the Joint FAO / WHO Expert Committee on Food Additives*. WHO Technical Report Series, No.888.

Khamta, Y., Pattarawarapan, M., Nangola, S. and Tayapiwatana, C. (2009). Development of immunochromatographic assay for the on-site detection of salbutamol. *J. Immunoassay Immunochemistry*, 30: 441–456.

Kirbi, A., Marinsek, J. and Flajs, V.C. (2005). Introduction of the HPLC method for the determination of quinolone residues in various muscle tissues. *Biomed Chromatogr*, 19: 259–265.

Karsourd, G.O., Boison, J.O., Nouws, J.F. and MacNeil, J.D. (1998). Bacterial inhibition tests used to screen for antibacterial veterinary drug residues slaughtered animals. *J. AOAC-Int.*, 81: 21–24.

Kumar, N., Das, S. and Manju, G. (2009). A kit for detection of β-lactam antibiotic group in milk using bacterial spore as biosensor (Patent Reg # IPR/115/del/2009).

Lai, W., Xu, Y., Fung, D.Y.C. and Xiong, Y. (2007). Development of a lateral-flow assay for rapid screening of the performance-enhancing sympathomimetic drug clenbuterol used in animal production; food safety assessments. *J. Clin. Nutr.*, 16: 106–110.

Lee, M.H., Lee, H.L. and Ryu, P.D. (2001). Public health risks: Chemical and antibiotic residues. *Asian-Australian J Animal Sci.*, 14(3): 402–413.

Lehotay, S.J., Mastovska, K. and Yun, S.J. (2005). Evaluation of two fast and easy methods for pesticide residue analysis in fatty food matrices. *J. AOAC Int.*, 88: 630.

Leitner, A., Z¨ollner, P. and Lindner, W. (2001). Determination of the metabolites of nitrofuran antibiotics in animal tissue by high-performance liquid chromatography–tandem mass spectrometry. *J Chromatogr A*, 939: 49–58.

Li, C., Wang, Z.H., Cao, X.Y., Beier, R.C., Zhang, S.X., Ding, S.Y., Li, X.W. and Shen, J.Z. (2008). Both incidence and severity of white rust disease reflect host resistance in *Brassica juncea* germplasm from Australia, China and India. *J. Chromatogr A*, 1209: 1–8.

Li, K., Liu, L., Xu, C.L. and Chu, X.G. (2007). Rapid Determination of chloramphenicol residues in aquaculture tissues by immunochromatographic assay. *J Anal Sciences*, 23: 1281–1284.

Liqiang, L., Chifang, P., Zhengyu, J. and Chuanlai, X. (2007). Development and evaluation of a rapid lateral flow immunochromatographic strip assay for screening 19-nortestosterone. *J Biomed Chromatogr*, 21: 861–866.

Mandal, P.K., Biswas, A.K., Choi, K. and Pal, U.K. (2011). Methods for rapid detection of foodborne pathogens: An overview. *Am. J. Food Technol.*, 6: 87–102.

Mandell, G.L. and Sande, M.A. (1985). Antimicrobial agents, in Goodman and Gilman's *The Pharmacological Basis of Therapeutics*. Macmillan Publishing Company, pp. 1115–1149.

Marti, A.M., Mooser, A.E. and Koch, H. (1990). Determination of benzimidazole anthelmintics in meat samples. *J. Chromatogr.*, 498: 145–157.

McLaughlin, L.G., Henion, J.D. and Kijak, P.J. (1994). Multi-residue confirmation of aminoglycoside antibiotics and bovine kidney by ion spray high-performance liquid chromatography/tandem mass spectrometry. *Biol. Mass Spectrom.*, 23: 417–429.

Miller, D.J.S. (1993). Present state and trends in the use of veterinary drugs. *Proceeding of the Euro Residue II Conference*, Veldhoven, The Netherlands, 3–5 May.

Mitchell, J.M., Griffith, M.W., McEwen, S.A., MeNabb, W.B. and Yee, A.J. (1998). Antimicrobial drug residues in milk and meat: Causes, concerns, prevalence, regulations, test and test performance. *J. Food Prot.*, 61: 742–56.

Mohamed, R., Richoz-Payot, J., Gremaud, E., Mottier, P., Yilmaz, E., Tabet, J.C. and Guy, P.A. (2007). *Anal Chem.*, 79: 9557.

Mortier, L., Daeseleire, E. and Delahaut, P. (2003). Simultaneous detection of five coccidiostats in eggs by liquid chromatography–tandem mass spectrometry. *Anal Chem. Acta*, 483: 27–37.

Nightingale, C.H., Quintiliani, R. and Nicolau, D.P. (1994). Intelligent dosing of antimicrobials. *In*: Remington, J.S. and Swartz, M.N. (*Eds.*). *"Current Clinical Topics in Infectious Diseases"*. Cambridge, M.A., Blackwell Scientific Publications, pp. 252–265.

NRCIM. (1999). The use of drugs in food animals: Benefits and risks. *Natural Research Council and Institute of Medicine*, National Academy Press, Washington, DC.Nelson, G.L. (1987). Regulatory aspects of fire toxicology. *Toxicology*, 47: 181–199.

Oka, H., Ikai, Y., Haggan, A.S., Ito, Y., Hayakawa, J., Harada, K., Suzuki, M., Odani, H. and Maeda, K. (1997). Improvement of chemical analysis of antibiotics XXIII. Identification of residual tetracyclines in bovine tissues by electrospray high-performance liquid chromatography-tandem mass spectrometry. *J. Chromatogr B.*, 693: 337–344.

Paige, J.C. (1994). Analysis of tissue residues. *FDA Vet.*, 9(6): 4–6.

Paige, J.C., Tollefson, L. and Miller, M. (1997). Public health and drug residues in animal tissues. *Vety. Human Toxicol.*, 30: 162–169.

Pellegrini, G.E., Carpico, G. and Coni, E. (2004). Electrochemical sensor for the detection and presumptive identification of quinolone and tetracycline residues in milk. *Anal. Chim. Acta*, 520: 13–18.

Pleasance, S., Kelly, J., LeBlanc, M.D., Quilliam, M.A., Boyd, R.K., Kitts, D.D., McErlane, K., Bailey, M.R. and North, D.H. (1992). Determination of erythromycin A in salmon tissue by liquid chromatography with ionspray mass spectrometry. *Biol. Mass Spectrom.*, 21: 675–687.

Prasad, R., Rao, G.S., Biswas, A.K., Sarkar, S.N., Kumar, D. and Malik, J.K. (2005). Tissue residues of sulfamethoxazole and trimethoprim after oral administration of sulfamethoxazole-trimethoprim combination in broiler chickens. Presented in National Seminar on "Veterinary pharmacology in post WTO era" and 5th Annual Conference of ISVPT. (pp69), 24–26 November.

Qian, S. and Bau, H.H. (2004). Analysis of lateral flow biodetectors: Competitive format. *Anal. Biochem.*, 326: 211–224.

Rabbolini, S., Verardo, E., Da Col, M., Gioacchini, A.M. and Traldi, P. (1998). Negative ion electrospray ionization tandem mass spectrometry in the structural characterization of penicillins. *Rapid Commun. Mass Spectrom.*, 12: 1820–1826.

Reig, M., Batlle, N., Navarro, J.L. and Toldra, F. (2005). A modified HPLC method for the detection of 6-methyl-2-thiouracil in cattle urine. Proceedings of international congress in meat science and technology (p. 90), Baltimore, Maryland, August 7–12.

Reig, M., Mora, L., Navarro, J.L. and Toldra, F. (2006). A chromatography method for the screening and confirmatory detection of dexamethasone. *Meat Sci.*, 74: 676–680.

Roeder, A.M. and Roeder, M. (2000). Antibiotics in foods of animal origin. In: Francis F., Encyclopedia of Food Science and Technology. Vol. 1, 2nd Ed. John Wiley & Sons Inc., New York: pp. 54–63.

Rotman, B. (2001). Using living spores for real-time biosensing. *Gen. Eng. News*, 21: 65.

Rubies A, Vaquerizo, R., Centrich, F., Compano, R., Granados, M. and Prat, M.D. (2007). Validation of a method for the analysis of quinolones residues in bovine muscle by liquid chromatography with electrospray ionisation tandem mass spectrometry detection. *Talanta*, 72: 269–276.

Sams, M.J., Strutt, P.R., Barnes, K.A., Damant, A.P. and Rose, M.D. (1998). Determination of dimetridazole, ronidazole and their common metabolite in poultry muscle and eggs by high performance liquid chromatography with UV detection and confirmatory analysis by atmospheric pressure chemical ionisation mass spectrometry. *Analyst*, 123: 2545–2549.

Schmidt, C.W. (2002). More at Stake than Steak. *Env. Health Pers.*, 110(7): A397–A402.

Smith, K.E., Besser, J.M., Hedberg, C.W., Leaner, F.T., Bender, J.B., Wickland, J.H., Johnson, B.P., Moore, K.A. and Osterholm, M.T. (1999). Quinolone-resistant *Campylobacter jejuni* infections in Minnesota, 1992-1998. *New England J. Medicine*, 340: 1525–1532.

Schneider, R.P., Ericson, J.F., Lynch, M.J. and Fouda, H.G. (1993). Confirmation of danofloxacin residues in chicken and cattle liver by microbore high-performance liquid chromatography electrospray ionization tandem mass spectrometry. *Biol. Mass Spectrom.*, 22: 595–599.

Setford, S.J., Van Es, R.M., Blankwater, Y.J. and Kroger, S. (1999). Receptor binding protein amperometric affinity sensor for rapid β-lactam quantification in milk. *Anal. Chim. Acta*, 398: 13–22.

Stolker, A.A.M. and Brinkman, U.A.T. (2005). Analytical strategies for residue analysis of veterinary drugs and growth-promoting agents in food-producing animals—a review. *J Chromatog A*, 1067: 15–53

Sua´rez, G., Jin, Y.H., Auerswald, J., Berchtold, S., Knapp, H.F., Dieserens, J.M., Leterrier, Y., Manson, J.A. and Voirin, G. (2009). Lab-on-a-chip for multiplexed biosensing of residual antibiotics in milk. *Lab. Chip.*, 9: 1625–1630.

Tanase, S., Tsuchiya, H., Yao, J., Ohmoto, S., Takagi, N. and Yoshida, S. (1998). Reverse-phase ion-pair chromatographic analysis of tetracycline antibiotics: Application to discoloured teeth. *J. Chromatogr B.*, 706: 279–285.

Tieming, H., Chifang, P., Chuanlai, X. and Liqiang, L. (2006). Development of colloidal gold-based immunochromatographic assay for the rapid detection of medroxyprogesterone acetate residues. *J. Food Agric. Immunol.*, 17: 183–190.

Van Poucke, L.S.G., Depourcq, G.C.I. and Van Peteghem, C.H. (1991). A quantitative method for the detection of sulphonamide residues in meat and milk samples with a high-performance thin-layer chromatographic method. *J. Chromatogr.*, 29: 423–427.

Van Herwijnen, R. (2006). The use of lateral flow device to detect food allergens. *In*: Koppelman, J. and Helfe, S.L., Detecting Allergens in Food. Woodhead Publishing, Cambridge. pp. 175–181.

Wang, S., Zhang, H.Y., Wang, L., Duan, Z.J. and Kennedy, I. (2006). Analysis of sulphonamide residues in edible animal products: A review. *Food Additives Cont.*, 23: 362–384.

Weber, C.C., Link, N., Fux, C., Zisch, A.H., Weber, W. and Fussenegger, M. (2005). Broad spectrum protein biosensors for class specific detection of antibiotics. *Biotechnol. Bioeng.*, 89: 9–17.

Wilson, R.T., Groneck, J.M., Henry, A.C. and Rowe, L.D. (1991). Multiresidue assay for benzimidazole anthelmintics by liquid chromatography and confirmation by gas chromatography/selected-ion monitoring electron impact mass spectrometry. *J. Assoc. Off Anal. Chem.*, 74: 56–67.

Zhang, G.P., Wang, X.N., Yang, J.F., Yang, Y.Y., Xing, G.X., Zhao, D., Li, Q.M., Chai, S.J. and Guo, J.Q. (2006). Development of an immunochromatographic lateral flow test strip for detection of β-adrenergic agonist Clenbuterol residues. *J. Immunol. Methods*, 312: 27–33.

Zhang, M.Z., Wang, M.Z., Chen, Z.L., Fang, J.H., Fang, M.M., Liu, J. and Yu, X.P. (2009). Development of a colloidal gold-based lateral-flow immunoassay for the rapid simultaneous detection of clenbuterol and ractopamine in swine urine. *Anal. Bioanal. Chem.*, 395: 2591–2599.

Zhao, Y., Zhang, G., Liu, Q., Teng, M., Yang, J. and Wang, J. (2008). Development of a lateral flow colloidal gold immunoassay strip for the rapid detection of enrofloxacin residues. *J. Agric. Food Chem.*, 56: 12138–12142.

Zhao, S.J., Li, X.L., Ra, Y.K., Li, C., Jiang, H.Y., Li, J.C., Qu, Z.N., Zhang, S.X., He, F.Y., Wan, Y.P., Feng, C.W., Zheng, Z.R. and Shen, J.Z. (2009). Development and optimizing an immunoaffinity cleanup for determination of quinolones from chicken muscles. *J. Agric. Food Chem.*, 57: 365–371.

Zhu, Y., Li, L., Wang, Z., Chen, Y., Zhao, Z., Zhu, L., Wu, X., Wan, Y., He, F. and Shen, J. (2008). Development of an immunochromatography strip for the rapid detection of 12 fluoroquinolones in chicken muscle and liver. *J. Agric. Food Chem.*, 56: 5469–5474.

26

Production of Designer Eggs

Z.F. Bhat, Sunil Kumar and Pavan Kumar

ABSTRACT

Research shows that dietary cholesterol has only a small effect on blood cholesterol and that the consumption of eggs (up to an intake of one egg per day) has no detectable effect on heart disease risk in healthy people. Consuming an average of one egg per day is acceptable for many people but not for everyone. Although specific recommendations for strict limitations on egg or egg yolk intake/week have been dropped, the American Heart Association still recommends that the intake of dietary cholesterol be limited to an average of no more than 300 mg/day. An egg a day fits within this limitation only if the individual's overall diet is otherwise low in cholesterol. Eggs contain certain components that may have health benefits that go beyond basic nutrition and it is possible to modify some aspects of the nutritional composition of eggs by feeding the hens special diets. The cholesterol content of the egg has been lowered by either reducing the mass of the yolk relative to the egg i.e., a non-specific reduction or by using feed additives such as beta-cyclodextrin that have a specific effect on the cholesterol content of the yolk. Dietary manipulation can be used to either increase the amount of the n-3 PUFAs directly (using fish oil) or indirectly by increasing the levels of the precursor n-3 PUFA by feeding alpha-linolenic acid (ALA) in the form of linseeds, flaxseeds or similar. Apart from reducing the rate of PUFA oxidation during storage and cooking, enriching the level of antioxidant such as vitamin E in the egg can also enhance levels of antioxidants in the consumer.

1. INTRODUCTION

Consumers have started viewing food from a radically different vantage point which has evolved into an exciting area of the food and nutrition sciences known as functional foods. Functional foods can be defined as those providing health benefits beyond basic nutrition and include whole, fortified, enriched or enhanced foods which have a potentially beneficial effect on health when consumed as part of a varied diet on a regular basis at effective levels (Hasler,

2000). Besides these new food items designed specifically to enhance health, however, functional foods can also include those traditional, familiar foods (like eggs) for which recent research findings have highlighted new health benefits or dispelled old dogma about potential adverse health effects (Hasler, 2000). Data from free-living populations show that egg consumption is not associated with higher cholesterol levels and as a whole, the epidemiologic literature does not support the idea that egg consumption is a risk factor for coronary disease (Kritchevsky, 2004).

Eggs have been described as nature's original functional food (Hasler, 2000) packed with thirteen important vitamins and minerals (Shallo, 2001). In its efficient ovoid container, eggs are naturally a functional food providing valuable nutrients, from the high-quality protein to significant levels of beneficial vitamins, antioxidants, and other healthful compounds. Egg components have been attributed diverse biological activities, including antimicrobial and antiviral activity, protease inhibitory action, vitamin-binding properties, anticancer activity, and immunomodulatory activity (Li-Chan *et al.*, 1995). In addition, bioactive components of the egg are of importance in advanced research, and represent promising leads for treatment of various diseases. This chapter introduces the concept of designer egg and the contribution that eggs can make to the human diet and considers the additional benefits that can be derived from modifying the egg's nutritional profile and in particular the egg's fats and antioxidants. The possibility of manipulating the nutrient composition of eggs was considered as long ago as 1934 (Cruikshank, 1934) and modification of the polyunsaturated fatty acid composition has been pursued since the early sixties (Wheeler *et al.*, 1959). Attempts to modify the egg's fat component have tended to focus on the means by which the cholesterol level in the egg may be reduced or the ratio of n-3 to n-6 polyunsaturated fatty acids (PUFAs) manipulated.

2. EGG NUTRITION

For years, eggs have been held up as a powerhouse of nutrition due to their exceptional nutrition profile as a nutrient-dense food containing high quality protein and a substantial amount of many essential vitamins and minerals (Applegate, 2000). It is well known that eggs contain a rich array of nutrients. They have been promoted for their high quality protein, yet compared to other animal protein sources, eggs are the most inexpensive. Egg white is considered an ideal protein because it contains all the essential amino acids in proper proportion for human nutrition. A typical egg would contribute 3–4% of an adult's average energy requirement and approximately 6.5 g of protein. The major albumen proteins are ovalbumin (54%), ovotransferrin (14%), ovomucoid (11%), ovomucin (3.5%), lysozyme (3.5%) and globulins (8%) (Li *et al.*, 1995). It is notable that the quality of protein in eggs is considered to be sufficiently high to be used as the standard for measuring the quality of other food proteins. Similarly eggs are a source of the B vitamins and vitamin A in particular and contain a significant amount of phosphorus, iodine, selenium and iron. Of the total fat in eggs, more than half is the unsaturated variety. Unlike many other foods that are high in cholesterol, such as fatty-meats and full-fat dairy products, eggs are relatively low in saturated fat and calories. Thus, eggs provide a high nutrient density in proportion to their energy content. The nutrient density of eggs makes them a valuable contributor to the overall nutritional balance of the diet and, as an economical source of high quality protein, an important component in the diets of the elderly, low-income families, growing children and people limiting calories for weight loss purposes. Eggs are a significant source of iron, riboflavin, folate and vitamins B12, D, and E. The iron in egg yolks, like the iron in meat, is highly bioavailable, egg yolks may therefore be valuable in the diets of individuals who may need more iron, such as infants (Makrides *et al.*, 2002). An egg's contribution of vitamin D is noteworthy because the egg is one of the very few foods that supply this nutrient. Meyer *et al.*, 2003 ranked eggs at third most important provider of n-3 PUFA (6%), behind seafood (71%) and meat (20%) in an Australian survey. In another Australian study, McNaughton and Marks (2002) reported that eggs were a relatively rich source of selenium (9.0–41.4 μg/100g)-again coming third to seafood and meat. In addition to their nutritional value, eggs also have other desirable properties. They are inexpensive, convenient, easy to prepare, and easy to chew, and they play important roles in a wide variety of recipes.

2.1. A Shift in the Egg Paradigm

Despite of many beneficial nutrients, eggs have been linked with adverse effects on blood cholesterol. A significant barrier to increasing egg consumption in western countries is the perception that egg consumption is associated with a rise in blood cholesterol levels (Yaffee *et al.*, 1991) and as a consequence is deleterious to health and life expectancy. In this respect, expert opinion on the role of dietary cholesterol in the development of heart diseases has changed recently, indicating that, for the general population, dietary cholesterol makes no significant contribution to atherosclerosis and risk of cardiovascular disease (McNamara, 2000). The specific recommendations for strict limitations on egg or egg yolk intake/week have been dropped by American Heart Association although; it still recommends that the intake of dietary cholesterol be limited to an average of no more than 300 mg/day (AHA, 2000). Experimental feeding studies have shown that the consumption of an additional egg per day will lead to a modest increase in average serum cholesterol (McNamara, 2000) and population-based studies examining the association between egg consumption and serum cholesterol levels show either no association, or, paradoxically, an inverse association. In a meta-analysis of 224 studies covering more than 30 years of research, it was demonstrated that dietary cholesterol has only a limited effect on plasma cholesterol (Howell *et al.*, 1997). Moreover, in the largest epidemiologic study conducted to date on the relationship between egg consumption and coronary heart disease, involving more than 118,000 men and women from both the Nurses' Health Study and the Health Professional Follow-Up Study, consumption of up to one egg per day did not have a substantial overall impact on the risk of coronary heart disease and stroke (Hu, 1999). Moreover, research demonstrated that persons who reported eating more than four eggs per week had significantly lower mean serum cholesterol than those who reported eating one or fewer eggs per week (Kerver, 2000). Furthermore, results of many other studies (Ginsberg *et al.*, 1994, McNamara, 1995; Simopoulos, 2000) have shown that relatively large numbers of eggs can be consumed without any significant changes to plasma cholesterol or other lipid components. This should improve the image of eggs. In this respect, eggs, which are consumed regularly by most of the population, when enriched with DHA, vitamin E, lutein and selenium, are capable of substantially improving the diet.

The time has come to change consumer attitudes about how eggs can contribute to overall health rather than incorrectly positioning them as a risk factor in heart disease. Scientists, medical professionals and communicators are challenged with changing consumer attitudes and behaviors about eggs. Beliefs and behaviors can stem from cultural influences rather than from nutrition science. The association of eggs with heart disease stemmed from observational studies, but its perpetuation is driven by more culture and public health than by science (Applegate, 2000). The time is right to change the egg message. Furthermore, it is time that eggs be recognized as a nutritious food which also has health benefits beyond basic nutrition. The concept of eggs as a 'functional food' is new to many and requires a change in the perception of role of eggs in the diet. And finally, the evidence that eating eggs is unrelated to heart disease risk needs to be widely disseminated to health professionals and the public so that everyone can benefit from including eggs in the diet (Applegate, 2000).

3. FUNCTIONAL FOODS

Functional foods have been defined as foods that may provide health benefits beyond basic nutrition and are similar in appearance to conventional food that is intended to be consumed as part of a normal diet, but has been modified to subserve physiological roles beyond the provision of simple nutrient requirements (Bech-Larsen and Grunert, 2003). Thus, a food marketed as functional contains added, technologically developed ingredients with a specific health benefit (Niva, 2007). The concept of functional food was first promoted in 1984 by Japanese scientists (Harris, 2000) who studied the relationships between nutrition, sensory satisfaction, fortification and modulation of physiological systems and came up with specific health-related food category called FOSHU (Food for Specified Health Uses) with specific health claims (Burdock *et al.*, 2006; Stanton *et al.*, 2005; Menrad, 2003; Kwak & Jukes, 2001; Roberfroid, 2000b). Although the term "functional food" has already been defined several times (Roberfroid, 2002), so far there is no unitary accepted definition for this group of food (Alzamora *et al.*, 2005). The amount of intake and form of the functional food should be as it is normally expected for dietary purposes. Therefore, it could not be in the form of pill or capsule just as normal food form (Diplock *et al.*, 1999) but FOSHU products in Japan can also take the form of capsules and tablets, although a great majority of products are still in more conventional forms (Ohama *et al.*, 2006).

3.1. Demand of Functional Foods

Consumer demands in the field of food production have changed considerably in the last few decades. It was recognized that there is a demand for these products as different demographical studies revealed that the medical service of the aging population is rather expensive (Mark-Herbert, 2004; Menrad, 2003; Side, 2006). Consumers more and more believe that foods contribute directly to their health (Mollet & Rowland, 2002; Young, 2000). Foods are not intended to only satisfy hunger and to provide necessary nutrients for humans now a day but also to prevent nutrition-related diseases and improve physical and mental well-being of the consumers (Menrad, 2003; Roberfroid, 2000b). There is an increasing demand of functional foods in developed countries. This rapidly growing market of functional or health-enhancing foods has emerged as a response to demographics, patterns of health and disease, innovation in food and health related research, and globalization and can be explained by certain factors like increasing cost of healthcare, the steady increase in life expectancy, and the desire of older people for improved quality of their later years (Kotilainen *et al.*, 2006; Roberfroid, 2000a,b). This sizeable demand of functional foods provides important export opportunities for developing countries.

3.2. Egg as a Functional Food

Meeting consumer demands is a constant challenge for the animal food industry. Many consumers desire somewhat distinct products with respect to safety, healthfulness, freshness, taste, color, etc. To tap into this market, companies have developed several designer and speciality eggs which have appeared on store shelves.

The potential of hen egg as a functional food dates back to 300 A.D. when an oral drench of up to 12 whole eggs was recommended for the treatment of diarrhea (Losch *et al.*, 1986). Moreover, hen egg yolk has been promoted as a novel source of polyclonal antibodies suitable for providing passive protection against gastrointestinal infection (Kuroki *et al.*, 1993; Peralta *et al.*, 1994; O'Farrelly *et al.*, 1992; Ikemori *et al.*, 1992; Yolken *et al.*, 1988; Yokoyama *et al.*, 1992; Deignan *et al.*, 2001). Egg yolk from hens immunized with a bacterial pathogen has high levels of specific antibodies. Oral administration of specific antipathogen antibodies, purified from egg yolk from immunized hens, has been shown to prevent infection in several *in vivo* studies (Yokoyama *et al.*, 1992; Kuroki *et al.*, 1993; Deignan *et al.*, 2001). It has been suggested that specific egg-yolk antibodies from immunized hens bind potential pathogens in the gut lumen and prevent their adherence to the gut mucosa, thereby preventing the first step in the infection process. *In vitro* studies have confirmed the ability of specific egg-yolk antibodies to reduce bacterial adherence to epithelial cells (Peralta *et al.*, 1994). However, *in vivo* study has shown that oral administration of whole egg yolk from unimmunized hens also provided protection against bacterial challenge (O'Farrelly *et al.*, 1992) which suggests that egg yolk from unimmunized hens may contain protective factors other than antibodies (Deignan *et al.*, 2001).

Eggs are of particular interest from a functionality point of view, because they are relatively rich in fatty acids and the associated fat-soluble compounds. The type and ratio of fatty acids is an important determinant of human health. Eggs contain a number of beneficial nutrients, some of which have functions that are currently being studied. Eggs are one of the best sources of a nutritionally important substance called choline (NAS, 2000) which has been identified as contributing to fetal memory and brain development and are being investigated for possible beneficial effects on cognitive function (Hasler, 2000). Choline is found in the form of phosphatidylcholine and sphingomyelin and is an important part of cellular compounds such as the neurotransmitter acetylcholine and lecithin, a naturally occurring emulsifier present in cell membranes and bile. According to the National Academy of Sciences, the recommended daily intakes of choline range from 425 to 550 milligrams for adults, including pregnant and lactating women and one large egg contains approximately 300 milligrams choline (Zeisel, 2000). Eggs are a source of highly bioavailable forms of the carotenoids, lutein and zeaxanthin (Handelman *et al.*, 1999; Applegate, 2000). These antioxidant-like compounds have been shown to help in the prevention of macular degeneration, a leading cause of blindness in the elderly, and have been associated with lower risk of cataract extraction (Hasler, 2000; Moeller *et al.*, 2000). Recent research demonstrated the link between these dietary compounds and the macular pigment of the retina of the eye (Landrum & Bone, 2001). Dwyer *et al.* (2001) reported that increased amounts of dietary lutein from green leafy vegetables and egg yolks could be protective against atherosclerosis by slowing the progression of atherosclerotic lesions in humans and animals.

Analyses of recent literature conclude that eggs ideally fit the requirements of a functional food. For example, the levels of certain nutrients (vitamin E and DHA) could be increased in the egg to such an extent that consumption of a single egg could deliver these nutrients in amounts comparable or higher than daily requirement (Surai, 2000). Commercially, it is possible to produce designer eggs enriched with many different nutrients simultaneously or with a single nutrient depending on the consumer demand. As a result, price for the production of such eggs could substantially vary. These eggs are marketed as specialty eggs of added value because of their unique features and usually command a higher price than conventional eggs in the market place. The idea of egg enrichment with omega-3 fatty acids simultaneously with antioxidants and other vitamins has recently been used to produce VITA Eggs by Freshlay Foods (Devon, UK) enriched with omega-3 fatty acids, Se, vitamins D, E, B_{12} and folic acid. Eggs enriched in omega-3 and vitamin E produced by Belovo under the trade name of Columbus first appeared in Belgium in 1997, and since then they have been sold in the UK (from, 1998), The Netherlands (from 1999), India, Japan and South Africa (from, 2000). Currently, production of Columbus egg exceeds 50 millions/ year in Europe. Similar eggs are produced by Pilgrim's Pride Company, Gold Circle Farms and OmegaTech in the USA (Surai & Sparks, 2001).

4. DESIGNER EGG

In addition to the naturally containing essential health promoting and functional nutrients, the nutrient content of eggs can be modified to provide nutrients above and beyond what is normally found in generic shell eggs. The possibility of manipulating the nutrient composition of eggs was considered as long ago as 1934 (Cruikshank, 1934). Modifying egg quality can be accomplished by inducing metabolic changes in the hen that can result in synthesis of compounds that essentially end up in her egg. Another way is to change the characteristics of membrane transport to facilitate movement of compounds into the egg. One of the more common acceptable ways to modify egg quality is to manipulate the diet of the hen such that the desired compounds eventually find their way into the egg. If a food, such as the egg, is nutritionally modified it is referred to as a designer or functional food (American Dietetic Association, 1995). Thus "designer eggs" are those in which the content has been modified from the standard egg. The contents of the chicken egg can be changed in such ways as to be more healthful and appealing to a segment of consumers who are willing to pay for those changes in the egg. Poultry researchers have been dedicating a considerable amount of their efforts in recent decades to studies with hens in an attempt to lower egg yolk cholesterol to satisfy concerns of health-conscious consumers. Research has shown that not only cholesterol, but the egg content of several other nutrients can be manipulated to produce today's so-called 'designer' eggs.

The nutritive value of an egg is potentially affected by geography and by factors that influence egg size or weight or the proportion of the constituents. Thus, it is well documented, for example, that egg size tends to increase with increasing age of the hen (Cook & Briggs, 1977). So, while the proportion of yolk tends to be greater in small eggs rather than large ones, the absolute amount of yolk increases with increasing egg size (Cook & Briggs, 1977). It can be important therefore, when considering the nutritive value of eggs, to ensure that the weight of the egg is known. Similarly some constituents of the egg are sensitive to manipulation through the diet of the hen (Surai *et al.*, 2000). However, while the concentration of many of the nutrients found in an egg can be influenced by dietary manipulation of the laying hens ration, consideration should be given to the efficacy with which the nutrient is taken up by the egg and also the potential benefits to the consumer of manipulating the nutrient level.

4.1. Modifications in the Nutritional Profile of a Designer Egg

Modification of the nutrient content of the egg by modifying the diet of the hen has led to "designer eggs" that serve as functional foods and nutraceuticals

(Stadelman, 1999). Through special feeding practices, it has been possible to produce eggs with even less cholesterol and saturated fat and with substantially increased amounts of omega-3 polyunsaturated fatty acids, carotenoids, vitamins, minerals, antibodies and even bioactive peptides (Leskanich & Noble, 1997; Surai & Sparks, 2001; Watson, 2002; Leeson & Caston, 2003; Yaroshenko *et al.*, 2003; Grashorn & Steinberg, 2002; Kang *et al.*, 2003; Kim *et al.*, 2004; Davalos *et al.*, 2004; Chowdhury *et al.*, 2005). Hens fed on all grain diet, free of animal fat and byproducts, lay eggs marketed as vegetarian eggs. Vegetarian eggs may differ slightly in nutrient content from generic shell eggs depending on the nutrient content and quality of the vegetarian feed. These types of designer egg appeals to individuals who are lacto-ovo-vegetarians. Eggs with modified fat content have become very popular and are available in many parts of the country. Modified fat content eggs are marketed as containing less cholesterol, less saturated fat, higher amounts of omega-3 fatty acids, higher vitamin E content, and high amounts of iodine when compared to generic shell eggs (Shallo, 2001).

5. OMEGA-3-ENRICHED EGGS

Commercial table eggs contain a high proportion of n-6 PUFA (mainly 18:2n-6) but are a poor source of n-3 fatty acids. The early generation of modified eggs was enriched with the n-3 PUFA and to protect the PUFA from oxidation, vitamin E content was also increased. Studies have shown that diets high in omega-3 fatty acids can reduce heart disease risk. Hens fed diets high in omega-3 fatty acids produce eggs with high omega-3 fatty acid content in the yolks. The type of omega-3 fatty acid in the feed can be linolenic acid from flaxseed or docosa-hexanoic acid (DHA) or eicosapentaenoic acid (EPA) from marine algae or fish oil. Individuals who do not eat or do not have access to fish products can consume an adequate amount of omega-3 fatty acids by incorporating omega-3 eggs into their diet. Leskanich and Noble (1997) suggested two feeding strategies to increase the concentration of n-3 fatty acids *i.e.*, either by feeding a diet containing appropriate levels of flaxseeds, linseeds or the oils from these seeds that enhanced the concentration of alpha-linolenic acid (ALA) and, to a lesser extent docosa-hexanoic acid (DHA, one of the omega-3 fatty acids) in the yolk or by feeding a diet rich in fish oil which also significantly enhanced the levels of DHA in the egg. The second feeding strategy often used fish oil from menhaden, tuna or herring although algal forms are also available and the potential advantage is that most of the beneficial health properties of the n-3 PUFAs are associated with DHA and the conversion of ALA to DHA is not always efficient, particularly in the young and elderly receiving diets high in n-6 PUFAs. However, eggs enriched with PUFAs may be associated with off-odours and in particular fishy taints which could be minimised by feeding ≤1.5% of high quality fish oil or ≤5% flaxseed and protecting the PUFAs both in the diet and in the egg from oxidation (Surai & Sparks, 2001). However, due to manipulations in diet interactions in many egg constituents can occur. As an example, Szymczyk and Pisulewski (2003) reported that dietary CLA depressed the concentration of linoleic and linolenic acids significantly without affecting the concentration of cholesterol.

5.1. Health Benefits of Omega-3 Enriched Eggs

The major advantage to the consumer of omega-3 eggs is an enrichment of plasma lipids with these fatty acids (Ferrier *et. al.*, 1995; Farrell, 1998; Farrell & Gibson, 1991) which have beneficial effects in prevention and management of human diseases like coronary heart disease, embolism, hypertension, allergic problems, thrombosis, chronic obstructive pulmonary disease, type 2 diabetes, prostate and breast cancer, renal disease, rheumatoid arthritis, autoimmune disease, ulcerative colitis, crohn's disease and depression besides their positive effects on longevity, immunological functions, fetal brain and visual development and improvement of learning ability (Surai & Sparks, 2001). There are also reports indicating other beneficial effects, including decreased plasma triglyceride concentration (Jiang & Sim, 1993; 'Ferrier *et al.*, 1992) and lowered systolic and diastolic blood pressures (Oh *et al.*, 1991; Farrell, 1993; Farrell, 1994) and platelet aggregation (Van Elswyk, 1998). In some experiments, even total plasma cholesterol level was reduced (Lewis *et al.*, 2000) due to consumption of modified eggs.

Many studies have been undertaken regarding the effects of consuming n-3 enriched eggs and have reported positive effects in healthy adults, the very young and the aged with increase in PUFA content of the plasma and erythrocyte membrane, enhanced erythrocyte DHA concentration and accelerated maturation in visual acuity (Makrides *et al.*, 2002; Sindelar *et al.*, 2004; Payet *et al.*, 2004; Hoffman *et al.*, 2004). In a study, Farrell (1998) enriched eggs with omega-3 fatty acids and fed human volunteers the enriched eggs for twenty-two weeks. The subjects consumed seven eggs per week for the first twenty weeks and approximately 10 eggs per week for the last two weeks. The control group received the same number of generic eggs. Consumption of one enriched egg per day significantly increased levels of EPA, DHA and total omega-3 fatty acids in blood. High density lipoprotein (HDL) levels gradually rose in the subjects throughout the 24 week feeding study while total blood cholesterol levels and triacylglycerol levels did not increase. The author concluded that eggs enriched with EPA and DHA could contribute to the dietary intake of omega-3 polyunsaturated fatty acids and could be an alternative source of omega-3 fatty acids to fish. In a similar controlled trial, Surai *et al.*, (2000) fed human volunteers one designer egg (having 26 times more vitamin E, 16 times higher carotenoid content, over 7 times greater selenium content, and 6 times higher DHA content than generic eggs) each day for eight weeks. Plasma samples were collected before and after the eight-week feeding period. Alpha-tocopherol (vitamin E), lutein and DHA significantly increased in the plasma of subjects fed designer eggs when compared to subjects fed generic shell eggs. The authors concluded that DHA, vitamin E and lutein from eggs may improve the diet and could result in beneficial to health.

5.2. Organoleptic Quality of Omega-3-Enriched Eggs

Although in some cases panelists are able to detect off-flavours, the organoleptic quality of omega-3 eggs tends to be similar to regular table eggs (Van Elswyk *et al.*, 1992; Caston *et al.*, 1994; Ahn *et al.*, 1995; Surai & Sparks, 2001). Scheideler *et al.* (1997) tested the effects of flax seed or menhaden fish oil with supplemented vitamin E on consumer acceptance, oxidative products and the yolk color of fresh and stored eggs. In general, they found that consumer acceptability of eggs from hens fed flax seed was similar to eggs from diets enriched with menhaden fish oil and not greatly different from eggs produced by hens fed standard diets. A 'fishy' or fish-product related flavour was detected in eggs from hens on diet containing 15–20% flax seed (Ferrier *et al.*, 1994; Jiang & Sim, 1994). Data of Leeson *et al.* (1998) also suggest that high (>10%) levels of flax seed will result in some decrease in overall egg acceptability as assessed by aroma and flavour. Studies suggest that the use of combinations of anti-oxidants in the hen's diet could help to suppress these off-flavours (Farrell, 1998), however, high levels of vitamin E did not prevent the formation of off-flavours in hens fed on the high (>10%) flax diet (Leeson *et al.*, 1998). In general, therefore, fishy taints in eggs are not detectable provided that the hens are fed 5% (or less) flaxseed or low levels of high quality oil (Scheideler *et al.*, 1997; Cloughley *et al.*, 1997; Maurice, 1994; Marshall *et al.*, 1994). In this respect, a dried DHA-enriched marine micro-algal product showed promising results (Herber & Van Elswyk, 1996; Farrell, 1993; Surai & Sparks, 2001). The fishy taint can result from rancidity of the n-3 enriched diet; therefore, the quality of oils used is important in eliminating off-flavour in eggs. While the oil can be protected with anti-oxidants or micro-encapsulated once the oil is oxidized the addition of anti-oxidants will not reverse this situation. Cooking characteristics of omega-3 eggs, including emulsification capacity, hardiness and springiness of sponge cakes prepared using these eggs, were the same as in ordinary eggs (Leskanich & Noble, 1997; Surai & Sparks, 2001).

5.3. Commercial Production of Omega-3 Enriched Eggs

There are various egg types on supermarket shelves in different countries. For example, in the UK, the designer egg available through the supermarkets is the 'Columbus' egg produced by a Belgium company, Belovo. These eggs, enriched in n-3 fatty acids and vitamin E, first appeared in Belgium in 1997, and since then they have been sold in the UK (1998), The Netherlands (1999) and India, Japan and South Africa (2000). These eggs are characterized by a balanced nutritional lipid composition (C18, omega-6: omega-3=1:1) and a favourable structural lipid ratio (long-chain PUFA, omega-6: omega-3=1:3). Pilgrim's Pride Company, producer of poultry products in North America and Mexico introduced the so-called EggsPlus with an increased level of vitamin E and omega-3 fatty acids. Similar eggs are produced by Gold Circle Farms (containing 150 mg DHA and 6 mg vitamin E, CO, USA) and OmegaTech (USA) produces the so called Gold Circle Farms Eggs. Omega Tech launched the sale of DHA eggs in Germany, Spain, Portugal, Belgium, Norway and Andorra.

The production of designer eggs involves extra cost and they are usually sold at premium prices. The prices could differ substantially and will ultimately depend in part on the sources of nutrients used in the dietary formulations. It appears that consumers are ready to pay extra for designer eggs provided that they understand the possible health-promoting properties (Surai & Sparks, 2001).

6. LOW-CHOLESTEROL EGGS

During the last few decades the egg industry has been under intense attack by anti-cholesterol advocates and the egg has been blamed, blasted and defamed as one of the major culprits in coronary heart disease (CHD). Yaffee *et al.* (1991) concluded that the public's perception of eggs as a major source of dietary cholesterol was a significant contributing factor to the overall decline in egg consumption since the late 1960s. The public's negative perception about eggs, however, is changing and the eggs are now being accepted as the nutritious food they are. In order for a company to claim a reduce amount of a nutrient the product must have 25 percent less than the normal product standards for that nutrient. Eggs from commercial egg-type laying hens typically contain about 200 mg cholesterol per egg (Beyer & Jensen, 1989). The cholesterol in an egg is found only in the yolk. Over 95% of the yolk cholesterol is associated with yolk triglyceride-rich lipoproteins (Griffin, 1992).

It is relatively difficult to manipulate the cholesterol content of the yolk, unlike some of the other egg constituents, and as such to produce so-called low-cholesterol eggs. It has been suggested that the relative resistance of egg composition to alterations in diet apparently reflects the nutritional and structural requirements for avian embryonic development (Kuksis, 1992). There are normally two strategies that are adopted for producing these eggs. Either by reducing the amount of cholesterol per egg, for example by reducing the size of the yolk or by altering the lipid profile of the yolk. Efforts to reduce egg yolk cholesterol content have involved the investigation of the effects of diet, pharmacological intervention and genetic selection (Griffin, 1992).

Research into lowering egg cholesterol has centered mostly around diet and pharmacological intervention (drugs). Drugs have been successful in lowering egg cholesterol by as much as 50%. Drugs lower cholesterol in the egg by either inhibiting the synthesis of cholesterol in the hen or by inhibiting the transfer of cholesterol from the blood to the developing yolk on the ovary. The use of various pharmacological agents to lower egg yolk cholesterol will be limited if such agents or their metabolites are excreted into the egg (Hargis, 1988).

6.1. Cholesterol Reduction Through Genetic Selection

Genetic selection of hens for lowered cholesterol has not been successful in lowering the egg cholesterol content. Genetic selection studies have indicated sufficient genetic variability in yolk cholesterol levels to make a genetic selection program feasible. Selection for lower egg cholesterol has not, however, been successful by this means and has resulted in only a slight reduction in egg cholesterol concentration. Unfortunately, this reduction is associated with a decline in egg production (Hargis, 1988). There have been claims that eggs from the Araucana chicken contain less cholesterol when compared to eggs of other breeds. However, there is no scientific evidence to support this claim. Several well-designed scientific studies have compared the cholesterol content of Araucana eggs to eggs from other breeds and suggest that Arauanca chickens simply lay a smaller egg so that cholesterol content, on a per egg basis, is lower.

6.2. Cholesterol Reduction Through Chromium Supplementation

Chromium supplementation to laying hen diets at concentrations of less than 1 ppm have been shown to lower egg cholesterol and also improve egg interior quality. McCarty (1991) has shown that the organic form of chromium is superior to the inorganic form and has been shown to have a serum cholesterol lowering effect (Evans, 1989; Lien *et al.*, 1993). Lien *et al.* (1996) fed commercial egg type White Leghorn hens 100 g/day of a corn-soybean meal based diet supplemented with either 0, 200, 400 or 800 ppb organic chromium (picolinate). These authors reported a 39% reduction in serum cholesterol with the 800 ppb level of chromium of supplementation. Compared to egg yolks from hens fed the control ration, yolks from hens fed 200, 400 or 800 ppb chromium contained 14, 29 and 34% less cholesterol, respectively. Nakaue and Hu (1997) used 22- and 75-week old commercial egg-type laying hens fed diets supplemented with 0, 200 or 800 ppb chromium in an experiment designed to evaluate the cholesterol lowering ability of chromium. The younger group of hens laid eggs with significantly lower yolk cholesterol levels at both levels of chromium supplementation. A numerical decrease in yolk cholesterol occurred in the older group of hens fed the chromium supplemented diets.

6.3. Pharmacological Ways of Cholesterol Reduction

Researchers continue to search for pharmacological ways to efficiently produce low cholesterol eggs. Elkin *et al.* (1997) reported that the amount of cholesterol in an egg can be reduced by as much as 45–50% by the drug atorvastatin without seriously impairing egg production. Some researchers have found that when n-dihydroxyphenylalanine (L-DOPA) in the bloodstream of laying hens is elevated, the hens produce eggs with reduced cholesterol content, a lower ratio of saturated to unsaturated fatty acids and increased total protein. The L-DOPA can be administered to the hens as a dietary supplement. Chowdhury *et al.* (2005) studied the effects of dietary tamarind and reported quadratically decreasing serum cholesterol concentrations with increasing dietary levels of tamarind however, the concentrations of cholesterol in the yolk were unaffected by these treatments. In contrast to this, Kim *et al.* (2004) administered 0.06%

pravastatin and reported reduced egg cholesterol levels by almost 20%, however, the concentrations of cholesterol in the plasma were unaffected. Only pravastatin (3-hydroxy-3-methlyglutaryl coenzyme A reductase) could specifically reduce egg cholesterol levels whereas, lovastatin and simvastatin reduced plasma cholesterol. Park *et al.* (2005) also reported significant reduction (by 4 mg/g) of the cholesterol content of the yolk relative to the control by feeding betacyclodextrin at 6%.

7. VITAMIN ENRICHED EGGS

Nabor (1993) reviewed the successful attempts to modify vitamin composition of eggs by dietary vitamin supplementation and concluded that egg vitamin content is highly variable and dependent primarily on the vitamin concentration of the hen's diet. Vitamin A content of eggs responds slowly to changes in dietary vitamin A content, whereas, the riboflavin content responds rapidly to dietary changes in riboflavin concentration. Vitamin D, pantothenic acid, folic acid, biotin and B_{12} respond greatly to increases in the dietary levels of these vitamins. Transfer efficiency to the egg is very high for vitamin A and high for riboflavin, pantothenic acid, biotin and B_{12}. Transfer efficiency is medium for vitamin D_3 and vitamin E and low for vitamin K, thiamin and folic acid. Vitamin content of eggs can be increased over certain ranges of diet fortification and with varying efficiency of vitamin transfer. If development of designer eggs containing higher concentrations of certain vitamins is ever the objective of a commercial enterprise, vitamin transfer efficiency and vitamin cost would be two of the major considerations used in determining the economic feasibility of marketing such eggs (Nabor, 1993).

The development of designer eggs containing increased concentrations of vitamin E, β-carotene and vitamin A has recently been of interest to researchers due to their proposed strong protective association with certain human cancers (Jiang *et al.*, 1994). These three nutrients are known to function as antioxidants, quenching singlet oxygen and free radicals (Bendich & Olson, 1989; Krinsky, 1989; Knekt *et al.*, 1991).

7.1. Vitamin A Enriched Eggs

Vitamins in the egg would be expected to vary with their concentration in the feed. However, for some vitamins, as vitamin A, the liver acts as a reservoir so that the concentration in the yolk is buffered against large changes in the diet (Naber, 1979). Naber and Squires (1991) performed experiments with this vitamin and laying hens. At low dietary levels, liver storage was minimal but egg yolk increased sharply. At higher dietary levels liver storage increased markedly while egg yolk level continued to increase at a moderate rate. Over the range of 2600 to 22000 IU/kg diet liver storage of the vitamin approached a 200 fold increase while egg yolk vitamin content doubled (NRC 4000 IU/kg, 117 IU/egg).

7.2. Vitamin B_{12} Enriched Eggs

Naber and Squires (1991) increased the feed dosage of vitamin B_{12} from 24 to 200 pg/kg (NRC 4 μg/kg) and observed a 5 fold increase in yolk vitamin level (normal level 0.6 pg/egg). Thus it is possible to modify the content of the egg with vitamin B_{12}.

7.3 Vitamin D Enriched Eggs

In general, the transfer efficiency of dietary fat-soluble vitamins to egg is lower than that of water-soluble vitamins (Naber, 1993). In contrast to the situation with vitamin A, there is a linear relationship of dietary vitamin D to vitamin D in egg yolk. Excess dietary vitamin D is readily absorbed, transferred to the ovary and secreted in egg yolk during its formation (Naber & Squires, 1991). The level of vitamin D in eggs was increased considerably in response to the increase of dietary level of this vitamin (Park *et al.*, 2005, Mattila *et al.*, 2004). In other experiments, the concentration of vitamin D_2 in the egg yolk was 14.5 IU/g when the concentration in the diet was 20 IU/g (Kawazoe *et al.*, 1996). Hence, egg yolk can be fortified with vitamin D in order to increase its nutritional value.

7.4. Folic Acid and Biotin Enriched Eggs

Sherwood *et al.* (1993) studied the supplementation of food with folic acid; they found that folates in egg yolk were 43 times more concentrated than in hen plasma; and the folate concentration in egg yolk was 100% above that in egg white.

A positive relationship between dietary biotin and the amount of biotin in the egg has been reported (Robel, 1991).

7.5. Vitamin K Enriched Eggs

Although there is no available data on transfer of dietary vitamin K to eggs, Griminger and Brubacher (1966) demonstrated a marked reduction in the prothrombin time in plasma of chicks hatched from eggs laid by hens given ten times more dietary vitamin K_1 than the control. This was an indirect proof that dietary vitamin K_1 transferred to the egg.

Park *et al.* (2005) studied the effect of enrichment of Vitamins D_3, K and Iron in eggs of laying hens and concluded that vitamins D_3 and K in eggs can be effectively enriched by feeding diets supplemented with 12,000 IU vitamin D3 and 7.5 mg vitamin K/kg of diet for 20 days. Compared to the control, the maximum levels achieved by these supplementation were 4.6 times greater for vitamin D_3 and 4.8 times for vitamin K, respectively.

Leeson and Caston (2003) studied the effect of dietary vitamins by increasing the concentration of the twelve vitamins (A, D_3, E, K, B_1, B_2, B_6, biotin, folic acid, niacin, pantothenic acid, vitamin B_{12}) in the hen's diet by between three and ten times the normal amount. Of the twelve vitamins examined only five (vitamin E, 25%; vitamin K, 108%; biotin, 60%; pantothenic acid 24% and vitamin B_{12}, 139%) were deposited in the egg in sufficient quantities to exceed 20% of the recommended daily intake. However, when compared with the control eggs the increases in vitamin K and biotin were not dissimilar. The authors concluded that while eggs could be enriched for vitamin E and B_{12}, there was little justification in enriching eggs with other vitamins.

8. MINERAL ENRICHED EGGS

The shell contains the majority of the minerals in an egg. There are approximately 2,200 mg of calcium and 20 mg of phosphorus in the shell. There has been very little success in changing the calcium and phosphorus content of the albumen and yolk. It is possible, however, to increase the content of chromium, selenium, iron, zinc, iodine and manganese. This has been done through dietary supplementation of the hen. These minerals are important in human health. There has been some interest, therefore, in promoting these eggs as designer eggs.

8.1. Chromium Enriched Eggs

Mineral deficiencies are causing more health problems than the overconsumption of fats, cholesterol and refined carbohydrates that are the targets of almost all of public health measures (Brown, 1996). There are atleast 256 perceived benefits of organic chromium. A portion of the need for effective fiber is, in fact, a need for organic chromium.

The chromium content in egg yolk and albumen was higher for chromium supplemented hens than for control ones, without differences from chromium source (inorganic or organic). The chromium content increased sharply the first two week (10 times) while decreased to two- three-fold the basal level thereafter (Kang *et al.*, 1996). Chromium content of eggs was also increased in other experiments (more than 5-fold) by supplementation of chromium to the hens diet; however, chromium content decreased with time of supplementation (30 to 75 d) (Meluzzi *et al.*, 1996).

8.2. Selenium Enriched Eggs

Like in UK, the consumption of selenium falls below the recommended daily allowance (RDA) in many countries (Yaroshenko *et al.*, 2003) and may be added to laying hen diets either as inorganic forms (selenite or selenate) or as organic form (selenomethionine). It is feasible to supply 50% of the RDA for selenium in one egg by supplementing the feed of the layer hen with 0.4 ppm selenium in the form of selenomethionine (Yaroshenko *et al.*, 2003; Surai, 2000).

8.3 Iron Enriched Eggs

Since one standard egg contains about 1 mg of iron, eggs could contribute significantly to the-dietary iron supply. Eggs might be a food of choice for those who require a generous supply of dietary iron and a limited calorie intake, and iron fortified-eggs will have an increased value. Morck and Austic (1981) fed graded levels of iron from 15 to 65 ppm to hens and observed a dose-response of hematocrit and hatchability. Biehl *et al.* (1997) observed marked improvement in weight gain, hemoglobin, and hematocrit when increasing levels (0 to 80 mg/kg) of added iron from analytical grade ferrous sulfate were fed to chicks. Organic sources of iron showed no clear advantage over the inorganic salt in several experiments (Biehl *et al.*, 1997). The iron content of the egg yolk was increased up to 125% and 140% with inorganic and organic iron sources.

8.4. Zinc Enriched Eggs

Eggs represent a good opportunity for supplying extra zinc to human diets. Zinc content of egg yolk is closely related to diet content. Adding 80 mg/kg of zinc to a laying hen diet containing 65 mg/kg increased egg yolk zinc levels from 0.84 to 1.62 mg/egg (Naber & Squires, 1991). Kidd *et al.* (1992) reported a study in which progeny from broiler breeders fed 72 mg Zn/kg diet had heavier embryonic bones and improved immune status, as a result of Zn accumulation in the egg. Flinchum *et al.* (1989) also observed similar results.

8.5. Iodine and Manganese Enriched Eggs

Iodine and manganese are other elements that vary according to the diet (Naber, 1979). Naber and Squires (1991) reported that egg iodine can be increased markedly by feeding potassium iodide, kelp or iodized linseed meal to laying hens that resulted in a high egg uptakes of iodine that were 10 times the basal level (up to 400 μg/egg) when 16 mg/kg of iodine from these sources were fed. Ternes and Leitsch (1997) reviewed that a lower enrichment of iodine in the diet (2.6 mg/kg) produced iodine enriched eggs (74 μg/egg compared to 26 μg/egg of generic eggs).

Manganese levels in egg yolk responded only fractionally to the increase in the dietary manganese supplement. Adding 60 mg/kg of manganese to a laying diet containing 37.5 mg/kg increased egg yolk manganese levels from 156 to 222 μg/egg (Naber & Squires, 1991). However, there is an impairment of iron absorption when high levels of manganese are fed (Baker & Halpin, 1991).

9. CAROTENOID ENRICHED EGGS

The type of pigment in the egg and its concentration are directly influenced by the dietary concentration of any particular pigment. The color of the yolk is a reflection of its pigment content and most of the carotenoids in egg yolk are hydroxyl compounds called xanthophylls. Lutein and zeaxanthin are two of the most common xanthophylls found in egg yolk. Lutein and zeaxanthin are high in pigmented feed ingredients such as yellow corn, alfalfa meal, corn gluten meal, dried algae meal, and marigold-petal meal. The high protein blue-green algae known as spirulina has also been shown to be a very efficient pigment source for poultry skin and egg yolk.

Besides a precursor for vitamin A or a pigment, scavenging free radicals, affecting innate, humoral and cell-mediated immune response are some of the effects of carotenoids that are taking the attention of poultry nutritionists. Canthaxanthin is a carotenoid commonly used in laying hen diets which is deposited in the yolk at a level linearly related to the level in the diet (Grashorn & Steinberg, 2002). Other carotenoids that have been considered for inclusion in eggs include lycopene (Kang *et al.*, 2003).

Research suggests that eggs may be beneficial in preventing macular degeneration, a major cause of blindness in the elderly. A recent study indicated that a higher intake of carotenoids reduce the risk of age-related macular degeneration. Lutein and zeaxanthin are most effective in preventing macular degeneration and are commonly found in dark-green leafy vegetables (such as spinach and collard greens) and are efficiently transferred to the yolk when these various feed ingredients are fed to laying hens. With a growing problem of macular degeneration in the elderly, the egg industry may want to seize this opportunity and perhaps there would be a market for eggs having a higher level of lutein and zeaxanthin.

10. PHARMACEUTICALS AND THERAPEUTIC ANTIBODIES

With a hen laying more than 300 eggs per year, and with the vast infrastructure of the layer industry housing more than 100,000 hens, the hen holds great promise as a low - cost, high-yield bioreactor of active molecules usable by the food, pharmaceutical, and cosmetic industries. Novel biotechnological techniques are being used to develop genetically modified chickens that produce compounds that can be harvested from the eggs. Gene transfer will eventually make it possible to use the laying hen as a 'bioreactor' for the production of pharmaceuticals and other proteins (Shuman, 1991). These compounds include insulin for the treatment of diabetes. While an appropriate technology for the production of transgenic chickens that deposited novel proteins in their eggs was slow to develop, the concept remained attractive throughout the 1990s and into the first decade of the twenty-first century (Kamihira *et al.*, 2004; Lillico *et al.*, 2005). A major technical hurdle has been the introduction of large DNA constructs that would direct high-level and tissue-specific deposition of proteins into eggs. This hurdle has now been overcome (van de Lavoir *et al.*, 2006a) and high levels of therapeutic monoclonal antibodies have been produced in egg white (Zhu *et al.*, 2005). There are two possible ways to achieve this product. The first way would be to express the pharmaceutical gene in the oviduct of the hen in order to have the resulting protein incorporated into the egg albumen. Another way would be to express the gene in the liver and manipulate the resulting protein so as to have it incorporated in the egg yolk. Thus a new manufacturing system capable to produce bioproteins at a cost which would be less than the current mammalian or bacterial cell culture techniques could possibly be developed. Many factors like high rate of egg production, relatively short generation interval of hens and high protein ratio in eggs make laying hens a more advantageous bioreactor than their mammalian counterparts such as, rabbits, mice, goats, sheep and cattle. As new biotechnological knowledge is gained, this type of designer egg will most likely gain in popularity and use, resulting in a range of antibodies for treatment of snake venoms to the countering of microorganisms which cause tooth decay (Yamamoto *et al.*, 1997). Thus it can safely be said that the future for designer eggs is exciting.

Passive immunization consists of oral administration of antibodies for disease prevention, and has emerged as an alternative to the use of antibiotics, without any possible resistance from the microorganisms. Passive immunization requires large amounts of antibodies. Avian IgY can be produced in larger amounts and lesser cost than mammalian IgG, and it has additional advantages like welfare of the antibody-producing animals, the collection of the antibody source is much easier and there are less cross-reactions when administered. Moreover, antigen binding and bacterial cell agglutinating capacity of the IgY remains after heating up to 70°C for 15 min, after digestion with alimentary tract proteases, freezing and freeze-drying repeated several times. Considering its availability to obtain in the technical scale, it seems a good idea to select yolk gammaglobulin, as a food additive to fortify the products ability to last longer and be safer for the consumer. Stefaniak and Kopec (1997) obtained egg yolk IgYs against several species of bacteria-human alimentary tract pathogens (*Escherichia* coli, *Klebsiella pneumoniae*, *Salmonella enterifidis*, *S. typhimurium*), with no evidence of loss of activity of the IgY from eggs refrigerated up to 21 days. The specific antibody enriched eggs can be directed either to human or livestock disease prevention, and several reports have been published recently confirming its efficacy in both cases.

Thus immunologists are taking advantage of the fact that the hen can develop antibodies against a large array of antigens and concentrate them in the egg as protection to the developing chick. Specific antigens are now being selected and injected into the hen that develops antibodies against them. The egg is collected and a specific antibody is isolated and purified. This type of designer egg which serves as a source of specific antibodies is still in its developmental infancy.

11. BIOACTIVE PEPTIDES

Many of the properties of food proteins, including egg white proteins, with physiological significance beyond the nutritional requirements of nitrogen for growth and maintenance, are attributed to protein-encrypted physiologically active peptides (Korhonen & Pihlanto, 2003) that are released during *in vivo* gastrointestinal digestion or *in vitro* enzyme treatment and processing as bioactive peptides or "food hormones". Some studies describe the production of antihypertensive hydrolysates and peptides from egg proteins. The sequences FRADHPFL (ovokinin) and RADHPFL (ovokinin 2–7), that have shown endothelium-dependent vasodilator activity, were obtained at first. Some egg white hydrolysates with ACE inhibitory activity have also been described.

It has been reported that certain egg white derived peptides can play a role in controlling the development of hypertension by exerting vasorelaxing effects (Davalos *et al.*, 2004). Davalos *et al.* (2004) reported that the enzymatic hydrolysis of crude albumen proteins with pepsin resulted in the production of peptides with strong antioxidant activities. The egg white peptide Tyr–Ala–Glu–Glu–Arg–Tyr–Pro–Ile–Leu, which was shown previously to possess ACE inhibitory activity, also exhibited a high radical—scavenging activity. These results would suggest that the combined antioxidant and ACE-inhibitory properties of albumen hydrolysates, or the corresponding peptides, would make a useful multifunctional preparation for the control of cardiovascular diseases, in particular hypertension. Miguel *et al.* (2004) examined peptides with ACE-inhibitory properties produced by enzymatic hydrolysis of crude egg white. Peptide sequences RADHPFL, YAEERYPIL, and IVF with IC_{50} values of 4.7, 6.2, and 33.11 μ mol/L, respectively, derived after 3h incubation with pepsin followed by ultrafiltration exhibited potent ACE-inhibitory activity and antihypertensive activity (Miguel *et al.*, 2005).

12. RESEARCH NEEDS

Designer eggs with new functional properties are highly demanding, however, still there is lack of knowhow for their commercial production. There is need for more research for commercial production and marketing of these new generation egg and egg products. Further, more research should be done in this fascinating area to improve designer egg quality and assess long-term effects of their consumption and ultimately to convince customers of the benefits of eating these eggs.

REFERENCES

AHA, (2000). AHA Dietary Guidelines. Revision: A statement for healthcare professionals from the nutrition committee of the American Heart Association, *Circulation*, 102: 2296–2311.

Ahn, D.U., Sunwoo, H.H., Wolfe, F.H. and Sim, J.S. (1995). Effects of dietary alpha-linolenic acid and strain of hen on the fatty acid composition, storage stability, and flavor characteristics of chicken eggs. *Poult Sci.*, 74: 1540–1547.

Alzamora, S.M., Salvatori, D., Tapia, S., Lo'pez-Malo, M.A., Welti-Chanes, J. and Fito, P. (2005). Novel functional foods from vegetable matrices impregnated with biologically active compounds. *J. Food Eng.*, 67: 205–214.

American Dietetic Association, (1995). Position of the american dietetic association. phytochemicals and functional foods. *J. Am. Dietetic Assoc.*, 95: 493–496.

Applegate, E. (2000). Introduction: Nutritional and functional roles of eggs in the diet. *J Am Col Nutr.*, 19(5): 495S–498S.

Baker, D.H. and Halpin, K.M. (1991). Manganese and iron interrelationship in the chick. *Poultry Sci.*, 70: 146–152.

Bech-Larsen, T. and Grunert, K.G. (2003). The perceived healthiness of functional foods-A conjoint study of Danish, Finnish and American consumers' perception of functional foods. *Appetite*, 40: 9–14.

Beyer, R.S. and Jensen, L.S. (1989). Cholesterol content in commercially produced eggs in Georgia. *Poultry Sci.*, 68: 1703–1706.

Biehl, R.R., Emmert, J.L. and Baker, D.H. (1997). Iron bioavailability in soybean meal as affected by supplemental phytase and 1 alpha-hydroxycholecalciferol. *Poultry Sci.*, 76(10): 1424–1427.

Burdock, G.A., Carabin, I.G. and Griffiths, J.C. (2006). The importance of GRAS to the functional food and nutraceutical industries. *Toxicology*, 221: 17–27.

Caston, L.J., Squires, E.J. and Leeson, S. (1994). Hen performance, egg quality, and the sensory evaluation of eggs from SCWL hens fed dietary flax. *Can. J. Anim. Sci.*, 74: 347–353.

Chowdhury, S.R., Sarker, D.K., Chowdhury, S.D., Smith, T.K., Roy, P.K. and Wahid, M.A. (2005). Effects of dietary tamarind on cholesterol metabolism in laying hens. *Poult. Sci.*, 84(1): 56–60.

Cloughley, J., Noble, R., Speake, B. and Sparks, N. (1997). Manipulation of docosahexaenoic (22:6n-3) acid in chicken's egg. *Prostaglandins, Leucotrienes* and *Essential Fatty Acids*, 57: 222.

Cook, F. and Briggs, G.M. (1977). Nutritive value of eggs. *In*: Egg Science and Technology, Stadelman, W.J. and Cotterill, O.J. (*Eds*.), AVI Publishing Co., West Port, CT, pp. 92–108.

Cruikshank, E.M. (1934). Studies in fat metabolism in the fowl. *Biochem. J.*, 28: 965–977.

Davalos, A., Miguel, M., Bartolome, B. and Lopez – Fandino, R. (2004). Antioxidant activity of peptides derived from egg white proteins by enzymatic hydrolysis. *J. Food Prot.*, 67: 1939–1944.

Deignan, T., Alwan, A., Malone, L., Kelly, J. and O'Farrelly, C. (2001). Hen egg yolk prevents bacterial adherence: A novel function for a familiar food. *J. Food Sci.*, 66(1): 158–161.

Diplock, A.T., Aggett, P.J., Ashwell, M., Bornet, F., Fern, E.B. and Roberfroid, M.B. (1999). Scientific concepts of functional foods in Europe: Concensus document. *British J. Nutr.*, 81: S1–S27.

Dwyer, J.H., Navab, M., Dwyer, K.M., Hassan, K., Sun, P., Shircore, A., Hama-Levy, S., Hough, G., Wang, X., Drake, T., Bairey Merz, C.N. and Fogelman, A.M. (2001). Oxygenated carotenoid lutein and progression of early atherosclerosis: The Los Angeles atherosclerosis study. *Circulation*, 103: 2922–2927.

Evans, G.W. (1989). The effect of chromium picolinate on insulin controlled parameters in humans. *Int. J. Biosoc. Med. Res.*, 11: 163.

Farrell, D.J. (1993). One's Designer Egg: Recent Advances. *In*-Recent Advances in Animal Nutrition in Australia, pp. 291–302.

Farrell, D.J. (1994). The fortification of hen's eggs with Omega-3 long chain fatty acids and their effect in humans. *In*: Egg Uses and Processing Technologies-New Developments, Sim, J.S. and Nakai, S. (*Eds*.), CAB International, Wallingford, pp. 386–401.

Farrell, D.J. (1998). Enrichment of hen eggs with n-3 long-chain fatty acids and evaluation of enriched eggs in humans. *Am. J. Clin. Nutr.*, 68: 538–44.

Farrell, D.J. and Gibson, R.A. (1991). The enrichment of eggs with omega-3 fatty acids and their effects in humans. Recent Advances in Animal Nutrition in Australia. pp. 256–270.

Federation of American Societies of Experimental Biology (1995). Report on nutrition monitoring in the United States. US Government Printing Office, Washington, DC.

Ferrier, L.K., Caston, L., Leeson, S., Squires, E.J., Celi, B., Thomas, L. and Holub, B.J. (1992). Changes in serum lipids and platelet fatty acid composition following consumption of eggs enriched in alpha-linolenic acid. *Food Res. Int.*, 25: 263–268.

Ferrier, L.K., Caston, L.J., Leeson, S., Squires, J., Weaver, B.J. and Holub, B.J. (1995). Alpha-linolenic acid and docosahexaenoic acid-enriched eggs from hens fed flaxseed: Influence on blood lipids and platelet phospholipid fatty acids in humans. *Am. J. Clin. Nutr.*, 62: 81–86.

Ferrier, L.K., Leeson, S., Holub, B.J., Caston, L. and Squires, D.J. (1994). High linolenic acid eggs and their influence on blood lipids in man. *In*: Egg Uses and Processing Technologies-New Developments, Sim, J.S. and Nakai, S. (*Eds.*), CAB International, Wallingford, pp. 362–373.

Flinchum, J.D., Nockels, C.F. and Moreng, R.E. (1989). Aged hens fed zinc-methionine had chicks with improved performance. *Poultry Sci.*, 68(S1): 55.

Ginsberg, H.N., Karnally, W., Siddiqyi, M., Holleran, S., Tall, A.R., Ramsey, S.C., Deckelbaum, R.J., Blaner, W.S. and Ramakrishnan, R. (1994). A dose–response study of the effects of dietary cholesterol on fasting and postprandial lipid and lipoprotein metabolism in healthy young men. *Arterio Thrombosis*, 14: 576–586.

Grashorn, M.A. and Steinberg, W. (2002). Deposition rates of canthaxanthin in egg yolks. *Archiv fur Geflügelkunde*, 66(6): 258–262.

Griffin, H.D. (1992). Manipulation of egg yolk cholesterol: A physiologist's view. *World's Poultry Sci. J.*, 48: 101–112.

Griminger, P. and Brubacher, G. (1966). The transfer of vitamin K_1 and menadione from the hen to the egg. *Poult. Sci.*, 31: 777–780.

Handelman, G.J., Nightingale, Z.D., Lichtenstein, A.H., Schaefer, E.J. and Blumberg, J.B. (1999). Lutein and zeaxanthin concentrations in plasma after dietary supplementation with egg yolk. *Am. J. Clin. Nutr.*, 70: 247–251.

Hargis, P.S. (1988). Modifying egg yolk cholesterol in domestic fowl-a review. *World's Poultry Sci. J.*, 44: 17–29.

Harris, C. (2000). Meat products are perfect as functional foods. Meat Processing January/February 19, NAS (National Academy of Sciences), Dietary reference intakes for thiamin, riboflavin, niacin, vitamin B6, folate, vitamin B_{12}, pantothenic acid, biotin, and choline, National Academy Press, Washington, D.C.

Hasler, C.M. (2000). The changing face of functional foods. *J. Am. Col. Nutr.*, 19: 499S–506S.

Herber, S.M. and Van Elswyk, M.E. (1996). Dietary marine algae maintains egg consumer acceptability while enhancing yolk colour. *Poultry Sci.*, 77: 493–496.

Hoffman, D.R., Theur, R.C., Castaneda, Y.A., Wheaton, D.H., Bosworth, R.G., O'Connor, A.R., Morale, S.E., Wiedemann, L.E. and Birch, E.E. (2004). Maturation of visual acuity is accelerated in breast-fed term infants fed baby food containing DHA-enriched egg yolk. *J. Nutr.*, 134(9): 2307–2313.

Howell, W.H., McNamara, D.J., Tosca, M.A., Smith, B.T. and Gaines, J.A. (1997). Plasma lipid and lipoprotein responses to dietary fat and cholesterol: A meta-analysis. *Am. J. Clin. Nutr.*, 65: 1747–1764.

Hu, F.B., Stampfer, M.J., Rimm, E.B., Manson, J.E., Ascherio, A., Colditz, G.A., Rosner, B.A., Spiegelman, D., Speizer, F.E., Sacks, F.M., Hennekens, C.H. and Willet, W.C. (1999). A prospective study of egg consumption and risk of cardiovascular disease in men and women. *JAMA*, 281:1387–1394.

Ikemori, Y., Kuroki, M., Peralta, R.C., Yokoyama, H. and Kodama, Y. (1992). Protection of neonatal calves against fatal enteric colibacillosis by administration of egg yolk powder from hens immunised with K99-pilated enterotoxigenic *Escherichia* coli (ETEC). *Am. J. Vet. Res.*, 53: 2005–2008.

Jiang, Y.H., McGeachin, R.B. and Bailey, C.A. (1994). α-tocopherol, β-carotene and retinol enrichment of chicken eggs. *Poultry Sci.*, 73: 1137–1143.

Jiang, Z. and Sim, J.S. (1993). Consumption of n-3 polyunsaturated fatty acid-enriched eggs and changes in plasma lipids of human subjects. *Nutrition*, 9: 513–518.

Jiang, Z. and Sim, J.S. (1994). Fatty acid modification of yolk lipids and cholesterol-lowering eggs. *In*: Egg Uses and Processing Technologies-New Developments, Sim, J.S. and Nakai, S. (*Eds.*), CAB International, Wallingford, pp. 349–361.

Kamihira, M., Ono, K., Esaka, K., Nishijima, K., Kigaku, R., Komatsu, H., Yamashita, T., Kyogoku, K. and Iijima, S. (2005). High-level expression of single-chain Fv-Fc fusion protein in serum and egg white of genetically manipulated chickens by using a retroviral vector. *J. Virol.*, 79: 10864–10874.

Kang, C.W., Lee, H.Y., Nam, K.T. and Yoo, I.J. (1996). The effect of dietary inorganic and organic chromium on their transference rates to egg and accumulation in body tissues of laying hens. *In*: Proceedings of the XX World's Poultry Congress, 1996, WPSA, pp. 53–57.

Kang, D.K., Kim, S.I., Cho, C.H., Yim, Y.H. and Kim, H.S. (2003). Use of lycopene, an antioxidant carotenoid, in laying hens for egg yolk pigmentation. *Asian-Australian Journal of Animal Sciences*, 16(12):1799–1803.

Kerver, J.M., Bianchi, L.J., McNamara, D.J. and Song, W.O. (2000). The nutritional contributions of eggs to the American diet in the 90's. *FASEB J.*, 14: A 220 (161.1).

Kidd, M.T., Anthony, N.B. and Lee, S.R. (1992). Progeny performance when dams and chicks are fed supplemental zinc. *Poultry Sci.*, 71(7): 1201–1206.

Kim, J.H., Hong, S.T., Lee, H.S. and Kim, H.J. (2004). Oral administration of pravastatin reduces egg cholesterol but not plasma cholesterol in laying hens. *Poult. Sci.*, 83(9): 1539–1543.

Knekt, P., Aromaa, A. and Maatela, J. (1991). Vitamin E and cancer prevention. *Amer. J. Clin. Nutr.*, 53: 283s–286s.

Korhonen, H. and Pihlanto, A. (2003). Food-derived bioactive peptides-opportunities for designing future foods. *Curr. Pharm Des.*, 9: 1297–1308.

Kotilainen, L., Rajalahti, R., Ragasa, C. and Pehu, E. (2006). Health enhancing foods: Opportunities for strengthening the sector in developing countries. Agriculture and Rural Development Discussion Paper 30, World Bank, Washington, DC.

Krinsky, N.I. (198)9. Carotenoids and cancer in animal models. *J. Nutr.*, 119:123–126.

Kritchevsky, S.B. (2004). A review of scientific research and recommendations regarding eggs. *J. Amer. Col. Nutr.*, 23(6): 596S–600S.

Kuksis, A. (1992). Yolk lipids. *Biochem. Biophys. Acta.*, 1124: 205–222.

Kuroki, M., Ikemori, Y., Yokoyama, H., Peralta, R.C., Icatlo, F. and Kodama, Y. (1993). Passive protection against bovine rotavirus-induced diarrhea in a murine model by specific immunoglobulins from egg yolk. *Vet. Microbiol.*, 37: 135–146.

Kwak, N.S. and Jukes, D.J. (2001). Functional foods. Part 1. The development of a regulatory concept. *Food Control*, 12: 99–107.

Landrum, J.T. and Bone, R.A. (2001). Lutein, Zeaxanthin, and the Macular Pigment. *Arch. Biochem. Biophys.*, 385: 28–40.

Latshaw, D.J. (1987). Eggs give clues about dietary selenium. Poultry International (Feb.): 50.

Leeson, S. and Caston, L.J. (2003). Vitamin enrichment of eggs. *J. Appl. Poult. Res.*, 12(1): 24–26.

Leeson, S., Caston, L. and MacLaurin, T. (1998). Organoleptic evaluation of eggs produced by laying hens fed diets containing graded levels of flaxseed and vitamin E. *Poultry Sci.*, 77: 1436–1440.

Leskanich, C.O. and Noble, R.C. (1997). Manipulation of the n-3 polyunsaturated fatty acid composition of avian egg and meat. *World's Poultry Sci. J.*, 53: 155–183.

Lewis, N.M., Seburg, S. and Flanagan, N.L. (2000). Enriched eggs a source of n-3 polyunsaturated fatty acids for humans. *Poultry Sci.*, 79: 971–974.

Li-Chan, E.C.Y., Powrie, W.D. and Nakai, S. (1995). The chemistry of eggs and egg products. *In*: Egg Science and Technology. *Eds.* Stadelman, W.J. and Cotterill, O.J., the Haworth press Inc., New York.

Lien, T.F., Chen, S.Y., Chen, C.L. and Wu, C.P. (1993). The effects of various levels of chromium picolinate on growth performances and serum traits in pigs. *J. Chin. Soc. Anim. Sci. Taiwan*, R.O.C. 22: 349.

Lien, T.F., Chen, S.Y., Shiau, S.P., Froman, D.P. and Hu, C.Y. (1996). Chromium picolinate reduces laying hen serum and egg yolk cholesterol. *The Professional Anim Scientist*, 12: 77–80.

Lillico, S.G., McGrew, M.J., Sherman, A. and Sang, H.M. (2005). Transgenic chickens as bioreactors for protein-based drugs. *Drug Discov. Today*. 10: 191–196.

Losch, U., Schranner, I., Wanke, R. and Jurgens, L. (1986). The chicken egg, an antibody source. *J. Vet. Med. Series B.*, 33: 609–619.

Makrides, M., Hawkes, J.S., Newmann, M.A. and Gibson, R.A. (2002). Nutritional effect of including egg yolk in the weaning diet of breast-fed and formula-fed infants: A randomized controlled trial. *Am. J. Clin. Nutr.*, 75: 1084–1092.

Mark-Herbert, C. (2004). Innovation of a new product category-functional foods. *Technovation.*, 24: 713–719.

Marshall, A.C., Kubena, K.S., Hinton, K.R., Hargis, P.S. and Van Elswyk, M.E. (1994). n-3 fatty acid enriched table eggs: A survey of consumer acceptability. *Poultry Sci.*, 73: 1334–1340.

Mattila, P., Valaja, J., Rossow, L., Venäläinen, E. and Tupasela, T. (2004). Effect of vitamin D2- and D3-enriched diets on egg vitamin D content, production, and bird condition during an entire production period. *Poult. Sci.*, 83: 433–440.

Maurice, D.V. (1994). Dietary fish oils. Feeding to produce designer eggs. *Feed Management*, 45: 29–32.

McCarty, M.F. (1991). The case for supplemental chromium and a survey of clinical studies with chromium picolinate. *J. Appl. Nutr.*, 43: 58.

McNamara, D.J. (1995). Dietary cholesterol and the optimal diet for reducing risk of atherosclerosis. *Can. J. Cardiol.*, 11(G): 123G–126G.

McNamara, D.J. (2000a). Dietary Cholesterol and Atherosclerosis. *Biochem. Byophys. Acta*, 2529: 310–320.

McNamara, D.J. (2000b). The impact of egg limitations on coronary heart disease risk: Do the numbers add up? *J. Am. Coll. Nutr.*, 19: 540S–548S.

McNaughton, S.A. and Marks, G.C. (2002). Selenium content of Australian foods: A review of literature values. *Journal of Food Composition and Analysis*, 15(2): 169–182.

Meluzzi, A., Simoncini, F., Sirri, F., Vandi, L. and Giordani, G. (1996). Feeding hens diets supplemented with heavy metals (chromium, nickel and lead). *Arch. Geflügelk.*, 60(3): 119–125.

Menrad, K. (2003). Market and marketing of functional food in Europe. *J. Food Eng.*, 56: 181–188.

Meyer, B.J., Mann, N.J., Lewis, J.L., Milligan, G.C., Sinclair, A.J. and Howe, P.R.C. (2003). Dietary intakes ad food sources of omega-6 and omega-3 polyunsaturated fatty acids. *Lipids*, 38(4): 391–398.

Miguel, M., Lopez-Fandino, R., Ramos, M. and Aleixandre, A. (2005). Short-term effect of egg white hydrolysate products on the arterial blood pressure of hypertensive rats. *Br. J. Nutr.*, 94: 731–737.

Miguel, M., Recio, I., Gomez-Ruiz, J.A., Ramos, M. and Lopez-Fandino, R. (2004). Angiotensin I-converting enzyme inhibitory activity of peptides derived from egg white proteins by enzymatic hydrolysis. *J. Food Protect.*, 67: 1914–1920.

Moeller, M.S., Jacques, P.F. and Blumberg, J.B. (2000). The potential role of dietary xanthophylls in cataract and age-related macular degeneration. *J. Am. Coll. Nutr.*, 19: 522S–527S.

Mollet, B. and Rowland, I. (2002). Functional foods: At the frontier between food and pharma. *Curr. Opin. Biotech.*, 13: 483–485.

Morck, T.A. and Austic, R.E. (1981). Iron requirements of White Leghorn Hens. *Poultry Sci.*, 60: 1497–1503.

Naber, E.C. (1979). The effect of nutrition of nutrition on the composition of eggs. *Poultry Sci.*, 58: 518–528.

Naber, E.C. (1993). Modifying vitamin composition of eggs: A review. *App. Poult. Sci.*, 4: 385–393.

Naber, E.C. and Squires, M.W. (1991). Changing the cholesterol and vitamin content of eggs for improved human nutrition. *In*: Qualify of Poultry Products Il. Eggs and Egg Products, Oosterwould, A. and de Vries, A.W. (*Eds.*). Spelderholt Centre for Poultry Research and Information Services, Beekbergen, pp. 249–256.

Nakaue, H.S. and Hu, C.Y. (1997). Chromium picolinate (CrP) effect on blood and egg cholesterol levels in young and old single comb white leghorn (SCWL) laying chickens. *Poultry Sci.*, 76(1): 61.

Niva, M. (2007). All foods affect health: Understandings of functional foods and healthy eating among health-oriented Finns. *Appetite*, 48: 384–393.

O'Farrelly, C., Branton, D. and Wanke, C.A. (1992). Oral ingestion of egg yolk immunoglobulin from hens immunized with an enterotoxigenic *Escherichia* coli strain prevents diarrhea in rabbits challenged with the same strain. *Infect Immun.*, 60: 2593–2597.

Oh, S.Y., Ryue, J., Hseih, C.H. and Bell, D.E. (1991). Eggs enriched in omega-3 fatty acids and alterations in lipid concentrations in plasma and lipoproteins and in blood pressure. *Am. J. Clin. Nutr.*, 54: 689–695.

Ohama, H., Ikeda, H. and Moriyama, H. (2006). Health foods and foods with health claims in Japan. *Toxicology*, 22: 95–111.

Park, B.S., Kang, H.K. and Jung, A. (2005). Influence of feeding beta-cyclodextrin to laying hens on the egg production and cholesterol of egg yolk. *Asian Austral J. Anim. Sci.*, 18(6): 835–840.

Payet, M., Esmail, M.H., Polichetti, E., Brun, L.E., Adjemout, G., Donnarel, L., Portugal, G.H. and Pieroni, G. (2004). Docosahexaenoic acid-enriched egg consumption induces accretion of arachidonic acid in erythrocytes of elderly patients. *Brit. J. Nutr.*, 91(5): 789–796.

Peralta, R.C., Yokoyama, H., Ikemori, Y., Kuroki, M. and Kodama, Y. (1994). Passive immunization against experimental salmonellosis in mice by orally administered hen egg-yolk antibodies specific for 14-kDa fimbriae of *Salmonella enteritidis*. *J. Med. Microbiol.*, 41: 29–35.

Robel, E.J. (1991). The value of supplemental biotin for increasing hatchability of turkey eggs. *Poultry Sci.*, 70: 1716–1722.

Roberfroid, M.B. (2000a). Concepts and strategy of functional food science: The European perspective. *Am. J. Clin. Nutr.*, 71: S1660–S1664.

Roberfroid, M.B. (2000b). An European consensus of scientific concepts of functional foods. *Nutrition*, 16: 689–691.

Roberfroid, M.B. (2002). Global view on functional foods: European perspectives. *Brit. J. Nutr.*, 88: S133–S138.

Scheideler, S.E., Froning, G. and Cuppett, S. (1997). Studies of consumer acceptance of high omega-3 fatty acid-enriched eggs. *J. Appl. Poultry Sci.*, 6: 147–154.

Shallo. H.E. (2001). Designer foods: Egg products. Egg nutrition center, 1050 17th St NW, Suite 560, Washington, DC 20036, http://www.fass.org/fass01/pdfs/Shallo.pdf.

Sherwood, T.A., Alphin, R.L., Saylor, W.W. and White, H.B. (1993). Folate metabolism and deposition in eggs by laying hens. *Arch. Biochem. Biophys.*, 3(07): 66–72.

Side, C. (2006). Overview on marketing functional foods in Europe. Functional food network general meeting, March 8–10, 2006, University of Turku, Educarium building, address Assistentinkatu 5, Turku, Finland. http://congress.utu.fi/ffnet/PROGRAMME%20Jan%2012.pdf.

Simopoulos, A.P. (2000). Human requirement for n-3 polyunsaturated fatty acids. *Poult. Sci.*, 79: 961–970.

Sindelar, C.A., Scheerger, S.B., Plugge, S.L., Eskridge, K.M., Wander, R.C. and Lewis, N.A. (2004). Serum lipids of physically active adults consuming omega-3 fatty acid-enriched eggs or conventional eggs. *Nutr. Res.*, 24(9): 731–739.

Stadelman, W.J. (1999). The incredibly functional egg. *Poultry Sci.*, 78: 807–811.

Stanton, C., Ross, R.P., Fitzgerald, G.F. and Sinderen, D.V. (2005). Fermented functional foods based on probiotics and their biogenic metabolites. *Curr. Opin. Biotech.*, 16: 198–203.

Stefaniak, T. and Kopec, W. (1997). The activity of the hen's egg gamma globulin preparations as the food additives against the human alimentary tract pathogens. *In*: Eggs and Egg Products Quality, Kijowski, J. and Pikul, J. (*Eds.*), WPSA, pp. 248–254.

Surai, P.F. (2000). Organic selenium and the egg: Lessons from nature. *Feed Compounder*, 20: 16–18.

Surai, P.F. and Sparks, N.H.C. (2001). Designer eggs: From improvement of egg composition to functional food. *Trends Food Sci. Technol.*, 12: 7–16.

Surai, P.F., MacPherson, A., Speake, B.K. and Sparks, N.H.C. (2000). Designer egg evaluation in a controlled trial. *Eur. J. Clin. Nutr.*, 54: 298–305.

S.W. Park, Namkung, H., Ahn1, H.J. and Paik, I.K. (2005). Enrichment of vitamins d3, k and iron in eggs of laying hens. *Asian-Aust J Anim Sci.*, 18(2): 226–229.

Szymczyk, B. and Pisulewski, P.M. (2003). Effects of dietary conjugated linoleic acid on fatty acid composition and cholesterol content of hen egg yolks. *Brit. J. Nutr.*, 90(1): 93–99.

Ternes, W. and Leitsch, S. (1997). Chemistry of egg yolk. *ln*: Eggs and Egg Products Quality, Kijowski, J. and Pikul, J. (*Eds.*), WPSA, pp. 127–144.

Van de Lavoir, M.C., Diamond, J.H., Leighton, P., Mather-Love, C., Heyer, B.S., Kerchner, A., Hooi, L., Gessaro, T., Swanberg, S., Delany, M.E. and Etches, R.J. (2006a). Germline transmission of genetically modified primordial germ cells. *Nature*, 441: 766–769.

Van Elswyk, M.E., Hatch, S.D., Stella, G.G., Mayo, P.K. and Kubena, K.S. (1998). Poultry-based alternatives for enhancing the n-3 fatty acid content of american diets. *In*: The Return of n-3 Fatty Acids into the Food Supply, Simopoulos AP (Ed), Karger, New York, pp. 24–37.

Van Elswyk, M.E., Sams, A.R. and Hargis, P.S. (1992). Composition, functionality, and sensory evaluation of eggs from hens fed dietary menhaden oil. *J. Food Sci.*, 57: 342–349.

Watson, R.R. (2002). Eggs and health promotion, 2002 ed, Iowa State Press, Ames, IA.

Wheeler, P., Peterson, D.W. and Michaels, G.D. (1959). Fatty acid distribution in egg yolk as influenced by type and level of dietary fat. *J. Nutr.*, 69: 253–260.

Yaffee, M., Schutz, H., Stone, J., Bokhari, S. and Zeidler, G. (1991). Consumer perception and utilization of eggs and egg products. *Poult Sci.*, 70(1): 88.

Yamamoto, T., Juneja, L.R., Hatta, H. and Kim, M. (1997). Hen eggs: Their basis and applied science. CRC Press, New York.

Yaroshenko, F.O., Dvorska, J.E., Surai, P.F. and Sparks, N.H.C. (2003). Selenium-enriched eggs as a source of selenium for human consumption. *Applied Biotechnology, Food Science and Policy*, 1(1): 13–23.

Yokoyama, H., Peralta, R.C., Diaz, R., Sendo, S., Ikemori, Y. and Kodama, Y. (1992). Passive protective effect of chicken egg yolk immunoglobulins against experimental enterotoxigenic *Escherichia coli* infection in neonatal piglets. *Infect Immun.*, 60: 998–1007.

Yolken, R.H., Leister, F., Siok-Bi, W., Miskuff, R. and Vonderfecht, S. (1988). Antibodies to rotaviruses in chickens' eggs: A potential source of antiviral immunoglobulins suitable for human consumption. *Pediatr*, 81: 291–195.

Young, Y. (2000). Functional foods and the european consumer. *In*: Functional Foods. II. Claims and Evidence, Buttriss, J. and Saltmarsh, M. (*Eds.*), The Royal Society of Chemistry, London, UK.

Zeisel, S.H. (2000). Choline: Needed for normal development of memory. *J. Amer. Col. Nutr.*, 19: 528S–531S.

Zhu, L., van de Lavoir, M.C., Albanese, J., Beenhouwer, D.O., Cardarelli, P.M., Cuison, S., Deng, D.F., Deshpande, S., Diamond, J.H., Green, L., Halk, E.L., Heyer, B.S., Kay, R.M., Kerchner, A., Leighton, P.A., Mather, C.M., Morrison, S.L., Nikolov, Z.L., Passmore, D.B., Pradas-Monne, A., Preston, B.T., Rangan, V.S. and Shi, M. (2005). Production of human monoclonal antibody in eggs of chimeric chickens. *Nat Biotechnol.*, 23: 1159–1169.

27

Application of Proteomics in Meat Science

Ashim Kumar Biswas and Prabhat Kumar Mandal

ABSTRACT

Proteomics allows studying large numbers of proteins, including their post-translational modifications. Proteomics has been, and still are, used in numerous studies on skeletal muscle. Meat quality depends on several factors such as genetics, handling of animals during production, transportation and slaughter and also the handling of meat during processing. Proteome analyses were used to detect, as early as possible, animals carrying these modifications. Proteomic tools could help in detection of changes in protein profiles during myogenesis in cattle, pigs and fowl using across different ontogenetic stages. This approach allows a better understanding of the key stages of myogenesis and helps identifying processes that are similar or divergent between species. Biological markers of meat quality, particularly tenderness in cattle, pigs and fowl are presented, including protein modifications during meat ageing in cattle, protein markers of PSE meat in turkeys and of post-mortem muscle metabolism in pigs. Finally, we discuss the interest of proteomics as a tool to understand better biochemical mechanisms underlying the effects of stress during the pre-slaughter period on meat quality traits. This includes water-holding capacity, tenderness, colour, growth of tissues (muscle, bone or adipose tissue) or meat processing ability. All these fields of interest for meat scientists are now easily accessible because of recent technical developments in 2DE and of facilitated access to MS technology. However, all meat quality studies must not necessarily include proteomics studies while those including them will not always lead to the most relevant results for meat quality improvement. Proteomic tools can give a new point of view on these problems and help the biologist to understand and finally resolve it.

1. INTRODUCTION

Product quality and safety are still highly relevant issues for the meat industry worldwide as they are directly linked to public health and welfare. Most consumers buy exactly what they want to eat, instead of only what is readily available or affordable. Consumers also frequently look for quality seals and trust marks on meat products, and expect manufacturers and retailers to provide products of high quality. All of these factors have underlined the need for reliable techniques to monitor developments in meat science and technology and to evaluate the meat quality and safety. In recent years, a number of "omics" technologies (genomics, proteomics, metabolomics, and others) have become available that hold promise for increasing the understanding of the complexities of food quality and pathogen behavior at the molecular level. Genomics is the study of genes and their function, transcriptomics refers to global analyses of gene expression, and proteomics is the study of the complete set of proteins produced by a species and their modifications, expression, involvement in metabolic pathways, and interactions. Omics-based tools enable researchers to explore complex biological processes in a quantitative and integrative manner via a systems biology approach. These methods of analysis are facilitating the identification of genes that are responsible for survival and persistence in specific environments, and revealing genes that are potential targets for interventions and that play a role in pathogenesis, stress responses, and biofilm formation.

Proteomics is the study of the proteome, the protein complement of the genome. Proteomics or, more appropriately functional proteomics refers to the branch of discovery of science focusing on proteins. The term 'proteome' is used to describe the complete set of proteins that is expressed, and modified following expression, by the entire genome in the lifetime of a cell. The term "proteomics" and "proteome" were coined by Marc Wilkins and colleagues in the early 1990s and mirror the terms "genomics" and "genome", which describe the entire collection of genes in an organism. Today, proteomics is a scientific discipline that promises to bridge the gap between our understanding of genome sequence and cellular behaviour. It can be viewed as more of a biological assay or tool for determining gene function.

2. PROTEOMICS AND THEIR CATEGORIES

2.1. Proteomic Analysis/Analytical Protein Chemistry)

This includes large-scale identification and characterization of proteins, including their posttranslational modifications, such as phosphorylation and glycosylation. Analysis is done with the aid of mass spectrometry or Edman degradation.

2.2. Expression Proteomics/Differential Display Proteomics)

Two-dimensional gels are used for global profiling of expressed proteins in cell lysates and tissues. This conventional approach is being challenged by non-2D gel methods, such as liquid-based isoelectric focusing (IEF) or ion-exchange chromatography/reversed-phage high-performance liquid chromatography (RP-HPLC). Proteins are typically identified by mass spectrometry (MS). In many situations, these methods are complemented by DNA-based array methods.

2.3. Cell-Mapping Proteomics/Cataloging of Protein-Protein Interactions)

Protein-protein interactions and intracellular signaling are determined by identification of protein complexes (obtained by affinity purification and protein identification by MS) or by direct DNA readout (*e.g.*, yeast two-two hybrid, phage display, ribosome display, and RNA-peptide fusions).

3. PROTEOMICS *VS.* GENOMICS

3.1. Large-Scale Genome Sequencing

One of the most biological achievements to emerge during the last 40 years has been the completion of draft DNA sequences of the human genome, published by the International Human Genome Sequencing Consortium funded project and by Celera Genomics. This has provided a blue print of the information needed to create a human being and revealed for the first time the organization of a vertebrate's DNA. The public project estimates that there are 31,000 protein encoding genes, where as Celera finds ~26,000, with many more still to be found. A current estimate suggests that the number of protein-encoding genes may be on the order of 60,000. Interestingly, the number of coding genes in the human sequence is not dramatically different from the numbers reported for phylogenetically remote organisms: 6,000 for a yeast cell, 13,000 for a fly, 18,000 for a worm, and 26,000 for a plant (Genomes Online Databases at http://wit.integratedgenomics.com/GOLD).

The number of genes reported for multicellular organisms is not highly accurate because of limitations of existing abinitiogene prediction methods used to identify genes. The existence of an open reading frame (ORF) in genomic data does not necessarily imply the existence of functional gene. In human DNA, gene prediction by abinitiomethods is difficult because of extensive alternative splicing, lower density exons, and high proportions of interspersed repetitive sequences. Given the unreliability of abinitiogene prediction software, all genes will need to be experimentally identified and annotated. Hence, verification of a gene product by proteomic analysis is an important first step in annotating the genome.

3.2. Disparity between mRNA Profiling and Protein Profiling

There is no simple correlation between changes in mRNA expression levels (transcriptomics) and those in protein levels (proteomics). The link between transcript levels and protein levels in a given cell or tissue is difficult. It is also understood that array-based gene expression monitoring or other gene expression methods for measuring mRNA abundances, alone are insufficient for analyzing the cell's protein complement. There is a marked disparity between the relative expression levels of mRNAs and those of their corresponding proteins. Differing stability of mRNAs and different efficiencies in translation can affect the generation of new proteins. Once formed, proteins differ significantly in stability and turnover rates.

Many proteins involved in signal transduction, transcription factor regulation, and cell-cycle control are rapidly turned over as a means of regulating their activities. Also mRNA levels tell us nothing about the regulatory status of the corresponding proteins, whose activities and functions are subject to many endogenous posttranslational modifications and other modifications by environmental agents. Therefore, the complication arises when considering the complementarity of genomics and proteomics. Despite the notion that one gene gives rise to one protein, in eukaryotic cells is more likely six to eight proteins per gene. Thus, there may be several hundred thousand human proteins after splice variants and essential post translational modifications are included. For example, 22 different forms of human α-1-antitrypsin have been observed in human plasma. Such biological complexities can be unraveled using proteomic studies to understand how cells modulate and integrate signals.

4. IDENTIFICATION AND ANALYSIS OF PROTEINS

Four key platform technologies are crucial to any proteomics strategy aimed at elucidating the function of an unknown gene.

- Sample preparation & handling
- Determination of partial amino acid sequence information
- Protein identification and quantification
- Cell mapping

4.1. Protein Separation Strategies

One of the rate-limiting steps in any proteomic analysis study is obtaining, and then handling, sufficient quantities of target protein (s) from its original biological source. The classical method of quantitative and qualitative expression proteomics combines protein separation by high-resolution 2D gel electrophoresis with MS or MS/MS identification of selected protein spots. Because, even the best 2D gels can routinely separate no more than 1500 proteins, this technique is limited to the most abundant proteins if a crude protein mixture (whole cell lysate) is used. 2D electrophoresis is limited by the amount of material that can be applied to the first-dimension immobilized pH gradient gel (~150 μg to low milligram quantities). Hence, 2D gels have limited 'scale up' capability. For this reason, it is often desirable to "trace enrich" for a particular subclass of proteins. By analyzing proteins in a cellular compartment or organelle, it is possible to reduce the complexity and differences in abundance of a subset of proteins within a cell.

4.2. Two-Dimensional SDS-PAGE

This separation method has become synonymous with proteomics and remains the single best method for resolving highly complex protein mixtures. Two dimensional SDS-PAGE is actually a combination of two different types of separations. In the first, the proteins are resolved on the basis of isoelectric point by IEF. In the second, focused proteins are then further resolved by electrophoresis on a polyacrylamide gel. Thus 2D-SDS-PAGE resolves proteins in the first dimension by isoelectric point and in the second dimension by molecular weight. Dedicated 2D-SDSPAGE systems are available that use immobilized pH gradient (IPG) strips and relatively foolproof hardware to facilitate the transfer of proteins from the IPG strip into the SDS-PAGE slab gel. The IPG strip is based on the use of immobilized pH gradients, in which polycarboxylic acid ampholytes are immobilized on supports to reproducibly create stable pH gradient. The use of narrow pH ranges facilitates the separation of proteins with highly similar isoelectric points. Proteins separated by 2D gels are visualized by conventional staining techniques, including silver, coomassie, amido black stains and fluorescent staining. Silver-staining and newer fluorescent dyes are the most sensitive.

4.3. Cell Mapping and Identification of Proteins in Complexes

One way to observe interacting proteins involved in a given biological process is to specifically enrich for these proteins. Typically, this requires knowledge of the activity of at least one protein in the multi-protein complex. Under non-denaturing conditions, interacting proteins can be enriched from complex protein mixtures (*e.g.*, cell lysates) using methods such as:

- *Coimmunoprecipitation* or “pull-down” techniques using antibodies directed against one of the component proteins.
- *Coprecipitation* using affinity-tagged recombinant proteins and antibodies directed against the “tag” epitope.
- *Protein-affinity-interaction chromatography* (*e.g.*, using recombinant glutathione S-transferase (GST)- fusion proteins and glutathione-affinity chromatography).
- *Isolation of intact multiprotein complexes* (*e.g.*, nuclear pore complex, ribosome complexes, spliceosomes).

4.4. Determination of Partial Amino Acid Sequence

Usually, the final step of most proteomic studies, independent of the purification method employed, utilizes either SDS-PAGE or 2D acrylamide gels to separate the proteins for identification and characterization. Following electroblot and transfer to an inert membrane, such as polyvinylidine difluoride (PVDF), intact proteins can be identified directly by amino- or carboxy-terminal amino acid sequence analysis or indirectly from peptides generated by in-gel or on-membrane digestion of the protein with a protease usually trypsin. MS-based methods usually identify a protein, not by analyzing it directly, but by analyzing the peptides derived from proteolytic digestion. Usually, a small number of peptides yield sufficient information to permit protein identification by peptide mass finger printing (PMF) and/or MS/MS of individual peptides.

In contrast to peptides, the molecular mass of intact proteins is usually insufficient to allow database identification. MALDI-MS is used to determine the accurate mass of a group of peptides derived from a protein by digestion with a sequence-specific protease, usually trypsin, thus generating a peptide mass map or peptide mass fingerprint. Because trypsin cleaves proteins at the amino acids arginine and lysine, the masses of tryptic peptides can be predicted theoretically for any entry in a protein sequence database. Electrospray ionization (ESI) and tandem mass spectrometry (TMS) is used to sequence the isolated peptides from a peptide mixture.

5. DIFFERENTIAL DISPLAY PROTEOMICS

A fundamental aspect of proteomic research is the determination of protein expression levels between two different states of a biological system (*e.g.*, relative quantification of protein levels), such as that encountered between a normal and diseased cells or tissues. This is referred to as differential display or comparative proteomics. This can be done in two ways such as running and comparing samples in 2D-SDS-PAGE or with LC-MS and isotope tags.

For differences in the protein-expression profiling, we compare 2D-gels from two different samples for differences in the occurrence or intensity of protein spots. This approach provides a useful means of comparing proteomes. However, identification of protein is cumbersome and difficult by this procedure. Application of peptide mass fingerprinting and LC-MS-MS analysis now makes it possible to identify essentially any protein one can detect by staining the gel. Therefore, the critical task in comparative proteomics with 2D gels is identifying the features that differ between the gels. The LC-MS approach to proteome comparison is conceptually the opposite of the 2D gel approach.

Whereas, the 2D gel approach separates proteins and begins with an image comparison, the LC-MS approach separates peptides and ends with data mining to assess differences between samples. Two protein samples are treated with reagents to "tag" them. The tags are chemically identical, except that one contains heavy isotopes (*e.g.*, 2H, 15N, 13C, 18O etc.) and the other contains light isotopes. The samples are digested and the peptides are analyzed by LC-MS-MS. Analysis of the MS-MS data allows identification of the protein present. Examination of the full-scan spectra corresponding to each MS-MS scan then allows measurement of the ratio of the light and heavy isotope tagged peptides. This ratio corresponds to the ratio of that protein in the two samples. This approach provides a relative quantification of the level of a particular protein in two samples.

6. MAPPING PROTEIN MODIFICATIONS

Vast majority of all eukaryotic proteins are post translationally modified and more than 200 posttranslational modifications (PTMs) of amino acids have been reported so far. Practically all PTMs are associated with either an increase or a decrease in molecular mass. The two major PTMs of proteins are phosphorylation and glycosylation. Phosphorylation of proteins is a ubiquitous regulatory mechanism in both eukaryotes and prokaryotes. Intracellular phosphorylation is regulated by protein kinases (dephosphorylation is regulated by protein phosphatases), which are activated in response to extracellular signals and trigger cells to switch on or off many diverse processes such as metabolic pathways, kinase cascade activation, membrane transport, gene transcription, and motor mechanisms. Protein phosphorylation can be examined in several ways such as 'phosphopeptide mapping of 32_P labeled proteins and peptides', amino-terminal sequencing using Edman degradation procedure and mass spectrometry. While all the three methods are equally good, MS is an ideal tool in proteomics studies of PTM identification and characterization because of its high sensitivity.

7. APPLICATIONS OF PROTEOMICS IN MEAT SCIENCE

In meat science proteomics is a fairly new tool. Prediction of meat quality in the slaughter line on the day of slaughter is very much desired, from both a logistic and an economical point of view. The vast progress in the production of lean meat during the last decades has largely taken place at the expense of meat quality, causing serious problems in the tenderness and the water holding capacity (drip loss) of meat. Drip loss may vary 2–12%, which causes a direct economical loss because of reduced weight. Indirect losses are caused by reduced possibilities for processing of meat into high quality products such as ham.

Rapid on-site prediction of meat quality, preferably even before the meat is chilled, may allow for allocation of carcasses to different production lines or for branding of meat (premium-meat, etc.). Meat processors companies, are becoming more aware that "raw" meat proteins must have consistent and specific quality and functionality characteristics. The latest consumer demands have already led to product differentiation and greater pressure on the value of meat quality parameters, especially tenderness, juiciness and flavor of fresh and value-added meat products (Sosnicki *et al.*, 2003). Moreover, tenderization through proteolytic degradation, and water holding capacity are interrelated phenomena, as shown by a published model for possible causes of drip loss.

7.1. Proteome Mapping

Several proteome studies have been performed aiming at describing the different proteins that are present in a meat sample. This approach can be described as proteome mapping. A proteome mapping study of bovine *semitendinosus* (ST) muscle using a combination of 2-DE and mass spectrometry allowed the detection of roughly 500 reproducible protein spots (Bouley *et al.*, 2004) from these, a total of 129 protein spots corresponding to 75 different gene products were identified by matrix-assisted laser desorption ionization-time of flight (MALDI-TOF) MS. Approximately 25 percent of the identified proteins were involved in metabolic pathways, while 17, 16, and 14.5 percent were involved in cell structure, cell defense, and the contractile apparatus, respectively.

However, how these proteins are related to meat quality was not investigated in this study. A mapping strategy has also been used to compare the proteome expression of sarcoplasmic proteins between four different ovine muscles (Hamelin *et al.*, 2006). Muscle samples were taken within 30 min after slaughter from *Longissimus dorsi* (LD), *Vastus medialis* (VM), *Semi-membranosus* (SM), and *Tensor fasciae latae* (TL). After 2-DE and data analysis, 77 protein spots were found to be differentially expressed between the muscles, and 47 of these were identified by MALDI-TOF MS. The VM muscle had a different protein expression than the other three muscles. This muscle had the highest proportion of slow-twitch oxidative fibers and the highest abundance of enzymes involved in oxidative metabolism and oxidative stress-related proteins. The TL, on the contrary, had a reduced expression of most of these proteins, which reflects that this muscle had the highest rate of fast twitch glycolytic fibers.

7.2. Changes in Proteomes due to Genetic Factors

Genetic variations may cause phenotypic differences that can be studied using proteomics. Changes in expression of bovine skeletal muscle proteins induced by hypertrophy were studied by proteome analyses (Bouley *et al.*, 2005). The muscle hypertrophy was caused by a deletion in the *myostatin* gene, and samples from the ST muscle were taken from bulls that were heterozygote and homozygote for the *myostatin* deletion and from control bulls. The 2-DE analysis allowed the detection of 400 common protein spots, and statistical analyses revealed 28 protein spots that differed between the control and the homozygote bulls, while only one protein differed between the control and the heterozygotic bulls. Changes of proteins in the contractile apparatus and metabolic enzymes indicate that the *myostatin* deletion results in a shift towards a fast-twitch glycolytic muscle type, which is in accordance with the results from the muscle fibre type analyses performed on the same animals. This demonstrates that proteomics reflects the shifts in fiber type in the muscles.

In another study the effect of quantitative trait loci (QTL) for muscle hypertrophy on sarcoplasmic proteins expression in four ovine muscles was investigated (Hamelin *et al.*, 2006). Thirty minutes after slaughter, samples from LM, VM, SM, and TL were taken for analyses by 2-DE. Only three of the muscles (LM, SM, and TL) were hypertrophied due to the QTL, while the VM was normal. By use of MALDI-TOF MS, 63 spots that were expressed differentially between the genotypes were identified. These changes included increased levels of enzymes involved in the glycolytic metabolism, which indicates a switch towards a more glycolytic muscle type. Moreover, several chaperon proteins were also increased in abundance, which may indicate a requirement for more building blocks to promote myofibril assembly.

7.3. Proteome Changes due to Pre-Slaughter Factors

Proteomic tools have been applied to investigate proteome changes induced by different pre-slaughter conditions. On proteome study during compensatory growth in pigs, it has been found that the changes of meat are associated with tenderness (Kristesen *et al.*, 2004). Seven proteins were changed according to compensatory growth and among them several stress proteins and glycolytic proteins were decreased in abundance (Lametsch *et al.*, 2006). In the same study eight proteins were affected 48 h after slaughter including both structural proteins and enzymes. In another study, the effect of postmortem storage time (0, 12, and 72 h) and different pre-slaughter treatments like transported immediately before slaughter vs. transported 12 h before slaughter of pigs on the LD muscle proteome were analyzed (Morzel *et al.*, 2004). Most differences were found as a result of postmortem storage, and 37 spots varied with storage time. Ten of these increased in abundance during the storage period, such as fragments of actin, troponin T, and α-crystallin resulting from proteolysis. The pre-slaughter treatments affected 8 protein spots, while they had no effect on the proteolytic events taking place during the storage period.

7.4. Proteome Changes affects on Lean Colour

Lean colour is a complex trait influenced by genetic components, the handling of animals during production, transportation, and slaughtering, and the treatment during processing. The pigment responsible for most of this colour is the protein myoglobin. Meat myoglobin (Mb), is a complex sarcoplasmic heme protein, and comprised of 153 amino acids, and the primary structure of Mb depends up on species. Despite differences in the amino acid sequences, many structural and functional properties of Mb are conserved across livestock species. The primary structure of protein dictates the tertiary structure, which in turn influences interactions with ligands and biomolecules. It is known that Mb primary structure influences meat color stability via mechanisms such as autoxidation, heme retention, structural stability, thermostability, and oxygen affinity. The interactions of pH, temperature and post-mortem time also affect the biochemical dynamics of early Mb discolouration and hence darker meat colour.

Furthermore, variations in amino acid sequence of Mb influence meat color stability through species specific interactions with small biomolecules like lactate and aldehydes. For Mb, a heme protein with different redox states, this is extremely critical, because Mb stability and aforementioned molecular interactions govern meat color/color stability. Recently, 2-DE and MS have been used to investigate the biochemical mechanisms behind variation in meat colour Comparison of the sarcoplasmic proteome in pig SM muscle from two groups of animals having a light or dark meat colour revealed that 22 protein spots were differentially expressed (Sayd *et al.*, 2006). The animals, 12 animals in each group, were selected from samples of 1000 pigs based on extreme *L*-values. While the dark muscles had an increased abundance of mitochondrial proteins, indicating a more oxidative metabolism, the light muscles had an increased abundance of cytosolic proteins involved in glycolysis.

7.5. Proteome Changes Due to Post-Mortem Factors

7.5.1. *Tenderness*

Postmortem degradation of muscle proteins is an important factor in the meat tenderization process, as postmortem degradation of several structural proteins including troponin T, nebulin, titin, vinculin, desmin and dystrophin has been demonstrated using one-dimensional SDS–PAGE and immuno-blotting (Melody *et al.*, 2004; Geesink & Koohmaraie, 1992). This is supported by the fact that titin, a giant muscle protein spanning from the Z-line to M-line region, and nebulin, which runs parallel with the thin filaments to the Z-line both have been shown to degrade faster in tender meat than in tough meat. Moreover, electron microscopy studies of the myofibrils during tenderization have shown that the attachment of the sarcolemma to myofibrils and the junctions between myofibrils at the level of the Z-disk and M-line are disrupted postmortem, which suggests that proteolysis of the costamere proteins desmin, dystrophin, and vinculin, which attach the myofibrils to the sarcolemma, is taking place. Strong correlations have been found between postmortem degradation of troponin T and shear force.

However, whether the degradation of troponin T is an essential part of the tenderization process or just a marker for postmortem proteolytic activity is still unclear. Results of proteomic study revealed that changes of proteins occurred in muscle during postmortem storage. A total15 proteins were changed, some increasing and some were decreasing in abundance after slaughter. Several of these proteins were identified as fragments of structural proteins such as actin, myosin heavy chain and troponin T. Proteome analysis by two-dimensional gel electrophoresis (2DE) and matrix-assisted laser desorption/ionization time-of-flight mass spectrometry (MALDI-TOF MS) further extended the number of observed degrading proteins *L. dorsi* of pigs during the first 72 h postmortem to 103, which included actin, myosin heavy chain (HC), myosin light chain (LC) I and II, CapZ and cofilin (Lametsch *et al.*, 2004).

It indicated that using 2-DE the resolution allows for a better separation of the different actin fragments and demonstrates the potential of 2-DE. Some of the actin fragments observed in postmortem pork muscle were also significantly correlated with tenderness. In addition, several metabolic enzymes involved in energy metabolism were altered in the pork muscle. More recently, a tentative 2DE-based reference map encompassing 133 proteins was prepared by (Hwang *et al.*, 2005).

In a study of proteome changes related to tenderness in pork LD muscle 6 proteins were affected (Lametsch *et al.*, 2003). A great increase of fragments of actin during postmortem storage was observed. Three actin fragments as well as a myosin heavy chain fragment were correlated to shear force. Moreover, myosin light chain II and the glycolytic enzyme triose phosphate isomerase were correlated to tenderness. This has also been confirmed by (Hwang *et al.*, 2005). It has also been found that proteome analysis of SM muscle from normal hams and from PSE zones of defective hams demonstrated a reduced proteolysis of troponin T, C, I and α-crystallin in the defect muscles (Laville *et al.*, 2005).

During postmortem storage the calpain system is believed to be important for degradation of myofibrillar proteins and development of tenderness. In a proteome study in pork LD muscle several of the myofibrillar substrates for μ-calpain were identified (Lametsch *et al.*, 2004). Among the substrates that were degraded by *in vitro* incubation with μ-calpain were desmin, actin, myosin heavy chain, myosin light chain I, troponin T, tropomyosin α1, tropomyosin α4, thioredoxin and CapZ. Changes in metabolic protein composition in biopsies from live animals to postmortem samples collected shortly after slaughter in the cattle LD muscle revealed that 24 protein spots were changed (Jia *et al.*, 2006). This reflects the contribution of several factors such as transportation, lairage, stunning, exsanguination and dehiding on the LD muscle proteome. Identification of the proteins by MALDI-TOF/TOF MS revealed that a wide range of metabolic enzymes and stress proteins increased in abundance after slaughter.

Several of these proteins were glycolytic enzymes such as enolase, aldehyde dehydrogenase, phosphoglycerate kinase or enzymes involved in oxidative metabolism such as ATP-specific succinyl-CoA synthetase and isocitrate dehydrogenase. This supports an expected shift in energy metabolism in the muscle postmortem via the glycolytic pathway, and also an increase in aerobic energy metabolism the first hour after slaughter. Proteomic study on the changes in the muscle proteome between slaughter and 24 h storage in bovine LD and ST muscles showed that 5 proteins were changed in both muscles, namely cofilin, lactoylglutathione lyase, substrate protein of mitochondrial ATP-dependent proteinase SP-22, HSP27KDa and HSP20KDa. However, 15 proteins were changed in either LD or ST muscles. These differences reflect distinct metabolic and physiological functions of the different muscles.

7.5.2. *Water holding capacity* (WHC)

The changes in water holding capacity (WHC) in muscle postmortem involves lateral shrinkage of myofibrils as the principle mechanism driving water loss (Kristensen & Purslow, 2001). Lateral shrinkage of myofibrils, muscle fibers and entire muscle fascicles, generates drip channels between fascicles. This mechanism requires that lateral connections formed by intermediate filaments (IFs) through the costameres to extracellular structures are sufficiently strong to laterally displace fluid released by shrinkage of individual myofibrils to the extracellular spaces (Purslow *et al.*, 1997). Apart from involvement of IF there are many other factors, both genetic and non-genetic, that can influence WHC (Henckel *et al.*, 2000). Newly discovered polymorph expression products in this respect are the melanocortin-4 receptor, the γ-3 subunit of AMP activated proteinkinase, and the ryanodine receptor. Non-genetic factors include stress and nutrition status of the animal before slaughter, creatine phosphate (CP) and glycogen content, ultimate pH and fiber type composition of meat, stunning method and cooling regime.

7.6. Proteome Changes in Dry-Cured Hams

Proteome analyses have also been used to study protein changes in dry-cured hams. In one study, the myofibrillar proteins of raw meat and dry-cured hams after 6, 10 and 14 months of ripening were analyzed by 2-DE (Di Luccia *et al.*, 2005). Both actin, tropomyosin and myosin light chain disappeared during the ripening period, and were almost completely hydrolyzed after 12 months.

7.7. Application of Proteomic Tools in Meat Food Authentication

Meat articles are continually being adulterated, mainly from the desire of manufacturers to increase their profits even though they are required to label meat products very precisely and specify exactly all the applied ingredients on the packaging. Meat derived from different animal species carries a different price and consumers want to be certain what kind of meat they buy. Adulteration of meat articles poses a serious problem not only for economical reasons, but equally important, for religious and moral objections as well as due to allergies to meat derived from individual animal species. The commercial quality and appropriate labeling of meat and meat products control carried out in Poland in the second quarter of 2009 by the Commercial Inspection revealed that 24 percent of the examined batches of meat and meat articles were labeled in contravention of the existing regulations. The most frequent faults included, among others, lower input of meat fat raw materials, lower meat content, lower content of protein and higher content of starch.

The above mentioned faults indicate incorrect quantitative and qualitative selection of raw material for production; in other words, manufacturers used cheaper meat of lower quality. This means that when buying processed meat, especially homogenized sausages, finely comminuted sausages as well as canned meat products, consumers do not always get what they pay for. Similar proportions of meat product falsification have also been reported in other countries. Ayaz *et al.* (2006) conducted investigations of 100 products purchased on the Turkish market, including sausages, salami, frankfurters, bacons, and canned products, raw and cooked meats. The performed tests revealed that 22 percent of all the examined products were adulterated and contained meat species incompatible with specifications on the label. In 1995, Rao and Hsieh (2007) reported that in the United States 15.9 percent and 22.5 percent of fresh and cooked meat articles, respectively were falsified.

Similarly, in Great Britain, in 1999, 14.6 percent of meat products obtained from shops contained meat species (Rao & Hsieh, 2007) that were not specified on the package. Investigations carried out in Spain by Pascoal *et al.* (2004) showed that 30 percent of meat products purchased in supermarkets had incorrect labeling. The most common irregularity was the addition of turkey meat undeclared on the label.

Methods aiming at investigating meat and meat product authenticity may be based either on the analysis of protein composition or on the analysis of nucleic acids. Each species is characterized by its own individual set of genes (genome) which predetermines, in each cell, the amino acid composition of all proteins, their quantity and the rate of their synthesis. Thus, authenticity investigations of meat and meat products are based either on the analysis of protein composition, including all isoforms and modified proteins, or on the analysis of nucleic acids. In proteomic analysis for authentication of meat and meat products two-dimensional electrophoresis (2-DE) has been developed in recent years.

This technique makes it possible to separate simultaneously thousands of proteins in one gel. During the first dimension, proteins are separated with the assistance of the native technique (most frequently IEF) depending on their isoelectric point (pI) while the second dimension uses SDS-PAGE to separate proteins depending on their molecular weight (MW). The combination of the above mentioned techniques results in the development of 2-D maps on which proteins are visible as spots with positions determined by their pI and MW. This technique can be used to separate simultaneously up to 10,000 proteins as well as to identify and determine quantitatively spots that contain less than 1 ng protein (Lopez, 2007). Moreover, it allows analysis not only of intact proteins but, equally importantly, of changes in protein expression levels, isoforms and posttranslational modifications. Potentials of the 2-DE technique were first described by O'Farrell (1975).

However, the introduction in recent years of immobilized pH gradients (IPG) and the availability of ready-made IPG strips considerably enhanced possibilities of its utilization. The analysis of separations has been greatly facilitated by currently available computer software since they provide tools for the segmentation, alignment and matching of 2-DE images. It is also possible to distinguish, on the basis of protein separation extracted from muscle tissue, between wild cod and farmed cod (Martinez *et al.*, 2007). The performed analysis of parvalbumins with the assistance of 2-DE in combination with MALDI-TOF MS allowed identification of 10 closely related fish species from the *Merlucciidae* family (Carrera *et al.*, 2006) Mass spectrometry made it possible to differentiate, on the basis of all isoforms, parvalbumins isolated from 9 different hake species and two grenadier sub-species.

These investigations corroborate the fact that the 2-DE technique allows investigations of similarities and differences between proteins extracted from muscles of different animal species. Experiments aimed at verifying whether the 2-DE method is appropriate for species identification of poultry, pork and beef have confirmed that the obtained separations of chicken, duck, swine and cattle meats were species specific. Despite certain limitations associated with the application of the 2-DE method, this technique is quite versatile and most common in proteome analysis.

7.8. Application in Pathogen Detection

Microarray based comparative genomics research, which takes advantage of information available from whole genome sequences, is leading to an increased understanding of the evolution and pathogenesis of food borne pathogens and is providing critical information for the development of improved detection and genotyping methods. DNA microarray technology provides accurate measurements of gene expression for every gene in a genome and allows this expression to be analyzed in response to specific environmental variables. Further, this technology can be utilized to identify genes that are controlled by specific regulators by comparing gene expression in mutant and wild type bacteria.

However, the potential for the analysis of gene expression of pathogens in food environments has not yet been completely realized due to the substantial technical challenges associated with accurately measuring bacterial gene expression in complex matrices. Genomotyping involves the comparison of whole genomes of bacteria using DNA microarrays and has been utilized to identify potential genes associated with virulence, disease severity, and adaptation to different hosts and ecological niches. Used as diagnostic tools, DNA microarrays offer the capability to detect and characterize a broad spectrum of pathogens simultaneously in a relatively short period of time.

Various food grade bacteria with specific reference to lactic acid bacteria are used for production of fermented foods. Different lactic acid bacteria produce different proteins in the system during fermentation. Global analysis of proteome of useful lactic acid bacteria using proteomic approaches will help to identify various proteins expressed in the system. Systematic analysis of the proteins expressed will give an idea about the importance of various proteins during fermentation. Differential expression proteomics in between various useful microorganisms will help us to identify useful biomolecules specific for a particular microorganism. This will help us to identify organism specific cellular markers for future application in food product development.

8. FUTURE PROSPECTS

Proteomics has turned out to be a promising and powerful tool in food science, and in particular meat science over the last years yielding information on differences between muscles in a carcass. Furthermore, the combined approach of 2-DE and MS has demonstrated a great strength in studying metabolic alterations, postmortem proteolysis, and changes induced by environmental and processing conditions. As a result an increasing number of studies are emerging in the literature using proteomics as the key tool to unleash the molecular mechanisms behind different genetic backgrounds or processing techniques of meat. Although the genetic background is important, the major contribution to meat quality is caused by processing and environmental conditions. Thus proteomics could be the tool to reflect the important mechanisms and contributions to development of a satisfactory meat quality.

Proteomic approaches will help to identify useful low abundant proteins, which can be studied further for understanding their beneficial properties. Global genetic based analyses provide information regarding which genes an organism contains or which genes are expressed under specific conditions; however, examining the posttranslational protein output of an organism allows one to query the ultimate outcome of the organism's genetic and regulatory activities. Techniques that fall within the category of either proteomics or protein arrays are used for global analysis of cellular protein output under different conditions and potentially those relevant to food and food processing environments. The integration of both genetic and protein based approaches provides a global analysis of treatment or environment-related changes at the molecular level, thus presenting a more comprehensive view of cellular activities. The development of high-throughput analysis techniques will make possible multi-omics approaches to understanding complete biological systems, a field known as systems (integrated) biology.

Although omics technologies are becoming standard research tools that offer tremendous opportunities, there are also significant challenges. There is a need to properly manage the large quantity of complex raw data generated by these technologies in a manner such that it can be adequately analyzed, scrutinized, and compared for the benefit of the scientific community. There are various omics standardization activities underway, which are critical for the integration and interpretation of data from different data sources. Lastly, there is a need to bridge the gap between knowledge of the genome, proteome, and metabolome, and results obtained in relevant systems by studying the behavior of pathogens in foods and in the animal host, not only in model systems under laboratory conditions. The knowledge garnered from omics-based research in the coming years will play an important role in understanding how pathogens survive food safety barriers and interact with host species. Each new advance in our understanding will potentially give rise to improved and novel strategies for detection, identification, and control of food borne pathogens, as well as for diagnosis and control of infections.

REFERENCES

Ayaz, Y., Ayaz, N.D. and Erol, I. (2006). Detection of species in meat and meat products using enzyme linked immunosorbent assay. *J. Muscle Foods*, 17: 214–220.

Bouley, J., Meunier, B., Chambon, C., Smet, S.De, Hocquette, J.F. and Picard, B. (2005). Proteomic analysis of bovine skeletal muscle hypertrophy, *Proteomics*, 5: 490–500.

Carrera, M., Cañas, B., Piñeiro, C., Vázquez, J. and Gallardo, J.M. (2006). Identification of hake and grenadier species by proteomic analysis of the parvalbumin fraction. *Proteomics*, 6: 5278–5287.

Di Luccia, A., Picariello, G., Cacace, G., Scaloni, A., Faccia, M. and Liuzzi, V. (2005). Proteomic analysis of water soluble and myofibrillar protein changes occurring in dry-cured hams. *Meat Sci.*, 69: 479–491.

Geesink, G.H. and Koohmaraie, M. (1999). Postmortem proteolysis and calpain/calpastatin activity in callipyge and normal lamb biceps femoris during extended postmortem storage. *J. Anim. Sci.*, 77: 1490–1501.

Hamelin, M., Sayd, T., Chambon, C., Bouix, J., Bibe, B. and Milenkovic, D. (2006). Proteomic analysis of ovine muscle hypertrophy. *J. Anim. Sci.*, 84: 3266–3276.

Henckel, P., Karlsson, A., Oksbjerg, N. and Petersen, J.S. (2000). Control of postmortem pH decrease in pig muscles: Experimental design and testing of animal models. *Meat Sci.*, 55: 131–138.

Hwang, I.H., Park, B.Y., Kim, J.H., Cho, S.H. and Lee, J.M. (2005). Assessment of postmortem proteolysis by gel-based proteome analysis and its relationship to meat quality traits in pig longissimus, *Meat Sci.*, 69: 79–91.

Jia, X., Hollung, K., Therkildsen, M., Hildrum, K.I. and Bendixen, E. (2006). Proteome analysis of early post-mortem changes in two bovine muscle types: *M. longissimus dorsi* and *M. semitendinosis*, *Proteomics*, 6: 936–944.

Kristensen, K. and Purslow, P.P. (2001). The effect of ageing on the waterholding capacity of pork: Role of cytoskeletal proteins. *Meat Sci.*, 58: 17–23.

Kristensen, L., Therkildsen, M., Aaslyng, M.D., Oksbjerg, N. and Ertbjerg, P. (2004). Compensatory growth improves meat tenderness in gilts but not in barrows. *J. Anim. Sci.*, 82: 3617–3624.

Lametsch, R., Karlsson, A., Rosenvold, K., Anderson, H.J., Roepstorff, P. and Bendixen, E. (2003). Postmortem proteome changes of porcine muscle related to tenderness. *J. Agric Food Chem.*, 51: 6992–6997.

Lametsch, R., Roepstorff, P., Molle, H.S. and Bendixen, E. (2004). Identification of myofibrillar substrates for l-calpain. *Meat Sci.*, 68: 515–521.

Lametsch, R., Kristensen, L., Larsen, M.R., Therkildsen, M., Oksbjerg, N. and Ertbjerg, P. (2006). Changes in the muscle proteome after compensatory growth in pigs. *J. Anim. Sci.*, 84: 918–924.

Laville, E., Sayd, T., Sante-Lhoutellier, V., Morzel, M., Labas, R. and Franck, M. (2005). Characterisation of PSE zones in semimembranosus pig muscle. *Meat Sci.*, 70: 67–172.

López, J.L. (2008). Two-dimensional electrophoresis in proteome expression analysis. *J. Chrom. B.*, 849: 190–202.

Martinez, I., Slizyte, R. and Daukas, E. (2007). High resolution two-dimensional electrophoresis as a tool to differentiate wild from farmed cod (*Gadus morhua*) and to assess the protein composition of klipfish. *Food Chem.*, 102: 504–510.

Melody, J.L., Lonergan, S.M., Rowe, L.J., Huiatt, T.W., Mayes, M.S. and Huff-Lonergan, E. (2004). Early postmortem biochemical factors influence tenderness and water-holding capacity of three porcine muscles. *J Anim. Sci.*, 82: 1195–1205.

Morzel, M., Chambon, C., Hamelin, M., Sante´-Lhoutellier, V., Sayd, T. and Monin, G. (2004). Proteome changes during pork meat ageing following use of two different pre-slaughter handling procedures. *Meat Sci.*, 67: 689–696.

O'Farrell, P.H. (1975). High resolution two-dimensional electrophoresis of proteins. *J. Biol. Chem.*, 250: 4007–4021.

Pascoal, A., Prado, M., Castro, J., Cepeda, A. and Barros-Velázquez, J. (2004). Survey of authenticity of meat species in food products subjected to different technological processes, by means of PCR-RFLP analysis. *Eur. Food Res. Technol.*, 218: 306–312.

Purslow, P., Morrison, H. and Mielche, M. (1997). Are there intermediate filament proteins involved in determining meat quality? In Proceedings 43rd international congress of meat science and technology, (p. 598) Auckland, New Zealand, G1-25.

Rao, Q. and Hsieh, Y.H.P. (2007). Evaluation of commercial lateral flow feed test for rapid detection of beef and sheep content in raw and cooked meats. *Meat Sci.*, 76: 489–494.

Sayd, T., Morzel, M., Chambon, C., Franck, M., Figwer, P. and Larzul, C. (2006). Proteome analysis of the sarcoplasmic fraction of pig semimembranosus muscle: Implications on meat color development. *J. Agric. Food Chem.*, 54(7): 2732–2737.

28

Animal By-products Utilization

Prabhat Kumar Mandal

ABSTRACT

Slaughtering of animals in organised slaughterhouses enables collection and complete utilisation of by-products. Recovery of carcass by-products is feasible. Many of these by-products can be converted into high value products. When the animals die due to natural causes, the owner loses the regular income therefore he discards the carcass. A dead animal, then, becomes a social property in the country. In India the availability of fallen animals is very large and we cannot neglect this sector. Proper recovery and utilisation of fallen animals would bring wealth from waste and also improve the quality and availability of hides and skins. In this chapter scope for processing and utilization of major by-products from slaughter and fallen animals are discussed. This includes processing and utilization of blood, bones, hides and skin, gut and intestine, glandular and other by-products.

1. INTRODUCTION

It is said that "nothing except the last cry of the animal is lost in the modern slaughter house". This may be to some extent true in the developed countries having a proper system of byproducts utilization adjacent to modern slaughter houses. On the contrary, it is still a matter of concern in the developing countries. Unfortunately, due to lack of proper guidance and technical knowhow, improper national policies or their implementation, the approach towards the proper utilization of slaughter house byproducts is still unimaginative leading to wasteful losses.

In view of huge livestock population, large number of slaughter houses and developments in meat and meat products export trade the economic losses encountered through under-utilization of the valuable raw material should be given due consideration. These topics are discussed by eminent scientists and policy makers but so far no steps have been taken to improve and encourage the byproduct industry. Only modernization of abattoirs will not be sufficient for the establishment of meat industry in India, but due consideration must be given to have a well defined system of by-products processing attached to modern slaughter houses.

Animal by-products utilisation has been a remarkable economic and public health phenomenon but in spite of this, published information in this area is scanty. A limited number of publications of the Food and Agricultural Organisation (Rome) and the Commonwealth Science Council (London) are available. Moreover because of lack of awareness about the significant economic benefits, of wastage or under utilisation of animal by-products are common in several developing countries. According to one estimate, the annual loss sustained by India due to under utilisation of slaughterhouse by-products is of the order of Rs. 8000 million. The situation may not be any better in many developing countries.

It has become a basic need to promote animal by-products utilisation through publications, conducting workshops, seminars and giving publicity through mass media. This will not only help to solve pollution problems and energy crisis but also meet the challenges of the carcass by-products industries through innovations and find better opportunities in future. Any industry that processes carcass wastes into valuable products ought to be supported and encouraged. There is a general impression that meat is the main edible product and there is no by-product, which can be obtained in sufficient quantity from the abattoir.

With the help of the genetic engineering chemical synthesis of some hormones like insulin, FSH, LH, adrenaline etc. are possible. This has again raised a point of argument amongst various technologists. Even though these procedures are very efficient, accurate and high yielding, they have some limitations. First of all, by these processes only proteins can be synthesized. While some drugs and biochemical are protein-carbohydrate complexes; secondly, it requires sophisticated and expensive machineries which only developed countries can afford. Hence, for developing countries like India it will not be justifiable to look for such expensive processes overlooking the enormous raw materials available (Table 1) from over 3600 slaughter houses. It is therefore, felt that there is an urgent need for organized collection of by-products and processing them into value added products.

Table 1: Percentage Yield of Percentage Yield of Meat and by-products

Name of the product/by-product	*Percentage yield on live weight*
Meat	35.00
Bones (Green)	25.00
Hide	6.50
Blood	4.00
Horns and Hooves	0.60

Table 1: (*Contd...*)

Table 1: *(Contd...)*

Name of the product/by-product	*Percentage yield on live weight*
Small intestines	1.00
Large intestines	1.00
Stomach	2.00
Rumen digest	11.00
Liver	1.20
Heart	0.35
Kidneys	0.20
Lungs	1.20
Brain	0.17
Spleen	0.15
Pancreas	0.07
Bile	0.06
Tongue	0.30
Hair, teeth and other	1.50
Cutting losses etc	8.70
Total	**100.00**

2. UTILIZATION OF BLOOD

Blood contributes approximately 4–5% of the live weight of the animal. Dried blood is high in protein (80–90%) and rich in lysine and commands a good price. Efficient collection and processing of blood will ensure a profitable return. At the same time inefficient collection and processing is bound to create effluent disposal problems as blood causes an increase in BOD and its red colour is at once recognizable as a highly polluting waste material. There are many potential uses of blood products. The most common uses would include: as a nutrient for tissue culture media a s well as ingredient in blood agar media for bacteriological purposes. Many blood components are isolated and used in chemical analysis or as nutritive supplements. Blood plasma has also been used as diluents for semen from boars and bulls.

Table 2: Uses of animal blood in various industries

Food	Emulsifier, stabilizer, clarifier, colour additive, nutritional component.
Feed	Lysine supplement, vitamin stabilizer, milk substitute, nutritional component.
Fertilizer	Seed coating, soil pH stabilizer, mineral component.
Laboratory	Tissue culture media, tannin analysis, active carbon, haemin, blood agar, peptones, glycerophosphates, albumins, globulins, sphingomyelins, catalase.
Medicine	Agglutinin test, immunoglobulins, fractionation techniques, blood clotting factors, sutures, fibrinogen, fibrinolysin, fibrin products, serotonin, plasminogen, plasma extenders.
Industry	Adhesive. resin extender, finishes for leather and textiles, insecticide spray adjuvants, egg albumin substitute, foam fire extinguisher, porous concrete, ceramic and plastic manufacturer, plastic and cosmetic base formulations.

Industrial uses of blood are as adhesives, in film forming properties in the paper, lithographic, plywood, fibre, plastics and glue industries. It also finds use in insecticide and fungicide formulations, in foam fire extinguishers, in moulded ceramic products, in leather finishes, in porous concrete and as a stabilizer for biological material and drugs. Blood, as fertilizer, contributes to nitrogen source. It aids in humus formation and improves the soil structure. Blood is also useful in seed coating and regulating soil pH. The disadvantage of blood spread on the soil is that it will affect rats and vermin.

Blood in most of the abattoirs in developing countries is being wasted completely or under-utilized, causing environmental pollution problem. In some of the metropolitan cities of India, bovine blood is collected by pharmaceutical firms for utilization in haemostatic preparations as a source of iron and protein supplement. Blood is particularly rich in haem iron, which is biologically the most available form of iron. Table 2 shows different end uses of blood and Table 3 shows uses of blood components. Some Selected uses of blood products are described below:

Table 3: Uses of blood components

Raw material	*Product*	*Uses*
Whole blood	Foam compound	Fire foam extinguisher
	Blood char	Active carbon in Pharmaceuticals
	Blood meal	Animal feed supplement
	L. Histidine	Research and food additive
	Thrombin	Aid in coagulation treat wounds in brain & ulcers of liver surgery
Whole liquid blood	Blood agar	Bacteriology
Serum	Sterile serum	Tissue culture and virology in media
Sediment of RBC	Haemoglobin syrup	Haematinic tonic
Plasma	Blood clotting factor	Therapeutic and diagnostic media
Fibrin	Fibrinogen	Haemostatic
	Fibrin powder	Haemostatic
	Fibrin sheets	Wound dressing material
	Fibrin glue	Therapeutic uses
Serum	Albumin	As substitute to egg albumin
	Bovine serum albumin	Claimed-research RH factor-cosmetics, blood loss etc.
Foetal calf blood	Foetal calf serum	Tissue culture and vaccine production

2.1. Blood Plasma/Serum

To prepare serum, blood is collected carefully and allowed to clot and subjected to further processing in a cold room. Chilled clotted blood is cut into smaller lumps to bring about quick contraction of clots. Serum collected in first 12 hours is generally clear except for some suspended RBCs. It is centrifuged in a bucker centrifuge at 1000 rpm and filtered through seitz filter. The yield of serum is about 10–12% of the weight of whole blood. It keeps well at 4–5°C for one month and at –20 degree c for six months.

Dried blood serum is referred as blood albumen and is used as the cheapest substitute for dried egg albumen powder in the industry. To prepare it, clear yellow serum is added 0.05% phenol on weigh basis. It is sprayed or vacuum dried to soluble fine powder. The yield of albumen is 10–20% of the weight of the serum. It is stable in air tight containers in cool places for several months.

Blood plasma upon heating forms a gel. Blood plasma proteins such as albumin and globulins are good emulsifiers. When blood plasma is used in meat sausage products as an emulsifier, it will decrease the shrinkage and increase yield (approximately 4–5%) and the texture of the finished product will become more rigid due to the gelling properties of the plasma. Blood plasma can also be used in bakery products because of good foaming properties similar to egg albumin. Serum is used in the laboratory for inactivation of proteolytic enzymes, as a medium in virus propagation, in production of virus vaccines and in bacteriological media.

2.2. Fibrin

Large quantities of fibrin are rejected when haem is collected from blood. Methods for utilisation of fibrin have been developed by CLRI. Fibrin forms an insoluble network in blood coagulation. Fibrin contains two types of peptides, peptide A and peptide B. There is great scope for fibrin and fibrin based prosthetic materials in various surgical manifestations. Fibrin can be processed into various forms such as foam material, powder, sheets, and sponges and for hard bone replacement moulds (bioplasts). Fibrin could be processed in the form of foam material. This foam is a white sponge-like material and it can be cut easily into different shapes. When placed on a bleeding surface, it soaks up the blood, which then clots in its pores. One gram of foam soaks up as much as 30 gms of blood.

2.2.1. *Fibrin powder*

Fibrin powder is a yellowish light lyophilic powder. It presents a large surface and sticks to the wound surfaces. The beneficial effect of fibrin powder was studied. We observed the application of fibrin powder to fresh wounds had speeded up haemostasis. It also sticks to the wound surfaces. The powder could be sprinkled on the surface of the wound or may be blown with simple devices into less accessible places. Using fibrin powder can immediately control bleeding.

2.2.2. *Fibrin sheets/bioplasts*

Fibrin sheets can also be prepared and used for covering bleeding wounds and ulcers as a wound dressing material. This sheet not only covers the wound but also prevents infections. Fibrin bioplasts are moulded materials resembling plastic which are prepared from fibrin powder. They can be moulded even as bone joints and need not be removed from the body.

2.2.3. *Fibrin glue*

Fibrin glue can be prepared from the human/animal blood. There are numerous applications for fibrin glue *e.g.*, control of bleeding, adhesion of tissues together and scaling of tissues defects. Fibrin can be incorporated in the form of a coating in orthopaedic implants. Fibrin along with hydroxyapatite and tricalcium phosphate can be used as a bone implant material. Fibrin helps in haemostasis at the operative site and thereby puts the surgeon at ease. This also helps in preventing loss of blood. Fibrin can also be coupled with bone growth factors such as bone morphogenetic protein (BMP), osteogenin and transforming growth factor (TGF) could be used along with osteoconductive material at the orthopaedic defective site.

2.3. Blood Meal

Under the existing system of slaughter of meat animals in developing countries, it is always a better choice to process blood into blood meal for utilisation as a protein supplement in animal feed formulation (Table 4). Most of the protein feed stuffs does not contain required levels of lysine, an essential amino acid for growing monogastric animals. A good quality blood meal containing high levels of available lysine, should find a place in the market.

Table 4: Composition of blood meal

Crude protein	—	80%
Digestible protein	—	56%
Total Digestible	—	61%
Fat	—	1%
Fibre	—	1%
Ash	—	6%
Calcium	—	0.28%
Phosphorus	—	0.28%

3. UTILIZATION OF BONES

Bone is a highly specialized form of connective tissue, composed of interconnected cells in an intercellular substance and forming the skeleton or framework of the bodies of most vertebrates. Bone is an important carcass by-product with diverse applications and potential. Bone is a special raw material containing both organic and inorganic constituents in appreciable quantities resulting in the preparation of various value-added end products. Many types of industries based on bones are established. Traditionally bones have been used in handicrafts for the manufacture of inlay works, jewellery, fancy and decorative articles. But due to the advent of plastics, the use of bones in the industry has declined.

Bones constitute almost 15% of the weight of dressed carcass. The yield of bones may vary according to the breed, age, health status etc. On live weight basis, bones form 12–30% in sheep and goat. Fresh bones are heavy and contain comparatively high moisture, fat and protein content. The average composition of bone is given below. Defatted and dried bones contains organic and inorganic matter in a ratio of 1:2. Bone collagen called *ossein* is the main constituent of the organic matter and accounts for 23–36%. Inorganic matter consists of nearly 32.6% of calcium and 15.2% phosphorus, small quantities of sodium, potassium and magnesium and traces of copper, zinc, iron, cobalt, manganese etc. Almost 96% of red and yellow marrow is fat.

Composition of fresh bones:

Moisture	:	50%
Red and yellow marrow	:	15%
Organic matter	:	12%
Inorganic matter	:	23%

3.1. Bone Crushing

Bones are available in nature having different shapes and sizes and therefore, it is not convenient to use them as such for processing them into various end products. Commercially the most popular sizes in which bones are crushed are 19 mm, 16 mm and 10 mm. These sizes are used both for manufacture of ossein as well as glue. The material passing through 5 mm but resting on 2.5 mm sieve called bone grist, which is an ideal raw material for the manufacture of bone char. The material that passes through 2.5 mm sieve is called bone meal (Table 5).

Table 5: Yield of various products obtained after crushing of raw bones

Crushed bones	50–55%
Grist	10–12%
Bone meal (raw)	20–25%
Sinews	5–10%
Loss	5–10%

3.2. Mineral Based Products

A number of mineral based products are made from bones. They are Sterilized bone meal, Steam digested bone meal, Bone ash, Bone char and di-calcium phosphate.

3.2.1. *Sterilised bone meal*

The crushed bones are digested in the bone digester at a pressure of 2.8 kg/cm^2 (28 psi) for a period of 1–2 hours. The bones get sterilised and become soft.

These are then taken out, dried and powdered. Sterilised bone meal (SBM) has got calcium and phosphorus content of 21 and 10%, respectively.

3.2.2. *Steamed bone meal*

This product is different from sterilised bone meal as regards the protein content. In case of steam digested bone meal, most of the protein has been extracted. However, some quantities of protein remain intact in the sterilised bone meal. The availability of calcium and phosphorus in the steamed bone meal is 31 and 15%, respectively. The higher percentage of calcium and phosphorus is due to the absence of most of the protein from the bone.

3.3. Bone Ash

Bone ash contains exclusively the inorganic part of bone prepared by calcining the bones burning out all the organic matter (Table 6). The ashing of the bones can be done in rotary kilns like the ones used in cement industry. Conventional kilns used for fixing limestone and pottery are also used. The temperature required for complete ashing of bones is in the range of 600–800°C for 4–5 hours. Bone ash finds use both in agriculture and livestock feed. Bone ash is used for bone china in ceramic industry. Bone ash can also be used as starting material for the manufacture of dicalcium phosphate.

3.4. Bone Char

Bone char is also called as active carbon. Bone char is prepared by treating carbon granular bones at high temperatures (600–800°C) in a suitable furnace in the absence of air. Low quality bones, cooked bones and other waste bones of granular nature are used for the production of bone char. Active carbon is extensively used for decolourisation and adsorption purposes in sugar and pharmaceutical industries. The property of decolourising and adsorption of bone char is not due to the presence of carbon, but only due to phosphate. During the processing of carbonisation large quantities of gases are evolved which on scrubbing results in the production of bone tar, bone oil etc.

Table 6: Yield of different bone products from green bones

Product/End product	*% Yield*	*Qty. in kg*
Green bones	–	1000
Tallow/1000 kg green bones	5	50
Dry bones/1000 kg green bones	35	350
Crushed bone, bone meal and bone grist	93	325
Bone sinews/350 kg dry bones	7	25
Glue/25 kg bone sinews	30	7.5

Table 6: (*Contd...*)

Table 6: *(Contd...)*

Product/End product	*% Yield*	*Qty. in kg*
Glue/325 kg crushed bones	16	52
Ossein/325 kg crushed bones	30	97.5
Dicalcium phosphate/325 kg of crushed bone	60	195
Gelatin /97.5 kg ossein	50	49
Bone ash /273 kg deprotein bone	80	218
Bone char /325 kg crushed bones	50	162.5

4. UTILIZATION OF HIDE AND SKIN

Leather is an animal hide or skin converted by chemical treatment and processing known as tanning. Leather is resistant to decomposition or bacterial decay, particularly when it is wet and it has attained certain physical properties like tensile strength, flexibility, and resilience and abrasion resistance.

4.1. Processing of Hide/Skin

(*i*) *Washing and soaking*: Washing with water helps in removing salt used during curing, the water soluble proteins from blood and lymph, dung and other matter. It also helps the fibres to absorb water restoring the normal shape and dimensions.

(*ii*) *Fleshing*: The flesh adhered to the inner side of the skin is scraped by a serrated knife. This is done for better penetration of the chemicals during further processing.

(*iii*) *Liming*: Hides are immersed in vats containing saturated lime solution. This helps in separating keratinous tissues (hair and epidermal cells) and losses the hair follicles are removed by serrated knife.

(*iv*) *Washing and deliming*: The hides are washed to remove lime and hair. The residual lime is removed by soaking in weak acid.

(*v*) *Bating*: This is a treatment of delimed hides with proteolytic enzymes which makes them soft and pliable.

(*vi*) *Pickling*: The skins are soaked (pickled) in a solution of sulphuric acid and salt in order to condition them for tanning. Pickling brings the skin to a uniform degree of acidity and cleanses the goods.

4.2. Tanning of Leather

Tanning is the conversion of hide/skin in to insoluble and non-putrisible leather without destruction of original structure. It is done by treating the conditioned hides and skin with organic, inorganic or synthetic tanning agents. Among these, vegetable tanning material and basic chromium sulphate are being commonly used.

4.2.1. *Vegetable tanning*

In this process, extracts of infusion derived from tannin bearing plants or barks are used. In India plants like Avaram (*Cassia auriculata*) Babul (*Acacia Arabica*), Myrabalan (*Terminalia chebula*) are used. De-limed pelts (hide/skin) are immersed in progressively stronger tan liquors containing in pits. The pelts are moved at a fixed interval from one pit to another. The tannin particles diffuse into the pelt and impart a uniform colour. The spread of tannin is improved by the use of revolving drum and strong tanning liquor. Vegetable tanning is employed in producing sole belting, harness and saddle and other heavy leather.

4.2.2. *Chrome tanning*

This method yields soft, supple and strong leather which is permeable to air. This is most popular method of tanning. The chemical used is chromic sulphate [$Cr(OH)SO_4$]. Chrome tanning takes very less time 4–10 hours and retanning can be done if necessary. Disposal of chrome is a problem since it is toxic.

4.2.3. *Post tanning operations*

Before furnishing leather tanned leather undergo various operations like removal of water and tan liquor, splitting and shaving , neutralizing and dyeing, fat liquoring and stuffing, setting out, shammoying, staking and buffing. The sole leather is furnished by applying oil on grain surface and then rolling under a heavy roller. Chrome leathers are glazed after seasoning and drying. Glazing is done by a glass cylinder clamped to the end of a moving arm. The heat generated by the friction softens the wax. The finish and a continuous plastic coat of high lustre are produced on the grain surface. After glazing and staking, leathers are smooth plated on the grain side in a specially designed press. Pleasing grain platforms are produced by boarding. Vegetable tanned leathers are given decorative pattern by embossing with hydraulic press (Table 7).

Table 7: Physical properties of leather

Property	*Value*
Tensile strength	15.3–37.5
Elongation at break (%)	29.5–73.0
Stitch tear strength (N/cm)	1280–2275
Thickness (mm)	1.5–2.4
Bursting strength (kn/cm)	1.1–24.5
Apparent density (g/cm^2)	0.6–0.9
Real density (g/cm^2)	1.4–1.6
Heat resistance	Shrinks

5. UTILIZATION OF GUTS AND INTESTINE

Various parts of alimentary tract of food animals viz. cattle, buffalo, sheep, goat, pig and horses after processing are used as containers for the preparation of comminuted meat products. Intestines of animals are processed into value added products such as sausage casings, surgical sutures (catguts), guts for sports rackets, musical instruments strings. Amongst these uses, sausage casings are most profitable way of utilizing animal intestine. Sausage casings can be prepared from small and large intestines, oesophagus, urinary bladder, stomach and even rectum.

Different countries have preference to different types of sausages prepared using different types of casings.

5.1. Principles of Casing Preparation

To get good quality natural casings it is necessary to collect intestines from healthy animals which are properly slaughtered and passed both ante mortem and post mortem inspection. Sound raw materials are devoid of ulcers and damages due to infestation with parasites resulting in nodule formation. They should be removed carefully and handled using sanitary precautions. Bulks of the casings are produced from small intestines. Commercially, intestines of cattle, sheep, goats and pigs are used for making casings.

Quality casings are essentially prepared from the middle submucosa layer of intestine only. The submucosa layer is grayish-yellow or red in colour and composed of dense connective tissue network interwoven with an elastic network.. The casings are, therefore, made from submucous coat only because of its fine structure, having sufficient flexibility, elasticity and strength. Moreover, they are able to shrink and extend without cracking together with delicacy makes them suitable for sausage casings.

5.2. Preparation of casings

(i) *Collection of intestine*: After slaughter of food animals intestines are collected from the slaughter house cleaned of the intestinal contents and washed in clean water. They should be collected fresh and cleaned quickly to avoid decomposition by bacteria and damage to submucous layer. Fat and other adhering tissues are removed carefully. The cleaned intestines are dipped in a solution containing 0.2% sodium pyrophosphate and 1% sodium chloride for one minute with constant stirring and then removed.

(ii) *Sliming*: Removal of the other layers *i.e.*, outer serous, muscular and inner mucosal is called sliming. It is done by pulling the intestines through the gentle pressure with blunt knives over the surface of the table or a special device, so that above stated layers are pressed out leaving behind the submucosa layer. The process is repeated till the clean white coloured casings are obtained.

(iii) *Washing and grading*: The casings are washed by infusing with water and graded according to prescribed standards such as diameter, length, colour and physical defects such as leakages, nodules, etc.

(iv) *Curing and draining*: Casings are preserved by adding salt in wooden boxes with holes underneath to drain brine formed for about 3 days and then shaken to remove excess of salt.

(v) *Wet and dry salting*: In wet salting the cured casings are simply placed in saturated salt solution in wooden barrels, but in dry salting, the casings are rubbed with fine grade common salt of about 40% weight of the cured casings.

(vi) *Storage and transport*: Casings can be transported in plastic cover, bottles with salt or in tins/wooden boxes.

6. UTILIZATION OF OTHER BYPRODUCTS

6.1. Collagen Products from Animal Tissues

Collagen derived products are emerging as valuable products of biomedical technology. Technology for collagen-based biomaterials has been developed through research in Central Leather Research Institute, Chennai. Two such products are to reconstituted collagen sheet (RCS) and collagen sheet (CS). RCS and CS have been widely established as an effective dressing for open wounds. They are especially used for the treatment of both 1st and 2nd degree superficial and deep dermal non-infected burns; skin ulcers and open skin conditions resulted due to injuries and accidents. Both RCS and CS are products derived from by-products of meat industry.

The sheets consist of mainly type I collagen with small percentage of type II and type V collagens. Chemically, the bovine collagen is very similar to the human form. This is crucial, as the human immune system will reject everything that deviates too much from its own proteins. Because of all these reasons, collagen membranes are well pronounced for usage as an effective wound cover. Collagen as a natural biomaterial has unique properties like degrading in biologic environment, leaving no foreign residue, low antigenicity, ability to support cellular growth and is a haemostat. The scientists have been successful in bringing out this wonder material in different forms and shapes to handle problems in the health care industry, more specifically in the wound care segment. This commercial biomaterial is extracted from Bovine animal sources.

Collagen Sheet is prepared from the serosa layer of bovine intestine by a series of chemical and enzymatic treatments. It is further stabilised by cross-linking with basic chromium sulphate (BCS) or glutaraldehyde (GTA) and finally sterilised by ethylene oxide treatment or γ-irradiation. The sterilised membranes are sealed individually in a sterile fluid for extended shelf life. Reconstituted Collagen Sheet is prepared from the Achilles tendons of bovine animals. Sheets are prepared from the pure collagen in a sterile and dust free chamber. They are cross-linked with GTA and sealed in polypropylene sachets. The sachets are sterilised by, ethylene oxide fumigation or γ-irradiation for extended shelf life.

Both Collagen Sheet (CS) and Reconstituted Sheet (RCS) act as a temporary biological cover for second and third degree burns, open injury, wound ulcer etc. They are India's first natural non-immunogenic sterile temporary biological skin covers for burns and open raw wounds. Both are ideal skin substitutes for the management of first and second degree non infected burns, raw areas caused by release of burns contractures, traumatic loss of skin cover, skin donor sites, amputation sites, chronic skin ulcers, shallow pressure sores cases. In third degree burns they are useful as a temporary cover after tangential excision. Collagen membranes have been widely established as an effective dressing for wounds especially the burns because of its easy commercial availability in sterile condition in various sizes.

6.2. High Value Products from Keratinous Wastes

6.2.1. *Products from animal hair, poultry feathers, and hoofs and horns*

Technology are developed to convert keratinous wastes *viz*. animal hair, poultry feathers and horns and hoofs into leather fillers-cum-retanning material and chrome exhaustion aid. These materials could be suitably solubilised by controlled alkali hydrolysis to make keratin hydrolysate (KH). The hydrolysate made by alkali process has been found to be excellent leather filler—an auxiliary used to improve the physical characteristics of leathers such as chrome tanned, goat, sheep, buff calf and cow uppers. KH is used as a filler syntan-cum-retanning material and as a chrome exhaustion aid in leather processing. Use of biological wastes to produce KH for the leather industry is of immense practical relevance for the country.

KH is prepared from animal by-product keratinous wastes. These sources are rich in keratin protein. Conversion of these sources into hydrolysates would eliminate the environmental pollution posed by these materials. Poultry feathers are a waste material in India, and do constitute a nuisance. The situation with tannery hair is similar. Tannery hair is burnt off in liming operations using lime and sodium sulphide. The disposal of the burnt hair along with lime and sulphide is a major problem to a tanner. Horns and hoofs do have some value but still large quantity remains unutilised. Conversion of these materials into KH would reduce pollution at two levels in a tannery, firstly, utilisation of pollution causing burnt off hair and secondly removal of chromium salts from the waste liquors.

6.2.2. *Steamed horn and hoof meal*

Horns and hooves are very useful by-products. They are screened by hand picking of bones and other impurities. Cleaned horns and hooves are loaded into a digester/ autoclave. Water is filled to the digester and drained off after 30 minutes to remove the adhering debris and soil. After that, steam is passed into the digester at 80–100 lb-sq/in for 1–2 hrs and then the steam is released. The steam coming out from the digester contains sulphur dioxide which is a toxic gas and will injure living things. Utmost care should be taken while delivering the steam. The toxicity will be nullified by diluting it with water.

Overcooking generates lot of liquid and also the colour of the cooked material is black. If they are semi-cooked, uneven digestion of horns and hooves will result. Products of overcooking and less cooking are not desirable. The desirable colour is golden red. Only by practice one can get good product. Steamed horns and hooves are dried under shade. After drying it can be crushed into different sizes as per the requirement. Steamed horn and hoof meal is a very good organic fertiliser, because it will slowly release nitrogen into the soil Soil (Table 8).

Table 8: Animal by-products in our daily lives

Intestines	*Fats and fatty acids*	*Bones, horns and hooves*
Sausage casings	Explosives	Syringes
Instrument strings	Solvents	Piano keys
Surgical sutures	Chewing gum	Marshmallow
Tennis racket strings	Paints	Pet food ingredients
	Industrial lubricants	Bandage strips
	Cosmetics, Shampoo	Bone charcoal products
	Dog food	Gelatin
	Mink oil	Adhesive tape
	Oleo margarine	Phonograph records
	Ceramics	Combs and Toothbrushes
	Medicines	Buttons
	Soaps	Jewelry
From hide and wool	Creams and Lotions	Bone meal
Lanolin	Tires, Rubber products	Emery boards and Cloth
Clothing	Paraffin	Ice cream
Drum heads	Biodegradable detergents	Horn and Bone handles
Luggage	Antifreeze	Wallpaper and Wallpaper paste
Yarns	Crayons	Dog biscuits
Artist's brushes	Floor wax	Steel ball bearings
Sports equipment	Chemicals	Fertilizer
Fabrics	Chemicals	Neatsfoot oil
Pelt products	Insecticides	Adhesives
Insulation	Candles	Plywood and Paneling
Textiles	Herbicides	Shampoo and Conditioner

Table 8: (*Contd...*)

Table 8: *(Contd...)*

Intestines	*Fats and fatty acids*	*Bones, horns and hooves*
Tennis balls	Shaving cream	Dice
Carpet	–	Collagen cold cream
Footwear	–	Crochet needles
Woolen goods	–	Cellophane products
Baseballs	–	Glycerine
Upholstery	–	Photographic film
Hide glue	–	Laminated wood products

6.2.3. *Fire foam compounds*

Here also both horn and hoof meal and poultry feathers are suitable for making fire foam compounds. Keratins under controlled condition of hydrolysis can be solubilised and brought to the leave of molecular weight profile suitable as fire foam compound. We get keratin hydrolysate with very high foaming stability. The critical requirement is 4.5 litres of fire foam compound from 1 kg. raw material and it should have the specific gravity of 1.22. The foaming capacity of the product depends upon the reaction time and the pressure. The molecular weight distribution of the product at maximum foaming is between 5.5×10^3 and 1.5×10^5 (average molecular weight 2.93×10^4).

6.2.4. *Food flavour (Soup concentrate)*

Both horn and hooves and poultry feather are suitable for making food flavour. The concentrate obtained by acid hydrolysis of the raw material has a meaty flavour and can be added in soups and other food preparations.

6.2.5. *Hair care preparation*

Keratinous wastes by acid hydrolysis can be converted into hydrolysates, which can be used as hair conditioners either independently or with other ingredients. The requirement for the hair care preparation is that the molecular weight of the hydrolysate should be low enough to penetrate into the hair shaft. The molecular weight should be somewhat high enough to resist their removal in subsequent washing. It should achieve substantive absorption by hair normally 200 mg of the hydrolysate is retained by 100 g hair. The ideal molecular weight of the hydrolysate is between 500 and 2000 Dalton.

6.2.6. *Feed supplement*

Keratins contain amino acids. Poultry feather can be steam digested (40–60 psi) and converted into hydrolysed feather meal. Both dry rendering and wet rendering technique can be adopted to convert poultry feathers into hydrolysed feather meal. The following essential amino acids are made available in the hydrolysed feather meal required in chick feed rations.

Roughly up to 5% of poultry rations can be substituted with hydrolysed feather meal.

7. BIO-PHARMACEUTICAL UTILIZATION OF BYPRODUCTS

7.1. Blood

Blood, a viscous scarlet colored fluid confined to the circulatory system of the animal's body is a valuable potential human food of high protein content (80% DM) with excellent amino acid balance and nutritional value. Biochemicals and Pharmaceuticals prepared from blood are given below.

- *Histidine*: An amino acid prepared from blood corpuscle paste is required for haemopoesis. It is used in control of allergic disorders, cardio vascular and gastrointestinal disturbances. By intramuscular injection histidine mono-hydrate chloride relives pain, nausea and digestive disturbances. Yield is about 1.4% pure histidine from blood corpuscle paste.
- *Leucine*: It is one of the essential amino acids required by all species of animals. As a pure amino acid leucine is used in biological research. About 5% leucine can be produced from RBC on dry weight basis.
- *Phenylalanine*: An essential amino acid that can be isolated in pure crystalline form from animal blood and yield is about 2% from RBC.
- *Lysine*: is an essential amino acid that can be isolated as picrate or hydrochloride from fibrin and blood corpuscle. About 1.2–1.5% of lysine hydrochloride is obtained from blood corpuscles.
- *Haemin*: Is formed by the removal of the globin portion of the hemoglobin molecule in which the iron complexed porphyrin has chlorine in place of hydroxyl group in the ferric iron. Haemin is used for coloring canned meat and has several biochemical properties. Yield is about 0.35–0.45% crude haemin from blood.
- *Protoporphyrine*: Is a diethyl ester moiety of porphyrin without iron. It can be prepared synthetically and is used mainly in research.
- *Oxyhaemoglobin*: is a tetrapyrole porphyrine with a closed ring structure. It can be prepared from fresh blood or RBC. This is a starting material for most of the works on the chemistry of hemoglobin and its derivatives.
- *Prothrombin*: A plasma protein in the blood and plays an active role in blood coagulation by conversion into thrombin. This is extensively used in research in clinical hematology and clinical medicine as a starting material for making thrombin.
- *Thrombin*: This is a factor formed from prothrombin when blood is shed. It is applied as a dry powder or in combination with glucose and sterile saline to control capillary or other surgical bleeding.

- *Fibrinogen:* Is the precursor of fibrin and can be isolated from bovine blood plasma, used in clinical hematology and veterinary therapeutics.

7.2. Pancreas

In India only Boots Drug company, Bombay is manufacturing insulin from pancreas meeting only 50% demand of our consumption. It is estimated that the annual requirement of insulin in our country is about 1500 mega units, whereas, the production was 761 MU in 1971and 660 MU in 1972 from the imported pancreatic tissue. For this purpose our country imports about 750 to 1000 tons of glands costing more than 2 million Rupees. Standard insulin contains 22 IU/mg of insulin. Scientists of V.P. Chest Institute, Delhi has developed a process for insulin extraction from sheep pancreas. While a team of workers from Bombay Veterinary College, Bombay (1980) isolated it from buffalo pancreas. Central Drug Research Institute (CDRI) has developed a process for the preparation of crystalline insulin.

7.3. Heparin

It is the physiological anticoagulant used to prevent clotting during transfusion of blood, post-operative thrombi and in acute thrombotic diseases. Liver, lungs and intestinal mucosa are rich source of heparin. Commercially it is isolated either from beef lungs or pig intestinal mucosa. Heparin is present in trace amounts in the body. Highest concentration (mg/kg of tissue) were found in lungs, (beef 190 mg, hog 340 mg, dog 330 mg) and muscle (beef 600 mg), but beef thymus. Spleen and lungs also contain large amounts (310, 230, 230 mg/kg, respectively. Heparin is a greenish brown power, fairly hygroscopic, readily soluble in water and in normal saline forming a colorless or straw colored solution.

7.4. Liver Extract

Liver extract was first produced in India in 1930 by Bengal Chemical and Pharmaceutical works Ltd. Calcutta. The quantity of liver obtained from slaughter house in the big cities is reported to be adequate to meet the countries requirements. Two types of liver extract are produced in India namely aqueous and proteolysed. Both types are used in the treatment of tropical anemia. Highly concentrated extracts are not produced in the country as the incidence of pernicious anemia is low. Aqueous liver extract is a brownish liquid at times may show slight turbidity or precipitate. Proteolysed liver extract is pale grayish brown powder or a brown paste with pleasant meaty odour. Nitrogen should not be less than 11.5%. The appropriate anti-anaemic potency of liver extract is expressed in terms of vitamin B_{12} content, which should be at least 0.1 micrograms of vit B_{12}/ml. Liver extract should be preserved in tightly closed light resistant containers preferably at temperature below 20°C.

7.5. Bile Salts

Two or three adult buffaloes or 20–30 sheep or goat can yield about a pound of liquid bile. Bile from Indian cattle contains only 7% total solids and 18 gm/lit of colic acid as compared to 9.6% total solids and 25 gm/lit colic acid in bile obtained from cattle in foreign countries. Bile salts are prepared by a number of manufacturers notable among these being the Bengal Immunity Co. Ltd, Calcutta and Retakes Brett Co Ltd, Bombay. For preparation of sodium tauroglycocholate, raw bile is evaporated to dryness with the addition of animal charcoal. The resultant pasty mass is extracted with alcohol and sodium tauroglycocholate is precipitated from the extract by adding ether.

7.6. Pituitary Gland

The pituitary gland is a small bean shaped gland situated in bony cavity at the base of the brain, it has anterior lobe, forming a firm and glandular structure, posterior lobe made up of nerve cells and an intermediate lobe—which is present in some mammals. Six active principles have been isolated from anterior lobe: two gonadotropic hormones namely follicle stimulating hormone (FSH) and interstitial cell stimulating hormone (ICSH); lactogenic hormone (prolactein); growth hormone (G.H); thyrotropic hormone (TH) and adrenocorticotropic hormone (ACTH).

Prolactin initiates and maintains lactation and is responsible for the nesting instinct of mammals. Prolactin when administered in postnatal stage, stimulates lactation and also advocated in menstrual disturbances.

Growth hormone controls the growth and development of skeleton. It is prepared from desiccated beef pituitary anterior lobe. The material is extracted with calcium hydroxide: ammonium sulphate is added to the extract and the precipitate obtained is treated with cysteine. Fractionation is effected by pH variation.

ACTH can be prepared from sheep and pig pituitary by extraction with acidified acetone. The extractable material is precipitated by increasing the concentration of acetone in the extract to 92%. The precipitate is purified by treatment with 5% lead acetate at ph. 7.0 and also with TCA.ACTH is of value in allergies, gout, rheumatoid arthritis, leukemia and pneumonia.

The posterior pituitary lobe contains two active principles oxytocin, (pitocin, α-hypophamine) and vasopression (pitrissin, β-hypophamine). Oxitocin stimulates uterian contractions while vassopressin raises the blood pressure. Extracts of posterior pituitary are used for the treatment of diuresis in diabetes insipidus, for controlling hemorrhages after child birth and in many surgical operations, to induce labor.

7.7. Thyroid

This is one of the largest and most easily accessible gland which secrets thyroxine. It consists of two lobes situated on either side of the trachea and in many species joined by narrow isthmus. Desiccated thyroid used in the treatment of several clinical conditions such as myxedema and cretinism. It is also widely used in the treatment of menstrual disturbances, sterility, abortion, psoriasis and other skin affections. The purified desiccated thyroid gland contains not less than 0.17% and not more than 0.23% iodine and is composed of iodine containing protein thyroglobulin.

7.8. Adrenal Gland

The adrenal or suprarenal gland consists of two pear shaped bodies, each resting upon the upper border of kidney. This gland is made up of cortex and medulla. Adrenaline or epinephrine is formed in the medulla and also in certain extra modularly nervous tissue. Adrenaline is extracted from the gland by water or alcohol. By precipitation with neutral lead acetate impurities are removed from extract and separation is carried out by precipitation with ammonia. The adrenaline content of the fresh gland is about 0.2% (1% in dried tissue) and the commercial yield is reported to be 0.1% of the fresh gland. It is extensively used for its vasoconstriction property. Applied locally or on congested mucous membrane and to bleeding surface. It has relaxing effect on bronchial muscle hence, used in treatment of asthma.

7.9. Pepsin

It is a proteolytic enzyme extracted from pig's abdominal mucosa with cold water and precipitated by alcohol. Pepsin can be dried with chilled acetone. A commercial method of pepsin production from buffalo, sheep and goat abomasum lining has been developed by CDRI, Lucknow. According to this process about 200 gm of lining obtained from each buffalo stomach will give 7 to 8 gm of crystalline pepsin. Pepsin is used for digestive disorders and has some industrial use mainly in breweries. Rennet enzyme extracted from 4th stomach of young un-weaned calf is principally used in cheese industry. Today it is also used in the manufacture of Junket (flavoured milk curdled with rennet).

7.10. Spinal Cord and Brain

These are rich source of cholesterol, sphingomyelin, cerebrosides etc. A method for the isolation of cholesterol from spiral cord has been standardized by CDRI, Lucknow. The V.P Chest Institute, New Delhi is manufacturing cholesterol from buffalo brain. An indigenous technology has been reported for isolation of cholesterol from spinal cord with an yield of 3.5–4%. The buffalo spinal cord found to contain more cholesterol (7%) than goat and sheep spinal cord (5%). The cholesterol recovered from these sources was comparable in all aspects with pure cholesterol available in the market. Cholesterol is used in cosmetic industry, in biochemical and clinical research as reference standard.

7.11. Testes

An enzyme hyaluronidase is extracted commercially from ram or bull testes. Sheep testes are considered to be rich source of this enzyme. Hyaluronidase is also designated as the spreading factor as it reduces the viscosity by destroying tissue hyaluronic acid, thus permitting greater spreading of materials in the tissue. This biological property forms the basis of the clinical use of hyaluranidase to promote absorption of fluids administered sub-cutaneously or on the diffusion of drugs such as local anaesthetics injected in to the tissues (Table 9).

7.12. Heart

Heart and brain infusions are used in preparing culture media. Cardiolipin and cytochrome C are also extracted from heart and have diagnostic value in medicine and biochemical research.

Table 9: By-products utilization in bio-pharmaceutical industry

Name of the product	*Name of the end product*	*Uses*
1. Intestines	Collagen sheet	Burns/Wound dressing
	Heparin	Anticoagulant
2. Bones	Gelatin	Gelatin capsule
		Food industry
3. Blood	Hydrolyzed RBC	Syrup
	Haemoglobin	
	Blood albumen	Protein base
4. Glands and organs		
Pancreas	Insulin	life savor for diabetics
Liver	Liver extract	Tonic in anemia
Bile	Bile paste	Digestive troubles
Lung	Heparin	Anticoagulant
Adrenals	Adrenaline	Cardio accelerator
Pituitary	Pituitary hormones	Hormone therapy
Testes	Sex hormones	Hormone therapy
Thymus (Calf)	Hyaluronidase	spreading agent for drugs
	Histones and DNA	drugs

8. SCOPE FOR BY-PRODUCTS UTILISATION INDUSTRIES

In developing countries two tier technology for utilisation of animal by-products will be most appropriate. As per this technology, the raw materials should be conserved at the place of their availability. Once when sufficient quantity of conserved raw material is available, it could be transported to central place for processing thus ensuring economical viability. Necessary facility could be established in the central place for processing conserved raw materials into value added end products.

It is well known that raw material such as soft tissues; blood, ruminal content etc. are perpetually wasted in most of the developing countries. Various simple, cost effective, appropriate, two-tier technologies for the conservation of small quantities of raw materials have now been developed. More than 15 different industries like food, pharmaceutical, leather, biochemical, sports, dairy, cosmetic, animal feed, detergent, handicraft and fat refining industries are associated in processing various carcass by-products. A variety of economically useful products are produced all over the world. Animal By-products in our daily lives is given in table.

A good number of products like sausage casings, gelatine, glue, surgical sutures, sport guts, collagen sheets, bone meal, blood meal, horn and hoof meal, tallow, keratin hydrolysate, ossein, pet foods, animal feeds, fertiliser and some bio-chemicals are produced in the country. Further, many of the by-products like guts, bladders, horns, hooves and bones are sources of valuable foreign exchange and the demand for these items is increasing every year.

The scope for by-products utilisation and establishment of related industries in India is excellent. This is a very important area as it may play a major role in the availability of the raw material on continuous basis and also solve the disposal problem of a large quantity of bio-waste emanating from various sources such as slaughterhouses, meat processing plant, poultry dressing unit, fish processing plant, fallen carcass utilisation centres and tanneries. For developing full fledge packages following factors are important.

- Techno-economic viabilities of several processes.
- Engineering aspects of unit process and operation of carcass by-products process.
- Development of component engineering organisations to implement these technologies on turnkey basis.
- Development of by-products processing plants, which can fully utilise the entire wastes.

9. SUGGESTIONS

The following suggestions would be useful for the development of animal by-products industry in the country.

(*i*) A carcass by-products development authority with an immediate objective of developing by-product industries may be established. Such an authority could arrange to conduct nation wide survey and generate reliable databases for the assessment of economic potential and identify suitable technologies for utilisation of carcass by-products.

(*ii*) Awareness among public about the socio-economic benefits on the profitable utilisation of carcass by-products into value added end products could be created through audio visual aids, demonstration and training programming etc.

(*iii*) Encouragement should be provided to the local entrepreneur for setting up by-product utilisation industries by providing incentives like grants, subsidies, tax benefit etc.

(*iv*) Demonstration cum training centre for by-product conservation and utilisation could be established.

(*v*) Carcass by-product complex could be established at appropriate place in and around where large quantities and carcass by-products are available.

(*vi*) Provides suitable quality control and products specification measures by centralising the activity at the district level and develop marketing link with the user industry.

(*vii*) Meet the user industry requirement by providing the quality control materials obtained from the processed animal by-products.

REFERENCES

Divakaran, S. (1982). Processing and utilization of animal blood, FAO

Ockerman, W.H. and Hansen, C.L. (1988). Animal by-products processing, VCH Pub.

Kumar, Mahendra (1987). Processing of animal byproducts in developing countries—a manual" Commonwealth Science Council London, UK.

Kumar, Mahendra. (1989). Handbook of rural technology for the processing of animal byproduct (FAO agricultural service bulletin No.79) Food and Agricultural Organisation of United Nation, Rome, Italy.

Mandal, P.K., Pal, U.K., Das, C.D. and Rao, V.K. (2006). Use of Animal By-products in Pharmaceutical and Biomedical Industry. National Symposium on Prospects and Challenges in Indian Meat Industry" 27–29 July, 2006, Chennai, pp. 225–231.

Mann, I. (1962). Processing and utilisation of animal byproducts Food and Agricultural Organisation of United Nations, Rome, 1962.

Rao, V.K., Pal, U.K., Mandal, P.K. and Das, C.D. (2003). Technologies for by-products utilization—indian scenario. Compendium of National Workshop on Identification of Technologies & Equipments for meat and Milk products, Sept. 5–6, 2003, IVRI, pp. 41–47.

Sastry, T.P., Sehgal, P.K., Gupta, K.B. and Kumar, Mahendra., *Leather Science*, 33(12): 345–359.

Sehgal, P.K., Sastry, T.P. and Kumar, Mahendra. (1986). *Leather Science*, 33(12): 333–344.

Sharma, B.D. (2003). Modern abattoir practices and animal by-products technology, JP Brothers, New Delhi.

Subject Index

A

Accelerated ripening 209
Acetaldehyde 204
Active packaging 354
Additives 371, 377
Adrenal 604
Advantages of enrobed products 255
Advantages of nanotechnology 483
Alginate coatings 263
Allergen reduction 404
Animal products 1, 2, 18
Antimicrobial packaging 354, 364
Antimicrobials 185, 187
Antioxidants 185, 186, 404
Applications of proteomics 575
Avidin 190

B

Back pressure valve 54
Bacteriocins 154, 185, 192, 204, 497
Bacteriocins classification 193
Bacteriophages 162, 195
Baking quality 133
Batch pasteurization 53
Beta-Star test 526
Beverages 344, 347
Bile salts 603
Bioactive components 349
Bioactive peptides 204, 433, 550
Bioavailability 4
Biocontrol 161
Biogenic amines 210
Biopreservation 164
Biopreservatives 185, 490, 491
Biosensor 165, 522
Biosensor techniques 319
Biotechnology 152, 153
Biothickener 207
Blending 81
Blood plasma 589
Blood proteins 586
Boil-in-bag 41
Bone char 592
Bone meal 592
Booster pump 54
Byproducts utilization 586

C

Capillary electrophoretic (CE) methods 327
Carbohydrate 371, 372
Carbon dioxide absorbers and emitters 360
Caroteinoid enriched eggs 558
Carrageenan 377
Caseinates 456, 457
Casings 596
Celery 490, 493
Cellophane 33
Ceratonia siliqua 490, 493
Cereal based fermented dairy beverages 351
Certification 487
CHARM I and II tests 526

Chitinases 191
Chitosan 190, 496
Cholesterol 2
Chopping 80
Clove oil 492
Coagulant 137
Collagen 592
Colostrum 13
Comminuted meat products 79, 80
Conjugated 213
Conjugated linoleic acid 5, 350, 404, 405
Conjugation 152, 153
Continuous pasteurization 53
Continuous production 144
Cooking media 260
Cooking methods 261
Cooling time 59, 62
Copolymers 44
Corrective action 182
Critical control points (CCPs) 181
Critical limits 181
Curing, fermentation 404
Curry leaf 494
Curry leaves 276

D

Dairy beverages 344, 348
Dairy ingredients 456, 457
Deep fat frying 256
Designer eggs 548
Designer milk 166
Developing countries 174, 175
Diacetyl 204
Dielectric heating 217
Dietary supplementation 408
Differential display proteomics 570
Direct acidification 135
Drumstick leaves 276
Dry sausages 89
D-value 56

E

Effect of bacteriocins 194
Egg and health 10
Egg as functional food 547
Egg nutrition 544
Electrical stimulation 18, 25
Emulsifiers 376
Emulsion 82
Emulsion type meat products 84
Enrobed meat products 456, 467
Enrobed products 21
Enrobing 254, 255
Enrobing ingredients 257
Essential oil 187, 490, 491
Ethanol emitters 365
Ethylene-vinyl acetate (EVA) 44
Exopolysaccharides 204
Extenders 84
Extraction 518

F

Fat 371, 372
Fat and cholesterol reduction 404
Fat leakage 136
Fat-mimetics 375
Fat-replacer 371, 372
Fat-substitutes 376
FDV 72
Fenugreek extracts 498
Fermented foods 152, 153, 344, 351
Fermented meat products 12, 22
Fermented milk products 344
Fiber enrichment 404
Fibrin 588
Flaking 80
Flash pasteurization 54
Flavonoids 495
Flavour and odour absorbers 365
Flow diagram 179
Foam heater 62
Food safety 102, 111, 154
Food safety management system 174, 176
Freezing and thawing 396
Fresh meat 102, 354
Freshness indicators 366
Functional dairy beverages 344, 352
Functional foods 1, 7, 271, 404, 405, 547
Functional meat products 26, 93
Functional starters 208
Further processing 79

G

Gamma irradiation 119
Gas chromatography 322
Gas packaging 357
Gelatin coating 263
Gelation properties 388
Genetic engineering 156
Genetic manipulations 404
Genetic selection 553
Genomics 570
Goat milk 13
Good manufacturing practice 118
Gooseberry 280
GRAS 376
Grinding 80

H

HACCP 102, 110, 174, 175
Hazard 484
Hazard analysis 174, 175
Health benefits 197, 404, 406, 543, 544
Heating time 62
Herbs 404, 405
HHP and combination treatment 400
Hide and skin 594
High density polyethylene 49
High pressure processing 216, 224
High pressure technology 18, 31
Holding tube 64
HPLC 322
HPLC purification 532
HPTLC 532
HTST 53
Human health 1, 2
Hydrogen peroxide 190

I

Identification 570
IDEXX SNAP test 525
Immunoaffinity chromatography 519
Immunoassays 522
Infrared heating 30, 236
Intelligent packaging 362
Intermediate moisture foods 22
Irradiation 18, 28, 216, 239

K

Keratin 598

L

Lactic acid 203, 204
Lactic acid bacteria 203, 204
Lactic fermented products 345
Lactoferrin 190
Lactoperoxidase 190
Lactose 457
Lactose negative starters 205
Laminates 38
Lassi 345
Lateral Flow Assay (LFA) 317, 527
LC-MS-MS methods 331
Leafy green vegetable 495
Limitation 401
Linoleic acid (CLA) 206
Lipid oxidation 271, 393
Liver extract 602
Low calorie sugar 211
Low cholesterol eggs 553
Low density poly ethylene 49
Low fat meat products 27, 456, 461
LTLT 53
Luminex xMAP^R technology 305, 322
Lysozyme 190

M

Malondialdehyde 272
Mannitol 207
Master pack 358
Meat 371, 372
Meat as functional food 404
Meat colour 392
Meat preservation 102, 114
Meat production 177
Meat products 24, 456, 457
Meat quality 103, 389
Mechanism of action 194
Meltability 136
Membrane processing 31
Membrane technology 216, 242
Methods of preparation of soup 289, 291
Microarray technology 321

Microbiological quality 394
Microbiological quality of meat 110
Microwave (MW) 217
Microwave heating 18, 29
Milk based functional foods 12
Milk pasteurization 52
Milk powder 456
Milk products in human health 11
Milk proteins 456, 457
Milk type 133
Mineral enriched eggs 557
Minerals 404, 411
Mint leaves 277
Modifications 549
Modified starch 376
Moisture absorbers 363
Molecular imprinted polymers 305, 321
Molecular techniques 152, 153
Monitoring system 181
Monosodium glutamate 289, 292
Mozzarella 133, 134
mRNA profiling 571
Mycotoxins 305, 306
Mylar 42

N

Nanoencapsulation 478
Nano-packaging 479
Nanoparticles 474
Nanoparticles in food packaging 479
Nanoproducts 484
Nanosensors 476
Nanotechnology 472, 473
Natural antioxidant 270, 271, 490, 491
Natural preservatives 185, 186, 494
NFDM 456, 457
Non-fat dry milk 456, 463
Non-meat protein 382
Non-thermal processing 224
Novel texture 396
Nuggets 456, 460
Nutritional requirements 2
Nutritional value of egg 9
Nutritive value of soup 289, 295
Nylons 40

O

Oatrim 376
Ohmic heating 18, 29, 222
Olestra 375
Omega-3 enriched eggs 550
Omega-3 fatty acids 409
Organic acids 187
Organoleptic quality 552
Oscillating magnetic fields 247
Oxygen scavengers 354, 361

P

Packaging 143, 354, 355
Packaging evolution 33
Packaging of enrobed products 261
Packaging of meat 115
Packaging of soup 289, 296
Packaging properties 33
Pancreas 588
Pasteurizer trouble shooting 76
Pathogens 161
Patties 461
PCR 152, 153
Pepsin 604
Pharmaceuticals 559
Pharmacological ways 554
PHE 53
Phytoalexins 187
Plant sterols and stanols 427
Plasma treatment 216
Plasmid DNA 155
Pliofilm 42
Polyamides 40
Polycarbonates 41
Polyester 33, 35
Polyethylene 33, 35
Polyethylene terephthalate 48
Polypropylene 49
Polystyrene 33, 39
Polytetrafluoroethylene 40
Polyvinyl chloride 33, 38
Polyvinylidene chloride 33, 35
Pomegranate extract 490, 494
Post production factors 103
Prebiotics 405, 426
Prerequisite program (PRP) 180

Principles 388
Probiotic dairy beverages 348
Probiotics 204, 404, 426
Processed meats 457
Processing of meat 102, 117
Processing strategies 414
Product description 178
Production factor 103
Propolis 496
Protein modifications 569, 575
Protein profiling 571
Proteins 372
Proteolytic enzymes 391
Proteome mapping 575
Proteomics 569, 570
PUFA 372
Pulse electrified field (PEF) 30, 228

Q

Q_{10}-value 58
Quality assurance 102, 111
Quality changes in soup 297
Quality monitoring 116

R

Rabdi 351
Radio frequency (RF) 217
Record keeping 182
Red grapes 280
Regeneration efficiency 70
Regulations 196, 387, 305, 336, 395, 484
Residues and health risk 332, 534
Residues in meat 102, 123
Restructured meat 21, 274, 464
Retort pouches 92
Rosemary extract 490, 496
Rubber hydrochloride 42

S

Sausages 80, 456, 458
Se-enriched eggs 560
Shelf life of soup 289, 299
Shred 145
Shrink packaging 356
Simmering method 261
Skim milk powder 456
Skin packaging 356
Smart packaging 481
Sodium reduction 404
Solid-phase extraction 518
Soup composition 290
Soy proteins 84
Spices 404, 405
Spices and condiments 292
Starter culture 135, 152, 153
Storage 133, 136
Stress and meat quality 106
Stretch 140
Structured lipids 378
Stunning and Slaughter 107
Super-critical fluid extraction 371, 379
Synthetic antioxidants 270, 275
Systems of enrobing 260

T

Tanning 594
TBA value 257
Team assembly 177
Temperature indicator 366
Tenderness 83, 389
Terelyn 42
Texture 387
Therapeutic antibodies 559
Thermal processing 217
Thickeners 293
Thin-layer chromatography (TLC) 326
Thyroid 604
Tomato 280
Transgenic cows 166
Trichinella spiralis of pork 395

U

UHT 54
Ultrasound treatment 30, 233
Unsaturated fatty acids 5
UV processing 29, 216, 244

V

Vacuum packaging 355
Value added meat products 19
Vegetable proteins 405
Velocity profile 71

Verification procedure 182
Vitamin enriched eggs 555
Vitamins 203, 206, 404

W

Water holding capacity 380
Whey proteins 456, 458
Whey-based beverages 347
WPC 456, 458

Z

Z-trim 376
Z-value 57